Circuit Design

The Newnes Know It All Series

PIC Microcontrollers: Know It All
Lucio Di Jasio, Tim Wilmshurst, Dogan Ibrahim, John Morton, Martin Bates, Jack Smith, D.W. Smith, and
Chuck Hellebuyck
ISBN: 978-0-7506-8615-0

Embedded Software: Know It All
Jean Labrosse, Jack Ganssle, Tammy Noergaard, Robert Oshana, Colin Walls, Keith Curtis,
Jason Andrews, David J. Katz, Rick Gentile, Kamal Hyder, and Bob Perrin
ISBN: 978-0-7506-8583-2

Embedded Hardware: Know It All
Jack Ganssle, Tammy Noergaard, Fred Eady, Lewin Edwards, David J. Katz, Rick Gentile, Ken Arnold,
Kamal Hyder, and Bob Perrin
ISBN: 978-0-7506-8584-9

Wireless Networking: Know It All
Praphul Chandra, Daniel M. Dobkin, Alan Bensky, Ron Olexa, David Lide, and Farid Dowla
ISBN: 978-0-7506-8582-5

RF & Wireless Technologies: Know It All
Bruce Fette, Roberto Aiello, Praphul Chandra, Daniel Dobkin, Alan Bensky, Douglas Miron, David Lide,
Farid Dowla, and Ron Olexa
ISBN: 978-0-7506-8581-8

Electrical Engineering: Know It All
Clive Maxfield, Alan Bensky, John Bird, W. Bolton, Izzat Darwazeh, Walt Kester, M. A. Laughton,
Andrew Leven, Luis Moura, Ron Schmitt, Keith Sueker, Mike Tooley, DF Warne, Tim Williams
ISBN: 978-1-85617-528-9

Audio Engineering: Know It All
Douglas Self, Richard Brice, Don Davis, Ben Duncan, John Linsley Hood, Morgan Jones, Eugene Patronis,
Ian Sinclair, Andrew Singmin, John Watkinson
ISBN: 978-1-85617-526-5

Circuit Design: Know It All
Darren Ashby, Bonnie Baker, Stuart Ball, John Crowe, Barrie Hayes-Gill, Ian Grout, Ian Hickman,
Walt Kester, Ron Mancini, Robert A. Pease, Mike Tooley, Tim Williams, Peter Wilson, Bob Zeidman
ISBN: 978-1-85617-527-2

Test and Measurement: Know It All
Jon Wilson, Stuart Ball, GMS de Silva,Tony Fischer-Cripps, Dogan Ibrahim, Kevin James, Walt Kester,
M. A. Laughton, Chris Nadovich, Alex Porter, Edward Ramsden, Stephen Scheiber, Mike Tooley,
D. F. Warne, Tim Williams
ISBN: 978-1-85617-530-2

Mobile Wireless Security: Know It All
Praphul Chandra, Alan Bensky, Tony Bradley, Chris Hurley, Steve Rackley, John Rittinghouse,
James Ransome, Timothy Stapko, George Stefanek, Frank Thornton, Chris Lanthem, Jon Wilson
ISBN: 978-1-85617-529-6

For more information on these and other Newnes titles visit: www.newnespress.com

Circuit Design

Darren Ashby
Bonnie Baker
Stuart Ball
J. Crowe
Barrie Hayes-Gill
Ian Hickman
Walt Kester
Ron Mancini
Ian Grout
Robert A. Pease
Mike Tooley
Tim Williams
Peter Wilson
Bob Zeidman

ELSEVIER

AMSTERDAM • BOSTON • HEIDELBERG • LONDON
NEW YORK • OXFORD • PARIS • SAN DIEGO
SAN FRANCISCO • SINGAPORE • SYDNEY • TOKYO

Newnes is an imprint of Elsevier

Newnes

Newnes is an imprint of Elsevier
30 Corporate Drive, Suite 400, Burlington, MA 01803, USA
Linacre House, Jordan Hill, Oxford OX2 8DP, UK

Library of Congress Cataloging-in-Publication Data
Application submitted

British Library Cataloguing-in-Publication Data
A catalogue record for this book is available from the British Library.

ISBN: 978-1-85617-527-2

For information on all Newnes publications
visit our Web site at www.elsevierdirect.com

Printed in the United States of America
08 09 10 10 9 8 7 6 5 4 3 2 1

Contents

About the Authors

Darren Ashby (Chapters 1, 19, 26, and 35) author of *Electrical Engineering 101*, is a self-described "techno geek with pointy hair." He considers himself a Jack of all trades, master of none. He figures his common sense came from his dad and his book sense from his mother. Raised on a farm and graduating from Utah State University seemingly ages ago, he has nearly 20 years of experience in the real world as a technician, engineer and a manager. He has worked in diverse areas of compliance, production, testing and his personal favorite, R&D.

He jumped at a chance some years back to teach a couple of semesters at his alma mater. For about two years, he wrote regularly for the online magazine "chipcenter.com." He is currently the Director of electronics R&D at a billion dollar consumer product company. His passions are boats, snowmobiles, motorcycles and pretty much anything with a motor in it. When not at his day job, he spends most his time with his family and a promising R&D consulting/manufacturing firm he started a couple of years ago.

He lives with his beautiful wife, four strapping boys and cute little daughter next to the mountains in Richmond, Utah.

Bonnie Baker (Chapters 18, 19, 25 and 38) author of *A Baker's Dozen*, also writes the monthly "Baker's Best" for *EDN* magazine. She has been involved with analog and digital designs and systems for over 20 years. Bonnie started as a manufacturing product engineer supporting analog products at Burr-Brown. From there, Bonnie moved up to IC design, analog division strategic marketer, and then corporate applications engineering manager. In 1998, she joined Microchip Technology and served as their analog division analog/mixed-signal applications engineering manager and staff architect engineer for one of their PICmicro divisions. This expanded her background to not only include analog applications, but microcontroller solutions as well. At present, she has returned to the Precision Analog fold at Texas Instruments in Tucson, Arizona.

Bonnie holds a Masters of Science in Electrical Engineering from the University of Arizona (Tucson, AZ) and a bachelor's degree in music education from Northern Arizona University (Flagstaff, AZ). In addition to her fascination with analog design, Bonnie has a drive to share her knowledge and experience and has written almost 300 articles, design notes, and application notes and she is a frequent presenter at technical conferences and shows.

Stuart Ball, P.E., (Chapters 20, 34) author of *Analog Interfacing to Embedded Microprocessors*, is an electrical engineer with over 20 years of experience in electronic and embedded systems. He is currently employed with Seagate Technologies, a manufacturer of computer hard disc drives.

Bruce Carter (Chapter 19) a contributor to *Electrical Engineering 101*, is currently an Engineer for the Test and Measurement group of Texas Instruments. Carter earned a BS in Engineering Physics from Texas Tech University, and a BS in Electrical Engineering from the University of Texas. He authored several technical articles, including four chapters in Op-Amps for Everyone. New edition publishing soon.

John Crowe (Chapter 13) co-author of *Introduction to Digital Electronics*, is Reader in Biomedical Informatics in the School of Electrical & Electronic Engineering, University of Nottingham, UK. His contribution to this book is based upon material used in a Digital Electronics module delivered to 1st and 2nd year undergraduate students.

His research concerns the development of novel biomedical instrumentation such as fetal heart rate monitors and integrated optical and electronics ASICs for imaging skin blood flow.

Ian Grout (Chapters 7, 8, 9, 10, 11, and 12) the author of *Digital Systems Design* received his B.Eng in Electronic Engineering (1991) and PhD (1994) from Lancaster University (UK). He has worked in both industry and the academic field in microelectronic circuit and electronics design and test. He currently works in the areas of mixed-signal integrated circuit (IC) design for testability (DfT) and digital electronic circuit design using programmable logic. The author is currently a lecturer within the Department of Electronic and Computer Engineering at the University of Limerick (Ireland). He currently teaches programmable logic and integrated circuit design and test principles within the university and has worked in Limerick since 1998. Prior to this he was a lecturer in the Engineering Department at Lancaster University (UK).

Barrie Hayes-Gill (Chapter 13) co-author of *Introduction to Digital Electronics*, is Associate Professor in Integrated Circuit Design and Electronic Instrumentation in the School of Electrical & Electronic Engineering, University of Nottingham, UK. He has lectured inintegrated circuit design both within the University of Nottingham and at international locations around the World.

His research and industrial work concerns the development of compact and low noise instrumentation for medical devices and instrumentation where he deploys off-the-shelf electronic components and semi-custom and full custom integrated circuits for integrated optical sensors. He has published widely with over 150 publications and 10 patents on medical devices and VLSI systems. In addition to his University post he is also an Executive Directorat Monica Healthcare Ltd.

Ian Hickman, Eur. Ing. D. I. H. May B.Sc.Hons, C.Eng., MIEE, MIEEE (Chapters 2, 4, 5, 23, 24) is the author of *Analog Electronics*. He has been interested in electronics since the late 1940s, and professionally involved in it since 1954. Starting with a crystal set, his interests over the years have covered every aspect of electronics, though mainly concentrating on analog. Now retired, Ian was a consultant to Electronics World for many years. He is a Member of the Institution of Engineering and Technology: and a Life Member of the Institute of Electrical & Electronics Engineers. He has also written several books including *Practical RF Handbook,* Hickman's *Analog and RF Circuits,* and *Analog Circuits Cookbook,* to name just a few.

Walt Kester (Chapters 16, 17) is the author of *Mixed Signal and DSP Design Techniques*. He is a corporate staff applications engineer at Analog Devices. For over 35 years at Analog Devices, he has designed, developed, and given applications support for high-speed ADCs, DACs, SHAs, op-amps, and analog multiplexers. Besides writing many papers and articles, he prepared and edited eleven major applications books, which form the basis for the Analog Devices world-wide technical seminar series including the topics of op-amps, data conversion, power management, sensor signal conditioning, mixed-signal, and practical analog design techniques. He also is the editor of *The Data Conversion Handbook*, a 900+ page comprehensive book on data conversion published in 2005 by Elsevier. Walt has a BSEE from NC State University and MSEE from Duke University.

Thomas Kugelstadt (Chapter 22) was a contributor to *Op Amps for Everyone*. He is a senior application engineer at Texas Instruments. He is writing many technical articles on various subjects, often system related.

He also provides freelance writing services if your company were ever interested in a technical subject but experienced difficulties finding a writer.

Ron Mancini (Chapter 26) the editor of *Op Amps for Everyone* has spent nearly fifty years in electronics. Recently retired, he was a Staff Scientist at Texas Instruments for many years. He was also a regular columnist for *EDN*.

Richard Palmer (Chapter 26) was a contributor to *Op Amps for Everyone*.

Robert A. Pease (Chapters 1, 3, 6) author of *Troubleshooting Analog Circuits*, attended Mt. Hermon School, and graduated from MIT in 1961 with a BSEE. He worked at Philbrick Researches up to 1975 and designed many Op-Amps and Analog Computing Modules.

Pease joined National Semiconductor in 1976. He has designed about 24 analog ICs including power regulators, voltage references, and temp sensors. He has written 65+ magazine articles and holds about 21 US patents. Pease is the self-declared Czar of Bandgaps since 1986. He enjoys hiking and trekking in Nepal, and ferroequinology. His position at NSC is Staff Scientist. He is a Senior Member of the IEEE.

Pease is a columnist in *Electronic Design* magazine, with over 240 columns published. The column, PEASE PORRIDGE, covers a wide range of technical topics. Pease also has posted many technical and semi-technical items on his main website: http://www.national.com/rap Many of Pease's recent columns are accessible there.

Pease was inducted into the E.E. Hall Of Fame in 2002. Refer to: http://www.elecdesign.com/Articles/Index.cfm?ArticleID=17269&Extension=pdf See Pease's other web site at http://www.transtronix.com

Mike Tooley (Chapters 1, 21, and 33) author of *Electronic Circuits*, is a technical author and consultant. He was formerly Vice Principal at Brooklands College in Surrey, England, where he was responsible for the delivery of learning to over 10,000 Further and Higher Education students increasingly by flexible, open and online distance learning. Mike is the well-known author of several popular engineering and related text books, including widely adopted course texts for BTEC, GCE A-level and GCSE qualifications in Engineering. Mike's hobbies include astronomy, amateur radio, aviation, computing and electronic circuit design and construction.

Tim Williams (Chapters 14, 15, 19, 35, 36, 37, 39, 40,41, 42, 43, and Appendix A) is the author of *The Circuit Designer's Companion, 2ⁿᵈ Edition*. He works at Elmac Services, which provides consultancy and training on all aspects of EMC, including design, testing and the application of standards, to companies manufacturing electronic products and concerned about the implications of the EMC Directive.

Tim Williams gained a BSc in Electronic Engineering from Southampton University in 1976. He has worked in electronic product design in various industry sectors including process instrumentation and audio visual control. He was design group leader at Rosemount Ltd before leaving in 1990 to start Elmac Services.

He is also the author of "EMC for Product Designers" (now in its fourth edition, Elsevier 2006), and has presented numerous conference papers and seminars. He is also author of "EMC for Systems & Installations" with Keith Armstrong. He is an EMC technical assessor for UKAS and SWEDAC.

Peter Wilson (Chapters 25, 30, 31, 32) author of Design Recipes for FPGAs, is Senior Lecturer in Electronics at the University of Southampton. He holds degrees from Heriot-Watt University, an MBA from Edinburgh Business School and a PhD from the University of Southampton. He worked in the Avionics and Electronics Design Automation Industries for many years at Ferranti, GEC-Marconi and Analogy prior to rejoining academia . He has published widely in the areas of FPGA design, modeling and simulation, VHDL, VHDL-AMS, magnetics and power electronics. He is a Senior Member of the IEEE, member of the IET, and a Chartered Engineer.

Bob Zeidman (Chapters 27, 28, 29) author of *Designing with FPGAs and CPLDs*, is the president of Zeidman Consulting (www.ZeidmanConsulting.com), a premiere contract research and development firm in Silicon Valley. He is also the president of Zeidman Technologies (www.zeidman.biz), a developer of tools for embedded systems hardware and software development, and president of Software Analysis and Forensic Engineering Corporation (www.SAFE-corp.biz), the leading provider of software intellectual property analysis tools. Bob has designed ASICs, FPGAs, and PC boards for RISC-based parallel processor systems, laser printers, network switches and routers, and other systems for clients including Apple Computer, Cisco Systems, Mentor Graphics, and Ricoh. He is the inventor of SynthOSTM, a tool for synthesizing software from a high-level description, and CodeSuite®, a tool for measuring software source code correlation. His publications include papers on hardware and software design methods and three textbooks: Designing with FPGAs and CPLDs, Verilog Designer's Library, and Introduction to Verilog. Bob has taught courses at conferences throughout the world. He holds several patents and earned bachelor's degrees in physics and electrical engineering at Cornell University and a master's degree in electrical engineering at Stanford University.

CHAPTER 1

The Fundamentals

Mike Tooley
Darren Ashby
Robert Pease

1.1   Electrical Fundamentals

This chapter has been designed to provide you with the background knowledge required to help you understand the concepts introduced in the later chapters. If you have studied electrical science, electrical principles, or electronics then you will already be familiar with many of these concepts. If, on the other hand, you are returning to study or are a newcomer to electronics or electrical technology this chapter will help you get up to speed.

1.1.1   Fundamental Units

You will already know that the units that we now use to describe such things as length, mass and time are standardized within the International System of Units (SI). This SI system is based upon the seven *fundamental units* (see Table 1.1).

1.1.2   Derived Units

All other units are derived from these seven fundamental units. These *derived units* generally have their own names and those commonly encountered in electrical circuits are summarized in Table 1.2, together with the corresponding physical quantities.

(Note that 0K is equal to $-273°C$ and an *interval* of 1K is the same as an *interval of 1°C*.)

If you find the exponent notation shown in Table 1.2 a little confusing, just remember that V^{-1} is simply 1/V, s^{-1} is 1/s, m^{-2} is $1/m^{-2}$, and so on.

Table 1.1: SI units

Quantity	Unit	Abbreviation
Current	ampere	A
Length	meter	m
Luminous intensity	candela	cd
Mass	kilogram	kg
Temperature	Kelvin	K
Time	second	s
Matter	mol	mol

Table 1.2: Electrical quantities

Quantity	Derived unit	Abbreviation	Equivalent (in terms of fundamental units)
Capacitance	farad	F	$A\ s\ V^{-1}$
Charge	coulomb	C	$A\ s$
Energy	joule	J	$N\ m$
Force	newton	N	$kg\ m\ s^{-1}$
Frequency	hertz	Hz	s^{-1}
Illuminance	lux	lx	$lm\ m^{-2}$
Inductance	henry	H	$V\ s\ A^{-1}$
Luminous flux	lumen	lm	$cd\ sr$
Magnetic flux	weber	Wb	$V\ s$
Potential	volt	V	$W\ A^{-1}$
Power	watt	W	$J\ s^{-1}$
Resistance	ohm	Ω	$V\ A^{-1}$

Example 1.1

The unit of flux density (the tesla) is defined as the magnetic flux per unit area. Express this in terms of the fundamental units.

Solution

The SI unit of flux is the weber (Wb). Area is directly proportional to length squared and, expressed in terms of the fundamental SI units, this is square meters (m^2). Dividing the flux (Wb) by the area (m^2) gives Wb/m^2 or $Wb\ m^{-2}$. Hence, in terms of the fundamental SI units, the tesla is expressed in $Wb\ m^{-2}$.

Example 1.2

The unit of electrical potential, the volt (V), is defined as the difference in potential between two points in a conductor, which when carrying a current of one amp (A), dissipates a power of one watt (W). Express the volt (V) in terms of joules (J) and coulombs (C).

Solution

In terms of the derived units:

$$\text{Volts} = \frac{\text{Watts}}{\text{Amperes}} = \frac{\text{Joules/seconds}}{\text{Amperes}}$$

$$= \frac{\text{Joules}}{\text{Amperes} \times \text{seconds}} = \frac{\text{Joules}}{\text{Coulombs}}$$

Note that: Watts = Joules/seconds and also that Amperes × seconds = Coulombs.

Alternatively, in terms of the symbols used to denote the units:

$$V = \frac{W}{A} = \frac{J/s}{A} = \frac{J}{As} = \frac{J}{C} = JC^{-1}$$

One volt is equivalent to one joule per coulomb.

1.1.3 Measuring Angles

You might think it strange to be concerned with angles in electrical circuits. The reason is simply that, in analog and AC circuits, signals are based on repetitive waves (often sinusoidal in shape). We can refer to a point on such a wave in one of two basic ways, either in terms of the time from the start of the cycle or in terms of the angle

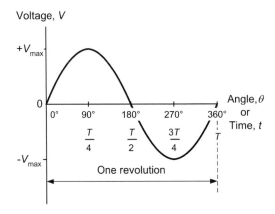

Figure 1.1: One cycle of a sine wave voltage

(a cycle starts at 0° and finishes as 360°—see Figure 1.1). In practice, it is often more convenient to use angles rather than time; however, the two methods of measurement are interchangeable and it's important to be able to work in either of these units.

In electrical circuits, angles are measured in either degrees or radians (both of which are strictly dimensionless units). You will doubtless already be familiar with angular measure in degrees where one complete circular revolution is equivalent to an angular change of 360°. The alternative method of measuring angles, the *radian*, is defined somewhat differently. It is the angle subtended at the center of a circle by an arc having length that is equal to the radius of the circle (see Figure 1.2).

You may sometimes find that you need to convert from radians to degrees, and vice versa. A complete circular revolution is equivalent to a rotation of 360° or

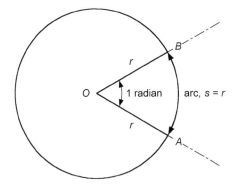

Figure 1.2: Definition of the radian

2π radians (note that π is approximately equal to 3.142). Thus, one radian is equivalent to $360/2\pi$ degrees (or approximately $57.3°$). Try to remember the following rules that will help you to convert angles expressed in degrees to radians and vice versa:

- From degrees to radians, divide by 57.3.

- From radians to degrees, multiply by 57.3.

Example 1.3

Express a quarter of a cycle revolution in terms of:

(a) degrees;

(b) radians.

Solution

(a) There are $360°$ in one complete cycle (i.e., one full revolution. Hence, there are $(360/4)°$ or $90°$ in one quarter of a cycle).

(b) There are 2π radians in one complete cycle. Thus, there are $2\pi/4$ or $\pi/2$ radians in one quarter of a cycle.

Example 1.4

Express an angle of $215°$ in radians.

Solution

To convert from degrees to radians, divide by 57.3. So, $215°$ is equivalent to $215/57.3 = 3.75$ radians.

Example 1.5

Express an angle of 2.5 radians in degrees.

Solution

To convert from radians to degrees, multiply by 57.3. Hence, 2.5 radians is equivalent to $2.5 \times 57.3 = 143.25°$.

1.1.4 Electrical Units and Symbols

Table 1.3 shows the units and symbols that are commonly encountered in electrical circuits. It is important to get to know these units and also be able to recognize their abbreviations and symbols. You will meet all of these units later in this chapter.

Table 1.3: Electrical units

Unit	Abbrev.	Symbol	Notes
ampere	A	I	Unit of electric *current* (a current of 1A flows when a charge of 1C is transported in a time interval of 1s)
coulomb	C	Q	Unit of electric *charge* or quantity of electricity
farad	F	C	Unit of *capacitance* (a capacitor has a capacitance of 1F when a potential of 1V across its plates produced a charge of 1C)
henry	H	L	Unit of *inductance* (an inductor has an inductance of 1H when an applied current changing at 1 A/s produces a potential difference of 1V across its terminals)
hertz	Hz	f	Unit of *frequency* (a signal has a frequency of 1 Hz if one cycle occurs in an interval of 1s)
Joule	J	W	Unit of *energy*
Ohm	Ω	R	Unit of *resistance*
second	s	t	Unit of *time*
siemen	S	G	Unit of *conductance* (the reciprocal of resistance)
tesla	T	B	Unit of *magnetic flux density* (a flux density of 1T is produced when a flux of 1 Wb is present over an area of 1 square meter)
volt	V	V	Unit of electric *potential* (e.m.f. or p.d.)
watt	W	P	Unit of *power* (equivalent to 1J of energy consumed in 1s)
Weber	Wb	φ	Unit of *magnetic flux*

1.1.5 Multiples and Sub-Multiples

Unfortunately, many of the derived units are either too large or too small for convenient everyday use, but we can make life a little easier by using a standard range of multiples and sub-multiples (see Table 1.4).

Table 1.4: Multiples and sub-multiples

Prefix	Abbreviation	Multiplier
tera	T	10^{12} ($= 1{,}000{,}000{,}000{,}000$)
giga	G	10^{9} ($= 1{,}000{,}000{,}000$)
mega	M	10^{6} ($= 1{,}000{,}000$)
kilo	K	10^{3} ($= 1{,}000$)
(none)	(none)	10^{0} ($= 1$)
centi	c	10^{-2} ($= 0.01$)
milli	m	10^{-3} ($= 0.001$)
micro	μ	10^{-6} ($= 0.000001$)
nano	n	10^{-9} ($= 0.000000001$)
pico	p	10^{-12} ($= 0.000000000001$)

Example 1.6

An indicator lamp requires a current of 0.075A. Express this in mA.

Solution

You can express the current in mA (rather than in A) by simply moving the decimal point three places to the right. Hence, 0.075A is the same as 75 mA.

Example 1.7

A medium-wave radio transmitter operates on a frequency of 1,495 kHz. Express its frequency in MHz.

Solution

To express the frequency in MHz rather than kHz, we need to move the decimal point three places to the left. Hence, 1,495 kHz is equivalent to 1.495 MHz.

Example 1.8

Express the value of a 27,000 pF in μF.

Solution

To express the value in μF rather than pF we need to move the decimal point six places to the left. Hence, 27,000 pF is equivalent to 0.027 μF (note that we have had to introduce an extra zero before the 2 and after the decimal point).

1.1.6   Exponent Notation

Exponent notation (or *scientific notation*) is useful when dealing with either very small or very large quantities. It's well worth getting to grips with this notation as it will allow you to simplify quantities before using them in formulae.

Exponents are based on *powers of ten*. To express a number in exponent notation the number is split into two parts. The first part is usually a number in the range 0.1 to 100 while the second part is a multiplier expressed as a power of ten.

For example, 251.7 can be expressed as 2.517×100, i.e., 2.517×10^2. It can also be expressed as $0.2517 \times 1,000$, i.e., 0.2517×10^3. In both cases the exponent is the same as the number of noughts in the multiplier (i.e., 2 in the first case and 3 in the second case). To summarize:

$$251.7 = 2.517 \times 10^2 = 0.2517 \times 10^3$$

As a further example, 0.01825 can be expressed as 1.825/100; that is, 1.825×10^{-2}. It can also be expressed as 18.25/1,000, i.e., 18.25×10^{-3}. Again, the exponent is the same as the number of zeros but the minus sign is used to denote a fractional multiplier. To summarize:

$$0.01825 = 1.825 \times 10^{-2} = 18.25 \times 10^{-3}$$

Example 1.9

A current of 7.25 mA flows in a circuit. Express this current in amperes using exponent notation.

Solution

$$1 \text{ mA} = 1 \times 10^{-3} \text{A}; \quad \text{thus,} \quad 7.25 \text{ mA} = 7.25 \times 10^{-3} \text{A}$$

Example 1.10

A voltage of 3.75×10^{-6}V appears at the input of an amplifier. Express this voltage in (a) V, and (b) mV, using exponent notation.

Solution

(a) 1×10^{-6}V $= 1\ \mu$V so 3.75×10^{-6}V $= 3.75\ \mu$V

(b)   There are 1,000 μV in 1 mV so we must divide the previous result by 1,000 in order to express the voltage in mV. So 3.75 μV $= 0.00375$ mV.

1.1.7   Multiplication and Division Using Exponents

Exponent notation really comes into its own when values have to be multiplied or divided. When multiplying two values expressed using exponents, you simply need to add the exponents. Here's an example:

$$(2 \times 10^2) \times (3 \times 10^6) = (2 \times 3) \times 10^{(2+6)} = 6 \times 10^8$$

Similarly, when dividing two values which are expressed using exponents, you only need to subtract the exponents. As an example:

$$(4 \times 10^6) \times (2 \times 10^4) = 4/2 \times 10^{(6-4)} = 2 \times 10^2$$

In either case it's important to remember to specify the units, multiples and sub-multiples in which you are working (e.g., A, kΩ, mV, μF, etc.).

Example 1.11

A current of 3 mA flows in a resistance of 33 kΩ. Determine the voltage dropped across the resistor.

Solution

Voltage is equal to current multiplied by resistance. Thus:

$$V = I \times R = 3 \text{ mA} \times 33 \text{ k}\Omega$$

Expressing this using exponent notation gives:

$$V = (3 \times 10^{-3}) \times (33 \times 10^3)\text{V}$$

Separating the exponents gives:

$$V = 3 \times 33 \times 10^{-3} \times 10^3 \text{V}$$

Thus, $V = 99 \times 10^{(-3+3)} = 99 \times 10^0 = 99 \times 1 = 99\text{V}$

Example 1.12

A current of 45 µA flows in a circuit. What charge is transferred in a time interval of 20 ms?

Solution

Charge is equal to current multiplied by time (see the definition of the ampere). Thus:

$$Q = It = 45\,\mu\text{A} \times 20 \text{ ms}$$

Expressing this in exponent notation gives:

$$Q = (45 \times 10^{-6}) \times (20 \times 10^{-3}) \text{ coulomb}$$

Separating the exponents gives:

$$Q = 45 \times 20 \times 10^{-6} \times 10^{-3} \text{ coulomb}$$

Thus, $Q = 900 \times 10^{(-6-3)} = 900 \times 10^{-9} = 900 \text{ nC}$

Example 1.13

A power of 300 mW is dissipated in a circuit when a voltage of 1,500V is applied. Determine the current supplied to the circuit.

Solution

Current is equal to power divided by voltage. Thus:

$$I = P/V = 300 \text{ mW}/1,500\text{V amperes}$$

Expressing this in exponent notation gives:

$$I = (300 \times 10^{-3})/(1.5 \times 10^3)\text{A}$$

Separating the exponents gives:

$$I = (300/1.5) \times (10^{-3}/10^{3})\text{A}$$
$$I = 300/1.5 \times 10^{-3} \times 10^{-3}\text{A}$$

Thus, $I = 200 \times 10^{(-3-3)} = 200 \times 10^{-6} = 200 \; \mu\text{A}$

1.1.8 Conductors and Insulators

Electric current is the name given to the flow of *electrons* (or negative charge carriers). Electrons orbit around the nucleus of atoms just as the earth orbits around the sun (see Figure 1.3). Electrons are held in one or more *shells*, constrained to their orbital paths by virtue of a force of attraction toward the nucleus, which contains an equal number of *protons* (positive charge carriers). Since like charges repel and unlike charges attract, negatively charged electrons are attracted to the positively charged nucleus. A similar principle can be demonstrated by observing the attraction between two permanent magnets; the two North u.c. Poles of the magnets will repel each other, while a North and South u.c. Pole will attract. In the same way, the unlike charges of the negative electron and the positive proton experience a force of mutual attraction.

The outer shell electrons of a *conductor* can be reasonably easily interchanged between adjacent atoms within the *lattice* of atoms of which the substance is composed. This makes it possible for the material to conduct electricity. Typical examples of conductors are metals such as copper, silver, iron and aluminum. By contrast, the outer shell

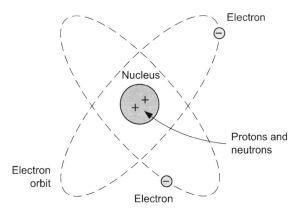

Figure 1.3: A single atom of helium (H$_\text{e}$) showing its two electrons in orbit around its nucleus

electrons of an *insulator* are firmly bound to their parent atoms and virtually no interchange of electrons is possible. Typical examples of insulators are plastics, rubber, and ceramic materials.

1.1.9   Voltage and Resistance

The ability of an energy source (e.g., a battery) to produce a current within a conductor may be expressed in terms of *electromotive force* (e.m.f.). Whenever an e.m.f. is applied to a circuit *a potential difference* (p.d.), or voltage, exists. Both e.m.f. and p.d. are measured in volts (V). In many practical circuits there is only one e.m.f. present (the battery or supply), whereas a voltage will be developed across each component present in the circuit.

The *conventional flow* of current in a circuit is from the point of more positive potential to the point of greatest negative potential (note that electrons move in the *opposite* direction!). *Direct current* results from the application of a direct e.m.f. (derived from batteries or a DC power supply). An essential characteristic of these supplies is that the applied e.m.f. does not change its polarity (even though its value might be subject to some fluctuation).

For any conductor, the current flowing is directly proportional to the e.m.f. applied. The current flowing will also be dependent on the physical dimensions (length and cross-sectional area) and material of which the conductor is composed.

The amount of current that will flow in a conductor when a given e.m.f. is applied is inversely proportional to its *resistance*. Therefore, resistance may be thought of as an opposition to current flow; the higher the resistance the lower the current that will flow (assuming that the applied e.m.f. remains constant).

1.1.10   Ohm's Law

Provided that temperature does not vary, the ratio of p.d. across the ends of a conductor to the current flowing in the conductor is a constant. This relationship is known as *Ohm's Law* and it leads to the relationship:

$$V/I = \text{a constant} = R$$

where V is the potential difference (or voltage drop) in volts (V), I is the current in amperes (A), and R is the resistance in ohms (Ω) (see Figure 1.4).

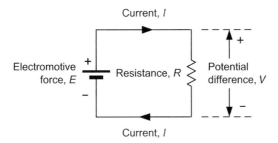

Figure 1.4: Simple circuit to illustrate the relationship between voltage (V), current (I) and resistance (R). Note that the direction of conventional current flow is from positive to negative.

The formula may be arranged to make V, I or R the subject, as follows:

$$V = I \times R,\; I = V/R \quad \text{and} \quad R = V/I$$

The triangle shown in Figure 1.5 should help you remember these three important relationships. However, it's worth noting that, when performing calculations of currents, voltages and resistances in practical circuits it is seldom necessary to work with an accuracy of better than $\pm 1\%$ simply because component tolerances are usually greater than this. Furthermore, in calculations involving Ohm's Law, it can sometimes be convenient to work in units of kΩ and mA (or MΩ and μA) in which case potential differences will be expressed directly in V.

$$\frac{V}{I} = R \qquad I = \frac{V}{R}$$

$$V = I \times R$$

Figure 1.5: Triangle showing the relationship between V, I and R

Example 1.14

A 12Ω resistor is connected to a 6V battery. What current will flow in the resistor?

Solution

Here we must use $I = V/R$ (where $V = 6V$ and $R = 12\Omega$):

$$I = V/R = 6V/12\Omega = 0.5A\,(\text{or}\,500\,\text{mA})$$

Hence a current of 500 mA will flow in the resistor.

Example 1.15

A current of 100 mA flows in a 56Ω resistor. What voltage drop (potential difference) will be developed across the resistor?

Solution

Here we must use $V = I \times R$ and ensure that we work in units of volts (V), amperes (A), and ohms (Ω).

$$V = I \times R = 0.1A \times 56\Omega = 5.6V$$

(Note that 100 mA is the same as 0.1A.)

This calculation shows that a p.d. of 5.6V will be developed across the resistor.

Example 1.16

A voltage drop of 15V appears across a resistor in which a current of 1 mA flows. What is the value of the resistance?

Solution

$$R = V/I = 15V/0.001A = 15,000\Omega = 15\,\text{k}\Omega$$

Note that it is often more convenient to work in units of mA and V, which will produce an answer directly in kΩ, that is:

$$R = V/I = 15V/1\,\text{mA} = 15\,\text{k}\Omega$$

1.1.11 Resistance and Resistivity

The resistance of a metallic conductor is directly proportional to its length and inversely proportional to its area. The resistance is also directly proportional to its *resistivity* (or *specific resistance*). Resistivity is defined as the resistance measured between the opposite faces of a cube having sides of 1 cm.

The resistance, *R*, of a conductor is given by the formula:

$$R = \rho \times l/A$$

where *R* is the resistance (ft), ρ is the resistivity (Ωm), *l* is the length (m), and *A* is the area (m^2).

Table 1.5 shows the electrical properties of some common metals.

Example 1.17

A coil consists of an 8m length of annealed copper wire having a cross-sectional area of 1 mm^2. Determine the resistance of the coil.

Solution

We will use the formula, $R = \rho\ l/A$.

Table 1.5: Properties of some common metals

Metal	Resistivity (at 20°C) (Ωm)	Relative conductivity (copper = 1)	Temperature coefficient of resistance (per °C)
Silver	1.626×10^{-8}	1.06	0.0041
Copper (annealed)	1.724×10^{-8}	1.00	0.0039
Copper (hard drawn)	1.777×10^{-8}	0.97	0.0039
Aluminum	2.803×10^{-8}	0.61	0.0040
Mild steel	1.38×10^{-7}	0.12	0.0045
Lead	2.14×10^{-7}	0.08	0.0040
Nickel	8.0×10^{-8}	0.22	0.0062

The value of ρ for annealed copper given in Table 1.5 is 1.724×10^{-8} Ωm. The length of the wire is 4m, while the area is 1 mm^2 or 1×10^{-6} m^2 (note that it is important to be consistent in using units of meters for length and square meters for area).

Hence, the resistance of the coil will be given by:

$$R = \frac{1.724 \times 10^{-8} \times 8}{1 \times 10^{-6}} = 13.724 \times 10^{(-8+6)}$$

Thus, $R = 13.792 \times 10^{-2}$ or 0.13792Ω

Example 1.18

A wire having a resistivity of 1.724×10^{-8} Ωm, length 20m and cross-sectional area 1 mm^2 carries a current of 5A. Determine the voltage drop between the ends of the wire.

Solution

First, we must find the resistance of the wire (as in Example 1.17):

$$R = \frac{\rho l}{A} = \frac{1.6 \times 10^{-8} \times 20}{1 \times 10^{-6}} = 32 \times 10^{-2} = 0.32\Omega$$

The voltage drop can now be calculated using Ohm's Law:

$$V = I \times R = 5A \times 0.32\Omega = 1.6V$$

This calculation shows that a potential of 1.6V will be dropped between the ends of the wire.

1.1.12   Energy and Power

At first you may be a little confused about the difference between energy and power. Simply put, energy is the ability to do work, while power is the rate at which work is done. In electrical circuits, energy is supplied by batteries or generators. It may also be stored in components such as capacitors and inductors. Electrical energy is converted into various other forms of energy by components such as resistors (producing heat), loudspeakers (producing sound energy), and light emitting diodes (producing light).

The unit of energy is the joule (J). Power is the rate of use of energy and it is measured in watts (W). A power of 1W results from energy being used at the rate of 1J per second. Thus:

$$P = W/t$$

where P is the power in watts (W), W is the energy in joules (J), and t is the time in seconds (s).

The power in a circuit is equivalent to the product of voltage and current. Hence:

$$P = I \times V$$

where P is the power in watts (W), I is the current in amperes (A), and V is the voltage in volts (V).

The formula may be arranged to make P, I or V the subject, as follows:

$$P = I \times P, I = P/V \quad \text{and} \quad V = P/I$$

The triangle shown in Figure 1.6 should help you remember these relationships.

The relationship, $P = I \times V$, may be combined with that which results from Ohm's Law ($V = I \times R$) to produce two further relationships. First, substituting for V gives:

$$P = I \times (I \times R) = I^2 R$$

Secondly, substituting for I gives:

$$P = (V/R) \times V = V^2/R$$

Example 1.19

A current of 1.5A is drawn from a 3V battery. What power is supplied?

Solution

Here we must use $P = I \times V$ (where $I = 1.5$A and $V = 3$V).

$$P = I \times V = 1.5\text{A} \times 3\text{V} = 4.5\text{W}$$

Hence, a power of 4.5W is supplied.

Example 1.20

A voltage drop of 4V appears across a resistor of 100Ω. What power is dissipated in the resistor?

Solution

Here we use $P = V^2/R$ (where $V = 4V$ and $R = 100Ω$).

$$P = V^{2/}R = (4V \times 4V)/100Ω = 0.16W$$

Hence, the resistor dissipates a power of 0.16W (or 160 mW).

Example 1.21

A current of 20 mA flows in a 1 kΩ resistor. What power is dissipated in the resistor?

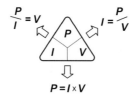

$$P = I \times V$$

Figure 1.6: Triangle showing the relationship between *P*, *I* and *V*

Solution

Here we use $P = I^2 \times R$ but, to make life a little easier, we will work in mA and kΩ (in which case the answer will be in mW).

$$P = I^2 \times R = (20 \text{ mA} \times 20 \text{ mA}) \times 1 \text{ kΩ} = 400 \text{ mW}$$

Thus, a power of 400 mW is dissipated in the 1 kΩ resistor.

1.1.13   Electrostatics

If a conductor has a deficit of electrons, it will exhibit a net positive charge. If, on the other hand, it has a surplus of electrons, it will exhibit a net negative charge. An imbalance in charge can be produced by friction (removing or depositing electrons using materials such as silk and fur, respectively), or induction (by attracting or repelling electrons using a second body, which is, respectively, positively or negatively charged).

1.1.14 Force Between Charges

Coulomb's Law states that, if charged bodies exist at two points, the force of attraction (if the charges are of opposite polarity), or repulsion (if the charges have the same polarity), will be proportional to the product of the magnitude of the charges divided by the square of their distance apart. Thus:

$$F = \frac{kQ_1Q_2}{r^2}$$

where Q_1 and Q_2 are the charges present at the two points (in coulombs), r the distance separating the two points (in meters), F is the force (in newtons), and k is a constant depending upon the medium in which the charges exist.

In vacuum or "free space":

$$k = \frac{1}{4\pi\varepsilon_0}$$

where ε_0 is the *permittivity of free space* (8.854×10^{-12} C/Nm^2).

Combining the two previous equations gives:

$$F = \frac{kQ_1Q_2}{4\pi \times 8.854 \times 10^{-12}r^2} \text{ Newtons}$$

1.1.15 Electric Fields

The force exerted on a charged particle is a manifestation of the existence of an electric field. The electric field defines the direction and magnitude of a force on a charged object. The field itself is invisible to the human eye, but can be drawn by constructing lines, which indicate the motion of a free positive charge within the field; the number of field lines in a particular region being used to indicate the relative strength of the field at the point in question.

Figures 1.7 and 1.8 show the electric fields between charges of the same and opposite polarity, while Figure 1.9 shows the field that exists between two charged parallel plates. You will see more of this particular arrangement when we introduce capacitors.

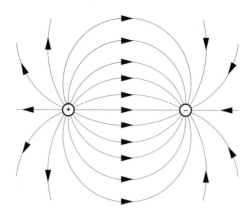

Figure 1.7: Electric field between two unlike electric charges

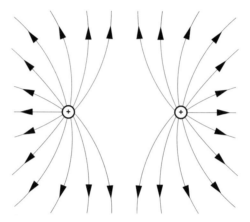

Figure 1.8: Electric field between two like electric charges (in this case both positive)

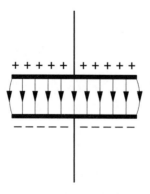

Figure 1.9: Electric field between two parallel plates

1.1.16 Electric Field Strength

The strength of an electric field (E) is proportional to the applied potential difference and inversely proportional to the distance between the two conductors. The electric field strength is given by:

$$E = V/d$$

where E is the electric field strength (V/m), V is the applied potential difference (V) and d is the distance (m).

Example 1.22

Two parallel conductors are separated by a distance of 25 mm. Determine the electric field strength if they are fed from a 600V DC supply.

Solution

The electric field strength will be given by:

$$E = V/d = 600/25 \times 10^{-3} = 24 \text{ kV/m}$$

1.1.17 Permittivity

The amount of charge produced on the two plates shown in Figure 1.9 for a given applied voltage will depend not only on the physical dimensions, but also on the insulating dielectric material that appears between the plates. Such materials need to have a very high value of resistivity (they must not conduct charge) coupled with an ability to withstand high voltages without breaking down.

A more practical arrangement is shown in Figure 1.10. In this arrangement the ratio of charge, Q, to potential difference, V, is given by the relationship:

$$\frac{Q}{V} = \frac{\varepsilon A}{d}$$

where A is the surface area of the plates (in m), d is the separation (in m), and ε is a constant for the dielectric material known as the *absolute permittivity* of the material (sometimes also referred to as the *dielectric constant*).

The absolute permittivity of a dielectric material is the product of the permittivity of free space (ε_0) and the *relative permittivity* (ε_r) of the material. Thus:

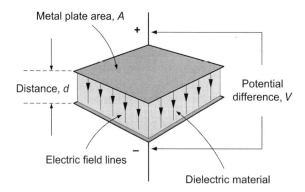

Figure 1.10: Parallel plates with an insulating dielectric material

$$\varepsilon = \varepsilon_0 \times \varepsilon \quad \text{and} \quad \frac{Q}{V} = \frac{\varepsilon_0 \varepsilon_r A}{d}$$

The *dielectric strength* of an insulating dielectric is the maximum electric field strength that can safely be applied to it before breakdown (conduction) occurs. Table 1.6 shows values of relative permittivity and dielectric strength for some common dielectric materials.

Table 1.6: Properties of some common insulating dielectric materials

Dielectric material	Relative permittivity (free space = 1)	Dielectric strength (kV/mm)
Vacuum, or free space	1	∞
Air	1	3
Polythene	2.3	50
Paper	2.5 to 3.5	14
Polystyrene	2.5	25
Mica	4 to 7	160
Pyrex glass	4.5	13
Glass ceramic	5.9	40
Polyester	3.0 to 3.4	18
Porcelain	6.5	4
Titanium dioxide	100	6
Ceramics	5 to 1,000	2 to 10

1.1.18   Electromagnetism

When a current flows through a conductor, a magnetic field is produced in the vicinity of the conductor. The magnetic field is invisible, but its presence can be detected using a compass needle (which will deflect from its normal North-South position). If two current-carrying conductors are placed in the vicinity of one another, the fields will interact with one another and the conductors will experience a force of attraction or repulsion (depending upon the relative direction of the two currents).

1.1.19   Force Between Two Current-Carrying Conductors

The mutual force that exists between two parallel current-carrying conductors will be proportional to the product of the currents in the two conductors and the length of the conductors but inversely proportional to their separation. Thus:

$$F = \frac{k I_1 I_2 l}{d}$$

where I_1 and I_2 are the currents in the two conductors (in amps), l is the parallel length of the conductors (in meters), d is the distance separating the two conductors (in meters), F is the force (in newtons), and k is a constant depending upon the medium in which the charges exist.

In vacuum or "free space",

$$k = \frac{\mu_0}{2\pi}$$

where μ_0 is a constant known as the *permeability of free space* ($4\pi \times 10^{-7}$ or 12.57×10^{-7} H/m).

Combining the two previous equations gives:

$$F = \frac{\mu_0 I_1 I_2 l}{2\pi d}$$

or,

$$F = \frac{4\pi \times 10^{-7} I_1 I_2 l}{2\pi d}$$

or,

$$F = \frac{2 \times 10^{-7} I_1 I_2 l}{d} \text{ Newtons}$$

1.1.20 Magnetic Fields

The field surrounding a straight current-carrying conductor is shown in Figure 1.11. The magnetic field defines the direction of motion of a free North Pole within the field. In the case of Figure 1.11, the lines of flux are concentric and the direction of the field determined by the direction of current flow) is given by the right-hand rule.

1.1.21 Magnetic Field Strength

The strength of a magnetic field is a measure of the density of the flux at any particular point. In the case of Figure 1.11, the field strength will be proportional to the applied current and inversely proportional to the perpendicular distance from the conductor. Thus:

$$B = \frac{kI}{d}$$

where B is the magnetic flux density (in tesla), I is the current (in amperes), d is the distance from the conductor (in meters), and k is a constant.

Assuming that the medium is vacuum or "free space," the density of the magnetic flux will be given by:

$$B = \frac{\mu_0 I}{2\pi d}$$

where B is the *flux density* (in tesla), μ_0 is the permeability of free space ($4\pi \times 10^{-7}$ or 12.57×10^{-7}), I is the current (in amperes), and d is the distance from the center of the conductor (in meters).

The flux density is also equal to the total flux divided by the area of the field. Thus:

$$B = \Phi/A$$

where Φ is the flux (in webers) and A is the area of the field (in square meters).

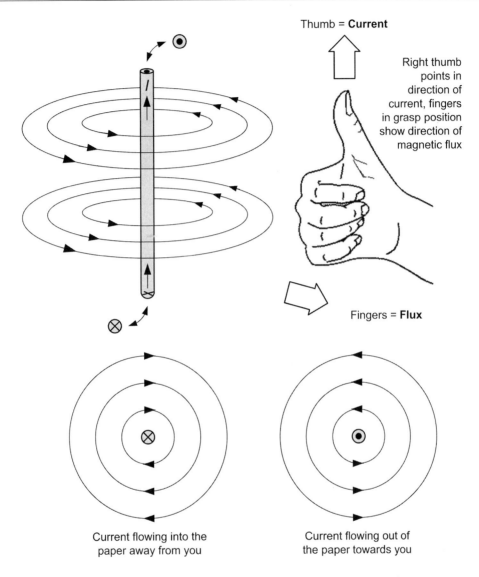

Thumb = **Current**

Right thumb points in direction of current, fingers in grasp position show direction of magnetic flux

Fingers = **Flux**

Current flowing into the paper away from you

Current flowing out of the paper towards you

Figure 1.11: Magnetic field surrounding a straight conductor

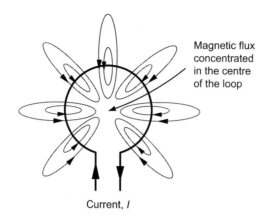

Current, *I*

Figure 1.12: Forming a conductor into a loop increases the strength of the magnetic field in the center of the loop

In order to increase the strength of the field, a conductor may be shaped into a loop (Figure 1.12) or coiled to form a solenoid (Figure 1.13). Note, in the latter case, how the field pattern is exactly the same as that which surrounds a bar magnet.

Example 1.23

Determine the flux density produced at a distance of 50 mm from a straight wire carrying a current of 20A.

Solution

Applying the formula $B = \mu_0 I/2\pi\, d$ gives:

$$B = \frac{12.57 \times 10^{-7} \times 20}{2 \times 3.142 \times 50 \times 10^{-3}} = \frac{251.4 \times 10^{-7}}{314.2 \times 10^{-3}}$$

from which:

$$B = 0.8 \times 10^{-4} \text{ tesla}$$

Thus, $B = 80 \times 10^{-6}$ T or $B = 80$ μT.

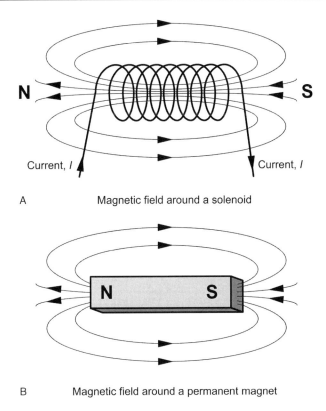

A   Magnetic field around a solenoid

B   Magnetic field around a permanent magnet

Figure 1.13: The magnetic field surrounding a solenoid coil resembles that of a permanent magnet

Example 1.24

A flux density of 2.5 mT is developed in free space over an area of 20 cm^2. Determine the total flux.

Solution

Rearranging the formula $B = \Phi/A$ to make Φ the subject gives $\Phi = B \times A$ thus:

$$\Phi = (2.5 \times 10^{-3}) \times (20 \times 10^{-4}) = 50 \times 10^{-7} \text{webers}$$

from which $B = 5\ \mu\text{Wb}$

1.22 Magnetic Circuits

Materials such as iron and steel possess considerably enhanced magnetic
properties. They are employed in applications where it is necessary to increase
the flux density produced by an electric current. In effect, magnetic materials
allow us to channel the electric flux into a "magnetic circuit," as shown in
Figure 1.14.

In the circuit of Figure 1.14(B), the *reluctance* of the magnetic core is analogous
to the resistance present in the electric circuit shown in Figure 1.14(A). We can
make the following comparisons between the two types of circuit (see
Table 1.7).

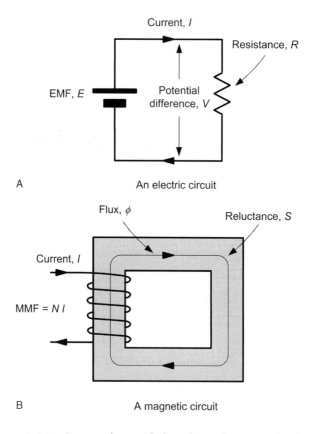

Figure 1.14: Comparison of electric and magnetic circuits

Table 1.7: Comparison of electric and magnetic circuits

Electric circuit Figure 1.14(A)	Magnetic circuit Figure 1.14(A)
Electromotive force, e.m.f. $= V$	Magnetomotive force, m.m.f. $= N \times I$
Resistance $= R$	Reluctance $= S$
Current $= I$	Flux $= \Phi$
e.m.f. $=$ current $\times$ resistance	m.m.f. $=$ flux $\times$ reluctance
$V = I \times R$	$N I = S \, \Phi$

In practice, not all of the magnetic flux produced in a magnetic circuit will be concentrated within the core and some "leakage flux" will appear in the surrounding free space (as shown in Figure 1.15). Similarly, if a gap appears within the magnetic circuit, the flux will tend to spread out as shown in Figure 1.16. This effect is known as *fringing*.

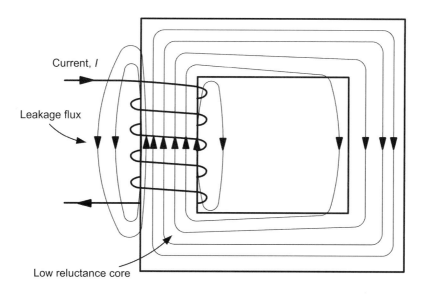

Figure 1.15: Leakage flux in a magnetic circuit

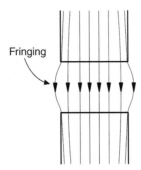

Fringing

Figure 1.16: Fringing of the magnetic flux at an air gap in a magnetic circuit

1.1.23 Reluctance and Permeability

The reluctance of a magnetic path is directly proportional to its length and inversely proportional to its area. The reluctance is also inversely proportional to the *absolute permeability* of the magnetic material. Thus:

$$S = \frac{l}{\mu A}$$

where S is the reluctance of the magnetic path, l is the length of the path (in meters), A is the cross-sectional area of the path (in square meters), and μ is the absolute permeability of the magnetic material.

The absolute permeability of a magnetic material is the product of the permeability of free space (μ_0) and the *relative permeability* of the magnetic medium (μ_0). Thus:

$$\mu = \mu_0 \times \mu \text{ and } S = \frac{l}{\mu_0 \mu_r A}$$

The permeability of a magnetic medium is a measure of its ability to support magnetic flux and it is equal to the ratio of flux density (B) to *magnetizing force* (H). Thus:

$$\mu = \frac{B}{H}$$

where B is the flux density (in tesla) and H is the magnetizing force (in ampere/meter). The magnetizing force (H) is proportional to the product of the number of turns and current but inversely proportional to the length of the magnetic path.

$$H = \frac{NI}{l}$$

where H is the magnetizing force (in amperes/ meters), N is the number of turns, I is the current (in amperes), and l is the length of the magnetic path (in meters).

1.1.24 B-H Curves

Figure 1.17 shows four typical *B-H* (flux density plotted against permeability) curves for some common magnetic materials. If you look carefully at these curves you will notice that they flatten off due to magnetic *saturation* and that the slope of the curve (indicating the value of μ corresponding to a particular value of H) falls

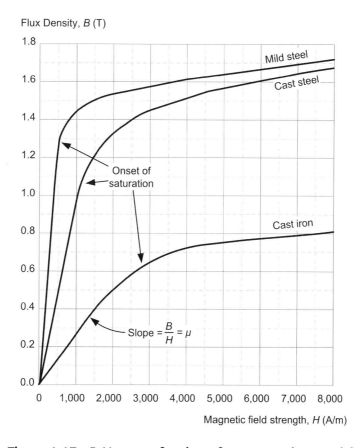

Figure 1.17: *B-H* **curves for three ferromagnetic materials**

as the magnetizing force increases. This is important since it dictates the acceptable working range for a particular magnetic material when used in a magnetic circuit.

Example 1.25

Estimate the relative permeability of cast steel (see Figure 1.18) at (a) a flux density of 0.6T, and (b) a flux density of 1.6T.

Solution

From Figure 1.18, the slope of the graph at any point gives the value of μ at that point. We can easily find the slope by constructing a tangent at the point in question and then finding the ratio of vertical change to horizontal change.

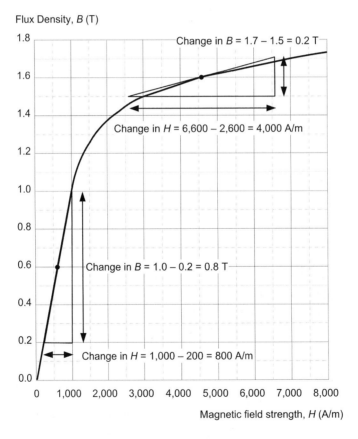

Figure 1.18: *B-H curve for a sample of cast steel*

(a)   The slope of the graph at 0.6T is $0.6/800 = 0.75 \times 10^{-3}$

Since $\mu = \mu_0 \times \mu_r$, $\mu_r = \mu/\mu_0 = 0.75 \times 10^{-3}/12.57 \times 10^{-7}$, thus $\mu_r = 597$ at 0.6T.

(b)   The slope of the graph at 1.6T is $0.2/4,000 = 0.05 \times 10^{-3}$

Since $\mu = \mu_0 \times \mu_r$, $\mu_r = \mu/\mu_0 = 0.05 \times 10^{-3} / 12.57 \times 10^{-7}$, thus $\mu_r = 39.8$ at 1.6T.

(This example clearly shows the effect of saturation on the permeability of a magnetic material!)

Example 1.26

A coil of 800 turns is wound on a closed mild steel core having a length 600 mm and cross-sectional area 500 mm^2. Determine the current required to establish a flux of 0.8 mWb in the core.

Solution

Now $B = \Phi/A = (0.8 \times 10^{-3})/(500 \times 10^{-6}) = 1.6$T

From Figure 1.17, a flux density of 1.6T will occur in mild steel when H = 3,500 A/m. The current can now be determined by re-arranging $H = N\,I/l$ as follows:

$$I = \frac{H \times l}{N} = \frac{3,500 \times 0.6}{800} = 2.625\text{A}$$

1.1.25   Circuit Diagrams

Finally, and just in case you haven't seen them before, we will end this section with a brief word about circuit diagrams. We are introducing the topic here because it's quite important to be able to read and understand simple electronic circuit diagrams before you can make sense of some of the components and circuits that you will meet later on.

Circuit diagrams use standard symbols and conventions to represent the components and wiring used in an electronic circuit. Visually, they bear very little relationship to the physical layout of a circuit but, instead, they provide us with a "theoretical" view of the circuit. In this section we show you how to find your way around simple circuit diagrams.

To be able to understand a circuit diagram you first need to be familiar with the symbols that are used to represent the components and devices. It's important to be aware that there are a few (thankfully quite small) differences between the symbols used in circuit diagrams of American and European origin.

As a general rule, the input to a circuit should be shown on the left of a circuit diagram and the output shown on the right. The supply (usually the most positive voltage) is normally shown at the top of the diagram and the common, 0V, or ground connection is normally shown at the bottom. This rule is not always obeyed, particularly for complex diagrams where many signals and supply voltages may be present.

Note also that, in order to simplify a circuit diagram (and avoid having too many lines connected to the same point) multiple connections to common, 0V, or ground may be shown using the appropriate symbol. The same applies to supply connections that may be repeated (appropriately labeled) at various points in the diagram.

A very simple circuit diagram (a simple resistance tester) is shown in Figure 1.20. This circuit may be a little daunting if you haven't met a circuit like it before but you can still glean a great deal of information from the diagram even if you don't know what the individual components do.

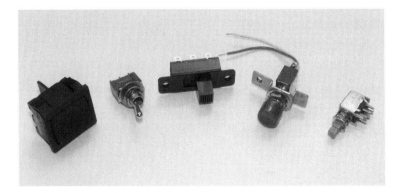

Figure 1.19: Various types of switches. From left to right: a mains rocker switch, an SPDT miniature toggle (changeover) switch, a DPDT side switch, an SPDT push-button (wired for use as an SPST push-button), a miniature PCB mounting DPDT push-button (with a latching action).

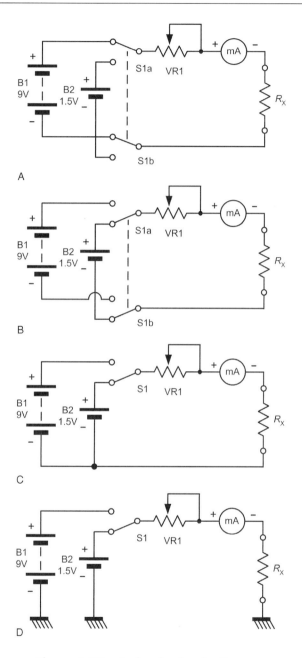

A

B

C

D

Figure 1.20: A simple circuit diagram

The circuit uses two batteries, B1 (a 9V multi-cell battery) and B2 (a 1.5V single-cell battery). The two batteries are selected by means of a double-pole, double-throw (DPDT) switch. This allows the circuit to operate from either the 9V battery (B1) as shown in Figure 1.20(A) or from the 1.5V battery (B2) as shown in Figure 1.20(B) depending on the setting of S1.

A variable resistor, VR1, is used to adjust the current supplied by whichever of the two batteries is currently selected. This current flows first through VR1, then through the milliammeter, and finally through the unknown resistor, R_X. Notice how the meter terminals are labeled showing their polarity (the current flows *into* the positive terminal and *out of* the negative terminal).

The circuit shown in Figure 1.20(C) uses a different type of switch but provides exactly the same function. In this circuit a single-pole, double-throw (SPDT) switch is used and the negative connections to the two batteries are "commoned" (i.e., connected directly together).

Finally, Figure 1.20(D) shows how the circuit can be redrawn using a common "chassis" connection to provide the negative connection between R_X and the two batteries. Electrically this circuit is identical to the one shown in Figure 1.20(C).

1.2   Passive Components

This section introduces several of the most common types of electronic component, including resistors, capacitors and inductors. These are often referred to as *passive components* as they cannot, by themselves, generate voltage or current. An understanding of the characteristics and application of passive components is an essential prerequisite to understanding the operation of the circuits used in amplifiers, oscillators, filters and power supplies.

1.2.1   Resistors

The notion of resistance as opposition to current was discussed in the previous section. Conventional forms of resistor obey a straight line law when voltage is plotted against current (see Figure 1.21) and this allows us to use resistors as a means of converting current into a corresponding voltage drop, and vice versa (note that doubling the applied current will produce double the voltage drop, and so on). Therefore, resistors provide us with a means of controlling the currents and voltages present in electronic

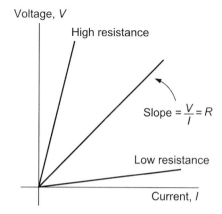

Figure 1.21: Voltage plotted against current for three different values of resistor

circuits. They can also act as *loads* to simulate the presence of a circuit during testing (e.g., a suitably rated resistor can be used to replace a loudspeaker when an audio amplifier is being tested).

The specifications for a resistor usually include the value of resistance expressed in ohms (Ω), kilohms (kΩ) or megohms (MΩ), the accuracy or tolerance (quoted as the maximum permissible percentage deviation from the marked value), and the power rating (which must be equal to, or greater than, the maximum expected power dissipation).

Other practical considerations when selecting resistors for use in a particular application include temperature coefficient, noise performance, stability and ambient temperature range. Table 1.8 summarizes the properties of five of the most common types of resistor. Figure 1.22 shows a typical selection of fixed resistors with values from 15Ω to 4.7 kΩ.

1.2.2 Preferred Values

The value marked on the body of a resistor is not its *exact* resistance. Some minor variation in resistance value is inevitable due to production tolerance. For example, a resistor marked 100Ω and produced within a tolerance of $\pm10\%$ will have a value which falls within the range 90Ω to 110Ω. A similar component with a tolerance of $\pm1\%$ would have a value that falls within the range 99Ω to 101Ω. Thus, where accuracy is important it is essential to use close tolerance components.

Table 1.8: Characteristics of common types of resistors

Property	Resistor type					
	Carbon film	Metal film	Metal oxide	Ceramic wirewound	Vitreous wirewound	Metal clad
Resistance range (Ω)	10 to 10M	1 to 1M	10 to 10 M	0.47 to 22k	0.1 to 22k	0.05 to 10k
Typical tolerance (%)	±5	±1	±2	±5	±5	±5
Power rating (W)	0.25 to 2	0.125 to 0.5	0.25 to 0.5	4 to 17	2 to 4	10 to 300
Temperature coefficient (ppm/°C)	−250	+50 to +100	+250	+250	+75	+50
Stability	Fair	Excellent	Excellent	Good	Good	Good
Noise performance	Fair	Excellent	Excellent	n.a.	n.a.	n.a.
Ambient temperature range (°C)	−45 to +125	−45 to +125	−45 to +125	−45 to +125	−45 to +125	−55 to +200
Typical applications	General-Purpose	Amplifiers, test equipment, etc., requiring low-noise high-tolerance components		Power supplies, loads, medium and high-power applications		Very high power applications

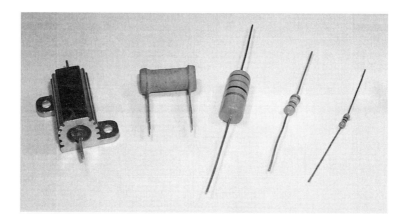

Figure 1.22: A selection of resistors including high-power metal clad, ceramic wirewound, carbon and metal film types with values ranging from 15Ω to 4.7 kΩ

Resistors are available in several series of fixed-decade values, the number of values provided with each series being governed by the tolerance involved. In order to cover the full range of resistance values using resistors having a ±20% tolerance it will be necessary to provide six basic values (known as the *E6 series*). More values will be required in the series, which offers a tolerance of ±10%, and consequently, the *E12 series* provides twelve basic values. The *E24 series* for resistors of ±5% tolerance provides no fewer than 24 basic values and, as with the E6 and E12 series, decade multiples (i.e., ×1, ×10, ×100, ×1 k, ×10 k, ×100k and ×1M) of the basic series. Figure 1.23 shows the relationship between the E6, E12 and E24 series.

1.2.3   Power Ratings

Resistor power ratings are related to operating temperatures and resistors should be derated at high temperatures. Where reliability is important resistors should be operated at well below their nominal maximum power dissipation.

Example 1.27

A resistor has a marked value of 220Ω. Determine the tolerance of the resistor if it has a measured value of 207Ω.

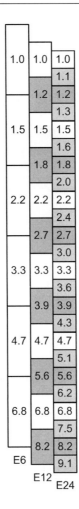

Figure 1.23: The E6, E12, and E24 series

Solution

The difference between the marked and measured values of resistance (the error) is $(220\Omega - 207\Omega) = 13\Omega$. The tolerance is given by:

$$\text{Tolerance} = \frac{\text{error}}{\text{marked value}} \times 100\%$$

The tolerance is thus, $(13/220) \times 100 = 5.9\%$.

Example 1.28

A 9V power supply is to be tested with a 39Ω load resistor. If the resistor has a tolerance of 10% find:

(a)   the nominal current taken from the supply;

(b)   the maximum and minimum values of supply current at either end of the tolerance range for the resistor.

Solution

(a)   If a resistor of *exactly* 39Ω is used the current will be:

$$I = V/R = 9V/39\Omega = 231 \text{ mA}$$

(b)   The lowest value of resistance would be $(39\Omega - 3.9\Omega) = 35.1\Omega$. In which case the current would be:

$$I = V/R = 9V/35.1\Omega = 256.4 \text{ mA}$$

At the other extreme, the highest value would be $(39\Omega + 3.9 \ \Omega) = 42.9\Omega$.

In this case, the current would be:

$$I = V/R = 9V/42.9\Omega = 209.8 \text{ mA}$$

The maximum and minimum values of supply current will thus be 256.4 mA and 209.8 mA, respectively.

Example 1.29

A current of 100 mA ($\pm 20\%$) is to be drawn from a 28V DC supply. What value and type of resistor should be used in this application?

Solution

The value of resistance required must first be calculated using Ohm's Law:

$$R = V/I = 28V/100 \text{ mA} = 280\Omega$$

The nearest preferred value from the E12 series is 270Ω (which will actually produce a current of 103.7 mA (i.e., within $\pm 4\% >$ of the desired value). If a resistor of $\pm 10\%$ tolerance is used, current will be within the range 94 mA to 115 mA (well within the $\pm 20\%$ accuracy specified).

The power dissipated in the resistor (calculated using $P = I \times V$) will be 2.9W and thus a component rated at 3W (or more) will be required. This would normally be a vitreous enamel coated wirewound resistor (see Table 1.8).

1.2.4   Resistor Markings

Carbon and metal oxide resistors are normally marked with color codes which indicate their value and tolerance. Two methods of color-coding are in common use; one involves four colored bands (see Figure 1.24), while the other uses five color bands (see Figure 1.25).

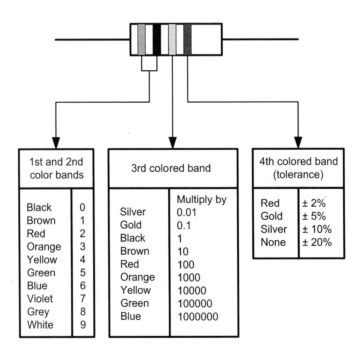

1st and 2nd color bands		3rd colored band		4th colored band (tolerance)	
			Multiply by	Red	± 2%
Black	0	Silver	0.01	Gold	± 5%
Brown	1	Gold	0.1	Silver	± 10%
Red	2	Black	1	None	± 20%
Orange	3	Brown	10		
Yellow	4	Red	100		
Green	5	Orange	1000		
Blue	6	Yellow	10000		
Violet	7	Green	100000		
Grey	8	Blue	1000000		
White	9				

Figure 1.24: Four-band resistor color code

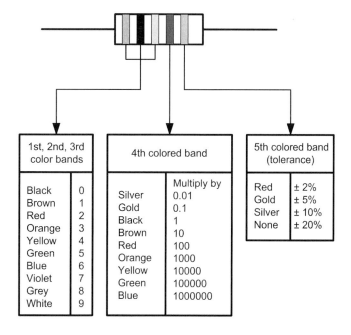

Figure 1.25: Five band resistor color code

Example 1.30

A resistor is marked with the following colored stripes: brown, black, red, silver. What is its value and tolerance?

Solution

See Figure 1.26.

Example 1.31

A resistor is marked with the following colored stripes: red, violet, orange, gold. What is its value and tolerance?

Solution

See Figure 1.27.

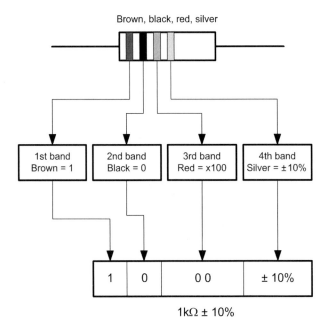

Figure 1.26: See Example 1.30

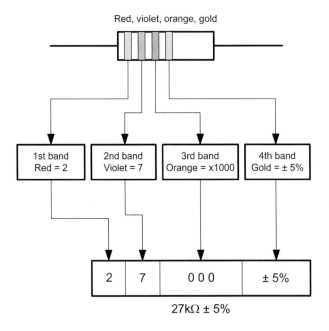

Figure 1.27: See Example 1.31

Example 1.32

A resistor is marked with the following colored stripes: green, blue, black, gold. What is its value and tolerance?

Solution

See Figure 1.28.

Example 1.33

A resistor is marked with the following colored stripes: red, green, black, black, brown. What is its value and tolerance?

Solution

See Figure 1.29.

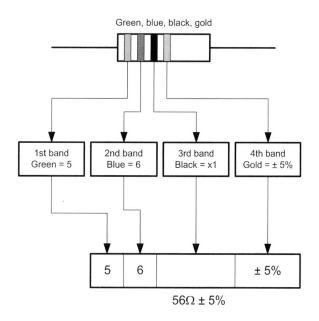

Figure 1.28: See Example 1.32

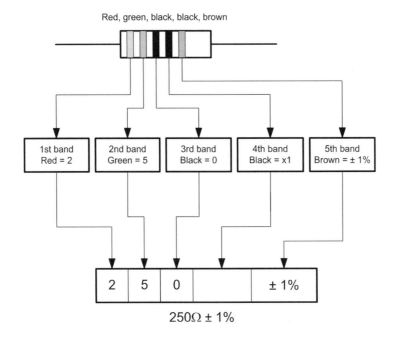

Figure 1.29: See Example 1.33

Example 1.34

A 2.2 kΩ of ±2% tolerance is required. What four-band color code does this correspond to?

Solution

Red (2), red (2), red (2 zeros), red (2% tolerance). Thus, all four bands should be red.

1.2.5   BS 1852 Coding

Some types of resistor have markings based on a system of coding defined in BS 1852. This system involves marking the position of the decimal point with a letter to indicate the multiplier concerned as shown in Table 1.9. A further letter is then appended to indicate the tolerance as shown in Table 1.10.

Table 1.9: BS 1852 resistor multiplier markings

Letter	Multiplier
R	1
K	1,000
M	1,000,000

Table 1.10: BS 1852 resistor tolerance markings

Letter	Multiplier
F	±1%
G	±2%
J	±5%
K	±10%
M	±20%

Example 1.35

A resistor is marked coded with the legend 4R7K. What is its value and tolerance?

Solution

$4.7\Omega \pm 10\%$

Example 1.36

A resistor is marked coded with the legend 330RG. What is its value and tolerance?

Solution

$330\Omega \pm 2\%$

Example 1.37

A resistor is marked coded with the legend R22M. What is its value and tolerance?

Solution

$0.22\Omega \pm 20\%$

1.2.6 Series and Parallel Combinations of Resistors

In order to obtain a particular value of resistance, fixed resistors may be arranged in either series or parallel as shown in Figures 1.30 and 1.31.

The effective resistance of each of the series circuits shown in Figure 1.30 is simply equal to the sum of the individual resistances. So, for the circuit shown in Figure 1.30(A):

$$R = R_1 + R_2$$

while for Figure 1.30(B):

$$R = R_1 + R_2 + R_3$$

Turning to the parallel resistors shown in Figure 1.31, the reciprocal of the effective resistance of each circuit is equal to the sum of the reciprocals of the individual resistances. Hence, for Figure 1.31(A):

$$\frac{1}{R} = \frac{1}{R_1} + \frac{1}{R_2}$$

while for Figure 1.32(B):

$$\frac{1}{R} = \frac{1}{R_1} + \frac{1}{R_2} + \frac{1}{R_3}$$

In the former case, the formula can be more conveniently rearranged as follows:

$$R = \frac{R_1 \times R_2}{R_1 + R_2}$$

You can remember this as the *product* of the two resistance values *divided by* the *sum* of the two resistance values.

Example 1.38

Resistors of 22Ω, 47Ω, and 33Ω are connected (a) in series and (b) in parallel. Determine the effective resistance in each case.

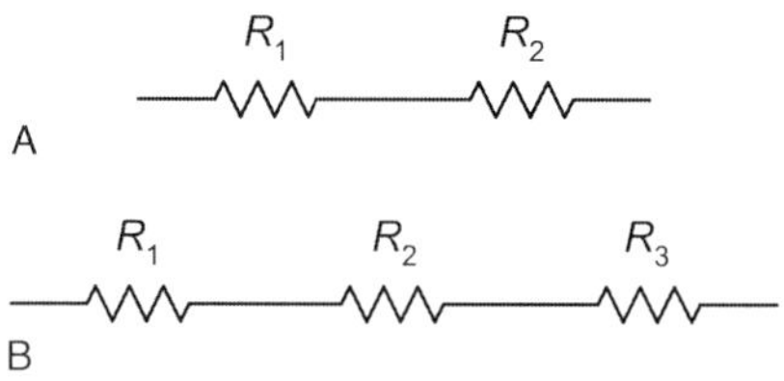

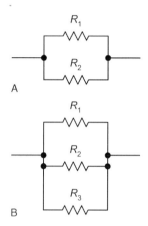

Figure 1.30: Resistors in series

Figure 1.31: Resistors in parallel

Solution

(a)   In the series circuit $R = R_1 + R_2 + R_3$, thus $R = 22\Omega + 47\Omega + 33\Omega = 102\Omega$

(b)   In the parallel circuit:

$$\frac{1}{R} = \frac{1}{R_1} + \frac{1}{R_2} + \frac{1}{R_3}$$

Thus,

$$\frac{1}{R} = \frac{1}{22\Omega} + \frac{1}{47\Omega} + \frac{1}{33\Omega}$$

Or,

$$\frac{1}{R} = 0.045 + 0.021 + 0.03$$

from which,

$$\frac{1}{R} = 0.096 = 10.42\Omega$$

Example 1.39

Determine the effective resistance of the circuit shown in Figure 1.32.

Solution

The circuit can be progressively simplified as shown in Figure 1.33. The stages in this simplification are:

(a) R_3 and R_4 are in series and they can replaced by a single resistance (R_A) of $(12\Omega + 27\Omega) = 39\Omega$.

(b) R_A appears in parallel with R_2. These two resistors can be replaced by a single resistance (R_B) of $(39\Omega + 47\Omega)/(39\Omega + 47\Omega) = 21.3\Omega$.

(c) R_B appears in series with R_1. These two resistors can be replaced by a single resistance (R) of $(21.3\Omega + 4.7\Omega) = 26\Omega$.

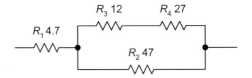

Figure 1.32: See Example 1.39

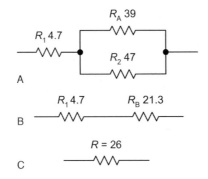

Figure 1.33: See Example 1.39

Example 1.40

A resistance of 50Ω rated at 2W is required. What parallel combination of preferred value resistors will satisfy this requirement? What power rating should each resistor have?

Solution

Two 100Ω resistors may be wired in parallel to provide a resistance of 50Ω as shown below:

$$R = \frac{R_1 \times R_2}{R_1 + R_2} = \frac{100 \times 100}{100 + 100} = \frac{10,000}{200} = 50\Omega$$

Note, from this, that when two resistors of the same value are connected in parallel the resulting resistance will be half that of a single resistor.

Having shown that two 100Ω resistors connected in parallel will provide us with a resistance of 50Ω we now need to consider the power rating. Since the resistors are identical, the applied power will be shared equally between them. Hence, each resistor should have a power rating of 1W.

1.2.7   Resistance and Temperature

Figure 1.34 shows how the resistance of a metal conductor (e.g., copper) varies with temperature. Since the resistance of the material increases with temperature, this characteristic is said to exhibit a *positive temperature coefficient* (*PTC*). Not all materials have a PTC characteristic. The resistance of a carbon conductor falls with temperature and it is therefore said to exhibit a *negative temperature coefficient* (*NTC*).

The resistance of a conductor at a temperature, *t*, is given by the equation:

$$R_\mathrm{t} = R_0(1 + \alpha t + \beta t^2 + \gamma t^3 \dots)$$

where α, β, γ, etc. are constants and R_0 is the resistance at 0°C.

The coefficients, β, γ, etc. are quite small and since we are normally only dealing with a relatively restricted temperature range (e.g., 0°C to 100°C), we can usually approximate the characteristic shown in Figure 1.34 to the straight line law shown in Figure 1.35. In this case, the equation simplifies to:

$$R_t = R_0(1 + \alpha t)$$

where α is known as the *temperature coefficient* of resistance. Table 1.11 shows some typical values for α (note that α is expressed in $\Omega/\Omega/°C$ or just $/°C$).

Example 1.41

A resistor has a temperature coefficient of $0.001/°C$. If the resistor has a resistance of $1.5\ k\Omega$ at $0°C$, determine its resistance at $80°C$.

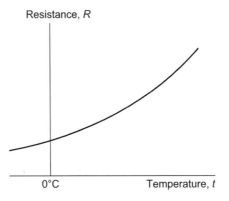

Figure 1.34: Variation of resistance with temperature for a metal conductor

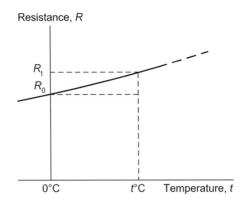

Figure 1.35: Straight line approximation of Figure 1.34

Solution

Now:

$$R_t = R_0(1 + \alpha t)$$

thus,

$$R_t = 1.5 \text{ k}\Omega \times (1 + (0.001 \times 80))$$

Hence,

$$R_t = 1.5 \times 1.08 = 1.62 \text{ k}\Omega$$

Example 1.42

A resistor has a temperature coefficient of 0.0005/°C. If the resistor has a resistance of 680Ω at 20°C, what will its resistance be at 80°C?

Solution

First we must find the resistance at 0°C. Rearranging the formula for R_t gives:

$$R_0 = \frac{R_t}{1 + \alpha t} = \frac{680}{1 + (0.0005 \times 20)} = \frac{680}{1 + 0.01}$$

Hence,

$$R_0 = \frac{680}{1 + 0.01} = 673.3\Omega$$

Now,

$$R_t = R_0(1 + \alpha t)$$

thus,

$$R_{90} = 673.3 \times (1 + (0.0005 \times 90))$$

Hence,

$$R_{90} = 673.3 \times 1.045 = 704\Omega$$

Example 1.43

A resistor has a resistance of 40Ω at 0°C and 44Ω at 100°C. Determine the resistor's temperature coefficient.

Solution

First we need to make α the subject of the formula:

$$R_t = R_0(1 + \alpha t)$$

Now,

$$\alpha = \frac{1}{t}\left(\frac{R_t}{R_o} - 1\right) = \frac{1}{100}\left(\frac{44}{40} - 1\right)$$

from which,

$$\alpha = \frac{1}{100}(1.1 - 1) = \frac{1}{100} \times 0.1 = 0.001/°C$$

Table 1.11: Temperature coefficient of resistance

Material	Temperature coefficient of resistance, α (/°C)
Platinum	+0.0034
Silver	+0.0038
Copper	+0.0043
Iron	+0.0065
Carbon	−0.0005

1.2.8 Thermistors

With conventional resistors we would normally require resistance to remain the same over a wide range of temperatures (i.e., α should be zero). On the other hand, there are applications in which we could use the effect of varying resistance to detect a temperature change. Components that allow us to do this are known as *thermistors*.

The resistance of a thermistor changes markedly with temperature and these components are widely used in temperature sensing and temperature compensating applications. Two basic types of thermistor are available, NTC and PTC (see Figure 1.36).

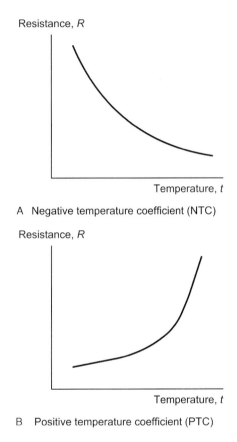

Figure 1.36: Characteristics of (A) NTC and (B) PTC thermistors

Typical NTC thermistors have resistances that vary from a few hundred (or thousand) ohms at 25°C to a few tens (or hundreds) of ohms at 100°C. PTC thermistors, on the other hand, usually have a resistance-temperature characteristic that remains substantially flat (typically at around 100Ω) over the range 0°C to around 75°C.

Above this, and at a critical temperature (usually in the range 80°C to 120°C) their resistance rises very rapidly to values of up to, and beyond, 10 kΩ (see Figure 1.36).

A typical application of PTC thermistors is over-current protection. Provided the current passing through the thermistor remains below the threshold current, the effects of self-heating will remain negligible and the resistance of the thermistor will remain low (i.e., approximately the same as the resistance quoted at 25°C). Under fault conditions, the current exceeds the threshold value by a considerable margin and the thermistor starts to self-heat. The resistance then increases rapidly and, as a consequence, the current falls to the rest value. Typical values of threshold and rest currents are 200 mA and 8 mA, respectively, for a device which exhibits a nominal resistance of 25Ω at 25°C.

1.2.9   Light-Dependent Resistors

Light-dependent resistors (LDR) use a semiconductor material (i.e., a material that is neither a conductor nor an insulator) whose electrical characteristics vary according to the amount of incident light. The two semiconductor materials used for the manufacture of LDRs are cadmium sulphide (CdS) and cadmium selenide (CdSe). These materials are most sensitive to light in the visible spectrum, peaking at about 0.6 μm for CdS and 0.75 μm for CdSe. A typical CdS LDR exhibits a resistance of around 1 MΩ in complete darkness and less than 1 kΩ when placed under a bright light source (see Figure 1.37).

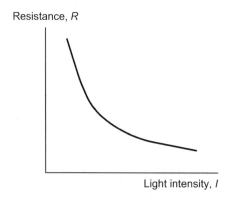

Figure 1.37: Characteristic of a light-dependent resistor (LDR)

1.2.10 Voltage Dependent Resistors

The resistance of a voltage dependent resistor (VDR) falls very rapidly when the voltage across it exceeds a nominal value in either direction (see Figure 1.38). In normal operation, the current flowing in a VDR is negligible; however, when the resistance falls, the current will become appreciable and a significant amount of energy will be absorbed. VDRs are used as a means of "clamping" the voltage in a circuit to a predetermined level. When connected across the supply rails to a circuit (either AC or DC) they are able to offer a measure of protection against voltage surges.

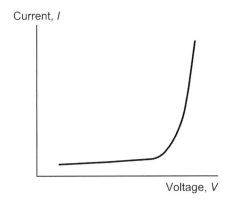

Figure 1.38: Characteristic of a voltage dependent resistor (VDR)

1.2.11 Variable Resistors

Variable resistors are available in several forms including those which use carbon tracks and those which use a wirewound resistance element. In either case, a moving slider makes contact with the resistance element. Most variable resistors have three (rather than two) terminals and as such are more correctly known as *potentiometers*. Carbon potentiometers are available with linear or semi-logarithmic law tracks (see Figure 1.39) and in rotary or slider formats. Ganged controls, in which several potentiometers are linked together by a common control shaft, are also available. Figure 1.40 shows a selection of variable resistors.

You will also encounter various forms of preset resistors that are used to make occasional adjustments (e.g., for calibration). Various forms of preset resistor are commonly used including open carbon track skeleton presets and fully encapsulated carbon and multiturn cermet types, as shown in Figure 1.41.

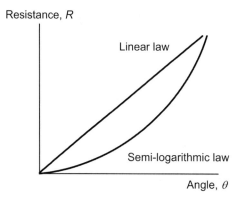

Figure 1.39: Characteristics for linear and semi-logarithmic law variable resistors

Figure 1.40: A selection of common types of carbon and wirewound variable
resistors/potentiometers

Figure 1.41: A selection of common types of standard and miniature preset
resistors/potentiometers

1.2.12 Capacitors

A capacitor is a device for storing electric charge. In effect, it is a reservoir into which charge can be deposited and then later extracted. Typical applications include reservoir and smoothing capacitors for use in power supplies, coupling AC signals between the stages of amplifiers, and decoupling supply rails (i.e., effectively grounding the supply rails as far as AC signals are concerned).

A capacitor can consist of nothing more than two parallel metal plates as shown in Figure 1.10. To understand what happens when a capacitor is being charged and discharged take a look at Figure 1.42. If the switch is left open (position A), no charge will appear on the plates and in this condition there will be no electric field in the space between the plates nor will there be any charge stored in the capacitor.

When the switch is moved to position B, electrons will be attracted from the positive plate to the positive terminal of the battery. At the same time, a similar number of electrons will move from the negative terminal of the battery to the negative plate. This sudden movement of electrons will manifest itself in a momentary surge of current (conventional current will flow from the positive terminal of the battery toward the positive terminal of the capacitor).

Eventually, enough electrons will have moved to make the e.m.f. between the plates the same as that of the battery. In this state, the capacitor is said to be *fully charged* and an electric field will be present in the space between the two plates.

If at some later time the switch is moved back to position A, the positive plate will be left with a deficiency of electrons while the negative plate will be left with a surplus of electrons. Furthermore, since there is no path for current to flow between the two plates the capacitor will remain charged and a potential difference will be maintained between the plates.

Now assume that the switch is moved to position C. The excess electrons on the negative plate will flow through the resistor to the positive plate until a neutral state once again exists (i.e., until there is no excess charge on either plate). In this state the capacitor is said to be *fully discharged* and the electric field between the plates will rapidly collapse. The movement of electrons during the discharging of the capacitor will again result in a momentary surge of current (current will flow from the positive terminal of the capacitor and into the resistor).

Figure 1.43 shows the direction of current flow in the circuit of Figure 1.42 during charging (switch in position B) and discharging (switch in position C). It should be

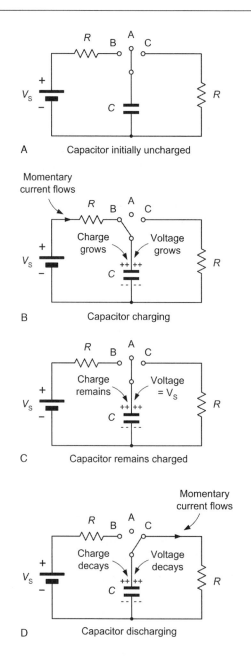

Figure 1.42: Capacitor charging and discharging

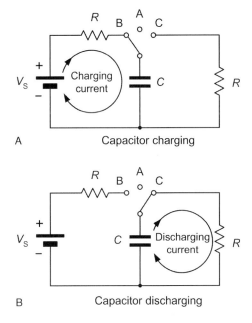

Figure 1.43: Current flow during charging and discharging

noted that current flows momentarily in both circuits even though you may think that the circuit is broken by the gap between the capacitor plates!

1.2.13 Capacitance

The unit of capacitance is the farad (F). A capacitor is said to have a capacitance of 1F if a current of 1A flows in it when a voltage changing at the rate of 1 V/s is applied to it. The current flowing in a capacitor will thus be proportional to the product of the capacitance, C, and the rate of change of applied voltage. Hence:

$$i = C \times (\text{rate of change of voltage})$$

Note that we've used a small i to represent the current flowing in the capacitor. We've done this because the current is changing and doesn't remain constant.

The rate of change of voltage is often represented by the expression *dv/dt* where *dv* represents a very small change in voltage and *dt* represents the corresponding small change in time. Expressing this mathematically gives:

$$i = C\frac{dV}{dt}$$

Example 1.44

A voltage is changing at a uniform rate from 10V to 50V in a period of 0.1s. If this voltage is applied to a capacitor of 22 µF, determine the current that will flow.

Solution

Now the current flowing will be given by:

$$i = C \times (\text{rate of change of voltage})$$

Thus,

$$i = C\left(\frac{\text{change in voltage}}{\text{change in time}}\right) = 22 \times 10^{-6} \times \left(\frac{50 - 10}{0.1}\right)$$

From which,

$$i = 22 \times 10^{-6} \times \left(\frac{40}{0.1}\right) = 22 \times 10^{-6} \times 400$$

so,

$$i = 8.8 \times 10^{-3} = 8.8 \text{ mA}$$

1.2.14   Charge, Capacitance and Voltage

The charge or quantity of electricity that can be stored in the electric field between the capacitor plates is proportional to the applied voltage and the capacitance of the capacitor. Thus:

$$Q = CV$$

where Q is the charge (in coulombs), C is the capacitance (in farads), and V is the potential difference (in volts).

Example 1.45

A 10 μF capacitor is charged to a potential of 250V. Determine the charge stored.

Solution

The charge stored will be given by:

$$Q = CV = 10 \times 10^{-6} \times 250 = 2.5 \text{ mC}$$

1.2.15 Energy storage

The energy stored in a capacitor is proportional to the product of the capacitance and the square of the potential difference. Thus:

$$W = \tfrac{1}{2}CV^2$$

where W is the energy (in joules), C is the capacitance (in farads), and V is the potential difference (in volts).

Example 1.46

A capacitor of 47 μF is required to store 4J of energy. Determine the potential difference that must be applied to the capacitor.

Solution

The foregoing formula can be rearranged to make V the subject as follows:

$$V = \sqrt{\frac{E}{0.5C}} = \sqrt{\frac{2E}{C}} = \sqrt{\frac{2 \times 4}{47 \times 10^{-6}}}$$

from which,

$$V = \sqrt{\frac{8}{47 \times 10^{-6}}} = \sqrt{0.170 \times 10^6} = 0.412 \times 10^3 = 412\text{V}$$

1.2.16 Capacitance and Physical Dimensions

The capacitance of a capacitor depends upon the physical dimensions of the capacitor (i.e., the size of the plates and the separation between them) and the dielectric material between the plates. The capacitance of a conventional parallel plate capacitor is given by:

$$C = \frac{\varepsilon_0 \, \varepsilon_r \, A}{d}$$

where C is the capacitance (in farads), ε_0 is the permittivity of free space, ε_r is the *relative permittivity* of the dielectric medium between the plates), and d is the separation between the plates (in meters).

Example 1.47

A capacitor of 1 nF is required. If a dielectric material of thickness 0.1 mm and relative permittivity 5.4 is available, determine the required plate area.

Solution

Rearranging the formula:

$$C = \frac{\varepsilon_0 \, \varepsilon_r \, A}{d}$$

to make A the subject gives:

$$A = \frac{C \, d}{\varepsilon_0 \, \varepsilon_r} = \frac{1 \times 10^{-9} \times 0.1 \times 10^{-3}}{8.854 \times 10^{-12} \times 5.4}$$

from which:

$$A = \frac{0.1 \times 10^{-12}}{47.8116 \times 10^{-12}}$$

thus,

$$A = 0.00209 \text{ m}^2 \quad \text{or} \quad 20.9 \text{ cm}^2$$

In order to increase the capacitance of a capacitor, many practical components employ multiple plates (see Figure 1.44). The capacitance is then given by:

Figure 1.44: A multi-plate capacitor

$$C = \frac{\varepsilon_0 \, \varepsilon_r (n - 1) \, A}{d}$$

where C is the capacitance (in farads), ε_0 is the permittivity of free space, ε_r is the *relative permittivity* of the dielectric medium between the plates), and d is the separation between the plates (in meters) and n is the total number of plates.

Example 1.48

A capacitor consists of six plates each of area 20 cm^2 separated by a dielectric of relative permittivity 4.5 and thickness 0.2 mm. Determine the value of capacitance.

Solution

Using:

$$C = \frac{\varepsilon_0 \, \varepsilon_r (n - 1) \, A}{d}$$

gives:

$$C = \frac{8.854 \times 10^{-12} \times 4.5 \times (6 - 1) \times 20 \times 10^{-4}}{0.2 \times 10^{-3}}$$

from which,

$$C = \frac{3,984.3 \times 10^{-16}}{0.2 \times 10^{-3}} = 19.921 \times 10^{-13} = 190 \times 10^{-12}$$

Thus,

$$C = 190 \times 10^{-12} \text{F} \quad \text{or} \quad 1.992 \text{ nF}$$

1.2.17   Capacitor Specifications

The specifications for a capacitor usually include the value of capacitance (expressed in microfarads, nanofarads or picofarads), the voltage rating (i.e., the maximum voltage which can be continuously applied to the capacitor under a given set of conditions), and the accuracy or tolerance (quoted as the maximum permissible percentage deviation from the marked value).

Other practical considerations when selecting capacitors for use in a particular application include temperature coefficient, leakage current, stability and ambient temperature range.

Table 1.12 summarizes the properties of five of the most common types of capacitor. Note that electrolytic capacitors require the application of a polarizing voltage in order to the chemical action on which they depend for their operation.

Table 1.12: Characteristics of common types of capacitor

Property	Capacitor type				
	Ceramic	Electrolytic	Polyester	Mica	Polystyrene
Capacitance range (F)	2.2p to 100n	100n to 10m	10n to 2.2μ	0.47 to 22k	10p to 22n
Typical tolerance (%)	±10 and ±20	−10 to +50	±10	±1	±5
Typical voltage rating (W)	50V to 200V	6.3V to 400V	100V to 400V	350V	100V
Temperature coefficient (ppm/°C)	+100 to −4700	+1000 typical	+100 to +200	+50	+250
Stability	Fair	Poor	Good	Excellent	Good
Ambient temperature range (°C)	−85 to +85	40 to +80	−40 to +100	−40 to +125	−40 to +100
Typical applications	High-frequency and low-cost	Smoothing and decoupling	General-Purpose	Tuned circuits and oscillators	General-Purpose

The polarizing voltages used for electrolytic capacitors can range from as little as 1V to several hundred volts depending upon the working voltage rating for the component in question.

Figure 1.45 shows some typical nonelectrolytic capacitors (including polyester, polystyrene, ceramic and mica types), while Figure 1.46 shows a selection of electrolytic (polarized) capacitors. An air-spaced variable capacitor is shown later in Figure 1.54.

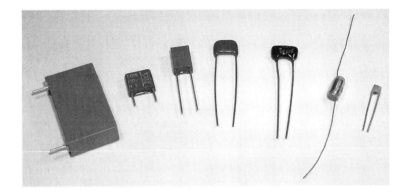

Figure 1.45: A typical selection of nonelectrolytic capacitors (including polyester, polystyrene, ceramic and mica types) with values ranging from 10 pF to 470 nF and working voltages from 50V to 250V

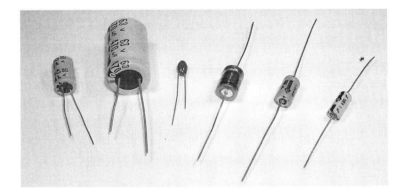

Figure 1.46: A typical selection of electrolytic (polarized) capacitors with values ranging from 1 μF to 470 μF and working voltages from 10V to 63V

1.2.18 *Capacitor Markings*

The vast majority of capacitors employ written markings which indicate their values, working voltages, and tolerance. The most usual method of marking resin dipped polyester (and other) types of capacitor involves quoting the value (μF, nF or pF), the tolerance (often either 10% or 20%), and the working voltage (often using _ and ∼ to indicate DC and AC, respectively). Several manufacturers use two separate lines for their capacitor markings and these have the following meanings:

First line: capacitance (pF or μF) and tolerance (K = 10%, M = 20%)

Second line: rated DC voltage and code for the dielectric material

A three-digit code is commonly used to mark monolithic ceramic capacitors. The first two digits of this code correspond to the first two digits of the value, while the third digit is a multiplier which gives the number of zeros to be added to give the value in picofarads. Other capacitors may use a color code similar to that used for marking resistor values (see Figure 1.48).

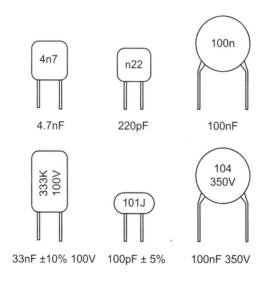

Figure 1.47: Examples of capacitor markings

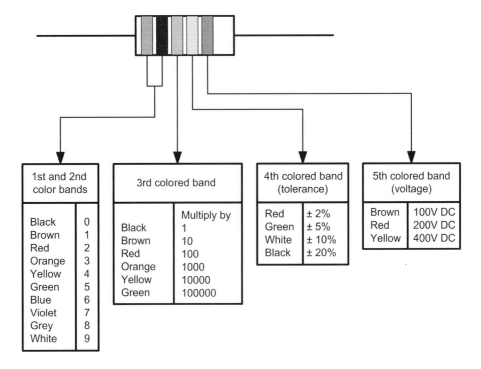

Figure 1.48: Capacitor color code

Example 1.49

A monolithic ceramic capacitor is marked with the legend "103K". What is its value?

Solution

The value (pF) will be given by the first two digits (10) followed by the number of zeros indicated by the third digit (3). The value of the capacitor is thus 10,000 pF or 10 nF. The final letter (K) indicates that the capacitor has a tolerance of 10%.

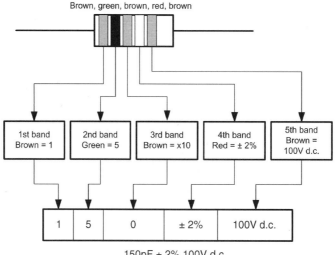

Brown, green, brown, red, brown

| 1st band
Brown = 1 | 2nd band
Green = 5 | 3rd band
Brown = x10 | 4th band
Red = ± 2% | 5th band
Brown =
100V d.c. |

| 1 | 5 | 0 | ± 2% | 100V d.c. |

150pF ± 2% 100V d.c.

Figure 1.49: See Example 1.50

Example 1.50

A tubular capacitor is marked with the following colored stripes: brown, green, brown, red, brown. What is its value, tolerance, and working voltage?

Solution

See Figure 1.49.

1.2.19 Series and Parallel Combination of Capacitors

In order to obtain a particular value of capacitance, fixed capacitors may be arranged in either series or parallel (Figures 1.50 and 1.51). The reciprocal of the effective capacitance of each of the series circuits shown in Figure 1.50 is equal to the sum of the reciprocals of the individual capacitances. Hence, for Figure 1.50(A):

$$\frac{1}{C} = \frac{1}{C_1} + \frac{1}{C_2}$$

while for Figure 1.50(B):

$$\frac{1}{C} = \frac{1}{C_1} + \frac{1}{C_2} + \frac{1}{C_3}$$

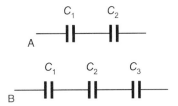

Figure 1.50: Capacitors in series

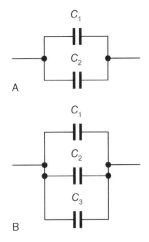

Figure 1.51: Capacitors in parallel

In the former case, the formula can be more conveniently rearranged as follows:

$$C = \frac{C_1 \times C_2}{C_1 + C_2}$$

You can remember this as the *product* of the two capacitor values *divided by* the *sum* of the two values—just as you did for two resistors in parallel.

For a parallel arrangement of capacitors, the effective capacitance of the circuit is simply equal to the sum of the individual capacitances. Hence, for Figure 1.51(A):

$$C = C_1 + C_2$$

while for Figure 1.51(B):

$$C = C_1 + C_2 + C_3$$

Example 1.51

Determine the effective capacitance of the circuit shown in Figure 1.52.

Solution

The circuit of Figure 1.52 can be progressively simplified as shown in Figure 1.53. The stages in this simplification are:

(a) C_1 and C_2 are in parallel and they can be replaced by a single capacitor (C_A) of (2 nF + 4 nF) = 6 nF.

(b) C_A appears in series with C_3. These two resistors can be replaced by a single capacitor (C_B) of (6 nF × 2 nF)/(6 nF + 2 nF) = 1.5 nF.

(c) C_B appears in parallel with C_4. These two capacitors can be replaced by a single capacitance (C) of (1.5 nF + 4 nF) = 5.5 nF.

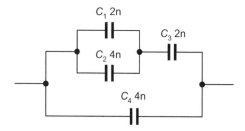

Figure 1.52: See Example 1.51

Example 1.52

A capacitance of 50 µF (rated at 100V) is required. What series combination of preferred value capacitors will satisfy this requirement? What voltage rating should each capacitor have?

Solution

Two 100 μF capacitors wired in series will provide a capacitance of 50 μF, as follows:

$$C = \frac{C_1 \times C_2}{C_1 + C_2} = \frac{100 \times 100}{100 + 100} = \frac{10,000}{200} = 50 \ \mu F$$

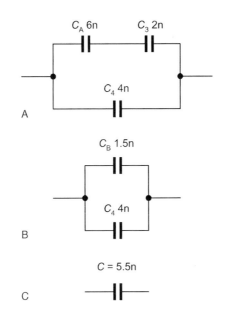

Figure 1.53: See Example 1.51

Since the capacitors are of equal value, the applied DC potential will be shared equally between them. Thus each capacitor should be rated at 50V. Note that, in a practical circuit, we could take steps to ensure that the DC voltage was shared equally between the two capacitors by wiring equal, high-value (e.g., 100 kΩ) resistors across each capacitor.

1.2.20 Variable Capacitors

By moving one set of plates relative to the other, a capacitor can be made variable. The dielectric material used in a variable capacitor can be either air (see Figure 1.54) or plastic (the latter tend to be more compact). Typical values for variable capacitors tend to range from about 25 pF to 500 pF. These components are commonly used for tuning radio receivers.

Figure 1.54: An air-spaced variable capacitor. This component (used for tuning an AM radio) has two separate variable capacitors (each of 500 pF maximum) operated from a common control shaft.

1.2.21   Inductors

Inductors provide us with a means of storing electrical energy in the form of a magnetic field. Typical applications include chokes, filters and (in conjunction with one or more capacitors) frequency selective circuits. The electrical characteristics of an inductor are determined by a number of factors including the material of the core (if any), the number of turns, and the physical dimensions of the coil. Figure 1.55 shows the construction of a typical toroidal inductor wound on a ferrite (high permeability) core.

In practice every coil comprises both inductance (L) and a small resistance (R). The circuit of Figure 1.56 shows these as two discrete components. In reality the inductance and the resistance (we often refer to this as a *loss resistance* because it's something that we don't actually want) are both distributed throughout the component but it is convenient to treat the inductance and resistance as separate components in the analysis of the circuit.

Figure 1.55: A practical coil contains inductance and resistance

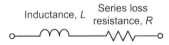

Figure 1.56: A practical coil contains inductance and a small amount of series loss resistance

To understand what happens when a changing current flows through an inductor, take a look at the circuit shown in Figure 1.57(A). If the switch is left open, no current will flow and no magnetic flux will be produced by the inductor. If the switch is closed, as shown in Figure 1.57(B), current will begin to flow as energy is taken from the supply in order to establish the magnetic field. However, the change in magnetic flux resulting from the appearance of current creates a voltage (an *induced e.m.f.*) across the coil which opposes the applied e.m.f. from the battery.

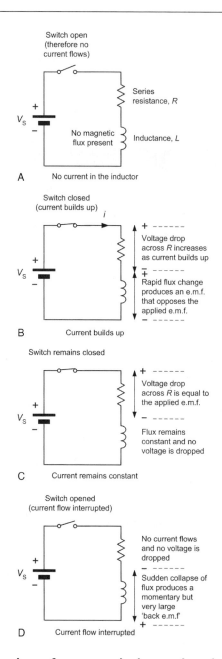

Figure 1.57: Flux and e.m.f. generated when a changing current is applied to an inductor

The induced e.m.f. results from the changing flux and it effectively prevents an instantaneous rise in current in the circuit. Instead, the current increases slowly to a maximum at a rate which depends upon the ratio of inductance (L) to resistance (R) present in the circuit.

After a while, a steady-state condition will be reached in which the voltage across the inductor will have decayed to zero and the current will have reached a maximum value determined by the ratio of V to R (i.e., Ohm's Law). This is shown in Figure 1.57(C).

If, after this steady-state condition has been achieved, the switch is opened, as shown in Figure 1.57(D), the magnetic field will suddenly collapse and the energy will be returned to the circuit in the form of an induced *back e.m.f.*, which will appear across the coil as the field collapses. For large values of magnetic flux and inductance this back e.m.f. can be extremely large!

1.2.22 Inductance

Inductance is the property of a coil which gives rise to the opposition to a change in the value of current flowing in it. Any change in the current applied to a coil/inductor will result in an induced voltage appearing across it. The unit of inductance is the henry (H) and a coil is said to have an inductance of 1H if a voltage of 1V is induced across it when a current changing at the rate of 1 A/s is flowing in it.

The voltage induced across the terminals of an inductor will thus be proportional to the product of the inductance (L) and the rate of change of applied current. Hence:

$$e = -L \times (\text{rate of change of current})$$

Note that the minus sign indicates the polarity of the voltage, i.e., opposition to the change.

The rate of change of current is often represented by the expression di/dt where di represents a very small change in current and dt represents the corresponding small change in time. Using mathematical notation to write this we arrive at:

$$e = -L\frac{di}{dt}$$

You might like to compare this with the similar relationship that we obtained for the current flowing in a capacitor shown in Section 1.2.13.

Example 1.53

A current increases at a uniform rate from 2A to 6A in a period of 250 ms. If this current is applied to an inductor of 600 mH, determine the voltage induced.

Solution

Now the induced voltage will be given by:

$$e = -L \times (\text{rate of change of current})$$

Thus,

$$e = -L\left(\frac{\text{change in current}}{\text{change in time}}\right) = -60 \times 10^{-3} \times \left(\frac{6-2}{250 \times 10^{-3}}\right)$$

From which,

$$e = -600 \times 10^{-3} \times \left(\frac{4}{0.25}\right) = -0.6 \times 10^{-3} \times 16$$

so,

$$\varepsilon = -9.6\text{V}$$

1.2.23 Energy Storage

The energy stored in an inductor is proportional to the product of the inductance and the square of the current flowing in it. Thus:

$$W = \frac{1}{2}LI^2$$

where W is the energy (in joules), L is the capacitance (in henries), and I is the current flowing in the inductor (in amps).

Example 1.54

An inductor of 20 mH is required to store 2.5J of energy. Determine the current that must be applied.

Solution

The foregoing formula can be rearranged to make I the subject as follows:

$$I = \sqrt{\frac{E}{0.5L}} = \sqrt{\frac{2E}{L}} = \sqrt{\frac{2 \times 2.5}{20 \times 10^{-3}}}$$

From which

$$I = \sqrt{\frac{5}{20 \times 10^{-3}}} = \sqrt{0.25 \times 10^{-3}} = \sqrt{250} = 15.81\text{A}$$

1.2.24 Inductance and Physical Dimensions

The inductance of an inductor depends upon the physical dimensions of the inductor (e.g., the length and diameter of the winding), the number of turns, and the permeability of the material of the core. The inductance of an inductor is given by:

$$L = \frac{\mu_0 \, \mu_r \, n^2 \, A}{l}$$

where L is the inductance (in henries), μ_0 is the permeability of free space, μ_r is the relative permeability of the magnetic core, l is the mean length of the core (in meters), and A is the cross-sectional area of the core (in square meters).

Example 1.55

An inductor of 100 mH is required. If a closed magnetic core of length 20 cm, cross-sectional area 15 cm^2 and relative permeability 500 is available, determine the number of turns required.

Solution

First we must rearrange the formula:

$$L = \frac{\mu_0 \, \mu_r \, n^2 \, A}{l}$$

in order to make n the subject:

$$n = \sqrt{\frac{L \times l}{\mu_0 \, \mu_r \, n^2 \, A}} = \sqrt{\frac{100 \times 10^{-3} \times 20 \times 10^{-2}}{12.57 \times 10^{-7} \times 500 \times 15 \times 10^{-4}}}$$

From which:

$$n = \sqrt{\frac{2 \times 10^{-2}}{94,275 \times 10^{-11}}} = \sqrt{21,215} = 146$$

Hence, the inductor requires 146 turns of wire.

1.2.25   Inductor Specifications

Inductor specifications normally include the value of inductance (expressed in henries, millihenries or microhenries), the current rating (i.e., the maximum current which can be continuously applied to the inductor under a given set of conditions), and the accuracy or tolerance (quoted as the maximum permissible percentage deviation from the marked value). Other considerations may include the temperature coefficient of the inductance (usually expressed in parts per million, p.p.m., per unit temperature change), the stability of the inductor, the DC resistance of the coil windings (ideally zero), the Q-factor (quality factor) of the coil, and the recommended working frequency range. Table 1.13

Table 1.13: Characteristics of common types of inductor

Property	Inductor type			
	Air cored	Ferrite cored	Ferrite pot cored	Iron cored
Core material	Air	Ferrite rod	Ferrite pot	Laminated steel
Inductance range (H)	50n to 100μ	10μ to 1m	1m to 100m	20m to 20
Typical DC resistance (Ω)	0.05 to 5	0.1 to 10	5 to 100	10 to 200
Typical tolerance (%)	±5	±10	±10	±20
Typical Q-factor	60	80	40	20
Typical frequency range (Hz)	1M to 500M	100k to 100M	1k to 10M	50 to 10k
Typical applications	Tuned circuits and filters	Filters and HF transformers	LF and MF filters and transformers	Smoothing chokes and filters

summarizes the properties of four common types of inductor. Some typical small inductors are shown in Figure 1.58. These have values of inductance ranging from 15 μH to 1 mH.

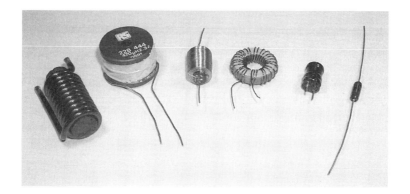

Figure 1.58: A selection of small inductors with values ranging from 15 μH to 1 mH

1.2.26   Inductor Markings

As with capacitors, the vast majority of inductors use written markings to indicate values, working current, and tolerance. Some small inductors are marked with colored stripes to indicate their value and tolerance (in which case the standard color values are used and inductance is normally expressed in microhenries).

1.2.27   Series and Parallel Combinations of Inductors

In order to obtain a particular value of inductance, fixed inductors may be arranged in either series or parallel as shown in Figs 1.59 and 1.60.

The effective inductance of each of the series circuits shown in Figure 1.59 is simply equal to the sum of the individual inductances. So, for the circuit shown in Figure 1.59(A):

$$L = L_1 + L_2$$

while for Figure 1.59(B):

$$L = L_1 + L_2 + L_3$$

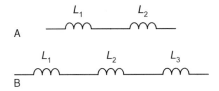

Figure 1.59: Inductors in series

Turning to the parallel inductors shown in Figure 1.60, the reciprocal of the effective inductance of each circuit is equal to the sum of the reciprocals of the individual inductances. Hence, for Figure 1.60(A):

$$\frac{1}{L} = \frac{1}{L_1} + \frac{1}{L_2}$$

while for Figure 1.60(B):

$$\frac{1}{L} = \frac{1}{L_1} + \frac{1}{L_2} + \frac{1}{L_3}$$

In the former case, the formula can be more conveniently re-arranged as follows:

$$L = \frac{L_1 \times L_2}{L_1 + L_2}$$

You can remember this as the *product* of the two inductance values *divided by* the *sum* of the two inductance values.

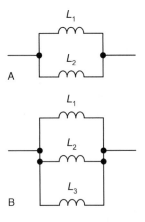

Figure 1.60: Inductors in parallel

Example 1.56

An inductance of 5 mH (rated at 2A) is required. What parallel combination of preferred value inductors will satisfy this requirement?

Solution

Two 10 mH inductors may be wired in parallel to provide an inductance of 5 mH as shown below:

$$L = \frac{L_1 \times L_2}{L_1 + L_2} = \frac{10 \times 10}{10 + 10} = \frac{100}{20} = 5 \text{ mH}$$

Since the inductors are identical, the applied current will be shared equally between them. Hence, each inductor should have a current rating of 1A.

Example 1.57

Determine the effective inductance of the circuit shown in Figure 1.61.

Solution

The circuit can be progressively simplified as shown in Figure 1.62. The stages in this simplification are as follows:

(a) L_1 and L_2 are in series and they can be replaced by a single inductance (L_A) of $(60 + 60) = 120$ mH.

(b) L_A appears in parallel with L_2. These two inductors can be replaced by a single inductor (L_B) of $(120 \times 120)/(120 + 120) = 60$ mH.

(c) L_B appears in series with L_4. These two inductors can be replaced by a single inductance (L) of $(60 + 50) = 110$ mH.

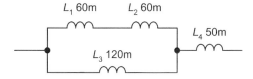

Figure 1.61: See Example 1.57

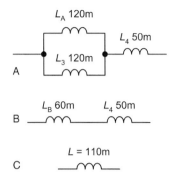

Figure 1.62: See Example 1.57

1.2.28   Variable Inductors

A ferrite cored inductor can be made variable by moving its core in or out of the former onto which the coil is wound. Many small inductors have threaded ferrite cores to make this possible (see Figure 1.63). Such inductors are often used in radio and high-frequency applications where precise tuning is required.

Figure 1.63: An adjustable ferrite cored inductor

1.2.29   Surface Mounted Components (SMC)

Surface-mount technology (SMT) is now widely used in the manufacture of printed circuit boards for electronic equipment. SMT allows circuits to be assembled in a much smaller space than would be possible using components with conventional wire leads and pins that are mounted using through-hole techniques. It is also possible to mix the two technologies, i.e., some through-hole mounting of components and some surface mounted components present on the same circuit board. The following combinations are possible:

- *Surface mounted components (SMC)* on both sides of a printed circuit board.

- SMC on one side of the board and conventional *through-hole components (THC)* on the other.

- A mixture of SMC and THC on both sides of the printed circuit board.

Surface mounted components are supplied in packages that are designed for mounting directly on the surface of a PCB. To provide electrical contact with the PCB, some SMC have contact pads on their surface. Other devices have contacts which extend beyond the outline of the package itself but which terminate on the surface of the PCB rather than making contact through a hole (as is the case with a conventional THC). In general, passive components (such as resistors, capacitors and inductors) are configured leadless for surface mounting, while active devices (such as transistors and integrated circuits) are available in both surface mountable types as well as lead as well as in leadless terminations suitable for making direct contact to the pads on the surface of a PCB.

Most surface mounted components have a flat rectangular shape rather than the cylindrical shape that we associate with conventional wire leaded components. During manufacture of a PCB, the various SMC are attached using re-flow soldering paste (and in some cases adhesives) which consists of particles of solder and flux together with binder, solvents and additives. They need to have good "tack" in order to hold the components in place and remove oxides without leaving obstinate residues.

The component attachment (i.e., soldering!) process is completed using one of several techniques including convection ovens in which the PCB is passed, using a conveyor belt, through a convection oven which has separate zones for preheating, flowing and cooling, and infra-red reflow in which infrared lamps are used to provide the source of heat.

Surface mounted components are generally too small to be marked with color codes. Instead, values may be marked using three digits. For example, the first two digits marked on a resistor normally specify the first two digits of the value while the third digit gives the number of zeros that should be added.

Example 1.58

In Figure 1.65, R88 is marked "102". What is its value?

Solution

R88 will have a value of 1,000Ω (i.e., 10 followed by two zeros).

Figure 1.64: Conventional components mounted on a printed circuit board. Note that components such as C38, R46, etc. have leads that pass through holes in the printed circuit boards

Figure 1.65: Surface mounted components (note the appearance of capacitors C35, C52, and C53, and resistors, R87, R88, R91, etc.)

1.3   DC Circuits

In many cases, Ohm's Law alone is insufficient to determine the magnitude of the voltages and currents present in a circuit. This section introduces several techniques that simplify the task of solving complex circuits. It also introduces the concept of exponential growth and decay of voltage and current in circuits containing capacitance and resistance and inductance and resistance. It concludes by showing how humble *C-R* circuits can be used for shaping the waveforms found in electronic circuits. We start by introducing two of the most useful laws of electronics.

1.3.1 Kirchhoff's Laws

Kirchhoff's Laws relate to the algebraic sum of currents at a junction (or *node*) or voltages in a network (or *mesh*). The term "algebraic" simply indicates that the polarity of each current or voltage drop must be taken into account by giving it an appropriate sign, either positive (+) or negative (−).

Kirchhoff's Current Law states that the algebraic sum of the currents present at a junction (node) in a circuit is zero (see Figure 1.66).

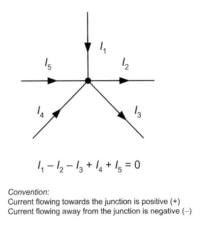

$$I_1 - I_2 - I_3 + I_4 + I_5 = 0$$

Convention:
Current flowing towards the junction is positive (+)
Current flowing away from the junction is negative (−)

Figure 1.66: Kirchhoff's Current Law

Example 1.59

In Figure 1.67, use Kirchhoff's Current Law to determine:

(a)   the value of current flowing between A and B, and

(b)   the value of I_3.

Solution

(a) I_1 and I_2 both flow toward Node A so, applying our polarity convention, they must both be positive. Now, assuming that a current I_5 flows between A and B and that this

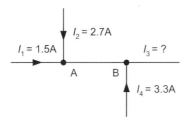

Figure 1.67: See Example 1.59

current flows away from the junction (obvious because I_1 and I_2 both flow toward the junction), we arrive at the following Kirchhoff's Current Law equation:

$$+I_1 + I_2 - I_5 = 0$$

From which:

$$I_5 = I_1 + I_2 = 1.5 + 2.7 = 4.2\text{A}$$

(b) Moving to Node B, let's assume that I_3 flows outward, so we can say that:

$$+I_4 + I_5 - I_3 = 0$$

From which:

$$I_3 = I_4 + I_5 = 3.3 + 4.2 = 7.5\text{A}$$

Kirchhoff's Voltage Law states that the algebraic sum of the potential drops in a closed network (or "mesh") is zero (see Figure 1.68).

Example 1.60

In Figure 1.69, use Kirchhoff's Voltage Law to determine:

(a) the value of V_2, and

(b) the value of E_3.

Solution

(a) In Loop A, and using the conventions shown in Figure 1.68, we can write down the Kirchhoff's Voltage Law equations:

$$E_1 - V_2 - E_2 = 0$$

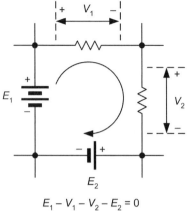

$$E_1 - V_1 - V_2 - E_2 = 0$$

Convention:
Move clockwise around the circuit starting with the positive
terminal of the largest e.m.f.
Voltages acting in the same sense are positive (+)
Voltages acting in the opposite sense are negative (−)

Figure 1.68: Kirchhoff's Voltage Law

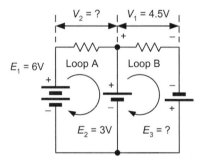

Figure 1.69: See Example 1.60

From which:

$$V_2 = E_1 - E_2 = 6 - 3 = 3\text{V}$$

(b)   Similarly, in Loop B, we can say that:

$$E_2 - V_2 + E_3 = 0$$

From which:

$$E_3 = V_2 - E_2 = 4.5 - 3 = 1.5\text{V}$$

Example 1.61

Determine the currents and voltages in the circuit of Figure 1.70.

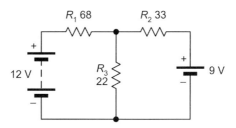

Figure 1.70: See Example 1.61

Solution

In order to solve the circuit shown in Figure 1.70, it is first necessary to mark the currents and voltages on the circuit, as shown in Figures 1.71 and 1.72.

By applying Kirchhoff's Current Law at Node A that we've identified in Figure 1.70:

$$+I_1 + I_2 - I_3 = 0$$

Therefore:

$$I_1 = I_3 - I_2 \qquad \text{(i)}$$

By applying Kirchhoff's Voltage Law in Loop A we obtain:

$$12 - V_1 - V_3 = 0$$

From which:

$$V_1 = 12 - V_3 \qquad \text{(ii)}$$

By applying Kirchhoff's Voltage Law in Loop B we obtain:

$$9 - V_2 - V_3 = 0$$

From which:

$$V_2 = 9 - V_3 \qquad \text{(iii)}$$

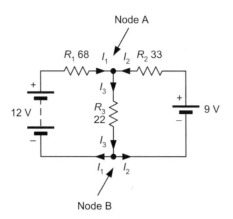

Figure 1.71: See Example 1.61

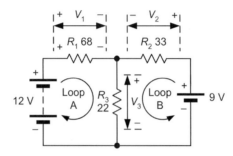

Figure 1.72: See Example 1.61

Next we can generate three further relationships by applying Ohm's Law:

$$V_1 = I_1 R_1 \quad \text{from which} \quad I_1 = \frac{V_1}{R_1}$$

$$V_2 = I_2 R_2 \quad \text{from which} \quad I_2 = \frac{V_2}{R_2}$$

and,

$$V_3 = I_3 R_3 \quad \text{from which} \quad I_3 = \frac{V_3}{R_3}$$

Combining these three relationships with the Current Law equation (i) gives:

$$\frac{V_1}{R_1} = \frac{V_3}{R_3} - \frac{V_2}{R_2}$$

from which:

$$\frac{V_1}{110} = \frac{V_3}{22} - \frac{V_2}{33}$$

Combining (ii) and (iii) with (iv) gives:

$$\frac{(12 - V_3)}{110} = \frac{V_3}{22} - \frac{(9 - V_3)}{33}$$

Multiplying both sides of the expression by 330 gives:

$$\frac{330(12 - V_3)}{110} = \frac{330V_3}{22} - \frac{330(9 - V_3)}{33}$$

$$3(12 - V_3) = 15\,V_3 - 10(9 - V_3)$$

From which:

$$36 - 3\,V_3 = 15\,V_3 - 90 + V_3$$
$$36 + 90 = 15\,V_3 + 10\,V_3 + 3\,V_3$$

and:

$$126 = 28V_3 \quad \text{so} \quad V_3 = 126/28 = 4.5\text{V}$$

From (ii):

$$V_1 = 12 - V_3 \quad \text{so} \quad V_1 = 12 - 4.5 = 7.5\text{V}$$

From (iii):

$$V_2 = 9 - V_3 \quad \text{so} \quad V_2 = 9 - 4.5 = 4.5\text{V}$$

Using the Ohm's Law equations that we met earlier gives:

$$I_1 = \frac{V_1}{R_1} = \frac{7.5}{110} = 0.068\text{A} = 68 \text{ mA}$$

$$I_2 = \frac{V_2}{R_2} = \frac{4.5}{33} = 0.136\text{A} = 136 \text{ mA}$$

$$I_3 = \frac{V_3}{R_3} = \frac{4.5}{22} = 0.204\text{A} = 204 \text{ mA}$$

Finally, it's worth checking these results with the Current Law equation (i):

$$+I_1 + I_2 - I_3 = 0$$

Inserting our values for I_1, I_2 and I_3 gives:

$$+0.068 + 0.136 - 204 = 0$$

Since the left and right hand sides of the equation are equal we can be reasonably confident that our results are correct.

1.3.2   The Potential Divider

The potential divider circuit (see Figure 1.73) is commonly used to reduce voltages in a circuit. The output voltage produced by the circuit is given by:

$$V_{\text{out}} = V_{\text{in}} \frac{R_2}{R_1 + R_2}$$

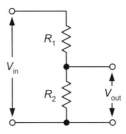

Figure 1.73: Potential divider circuit

It is, however, important to note that the output voltage (V_{out}) will fall when current is drawn from the arrangement.

Figure 1.74 shows the effect of *loading* the potential divider circuit. In the loaded potential divider (Figure 1.74) the output voltage is given by:

$$V_{\text{out}} = V_{\text{in}} \frac{R_{\text{p}}}{R_1 + R_{\text{p}}}$$

where:

$$R_{\text{p}} = \frac{R_2 \times R_{\text{L}}}{R_2 + R_{\text{L}}}$$

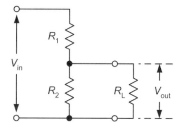

Figure 1.74: Loaded potential divider circuit

Example 1.62

The potential divider shown in Figure 1.75 is used as a simple *voltage calibrator*. Determine the output voltage produced by the circuit:

(a) when the output terminals are left open-circuit (i.e., when no load is connected); and

(b) when the output is loaded by a resistance of 10 kΩ.

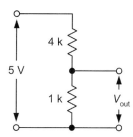

Figure 1.75: See Example 1.62

Solution

(a) In the first case we can simply apply the formula:

$$V_{out} = V_{in} \frac{R_2}{R_1 + R_2}$$

where $V_{in} = 5$V, $R_1 = 4$ kΩ and $R_2 = 1$ kΩ.

Hence:

$$V_{\text{out}} = 5 \times \frac{1}{4 + 1} = 1\text{V}$$

(b)   In the second case we need to take into account the effect of the 10 kΩ resistor connected to the output terminals of the potential divider.

First we need to find the equivalent resistance of the parallel combination of R_2 and R_L:

$$R_{\text{p}} = \frac{R_2 \times R_L}{R_2 + R_L} = \frac{1 \times 10}{1 + 10} = \frac{10}{11} = 0.909\text{ k}\Omega$$

Then we can determine the output voltage from:

$$V_{\text{out}} = V_{\text{in}} \frac{R_{\text{p}}}{R_1 + R_{\text{p}}} = 5 \times \frac{0.909}{4 + 0.909} = 0.925\text{V}$$

1.3.3   The Current Divider

The current divider circuit (see Figure 1.76) is used to divert a known proportion of the current flowing in a circuit. The output current produced by the circuit is given by:

$$I_{\text{out}} = I_{\text{in}} \frac{R_1}{R_1 + R_2}$$

It is, however, important to note that the output current (I_{out}) will fall when the load connected to the output terminals has any appreciable resistance.

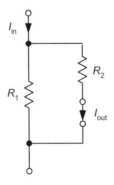

Figure 1.76: Current divider circuit

Example 1.63

A moving coil meter requires a current of 1 mA to provide full-scale deflection. If the meter coil has a resistance of 100Ω and is to be used as a milliammeter reading 5 mA full-scale, determine the value of parallel shunt resistor required.

Solution

This problem may sound a little complicated so it is worth taking a look at the *equivalent circuit* of the meter (Figure 1.77) and comparing it with the current divider shown in Figure 1.76.

We can apply the current divider formula, replacing I_{out} with I_m (the meter full-scale deflection current) and R_2 with R_m (the meter resistance). R_1 is the required value of shunt resistor, R_s, Hence:

$$I_{out} = I_{in} \frac{R_s}{R_s + R_m}$$

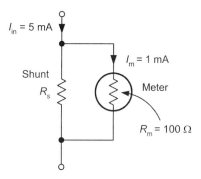

Figure 1.77: See Example 1.63

Rearranging the formula gives:

$$I_m \times (R_s + R_m) = I_{in} \times R_s$$

thus,

$$I_m R_s + I_m R_m = I_{in} R_s$$

or,

$$I_{in} R_s - I_m R_s = I_m R_m$$

from which,

$$R_s(I_{in} - I_m) = I_m R_m$$

So,

$$R_s = \frac{I_m R_m}{I_{in} - I_m}$$

Now $I_{in} = 1$ mA, $R_m = 100\Omega$ and $I_{in} = 5$ mA, thus:

$$R_s = \frac{1 \times 100}{5 - 1} = \frac{100}{4} = 25\Omega$$

1.3.4   The Wheatstone Bridge

The Wheatstone bridge forms the basis of a number of useful electronic circuits including several that are used in instrumentation and measurement.

The basic form of Wheatstone bridge is shown in Figure 1.78. The voltage developed between A and B will be zero when the voltage between A and Y is the same as that between B and Y. In effect, R_1 and R_2 constitute a potential divider as do R_3 and R_4.

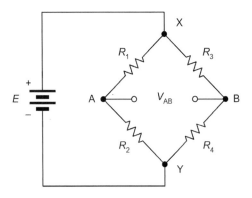

Figure 1.78: Basic Wheatstone bridge circuit

The bridge will be *balanced* (and $V_{AB} = 0$) when the ratio of $R_1 : R_2$ is the same as the ratio $R_3 : R_4$.

Hence, at balance:

$$\frac{R_1}{R_2} = \frac{R_3}{R_4}$$

A practical form of Wheatstone bridge that can be used for measuring unknown resistances is shown in Figure 1.79.

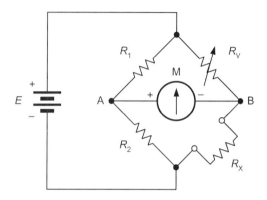

Figure 1.79: See Example 1.64

In this practical form of Wheatstone bridge, R_1 and R_2 are called the *ratio arms* while one arm (that occupied by R_3 in Figure 1.78) is replaced by a calibrated variable resistor. The unknown resistor, R_x, is connected in the fourth arm. At balance:

$$\frac{R_1}{R_2} = \frac{R_v}{R_x} \quad \text{thus} \quad R_x = \frac{R_2}{R_1} \times R_v$$

Example 1.64

A Wheatstone bridge is based on the circuit shown in Figure 1.79. If R_1 and R_2 can each be switched so that they have values of either $100\,\Omega$ or $1\,k\Omega$ and R_v is variable between $10\,\Omega$ and $10\,k\Omega$, determine the range of resistance values that can be measured.

Solution

The maximum value of resistance that can be measured will correspond to the largest ratio of R_2:R_1 (i.e., when R_2 is 1 kΩ and R_1 is 100Ω) and the highest value of RV (i.e., 10 kΩ). In this case:

$$R_x = \frac{1,000}{100} \times 10,000 = 100,000 = 100 \text{ k}\Omega$$

The minimum value of resistance that can be measured will correspond to the smallest ratio of R_2:R_1 (i.e., when R_1 is 100Ω and R_1 is 1 kΩ) and the smallest value of RV (i.e., 10Ω). In this case:

$$R_x = \frac{100}{1,000} \times 10 = 0.1 \times 10 = 1\Omega$$

Hence the range of values that can be measured extends from 1Ω to 100 kΩ.

1.3.5 Thévenin's Theorem

Thévenin's Theorem allows us to replace a complicated network of resistances and voltage sources with a simple equivalent circuit comprising a single *voltage source* connected in series with a single resistance (see Figure 1.70).

The single voltage source in the Thévenin equivalent circuit, V_{oc}, is simply the voltage that appears between the terminals when nothing is connected to it. In other words, it is the *open-circuit* voltage that would appear between A and B.

The single resistance that appears in the Thévenin equivalent circuit, R, is the resistance that would be seen *looking into* the network between A and B when all of the voltage sources (assumed perfect) are replaced by *short-circuit* connections. Note that if the voltage sources are not perfect (i.e., if they have some internal resistance) the equivalent circuit must be constructed on the basis that each voltage source is replaced by its own internal resistance.

Once we have values for V_{oc} and R, we can determine how the network will behave when it is connected to a load (i.e., when a resistor is connected across the terminals A and B).

Example 1.65

Figure 1.81 shows a Wheatstone bridge. Determine the current that will flow in a 100Ω load connected between terminals A and B.

Solution

First we need to find the Thévenin equivalent of the circuit. To find V_{oc} we can treat the bridge arrangement as two potential dividers.

The voltage across R_2 will be given by:

$$V = 10 \times \frac{R_2}{R_1 + R_2} = 10 \times \frac{600}{500 + 600} = 5.454\text{V}$$

Hence, the voltage at A relative to Y, V_{AY}, will be 5.454V.

The voltage across R_4 will be given by:

$$V = 10 \times \frac{R_4}{R_3 + R_4} = 10 \times \frac{400}{500 + 400} = 4.444\text{V}$$

Hence, the voltage at B relative to Y, V_{BY}, will be 4.444V.

The voltage V_{AB} will be the difference between V_{AY} and V_{BY}. This, the open-circuit output voltage, V_{AB}, will be given by:

$$V_{AB} = V_{AY} - V_{BY} = 5.454 - 4.444 = 1.01\text{V}$$

Next we need to find the Thévenin equivalent resistance looking in at A and B. To do this, we can redraw the circuit, replacing the battery (connected between X and Y) with a short circuit, as shown in Figure 1.82.

The Thévenin equivalent resistance is given by the relationship:

$$R = \frac{R_1 \times R_2}{R_1 + R_2} + \frac{R_3 \times R_4}{R_3 + R_4} = \frac{500 \times 600}{500 + 600} + \frac{500 \times 400}{500 + 400}$$

From which:

$$R = \frac{300,000}{1,100} + \frac{200,000}{900} = 272.7 + 222.2 = 494.9\Omega$$

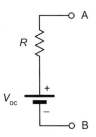

Figure 1.80: Thévenin equivalent circuit

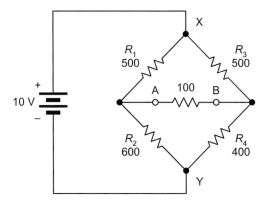

Figure 1.81: See Example 1.65

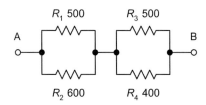

Figure 1.82: See Example 1.65

The Thévenin equivalent circuit is shown in Figure 1.83. To determine the current in a 100Ω load connected between A and B, we can simply add a 100Ω load to the Thévenin equivalent circuit, as shown in Figure 1.84. By applying Ohm's Law in Figure 1.84 we get:

$$I = \frac{V_{oc}}{R + 100} = \frac{1.01}{494.9 + 100} = \frac{1.01}{594.9} = 1.698 \text{ mA}$$

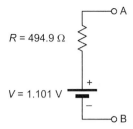

Figure 1.83: Thévenin equivalent of Figure 1.81

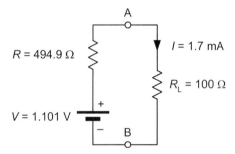

Figure 1.84: Determining the current when the Thévenin equivalent circuit is loaded

1.3.6 Norton's Theorem

Norton's Theorem provides an alternative method of reducing a complex network to a simple equivalent circuit. Unlike Thévenin's Theorem, Norton's Theorem makes use of a current source rather than a voltage source. The Norton equivalent circuit allows us to replace a complicated network of resistances and voltage sources with a simple equivalent circuit comprising a single constant current source connected in parallel with a single resistance (see Figure 1.85).

The constant current source in the Norton equivalent circuit, I_{sc}, is simply the *short-circuit* current that would flow if A and B were to be linked directly together. The resistance that appears in the Norton equivalent circuit, R, is the resistance that would be seen *looking into* the network between A and B when all of the voltage sources are replaced by *short-circuit* connections. Once again, it is worth noting that, if the voltage sources have any appreciable internal resistance, the equivalent circuit must be constructed on the basis that each voltage source is replaced by its own internal resistance.

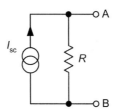

Figure 1.85: Norton equivalent circuit

As with the Thévenin equivalent, we can determine how a network will behave by obtaining values for I_{sc} and R.

Example 1.66

Three temperature sensors having the following characteristics shown in Table 1.14 are connected in parallel as shown in Figure 1.86:

Determine the voltage produced when the arrangement is connected to a moving-coil meter having a resistance of 1 kΩ.

Table 1.14: Temperature sensor characteristics

Sensor	A	B	C
Output voltage (open circuit)	20 mV	30 mV	10 mV
Internal resistance	5 kΩ	3 kΩ	2 kΩ

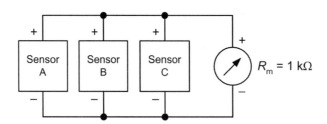

Figure 1.86: See Example 1.66

Solution

First we need to find the Norton equivalent of the circuit. To find I_{sc} we can determine the short-circuit current from each sensor and add them together.

For sensor *A*:

$$I = \frac{V}{R} = \frac{20 \text{ mV}}{5 \text{ k}\Omega} = 4 \text{ }\mu A$$

For sensor *B*:

$$I = \frac{V}{R} = \frac{30 \text{ mV}}{3 \text{ k}\Omega} = 10 \text{ }\mu A$$

For sensor *C*:

$$I = \frac{V}{R} = \frac{10 \text{ mV}}{2 \text{ k}\Omega} = 5 \text{ }\mu A$$

The total current, I_{sc}, will be given by:

$$I_{sc} = 4 \text{ }\mu A + 10 \text{ }\mu A + 5 \text{ }\mu A = 19 \text{ }\mu A$$

Next we need to find the Norton equivalent resistance. To do this, we can redraw the circuit showing each sensor replaced by its internal resistance, as shown in Figure 1.87.

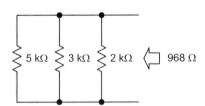

Figure 1.87: Determining the equivalent resistance in Figure 1.86

The equivalent resistance of this arrangement (think of this as the resistance seen *looking into* the circuit in the direction of the arrow shown in Figure 1.87) is given by:

$$\frac{1}{R} = \frac{1}{R_1} + \frac{1}{R_2} + \frac{1}{R_3} = \frac{1}{5,000} + \frac{1}{3,000} + \frac{1}{2,000}$$

where $R_1 = 5\text{k}\Omega$, $R_2 = 3\text{k}\Omega$, $R_3 = 2\text{k}\Omega$, hence:

$$\frac{1}{R} = \frac{1}{R_1} + \frac{1}{R_2} + \frac{1}{R_3} = \frac{1}{5,000} + \frac{1}{3,000} + \frac{1}{2,000}$$

or,

$$\frac{1}{R} = 0.0002 + 0.00033 + 0.0005 = 0.00103$$

from which:

$$R = 968\Omega$$

The Norton equivalent circuit is shown in Figure 1.88. To determine the voltage in a 1 kΩ moving coil meter connected between A and B, we can make use of the Norton equivalent circuit by simply adding a 1 kΩ resistor to the circuit and applying Ohm's Law, as shown in Figure 1.89.

The voltage appearing across the moving coil meter in Figure 1.90 will be given by:

$$V = I_{\text{sc}} \times \frac{R \times R_{\text{m}}}{R + R_{\text{m}}} = 19 \ \mu\text{A} \times \frac{1,000 \times 968}{1,000 + 968}$$

hence:

$$V = 19 \ \mu\text{A} \times 492\Omega = 9.35 \ \text{mV}$$

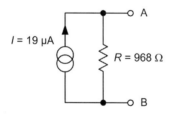

Figure 1.88: Norton equivalent of the circuit in Figure 1.86

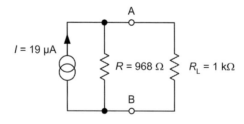

Figure 1.89: Determining the output voltage when the Norton equivalent circuit is loaded with 1 kΩ

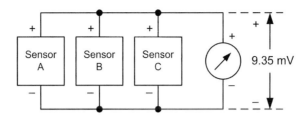

Figure 1.90: The voltage drop across the meter is found to be 9.35 mV

1.3.7 C-R Circuits

Networks of capacitors and resistors (known as *C-R* circuits) form the basis of many timing and pulse shaping circuits and are thus often found in practical electronic circuits.

1.3.8 Charging

A simple *C-R* circuit is shown in Figure 1.91. In this circuit *C* is charged through *R* from the constant voltage source, V_s. The voltage, v_c, across the (initially uncharged) capacitor voltage will rise exponentially as shown in Figure 1.92. At the same time, the current in the circuit, *i*, will fall, as shown in Figure 1.93.

The rate of growth of voltage with time (and decay of current with time) will be dependent upon the product of capacitance and resistance. This value is known as the *time constant* of the circuit. Hence:

Time constant, $t = C \times R$

where *C* is the value of capacitance (F), *R* is the resistance (F), and *t* is the time constant (s).

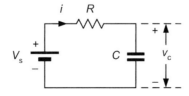

Figure 1.91: A C-R circuit in which C is charged through R

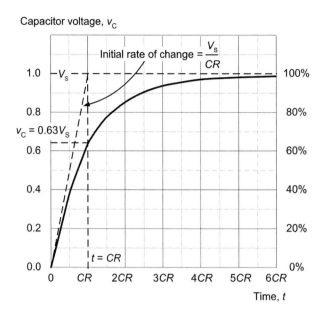

Figure 1.92: Exponential growth of capacitor voltage, v_c, in Figure 1.92

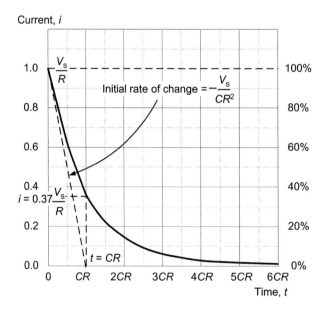

Figure 1.93: Exponential decay of current, i, in Figure 1.91

The voltage developed across the charging capacitor, v_c, varies with time, t, according to the relationship:

$$v_c = V_s \left(1 - e^{-\frac{t}{CR}} \right)$$

where v_c is the capacitor voltage, V_s is the DC supply voltage, t is the time, and CR is the time constant of the circuit (equal to the product of capacitance, C, and resistance, R).

The capacitor voltage will rise to approximately 63% of the supply voltage, V_s, in a time interval equal to the time constant.

At the end of the next interval of time equal to the time constant (i.e., after an elapsed time equal to $2CR$) the voltage will have risen by 63% of the remainder, and so on. In theory, the capacitor will *never* become fully charged. However, after a period of time equal to $5CR$, the capacitor voltage will to all intents and purposes be equal to the supply voltage. At this point, the capacitor voltage will have risen to 99.3% of its final value and we can consider it to be fully charged.

During charging, the current in the capacitor, i, varies with time, t, according to the relationship:

$$i = \frac{V_s}{R} \, e^{-\frac{t}{CR}}$$

where V_s is the DC supply voltage, t is the time, R is the series resistance and C is the value of capacitance.

The current will fall to approximately 37% of the initial current in a time equal to the time constant. At the end of the next interval of time equal to the time constant (i.e., after a total time of $2CR$ has elapsed) the current will have fallen by a further 37% of the remainder, and so on.

Example 1.67

An initially uncharged 1 μF capacitor is charged from a 9V DC supply via a 3.3 MΩ resistor. Determine the capacitor voltage 1s after connecting the supply.

Solution

The formula for exponential growth of voltage in the capacitor is:

$$v_c = V_s \left(1 - e^{-\frac{t}{CR}}\right)$$

Here we need to find the capacitor voltage, v_c, when $V_s = 9\text{V}$, $t = 1\text{s}$, $C = 1$ μF and $R = 3.3$ MΩ. The time constant, CR, will be given by:

$$CR = 1 \times 10^{-6} \times 3.3 \times 10^6 = 3.3\text{s}$$

Thus:

$$v_c = 9\left(1 - e^{-\frac{t}{3.3}}\right)$$

and,

$$v_c 9(1 - 0.738) = 9 \times 0.262 = 2.358\text{V}$$

Example 1.68

A 100 μF capacitor is charged from a 350V DC supply through a series resistance of 1 kΩ. Determine the initial charging current and the current that will flow 50 ms and 100 ms after connecting the supply. After what time is the capacitor considered to be fully charged?

Solution

At $t = 0$ the capacitor will be uncharged ($v_c = 0$) and all of the supply voltage will appear across the series resistance. Thus, at $t = 0$:

$$i = \frac{V_s}{R} = \frac{350}{1,000} = 0.35\text{A}$$

When $t = 50$ ms, the current will be given by:

$$i = \frac{V_s}{R} e^{-\frac{t}{CR}}$$

Where $V_s = 350$V, $t = 50$ ms, C $= 100$ μF, $R = 1$ kΩ. Hence:

$$i = \frac{350}{1,000}\, e^{-\frac{0.05}{0.1}} = 0.35\, e^{-0.5} = 0.35 \times 0.607 = 0.21\text{A}$$

When $t = 100$ ms (using the same equation but with $t = 0.1$s) the current is given by:

$$i = \frac{350}{1,000}\, e^{\frac{0.1}{0.1}} = 0.35\, e^{-1} = 0.35 \times 0.368 = 0.129\text{A}$$

The capacitor can be considered to be fully charged when $t = 5CR = 5 \times 100 \times 10^{-6} \times 1 \times 10^{3} = 0.5$s. Note that, at this point the capacitor voltage will have reached 99% of its final value.

Discharge

Having considered the situation when a capacitor is being charged, let's consider what happens when an already charged capacitor is discharged.

Figure 1.94: *C-R* circuits are widely used in electronics. In this oscilloscope, for example, a rotary switch is used to select different *C-R* combinations in order to provide the various timebase ranges (adjustable from 500 ms/cm to 1 μs/cm). Each *C-R* time constant corresponds to a different timebase range.

When the fully charged capacitor from Figure 1.89 is connected as shown in Figure 1.95, the capacitor will discharge through the resistor, and the capacitor voltage, v_C, will fall exponentially with time, as shown in Figure 1.96.

The current in the circuit, i, will also fall, as shown in Figure 1.97. The rate of discharge (i.e., the rate of decay of voltage with time) will once again be governed by the time constant of the circuit, $C \times R$.

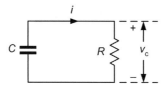

Figure 1.95: A C-R circuit in which C is initially charged and then discharges through R

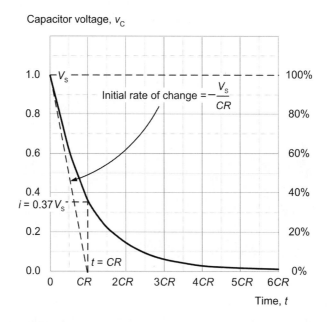

Figure 1.96: Exponential decay of capacitor voltage, v_c, in Figure 1.95

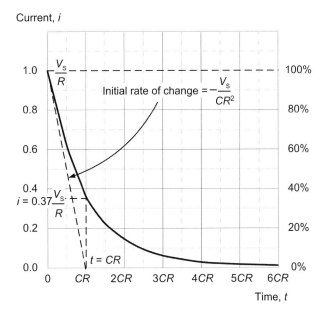

Figure 1.97: Exponential decay of current, *i*, in Figure 1.95

The voltage developed across the discharging capacitor, v_C, varies with time, t, according to the relationship:

$$v_c = V_s \, e^{-\frac{t}{CR}}$$

where V_s, is the supply voltage, t is the time, C is the capacitance, and R is the resistance.

The capacitor voltage will fall to approximately 37% of the initial voltage in a time equal to the time constant. At the end of the next interval of time equal to the time constant (i.e., after an elapsed time equal to $2CR$) the voltage will have fallen by 37% of the remainder, and so on.

In theory, the capacitor will *never* become fully discharged. However, after a period of time equal to $5CR$, the capacitor voltage will to all intents and purposes be zero.

At this point the capacitor voltage will have fallen below 1% of its initial value. At this point we can consider it to be fully discharged.

As with charging, the current in the capacitor, i, varies with time, t, according to the relationship:

$$i = \frac{V_s}{R} e^{-\frac{t}{CR}}$$

where V_s, is the supply voltage, t is the time, C is the capacitance, and R is the resistance. The current will fall to approximately 37% of the initial value of current, V_s/R, in a time equal to the time constant.

At the end of the next interval of time equal to the time constant (i.e., after a total time of $2CR$ has elapsed) the voltage will have fallen by a further 37% of the remainder, and so on.

Example 1.69

A 10 μF capacitor is charged to a potential of 20V and then discharged through a 47 kΩ resistor. Determine the time taken for the capacitor voltage to fall below 10V.

Solution

The formula for exponential decay of voltage in the capacitor is:

$$v_c = V_s e^{-\frac{t}{CR}}$$

where $V_s = 20V$ and $CR = 10$ μF $\times$ 47 kΩ $= 0.47s$.

We need to find t when $v_C = 10V$. Rearranging the formula to make t the subject gives:

$$t = -CR \times \ln\left(\frac{v_C}{V_s}\right)$$

thus,

$$t = -0.47 \times \ln\left(\frac{10}{20}\right) = -0.47 \times \ln(0.5)$$

or,

$$t = -0.47 \times -693 = 0.325s$$

In order to simplify the mathematics of exponential growth and decay, Table 1.15 provides an alternative tabular method that may be used to determine the voltage and current in a C-R circuit.

Table 1.15: Exponential growth and decay

t/CR or t /(L/R)	k (growth)	k (decay)
0.0	0.0000	1.0000
0.1	0.0951	0.9048
0.2	0.1812	0.8187 (1)
0.3	0.2591	0.7408
0.4	0.3296	0.6703
0.5	0.3935	0.6065
0.6	0.4511	0.5488
0.7	0.5034	0.4965
0.8	0.5506	0.4493
0.9	0.5934	0.4065
1.0	0.6321	0.3679
1.5	0.7769	0.2231
2.0	0.8647 (2)	0.1353
2.5	0.9179	0.0821
3.0	0.9502	0.0498
3.5	0.9698	0.0302
4.0	0.9817	0.0183
4.5	0.9889	0.0111
5.0	0.9933	0.0067

Notes: (1) See Example 1.70
(2) See Example 1.74
k is the ratio of the value at time, t, to the final value (e.g., v_c/V_s)

Example 1.70

A 150 μF capacitor is charged to a potential of 150V. The capacitor is then removed from the charging source and connected to a 2 MΩ resistor. Determine the capacitor voltage 1 minute later.

Solution

We will solve this problem using Table 1.15 rather than the exponential formula.

First we need to find the time constant:

$$C \times R = 150\,\mu F \times 2\ M\Omega = 300s$$

Next we find the ratio of *t* to *CR*:

After 1 minute, $t = 60s$ therefore the ratio of *t* to *CR* is 60/300 or 0.2. Table 1.15 shows that when $t/CR = 0.2$, the ratio of instantaneous value to final value (*k* in Table 1.15) is 0.8187.

Thus,

$$v_c/V_s = 0.8187$$

or,

$$v_c = 0.8187 \times V_s = 0.8187 \times 150V = 122.8V$$

1.3.9   *Waveshaping with C-R Networks*

One of the most common applications of *C-R* networks is in waveshaping circuits. The circuits shown in Figures 1.98 and 1.100 function as simple square-to-triangle and square-to-pulse converters by, respectively, *integrating* and *differentiating* their inputs.

The effectiveness of the simple integrator circuit shown in Figure 1.98 depends on the ratio of time constant, $C \times R$, to periodic time, *t*. The larger this ratio is, the more effective the circuit will be as an integrator. The effectiveness of the

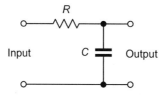

Figure 1.98: A C-R integrating circuit

circuit of Figure 1.98 is illustrated by the input and output waveforms shown in Figure 1.99.

Similarly, the effectiveness of the simple differentiator circuit shown in Figure 1.100 also depends on the ratio of time constant $C \times R$, to periodic time, t. The smaller this ratio is, the more effective the circuit will be as a differentiator.

The effectiveness of the circuit of Figure 1.100 is illustrated by the input and output waveforms shown in Figure 1.101.

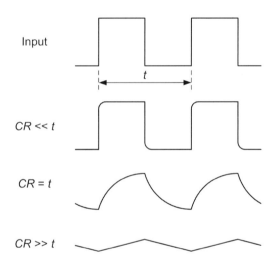

Figure 1.99: Typical input and output waveforms for the integrating circuit shown in Figure 1.98

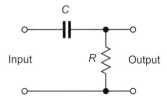

Figure 1.100: A *C-R* differentiating circuit

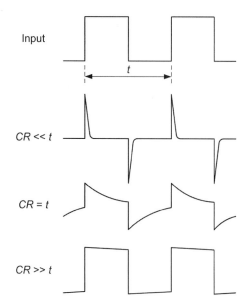

Figure 1.101: Typical input and output waveforms for the integrating circuit shown in Figure 1.98

Example 1.71

A circuit is required to produce a train of alternating positive and negative pulses of short duration from a square wave of frequency 1 kHz. Devise a suitable *C-R* circuit and specify suitable values.

Solution

Here we require the services of a differentiating circuit along the lines of that shown in Figure 1.100. In order that the circuit operates effectively as a differentiator, we need to make the time constant, $C \times R$, very much less than the periodic time of the input waveform (1 ms).

Assuming that we choose a medium value for R of, say, 10 kΩ, the maximum value which we could allow C to have would be that which satisfies the equation:

$$C \times R = 0.1t$$

where $R = 10 \text{ k}\Omega$ and $t = 1$ ms. Thus:

$$C = \frac{0.1t}{R} = \frac{0.1 \times 1 \text{ ms}}{10 \text{ k}\Omega} = 0.1 \times 10^{-3} \times 10^{-4} = 1 \times 10^{-8}\text{F}$$

or,

$$C = 10 \times 10^{-9} \text{ F} = 10 \text{ nF}$$

In practice, any value equal or less than 10 nF would be adequate. A very small value (say less than 1 nF) will, however, generate pulses of a very narrow width.

Example 1.72

A circuit is required to produce a triangular waveform from a square wave of frequency 1 kHz. Devise a suitable *C-R* arrangement and specify suitable values.

Solution

This time we require an integrating circuit like that shown in Figure 1.98. In order that the circuit operates effectively as an integrator, we need to make the time constant, $C \times R$, very much less than the periodic time of the input waveform (1 ms).

Assuming that we choose a medium value for R of, say, 10 kΩ, the minimum value which we could allow C to have would be that which satisfies the equation:

$$C \times R = 10t$$

where $R = 10 \text{ k}\Omega$ and $t = 1$ ms. Thus:

$$C = \frac{10t}{R} = \frac{10 \times 1 \text{ ms}}{10 \text{ k}\Omega} = 10 \times 10^{-3} \times 10^{-4} = 1 \times 10^{-6}\text{F}$$

or,

$$C = 1 \times 10^{-6} \text{ F} = 1 \text{ μF}$$

In practice, any value equal or greater than 1 μF would be adequate. A very large value (say more than 10 μF) will, however, generate a triangular wave which has a very small amplitude. To put this in simple terms, although the waveform might be what you want there's not a lot of it!

1.3.10   L-R Circuits

Networks of inductors and resistors (known as *L-R* circuits) can also be used for timing and pulse shaping. In comparison with capacitors, however, inductors are somewhat more difficult to manufacture and are consequently more expensive.

Inductors are also prone to losses and may also require screening to minimize the effects of stray magnetic coupling. Inductors are, therefore, generally unsuited to simple timing and waveshaping applications.

Figure 1.102 shows a simple *L-R* network in which an inductor is connected to a constant voltage supply. When the supply is first connected, the current, *i*, will rise exponentially with time, as shown in Figure 1.103. At the same time, the inductor voltage V_L, will fall, as shown in Figure 1.104). The rate of change of current with time will depend upon the ratio of inductance to resistance and is known as the *time constant*. Hence:

Time constant, $t = L/R$

where L is the value of inductance (H), R is the resistance (Ω), and t is the time constant (s).

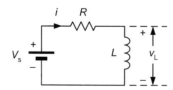

Figure 1.102: A C-R circuit in which C is initially charged and then discharges through R

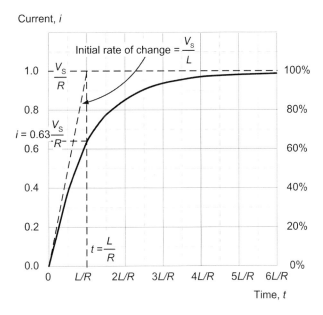

Figure 1.103: Exponential growth of current, *i*, in Figure 1.102

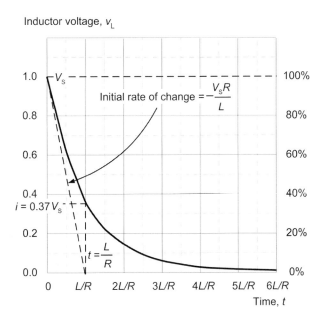

Figure 1.104: Exponential decay of voltage, v_L, in Figure 1.102

The current flowing in the inductor, i, varies with time, t, according to the relationship:

$$i = \frac{V_s}{R}\left(1 - e^{-\frac{tR}{L}}\right)$$

where V_s is the DC supply voltage, R is the resistance of the inductor, and L is the inductance.

The current, i, will initially be zero and will rise to approximately 63% of its maximum value (i.e., V_s/R) in a time interval equal to the time constant. At the end of the next interval of time equal to the time constant (i.e., after a total time of $2L/R$ has elapsed) the current will have risen by a further 63% of the remainder, and so on.

In theory, the current in the inductor will never become equal to V_s/R. However, after a period of time equal to $5L/R$, the current will to all intents and purposes be equal to V_s/R. At this point, the current in the inductor will have risen to 99.3% of its final value.

The voltage developed across the inductor, v_L, varies with time, t, according to the relationship:

$$v_L = V_s\, e^{-\frac{tR}{L}}$$

where V_s is the DC supply voltage, R is the resistance of the inductor, and L is the inductance.

The inductor voltage will fall to approximately 37% of the initial voltage in a time equal to the time constant.

At the end of the next interval of time equal to the time constant (i.e., after a total time of $2L/R$ has elapsed) the voltage will have fallen by a further 37% of the remainder, and so on.

Example 1.73

A coil having inductance 6H and resistance 24Ω is connected to a 12V DC supply. Determine the current in the inductor 0.1s after the supply is first connected.

Solution

The formula for exponential growth of current in the coil is:

$$i = \frac{V_s}{R}\left(1 - e^{-\frac{tR}{L}}\right)$$

where $V_s = 12\text{V}$, $L = 6\text{H}$ and $R = 24\Omega$.

We need to find i when $t = 0.1\text{s}$

$$i = \frac{12}{24}\left(1 - e^{-\frac{0.1 \times 24}{6}}\right) = 0.5\left(1 - e^{-0.4}\right) = 0.5(1 - 0.67)$$

thus,

$$i = 0.5 \times 0.33 = 0.165\text{A}$$

In order to simplify the mathematics of exponential growth and decay, Table 1.15 provides an alternative tabular method that may be used to determine the voltage and current in an *L-R* circuit.

Example 1.74

A coil has an inductance of 100 mH and a resistance of 10Ω. If the inductor is connected to a 5V DC supply, determine the inductor voltage 20 ms after the supply is first connected.

Solution

We will solve this problem using Table 1.15 rather than the exponential formula.

First we need to find the time constant:

$$L/R = 0.1\text{H}/10\Omega = 0.01\text{s}$$

Next we find the ratio of t to L/R.

When $t = 20$ ms the ratio of t to L/R is 0.02/0.01 or 2. Table 1.15 shows that when $t/(L/R) = 2$, the ratio of instantaneous value to final value (k) is 0.8647. Thus:

$$v_L/V_s = 0.8647$$

or,

$$v_L = 0.8647 \times V_s = 0.8647 \times 5V = 4.32V$$

1.4   Alternating Voltage and Current

This section introduces basic alternating current theory. We discuss the terminology used to describe alternating waveforms and the behavior of resistors, capacitors, and inductors when an alternating current is applied to them. The chapter concludes by introducing another useful component, the transformer.

1.4.1   Alternating Versus Direct Current

Direct currents are currents which, even though their magnitude may vary, essentially flow only in one direction. In other words, direct currents are *unidirectional*. Alternating currents, on the other hand, are *bidirectional* and continuously reverse their direction of flow. The polarity of the e.m.f. which produces an alternating current must consequently also be changing from positive to negative, and vice versa.

Alternating currents produce alternating potential differences (voltages) in the circuits in which they flow. Furthermore, in some circuits, alternating voltages may be superimposed on direct voltage levels (see Figure 1.105). The resulting voltage may be unipolar (i.e., always positive or always negative) or bipolar (i.e., partly positive and partly negative).

1.4.2   Waveforms and Signals

A graph showing the variation of voltage or current present in a circuit is known as a *waveform*. There are many common types of waveform encountered in electrical circuits including sine (or sinusoidal), square, triangle, ramp or sawtooth (which may be either positive or negative going), and pulse.

Complex waveforms, like speech and music, usually comprise many components at different frequencies. *Pulse waveforms* are often categorized as either repetitive

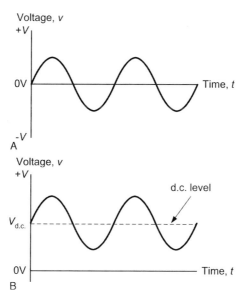

Figure 1.105: (A) Bipolar sine wave; (B) unipolar sine wave (superimposed on a *DC level*)

or nonrepetitive (the former comprises a pattern of pulses that repeats regularly while the latter comprises pulses which constitute a unique event). Some common waveforms are shown in Figure 1.106.

Signals can be conveyed using one or more of the properties of a waveform and sent using wires, cables, optical and radio links. Signals can also be processed in various ways using amplifiers, modulators, filters, etc. Signals are also classified as either *analog* (continuously variable) or *digital* (based on discrete states).

1.4.3 Frequency

The frequency of a repetitive waveform is the number of cycles of the waveform which occur in unit time. Frequency is expressed in hertz (Hz) and a frequency of 1 Hz is equivalent to one cycle per second. Hence, if a voltage has a frequency of 400 Hz, 400 cycles of it will occur in every second.

The equation for the voltage shown in Figure 1.105(A) at a time, *t*, is:

$$v = V_{max}\sin(2\pi f t)$$

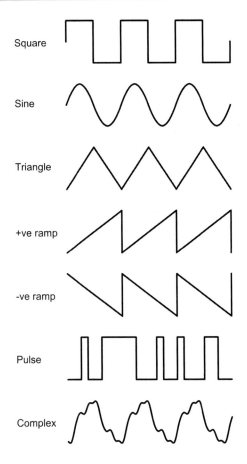

Figure 1.106: Common waveforms

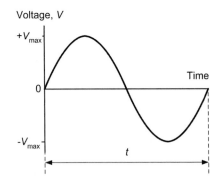

Figure 1.107: One cycle of a sine wave voltage showing its *periodic time*

while that in Figure 1.105(B) is:

$$v = V_{DC} + V_{max}\sin(20\pi ft)$$

where v is the instantaneous voltage, V_{max} is the maximum (or peak) voltage of the sine wave, V_{DC}, is the DC offset (where present), and f is the frequency of the sine wave.

Example 1.75

A sine wave voltage has a maximum value of 20V and a frequency of 50 Hz. Determine the instantaneous voltage present (a) 2.5 ms and (b) 15 ms from the start of the cycle.

Solution

We can find the voltage at any instant of time using:

$$v = V_{max} \sin(20\pi ft)$$

where $V_{max} = 20$V and $f = 50$ Hz.

In (a), $t = 2.5$ ms, hence:

$$v = 20 \sin(20\pi \times 50 \times 0.0025) = 20 \sin(0.785)$$
$$= 20 \times 0.707 = 14.14\text{V}$$

In (b), $t = 15$ ms, hence:

$$v = 20 \sin(20\pi \times 50 \times 0.0015) = 20 \sin(4.71)$$
$$= 20 \times -1 = -20\text{V}$$

1.4.4   Periodic Time

The periodic time (or *period*) of a waveform is the time taken for one complete cycle of the wave (see Figure 1.107). The relationship between periodic time and frequency is thus:

$$t = 1/f \quad \text{or} \quad f = 1/t$$

where t is the periodic time (in s) and f is the frequency (in Hz).

Example 1.76

A waveform has a frequency of 400 Hz. What is the periodic time of the waveform?

Solution

$$t = 1/f = 1/400 = 0.0025\text{s (or 2.5 ms)}$$

Example 1.77

A waveform has a periodic time of 40 ms. What is its frequency?

Solution

$$f = \frac{1}{t} = \frac{1}{40 \times 10^{-3}} = \frac{1}{0.04} = 25\text{Hz}$$

1.4.5   Average, Peak, Peak-Peak, and r.m.s. Values

The *average value* of an alternating current which swings symmetrically above and below zero will be zero when measured over a long period of time. Hence, average values of currents and voltages are invariably taken over one complete half-cycle (either positive or negative) rather than over one complete full-cycle (which would result in an average value of zero).

The *amplitude* (or *peak value*) of a waveform is a measure of the extent of its voltage or current excursion from the resting value (usually zero).

The *peak-to-peak value* for a wave which is symmetrical about its resting value is twice its peak value (see Figure 1.108).

The *r.m.s.* (or *effective*) *value* of an alternating voltage or current is the value which would produce the same heat energy in a resistor as a direct voltage or current of the same magnitude. Since the r.m.s. value of a waveform is very much dependent upon its shape, values are only meaningful when dealing with a waveform of known shape. Where the shape of a waveform is not specified, r.m.s. values are normally assumed to refer to sinusoidal conditions.

For a given waveform, a set of fixed relationships exist between average, peak, peak-peak, and r.m.s. values. The required multiplying factors are summarized for sinusoidal voltages and currents in Table 1.16.

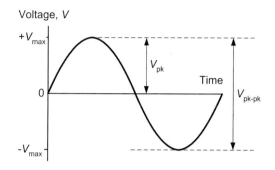

Figure 1.108: One cycle of a sine wave voltage showing its
peak **and** *peak-peak* **values**

Table 1.16: Multiplying factors for average, peak, peak-peak and r.m.s. values

Given	Wanted quantity			
Quantity	Average	Peak	Peak–peak	r.m.s.
Average	1	1.57	3.14	1.11
Peak	0.636	1	2	0.707
Peak-peak	0.318	0.5	1	0.353
r.m.s.	0.9	1.414	2.828	1

Example 1.78

A sinusoidal voltage has an r.m.s. value of 240V. What is the peak value of the voltage?

Solution

The corresponding multiplying factor (found from Table 1.16) is 1.414. Hence:

$$V_{pk} = 1.414 \times V_{r.m.s.} = 1.414 \times 240 = 339.4V$$

Example 1.79

An alternating current has a peak-peak value of 50 mA. What is its r.m.s. value?

Solution

The corresponding multiplying factor (found from Table 1.16) is 0.353. Hence:

$$I_{\text{r.m.s.}} = 0.353 \times V_{\text{pk-pk}} = 0.353 \times 0.05 = 0.0177\text{A (or } 17.7\text{mA)}.$$

Example 1.80

A sinusoidal voltage 10V pk-pk is applied to a resistor of 1 kΩ What value of r.m.s. current will flow in the resistor?

Solution

This problem must be solved in two stages. First we will determine the peak-peak current in the resistor and then we shall convert this value into a corresponding r.m.s. quantity.

Since $I = \dfrac{V}{R}$ we can infer that:

$$I_{\text{pk-pk}} = \frac{V_{pk\text{-}pk}}{R}$$

From which,

$$I_{\text{pk-pk}} = \frac{10}{1,000} = 0.01 = 10 \text{ mA pk-pk}$$

The required multiplying factor (peak-peak to r.m.s.) is 0.353. Thus:

$$I_{\text{r.m.s.}} = 0.353 \times I_{\text{pk-pk}} = 0.353 \times 10 = 3.53 \text{ mA}$$

1.4.6   Reactance

When alternating voltages are applied to capacitors or inductors the magnitude of the current flowing will depend upon the value of capacitance or inductance and on the frequency of the voltage. In effect, capacitors and inductors oppose the flow of current in much the same way as a resistor. The important difference being that the effective resistance (or reactance) of the component varies with frequency (unlike the case of a resistor where the magnitude of the current does not change with frequency).

1.4.7   Capacitive Reactance

The reactance of a capacitor is defined as the ratio of applied voltage to current and, like resistance, it is measured in Ohms. The reactance of a capacitor is inversely proportional to both the value of capacitance and the frequency of the applied voltage. Capacitive reactance can be found by applying the following formula:

$$X_C = \frac{1}{2\pi f C}$$

where X_c is the reactance (in ohms), f is the frequency (in hertz), and C is the capacitance (in farads).

Capacitive reactance falls as frequency increases, as shown in Figure 1.109. The applied voltage, V_c, and current, I_c, flowing in a pure capacitive reactance will differ in phase by an angle of 90° or $\pi/2$ radians (the *current leads the voltage*). This relationship is illustrated in the current and voltage waveforms (drawn to a common time scale) shown in Figure 1.110 and as a *phasor diagram* shown in Figure 1.111.

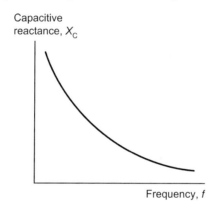

Figure 1.109: Variation of reactance with frequency for a capacitor

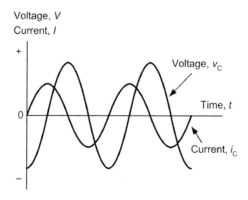

Figure 1.110: Voltage and current waveforms for a pure capacitor (the current leads the voltage by 90°)

Figure 1.111: Phasor diagram for a pure capacitor

Example 1.81

Determine the reactance of a 1 μF capacitor at (a) 100 Hz and (b) 10 kHz.

Solution

This problem is solved using the expression:

$$X_C = \frac{1}{2\pi fC}$$

(a) At 100 Hz:

$$X_C = \frac{1}{2\pi \times 100 \times 1 \times 10^{-6}} = \frac{0.159}{10^{-4}} = 1.59 \times 10^3$$

or,

$$X_C = 1.59 \text{ k}\Omega$$

(b) At 10 kHz:

$$X_C = \frac{1}{2\pi \times 1 \times 10^4 \times 1 \times 10^{-6}} = \frac{0.159}{10^{-2}} = 0.159 \times 10^2$$

or,

$$X_C = 15.9\Omega$$

Example 1.82

A 100 nF capacitor is to form part of a filter connected across a 240V 50 Hz mains supply. What current will flow in the capacitor?

Solution

First we must find the reactance of the capacitor:

$$X_C = \frac{1}{2\pi \times 50 \times 100 \times 10^{-9}} = 31.8 \times 10^3 = 31.8 \text{ k}\Omega$$

The r.m.s. current flowing in the capacitor will thus be:

$$I_C = \frac{V_C}{X_C} = \frac{240}{31.8 \times 10^3} = 7.5 \times 10^{-3} = 7.5 \text{ mA}$$

1.4.8 Inductive Reactance

The reactance of an inductor is defined as the ratio of applied voltage to current and, like resistance, it is measured in ohms. The reactance of an inductor is directly proportional to both the value of inductance and the frequency of the applied voltage. Inductive reactance can be found by applying the formula:

$$X_L = 2\pi f L$$

where X_L is the reactance in Ω, f is the frequency in Hz, and L is the inductance in H.

Inductive reactance increases linearly with frequency as shown in Figure 1.112. The applied voltage, V_1, and current, I_L, developed across a pure inductive reactance will differ in phase by an angle of 90° or $\pi/2$ radians (the *current lags the voltage*). This relationship is illustrated in the current and voltage waveforms (drawn to a common time scale) shown in Figure 1.113 and as a *phasor diagram* shown in Figure 1.114.

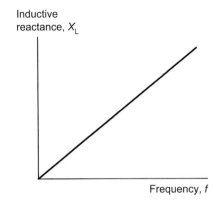

Figure 1.112: Variation of reactance with frequency for an inductor

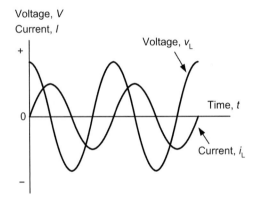

**Figure 1.113: Voltage and current waveforms for a pure inductor
(the voltage leads the current by 90°)**

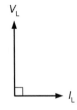

Figure 1.114: Phasor diagram for a pure inductor

Example 1.83

Determine the reactance of a 10 mH inductor at (a) 100 Hz and (b) at 10 kHz.

Solution

(a) at 100 Hz:

$$X_L = 2\pi \times 100 \times 10 \times 10^{-3} = 6.28\Omega$$

(b) At 10 kHz:

$$X_L = 2\pi \times 10 \times 10^3 \times 10 \times 10^{-3} = 628\Omega$$

Example 1.84

A 100 mH inductor of negligible resistance is to form part of a filter which carries a current of 20 mA at 400 Hz. What voltage drop will be developed across the inductor?

Solution

The reactance of the inductor will be given by:

$$X_L = 2\pi \times 400 \times 100 \times 10^{-3} = 251\Omega$$

The r.m.s. voltage developed across the inductor will be given by:

$$V_L = I_L \times X_L = 20 \text{ mA} \times 251\Omega = 5.02\text{V}$$

In this example, it is important to note that we have assumed that the DC resistance of the inductor is negligible by comparison with its reactance. Where this is not the case, it will be necessary to determine the *impedance* of the component and use this to determine the voltage drop.

1.4.9 Impedance

Figure 1.115 shows two circuits which contain both resistance and reactance. These circuits are said to exhibit impedance (a combination of resistance and reactance) which, like resistance and reactance, is measured in ohms.

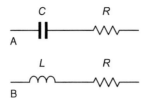

Figure 1.115: (A) *C* and *R* in series (B) *L* and *R* in series (note that both circuits exhibit an *impedance*)

The impedance of the circuits shown in Figure 1.115 is simply the ratio of supply voltage, V_S, to supply current, I_S. The impedance of the simple *C-R* and *L-R* circuits shown in Figure 1.115 can be found by using the impedance triangle shown in Figure 1.116. In either case, the impedance of the circuit is given by:

$$Z = \sqrt{R^2 + X^2}$$

and the phase angle (between V_S and I_S) is given by:

$$\phi = \tan^{-1}\left(\frac{X}{R}\right)$$

where Z is the impedance (in ohms), X is the reactance, either capacitive or inductive (expressed in ohms), R is the resistance (in ohms), and ϕ is the phase angle in radians.

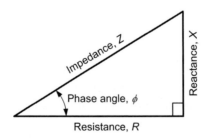

Figure 1.116: The impedance triangle

Example 1.85

A 2 μF capacitor is connected in series with a 100Ω resistor across a 115V 400 Hz AC supply. Determine the impedance of the circuit and the current taken from the supply.

Solution

First we must find the reactance of the capacitor, X_C:

$$X_C = \frac{1}{2\pi f C} = \frac{1}{6.28 \times 400 \times 2 \times 10^{-6}} = \frac{10^6}{5,024} = 199\Omega$$

Now we can find the impedance of the *C-R* series circuit:

$$Z = \sqrt{R^2 + X^2} = \sqrt{199^2 + 100^2} = \sqrt{49,601} = 223\Omega$$

The current taken from the supply can now be found:

$$I_S = \frac{V_S}{Z} = \frac{115}{223} = 0.52\text{A}$$

1.4.10 Power Factor

The power factor in an AC circuit containing resistance and reactance is simply the ratio of true power to apparent power. Hence:

$$\text{power factor} = \frac{\text{true power}}{\text{apparent power}}$$

The *true power* in an AC circuit is the power which is actually dissipated in the resistive component. Thus:

$$\text{true power} = I_S^2 \times R \ (\text{watts})$$

The *apparent power* in an AC circuit is the power which is apparently consumed by the circuit and is the product of the supply current and supply voltage (note that this is not the same as the power which is actually dissipated as heat). Hence:

$$\text{apparent power} = I_S^2 \times V_S (\text{volt - amperes})$$

Hence:

$$\text{power factor} = \frac{I_S^2 \times R}{I_S \times V_S} = \frac{I_S^2 \times R}{I_S \times (I_S \times Z)} = \frac{R}{Z}$$

From Figure 1.116,

$$\frac{R}{Z} = \cos\phi$$

Hence, the power factor of a series AC circuit can be found from the cosine of the phase angle.

Example 1.86

A *choke* (a form of inductor) having an inductance of 150 mH and resistance of 250Ω is connected to a 115V 400 Hz AC supply. Determine the power factor of the choke and the current taken from the supply.

Solution

First we must find the reactance of the inductor,

$$X_L = 2\pi \times 400 \times 0.15 = 376.8\Omega$$

We can now determine the power factor from:

$$\text{power factor} = \frac{R}{Z} = \frac{250}{376.8} = 0.663$$

The impedance of the choke, Z, will be given by:

$$Z = \sqrt{R^2 + X^2} = \sqrt{376.8^2 + 250^2} = 452\Omega$$

Finally, the current taken from the supply will be:

$$I_S = \frac{V_S}{Z} = \frac{115}{452} = 0.254A$$

1.4.11 L-C Circuits

Two forms of *L-C* circuits are illustrated in Figure 1.117. Figure 1.117(A) is a *series resonant* circuit, while Figure 1.117(B) constitutes a *parallel resonant* circuit. The impedance of both circuits varies in a complex manner with frequency.

The impedance of the series circuit in Figure 1.117(A) is given by:

$$Z = \sqrt{X_L^2 - X_C^2}$$

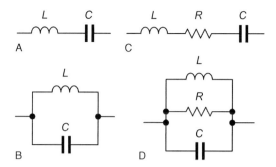

Figure 1.117: Series resonant and parallel resonant *L-C* and *L-C-R* circuits

where Z is the impedance of the circuit (in ohms), and X_L and X_C are the reactances of the inductor and capacitor respectively (both expressed in ohms).

The phase angle (between the supply voltage and current) will be $+\pi/2$ rad (i.e., $+90°$) when $X_L > X_C$ (above resonance) or $-\pi/2$ rad (or $-90°$) when $X_C > X_L$ (below resonance).

At a particular frequency (known as the *series resonant frequency*) the reactance of the capacitor, X_C, will be equal in magnitude (but of opposite sign) to that of the inductor, X_L. Due to this effective cancellation of the reactance, the impedance of the series resonant circuit will be zero at resonance. The supply current will have a maximum value at resonance (infinite in the case of a perfect series resonant circuit supplied from an ideal voltage source!).

The impedance of the parallel circuit in Figure 1.117B is given by:

$$Z = \frac{X_L \times X_C}{\sqrt{X_L{}^2 - X_C{}^2}}$$

where Z is the impedance of the circuit (in Ω), and X_L and X_C are the reactances of the inductor and capacitor, respectively (both expressed in Ω).

The phase angle (between the supply voltage and current) will be $+\pi/2$ rad (i.e., $+90°$) when $X_L > X_C$ (above resonance) or $\pi/2$ rad (or $-90°$) when $X_C > X_L$ (below resonance).

At a particular frequency (known as the *parallel resonant frequency*) the reactance of the capacitor, X_C, will be equal in magnitude (but of opposite sign) to that of the inductor, X_L. At resonance, the denominator in the formula for impedance becomes zero and thus the circuit has an infinite impedance at resonance. The supply current will have a minimum value at resonance (zero in the case of a perfect parallel resonant circuit).

1.4.12   L-C-R Circuits

Two forms of *L-C-R* network are illustrated in Figs 1.117(C)and 1.117(D); Figure 1.117 (C) is series resonant while Figure 1.117(D) is parallel resonant. As in the case of their simpler *L-C* counterparts, the impedance of each circuit varies in a complex manner with frequency.

The impedance of the series circuit of Figure 1.117(C) is given by:

$$Z = \sqrt{R^2 + (X_L - X_C)^2}$$

where Z is the impedance of the series circuit (in ohms), R is the resistance (in Ω), X_L is the inductive reactance (in Ω) and X_C is the capacitive reactance (also in Ω). At resonance the circuit has a minimum impedance (equal to R).

The phase angle (between the supply voltage and current) will be given by:

$$\phi = \tan^{-1}\left(\frac{X_L - X_C}{R}\right)$$

The impedance of the parallel circuit of Figure 1.117(D) is given by:

$$Z = \frac{R \times X_L \times X_C}{\sqrt{(X_L^2 - X_C^2) + R^2(X_L - X_C)^2}}$$

where Z is the impedance of the series circuit (in ohms), R is the resistance (in Ω), X_L is the inductive reactance (in Ω) and X_C is the capacitive reactance (also in Ω). At resonance the circuit has a minimum impedance (equal to R).

The phase angle (between the supply voltage and current) will be given by:

$$\phi = \tan^{-1}\frac{R(X_L - X_C)}{X_L \times X_C}$$

1.4.13 Resonance

The frequency at which the impedance is minimum for a series resonant circuit or maximum in the case of a parallel resonant circuit is known as the resonant frequency. The resonant frequency is given by:

$$f = \frac{1}{2\pi\sqrt{LC}}$$

where f_0 is the resonant frequency (in hertz), L is the inductance (in henries) and C is the capacitance (in farads).

Typical impedance-frequency characteristics for series and parallel tuned circuits are shown in Figs 1.118 and 1.119.

The series *L-C-R* tuned circuit has a minimum impedance at resonance (equal to R) and thus maximum current will flow. The circuit is consequently known as an *acceptor circuit*.

The parallel *L-C-R* tuned circuit has a maximum impedance at resonance (equal to R) and thus minimum current will flow. The circuit is consequently known as a *rejector circuit*.

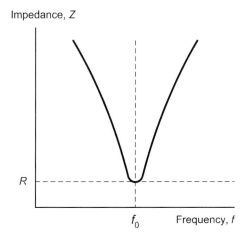

Figure 1.118: Impedance versus frequency for a series *L-C-R* acceptor circuit

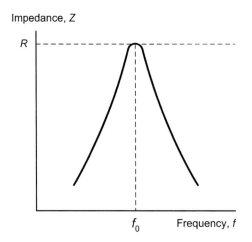

Figure 1.119: Impedance versus frequency for a parallel *L-C-R rejector circuit*

1.4.14 Quality Factor

The quality of a resonant (or tuned) circuit is measured by its *Q-factor*. The higher the *Q*-factor, the sharper the response (narrower bandwidth), conversely the lower the *Q*-factor, the flatter the response (wider bandwidth), see Figure 1.120. In the case of the series tuned circuit, the *Q*-factor will increase as the resistance, *R*, decreases. In the case of the parallel tuned circuit, the *Q*-factor will increase as the resistance, *R*, increases. The response of a tuned circuit can be modified by incorporating a resistance of

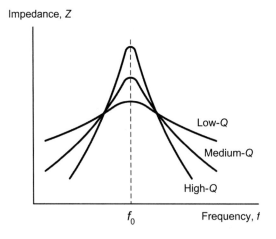

Figure 1.120: Effect of Q-factor on the response of a parallel resonant circuit (the response is similar, but inverted, for a series resonant circuit)

appropriate value either to "dampen" (low-Q) or "sharpen" (high-Q) the response. The relationship between bandwidth and Q-factor is:

$$\text{Bandwidth} = f_2 - f_1 = \frac{f_0}{Q} \text{ and } Q = \frac{2\pi f_0 L}{R}$$

where f_2 and f_1 are respectively the upper and lower cut-off (or *half-power*) frequencies (in Hertz), f_0 is the resonant frequency (in hertz), and Q is the Q-factor (see Figure 1.121).

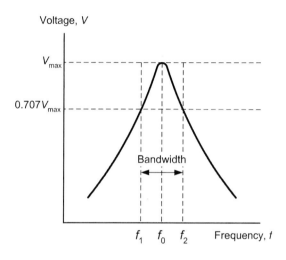

Figure 1.121: Bandwidth of a tuned circuit

Example 1.87

A parallel *L-C* circuit is to be resonant at a frequency of 400 Hz. If a 100 mH inductor is available, determine the value of capacitance required.

Solution

Rearranging the formula:

$$f = \frac{1}{2\pi\sqrt{LC}}$$

to make C the subject gives:

$$C = \frac{1}{f_0^2 (2\pi)^2 L}$$

Thus:

$$C = \frac{1}{400^2 \times 39.4 \times 100 \times 10^{-3}} = 1.58 \times 10^{-6} = 1.58 \ \mu F$$

This value can be made from preferred values using a 2.2 µF capacitor connected in series with a 5.6 µF capacitor.

Example 1.88

A series *L-C-R* circuit comprises an inductor of 20 mH, a capacitor of 10 nF, and a resistor of 100Ω. If the circuit is supplied with a sinusoidal signal of 1.5V at a frequency of 2 kHz, determine the current supplied and the voltage developed across the resistor.

Solution

First we need to determine the values of inductive reactance, X_L, and capacitive reactance X_C:

$$X_L = 20\pi \ fL = 6.28 \times 2 \times 10^3 \times 20 \times 10^{-3} = 251\Omega$$

$$X_C = \frac{1}{2\pi fC} = \frac{1}{6.28 \times 2 \times 10^3 \times 100 \times 10^{-9}} = 796.2\Omega$$

The impedance of the series circuit can now be calculated:

$$Z = \sqrt{R^2 + (X_L - X_C)^2} = \sqrt{100^2 + (251.2 - 796.2)^2}$$

From which:

$$Z = \sqrt{10,000 + 297,025} = \sqrt{307,025} = 554\Omega$$

The current flowing in the series circuit will be given by:

$$I = \frac{V}{Z} = \frac{1.5}{554} = 0.0027 = 2.7 \ mA$$

The voltage developed across the resistor can now be calculated using:

$$V = IR = 2.7mA \times 100\Omega = 270 \ mV$$

1.4.15   Transformers

Transformers provide us with a means of coupling AC power or signals from one circuit to another. Voltage may be *stepped-up* (secondary voltage greater than primary voltage) or *stepped-down* (secondary voltage less than primary voltage). Since no increase in power is possible (transformers are passive components like resistors, capacitors and inductors) an increase in secondary voltage can only be achieved at the expense of a corresponding reduction in secondary current, and vice versa (in fact, the secondary power will be very slightly less than the primary power due to losses within the transformer). Typical applications for transformers include stepping-up or stepping-down mains voltages in power supplies, coupling signals in AF amplifiers to achieve impedance matching and to isolate DC potentials associated with active components.

The electrical characteristics of a transformer are determined by a number of factors including the core material and physical dimensions. The specifications for a transformer usually include the rated primary and secondary voltages and current the required power rating (i.e., the maximum power, usually expressed in volt-amperes, VA) which can be continuously delivered by the transformer under a given set of conditions, the frequency range for the component (usually stated as upper and lower working frequency limits), and the *regulation* of a transformer (usually expressed as a percentage of full-load). This last specification is a measure of the ability of a transformer to maintain its rated output voltage under load.

Table 1.17 summarizes the properties of three common types of transformer. Figure 1.124 shows the construction of a typical iron-cored power transformer.

1.4.16   Voltage and Turns Ratio

The principle of the transformer is illustrated in Figure 1.125. The primary and secondary windings are wound on a common low-reluctance magnetic core. The alternating flux generated by the primary winding is therefore coupled into the secondary winding (very little flux escapes due to leakage). A sinusoidal current flowing in the primary winding produces a sinusoidal flux. At any instant the flux in the transformer is given by the equation:

$$\phi = \phi_{max} \sin(2\pi f t)$$

Table 1.17: Characteristics of common types of transformers

Property	Transformer core type			
	Air cored	Ferrite cored	Iron cored (audio)	Iron cored (power)
Core material/ construction	Air	Ferrite ring or pot	Laminated steel	Laminated steel
Typical frequency range (Hz)	30M to 1G	10k to 10M	20 to 20k	50 to 400
Typical power rating (VA)	(see note)	1 to 200	0.1 to 50	3 to 500
Typical regulation	(see note)	(see note)	(see note)	5% to 15%
Typical applications	RF tuned circuits and filters	Filters and HF transformers, switched mode power supplies	Smoothing chokes and filters, audio matching	Power supplies

Note: Not usually important for this type of transformer.

Figure 1.122: A selection of transformers with power ratings from 0.1 VA to 100 VA

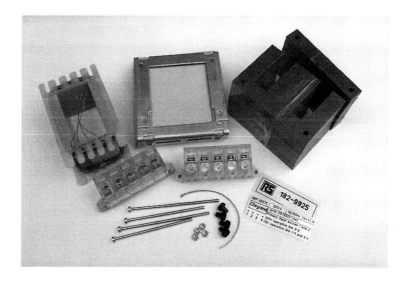

Figure 1.123: Parts of a typical iron-cored power transformer prior to assembly

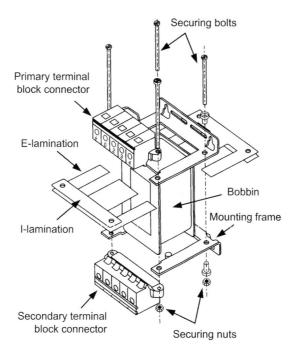

Figure 1.124: Construction of a typical iron-cored transformer

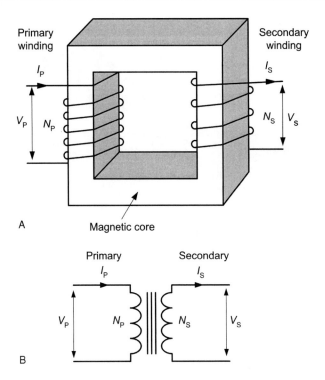

A

Magnetic core

B

Figure 1.125: The transformer principle

where ϕ_{max} is the maximum value of flux (in webers), f is the frequency of the applied current (in hertz), and t is the time in seconds.

The r.m.s. value of the primary voltage, V_P, is given by:

$$V_P = 4.44 f N_P \phi_{max}$$

Similarly, the r.m.s. value of the secondary voltage, V_S, is given by:

$$V_S = 4.44 f N_S \phi_{max}$$

Now:

$$\frac{V_P}{V_S} = \frac{N_P}{N_S}$$

where N_P/N_S is the *turns ratio* of the transformer.

Assuming that the transformer is loss-free, primary and secondary powers (P_P and P_S, respectively) will be identical. Hence:

$$P_P = P_S \quad \text{thus} \quad V_P \times I_P = V_S \times I_P$$

Hence,

$$\frac{V_P}{V_S} = \frac{I_S}{I_P} \quad \text{and} \quad \frac{I_S}{I_P} = \frac{N_P}{N_S}$$

Finally, it is sometimes convenient to refer to a *turns-per-volt* rating for a transformer. This rating is given by:

$$\text{turns-per-volt} = \frac{N_P}{V_P} = \frac{N_P}{V_S}$$

Example 1.89

A transformer has 2,000 primary turns and 120 secondary turns. If the primary is connected to a 220V r.m.s. AC mains supply, determine the secondary voltage.

Solution

Rearranging:

$$\frac{V_P}{V_S} = \frac{N_P}{N_S}$$

gives:

$$V_S = \frac{N_S \times V_P}{N_P} = \frac{120 \times 220}{2,000} = 13.2V$$

Example 1.90

A transformer has 1,200 primary turns and is designed to operate with a 200V AC supply. If the transformer is required to produce an output of 10V, determine the number of secondary turns required. Assuming that the transformer is loss-free, determine the input (primary) current for a load current of 2.5A.

Figure 1.126: Resonant air-cored transformer arrangement. The two inductors are tuned to resonance at the operating frequency (145 MHz) by means of the two small preset capacitors

Figure 1.127: This small 1:1 ratio toroidal transformer forms part of a noise filter connected in the input circuit of a switched mode power supply. The transformer is wound on a ferrite core and acts as a choke, reducing the high-frequency noise that would otherwise be radiated from the mains supply wiring

Solution

Rearranging:

$$\frac{V_P}{V_S} = \frac{N_P}{N_S}$$

gives:

$$N_S = \frac{N_P \times V_S}{V_P} = \frac{1,200 \times 10}{200} = 60 \text{ turns}$$

Rearranging:

$$\frac{I_S}{I_P} = \frac{N_P}{N_S}$$

gives:

$$N_S = \frac{N_S \times I_S}{N_P} = \frac{200 \times 2.5}{1,200} = 0.42\text{A}$$

1.5 Circuit Simulation

Computer simulation provides you with a powerful and cost-effective tool for designing, simulating, and analyzing a wide variety of electronic circuits. In recent years, the computer software packages designed for this task have not only become increasingly sophisticated but also have become increasingly easy to use. Furthermore, several of the most powerful and popular packages are now available at low cost either in evaluation, "lite" or student versions. In addition, there are several excellent freeware and shareware packages.

Whereas early electronic simulation software required that circuits were entered using a complex *netlist* that described all of the components and connections present in a circuit, most modern packages use an on-screen graphical representation of the circuit on test. This, in turn, generates a netlist (or its equivalent) for submission to the computational engine that actually performs the circuit analysis using mathematical models and *algorithms*. In order to describe the characteristics and behavior of components such as diodes and transistors, manufacturers often provide models in the form of a standard list of parameters.

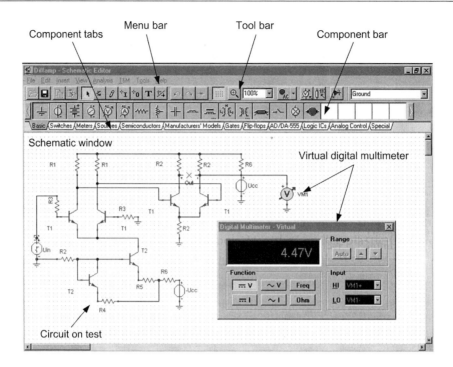

Figure 1.128: Using Tina Pro to construct and test a circuit prior to detailed analysis

Most programs that simulate electronic circuits use a set of algorithms that describe the behavior of electronic components. The most commonly used algorithm was developed at the Berkeley Institute in the United States and it is known as *SPICE* (Simulation Program with Integrated Circuit Emphasis).

Results of circuit analysis can be displayed in various ways, including displays that simulate those of real test instruments (these are sometimes referred to as *virtual instruments*). A further benefit of using electronic circuit simulation software is that, when a circuit design has been finalized, it is usually possible to export a file from the design/simulation software to a PCB layout package. It may also be possible to export files for use in screen printing or CNC drilling. This greatly reduces the time that it takes to produce a finished electronic circuit.

1.5.1 Types of Analysis

Various types of analysis are available within modern SPICE-based circuit simulation packages. These usually include:

1.5.1.1 DC Analysis

DC analysis determines the DC operating point of the circuit under investigation. In this mode any wound components (e.g., inductors and transformers) are short-circuited and any capacitors that may be present are left open-circuit. In order to determine the initial conditions, a DC analysis is usually automatically performed prior to a transient analysis. It is also usually performed prior to an AC small-signal analysis in order to obtain the linearized, small-signal models for nonlinear devices. Furthermore, if specified, the DC small-signal value of a transfer function (ratio of output variable to input source), input resistance, and output resistance is also computed as a part of the DC solution. The DC analysis can also be used to generate DC transfer curves in which a specified independent voltage or current source is stepped over a user-specified range and the DC output variables are stored for each sequential source value.

1.5.1.2 AC Small-Signal Analysis

The AC small-signal analysis feature of SPICE software computes the AC output variables as a function of frequency. The program first computes the DC operating point of the circuit and determines linearized, small-signal models for all of the nonlinear devices in the circuit (e.g., diodes and transistors). The resultant linear circuit is then analyzed over a user-specified range of frequencies. The desired output of an AC small-signal analysis is usually a transfer function (voltage gain, transimpedance, etc.). If the circuit has only one AC input, it is convenient to set that input to unity and zero phase, so that output variables have the same value as the transfer function of the output variable with respect to the input.

1.5.1.3 Transient Analysis

The transient analysis feature of a SPICE package computes the transient output variables as a function of time over a user-specified time interval. The initial conditions are automatically determined by a DC analysis. All sources that are not time dependent (for example, power supplies) are set to their DC value.

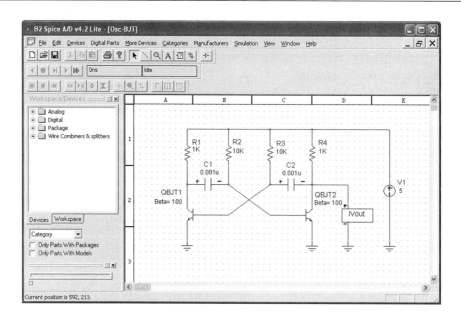

Figure 1.129: An astable multivibrator circuit being simulated using B2 Spice

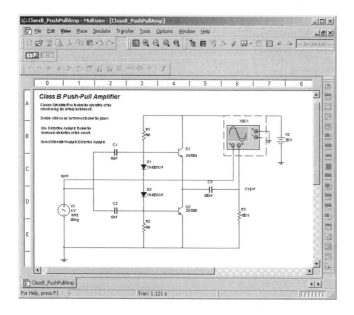

Figure 1.130: A Class B push-pull amplifier circuit being simulated by Multisim

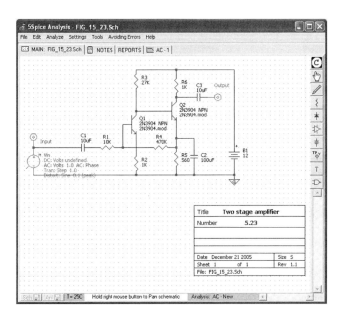

Figure 1.131: High-gain amplifier being analyzed using the 5Spice Analysis package

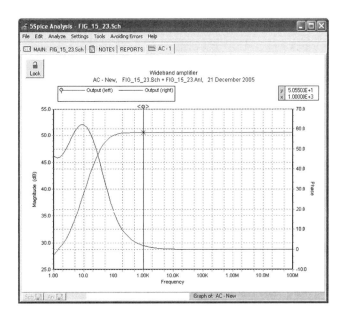

Figure 1.132: Gain and phase plotted as a result of small-signal AC analysis
of the circuit in Figure 1.131

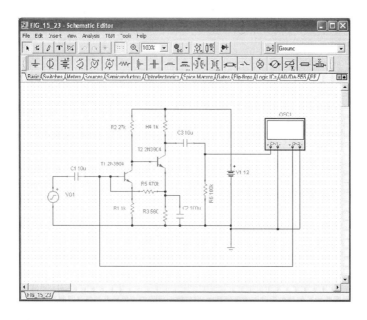

Figure 1.133: High-gain amplifier being analyzed using
the Tina Pro package

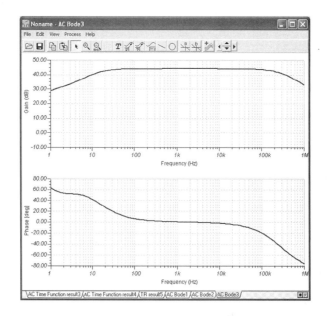

Figure 1.134: Gain and phase plotted as a result of small-signal AC analysis
of the circuit in Figure 1.133

1.5.1.4   Pole-zero Analysis

The pole-zero analysis facility computes the poles and/or zeros in the small-signal AC transfer function. The program first computes the DC operating point and then determines the linearized, small-signal models for all the nonlinear devices in the circuit. This circuit is then used to find the poles and zeros of the transfer function.

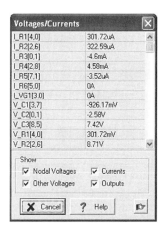

Figure 1.135: Results of DC analysis of the circuit shown in Figure 1.133

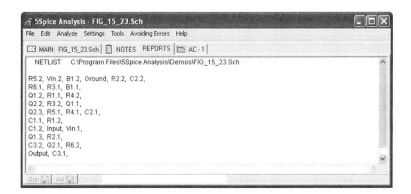

Figure 1.136: Computer generated netlist for the circuit shown in Figure 1.131

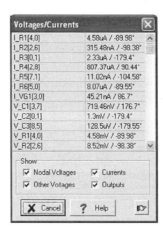

Figure 1.137: Results of AC analysis of the circuit shown in Figure 1.133

Two types of transfer functions are usually supported. One of these determines the voltage transfer function (i.e., output voltage divided by input voltage) and the other usually computes the output *transimpedance* (i.e., output voltage divided by input current) or *transconductance* (i.e., output current divided by input voltage). These two transfer functions cover all the cases and one can make it possible to determine the poles/zeros of functions like impedance ratio (i.e., input impedance divided by output impedance) and voltage gain. The input and output ports are specified as two pairs of nodes. Note that, for complex circuits it can take some time to carry out this analysis and the analysis may fail if there is an excessive number of poles or zeros.

1.5.1.5   Small-Signal Distortion Analysis

The distortion analysis facility provided by SPICE-driven software packages computes steady-state harmonic and inter-modulation products for small input signal magnitudes. If signals of a single frequency are specified as the input to the circuit, the complex values of the second and third harmonics are determined at every point in the circuit. If there are signals of two frequencies input to the circuit, the analysis finds out the complex values of the circuit variables at the sum and difference of the input frequencies, and at the difference of the smaller frequency from the second harmonic of the larger frequency.

1.5.1.6 Sensitivity Analysis

Sensitivity analysis allows you to determine either the DC operating-point sensitivity or the AC small-signal sensitivity of an output variable with respect to all circuit variables, including model parameters. The software calculates the difference in an output variable (either a node voltage or a branch current) by perturbing each parameter of each device independently. Since the method is a numerical approximation, the results may demonstrate second order affects in highly sensitive parameters, or may fail to show very low but nonzero sensitivity. Further, since each variable is perturbed by a small fraction of its value, zero-valued parameters are not analyzed (this has the benefit of reducing what is usually a very large amount of data).

1.5.1.7 Noise Analysis

The noise analysis feature determines the amount of noise generated by the components and devices (e.g., transistors) present in the circuit that is being analyzed. When provided with an input source and an output port, the analysis

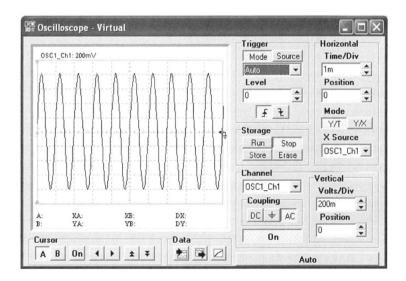

Figure 1.138: Using the virtual oscilloscope in Tina Pro to display the output voltage waveform for the circuit shown in Figure 1.133

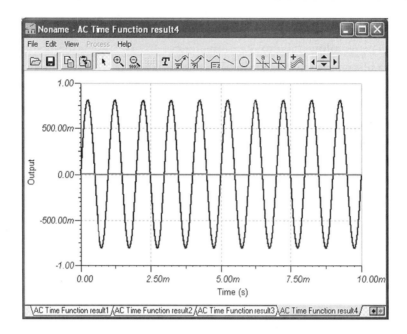

Figure 1.139: Alternative waveform plotting facility provided in Tina Pro

calculates the noise contributions of each device (and each noise generator within the device) to the output port voltage. It also calculates the input noise to the circuit, equivalent to the output noise referred to the specified input source. This is done for every frequency point in a specified range. After calculating the spectral densities, noise analysis integrates these values over the specified frequency range to arrive at the total noise voltage/current (over this frequency range).

1.5.1.8   Thermal Analysis

Many SPICE packages will allow you to determine the effects of temperature on the performance of a circuit. Most analyses are performed at normal ambient temperatures (e.g., 27°C) but it can be advantageous to look at the effects of reduced or increased temperatures, particularly where the circuit is to be used in an environment in which there is a considerable variation in temperature.

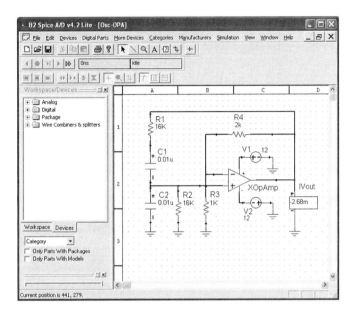

Figure 1.140: Analysis of a Wien Bridge oscillator using B2 Spice.

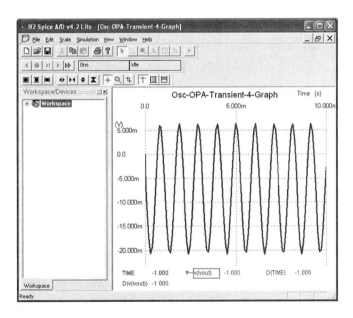

Figure 1.141: Transient analysis of the circuit in Figure 1.140 produced the output waveform plot

1.5.2 Netlists and Component Models

The following is an example of how a netlist for a simple *differential amplifier* is constructed (note that the line numbers have been included solely for explanatory purposes):

1. SIMPLE DIFFERENTIAL PAIR

2. VCC 7 0 12

3. VEE 8 0 -12

4. VIN 1 0 AC 1

5. RS1 1 2 1K

6. RS2 6 0 1K

7. Q1 3 2 4 MOD1

8. Q2 5 6 4 MOD1

9. RC1 7 3 10K

10. RC2 7 5 10K

11. RE 4 8 10K

12. MODEL MOD1 NPN BF = 50 VAF=50 IS=1.E-12 RB=100 CJC=.5PF TF=.6NS

13. TF V(5) VIN

14. AC DEC 10 1 100MEG

15. END

Lines 2 and 3 define the supply voltages. V_{CC} is +12V and is connected between node 7 and node 0 (signal ground). V_{EE} is –12V and is connected between node 8 and node 0 (signal ground). Line 4 defines the input voltage which is connected between node 1 and node 0 (ground) while lines 5 and 6 define 1 kΩ resistors (RS1 and RS2) connected between 1 and 2, and 6 and 0.

Lines 7 and 8 are used to define the connections of two transistors (Q1 and Q2). The characteristics of these transistors (both identical) are defined by MOD1 (see line 12). Lines 9, 10 and 11 define the connections of three further resistors (RC1, RC2 and RE respectively). Line 12 defines the transistor model. The device is NPN and has a current gain of 50. The corresponding circuit is shown in Figure 1.142.

Most semiconductor manufacturers provide detailed SPICE models for the devices that they produce. The following is a manufacturer's SPICE model for a 2N3904 transistor:

NPN (Is = 6.734f Xti = 3 Eg = 1.11 Vaf = 74.03 Bf = 416.4 Ne = 1.259
Ise = 6.734 Ikf = 66.78m Xtb = 1.5 Br = .7371 Nc = 2 Isc = 0 Ikr = 0
Rc = 1 Cjc = 3.638p Mjc = .3085 Vjc = .75 Fc = .5 Cje = 4.493p Mje = .2593
Vje = .75 Tr = 239.5n Tf = 301.2p Itf = .4 Vtf = 4 Xtf = 2 Rb = 10)

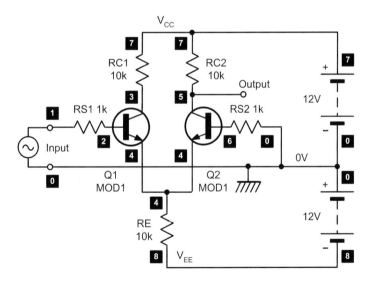

**Figure 1.142: Differential amplifier with the nodes marked for
generating a netlist**

1.5.3   Logic Simulation

As well as an ability to carry out small-signal AC and transient analysis of linear circuits (see Figure 1.130 and 1.143), modern SPICE software packages usually incorporate facilities that can be used to analyze logic and also "mixed-mode" (i.e., analog and digital) circuits. Several examples of digital logic analysis are shown in Figures 1.144, 1.145 and 1.146.

Figure 1.144 shows a four-stage shift register based on J-K bistables. The result of carrying out an analysis of this circuit is shown in Figure 1.145.

Finally, Figure 1.146 shows how a simple combinational logic circuit can be rapidly "assembled" and tested and its logical function checked. This circuit arrangement shows how the exclusive-OR function can be realized using only two-input NAND gates.

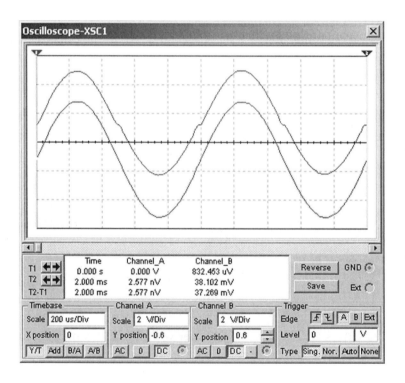

Figure 1.143: Cross-over distortion evident in the output waveform from the Class B amplifier shown in Figure 1.130

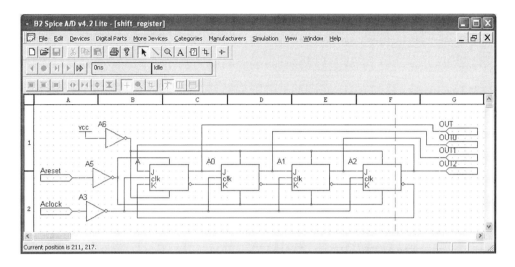

Figure 1.144: Four-stage circulating shift register simulated
using B2 Spice

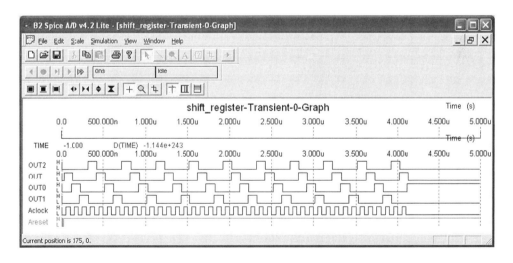

Figure 1.145: Waveforms for the four stage circulating shift register in
Figure 1.144

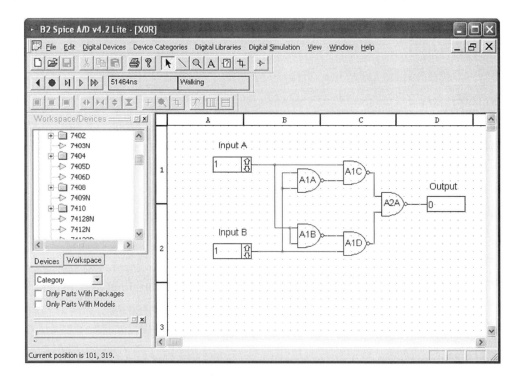

Figure 1.146: Using B2 Spice to check the function of a simple combinational logic circuit

1.6   Intuitive Circuit Design

Mechanical engineers have it easy. They can see what they are working on most of the time. As an EE, you do not usually have that luxury. You have to imagine how those pesky electrons are flittering around in your circuit. We are going to cover some basic comparisons that use things you are familiar with to create an intuitive understanding of a circuit. As a side benefit, you will be able to hold your own in a mechanical discussion as well. There are several reasons to do this:

- The typical person understands the physical world more intuitively than they understand the electrical one. This is because we interact with it using all of our senses, whereas the electrical world is still very magical, even to an educated engineer. This is because much of what happens inside a circuit cannot be seen, felt, or heard. Think about it. You flip on a light switch and the light goes on. You really don't consider how the electricity caused it to happen. Drag a heavy box across the floor, and you certainly understand the principle of friction.

- The rules for both disciplines are exactly the same. Once you understand one, you will understand the other. This is great, because you only have to learn the principles once. When you get a feel for what is happening inside a circuit, you can be an amazingly accurate troubleshooter. The human mind is an incredible instrument for simulation and, unlike a computer, it can make intuitive leaps to correct conclusions based on incomplete information. I believe that by learning these similarities you increase your mind's ability to put together clues to the operation and results of a given system, resulting in correct analysis. This will help your mind to "simulate" a circuit.

1.6.1   *Physical Equivalents of Electrical Components*

Before we move on to the physical equivalents, let's understand voltage, current and power. Voltage is the potential of the electron flow. Current is the amount of flow. Sometimes the best analogies are the old overused ones and that is true in this case. Think of it in terms of water in a squirt gun. Voltage is the amount of pressure in the gun. Pressure determines how far the water squirts, but a little pea shooter with a 30-foot shot and a dinky little stream won't get you soaked. Current is the size of the water stream, but a large stream that doesn't shoot far is not much help in a water fight. Voltage, current and power in electrical terms are related the same way. It is in fact a simple relationship; here is the equation:

$$voltage * current = power \qquad (1.1)$$

To get power you need both voltage and current. If either one of these are zero, you get zero power output. Let's discuss three basic components and look at how they relate to voltage and current.

1.6.2   The Resistor is Analogous to Friction

Think about what happens when you drag a heavy box across the floor. A force called *friction* resists the movement of the box. This friction is related to the speed of the box. The faster you try to move the box, the more the friction resists your work. It can be described by an equation.

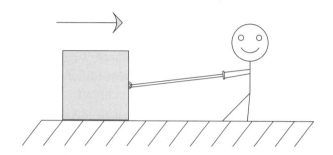

Figure 1.147: Friction resists smiley stick boy's efforts

$$friction = \frac{force}{speed} \tag{1.2}$$

Furthermore, the friction dissipates the energy loss in the system with heat. Let me rephrase that. Friction makes things get warm. Don't believe me? Try rubbing your hands together right now. Did you feel the heat? That is caused by friction. The function of a resistor in an electrical circuit is equal to friction. The resistor resists the flow of electricity just like friction resists the speed of the box. And, guess what, it heats up as it does so. An equation called *Ohm's Law* describes this relationship:

$$resistance = \frac{voltage}{current} \tag{1.3}$$

Do you see the similarity to the friction equation? They are exactly the same. The only real difference is the units you are working in.

1.6.3   The Inductor is Analogous to Mass

Let's stay with the box example for now. First let's eliminate friction, so as not to cloud our comprehension. The box is on a smooth track with virtually frictionless wheels.

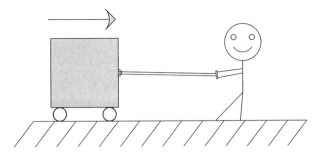

**Figure 1.148: Wheels eliminate friction but smiley has a hard time getting
it up to speed and stopping**

You notice that it takes some work to get the box going, but once moving, it
coasts along nicely. In fact, it takes work to get it to stop again. How much work
depends on how heavy the box is. This is known as the *law of inertia*. Newton
postulated this long before electricity was discovered, but it applies very well to
inductance. Mass resists a change in speed. Correspondingly, inductance resists
a change in current.

$$mass = \frac{force * time}{speed} \tag{1.4}$$

$$inductance = \frac{voltage * time}{current} \tag{1.5}$$

1.6.4   The capacitor is analogous to a spring

So what does a spring do? Take hold of a spring in your mind's eye. Stretch
it out and hold it, and then let it go. What happens? It snaps back into position.
A spring has a capacity to store energy. When a force is applied, it will hold that energy
till it is released. Capacitance is similar to the elasticity of the spring. (One note: the
spring constant that you may remember from physics texts is the inverse of the
elasticity.) I always thought it was nice that the word capacitor is used to represent
a component that has the capacity to store energy. (Technically an inductor can store
energy too. In a capacitor the energy is stored in the electric field that is generated
in and around the cap; in an inductor energy is stored in the magnetic field that is
generated. This energy stored in an inductor can be tapped very efficiently at high

currents. That is why most switching power supplies have an inductor in them as the primary passive component.)

$$spring = \frac{speed * time}{force} \qquad (1.6)$$

$$capacitance = \frac{current * time}{force} \qquad (1.7)$$

1.6.5   A Tank Circuit

Take the basic tank or LC circuit. What does it do? It oscillates. A perfect circuit would go on forever at the resonant frequency. How should this appear in our mechanical circuit? Think about the equivalents: an inductor and a capacitor, a spring and mass. In a thought experiment, hook the spring up to the box from the previous drawing. Now give it a tug. What happens? It oscillates.

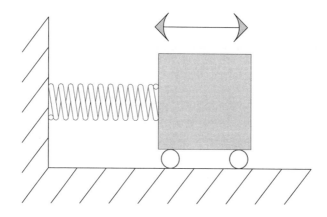

Figure 1.149: Get this started and it will keep bouncing until friction brings it to a halt

1.6.6   A Complex Circuit

Let's follow this reasoning for an LCR circuit. All we need to do is add a little resistance, or friction, to the mass-spring of the tank circuit. Let's tighten the wheels on our box a little too much so that they rub. What will happen after you give the box a tug? It will bounce back and forth a bit till it comes to a stop. The friction in the wheels slows it down. This friction component is called a *damper*, because it dampens the oscillation. What is it that a resistor does to an LC circuit? It dampens the oscillation.

There you have it, the world of electricity reduced to everyday items. Since these components are so similar, all the math tricks you may have learned apply as well to one system as they do to the other. Remember Fourier's theorems? They were discovered for mechanical systems long before anyone realized that they work for electrical circuits as well. Remember all that higher math you used to know or are just now learning about—Laplace transforms, integrals, derivatives, etc.? It all works the same in both worlds. You can solve a mechanical system using Laplace methods just the same as an electrical circuit.

Back in the '50s and '60s, the government spent mounds of dough using electrical circuits to model physical systems as described above. Why? You can get into all sorts of integrals, derivatives and other ugly math when modeling real-world systems. All that can get jumbled quickly after a couple of orders of complexity. Think about an artillery shell fired from a tank. How do you predict where it will land? You have the friction of the air, the mass of the shell, the spring of the recoil. Instead of trying to calculate all that math by hand, you can build a circuit with all the various electrical components representing the mechanical ones, hook up an oscilloscope, and fire away. If you want to test 1000 different weights of artillery at different altitudes, electrons are much cheaper than gunpowder.

Thumb Rules

- It takes voltage and current to make power.
- A resistor is like friction, it creates heat from current flow (resisting it), proportional to voltage across it.
- An inductor is like a mass.
- A capacitor is like a spring.
- The inductor is the inverse of the capacitor.

1.6.7   Learn an Intuitive Approach

1.6.7.1   Intuitive Signal Analysis

I'm not sure if this is actually taught in school. This is my name for it. It is something I learned on my own in college and the workplace. I didn't call it an actual discipline until I had been working for a while and had explained my methods to fellow engineers to help them solve their own dilemmas. I do think, however, that a lot of so-called bright people out there use this skill without really knowing it or putting a name to it. They seem to be able to point to something you have been working on for hours and say "your problem is there." They just seem to know intuitively what should happen. I believe this is a skill that can and should be taught.

There are three underlying principles needed to apply intuitive signal analysis. (Let's just call it *ISA*. After all, if I have any hope of this catching on in the engineering world, it has to have an acronym!)

First, you must drill the basics. For example, what happens to the impedance of a capacitor as frequency increases? It goes down. You should know that type of information off the top of your head. If you do, you can identify a high-pass or low-pass filter immediately. How about the impedance of an inductor—what does it do as frequency increases? What does negative feedback do to an op-amp, how does its output change? You do not necessarily need to know every equation by heart, but you do need to know direction of the change. As far as the magnitude of the change is concerned, if you have a general idea of the strength of the signal that is usually enough to zero in on the part of the circuit that is not doing what you want it to.

Second, you need experience and lots of it. You need to get a feel for how different components work. You need to spend a lot of time in the lab and you need to understand the basics of each component. You need to know what a given signal will do as it passes through a given component. Remember the physical equivalents of the basic components? These are the building blocks of your ability to visualize the operation of a circuit. You must imagine what is happening inside the circuit as the input changes. If you can visualize that, you can predict what the outputs will do.

Third, break the problem down. "How do you eat an elephant?" the knowledge seeker asked the wise old man. "One bite at a time," he replied. Pick a point to start and walk though it. Take the circuit and break it down into smaller chunks that can be handled easily. Draw arrows step by step that show the changes of signals in the circuit. "Does current go up here?" "Voltage at such and such point should be going down." These are the types of questions and answers you should be mumbling to yourself. Again, one thing you do not need to know is what the output will be precisely. You do not need to memorize every equation in the book to intuitively know your circuit, but you do need to know what effect changing a value of a component will have. For example, given a low-pass RC filter and an AC signal input, if you increase the value of the capacitor, what should happen to the amplitude of the output? Will it get smaller or larger? You should know immediately with something this basic that the answer is "smaller." You should also know that how much smaller depends on the frequency of the signal and the time constant of the filter. What happens as you increase current into the base of a transistor? Current through the collector increases. What happens to voltage across a resistor as current decreases? These are simple effects of components, but you would be surprised at how many engineers don't know the answers to these types of questions off the top of their head.

Spending a lot of time in the lab will help immensely to develop this skill. If you look at the response of a lot of different circuits many, many times, you will learn how they should act. When this knowledge is integrated, a wonderful thing happens. Your head becomes a circuit simulator. You will be able to sum up the effects caused by the various components in the circuit and intuitively understand what is happening. Let me show you an example.

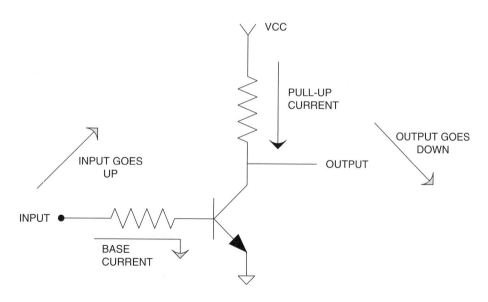

Figure 1.150: Use arrows to visualize what is happening to voltage and current

Now at this time you may not have a clue as to what a transistor is, so you may need to file this example away until you get past the transistor chapter, but be sure to come back to it so that the "Aha!" light bulb clicks on over your head. The analysis idea is what I am trying to get across, as you need it early on, but it creates a type of chicken-and-egg dilemma when it comes to an example. So for now, consider this example with the knowledge that the transistor is a device that moves current through the output that is proportional to the current through the base.

As voltage at the input above increases, base current increases. This causes the pullup current to increase, resulting in a larger voltage drop across the pull-up resistor. This means the voltage at the output must go *down* as the voltage at the input goes *up*. That is an example of putting it all together to really understand how a circuit works.

One way to develop this intuitive understanding is by use of computer simulators. It is easy to change a value and see what effect it has on the output and you can try several different configurations in a short amount of time. However, you have to be careful with these tools. It is easy to fall into a common trap, trusting the simulator so much that you will think there is something wrong with the real world when it doesn't work right in the lab. The real world is not at fault! It is the simulator that is missing something. I think it is best for the engineer to begin using simulators to model simple circuits. Don't jump into a complex model until you grasp what the basic components do—for example, modeling a step input into a RC circuit. With a simple model like this, change the values of R and C to see what happens. This is one way an engineer can develop the correct intuitive understanding of these two components. One word of warning though—don't spend all your time on the simulator. Make sure you get some good bench time too.

You will find this signal analysis skill very useful in diagnosing problems as well as in your design efforts. As your intuitive understanding increases, you will be able to leap to correct conclusions without all the necessary facts. You will know when you are modeling something incorrectly because the result just doesn't look right. Intuition is a skill no computer has, so make sure you take advantage of it!

Thumb Rules

- Drill the basics; know the basic formulas by heart.
- Get a lot of experience with basic circuits; the goal is to intuitively know how a signal will be affected by a component. Spend a lot of time on the bench!
- Break the problem down; draw arrows and notes on the schematic that indicate what the signal is doing.
- Determine what direction the signal is going; is it inversely related or directly related?
- Develop estimation abilities.
- Spend time on the bench with a scope and simple components.

1.6.8 "Lego" Engineering

1.6.8.1 Building Blocks

OK, so I came up with a fourth item. One of my instructors taught me a secret that I would like to pass on. Almost every discipline is easier to understand than you might think. The secret professors don't want you to know is there are usually about five or six basic principles or equations that lie at the bottom of the pile, so to speak. These fundamentals, once they are grasped, will allow you to derive the rest of the principles or equations in that field. They are like the old simple Legos®; you had five or six shapes to make everything. If you truly understand these few basic fundamentals in a given discipline, you will excel in that discipline. One other thing this instructor often said was that all the great discoveries were only one or two levels above these fundamentals. This means if you really know the basics well, you will excel at the rest. One thing you can be sure of is the human tendency to forget. All the higher level stuff is often left unused and will quickly be forgotten, but even an engineer-turned-manager like me uses the basics nearly every day.

Since this is a book on electrical engineering, let's list the fundamental equations for electrical circuits as I see them:

- Ohm's Law

- Voltage divider rule

- Capacitors impede changes in voltage

- Inductors impede changes in current

- Series and parallel resistors

- Thevenin's theorem

Let me touch on a couple of examples. You may say, "You didn't even list series and parallel capacitors. Isn't that a basic rule?" Well, you are right, it is fairly basic but it really isn't at the bottom of the pile. Series and parallel resistors are even more fundamental, because all that really happens when you add in the caps is the frequency of the signal is taken into account, other than that it is exactly the same equation! You would be better served to understand how a capacitor or inductor works and apply it to the basics than try to memorize too many equations. "What about Norton's theorem?" you might ask. Bottom line, it is just the flip side of Thevenin's theorem, so why learn two when one will do? I prefer to think of it terms of voltage so I set this to memory. You could work in terms of current and use Norton's theorem, but you would arrive at the same answer at the end of the day. So pick one and go with it.

You can always look the more advanced stuff up, but most of the time a solid application of the basics will force the problem at hand to submit to your engineering prowess. These six rules are things that you should memorize, understand and be able to do approximations of in your head. These are the rules that will make the intuition you are developing a powerful tool. It will unleash the simulation capability that you have right in your own brain.

If you really take this advice to heart, years down the road when you've advance to "management" and you have forgotten all the advanced stuff you used to know, you will still be able to solve engineering problems to the amazement of your engineers.

This can be generalized to all disciplines. Look at what you are trying to learn, figure out the few basic points being made from which you can derive the rest, and you will have discovered the basic "Legos" for that subject. Those are what you should know forward and backward to succeed in that field. Besides, Legos are fun, aren't they?

Thumb Rules

- There are a few rules in any discipline from which you can derive the rest.
- Learn these rules by heart; gain an intuitive understanding of them.
- Most significant discoveries are only a level or two above these basics.

1.7   Troubleshooting Basics

1.7.1   *If Only Everything Would Always Go Right...*

Why are we interested in troubleshooting? Because even the best engineers take on projects whose requirements are so difficult and challenging that the circuits don't work as expected—at least not the first time. I don't have data on switching regulators, but I read in an industry study that when disk drives are manufactured, the fraction that fails to function when power is first applied typically ranges from 20 to 70%. Of course, this fraction may occasionally fall as low as 1% and rise as high as 100%. But, on the average, production engineers and technicians must be prepared to repair 20, 40, or 60% of these complex units. Switching-regulated power supplies can also be quite complex. If you manufacture them in batches of 100, you shouldn't be surprised to find some batches with 12 pieces that require troubleshooting and other batches that have 46 such pieces. The troubleshooting may, as you well know, be tough with a new product whose bugs haven't been worked out. But it can be even tougher when the design is old and the parts it now uses aren't quite like the ones you used to be able to buy. Troubleshooting can be tougher still when there isn't much documentation describing how the product is supposed to work, and the designer isn't around any more. If there's ever a time when troubleshooting isn't needed, it's just a temporary miracle. You might try to duck your troubleshooting for a while. You might pretend that you can avoid the issue.

And, what if you decide that troubleshooting isn't necessary? You may find that your first batch of products has only three or four failures, so you decide that you don't need to worry. The second batch has a 12% failure rate, and most of the rejects have the same symptoms as those of the first batch. The next three batches have failure rates of 23, 49, and 76%, respectively. When you finally find the time to study the problems, you will find that they would have been relatively easy to fix if only you had started a couple of months earlier. That's what Murphy's Law can do to you if you try to slough off your troubleshooting chores...we have all seen it happen.

If you have a bunch of analog circuits that you have to troubleshoot, well, why don't you just look up the troubleshooting procedures in a book? The question is excellent, and the answer is very simple: Until now, almost nothing has been written about the troubleshooting of these circuits. The best previous write-up that I have found is a couple pages in a book by Jiri Dostal (Ref. 1.5). He gives some basic procedures for looking for trouble in a fairly straightforward little circuit: a voltage reference/regulator. As far as Dostal goes, he does quite well. But, he only offers a few pages of troubleshooting advice, and there is much to explain beyond what he has written.

Another book that has several good pages about the philosophy of troubleshooting is by John I. Smith (Ref. 1.6). Smith explains some of the foibles of wishing you had designed a circuit correctly when you find that it doesn't work "right." Unfortunately, it's out of print. Analog Devices sells a *Data Converter Handbook* (Ref. 1.7), and it has a few pages of good ideas and suggestions on what to look for when troubleshooting data converter and analog circuits.

What's missing, though, is general information. When I started writing about this troubleshooting stuff, I realized there was a huge vacuum in this area. So I have filled it up, and here we are.

You'll probably use general-purpose test equipment. What equipment can you buy for troubleshooting? For now, let me observe that if you have several million dollars worth of circuits to troubleshoot, you should consider buying a $100,000 tester. Of course, for that price you only get a machine at the low end of the line. And, after you buy the machine, you have to invest a lot of time in fixturing and software before it can help you. Yes, you can buy a $90 tester that helps locate short circuits on a PC board but, in the price range between $90 and $100,000, there isn't a lot of specialized troubleshooting equipment available. If you want an oscilloscope, you have to buy a general-purpose oscilloscope; if you want a DVM, it will be a general-purpose DVM.

Now, it's true that some scopes and some DVMs are more suitable for troubleshooting than others (and I will discuss the differences in the next chapter), but, to a large extent, you have to depend on your wits.

Your wits: Ah, very handy to use, your wits—but, then what? One of my favorite quotes from Jiri Dostal's book says that troubleshooting should resemble fencing more closely than it resembles wrestling. When your troubleshooting efforts seem like wrestling in the mud with an implacable opponent (or component), then you are probably not using the right approach. Do you have the right tools, and are you using them correctly? Do you know how a failed component will affect your circuit, and do you know what the most likely failure modes are? Ah, but do you know how to think about Trouble? That is this section's main lesson.

Even things that can't go wrong, do. One of the first things you might do is make a list of all the things that could be causing the problem. This idea can be good up to a point. I am an aficionado of stones about steam engines, and here is a story from the book Muster *Builders* of *Steam* (Ref. 1.8). A class of new 3-cylinder4-6-0 (four small pilot wheels in front of the drive wheels, six drive wheels, no little trailing wheels) steam engines had just been designed by British designer W. A. Stanier, and they were—perfect stinkers. They simply would not steam." So the engines' designers made a list of all the things that could go wrong and a list of all the things that could not be at fault; they set the second list aside.

The designers specified changes to be made to each new engine in hopes of solving the problem: "Teething troubles bring modifications, and each engine can carry a different set of modifications." The manufacturing managers "shuddered as these modified drawings seemed to pour in from Derby (Ed: site of the design facility—the Drawing Office), continually upsetting progress in the works." (Lots of fun for the manufacturing guys, eh?) In the end, the problem took a long time to find because it was on the list of "things that couldn't go wrong."

Allow me to quote the deliciously horrifying words from the text: "Teething troubles always present these two difficulties: that many of the clues are very subjective and that the 'confidence trick' applies. By the latter I mean when a certain factor is exonerated as trouble-free based on a sound premise, and everyone therefore looks elsewhere for the trouble: whereas in fact, the premise is not sound and the exonerated factor is guilty. In Stanier's case this factor was low super-heat. So convinced was he that a low degree of super-heat was adequate that the important change to increased superheater area was delayed far longer than necessary. There were some very sound men in the Experimental Section of the Derby Loco Drawing Office at that time, but they were young and their voice was only dimly heard. Some of their quite painstaking superhcatcr tcst results were disbelieved." But, of course, nothing like that ever happened to anybody you know—right?

1.7.2   Experts Have no Monopoly on Good Advice

Another thing you can do is ask advice only of "experts." After all, only an expert knows how to solve a difficult problem—right? Wrong! Sometimes, a major reason you can't find your problem is because you are too close to it—you are blinded by your familiarity. You may get excellent results by simply consulting one or two of your colleagues who are not as familiar with your design: they may make a good guess at a solution to your problem. Often a technician can make a wise (or lucky) guess as easily as can a savvy engineer. When that happens, be sure to remember who saved your neck. Some people are not just "lucky"; they may have a real knack for solving tricky problems, for finding clues, and for deducing what is causing the trouble. Friends like these can be more valuable than gold.

At National Semiconductor, we usually submit a newly designed circuit layout to a review by our peers. I invite everybody to try to win a Beverage of Their Choice by catching a real mistake in my circuit. What we *really* call this is a "Beercheck." It's fun because if I give away a few pitchers of brew, I get some of my dumb mistakes corrected—mistakes that I myself might not have found until a much-later, more-painful, and more-expensive stage. Furthermore, we all get some education. And, you can never predict who will find the little picky errors or the occasional real killer mistake. All technicians *and* engineers are invited.

1.7.3   Learn to Recognize Clues

There are four basic questions that you or I should ask when we are brought in to do troubleshooting on someone else's project:

- Did it *ever* work right?
- What are the symptoms that tell you it's not working right?
- When did it start working badly or stop working?
- What other symptoms showed up just before, just after, or at the same time as the failure?

As you can plainly see, the clues you get from the answers to these questions might easily solve the problem right away; if not, they may eventually get you out of the woods. So even if a failure occurs on your own project, you should ask these four questions—as explicitly as possible—of yourself or your technician or whoever was working on the project. For example, if your roommate called you to ask for a lift because the car had just quit in the middle of a freeway, you would ask whether anything else happened or if the car just died. If you're told that the headlights seemed to be getting dimmer and dimmer, that's a *clue*.

Figure 1.151: Peer review is often effective for wringing problems out of designs. Here, the author gets his comeuppance from colleagues who have spotted a problem *because* they are not as overly familiar with his circuit layout as he is. (Photo by Steve Allen.)

1.7.4 Ask Questions; Take Notes; Record Amount of Funny

When you ask these four questions, make sure to record the answers on paper—preferably in a notebook. As an old test manager I used to work with, Tom Milligan, used to tell his technicians, "When you are taking data, if you see something funny, Record Amount of Funny." That was such a significant piece of advice, we called it *Milligan's Law*. A few significant notes can save you hours of work. Clues are where you find them; they should be saved and savored.

Ask not only these questions but also any other questions suggested by the answers. For example, a neophyte product engineer will sometimes come to see me with a batch of ICs that have a terrible yield at some particular test. I'll ask if the parts failed any other tests, and I'll hear that nobody knows because the tester doesn't continue to test a part after it detects a failure. A more experienced engineer would have already retested the devices in the RUN ALL TESTS mode, and that is exactly what I instruct the neophyte to do.

Likewise, if *you* are asking another person for advice, you should have all the facts laid out straight, at least in your head, so that you can be clear and not add to the confusion.

I've worked with a few people who tell me one thing and a minute later start telling me the opposite. Nothing makes me lose my temper faster! Nobody can help you troubleshoot effectively if you aren't sure whether the circuit is running from +12V or ±12V and you start making contradictory statements.

And, if I ask when the device started working badly, don't tell me, "At 3:25 PM." I'm looking for clues, such as, "About two minutes after I put it in the 125°C oven," or, "Just after I connected the 4Ω load." So just as we can all learn a little more about troubleshooting, we can all learn to watch for the clues that are invaluable for fault diagnosis.

1.7.5   Methodical, Logical Plans Ease Troubleshooting

Even a simple problem with a resistive divider offers an opportunity to concoct an intelligent troubleshooting plan. Suppose you had a series string of 128 1 kΩ resistors. (See Figure 1.152.) If you applied 5V to the top of the string and 0V to the bottom, you would expect the midpoint-of the string to be at 2.5V. If it weren't 2.5V but actually 0V, you could start your troubleshooting by checking the voltage on each resistor, working down from the top, one by one. But that strategy would be absurd! Check the voltage at, say, resistor #96, the resistor which is halfway up from the midpoint to the top. Then, depending on whether that test is high, low, or reasonable, try at #112 or #80—at 5/8 or 7/8 of the span—then at #120 or #104 or #88 or #72, branching along in a sort of binary search—that would be much more effective. With just a few trials (about seven) you could find where a resistor was broken open or shorted to ground. Such branching along would take a lot fewer than the 64 tests you would need to walk all the way down the string.

Further, if an op-amp circuit's output were pegged, you would normally check the circuit's op-amp, resistors, or conductors. You wouldn't normally check the capacitors, unless you guessed that a shorted capacitor could cause the output to peg. Conversely, if the op-amp's V_{OUT} was a few dozen millivolts in error, you might start checking the resistors for their tolerances. You might not check for an open-circuited or wrong-value capacitor, *unless* you checked the circuit's output with a scope and discovered it oscillating!! So, in any circuit, you must study the data—your "clues"—until they lead you to the final test that reveals the true cause of your problem.

Thus, you should always first formulate a hypothesis and then invent a reasonable test or series of tests, the answers to which will help narrow down the possibilities of what is bad, and may in fact support your hypothesis. These tests should be performable. But you may define a test and then discover it is not performable or would be much too difficult to perform. Then I often think, "Well, if I could do that test, the answer would either come up 'good' or 'bad.' OK, so I can't easily run the test. But if I assume that I'd get one or the other of the answers, what would I do next to nail down the solution? Can I skip to the next test?"

For example, if I had to probe the first layer of metal on an IC with two layers of metal (because I had neglected to bring an important node up from the first metal to the second metal), I might do several other tests instead. I would do the other tests hoping that maybe I wouldn't have to do that probing, which is rather awkward even if I can "borrow" a laser to cut through all the layers of oxide. If I'm lucky, I may never have to go back and do that "very difficult or nearly impossible" test.

Of course, sometimes the actual result of a test is some completely unbelievable answer, nothing like the answers I expected. Then I have to reconsider—where were my assumptions wrong? Where was my thinking erroneous? Or, did I take my measurements correctly? Is my technician's data really valid? That's why troubleshooting is such a challenging business—almost never boring.

On the other hand, it would be foolish for you to plan everything and test nothing. Because if you did that, you would surely plan some procedures that a quick test would show are unnecessary. That's what they call "paralysis by analysis." All things being equal, I would expect the planning and testing to require equal time. If the tests are very complicated and expensive, then the planning should be appropriately comprehensive. If the tests are simple, as in the case of the 128 resistors in series, you could make them up as you go along. For example, the list above of resistors #80, 112, 120, 104, 88, or 72 are nominally binary choices. You don't have to go to exactly those places—an approximate binary search would be just fine.

1.7.6 You can make Murphy's Law Work for You

Murphy's Law is quite likely to attack even our best designs: "If anything can go wrong, it will." But, I can make Murphy's Law work *for* me. For example, according to this interpretation of Murphy's Law, if I drive around with a fire extinguisher, if I am prepared to put out any fire—will that make sure that I never have a fire in my car? When you first hear it, the idea sounds dumb. But, if I'm the kind of meticulous person who carries a fire extinguisher, I may also be neat and refuse to do the dumb things that permit fires to start.

Similarly, when designing a circuit I leave extra safety margins in areas where I cannot surely predict how the circuit will perform. When I design a breadboard, I often tell the technician, "Leave 20% extra space for *this* section because I'm not sure that it will work without modifications. And, please leave extra space around *this* resistor and *this* capacitor because I might have to change those values." When I design an IC, I leave little pads of metal at strategic points on the chip's surface, so that I can probe the critical nodes as easily as possible. To facilitate probing when working with 2-layer metal, I bring nodes up from the first metal through *vias* to the second metal. Sometimes I leave holes in my Vapox passivation to facilitate probing dice. The subject of testability has often been addressed for large digital circuits, but the underlying ideas of Design For Testability are important regardless of the type of circuit you are designing. You can avoid a lot of trouble by thinking about what can go wrong and how to keep it from going wrong before the ensuing problems lunge at you. By planning for every possibility, you can profit from your awareness of Murphy's Law. Now, clearly, you won't think of *every* possibility. (Remember, it was something that couldn't go wrong that caused the problems with Stanier's loco-motives.) But, a little forethought can certainly minimize the number of problems you have to deal with.

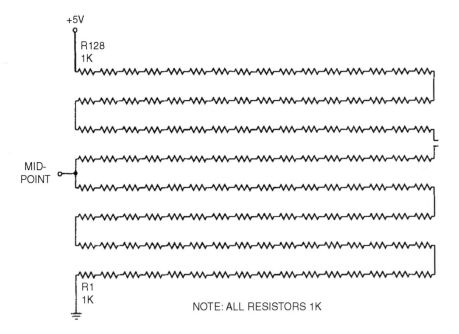

Figure 1.152: If you discovered that the midpoint was not at 2.5V, but at 0V, how would you troubleshoot this circuit? How would you search to detect a short, or an open?

1.7.7 *Consider Appointing a Czar for a Problem Area*

A few years ago we had so many nagging little troubles with band-gap reference circuits at National, that I decided (unilaterally) to declare myself "Czar of Band Gaps." The main rules were that all successful band-gap circuits should be registered with the Czar so that we could keep a log book of successful circuits; all unsuccessful circuits, their reasons for failure, and the fixes for the failures should likewise be logged in with the Czar so that we could avoid repeating old mistakes; and all new circuits should be submitted to the Czar to allow him to spot any old errors. So far, we think we've found and fixed over 50% of the possible errors, before the wafers were fabricated, and we're gaining. In addition, we have added Czars for start-up circuits and for trim circuits, and a Czarina for data sheet changes, and we are considering other czardoms. It's a bit of a game, but it's also a serious business to use a game to try to prevent expensive errors.

I haven't always been a good troubleshooter, but my "baptism of fire" occurred quite a few years ago. I had designed a group of modular data converters. We had to ship 525 of them, and some foolish person had bought only 535 PC (printed circuit) boards. When less than half of the units worked, I found myself in the trouble-shooting business because nobody else could imagine how to repair them. I discovered that I needed my best-triggering scope and my best DVM. I burned a lot of midnight oil. I got half-a-dozen copies each of the schematic and of the board layout. I scribbled notes on them of what the DC voltages ought to be, what the correct AC waveforms looked like, and where I could best probe the key waveforms. I made little lists of, "If this frequency is twice as fast as normal, look for Q17 to be damaged, but if the frequency is low, look for a short on bus B."

I learned where to look for solder shorts, hairline opens, cold-soldered joints, and intermittents. I diagnosed the problems and sent each unit back for repair with a neat label of what to change. When they came back, did they work? Some did—and some still had another level or two of problems. That's the Onion Syndrome: You peel off one layer, and you cry; you peel off another layer, and you cry some more... By the time I was done, I had fixed all but four of the units, and I had gotten myself one hell of a good education in troubleshooting.

After I found a spot of trouble, what did I do about it? First of all, I made some notes to make sure that the problem really was fixed when the offending part was changed. Then I sent the units to a good, neat technician who did precise repair work—much better than a slob like me would do. Lastly, I sent memos to the manufacturing and QC departments to make sure that the types of parts that had proven troublesome were not used again, and I confirmed the changes with ECOs (Engineering Change Orders). It is important to get the paperwork scrupulously correct, or the alligators will surely circle back to vex you again.

1.7.8   Sloppy Documentation Can End in Chapter XI

I once heard of a similar situation where an insidious problem was causing nasty reliability problems with a batch of modules. The technician had struggled to find the solution for several days. Finally, when the technician went out for lunch, the design engineer went to work on the problem. When the technician came back from lunch, the engineer told him, "I found the problem; it's a mismatch between Q17 and R18. Write up the ECO, and when I get back from lunch I'll sign it." Unfortunately, the good rapport between the engineer and the technician broke down: there was some miscommunication. The technician got confused and wrote up the ECO with an incorrect version of what should be changed. When the engineer came back from lunch, he initialed the ECO without really reading it and left for a two-week vacation.

When he came back, the modules had all been "fixed," potted, and shipped, and were starting to fail out in the field. A check of the ECO revealed the mistake—too late. The company went bankrupt. It's a true story and a painful one. Don't get sloppy with your paperwork; don't let it happen to you.

1.7.9   Failure Analysis?

One of the reasons you do troubleshooting is because you may be required to do a Failure Analysis on the failure. That's just another kind of paperwork. Writing a report is not always fun, but sometimes it helps clarify and crystallize your understanding of the problem. Maybe if a customer had forced my engineer friend to write exactly what happened and what he proposed to do about it, that disaster would not have occurred. When I have nailed down my little problem, I usually write down a scribbled quick report. One copy often goes to my boss, because he is curious why it's been taking me so long. I usually give a copy to friends who are working on similar projects. Sometimes I hang a copy on the wall, to warn *all* my friends. Sometimes I send a copy to the manufacturer of a component that was involved. If you communicate properly, you can work to avoid similar problems in the future.

Then there are other things *you* can do in the course of *your* investigation. When you find a bad component, don't just throw it in a wastebasket. Sometimes people call me and say, "Your ICs have been giving me this failure problem for quite a while." I ask, "Can you send me some of the allegedly bad parts?' And they reply, "Naw, we always throw them in the wastebasket..." Please don't do that, because often the ability to troubleshoot a component depends on having several of them to study. Sometimes it's even a case of "NTF"—"No Trouble Found." That happens more often than not. So if you tell me, "Pease, your lousy op-amps are failing in my circuit," and there's actually nothing wrong with the op-amps, but it's really a misapplication problem—I can't help you very well if the parts all went in the trash. Please save them, at least for a while. Label them, too.

Another thing you can do with these bad parts is to open them up and see what you can see inside. Sometimes on a metal-can IC, after a few minutes with a hacksaw, it's just as plain as day. For example, your technician says, "This op-amp failed, all by itself, and I was just sitting there, watching it, not doing anything." But when you look inside, one of the input's lead-bond wires has blown out, evaporated, and in the usage circuit, there are only a couple 10 kΩ resistors connected to it. Well, you can't blow a lead bond with less than 300 mA. Something must have bumped against that input lead and shorted it to a source that could supply half an ampere. There are many cases where looking inside the part is very educational. When a capacitor fails, or a trim-pot, I get my hammer and pliers and cutters and hack-saw and look inside just to see how nicely it was (or wasn't) built. To see if I can spot a failure mechanism—or a bad design. I'm just curious. But sometimes I learn a lot.

Now, when I have finished my inspection, and I am still mad as hell because I have wasted a lot of time being fooled by a bad component—what do I do? I usually WIDLARIZE it, and it makes me feel a lot better. How do you WIDLARIZE something? You take it over to the anvil part of the vice, and you beat on it with a hammer, until it is all crunched down to tiny little pieces, so small that you don't even have to sweep it off the floor. It sure makes you feel better. And you know that that component will never vex you again. That's not a joke, because sometimes if you have a bad pot or a bad capacitor, and you just set it aside, a few months later you find it slipped back into your new circuit and is wasting your time again. When you WIDLARIZE something, that is not going to happen. And the late Bob Widlar is the guy who showed me how to do it.

1.7.10   Troubleshooting by Phone—A Tough Challenge

These days, I do quite a bit of troubleshooting by telephone. When my phone rings, I never know if a customer will be asking for simple information or submitting a routine application problem, a tough problem, or an insoluble problem. Often I can give advice just off the top of my head because I know how to fix what is wrong. At other times, I have to study for a while before I call back. Sometimes, the circuit is so complicated that I tell the customer to mail or transmit the schematic to me. On rare occasions, the situation is so hard to analyze that I tell the customer to put the circuit in a box with the schematic and a list of the symptoms and ship it to me. Or, if the guy is working just a few miles up the road, I will sometimes drop in on my way home, to look at the actual problem.

Sometimes the problem is just a misapplication. Sometimes parts have been blown out and I have to guess what situation caused the overstress. Here's an example: In June, a manufacturer of dental equipment complained of an unacceptable failure rate on LM317 regulators. After a good deal of discussion, I asked, "Where did these failures occur?" Answer: North Dakota. "When did they start to occur?" Answer: In February. I put two and two together and realized that the climate in a dentist's office in North Dakota in February is about as dry as it can be, and is conducive to very high electrostatic potentials. The LM317 is normally safe against electrostatic discharges as high as 3 or 4 kV, but walking across a carpeted floor in North Dakota in February can generate *much* higher voltages than that. To make matters worse, the speed-control rheostat for this dental instrument was right out in the handle. The wiper and one end of the rheostat were wired directly to the LM317's ADJUST pin; the other end of the rheostat was connected to ground by way of a 1 kΩ resistor located back in the main assembly (see Figure 1.153). The speed-control rheostat was just wired up to a ct as a lightning rod that conducted the ESD energy right into the ADJUST pin.

The problem was easily solved by rewiring the resistor in series with the IC's ADJUST pin. By swapping the wires and connecting the rheostat wiper to ground (see Figure 1.154), much less current would take the path to the ADJUST pin and the diffused resistors on the chip would not be damaged or zapped by the current surges. Of course, adding a small capacitor from the ADJ pin to ground would have done just as well, but some customers find it easier to justify moving a component than adding one...

A similar situation occurs when you get a complaint from Boston in June, "Your op-amps don't meet spec for bias current." The solution is surprisingly simple: Usually a good scrub with soap and water works better than any other solvent to clean off the residual contaminants that cause leakage under humid conditions. (Fingerprints, for example. . .)

1.7.11 When Computers Replace Troubleshooters, Look Out

Now, let's think—*what* needs troubleshooting? Circuits? Television receivers? Cars? People? Surely doctors have a lot of troubleshooting to—they listen to symptoms and try to figure out the solution. What is the natural temptation? To let a computer do all the work! After all, a computer is quite good at listening to complaints and symptoms, asking wise questions, and proposing a wise diagnosis. Such a computer system is sometimes called an *Expert System*—part of the general field of Artificial Intelligence. But, I am still in favor of *genuine intelligence*. Conversely, people who rely on Artificial Intelligence are able to solve some kinds of problems, but you can never be sure if they can accommodate every kind of Genuine Stupidity as well as Artificial Stupidity. (That is the kind that is made up especially to prove that Artificial Intelligence works just great.)

I won't argue that the computer isn't a natural for this job; it will probably be cost-effective, and it won't be absent-minded. But, I am definitely nervous because if computers do all the routine work, soon there will be nobody left to do the thinking

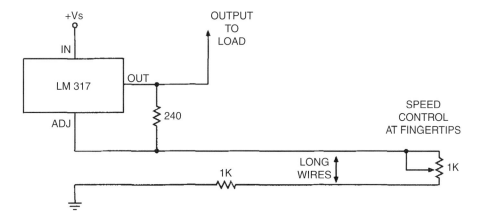

Figure 1.153: When you walk across a dry carpet and reach for the speed control, you draw an arc and most of the current from the wiper of the pot goes right into the LM3 17's ADJ pin

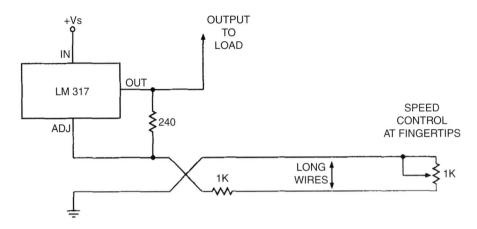

Figure 1.154: By merely swapping two wires, the ESD pulse is now sent to ground and does no harm

when the computer gives up and admits it is stumped. I sure hope we don't let the computers leave the smart troubleshooting people without jobs, whether the object is circuits or people.

My concern is shared by Dr. Nicholas Lembo, the author of a study on how physicians make diagnoses, which was published in the *New England Journal of Medicine*. He recently told the *Los Angeles Times*, "With the advent of all the

new technology, physicians aren't all that much interested (in bedside medicine) because they can order a $300 to $400 test to tell them something they could have found by listening." An editorial accompanying the study commented sadly: "The present trend…may soon leave us with a whole new generation of young physicians who have no confidence in their own ability to make worthwhile bedside diagnoses." Troubleshooting is still an art, and it is important to encourage those artists.

1.7.12 The Computer Is Your Helper…and Friend…???

I read in the San Francisco *Examiner* (Ref. 1.9) about a case when SAS, the Scandinavian airline, implemented an "Expert System" for its mechanics:

"Management knew something was wrong when the quality of the work started decreasing. It found the system was so highly mechanized that mechanics never questioned its judgment. So the mechanics got involved in its redesign. They made more decisions on the shop floor and used the computer to augment those decisions, increasing productivity and cutting down on errors. 'A computer can never take over everything,' said one mechanic. 'Now there are greater demands on my judgment, (my job) is more interesting.'" What can I add? Just be thoughtful. Be careful about letting the computers take over.

1.7.13 No Problems?? No Problem…Just Wait…

Now, let's skip ahead and presume we have all the necessary tools and the right receptive attitude. What else do we need? What is the last missing ingredient? That reminds me of the little girl in Sunday School who was asked what you have to do to obtain forgiveness of sin. She shyly replied, "First you have to sin." So, to do troubleshooting, first you have to have some trouble. But, that's usually not a problem; just wait a few hours, and you'll have plenty. Murphy's Law implies that if you are not prepared for trouble, you will get a lot of it. Conversely, if you have done all your homework, you may avoid most of the possible trouble.

I've tried to give you some insights on the philosophy of how to troubleshoot. Don't believe that you can get help on a given problem from only one specific person. In any particular case, you can't predict who might provide the solution. Conversely, when your buddy is in trouble and needs help, give it a try—you could turn out to be a hero. And, even if you don't guess correctly, when you do find out what the solution is, you'll have added another tool to *your* bag of tricks.

When you have problems, try to think about the right plan to attack and nail down the problem. So, if you do your "philosophy homework," it may make life easier and better for you. You'll be able not only to solve problems, but maybe even to avoid problems. That sounds like a good idea to me!

1.7.14   *Choosing the Right Equipment*

As discussed earlier, the most important thing you need for effective troubleshooting is your wits. In addition to those, however, you'll normally want to have some equipment. This section itemizes the equipment that is necessary for most general troubleshooting tasks; some you can buy off the shelf, and some you can build yourself.

Before you begin your troubleshooting task, you should know that the equipment you use has a direct bearing on the time and effort you must spend to get the job done. Also know that the equipment you need to do a good job depends on the kind of circuit or product you are working on. For example, a DVM may be unnecessary for troubleshooting some problems in digital logic. And, the availability and accessibility of equipment may present certain obstacles. If you only have a mediocre oscilloscope and your company can't go out and buy or rent or borrow a fancy full-featured scope, then you will have to make do.

If you lack a piece of equipment, be aware that you are going into battle with inadequate tools; certain clues may take you much longer than necessary to spot. In many cases when you spent too much time finding one small problem, the time was wasted simply because you were foolish or were unaware of a particular troubleshooting technique; but, in other cases, the time was wasted because of the lack of a particular piece of equipment. It's important for you to recognize this last-mentioned situation. Learning when you're wasting time because you lack the proper equipment is part of your education as a troubleshooter.

In addition to the proper tools, you also need to have a full understanding of how both your circuit and your equipment are supposed to work. I'm sure you've seen engineers or technicians work for many fruitless hours on a problem and then, when they finally find the solution, say, "Oh, I didn't know it was supposed to work that way." You can avoid this scenario by using equipment that you are comfortable and familiar with.

The following equipment is essential for most analog-circuit troubleshooting tasks. This list can serve as a guide to both those setting up a lab and those who just want to make sure that they have everything they need—that they aren't missing any tricks.

1. A dual-trace oscilloscope. It's best to have one with a sensitivity of 1 or 2 mV/cm and a bandwidth of at least 100 MHz. Even when you are working with slow op-amps, a wide-bandwidth scope is important because some transistors in "slow" applications can oscillate in the range of 80 or 160 MHz, and you should be able to see these little screams. Of course, when working with fast circuits, you may need to commandeer the lab's fastest scope to look for glitches. Sometimes a peak-to-peak automatic triggering mode is helpful and time-saving. Be sure you know how all the controls work, so you don't waste much time with setup and false-triggering problems.

2. Two or three scope probes. They should be in good condition and have suitable hooks or points. Switchable 1 ×/10× probes are useful for looking at both large and very small signals. You should be aware that 1× probes only have a 16- or 20-MHz bandwidth, even when used with a 100-MHz scope. When you use 10× probes, be sure to adjust the capacitive compensation of the probe by using the square-wave calibrator per Figure 1.155. Failure to do so can be a terrible time-wasting source of trouble.

   Ideally, you'll want three probes at your disposal, so that you can have one for the trigger input and one for each channel. For general-purpose troubleshooting, the probes should have a long ground wire, but for high-speed waveforms you'll need to change to a short ground wire (Figure 1.156) The shorter ground wires not only give you better frequency response and step response for your signal, but also better rejection of other noises around your circuit.

   In some high-impedance circuits, even a 10× probe's capacitance, which is typically 9 to 15 pF, may be unacceptable. For these circuits, you can buy an active probe with a lower input capacitance of 1.5 to 3 pF ($395 to $1800), or you can build your own (Figure 1.157).

   When you have to work with switching regulators, you should have a couple of current probes, so you can tell what those current signals are doing. Some current probes go down to DC; others are inherently AC coupled (and are much less expensive).

3. An analog-storage oscilloscope. Such a scope can be extremely useful, especially when you are searching for an intermittent or evanescent signal. The scope can trigger off an event that may occur only rarely and can store that event and the events that follow it. Some storage scopes are balky or tricky to apply, but it's often worth- while to expend the effort to learn how to use them. Digital-storage oscilloscopes (DSOs) let you do the same type of triggering and event storage as do the analog type, and some can display events that precede the trigger. They are sampled-data systems, however, so you must be sure to apply them correctly (Ref. 1.10). Once you learn how to use them, though, you'll appreciate the special features they offer, such as bright CRT displays, automatic pulse-parameter measurements, and the ability to obtain plots of waveforms.

4. A digital voltmeter (DVM). Choose one with at least five digits of resolution, such as the HP3455, the HP3456, the Fluke 8810A, or the Fluke 8842A. Be sure you can lock out the autorange feature, so that the unit can achieve its highest accuracy and speed. Otherwise, you'll be wasting time while the DVM autoranges. For many analog circuits, it's important to have a high-impedance ($\gg$10,000 MΩ) input that stays at high impedance up to 15 or 20V; the four DVMs mentioned above have this feature. There are many other fine DVMs that have 10 MR inputs above 2 or 3V; and, if a 10 MΩ input impedance is not a

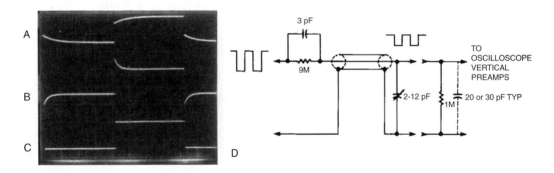

Figure 1.155: If an amplifier or a comparator is supposed to produce a square wave but the waveform looks like trace (A) or (B), how long should it take you to find the problem? No time at all! Just turn the screw that adjusts the 10 × probe's compensation, so the probe's response is flat at all frequencies (C). The schematic diagram of a typical 10 x oscilloscope probe is shown in (D).

problem, they are acceptable. The most important reason to use a high-input-impedance DVM is because sometimes it's necessary to put 33 kΩ or 100 kΩ resistors in series with the probe, right near the circuit-under-test, to prevent the DVM's input capacitance from causing the circuit to oscillate. If you're using a DVM with a 10 MΩ input impedance and you have a 100 kΩ resistor in series with the probe, the DVM's measurements would lose 1% of their accuracy. Fortunately, many good DVMs have less than 500 pA of input current, which would cause less than 50 μV of error in the case of 100 kΩ source resistance. A high-resolution DVM lets you detect 100- to 200-μV deviations in an 11V signal. You can best spot many semiconductor problems by finding these minor changes. A 4-digit DVM is a relatively poor tool; however, if you are desperate. you can detect small voltage changes if you refer the DVM's "low," or ground, side to a stable reference, such as a 10V bus. Then, with the DVM in the IV range, you can spot small deviations in an 11V signal. This measurement is more awkward and inconvenient than a ground-referenced measurement with a higher resolution meter would be, and this method can cause other problems as well. For instance, you can end up injecting noise generated by the DVM's A/D converter into the sensitive IO-V reference, thereby adversely affecting the performance of other circuits. In some cases, a little RC filtering may minimize this problem (see Figure 1.158) but you never can be sure how easy it will be to get the noise to an acceptable level . . .

Figure 1.156: When a fast square wave is supposed to be clean and fast-settling but looks like (A), don't repair the square-wave generator—just set aside the probe's 6-inch ground lead (C). If you ground the probe directly at the ground point near the tip (D) (special attachments that bring the ground out conveniently are available), your waveforms will improve considerably (B).

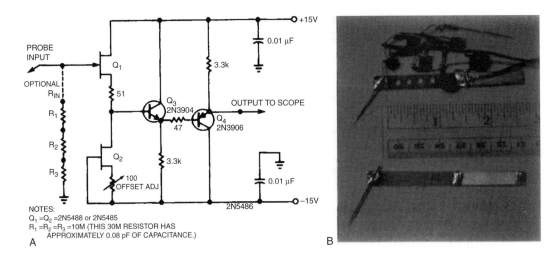

A

B

Figure 1.157: This probe circuit's input impedance is $10^{11}\Omega$ in parallel with 0.29 pF (A). Optimized primarily for its impedance characteristics and not its frequency response, the probe's bandwidth and slew rate are 90 *MHz* and 300V/μs, respectively. If the lack of physical rigidity of the TO-92 packaged FET makes it too wobbly to use as a probe, a piece of 1/16-in. glass epoxy board with the copper peeled off will add rigidity with only 0.08 pF of additional capacitance. The layout of the drilled-out beam shown in the top of Figure 1.157(B) adds only 0.06 pF to the input capacitance.

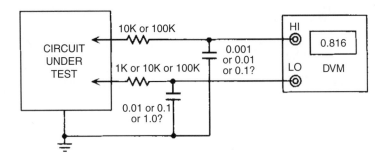

Figure 1.158: Even if it's battery-powered, a DVM is capable of Spitting out noise pulses into your delicate circuit. The RC filters shown here can help minimize this. Pick the values that work for your circuit.

5. Auxiliary meters. It may look silly to see a test setup consisting of two good DVM's, two little 3-digit DVMs monitoring a couple of voltage supplies. A couple more 3-digit DVMs monitoring current drain, and an analog meter monitoring something else. But, if you don't know exactly what you're looking for and you can borrow equipment, using lots of meters is an excellent way to attack a problem—even if you do have to wait until 5:15 PM to borrow all that equipment. When is an analog volt-meter better than a DVM? Well, the analog voltmeter usually has inferior accuracy and resolution, but when you watch an ordinary analog voltmeter your eye can detect a trend or rate-of-change that may be hard to spot on a DVM, especially in the presence of noise or jitter. As an example, if you suddenly connect an ordinary analog volt-ohmmeter across a 1 μF capacitor, your eye can resolve if the capacitor's value is 0.1 times or 10 times as large as it should be. You can't perform this kind of test with a DVM. Another advantage of analog meters is that they are passive devices: They don't inject noise into your circuit as digital meters can—even battery-powered ones. And they can have a lot less capacitance to ground.

6. A general-purpose function generator. While sine and square waves are popular test signals, I have often found triangular waveforms to be invaluable when searching for nonlinearities. Sometimes you need *two* function generators, one to sweep the operating point of the DUT, slowly, back and forth over its operating range, while you watch the response of the output to a small quick square wave—watching for oscillation or ringing or trouble.

7. Power supplies with stable outputs. They should have coarse and fine adjustment controls and adjustable current limits. Digital controls are usually not suitable; they don't let you continuously sweep the voltage up and down while you monitor the scope and watch for trends. In cases when the power supply's output capacitor causes problems, you may want a power supply whose output circuit, like that of an op-amp, includes no output capacitor. You can buy such a supply, or you can make it with an op-amp and a few transistors. The advantage of the supply shown in Figure 1.159 is that you can design it to slew fast when you want it to.

(For speed, use a quick LF356 rather than a slow LM741). Also, if a circuit latches and pulls its power supply down, the circuit won't destroy itself by discharging a big capacitor.

While we are on the topic of power, another useful troubleshooting tool is a set of batteries. You can use a stack of one, two, or four 9V batteries, ni-cads, gel cells,

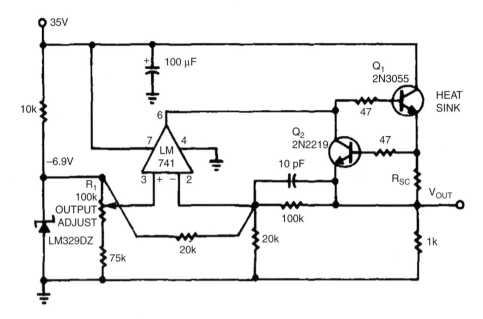

Figure 1.159: You can vary the output voltage of this DC power supply from 3 to 30V by adjusting R_1. R_{sc} should be between 3 and 100Ω; the short-circuit current is equal to about 20 mA + 600 mV/ R_{sc}

or whatever is suitable and convenient. Batteries are useful as an alternate power supply for low-noise preamplifiers: If the preamp's output doesn't get quieter when the batteries are substituted for the ordinary power supply, don't blame your circuit troubles on the power supply. You can also use these batteries to power low-noise circuits, such as those sealed in a metal box, without contaminating their signals with any external noise sources.

8.   A few RC substitution boxes. You can purchase the VIZ Model WC-412A, which I refer to affectionately as a "Twiddle-box" (Figure 1.160) from R & D Electronics, 1432 South Main Street, Milpitas CA 95035, (408) 262-7144. Or, inquire at VIZ, 175 Commerce Drive, Fort Washington, PA 19034, (800) 523-3696. You can set the unit in the following modes: R, C, series RC, parallel RC, open circuit, or short circuit. They are invaluable for running various "tests" that can lead to the right answer.

You may need component values beyond what the twiddle boxes offer; in our labs, we built a couple of home-brew versions (Figure 1.161). The circuit shown in Figure 1.161(A) provides variable low values of capacitance and is useful for fooling around with the damping of op-amps and other delicate circuits. You can make your own calibration labels to mark the setting of the capacitance and resistance values. The circuit shown in Figure 1.161(B) provides high capacitances of various types, for testing power supplies and damping various regulators.

9. An isolation transformer. If you are working on a line-operated switching regulator, this transformer helps you avoid lethal and illegal voltages on your test setup and on the body of your scope. If you have trouble obtaining an isolation transformer. you can use a pair of transformers (step-down, step-up) back-to-back (Figure 1.162). Or. If cost isn't an issue, you can use isolated probes. These probes let you display small signals that have common-mode voltages of hundreds of volts with respect to ground, and they won't require you to wear insulated gloves when adjusting your scope.

10. A variable autotransformer, often called a Variac™.[1] This instrument lets you change the line voltage and watch its effect on the circuit—a very useful trick. (Warning: A variable autotransformer is *not* normally an isolation transformer. You may need to cascade one of each, to get safe adjustable power.)

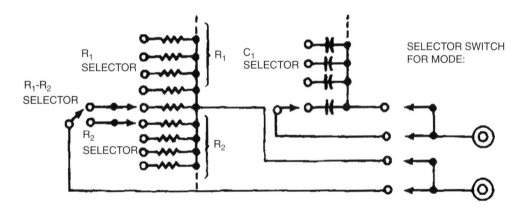

Figure 1.160: This general schematic is for a commercially available RC substitution box, the VIZ Model WC-4I2A. The unit costs around $139 in 1991 dollars and has resistor and capacitor values in the range of 15Ω to 10 MΩ and 100 pF to 0.2 μF, respectively. It can be configured to be an open circuit, a series RC, resistors, capacitors, a parallel RC, or a short circuit. See text for availability.

[1] Registered TM of GENRAD Corp., Concord, MA. Variacs can be purchased from JLM Electronics. 56 Somerset St., P.O. Box 10317, West Hartford, CT 06110. (203) 233-0600.

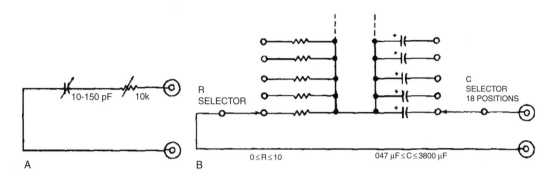

Figure 1.161: RC boxes based on these schematics extend component ranges beyond those available in off-the-shelf versions. You can house the series RC circuit in (A) in a 1 x 1 x 2-inch copper-clad box. Use the smaller plastic-film-dielectric tuning capacitors or whatever is convenient, and a small 1 -turn pot. Build the circuit in (B) with tantalum or electrolytic (for values of 1 μF and higher) capacitors, but remember to be careful about their polarities and how you apply them. Also, you might consider using mylar capacitors for smaller values. Sometimes it's very valuable to compare a mylar, a tantalum, and an aluminum electrolytic capacitor of the same value! Use 18-position switches to select R and C values. And, stay away from wirewound resistors; their inductance is too high.

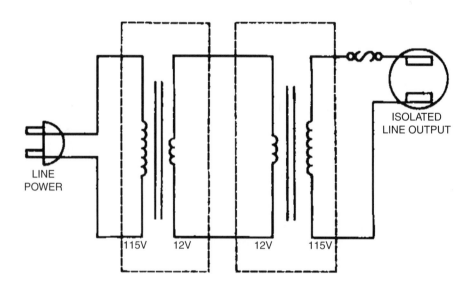

Figure 1.162: You can use this back-to-back transformer configuration to achieve line isolation similar to that of an isolation transformer.

11. A curve tracer. A curve tracer can show you that two transistors may have the same saturation voltage under a given set of conditions even though the slope of one may be quite different from the slope of the other. If one of these transistors works well and the other badly, a curve tracer can help you understand why. A curve tracer can also be useful for spotting nonlinear resistances and conductances in diodes, capacitors, light bulbs, and resistors. A curve tracer can test a battery by loading it down or recharging it. It can check semiconductors for breakdown. And, when you buy the right adapters or cobble them up yourself, you can evaluate the shape of the gain, the CMRR, and the PSRR of op-amps.

12. Spare repair parts for the circuit-under-test. You should have these parts readily available, so you can swap components to make sure they still work correctly.

13. A complete supply of resistors and capacitors. You should have resistors in the range from 0.1Ω to $100\ M\Omega$ and capacitors from 10 pF to 1 µF. Also, 10,100, and 1000 µF capacitors come in handy. Just because your circuit design doesn't include a 0.1Ω or a $100\ M\Omega$ resistor doesn't mean that these values won't be helpful in troubleshooting it. Similarly, you may not have a big capacitor in your circuit; but, if the circuit suddenly stops misbehaving when you put a 3800 µF capacitor across the power supply, you've seen a quick and dramatic demonstration that power-supply wobbles have a lot to do with the circuit's problems. Also, several feet of plastic-insulated solid wire (telephone wire) often come in handy. A few inches of this type of twisted-pair wire makes an excellent variable capacitor, sometimes called a *gimmick*. Gimmicks are cheap and easy to vary by simply winding or unwinding them. Their capacitance is approximately one picofarad per inch.

14. Schematic diagrams. It's a good idea to have several copies of the schematic of the circuit-under-test. Mark up one copy with the normal voltages, currents, and waveforms to serve as a reference point. Use the others to record notes and waveform sketches that relate to the specific circuit-under-test. You'll also need a schematic of any homemade test circuit you plan to use. Sometimes, measurements made with your homemade test equipment may not agree with measurements made by purchased test equipment. The results from each tester may not really be "wrong": They might differ because of some design feature, such as signal filtering. If you have all the schematics for your test equipment, you can more easily explain these incompatibilities. And, finally, the data sheets and schematics of any ICs used in your circuit will also come in handy.

15. Access to any engineering or production test equipment, if possible. Use this
     equipment to be sure that when you fix one part of the circuit, you aren't
     adversely affecting another part. Other pieces of equipment and testers also fall
     under the category of specialized test equipment; their usefulness will depend on
     your circuit. Three examples are a short-circuit-detector circuit, an AM transistor
     radio, and a grid-dip meter.

     A short-circuit-detector circuit.   This tool comes in handy when you have to
     repair a lot of large PC boards: It can help you find a short circuit between the
     ground bus and the power or signal busses. It's true that a sensitive DVM can also
     perform this function, but a short-circuit detector is much faster and more efficient.
     Also, this circuit turns itself off and draws no current when the probe is not
     connected. In the short-circuit-detector circuit shown in Figure 1.163, the LM10
     op-amp amplifies the voltage drop and feeds it to the LM331 voltage-to-frequency
     converter, which you set up to emit its highest pitch when $V_{in} = 0$ mV. When using
     this circuit, use a 50- to 100-mA current-limited power supply. To calibrate the
     circuit, first ground the detector's two probes and trim the OFFSET ADJUST for a
     high pitch. Then, move the positive test probe to V_s at A and trim the GAIN
     ADJUST for a low pitch. Figure 1.163 illustrates a case in which one of the five
     major power supply busses of the circuit-in-trouble has a solder short to ground.

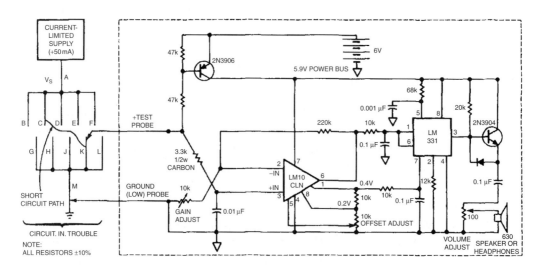

**Figure 1.163: You can use this short-circuit detector to find PC board shorts.
Simply slide the test probe along the various busses and listen for changes in pitch.**

To find the exact location of the short, you simply slide the positive input probe along the busses. In this example, if you slide it from A toward B or D, the pitch won't change because there is no change in voltage at these points—no current flowing along those busses. But, if you slide the probe along the path from A to C or from K to M, the pitch will change because the voltage drop is changing along those paths. It's an easy and natural technique to learn to follow the shifting frequency signals.

An AM radio.   What do you do when trouble is everywhere? A typical scenario starts out like this: You make a minor improvement on a linear circuit, and when you fire it up you notice a terrible oscillation riding on the circuit's output. You check everything about the circuit, but the oscillation remains. In fact, the oscillation is riding on the output, the inputs, on several internal nodes. and even on ground. You turn off the DVM, the function generator, the soldering iron, and finally even the *power supplies*, but the oscillation is still there.

Now you start looking around the lab to see who has started a new oscillator or switching regulator that is doubling as a medium-power transmitter. Aside from yelling, "Who has a new circuit oscillating at 87 kHz?" what can you do to solve the problem? One useful tool is an ordinary AM transistor radio. As we have all learned, FM radios reject many kinds of noise, but AM radios scoop up noise at repetition rates and frequencies that would surprise you.

How can a crummy little receiver with an audio bandwidth of perhaps 5 kHz detect noise in the kilohertz and megahertz regions? Of course, the answer is that many repetitive noise-pulse trains (whose repetition rates are higher than the audible spectrum but below the AM frequency band) have harmonics that extend into the vicinity of 600 kHz, where the AM receiver is quite sensitive. This sensitivity extends to signals with amplitudes of just a few microvolts per meter.

If you are skeptical about an AM radio's ability to detect these signals, tune its dial down to the low end, between stations. Then, hold it near a DVM or a computer or computer keyboard, and listen for the hash. Notice, too, that the ferrite stick antenna has definite directional sensitivity, so you can estimate where the noise is coming from by using either the null mode or pointing the antenna to get the strongest signal. So, the humble AM radio may be able to help you as you hike around the lab and smile pleasantly at your comrades until you find the culprit whose new switching regulator isn't working quite right but which he neglected to turn off when he went out to get a cup of coffee.

The grid-dip meter. On other occasions, the frequency and repetition rate of the noise are so high that an AM receiver won't be helpful in detecting the problem. What's the tool to use then? Back in the early days of radio, engineers found that if you ran a vacuum-tube oscillator and immersed it in a field of high-power oscillations at a comparable frequency, the tube's grid current would shift or dip when the frequencies matched. This tool became known as the "grid-dip meter." I can't say that I am an expert in the theory or usage of the grid-dip meter, but I do recall being impressed in the early days of monolithic ICs: A particular linear circuit was oscillating at 98 MHz, and the grid-dip meter could tickle the apparent rectified output error as I tuned the frequency dial back and forth.

That was 25 years ago, and, of course, Heathkit[2] has discontinued their old Grid Dip and Tunnel Dip meters in favor of a more modern design. The new one, simply dubbed HD-1250 "Dip Meter," uses transistors and tetrode FETs. At the bargain price of $89, every lab should have one. They'll help you ferret out the source of nasty oscillations as high as 250 MHz. The literature that comes with the HD-1250 dip-meter kit also lists several troubleshooting tips.

When grid-dip meters first became popular, the fastest oscilloscope you could buy had a bandwidth of only a few dozen megahertz. These days, it is possible to buy a scope with a bandwidth of many hundreds of megahertz, so there are fewer occasions when you might need a grid-dip meter. Still, there are times when it is exactly the right tool. For example, you can use its oscillator to activate passive tuned circuits and detect their modes of resonance. Also, in a small company where you can't afford to shell out the many thousands of dollars for a fast scope, the dip meter is an inexpensive alternative.

16. A few working circuits, if available. By comparing a bad unit to a good one, you can often identify problems. You can also use the good circuits to make sure that your specialized test equipment is working properly.

17. A sturdy, broad workbench. It should be equipped with a ground plane of metal that you can easily connect to the power ground. The purpose of this ground plane is to keep RF. 60-Hz, and all other noise from coupling into the circuit. Place insulating cardboard between the bench and the circuit-under-test. so that nothing tends to short to ground. Another way to prevent

---

[2] Heath Company, Benton Harbor, Michigan, 49022; (1-800-253-0570).

noise from interfering with the circuit is to use a broad sheet of single-sided copper-clad board. Placed copper-side down and with a ground wire soldered to the copper. it provides an alternate ground plane. To prevent electrostatic-discharge (ESD) damage to CMOS circuits, you'll need a wrist strap to ground your body through 1 MΩ.

18. Safety equipment. When working on medium- or high-power circuits that might explode with considerable power in the case of a fault condition, you should be wearing safety goggles or glasses with safety lenses. Keep a fire extinguisher nearby, too.

19. A suitable hot soldering iron. If you have to solder or unsolder heavy busses from broad PC-board traces, use a big-enough iron or gun. For small and delicate traces around ICs, a small tip is essential. And, be sure that the iron is hot enough. An easy way to delaminate a trace or pad, whether you want to or not, is to heat it for too long a time, which might happen if your iron weren't big enough or hot enough. (The old Heathkit warnings not to use a hot iron became obsolete along with the germanium transistor.) In some cases, a grounded soldering iron is required: in others, a portable (ungrounded or rechargeable) soldering iron is ideal. Make sure you know whether your iron is grounded or floating.

20. Tools for removing solder, such as solder wick or a solder sucker. You should be comfortable with whatever tools you are using; a well-practiced technique is sometimes critical for getting good results. If you are working on static-sensitive components, an antistatic solder-sucker is less likely to generate high voltages due to internal friction than is an ordinary solder-sucker. I have been cautioned that a large solder-sucker may cause problems when working on narrow PC traces: in that case, solder wick may be the better choice.

21. Hand tools. Among the tools you'll need are sharp diagonal nippers, suitable pliers, Screwdrivers, large cutters, wrenches, wire strippers, and a jack knife or Exacto™ knife.

22. Signal leads, connectors, cables, BNC adapters, wires, clip leads, ball hooks, and alligator clips—as needed. Scrimping and chintzing in this area can waste lots of time: shaky leads can fall off or short out.

23. Freeze mist and a hair dryer. The freeze mist available in aerosol cans lets you quickly cool individual components. A hair dryer lets you warm up a whole circuit. You'll want to know the dryer's output air temperature because that's the temperature to which you'll be heating the components.

NOTE: Ideally we should not use cooling sprays based on chlorofluorocarbons (CFCs), which are detrimental to the environment. I have a few cans that some people would say I shouldn't use. But what else should I do—send the can to the dump? Then it will soon enter the atmosphere, without doing anybody any good. I will continue to use up any sprays with CFC-based propellants that I already have, but when it is time to buy more, I'll buy environmentally safe ones.

24. A magnifying glass or hand lens. These devices are useful for inspecting boards, wires, and components for cracks, flaws, hairline solder shorts, and cold-soldered joints.

25. An incandescent lamp or flashlight. You should be able to see clearly what you are doing, and bright lights also help you to inspect boards and components.

26. A thermocouple-based thermometer. If your thermometer is floating and battery powered, you can connect the thermocouple to any point in the circuit and measure the correct temperature with virtually no electrical or thermal effect on the circuit. Figure 1.164 shows a thermocouple amplifier with designed-in cold-junction compensation.

   Some people have suggested that an LM35 temperature-sensor IC (Figure 1.165) is a simple way to measure temperature, and so it is. But, if you touch or solder an LM35 in its TO-46 package to a resistor or a device in a TO-5 or TO-3 case, the LM35 will increase the thermal mass and its leads will conduct heat away from the device whose temperature you are trying to measure. Thus, your measurements will be less accurate than if you had used a tiny thermocouple with small wires.

27. Little filters in neat metal boxes, to facilitate getting a good signal-to-noise ratio when you want to feed a signal to a scope. They should be set up with switch-selectable cut-off frequencies, and neat connectors. If in your business you need sharp roll-offs, well, you can roll your own. Maybe even with op-amps and batteries. *You* figure out what you need. Usually I just need a couple simple Rs and Cs, with an alligator clip to select the right ones.

28. Line adapters—those 3-wire-to-2-wire adapters for your 3-prong power cords. You need several of them. You only need them because too many scopes and function generators have their ground tied to the line-cord's neutral. You need some of these to avoid ground-loops. You also need a few spares because your

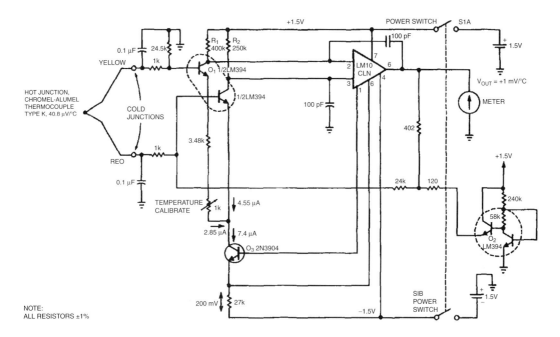

Figure 1.164: This thermocouple amplifier has inherent cold-junction compensation because of the two halves of Q_1 which run at a 1 6:1 current ratio. Their V_{BE}s are mismatched by 1 2 mV + 40.8 $\mu v/°$C. This mismatch exactly cancels out the 40.8 $\mu v/°$C of the cold junction. For best results, you should use four 100 kΩ resistors in series for R_1 and two 100 kΩ resistors in series with two 100 kΩ resistors in parallel for R_2—all resistors of the same type, from the same manufacturer. Q_2 and its surrounding components implement a correction for very cold temperatures and are not necessary for thermocouple temperatures above 0°C. Credit to Mineo Yamatake for his elegant circuit design.

buddies will steal yours. For that matter, keep a few spare cube taps. When they rewired our benches a few years ago, the electricians tried to give us five outlets per bench. I stamped my feet and insisted on ten per bench, and that's just barely enough, most of the time.

You've come to the end of my list of essential equipment for ordinary analog circuit troubleshooting. Depending on your circuit, you may not need all these items; and, of course, the list did not include a multitude of other equipment that you may find useful. Logic analyzers, impedance analyzers, spectrum analyzers, programmable current pumps, capacitance meters and testers, and pulse generators can all ease various troubleshooting tasks.

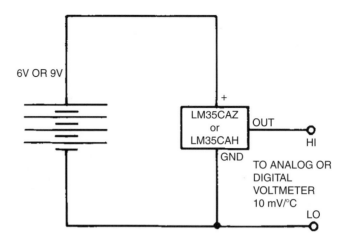

Figure 1.165: The LM35CAZ is a good, simple, convenient general-purpose temperature sensor. But beware of using it to measure the temperature of very small objects or in the case of extreme temperature gradients; it would then give you less accurate readings than a tiny thermocouple with small wires.

References

[1.1] Frederiksen Thomas M., *Intuitive IC Op Amps,* National Semiconductor Corp., 1984.
This classic paperback book was originally published in 1984. The book describes how op-amps work and how they are used, from a practical, commonsense perspective. It is currently out of print. However, you may be able to find it in university libraries or by browsing the Internet. As of March 2005, the book was also available from Rector Press. This book was written by the inventor of the most widely used op-amp in the world, the LM324. This book gave me the first hint that op-amps should be easy to use, not hard.

[1.2] Reliance Motion Control, Inc., "Pink Motor Book" *DC Motors Speed Controls Servo Systems, The Electro-Craft Engineering Handbook.*
I like to call this the pink motor book due to an interesting choice of color for the cover, and I highly recommend it for anyone who is working with DC motors. It's heavy on the equations, but a good source for understanding all the complexities of motors.

[1.3] Van Doren, Dr. Tom, *Grounding and Shielding of Electronic Systems,* University Missouri Rolla, Van Doren Company, Rt. 6, Box 319, Rolla, MO 65401.

[1.4] Weyrick, Robert C., *Yellow Control Theory: Fundamentals of Automatic Control,* McGraw Hill, ISBN 0-07-069493-1.
Good read, helped me understand control theory.

[1.5] Dostal, Jiri, *Operational Amplifiers,* Elsevier Scientific, The Netherlands, 1981: also, Elsevier Scientific, Inc., 655 Avenue of the Americas, NY, NY 10010. (212) 989-5800.

[1.6] Smith, John I., *Modern Operational Circuit Design,* John Wiley & Sons, New York, NY, 1971.

[1.7] *Data Converter Handbook,* Analog Devices Corp., P.O. Box 9106, Norwood MA 02062. 1984.

[1.8] Bulleid H. A. V., *Master Builders of Steam,* Ian Allan Ltd. London, UK, 1963, pp. 146–147.

[1.9] Caruso, Denise, "Technology designed by its users," The San Francisco, *Examiner,* E-15, Sunday, March 18, 1990.

[1.10] Collins, Jack, and David White, "Time-domain analysis of aliasing helps to alleviate DSO errors," *EDN,* September 15, 1988207.

CHAPTER 2

The Semiconductor Diode

Ian Hickman

The *semiconductor diode,* like its predecessor the *thermionic diode,* conducts current in one direction only. It is arguable that diodes in general are not really active devices at all, but simply nonlinear passive devices. The earliest semiconductor diode was a *point contact* device and was already in use before the First World War, being quite possibly contemporary with the earliest thermionic diodes. It consisted of a sharp pointed piece of springy wire (the "cat's whisker") pressed against a lump of mineral (the "crystal"), usually Galena, an ore containing sulfide of lead. The crystal detector was widely employed as the detector in the crystal sets that were popular in the early days of broadcasting. Given a long aerial and a good earth, the crystal set produced an adequate output for use with sensitive headphones, while with so few stations on the air in those days the limited selectivity of the crystal set was not too serious a problem. The crystal and cat's whisker variety of point contact diode was a very hit and miss affair, with the listener probing the surface of the crystal to find a good spot. Later, new techniques and materials were developed, enabling robust preadjusted point contact diodes useful at microwave frequencies to be produced. These were used in radar sets such as AI Mk.10, an airborne interceptor radar that was in service during (and long after!) the Second World War. Germanium point contact diodes are still manufactured and are useful where a diode with a low forward voltage drop at modest current (a milliampere or so) combined with very low reverse capacitance is required. However, for the last twenty years silicon has predominated as the preferred semiconductor material for both diode and transistor manufacture, while point contact construction gave way to junction technology even earlier.

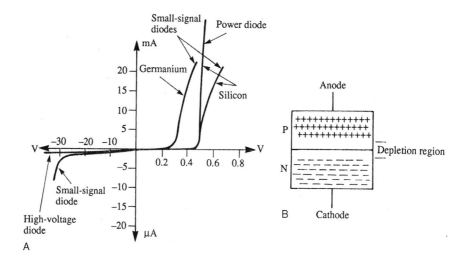

Figure 2.1: Semiconductor diodes. (A) *I/V* characteristics; (B) Diagrammatic representation of PN diode, showing majority carriers and depletion region.

Figure 2.1A shows the *I/V* characteristics of practical diodes. *Silicon* diodes are manufactured hundreds or thousands at a time, commencing with a thin wafer of single crystal silicon several inches in diameter, which is later scribed and separated to obtain the individual diodes. Silicon is one of those substances that crystallize in a cubic lattice structure; another is sodium chloride—common salt—but that is a compound, not an element like silicon. Silicon, in the form of silicon dioxide, is one of the most abundant elements in the earth's crust, occurring as quartz, in sandstone, etc. When reduced to elementary silicon, purified and grown from the melt as a single crystal, it is called *intrinsic* silicon and is a poor conductor of electricity, at least at room temperature. However, if a few of the silicon atoms in the atomic lattice are replaced with atoms of the pentavalent element phosphorus, (which has five valence electrons in its outer shell, unlike quadravalent silicon that has four outer electrons), then there are electrons "going begging," with no corresponding electron in a neighboring atom with which to form a bond pair. These spare electrons can move around in the semiconductor lattice, rather like the electrons in a metallic conductor. The higher the *doping level*, the more free electrons and the higher the conductivity of the silicon, which is described as n-type. This simply indicates that the current flow is by means of negative charge carriers, i.e., electrons. p-type silicon is obtained by doping the monocrystalline lattice with a sprinkling of trivalent atoms, such as boron. Where one of these is substituted in the lattice next to a silicon atom, the latter has one of its electrons "unpaired," a state of affairs described as a *hole*. If this is filled by an electron from a silicon atom to the right, then while that electron has moved to the left, the hole has effectively moved to the right. It turns out that the spare electrons in n-type silicon are more mobile than the holes in p-type, which explains why very high-frequency transistors are more easy to produce in NPN than PNP types—but that is jumping ahead.

To return to the silicon junction diode: the construction of this is as in Figure 2.1B, which indicates the lack of carriers (called a *depletion region*) in the immediate neighborhood of the junction. Here, the electrons from the N region have been attracted across to the P region to fill holes. This disturbance of the charge pattern that one would expect to find throughout n-type and p-type material represents a potential barrier, which prevents further migration of carriers across the junction. When the diode is reverse biased, the depletion region simply becomes more extensive.

The associated redistribution of charge represents a transient charging current, so that a reversed biased diode is inherently capacitive. If a forward bias voltage large enough to overcome the potential barrier is applied to the junction, about 0.6V in silicon, then current will flow; in the case of a large-area power diode, even a current of several amperes will only result in a small further increase in the voltage drop across the diode, as indicated in Figure 2.1A. The incremental or slope resistance r_d of a forward biased diode at room temperature is given approximately by $25/I_a$ ohms, where the current through the diode I_a is in milliamperes. Hence the incremental resistance at 10 μA is 2K5, at 100 μA is 250R, and so on, but it bottoms out at a few ohms at high currents, where the bulk resistance of the semiconductor material and the resistance of the leads and bonding pads, etc., come to predominate. This figure would apply to a small-signal diode: the minimum slope resistance of a high-current rectifier diode would be in the milliohm region.

Power diodes are used in power supply rectifier circuits and similar applications, while small *signal diodes* are widely used as detectors in radio-frequency circuits and for general purpose signal processing, as will appear in later chapters. Also worth mentioning are special purpose small-signal diodes such as the tunnel diode, backward diode, varactor diode, PIN diode, snap-off diode, Zener diode and Schottky diode; the last of these is also used as a rectifier in power circuits. However, the tunnel diode and its degenerate cousin the backward diode are only used in a few very specialized applications nowadays.

The *varactor diode* or varicap is a diode designed solely for reverse biased use. A special doping profile giving an abrupt or "hyperabrupt" junction is used. This results in a diode whose reverse capacitance varies widely according to the magnitude of the reverse bias. The capacitance is specified at two voltages, e.g., 1V and 15V, and may provide a capacitance ratio of 2:1 or 3:1 for types intended for use at UHF, up to 30:1 for types designed for tuning in AM radio sets. In these applications the peak-to-peak amplitude of the RF voltage applied to the diode is small compared with the reverse bias voltage, even at minimum bias (where the capacitance is maximum). So the varactor behaves like a normal mechanical air-spaced tuning capacitor except that it is adjusted by a DC voltage rather than a rotary shaft. Tuning varactors are designed to have a low series loss r_s so that they exhibit a high quality factor Q, defined as X_c/r_s, over the range of frequencies for that they are designed.

Another use for varactors is as frequency multipliers. If an RF voltage with a peak-to-peak amplitude of several volts is applied to a reverse biased diode, its capacitance will vary in sympathy with the RF voltage. Thus, the device is behaving as a nonlinear (voltage dependent) capacitor, and as a result the RF current will contain harmonic components that can be extracted by suitable filtering.

The p-type/intrinsic/n-type diode or PIN diode is a p-n junction diode, but fabricated so as to have a third region of intrinsic (undoped) silicon between the P and N regions. When forward biased by a direct current it can pass radio-frequency signals without distortion, down to some minimum frequency set by the lifetime of the current carriers—holes or electrons—in the intrinsic region. As the forward current is reduced, the resistance to the flow of the RF signal increases, but it does not vary over an individual cycle of the RF current. As the direct current is reduced to zero, the resistance rises toward infinity; when the diode is reverse biased, only a very small amount of RF current can flow, via the diode's reverse biased capacitance. The construction ensures that this is very small, so the PIN diode can be used as an electronically controlled RF switch or relay, on when forward biased and off when reverse biased. It can also be used as a variable resistor or attenuator by adjusting the amount of forward current. An ordinary p-n diode can also be used as an RF switch, but it is necessary to ensure that the peak RF current when "on" is much smaller than the direct current, otherwise waveform distortion will occur. It is the long lifetime of carriers in the intrinsic region (long compared with a single cycle of the RF), which enables the PIN diode to operate as an adjustable linear resistor, even when the peaks of the RF current exceed the direct current.

When a PN junction diode that has been carrying current in the forward direction is suddenly reverse biased, the current does not cease instantaneously. The charge has first to redistribute itself to re-establish the depletion layer. Thus for a very brief period, the reverse current flow is much greater than the small steady-state reverse leakage current. The more rapidly the diode is reverse biased, the larger the transient current and the more rapidly the charge is extracted. Snap-off diodes are designed so that the end of the reverse current recovery pulse is very abrupt, rather than the tailing off observed in ordinary PN junction diodes. It is thus possible to produce a very short sharp current pulse by rapidly reverse biasing a snap-off diode. This can be used for a number of applications, such as high-order harmonic generation (turning a VHF or UHF drive current into a microwave signal) or operating the sampling gate in a digital sampling oscilloscope.

Small-signal Schottky diodes operate by a fundamentally different form of forward conduction. As a result of this, there is virtually no stored charge to be recovered when they are reverse biased, enabling them to operate efficiently as detectors or rectifiers at very high frequencies. Zener diodes conduct in the forward direction like any other silicon diode, but they also conduct in the reverse direction, and this is how they are normally used. At low reverse voltages, a Zener diode conducts only a very small leakage current like any other diode. But when the voltage reaches the nominal Zener voltage, the diode current increases rapidly, exhibiting a low incremental resistance. Diodes with a low breakdown voltage, up to about 4V, operate in true Zener breakdown: this conduction mechanism exhibits a small negative temperature coefficient of voltage. Higher-voltage diodes, rated at 6V and up, operate by a different mechanism called *avalanche breakdown,* which exhibits a small positive temperature coefficient. In diodes rated at about 5V both mechanisms occur, resulting in a very low or zero temperature coefficient of voltage. However, the lowest slope resistance occurs in diodes of about 7V breakdown voltage.

Reference

"Current-feedback op-amps ease high-speed circuit design," P. Harold, *EDN,* July 1988.

CHAPTER 3

Understanding Diodes and Their Problems

Robert Pease

Even the simplest active devices harbor the potential for causing baffling troubleshooting problems. Consider the lowly diode. The task of a diode sounds simple: To conduct current when forward biased, and to block current when reverse biased, while allowing negligible leakage. That task sounds easy, but no diode is perfect, and diodes' imperfections are fascinating. Even these two-terminal devices are quite complex!

All diodes start conducting current exponentially at low levels, nanoamperes and up. An ideal diode may have an exponential characteristic with a slope $\Delta V/\Delta I$ of:

$$g = (38.6 \text{ mS/mA}) \times I_F,$$

where mS = millisiemens = millimhos, and I_F = forward current. And indeed transistors do have this slope of 38.6 mS/mA at their emitters, at room temperature. This corresponds to 60 mV per decade of current. But the slopes of the exponential curves of different real (two-terminal) diodes vary considerably. Some, like a 1N645, have a slope as good as 70 or 75 mV per decade. Others like 1N914s have a slope as poor as 113 mV per decade. Others have intermediate values such as 90 mV/decade. When you go shopping for a diode, the data sheets *never* tell you about this. To tell the truth, I didn't even really recognize this. When I wrote the first version of this, as published in *EDN*, I assumed that the slopes started out from 60 mV per decade and then got worse—shifted over to 120 mV per decade at higher current levels, and I said so. But I was wrong. And nobody ever contradicted me. Such a strange world!

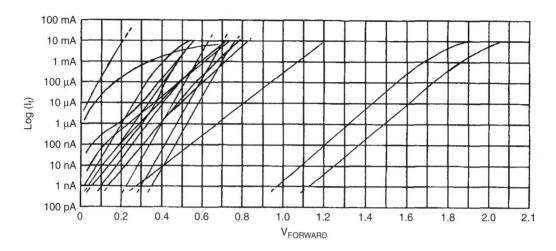

Figure 3.1: The diode made up of a transistor's emitter has high conductance over a wide range of currents. All the other diodes you can buy have inferior conductances, and they are just about all different . . . surprise.

Please refer to Figure 3.1, which shows just a few of the different curves you may get when you buy a diode. None of these slopes are characterized or guaranteed; if you change vendors, all bets are off. So, qualify each vendor of diodes carefully for each application. As the current level continues to increase, the conductance per milliampere gets even worse due to series resistance and high-level injection and other nonlinear factors. Therefore, at a large forward current, a diode's forward voltage, V_F, will be considerably larger than predicted by simple theory—and larger than desired. Of course, some rectifiers—depending on their ratings—can handle large currents from amperes to kiloamperes; but the V_Fs of all diodes, no matter what their ratings, err from the theoretical at high current levels.

These days, Schottky diodes have smaller V_Fs than ordinary p-n diodes. However, even germanium diodes and rectifiers still have their following because their low V_Fs are similar to the Schottky's. Just the other day I read about some new germanium Schottky diodes that have even lower V_Fs.

High-speed and ultra-high-speed (sometimes also called *high-efficiency*) silicon rectifiers are also available; they have been designed for fast switching-regulator and other high-frequency applications. They don't have quite as low V_Fs as Schottky diodes and are not quite as fast, but they are available with high reverse-voltage ratings and thus are useful for certain switch-mode circuit topologies that impress large flyback voltages on diodes.

When you reverse-bias these various diodes, ah, that is where you start to see even more wild dissimilarities. For example, the guaranteed reverse-current specification, I_{REV}, for many types of diodes is 25 nA max at 25°C. When you measure them, many of these devices actually have merely 50 or 100 pA of leakage. But the popular 1N914 and its close cousin, the 1N4148, actually *do* have about 10 or 15 nA of leakage at room temperature because of their gold doping. So although these diodes are inexpensive and popular, it's wrong to use them in low-leakage circuits since they're much leakier than other diodes with the same leakage specs.

Why, then, do some low-leakage diodes have the same mediocre 25-nA leakage spec as the 1N914? Diode manufacturers set the test and price at the level most people want to pay, because automatic test equipment can test at the 25-nA level—but no lower— without slowing down. If you want a diode characterized and tested for 100 pA or better, you have to pay extra for the slow-speed testing. Of course, high-conductance diodes such as Schottkys, germaniums, and large rectifiers have much larger reverse leakage currents than do signal diodes, but that's not normally a problem.

If you want a very-low-leakage diode, use a transistor's collector-base junction instead of a discrete diode (Ref. 3.1). The popular 2N930 or 2N3707 have low leakage, typically. Some 2N3904s do, too, but some of these are gold-doped and are leakier. The plastic-packaged parts are at least as good as the TO-18 hermetic ones. You can easily find such "diodes" having less than 1 pA leakage even at 7V, or 10 pA at 50V. Although this low leakage is not guaranteed, it's usually quite consistent. However, this c-b diode generally doesn't turn ON or OFF very quickly.

Another source of ultra-low-leakage diodes are the 2N4117A and the PN4117A, −18A, and −19A. These devices are JFETS with very small junctions, so leakages well below 0.1 pA are standard with 1.0 pA max, guaranteed—not bad for a $0.40 part. The capacitances are small, too.

3.1 Speed Demons

When a diode is carrying current, how long does it take to turn the current off? There's another wide-range phenomenon. Slow diodes can take dozens and hundreds of microseconds to turn off. For example, the collector-base junction of a 2N930 can take 30 μs to recover from 10 mA to less than 1 mA, and even longer to the nanoampere level. This is largely due to the recombination time of the carriers stored in the collector region of the transistor. Other diodes, especially gold-doped ones, turn off *much* faster—down into the nanosecond region. Schottky diodes are even faster, much faster than 1 ns. However, one of my friends pointed out that when you have a Schottky diode that turns off pretty fast, it is still in parallel with a p-n junction that may still turn off slowly at a light current level. If a Schottky turns off from 4 mA in less than 1 ns, there may still be a few microamperes that do not turn off for a microsecond. So if you want to use a Schottky as a precision clamp that will turn off very quickly, as in a settling detector (Ref. 3.2), don't be surprised if there is a small long "tail."

Switching regulators all have a need for diodes and high-current rectifiers and transistors to turn off quickly. If the rep rate is high and the current large and the diode turns off slowly, it can fail due to overheating. You *don't* want to try a 1N4002 at 20 or 40 kHz, as it will work very badly, if at all. Sometimes if you need a moderate amount of current at high speed, you can use several 1N914s in parallel. I've done that in an emergency, and it seemed to work well, but I can't be sure I can recommend it as the right thing to do for long-term reliability. The right thing is to engineer the right amount of speed for your circuit. High-speed, fast-recovery, and ultrafast diodes are available. The Schottky rectifiers are even faster, but not available at high voltage breakdowns. When you start designing switching regulators at these speeds, you really must know what you are doing. Or at least, wear safety goggles so you don't get hurt when the circuit blows up.

3.2   Turn 'em off—turn 'em on...

"Computer diodes" like the 1N914 are popular because they turn OFF quickly—in just a few nanoseconds—much faster than low-leakage diodes. What isn't well known is that these faster diodes not only turn OFF fast, they usually turn ON fast. For example, when you feed a current of 1.0 mA toward the anode of a 1N914 in parallel with a 40-pF capacitance (20 pF of stray capacitance plus a scope probe or something similar), the 1N914 usually turns ON in less than 1 ns. Thus, the V_F has only a few millivolts of overshoot.

But with some diodes—even 1N914s or 1N4148s from some manufacturers—the forward voltage may continue to ramp up past the expected DC level for 10 to 20 ns before the diode turns ON; this overshoot of 50 to 200 mV is quite surprising (Figure 3.2). Even more astonishing, the V_F overshoot may get worse at low repetition rates but can disappear at high repetition rates (Figure 3.2(B–D)).

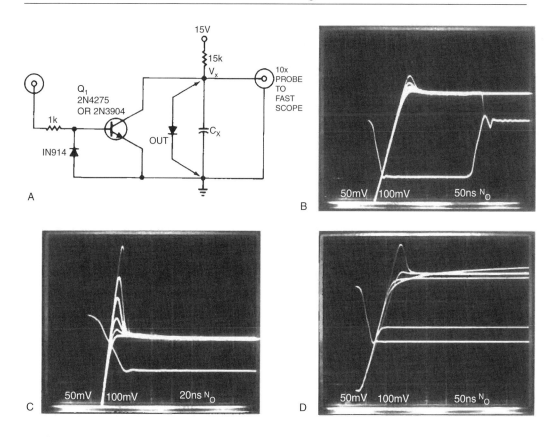

Figure 3.2: In this diode-evaluation circuit (A), transistor Q_1 simply resets V_x to ground periodically. When the transistor turns OFF, V_x rises to about 0.6V at which point the diode starts conducting. In (B), when dV_x/dt is 8 V/μps, this IN4148 overshoots as much as 140 mV at input frequencies below 10 kHz before it turns ON. At higher frequencies—120, 240, 480, 960, and 1920 kHz—as the repetition rate increases, the overshoot shrinks and disappears. Maximum overshoot occurs when $f_{in} < 7$ kHz. In (C), when dV_x/dt increases to 20 V/μs, this same IN4148 overshoots as much as 450 mV at 7 kHz but only 90 mV at 480 kHz and negligible amounts at frequencies above 2 MHz. In (D), various diode types have different turn-on characteristics. The superimposed, 120-kHz waveforms are all invariant with frequency except for the bad IN4148.

I spent several hours once discovering this particular peculiarity when a frequency-to-voltage converter suddenly developed a puzzling nonlinearity. The trickiest part of the problem with the circuit's diodes was that diodes from an earlier batch had not exhibited any slow-turn-on behavior. Further, some diodes in a batch of 100 from one manufacturer were as bad as the diodes in Figures 3.2(B) and 3.2(C). Other parts in that batch and other manufacturers' parts had substantially no overshoot.

When I confronted the manufacturers of these nasty diodes, they at first tried to deny any differences, but at length they admitted that they had changed some diffusions to "improve" the product. One man's "improvement" is another man's poison. Thus, you must always be alert for production changes that may cause problems. When manufacturers change the diffusions or the process or the masks, they may think that the changes are minor, but these changes could have a major effect on your circuit.

Many circuits, obviously, require a diode that can turn ON and catch, or clamp, a voltage moving much faster than 20 V/μs. Therefore, if you want any consistency in a circuit with fast pulse detectors (for example), you'll need to qualify and approve only manufacturers whose diodes turn ON consistently. So, as with any other unspecified characteristic, be sure to protect yourself against "bad" parts by first evaluating and testing and then specifying the performance you need. Also if you want to see fast turn-on of a diode circuit, with low overshoot, you must keep the inductance of the layout small. It only takes a few inches of wire for the circuit's inductance to make even a good fast rectifier look bad, with bad overshoot.

One "diode" that does turn ON and OFF quickly is a diode-connected transistor. A typical 2N3904 emitter diode can turn ON or OFF in 0.1 ns with negligible overshoot and less than 1 pA of leakage at 1V, or less than 10 pA at 4V. (This diode does, of course, have the base tied to the collector.) However, this diode can only withstand 5 or 6V of reverse voltage, and most emitter-base junctions start to break down at 6 or 8V. Still, if you can arrange your circuits for just a few volts, these diode-connected transistors make nice, fast, low-leakage diodes. Their capacitance is somewhat more than the 1N914's 1 pF.

3.3   Other Strange Things that Diodes Can Do to You. . .

If you keep LEDs in the dark, they make an impressive, low-leakage diode because of the high band-gap voltage of their materials. Such LEDs can exhibit less than 0.1 pA of leakage when forward biased by 100 mV or reverse biased by 1V.

Of course, you don't have to reverse-bias a diode a lot to get a leakage problem. One time I was designing a hybrid op-amp, and I specified that the diodes be connected in the normal parallel-opposing connection across the input of the second stage to avoid severe overdrive (Figure 3.3). I thought nothing more of these diodes until we had the circuit running—the op-amp's voltage gain was falling badly at 125°C. Why? Because the diodes were 1N914s, and their leakage currents were increasing from 10 nA at room temperature to about 8 μA at the high temperature. And, remember that the conductance of a diode at zero voltage is approximately (20 to 30 mS/mA) × $I_{LEAKAGE}$. That means each of the two diodes really measured only 6 kΩ.

Because the impedance at each input was only 6 kΩ, the op-amp's gain fell by a factor of four, even though the diodes may have only been forward or reverse biased by a millivolt. When we substituted collector-base junctions of transistors for the diodes, the gain went back up where it belonged.

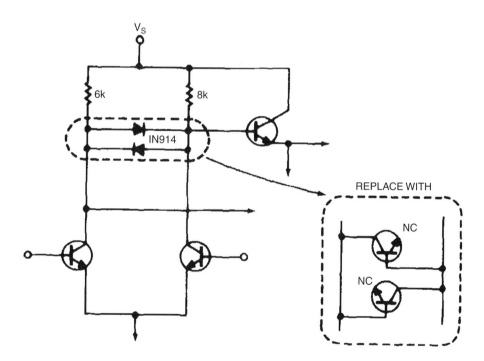

Figure 3.3: Even though the diodes in the first stage of this op-amp are forward or reverse biased by only a millivolt, the impedance of these diodes is much lower than the output impedance of the first stage or the input impedance of the second stage at high temperatures. Thus, the op-amp's gain drops disastrously.

Thus, you cannot safely assume that the impedance of a diode at zero bias is high if the junction's saturation current is large. For example, at 25°C a typical IN914 will leak 200 to 400 pA even with only 1 mV across it. Therefore, a 1N914 can prove unsuitable as a clamp or protection diode—even at room temperature—despite having virtually no voltage biased across it, in even simple applications such as a clamp across the inputs of a FET-input op-amp.

How can diodes fail? Well, if you were expecting a diode to turn ON and OFF, but instead it does something unexpected—of the sort I have been mentioning—that unexpected behavior may not be a *failure*, but it could sure cause *trouble*.

Further, you can kill a diode by applying excessive reverse voltage without limiting the current or by feeding it excessive forward current. When a diode fails, it tends to short out, becoming a small blob of muddy silicon rather than an open circuit. I did once see a batch of 1N4148s that acted like thermostats and went open circuit at 75°C, but such cases are rare these days.

One of the best ways to kill a diode is to ask it to charge up too big a capacitor during circuit turn-on. Most rectifiers have maximum ratings for how much current they can pass, on a repetitive and on a nonrecurring basis. I've always been favorably impressed by the big Motorola (Phoenix, AZ) books with all the curves of safe areas for forward current as a function of pulse time and repetition rate. These curves aren't easy to figure out at first, but after a while they're fairly handy tools.

Manufacturers can play tricks on you other than changing processes. If you expect a diode to have its arrow pointing toward the painted band (sometimes called the *cathode* by the snobbish) and the manufacturer put the painted band on the wrong end, your circuit won't work very well. Fortunately reverse-marked diodes are pretty rare these days. But just this morning, I heard an engineer call the "pointed" end of the diode an anode, which led to confusion and destruction. Sigh...

Once I built a precision test box that worked right away and gave exactly the right readings until I picked up the box to look at some waveforms. Then the leakage test shifted *way* off zero. Every time I lifted up the box, the meter gave an indication; I thought I had designed an altimeter. After some study, I localized the problem to an FD300 diode, whose body is a clear glass DO-35 package covered with black paint. This particular diode's paint had been scratched a little bit, so when I picked up the test box, the light shone under the fixture and onto the diode. Most of these diodes didn't exhibit this behavior; the paint wasn't scratched on most of them.

To minimize problems such as the ones I have listed, I recommend the following strategies:

- Have each manufacturer's components specifically qualified for critical applications. This is usually a full-time job for a components engineer, with help and advice from the design engineer and consultation with manufacturing engineers.

- Establish a good relationship with each manufacturer.

- Require that manufacturers notify you when, or preferably before, they make changes in their products.

- Keep an alternate source qualified and running in production whenever possible.

My boss may gripe if I say this too loudly, but it is well known that having two good sources is better than having one. The argument that "One source is better than two" falls hollow on my ears. Two may be better than seven or eight, but one is not better than two.

3.4 Zener, Zener, Zener...

Just about all diodes will break down if you apply too much reverse voltage, but zener diodes are *designed* to break down in a predictable and well-behaved way. The most common way to have problems with a zener is to starve it. If you pass too little current through a zener, it may get too noisy. Many zeners have a clean and crisp knee at a small reverse-bias current, but this sharp knee is not guaranteed below the rated knee current.

Some zeners won't perform well no matter how carefully you apply them. In contrast to high-voltage zeners, low-voltage (3.3 to 4.7V) zeners are poor performers and have poor noise and impedance specs and bad temperature coefficients—even if you feed them a lot of current to get above the knee, which is very soft. This is because "zeners" at voltages above 6V are really avalanche-mode devices and employ a mechanism quite different from (and superior to) the low-voltage ones, which are real zener diodes. At low-voltage levels, band-gap references such as LM336s and LM385s are popular, because their performance is good compared with low-voltage zeners.

Zener references with low temperature coefficients, such as the 1N825, are only guaranteed to have low temperature coefficients when operated at their rated current, such as 7.5 mA. If you adjust the bias current up or down, you can sometimes tweak the temperature coefficient, but some zeners aren't happy if operated away from their specified bias. Also, don't test your 1N825 to see what its "forward-conduction voltage" is because in the "forward" direction, the device's temperature-compensating diode may break down at 70 or 80V. This breakdown damages the device's junction, degrades the device's performance and stability, and increases its noise.

The LM329 is popular as a 6.9V reference because its TC is invariant of operating current, as it can run from any current from 1 to 10 mA. The LM399 is even more popular because of its built-in heater that holds the junction at $+85°C$. Consequently it can hold 1/2 or 1 ppm per $°C$. The LM329 and LM399 types also have good long-term stability, such as 5 or 10 ppm per 1000 hours, typically. The buried zeners in the LM129/LM199/LM169 also have better stability than most discrete references (1N825 or similar) when the references are turned on and off.

And before you subject a zener to a surge of current, check its derating curves for current vs. time, which are similar to the rectifiers' curves mentioned earlier. These curves will tell you that you can't bang an ampere into a 10V, 1W zener for very long.

In fact, most rectifiers are rated to be operated strictly within their voltage ratings, and if you insist on exceeding that reverse voltage rating and breaking them down, their reliability will be degraded. To avoid unreliability, you can redesign the circuit to avoid over-voltage, or you might add in an R-C-diode damper to soak up the energy; or you could shop for a controlled-avalanche rectifier. These rectifiers are rated to survive (safely and reliably) repetitive excursions into breakdown when you exceed their rated breakdown voltage. The manufacturers of these devices can give you a good explanation of how to keep out of trouble.

If you *do* need a zener to conduct a surge of current, check out the specially designed surge-rated zener devices—also called *transient-voltage suppressors*—from General Semiconductor Industries Inc. (Tempe, AZ). You'll find that their 1W devices, such as the 1N5629 through 1N5665A, can handle a surge of current better than most 10- or 50W-zeners. If you need a really high-current zener, a power transistor can help out (Figure 3.4). As mentioned earlier, a diode tends to fail by becoming a short circuit when overpowered, and zeners cannot absorb as much power as you would expect from short pulses. How dreadful; but, can IC designers serendipitously take advantage of this situation? Yes!

The V_{os} of an op-amp usually depends on the ratio of its first-stage load resistors. IC designers can connect several zeners across various small fractions of the load resistor.

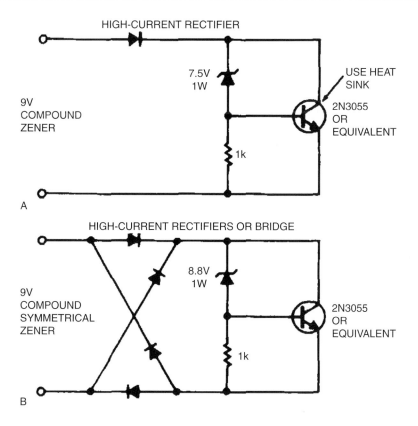

Figure 3.4: The power rating of this compound zener (A) is that of the power transistor. The second compound zener, (B) is almost the same as (A) but acts as a symmetrical, matched, double-ended compound zener.

When they measure the V_{os}, they can decide which zener to short out—or *zap*—with a 5-ms, 0.3- to 1.8A-pulse. The zener quickly turns into a low-impedance ($\approx 1\Omega$ short), so that part of the resistive network shorts out, and the V_{os} is improved.

In its LM108, National Semiconductor first used zener zapping, although Precision Monolithics (Santa Clara, CA) wrote about zener zapping first and used it extensively later on. Although zener zapping is a useful technique, you have to be sure that nobody discharges a large electrostatic charge into any of the pins that are connected to the zener zaps. If you like to zap zeners for fun and profit, you probably know that they really do make a cute lightning flash in the dark when you zap them. Otherwise, be careful not to hit zeners hard, if you don't want them to zap and short out.

These zener zaps are also becoming popular in digital ICs under the name of "vertical fuses" or, more correctly, "anti-fuses." If an IC designer uses platinum silicide instead of aluminum metallization for internal connections, the diode resists zapping.

3.5 Diodes that Glow in the Dark, Efficiently

Once I needed 100 LEDS, so I bought 200 LEDs from the cheapest supplier. I hoped to find some good ones and maybe just a few units that were weak or performed poorly, which I could use for worst-case testing. I lost out; every one of the 200 was of uniformly good intensity. In a variation on Murphy's Law, worst-case parts will typically appear only when you are depending on having uniform ones.

So long as you don't fry LEDs with your soldering iron or grossly excessive milliamps of current, LEDs are awfully reliable these days. I have a thermometer display on my wall, which has 650 inexpensive, plastic-packaged LEDs. These LEDs have amassed 40,000,000 device-hours with just one failure. The only problem I ever have with LEDs is trying to remember which lead is "plus"—I just measure the diode and re-derive it, every time.

3.6 Optoisolators

An optoisolator, also called a *photo-coupler* or *opto-coupler*, usually consists of an infrared LED and a sensitive phototransistor to detect the LED's radiation. In the course of working with the cheaper 4N28s, I've found it necessary to add circuitry to achieve moderate speeds. For example, if you tailor the biases per Figure 3.5, you can get a 4N28's response up toward 50 kHz; otherwise the devices can't make even 4 kHz reliably. The trick is decreasing the phototransistor's turn-off time by using a resistor from pin 4 to pin 6.

I've evaluated many different makes and lots of 4N28s and have found widely divergent responses. For example, the overall current gain at 8 mA can vary from 15 to 104%, even though the spec is simply 10% min. Further, the transfer efficiency from the LED to the photodiode varies over a range wider than 10:1, and the β of the transistor varies from 300 to 3000. Consequently, the transistor's speed of response, which is of course related to β and $f_{3\ dB}$, would vary over a 10:1 range.

If your circuit doesn't allow for gains and frequency responses that vary so wildly and widely, expect trouble. For example, two circuits, one an optoisolated switching regulator (Ref. 3.3) and the other a detector for 4- to 20-mA currents (Ref. 3.4), have enough degeneration so that any 4N28 you can buy will work. I used to have a group of

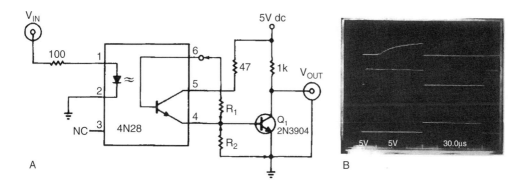

Figure 3.5: Adding R_1 and R_2 to the inexpensive 4N28 optoisolator lets it handle faster signals with less delay—5 µs vs. 60 µs. The scope photo's bottom trace is an input waveform, the top trace is the circuit's output without R_1 and R_2, and the center trace is the output with $R_1 = 2$ MΩ and $R_2 = 1$ kΩ.

several "worst-case" 4N28s from various manufacturers that I would try out in prototypes and problem circuits. Unfortunately, I don't have those marginal devices anymore, but they were pretty useful.

Also, the data sheets for optoelectronic components often don't have a clear V_F curve or list any realistic typical values; the sheets list only the worst-case values. Therefore, you may not realize that the V_F of an LED in an optoisolator is a couple hundred millivolts smaller than that of discrete red or infrared LEDs. Conversely, the V_F of high-intensity, or high-efficiency, red LEDs tends to be 150 mV larger than that of ordinary red LEDs. And the V_F of DEADs (a DEAD is a Darkness Emitting Arsenide Diode; that is, a defunct LED) is not even defined.

Once I was troubleshooting some interruptor modules. In these modules, a gap separated an infrared LED and a phototransistor. An interruptor—say a gear tooth—in the gap can thus block the light. I tested one module with a piece of paper and nothing happened—the transistor stayed ON. What was that again? It turned out that the single sheet of paper could diffuse the infrared light but not completely attenuate it. A thin sheet of cardboard or two sheets of paper would indeed block the light.

3.7   Solar Cells

Extraneous, unwanted light impinging on the p-n junction of a semiconductor is only one of many tricky problems you can encounter when you try to design and operate

Figure 3.6: With a solar-cell array, you can make electricity when the sun shines. (Photo copyright Peggi Willis.)

precision amplifiers—especially high-impedance amplifiers. Just like a diode's p-n junction, a transistor's collector-base junction makes a good photodiode, but a transistor's plastic or epoxy or metal package normally does a very good job of blocking out the light.

When light falls onto the p-n junction of any diode, the light's energy is converted to electricity and the diode forward biases itself. If you connect a load across the diode's terminals, you can draw useful amounts of voltage and current from it. For example, you could stack a large number of large-area diodes in series and use them for recharging a battery. The most unreliable part of this system is the battery. Even if you never abuse them, batteries don't like to be discharged a large number of cycles, and your battery will eventually refuse to take a charge. These days one reads all sorts of marvelous hype about battery-powered cars, but the writers always ignore the

Figure 3.7: Maintaining a healthy battery involves careful attention to charging, discharging, and temperature. (Photo copyright Peggi Willis.)

terrible expense of replacing the batteries after just a few hundred cycles. They seem to be pretending that if they ignore that problem, it will go away...

So much for the charms of solar-recharged batteries. It's much better to use a solar-powered night-light. Remember that one? A solar-powered night-light doesn't need a battery; it simply needs a 12,000-mile extension cord. To be serious, the most critical problem with solar cells is their packaging; most semiconductors don't have to sit out in the sun and the rain as solar cells do. And it's hard to make a reliable package when low cost is—as it is for solar cells—a major requirement.

In addition to packaging, another major trouble area with solar cells is their temperature coefficients. Just like every other diode, the V_F of a solar cell tends to decrease at 2 mV/°C of temperature rise. Therefore, as more and more sunlight shines on the solar cell, it puts out more and more current, but its voltage could eventually drop below the battery's voltage, whereupon charging stops. Using a reflector to get even more light onto the cell contributes to this temperature-coefficient problem. Cooling would help, but the attendant complications rapidly overpower the original advantage of solar cells' simplicity.

3.8   Assault and Battery

Lastly, I want to say a few things about batteries. The only thing that batteries have in common with diodes is that they are both two-terminal devices. Batteries are complicated electrochemical systems, and large books have been written about the characteristics of each type (Refs. 3.5–3.10). I couldn't possibly give batteries a full and fair treatment here, but I will outline the basics of troubleshooting them.

First, always refer to the manufacturer's data sheet for advice on which loads and what charging cycles will yield optimal battery life. When you recharge a nickel-cadmium battery, charge it with a constant current, not constant voltage. And be sure that the poor little thing doesn't heat up after it is nearly fully charged. Heat is the enemy of batteries as it is for semiconductors. If you're subjecting your battery to deep-discharge cycles, refer to the data sheet or the manufacturer's specifications and usage manual for advice. Some authorities recommend that you do an occasional deep discharge, all the way to zero; others say that when you do a deep discharge, some cells in the battery discharge before the others and then get reversed, which is not good for them. I cannot tell you who's correct.

Sometimes a NiCad cell will short out. If this happens during a state of low charge, the cell may stay shorted until you ZAP it with a brief burst of high current. I find that discharging a 470 μF capacitor charged to 12V into a battery does a good job of opening up a shorted cell. If 470 μF doesn't do it, I keep a 3800 μF to do the job.

When you recharge a lead-acid battery, charge it to a float voltage of 2.33V per cell. At elevated temperatures, you should decrease this float voltage by about 6 mV/°C; again, refer to the manufacturer's recommendations. When a lead-acid battery is deeply discharged (below 1.8V per cell), it should be recharged right away or its longevity will suffer due to sulfation.

Be careful when you draw excessive current from a lead-acid battery; the good strong ones can overheat or explode. Also be careful when charging them; beware of the accumulation of hydrogen or other gases that are potentially dangerous or explosive.

And, please dispose of all dead batteries in an environmentally sound way. Call your local solid-waste-disposal agency for their advice on when and where to dispose of batteries. Perhaps some can be recycled.

References

[3.1] Pease, Robert A., "Bounding, clamping techniques improve on performance," *EDN*, November 10,1983, p. 277.
[3.2] Pease, Bob, and Ed Maddox, "The Subtleties of Settling Time," *The New Lightning Empiricist,* Teledyne Philbrick, Dedham MA, June 1971.

[3.3] Pease, Robert A., "Feedback provides regulator isolation," *EDN*, November 24, 1983, p. 195.

[3.4] Pease, Robert A., "Simple circuit detects loss of 4-20 mA signal," *Instruments & Control Systems*, March 1982, p. 85.

[3.5] *Eveready Ni-Cad Battery Handbook*, Eveready, Battery Products Div., 39 Old Ridgebury Rd., Danbury, CT. (203) 794-2000.

[3.6] *Battery Application Manual*, Gates Energy Products, Box 861, Gainesville, FL 32602. (1-800-627-1700).
(Note: A sealed lead-acid and NiCd battery manual.)

[3.7] Perez, Richard, *The Complete Battery Book,* Tab Books, Blue Ridge Summit, PA, 1985.

[3.8] Small, Charles H., "Backup batteries," *EDN*, October 30, 1986, p. 123.

[3.9] Linden, David, Editor-in-Chief, *The Handbook of Batteries and Fuel Cells*, McGraw-Hill Book Co., New York, NY, 1984. (Note: The battery industry's bible.)

[3.10] Independent Battery Manufacturers Association, SLIG *System Buyer's Guide*, 100 Larchwood Dr., Largo, FL 33540. (813) 586-1408. (Note: Don't be put off by the title; this book is the best reference for lead-acid batteries.)

CHAPTER 4

Bipolar Transistors

Ian Hickman

Unlike semiconductor diodes, transistors did not see active service in the Second World War; they were born several years too late. In 1948 it was discovered that if a point contact diode detector were equipped with two cat's whiskers rather than the usual one, spaced very close together, the current through one of them could be influenced by a current through the other. The crystal used was germanium, one of the rare earths, and the device had to be prepared by discharging a capacitor through each of the cat's whiskers in turn to "form" a junction. Over the following years, the theory of conduction via junctions was elaborated as the physical processes were unraveled, and the more reproducible junction transistor replaced point contact transistors.

However, the point contact transistor survives to this day in the form of the standard graphical symbol denoting a bipolar junction transistor (Figure 4.1(A)). This has three separate regions, as in Figure 4.1(B), which shows (purely diagrammatically and not to scale) an NPN junction transistor. With the *base* (another term dating from point contact days) short-circuited to the *emitter*, no *collector* current can flow since the collector/base junction is a reverse biased diode, complete with depletion layer as shown. The higher the reverse bias voltage, the wider the depletion layer, which is found mainly on the collector side of the junction since the collector is more lightly doped than the base. In fact, the pentavalent atoms which make the collector n-type are found also in the base region. The base is a layer which has been converted to p-type by substituting so many trivalent (hole donating) atoms into the silicon lattice, e.g., by diffusion or ion bombardment, as to swamp the effect of the pentavalent atoms. So holes are the *majority carriers* in the base region, just as electrons are the majority carriers in the

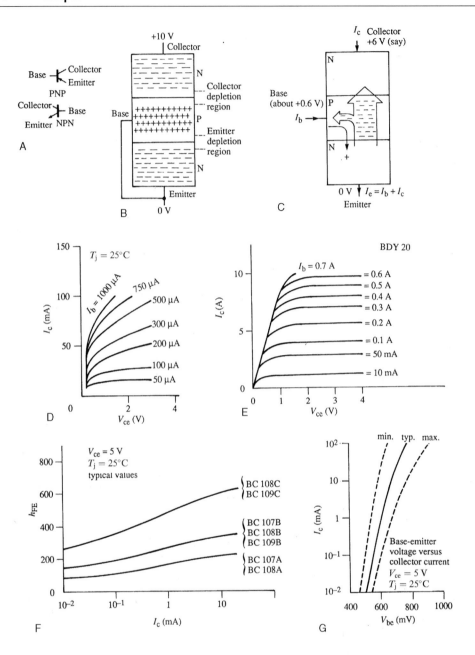

collector and emitter regions. The collector "junction" turns out then to be largely notional; it is simply that plane for which on one side (the base) holes or p-type donor atoms predominate, while on the other (the collector) electrons or n-type donor atoms predominate, albeit at a much lower concentration.

Figure 4.1(C) shows what happens when the base/emitter junction is forward biased. Electrons flow from the emitter into the base region and, simultaneously, holes flow from the base into the emitter. The latter play no useful part in transistor action: they contribute to the base current but not to the collector current. Their effect is minimized by making the n-type emitter doping a hundred times or more heavier than the base doping, so that the vast majority of current flow across the emitter/base junction consists of electrons flowing into the base from the emitter. Some of these electrons flow out of the base, forming the greater part of the base current. But most of them, being *minority carriers* (electrons in what should be a p-type region) are swept across the collector junction by the electric field gradient existing across the depletion layer. This is illustrated (in diagrammatic form) in Figure 4.1(C), while Figure 4.1(D) shows the collector characteristics of a small-signal NPN transistor and Figure 4.1(E) those of an NPN power transistor. It can be seen that, except at very low values, the collector voltage has comparatively little effect upon the collector current, for a given constant base current. For this reason, the bipolar junction transistor is often described as having a "pentode-like" output characteristic, by an analogy dating from the days of valves. This is a fair analogy as far as the collector characteristic is concerned, but there the similarity ends. The pentode's anode current is controlled by the g_1 (control grid) voltage, but there is, at least for negative values of control grid voltage, negligible grid

Figure 4.1: The bipolar transistor. (A) Bipolar transistor symbols. (B) NPN junction transistor, cut-off condition. Only majority carriers are shown. The emitter depletion region is very much narrower than the collector depletion region because of reverse bias and higher doping levels. Only a very small collector leakage current I_{cb} flows. (C) NPN small signal silicon junction transistor, conducting. Only minority carriers are shown. The DC common emitter current gain is $h_{FE} = I_c/I_b$, roughly constant and typically around 100. The AC small signal current gain is $h_{fe} = dI_c/dI_b = i_c/i_b$. (D) Collection current versus collector/ emitter voltage, for an NPN small signal transistor (BC 107/8/9). (E) Collector current versus collector/emitter voltage, for an NPN power transistor. (F) h_{FE} versus collector current for an NPN small signal transistor. (G) Collector current versus base/emitter voltage for an NPN small signal transistor. (Parts (D) to (G) reproduced by courtesy of Philips Components Ltd.)

current. By contrast, the base/emitter input circuit of a transistor looks very much like a diode, and the collector current is more linearly related to the base current than to the base/emitter voltage (Figure 4.1(F) and (G)). For a silicon NPN transistor, little current flows in either the base or collector circuit until the base/emitter voltage V_{be} reaches about +0.6V, the corresponding figure for a germanium NPN transistor being about +0.3V. For both types, the V_{be} corresponding to a given collector current falls by about 2 mV for each degree centigrade of temperature rise, whether this is due to the ambient temperature increasing or due to the collector dissipation warming the transistor up. The reduction in V_{be} may well cause an increase in collector current and dissipation, heating the transistor further and resulting in a further fall in V_{be}. It thus behooves the circuit designer, especially when dealing with power transistors, to ensure that this process cannot lead to thermal runaway and destruction of the transistor.

Although the base/emitter junction behaves like a diode, exhibiting an incremental resistance of $25/I_e$ at the emitter, most of the emitter current appears in the collector circuit, as has been described above.

The ratio I_c/I_b is denoted by the symbol h_{FE} and is colloquially called the DC *current gain* or static forward current transfer ratio. Thus, if a base current of 10 μA results in a collector current of 3 mA—typically the case for a high-gain general purpose audio-frequency NPN transistor such as a BC109—then $h_{FE} = 300$. As Figure 4.1(F) shows, the value found for h_{FE} will vary somewhat according to the conditions (collector current and voltage) at which it is measured. When designing a transistor amplifier stage, it is necessary to ensure that any transistor of the type to be used, regardless of its current gain, its V_{be}, etc., will work reliably over a wide range of temperatures: the no-signal DC conditions must be stable and well defined. The DC current gain h_{FE} is the appropriate gain parameter to use for this purpose. When working out the stage gain or AC small-signal amplification provided by the stage, h_{fe} is the appropriate parameter, this is the AC *current gain* $dI_c = dI_b$. Usefully, for many modern small-signal transistors there is little difference in the value of h_{FE} and h_{fe} over a considerable range of current, as can be seen from Figures 4.1(F) and 4.2(A) (allowing for the linear h_{FE} axis in one and the logarithmic h_{fe} axis in the other).

Once the AC performance of a transistor is considered, it is essential to allow for the effects of reactance. Just as there is capacitance between the various electrodes of a valve, so too there are unavoidable capacitances associated with the three electrodes of a transistor. The collector/base capacitance, though usually not the largest of these, is particularly important as it provides a path for AC signals from the collector circuit

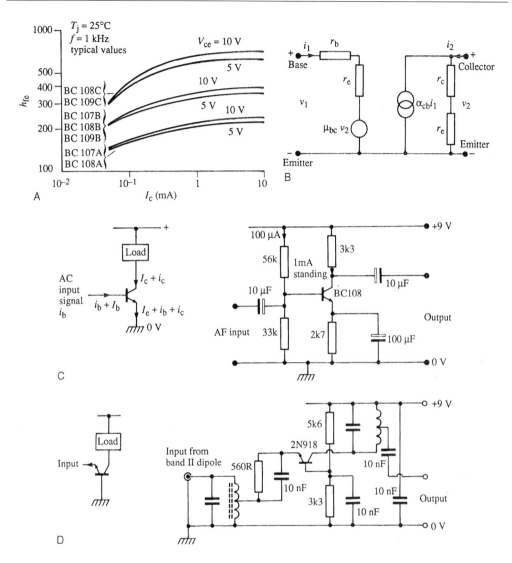

Figure 4.2: Small-signal amplifiers. (A) h_{fe} versus collector current for an NPN small-signal transistor of same type as in Figure 4.4(F) (reproduced by courtesy of Philips Components Ltd.). (B) Common emitter equivalent circuit. (C) Common emitter audio amplifier. I_b = base bias or standing current; I_c = collector standing current; i_c = useful signal current in load. (D) Common base RF amplifier.

(Continued)

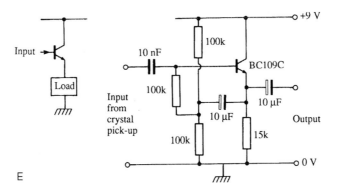

Figure 4.2: (Cont'd) (E) Common collector high-input-impedance audio amplifier.

back to the base circuit. In this respect, the transistor is more like a triode than a pentode, and as such the Miller effect will reduce the high-frequency gain of a transistor amplifier stage, and may even cause an RF stage to oscillate due to feedback of in-phase energy from the collector to the base circuit.

One sees many different theoretical models for the bipolar transistor, and almost as many different sets of parameters to describe it: *z, g, y,* hybrid, *s,* etc. Some equivalent circuits are thought to be particularly appropriate to a particular configuration, e.g., grounded base, while others try to model the transistor in a way that is independent of how it is connected. Over the years numerous workers have elaborated such models, each proclaiming the advantages of his particular equivalent circuit.

Just one particular set of parameters will be mentioned here, because they have been widely used and because they have given rise to the symbol commonly used to denote a transistor's current gain. These are the *hybrid* parameters which are generally applicable to any *two-port network,* i.e., one with an input circuit and a separate output circuit. Figure 4.3(A) shows such a two-port, with all the detail of its internal circuitry hidden inside a box—the well-known "black box" of electronics. The voltages and currents at the two ports are as defined in Figure 4.1(A), and I have used *v* and *i* rather than *V* and *I* to indicate small-signal alternating currents, not the standing DC bias conditions. Now v_1, i_1, v_2, and i_2 are all variables and their interrelation can be described in terms of four *h* parameters as follows:

$$v_1 = h_{11}i1 + h_{12}v_2 \tag{4.1}$$

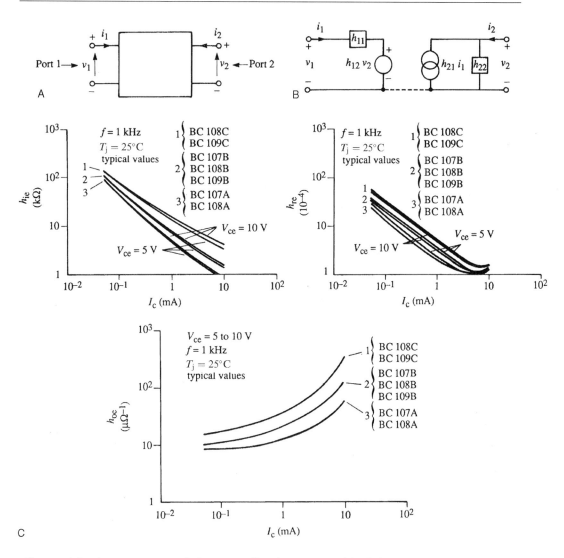

Figure 4.3: *h* parameters. (A) Generalized two-port black box. *v* and *i* are small-signal alternating qualities. At both ports, the current is shown as in phase with the voltage (at least at low frequencies), i.e., both ports are considered as resistances (impedances). (B) Transistor model using hybrid parameters. (C) *h* parameters of a typical small-signal transistor family. (Reproduced by courtesy of Philips Components Ltd.)

$$i_2 = h_{21}i_1 + h_{22}v_2 \tag{4.2}$$

Each of the h parameters is defined in terms of two of the four variables by applying either of the two conditions $i_1 = 0$ or $v_2 = 0$:

$$h_{11} = \left.\frac{v_1}{i_1}\right|_{v2=0} \tag{4.3}$$

$$h_{21} = \left.\frac{i_2}{i_1}\right|_{v2=0} \tag{4.4}$$

$$h_{12} = \left.\frac{v_1}{v_2}\right|_{i1=0} \tag{4.5}$$

$$h_{22} = \left.\frac{i_2}{v_2}\right|_{i1=0} \tag{4.6}$$

Thus h_{11} is the *input impedance* with the output port short-circuited as far as AC signals are concerned. At least at low frequencies, this impedance will be resistive and its units will be ohms. Next, h_{21} is the *current transfer ratio*, again with the output circuit short-circuited so that no output voltage variations result: being a pure ratio, h_{21} has no units. Like h_{11} it will be a complex quantity at high frequencies, i.e., the output current will not be exactly in phase with the input current. Third, h_{12} is the voltage feedback ratio, i.e., the voltage appearing at the input port as the result of the voltage variations at the output port (again this will be a complex number at high frequencies). Finally, h_{22} is the *output admittance*, measured—like h_{12}—with the input port open-circuit to signals. These parameters are called *hybrid* because of the mixture of units: impedance, admittance and pure ratios.

In equation (4.1) the input voltage v_1 is shown as being the result of the potential drop due to i_1 flowing through the input impedance plus a term representing the influence of any output voltage variation v_2 on the input circuit. When considering only small signals, to which the transistor responds in a linear manner, it is valid simply to add the two effects as shown. In fact the hybrid parameters are examples of partial differentials: these describe how a function of two variables reacts when first one variable is changed while the other is held constant, and then vice versa. Here, v_1 is a function of both i_1 and v_2—so $h_{11} = \partial v_1/\partial i_1$ with v_2 held constant (short-circuited), and $h_{12} = \partial v_1/\partial v_2$ with the other parameter i_1 held constant at zero (open-circuit). Likewise, i_2 is a function of both i_1 and v_2; the relevant parameters h_{21} and h_{22} are defined by equations (4.4) and (4.6), and i_2 is as defined in equation (4.2). Of course the interrelation of v_1, i_1, v_2 and i_2 could be specified in other ways: the above scheme is simply the one used with h parameters.

The particular utility of h parameters for specifying transistors arises from the ease of determining h_{11} and h_{21} with the output circuit short-circuited to signal currents. Having defined h parameters, they can be shown connected as in Figure 4.3(B). Since a transistor has only three electrodes, the dashed line has been added to show that one of them must be common to both the input and the output ports. The common electrode may be the base or the collector, but particularly important is the case where the input and output circuits have a common emitter.

Armed with the model of Figure 4.3(B) and knowing the source and load impedance, you can now proceed to calculate the gain of a transistor stage—provided you know the relevant values of the four h parameters (see Figure 4.3(C)). For example, for a *common emitter* stage you will need h_{ie} (the input impedance *h11* in the common emitter configuration), h_{fe} (the common emitter forward current transfer ratio or current gain corresponding to h_{21}), h_{re} (the common emitter voltage feedback ratio corresponding to h_{12}) and h_{oe} (the common emitter output admittance corresponding to h_{22}). You will generally find that the data sheet for the transistor you are using quotes maximum and minimum values for h_{fe} at a given collector current and voltage, and may well also include a graph showing how the typical or normalized value of h_{fe} varies with the standing collector current I_c. Sometimes, particularly with power transistors, only h_{FE} is quoted: this is simply the ratio I_c/I_b, often called the *DC current gain* or *static forward current transfer ratio*. As mentioned earlier, for most transistors this can often be taken as a fair guide or approximation for h_{fe} (for example, compare Figures 4.1(F) and 4.2(a)). From these it can be seen that over the range 0.1 to 10 mA collector current, the typical value of h_{FE} is slightly greater than that of h_{FE}, so the latter can be taken as a guide to h_{fe}, with a little in hand for safety. Less commonly you may find h_{oe} quoted on the data sheet, while h_{ie} and h_{re} are often simply not quoted at all. Sometimes a mixture of parameters is quoted; for example, data for the silicon NPN transistor type 2N930 quote h_{FE} at five different values of collector current, and low-frequency (1 kHz) values for h_{ib}, h_{rb}, h_{fe} and h_{ob}—all at 5V, 1 mA. The only data given to assist the designer in predicting the device's performance at high frequency are f_T and C_{obo}. The *transition frequency* f_T is the notional frequency at which $|h_{fe}|$ has fallen to unity, projected at −6dB per octave from a measurement at some lower frequency. For example, f_T (min.) for a 2N918 NPN transistor is 600 MHz measured at 100 MHz, i.e., its common emitter current gain h_{fe} at 100 MHz is at least 6. C_{obo} is the *common base output capacitance* measured at $I_c = 0$, at the stated V_{cb} and test frequency (10V and 140 kHz in the case of the 2N918).

If you were designing a *common base or common collector* stage, then you would need the corresponding set of *h* parameters, namely h_{ib}, h_{fb}, h_{rb} and h_{ob} or h_{ic}, h_{fc}, h_{rc} and h_{oc} respectively. These are seldom available—in fact, *h* parameters together with *z, v, i* and transmission parameters are probably used more often in the examination hall than in the laboratory. The notable exception are the *scattering* parameters *s*, which are widely used in radio-frequency and microwave circuit design. Not only are many UHF and microwave devices (bipolar transistors, silicon and gallium arsenide field-effect transistors) specified on the data sheet in *s* parameters, but *s* parameter test sets are commonplace in RF and microwave development laboratories. This means that if it is necessary to use a device at a different supply voltage and current from that at which the data sheet parameters are specified, they can be checked at those actual operating conditions.

The *h* parameters for a given transistor configuration, say grounded emitter, can be compared with the elements of an equivalent circuit designed to mimic the operation of the device. In Figure 4.2(B) r_e is the incremental slope resistance of the base/emitter diode; it was shown earlier that this is approximately equal to $25/I_e$ where I_e is the standing emitter current in milliamperes. Resistance r_c is the collector slope or incremental resistance, which is high. (For a small-signal transistor in a common emitter circuit, say a BC109 at 2 mA collector current, 15K would be a typical value: see Figure 4.3(C)). The base input resistance r_b is much higher than $25/I_e$, since most of the emitter current flows into the collector circuit, a useful approximation being $h_{fe} \times 25/I_e$. The ideal voltage generator μ_{bc} represents the voltage feedback h_{12} (h_{re} in this case), while the constant current generator α_{cb} represents h_{21} or h_{fe}, the ratio of collector current to base current. Comparing Figures 4.2(B) and 4.3(B), you can see that $h_{11} = r_b + r_e$, $h_{12} = \mu_{bc}$, $h_{21} = \alpha_{cb}$ and $h_{22} = 1/(r_e + r_c)$.

Not the least confusing aspect of electronics is the range of different symbols used to represent this or that parameter, so it will be worth clearing up some of this right here. The small-signal common emitter current gain is, as has already been seen, sometimes called α_{cb}, but more often h_{fe}; the symbol β is also used. The symbol α_{ce} or just α is used to denote the common base forward current transfer ratio h_{fb}: the term *gain* is perhaps less appropriate here, as i_c is actually slightly less than i_e, the difference being the base current i_b. It follows from this that $\beta = \alpha/(1 - \alpha)$. The symbols α and β have largely fallen into disuse, probably because it is not immediately obvious whether they refer to small-signal or DC gain: with h_{fe} and h_{FE}—or h_{fb} and h_{FB}—you know at once exactly where you stand.

When h parameters for a given device are available, their utility is limited by two factors: first, usually typical values only are given (except in the case of h_{fe}) and second, they are measured at a frequency in the audio range, such as 1 kHz. At higher frequencies the performance is limited by two factors: the inherent capacitances associated with the transistor structure, and the reduction of current gain at high frequencies.

In addition to their use as small-signal amplifiers, transistors are also used as switches. In this mode they are either reverse biased at the base, so that no collector current flows or conducting heavily so that the magnitude of the voltage drop across the collector load approaches that of the collector supply rail. The transistor is then said to be *bottomed*, its V_{ce} being equal to or even less than V_{be}. For this type of large-signal application, the small-signal parameters mentioned earlier are of little if any use. In fact, if (as is usually the case) one is interested in switching the transistor on or off as quickly as possible, it can more usefully be considered as a charge-controlled rather than a current-controlled device. Here again, although sophisticated theoretical models of switching performance exist, they often involve parameters (such as r_{bb}', the extrinsic or ohmic part of the base resistance) for which data sheets frequently fail to provide even a typical value. Thus one is usually forced to adopt a more pragmatic approach, based upon such data sheet values as are available, plus the manufacturer's application notes if any, backed up by practical in-circuit measurements.

Returning for the moment to small-signal amplifiers, Figure 4.2(C), (D) and (E) shows the three possible configurations of a single-transistor amplifier and indicates the salient performance features of each. Since the majority of applications nowadays tend to use NPN devices, this type has been illustrated. Most early transistors were PNP types; these required a radical readjustment of the thought processes of electronic engineers brought up on valve circuits, since with PNP transistors the "supply rail" was negative with respect to ground. The confusion was greatest in switching (logic) circuits, where one was used to the anode of a cut-off valve rising to the (positive) HT rail, this being usually the logic 1 state. Almost overnight, engineers had to get used to collectors flying up to –6V when cut off, and vice versa. Then NPN devices became more and more readily available and eventually came to predominate. Thus the modern circuit engineer has the great advantage of being able to employ either NPN or PNP devices in a circuit, whichever proves most convenient—and not infrequently both types are used together. The modern valve circuit engineer, by contrast, still has to make do without a thermionic equivalent of the PNP transistor.

A constant grumble of the circuit designer was for many years that the current gain h_{FE} of power transistors, especially at high currents, was too low. The transistor manufacturers' answer to this complaint was the *Darlington*, which is now available in a wide variety of case styles and voltage (and current) ratings in both NPN and PNP versions. The circuit designer had already for years been using the emitter current of one transistor to supply the base current of another, as in Figure 4.4(A). The Darlington compound transistor, now simply called the *Darlington*, integrates both transistors, two resistors to assist in rapid turn-off in switching applications, and usually (as in the case of the ubiquitous TIP120 series from Texas Instruments) an antiparallel diode between collector and emitter. Despite the great convenience of a power transistor with a value of h_{FE} in excess of 1000, the one fly in the ointment is the saturation or bottoming voltage. In a small-signal transistor (and even some power transistors) this may be as low as 200 mV, though usually one or two volts, but in a power Darlington it

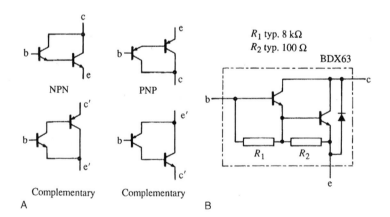

Figure 4.4: (A) Darlington connected discrete transistors. (B) Typical monolithic NPN Darlington power transistor. (Reproduced by courtesy of Philips Components Ltd.)

is often as much as 2 to 4V. The reason is apparent from Figure 4.4(B): the $V_{ce\ sat}$ of the compound transistor cannot be less than the $V_{ce\ sat}$ of the first transistor plus the V_{be} of the second.

Reference

"Current-feedback op-amps ease high-speed circuit design," P. Harold, *EDN*, July 1988.

CHAPTER 5

Field-Effect Transistors

Ian Hickman

An important milestone in the development of modern active semiconductor devices was the field-effect transistor, or FET for short. These did not become generally available until the 1960s, although they were described in detail and analyzed as early as 1952.

Figure 5.1(A) shows the symbols and Figure 5.1(B) and (C) the construction and operation of the first type introduced, the depletion mode *junction FET* or JFET. In this device, in contrast to the bipolar transistor, conduction is by means of majority carriers which flow through the *channel* between the *source* (analogous to an emitter or cathode) and the *drain* (collector or anode). The *gate* is a region of silicon of opposite polarity to the source cum channel cum drain. When the gate is at the same potential as the source and drain, its depletion region is shallow and current carriers (electrons in the case of the N-channel FET shown in Figure 5.1(C)) can flow between the source and the drain. The FET is thus a unipolar device, and minority carriers play no part in its action. As the gate is made progressively more negative, the depletion layer extends across the channel, depleting it of carriers and eventually pinching off the conducting path entirely when V_{gs} reaches $-V_p$, the pinch-off voltage. Thus for zero (or only very small) voltages of either polarity between the drain and the source, the device can be used as a passive voltage-controlled resistor. The JFET is, however, more normally employed in the active mode as an amplifier (Figure 5.1(D)) with a positive supply rail (for an N-channel JFET), much like an NPN transistor stage. Figure 5.1(E) shows a typical drain characteristic. Provided that the gate is reversed biased (as it normally will be) it draws no current.

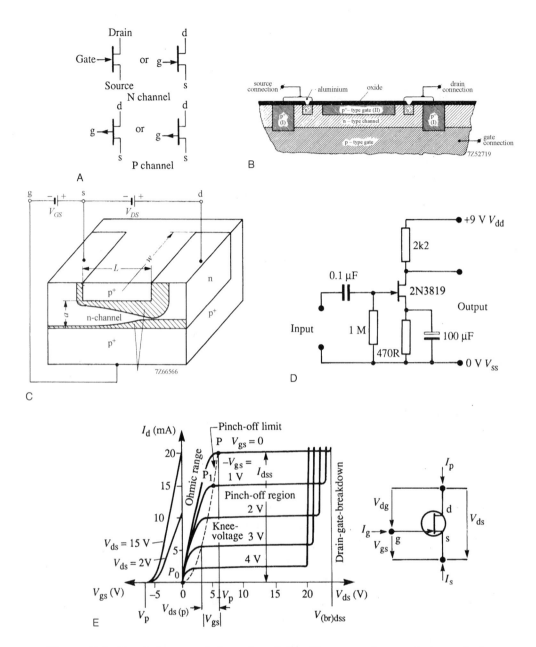

Figure 5.1: Depletion mode junction field-effect transistors. (A) Symbols.
(B) Structure of an N-channel JFET. (C) Sectional view of an N-channel JFET.
The P$^+$ upper and lower gate regions should be imagined to be connected in front
of the plane of the paper, so that the N-channel is surrounded by an annular gate
region. (D) JFET audio-frequency amplifier. (E) Characteristics of N-channel
JFET: pinch-off voltage $V_p = -6$V. (Parts (B), (C) and (E) reproduced by courtesy
of Philips Components Ltd.).

The positive excursions of gate voltage of an N-channel JFET, or the negative excursions in the case of a P-channel JFET, must be limited to less than about 0.5V to avoid turn-on of the gate/source junction; otherwise the benefit of a high input impedance is lost.

In the *metal-oxide semiconductor field-effect transistor* (MOSFET) the gate is isolated from the channel by a thin layer of silicon dioxide, which is a nonconductor; thus, the gate circuit never conducts regardless of its polarity relative to the channel. The channel is a thin layer formed between the substrate and the oxide. In the *enhancement* (normally off) MOSFET, a channel of semiconductor of the same polarity as the source and drain is induced in the substrate by the voltage applied to the gate (Figure 5.2(B)). In the *depletion* (normally on) MOSFET, a gate voltage is effectively built in by ions trapped in the gate oxide (Figure 5.2(C)). Figure 5.2(A) shows symbols for the four possible types, and Figure 5.2(D) summarizes the characteristics of N-channel types. Since it is much easier to arrange for positive ions to be trapped in the gate oxide than negative ions or electrons, P-channel depletion MOSFETs are not generally available. Indeed, for both JFETs and MOSFETs of all types, N-channel devices far outnumber P-channel devices. In consequence, one only chooses a P-channel device where it notably simplifies the circuitry or where it is required to operate with an N-channel device as a complementary pair.

Note that while the source and substrate are internally connected in many MOSFETS, in some (such as the Motorola 2N351) the substrate connection is brought out on a separate lead. In some instances it is possible to use the substrate, where brought out separately, as another input terminal. For example, in a frequency changer application, the input signal may be applied to the gate and the local oscillator signal to the substrate. In high-power MOSFETS, whether designed for switching applications or as HF/VHF/UHF power amplifiers, the substrate is always internally connected to the source.

In the N-channel *dual-gate* MOSFET (Figure 5.3) there is a second gate between gate 1 and the drain. Gate 2 is typically operated at +4V with respect to the source and serves the same purpose as the screen grid in a tetrode or pentode. Consequently the reverse transfer capacitance C_{rss} between drain and gate 1 is only about 0.01 pF, against 1 pF or thereabouts for small-signal JFETs, single-gate MOSFETs and most bipolar transistors designed for RF applications.

N-channel *power* MOSFETs for switching applications are available with drain voltage ratings up to 500V or more and are capable of passing 20A with a drain/source

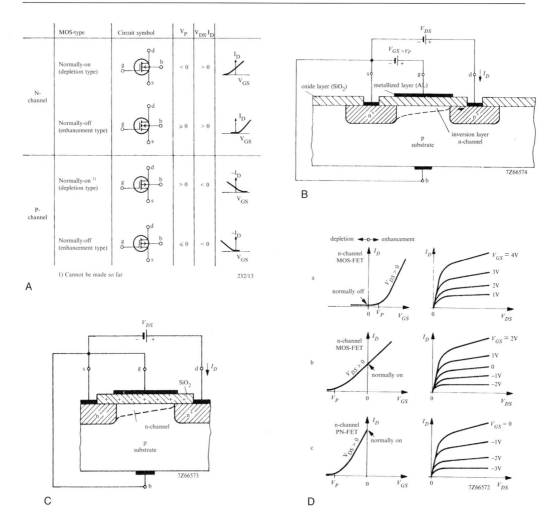

Figure 5.2: Metal-oxide semiconductor field-effect transistors. (A) MOSFET types.
Substrate terminal b (bulk) is generally connected to the source, often internally.
(B) Cross-section through an N-channel enhancement (normally off) MOSFET.
(C) Cross-section through an N-channel depletion (normally on) MOSFET.
(D) Examples of FET characteristics: (a) normally off (enhancement); (b) normally
on (depletion and enhancement); (c) pure depletion (JFETs only). (Parts (A) to
(D) reproduced by courtesy of Philips Components Ltd.).

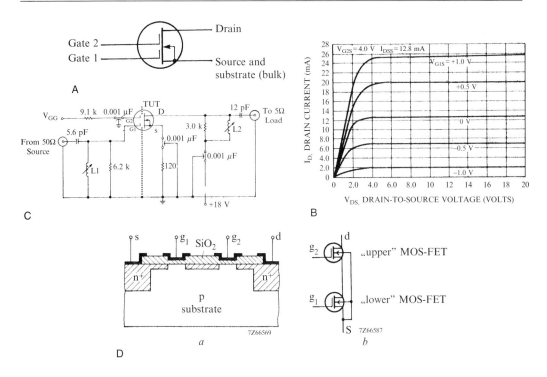

Figure 5.3: Dual-gate MOSFETS. (A) Dual-gate N-channel MOSFET symbol. Gate protection diodes, not shown, are fabricated on the chip in many device types. These limit the gate/source voltage excursion in either polarity, to protect the thin gate oxide layer from excessive voltages, e.g., static charges. (B) Drain characteristics (3N203/MPF203). (C) Amplifier with AGC applied to gate 2.50Ω source and load (3N203/MPF203). (D) Construction and discrete equivalent of a dual-gate N-channel MOSFET. Parts (B) and (C) reproduced by courtesy of Motorola Inc. Part (D) reproduced by courtesy of Philips Components Ltd.

voltage drop of only a few volts, corresponding to a drain/source resistance in the fully on condition of $r_{ds\ on}$ on of just a few hundred milliohms. Other devices with lower drain voltage ratings exhibit $r_{ds\ on}$ resistances as low as 0.010 ohms, and improved devices are constantly being developed and introduced. Consequently these figures will already doubtless be out of date by the time you read this. A very high drain voltage rating in a power MOSFET requires the use of a high-resistivity drain region, so that very low levels of $r_{ds\ on}$ cannot be achieved in high-voltage MOSFETs.

A development which provides a lower drain/source voltage drop in the fully on condition utilizes an additional p-type layer at the drain connection. This is indicated by

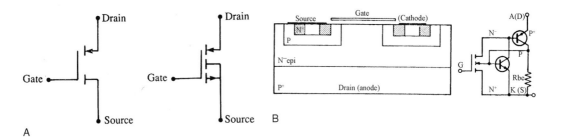

Figure 5.4: The gain enhanced MOSFET (GEMFET). (A) Symbols for GEMFET, COMFET (conductivity modulated FET) and other similar devices. (B) Structure and equivalent circuit of the GEMFET. (Reproduced by courtesy of Motorola Inc.).

the arrowhead on the symbol for this type of device (Figure 5.4A). The device is variously known as a *conductivity modulated power MOSFET* or *COMFET* (trademark of GE/RCA), as a gain enhanced MOSFET or GEMFET (trademark of Motorola), and so on. Like the basic MOSFET these are all varieties of *insulated gate* field-effect transistors (IGFETs). The additional heavily doped p-type drain region results in the injection of minority carriers (holes) into the main n-type drain region when the device switches on, supplementing the majority carrier electrons and reducing the drain region on voltage drop. However, nothing comes for free in this world, and the price paid here is a slower switch-off than a pure MOSFET; this is a characteristic of devices like bipolar transistors which use minority carrier conduction. An interesting result of the additional drain P layer is that the antiparallel diode inherent in a normal power MOSFET—and in Darlingtons—is no longer connected to the drain. Consequently COMFETS, GEMFETs and similar devices will actually block reverse drain voltages, i.e., N-channel types will not conduct when the drain voltage is negative with respect to source. Indeed, the structure has much in common with an insulated gate silicon controlled rectifier (SCR), to be covered later.

Reference

"Current-feedback op-amps ease high-speed circuit design," P. Harold, *EDN,* July 1988.

CHAPTER 6

Identifying and Avoiding Transistor Problems

Robert Pease

Although transistors—both bipolars and MOSFETs—are immune to many problems, you can still have transistor troubles. Robust design methods and proper assumptions regarding their performance characteristics will steer you past the shoals of transistor vexation and the rocks of transistor disasters.

Transistors are wonderful—they're so powerful and versatile. With a handful of transistors, you can build almost any kind of high-performance circuit: a fast op-amp, a video buffer, or a unique logic circuit.

On the other hand, transistors are uniquely adept at causing trouble. For example, a simple amplifier probably won't survive if you short the input to the power supplies or the output to ground. Fortunately, most op-amps include forgiving features, so that they can survive these conditions. When the µA741 and the LM101 op-amps were designed, they included extra transistors to ensure that their inputs and outputs would survive such abuse. But an individual transistor **is** vulnerable to damage by excessive forward or reverse current at its input, and almost every transistor is capable of melting. So it's up to us, the engineers, to design transistor circuits so that the transistors do not blow up, and we must troubleshoot these circuits when and if they do.

A simple and sometimes not-so-obvious problem is installing a transistor incorrectly. Because transistors have three terminals, the possibility of a wrong connection is considerably greater than with a mere diode. Small-signal transistors are often installed so close to a printed-circuit board that you can't see if the leads are crossed or shorted to a transistor's can or to a PC trace. In fact, I recall some boards in which the leads were often crossed and about every tenth transistor was the wrong gender—PNP where an NPN should have been, or vice versa. I've thought about it a lot, and I can't think of any circuits that work equally well whether you install a transistor of the opposite sex. So, mind your Ps and Qs, your Ps and Ns, your 2N1302s and 2N1303s, and your 2N3904s and 2N3906s.

In addition to installing a transistor correctly, you must design with it correctly. First of all, unless they are completely protected from the rest of the world, transistors require input protection. Most transistors can withstand dozens of milliamperes of forward base current but will die if you apply "only a few volts" of forward bias. One of my pet peeves has to do with adding protective components. MIL-HDBK-217 has always said that a circuit's reliability decreases when components are added. Yet when you add resistors or transistors to protect an amplifier's input or output, the circuit's reliability actually improves. It just goes to show that you can't believe everything you read in a military specification. (For detailed criticism of the notion of computing reliability per MIL-HDBK-217, see Ref. 6.1.)

Similarly, if you pump current *out of* the base of a transistor, the base-emitter junction will break down or "zener." This reverse current—even if it's as low as nanoamperes or very brief in duration—tends to degrade the low-current beta of the transistor, at least on a temporary basis. So in cases where accuracy is important, find a way to avoid reverse-biasing the inputs. Bob Widlar reminded me that the high-current beta of a transistor is generally not degraded by this zenering, so if you are hammering the V_{EB} of a transistor in a switch-mode regulator, that will not necessarily do it any harm, nor degrade its high-current beta.

Transistors are also susceptible to ESD—electrostatic discharge. If you walk across a rug on a dry day, charge yourself up to a few thousand volts, and then touch your finger to an NPN's base, it will probably survive because a forward-biased junction can survive a pulse of a few amperes for a small part of a microsecond. But, if you pull-up the emitter of a grounded-base NPN stage, or the base of a PNP, you risk reverse-biasing the base-emitter junction. This reverse bias can cause significant damage to the base-emitter junction and might even destroy a small transistor.

When designing an IC, smart designers add clamp diodes, so that any pin can survive a minimum of + and −2000V of ESD. Many IC pins can typically survive two or three times this amount. These ESD-survival design goals are based on the "human-body" model, in which the impedance equals about 100 pF in series with 1500Ω. With discrete transistors, whose junctions are considerably larger than the small geometries found in ICs, ESD damage may not be as severe. But in some cases, ESD damage can still happen. Delicate RF transistors such as 2N918s, 2N4275s, and 2N2369s sometimes blow up "when you just look at 'em" because their junctions are so small.

Other transistor-related problems arise when engineers make design assumptions. Every beginner learns that the V_{BE} of a transistor decreases by about 2 mV per degree Celsius and increases by about 60 mV per decade of current. Don't forget about the side effects of these rules, or misapply them at extreme temperatures. Don't make sloppy assumptions about V_{BE}s. For instance, it's not fair to ask a pair of transistors to have well-matched V_{BE}s if they're located more than 0.1 in. apart and there are heat sources, power sources, cold drafts, or hot breezes in the neighborhood. Matched pairs of transistors should be glued together for better results. Of course, for best results, monolithic dual transistors like the LM394 give the *best* matching.

I've seen people get patents on circuits that don't even work—based on misconceptions of the relationships between V_{BE} and current. It's fair to assume that two matched transistors with the same V_{BE} at the same small current will have about the same temperature coefficient of V_{BE}. But you wouldn't want to make any rash assumptions if the two transistors came from different manufacturers or from the same manufacturer at different times. Similarly, transistors from different manufacturers will have different characteristics when going into and coming out of saturation, especially when you're driving the transistors at high speeds. In my experience, a components engineer is a very valuable person to have around and can save you a lot of grief by preventing unqualified components from confusing the performance of your circuits.

Another assumption engineers make has to do with a transistor's failure mode. In many cases, people say that a transistor, like a diode, fails as a short circuit or in a low-impedance mode. But unlike a diode, the transistor is normally connected to its leads with relatively small lead-bond wires; so if there's a lot of energy in the power supply, the short circuit will cause large currents to flow, vaporizing the lead bonds. As the lead bonds fail, the transistor will ultimately fail as an *open* circuit.

6.1   More Beta—More Better?

It's nice to design with high-beta transistors, and, "if some is good, more's better." But, as with most things in life, too much can be disastrous. The h-parameter, h_{rb}, is equal to

$\Delta V_{BE}/\Delta V_{CB}$ with the base grounded. Many engineers have learned that as beta rises, so does h_{rb}. As beta rises and h_{rb} rises, the transistor's output impedance decreases; its Early voltage falls; its voltage gain decreases; and its common-emitter breakdown voltage, BV_{CEO}, may also decrease. (The Early voltage of a transistor is the amount of V_{CE} that causes the collector current to increase to approximately two times its low-voltage value, assuming a constant base drive. V_{Early} is approximately equal to $26\,mV \times (1/h_{rb})$). So, in many circuits there is a point where higher beta simply makes the gain lower, *not* higher.

Another way to effectively increase "beta" is to use the Darlington connection: but the voltage gain and noise may degrade, the response may get flaky, and the base current may decrease only slightly. When I was a kid engineer, I studied the ways that Tektronix made good use of the tubes and transistors in their mainframes and plug-ins. Those engineers didn't use many Darlingtons. To this day, I keep learning more and more reasons not to use Darlingtons or cascaded followers. For many years, it's been more important (in most circuits) to have matched betas than to have skyhigh betas. You can match betas yourself, or you can buy monolithic dual matched transistors like the LM394, or you can buy four or five matched transistors on one monolithic substrate, such as an LM3045 or LM3086 monolithic transistor array.

One of the nice things about bipolar transistors is that their transconductance, g_m, is quite predictable. At room temperature, $g_m = 38.6 \times I_C$. (This is much more consistent than the forward conductance of diodes, as mentioned in the previous chapter.) Since the voltage gain is defined as $A_V = g_m \times Z_L$, computing it is often a trivial task. You may have to adjust this simple equation in certain cases. For instance, if you include an emitter-degeneration resistor, R_e, the effective transconductance falls to $1/(R_e + g_m^{-1})$. A_V is also influenced by temperature changes, bias shifts in the emitter current, hidden impedances in parallel with the load, and the finite output impedance of the transistor. Remember—higher beta devices can have *much* worse output impedance than normal.

Also be aware that, although the transconductance of a well-biased bipolar transistor is quite predictable, beta usually has a wide range and is not nearly as predictable. So you have to watch out for adverse shifts in performance if the beta gets too low or too high and causes shifts in your operating points and biases.

6.2   Field-Effect Transistors

For a given operating current, field-effect transistors normally have much poorer g_m than bipolar transistors do. You'll have to measure your devices to see how much lower.

Additionally, the V_{GS} of FETs can cover a very wide range, thus making them harder to bias than bipolars.

JFETs (junction field-effect transistors) became popular 20 years ago because you could use them to make analog switches with resistances of 30Ω and lower. JFETs also help make good op-amps with lower input currents than bipolar devices, at least at moderate or cool temperatures. The BiFET™ process[1] made it feasible to make JFETs along with bipolars on a monolithic circuit. It's true that the characteristic of the best BiFET inputs are still slightly inferior to the best bipolar ones in terms of V_{OS} temperature coefficient, long-term stability, and voltage noise. But these BiFET characteristics keep improving because of improved processing and innovative circuit design. As a result, BiFETs are quite close to bipolar transistors in terms of voltage accuracy. and offer the advantage of low input currents. at room temperature.

JFETs can have a larger gate current when current flows through the source than when no current flows (which is called I_{gss}). When I discovered this, and discussed it with Joel Cohen at Crystalonics back 20 years ago, we called it the *Pease-Cohen Effect.* I thought it was caused by imperfect ohmic contacts, but other engineers showed that

Figure 6.1: Using equations to analyze circuits can sometimes help you define a problem. But if the equations are inapplicable, they do a lot more harm than good. (Photo copyright Peggi Willis.)

---

[1] A trademark of National Semiconductor Corporation.

it was actually caused by impact ionization, or "hot carriers." Either way, the gate current has a tendency to increase as a linear function of source current, with an exponential dependence on high drain-source voltages.

I recall working on a hybrid circuit that had some JFETs whose gate connection was *supposed to* be through the back of the die. I found that some of the dice didn't have proper metallurgical processing, which caused some strange behavior. Initially, the gate acted as if it really were connected to the metal under the die, and would act that way for a long period of time. Then, after a while the gate would act like an open circuit with as much as 1V of error between the actual gate and the bottom of the die. The amplifier's V_{OS} would grow as large as 1V! The gate would remain disconnected until a voltage transient restored the connection for another week! The intermittency was awful because nothing would speed up the 1-week cycle-to-failure time. So, we had to go back and add definite lead bonds to the gate's bond pad on the top of the chip, which we had been told was unnecessary. Ouch!

When designing hybrids, you need to make sure to connect the substrate of a chip to the correct DC level. The bottom of a FET chip is usually tied to the gate, but the connection may be through a large and unspecified impedance. You have to be a pretty good chemist or metallurgist to be sure that you don't have to add that bond to the gate's metal bonding-pad, on the top of the die, just to get a good gate connection.

The substrate of a discrete bipolar transistor's die is the collector. Most linear and digital IC substrates are tied to the negative supply. Exceptions include the LM117 and similar adjustable positive regulators—their substrate is tied to V_{out}. The LM196 voltage regulator's substrate is tied to the positive supply voltage, +Vs, as are the substrates of the MM74HC00 family of chips, the NSC LMC660 and LPC660 family, and most of the dielectrically isolated op-amps from Harris. So, be aware of your IC's substrate connection. If an LM101AH op-amp's metal can should happen to bump against ground or +V_S, you have a problem. Similarly, you shouldn't let an HA2525's case bump against ground or −Vs.

MOSFETs are widely used in digital ICs but are also very popular and useful in analog circuits, such as analog switches. The quad switches such as CD4016 and CD4066 are popular because of their low (typical) leakages and low price. Op amps with MOSFET inputs are starting to do well in the general-purpose op-amp market. MOSFETs used to have a bad reputation for excessive noise, but new IC devices, such as the LMC662 dual op-amp, demonstrate that clean processing can cure the problem, thus making MOSFETs competitive with BiFETs. They offer an advantage of a 1000:1 improvement in input current, decreased from 10 pA down to 10 fA. Just be careful not to let ESD near the inputs. Most MOSFET-input linear ICs *do* have protection diodes and may be able to withstand 600V, but they usually can't survive 2000V. If you work with unprotected MOSFETs, such as the 3N160, you must keep the pins securely shorted until the device is soldered into its PC board in which the protection diodes are already installed. I do all of that and wash the transistor package with both an organic solvent and soap and water. And, I keep the sensitive gate circuits entirely off the PC board by pulling the gate pin up in the air and using point-to-point wiring. Air, which is a superior dielectric, is also a good insulator (Ref. 6.2). So far, I haven't had any blown inputs or bad leakages—at least nothing as bad as 10 fA.

On the other hand, when using CMOS digital ICs, I always plug them into live sockets; I never use conductive foam; and I never wear a ground strap on my wrist. And I've almost never had any failures—with one exception. One time I shuffled across a carpeted floor and pointed an accusatory finger at a CMOS IC. There was a small *crack* of ESD—probably 5000V—followed by a *big* snap as the IC blew out and crowbarred the entire power supply. Since ESD testing is usually done with the power OFF, then if you did some tests with the power ON, you might get some messy failure modes like the one I just mentioned. Always be wary of any devices that manufacturers claim are safe from ESD.

One reader reminded me that in some cases, if you abuse CMOS ICs with ESD, they may not fail instantly, but they may become unreliable and fail at a later time. So, I must caution you that fooling around with CMOS ICs while you are not properly grounded might cause latent unreliability problems. If you *do* have to do troubleshooting of CMOS ICs while you are not grounded, if you decide to plug in CMOS ICs while the power busses are *hot*, just be aware that you might in some cases do some long-lasting harm to an occasional IC. But you have to use your judgment and trade off that possibility against the advantages of more free-swinging troubleshooting approaches.

6.3   Power Transistors may Hog Current

As you build a bipolar transistor bigger and bigger, you may be tempted to go to extremes and make a huge power transistor. But there are practical limitations. Soon, the circuit capacitances cause oppressive drive requirements, and removing the heat is difficult. Still, no matter how big you build power transistors, people will find a use for

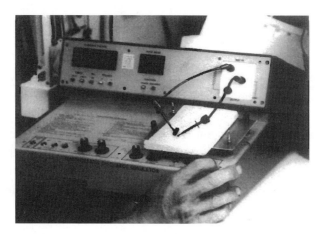

Figure 6.2: When you hit a component or circuit with a pulse of real ESD, you can never be sure what kind of trouble you'll get—unless you've already tested it with an ESD simulator. (Photo copyright Peggi Willis.)

them. Their most serious limitation on just building transistors bigger and bigger is secondary breakdown, which is what happens when you drive a transistor outside its "safe operating area." When you operate a power transistor at very high currents and low voltages, the distributed emitter resistance of the device—which includes the resistance of the emitter metal and the inherent emitter resistivity—can cause enough I × R drop to force the entire emitter and its periphery to share the current. Now, let's halve the current and double the voltage: The amount of dissipation is the same, but the I × R drop is cut in half. Now continue to halve the current and double the voltage. Soon you'll reach a point where the ballasting (Figure 6.3) won't be sufficient, and a hot spot will develop at a high-power point along the emitter. The inherent decrease of V_{BE} will cause an increase of current in one small area. Unless this current is turned OFF promptly, it will continue to increase unchecked. This "current hogging" will cause local overheating, and may cause the area to melt or crater—this is what happens in "secondary breakdown." By definition you have exceeded the secondary breakdown of the device. The designers of linear ICs use ballasting, cellular layouts, and thermal-limiting techniques, all of which can prevent harm in these cases (Ref. 3). Some discrete transistors are beginning to include these features.

Fortunately, many manufacturers' data sheets include permitted safe-area curves at various voltages and for various effective pulse-widths. So, it's possible to design reliable power circuits with ordinary power transistors. The probability of an unreliable

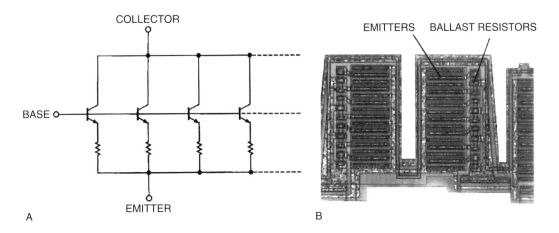

Figure 6.3: Ballast resistors, also known as sharing resistors, are often connected to the emitters of a number of paralleled transistors (A) to help the transistors share current and power. In an integrated circuit (B), the ballast resistors are often integrated with adjacent emitters. (Photo of National Semiconductor Corp's LM 138.)

design or trouble increases as the power level increases, as the voltage increases, as the adequacy of the heat sink decreases, and as the safety margins shrink. For example, if the bolts on a heat sink aren't tightened enough, the thermal path degrades and the part can run excessively hot.

High temperature *per se* does not cause a power transistor to fail. But, if the drive circuitry was designed to turn a transistor ON and only a base-emitter resistor is available to turn it OFF, then at a very high temperature, the transistor will turn itself ON and there will be no adequate way to turn it OFF. Then it may go into secondary breakdown and overheat and fail. However, overheating does not by itself cause failure. I once applied a soldering iron to a 3-terminal voltage regulator—I hung it from the tip of the soldering iron—and then ran off to answer the phone. When I came back the next day, I discovered that the TO-3 package was still quite hot— +300°C, which is normally recommended for only 10 seconds. When I cooled it off, the regulator ran fine and met spec. So, the old dictum that high temperature will necessarily degrade reliability is not always true. Still, it's a good practice to not get your power transistors that hot, and to have a base drive that can pull the base OFF if they do get hot.

You can also run into problems if you tighten the screws on the heat sink too tight, or if the heat sink under the device is warped, or if it has bumps or burrs or foreign matter on it. If you tighten the bolt too much, you'll overstress and warp the tab and die attach. Overstress may cause the die to pop right off the tab. The insulating washer under the power transistor can crack due to overstress or may fail after days or weeks or months. Even if you don't have an insulating washer, overtorqueing the bolts of plastic-packaged power transistors is one of the few ways a user can mistreat and kill these devices. Why does the number 10 inch-pounds max, 5 typ, stick in my head? Because that's the spec the Thermalloy man gave me for the 6/32 mounting bolts of TO-220 packages. For any other package, make sure you have the right spec for the torque. Don't hire a gorilla to tighten the bolts.

6.4   Apply the 5-Second Rule

Your finger is a pretty good heat detector—just be careful not to burn it with high voltages or very hot devices. A good rule of thumb is the 5-second rule: If you can hold your finger on a hot device for 5 seconds, the heat sink is about right, and the case temperature is about 85°C. If a component is hotter than that, too hot to touch, then dot your finger with saliva and apply it to the hot object for just a fraction of a second. If the moisture dries up quickly, the case is probably around 100°C; if it sizzles instantaneously, the case may be as hot as 140°C. Alternatively, you can buy an infrared imaging detector for a price of several thousand dollars, and you won't burn your fingers. You will get beautiful color images on the TV screen, and contour maps of isothermal areas. You will learn a lot from those pictures. About twice a year, I wish I could borrow or rent one.

6.5   Fabrication Structures Make a Difference

Another thing you should know when using bipolar power transistors is that there are two major fabrication structures: the epitaxial base, and the planar structure pioneered by Fairchild Semiconductor (Figure 6.5) (Ref. 6.4). (See my comments a couple paragraphs down concerning the obsolete single-diffused transistors.) Transistors fabricated with the epi-base structure are usually more rugged and have a wider safe operating area. Planar devices feature faster switching speeds and higher frequency response, but aren't as rugged as the epi-base types. You can compare the two types by looking at the data sheets for the Motorola 2N3771 and the Harris 2N5039. The 2N.5039 planar device has a current-gain bandwidth 10 times greater than the 2N3771 epi-base device. The 2N.5039

Figure 6.4: When using high-power amplifiers, there are there are certain problems you just never have if you use a big-enough heat sink. This heat sink's thermal resistance is lower than 0.5°C/W. (Photo copyright Peggi Willis.)

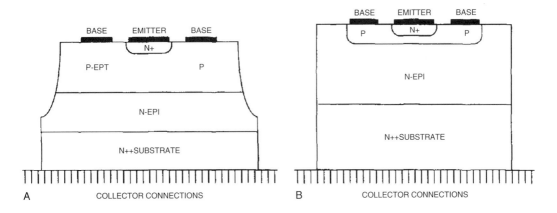

Figure 6.5: The characteristics of power transistors depend on their fabrication structure. The epitaxial-base structure (A) takes advantage of the properties of several different epitaxial layers to achieve good beta, good speed, low saturation, small die size, and low cost. This structure involves mesa etching, which accounts for the slopes at the die edges. Planar power transistors (B) can achieve very small geometries, small base-widths, and high-frequency responses, but they're less rugged than epitaxial-base types, in terms of Forward-Biased SOA.

also has a switching speed faster than the 2N3771 when used as a saturated switch, but the 2N3771 has a considerably larger safe area if used for switching inductive loads. You can select the characteristics you prefer, and order the type you need. . .

But be careful. If you breadboard with one type and then start building in production with the other, you might suddenly find that the bandwidth of the transistor has changed by a factor of 10 (or a factor of 0.1) or that the safe area doesn't match that of the prototypes. Also be aware that the planar power devices, like the familiar 2N2222 and 2N3904, are quite capable of oscillating at high frequencies in the dozens of megahertz when operated in the linear region, so you should plan to use beads in the base and/or the emitter, to quash the oscillation. The slower epi-base devices don't need that help very often.

When I first wrote these articles on troubleshooting back in 1988, you could still buy the older "single-diffused" transistors such as 2N3055H and the old 2N3771. I wrote all about how these devices had even more Safe Operating Area than the epi-base device, so you might want to order these if you wanted a "really gutsy" transistor for driving inductive loads. Unfortunately, these transistors were obsolescent and obsolete; they were slow (perhaps 0.5 MHz of fα), had a large die area, and were expensive. For example, although these transistors required only one diffusion, in some cases this diffusion had to run 20 hours. Because of all these technical reasons, sales shrank until, in the last two years, all the single-diffused power transistors have been discontinued.

So it's kind of academic to talk about the old single-diffused parts (see Figure 6.6), but I included a mention here just for historical interest. Also, I included it because if you looked in my old *EDN* write-up and then tried to buy the devices I recommended, you would meet with incredulity. You might begin to question the sanity of yourself, or the salesman, or of Pease. When I inquired into the availability of these parts, I talked to many sales people who had *no idea* what I was talking about. Finally, when I was able to talk to technical people, they explained why these transistors were not available—they admitted that I was not dreaming, but that the parts had been discontinued recently. These engineers at some of the major power-transistor manufacturers were quite helpful as they explained that newer geometries helped planar power transistors approach the safe area of the other older types without sacrificing the planar advantages of speed. Also, power MOSFETs had even wider amounts of SOA, and their prices have been dropping, and they were able to take over many new tasks where the planars did not have enough SOA. So the puzzle all fits together.

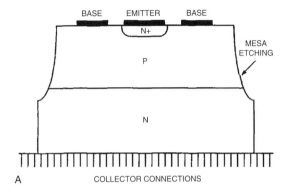

Figure 6.6: In the old single-diffused structure, n-type dopants were diffused simultaneously into the front and back of a thin p-type wafer. This structure produced rugged transistors with wider Safe Operating Areas than the more modern epitaxial-base transistor types, in terms of Forward-Biased SOA. However, this fabrication has been obsoleted.

There is still one tricky problem. Originally the old 2N3771 was a single-diffused part. If you wanted to buy an epi-base part, that was the MJ377 1. But now if you order a 2N3771, you get the epi-base part, which *does* meet and exceed the JEDEC 2N3771 specs. It just exceeds them a lot more than you would expect—like, the current-gain-bandwidth is 10 or $20\times$ higher. So, if you try to replace an *old* 2N377 1 with a *new* 2N377 1, please be aware that they are probably not very similar at all.

6.6   Power-Circuit Design Requires Expertise

For many power circuits, your transistor choice may not be as clear-cut as in the previous examples. So, be careful. Design in this area is not for the hotshot just out of school— there are many tricky problems that can challenge even the most experienced designers. For example, if you try to add small ballasting resistors to ensure current sharing between several transistors, you may still have to do some transistor matching. This matching isn't easy. You'll need to consider your operating conditions; decide what parameters, such as beta and V_{BE}, you'll match; and figure out how to avoid the mix-and-match of different manufacturers' devices. Such design questions are not trivial. When the performance or reliability of a power circuit is poor, it's probably not the fault of a bad transistor. Instead, it's quite possibly the fault of a bad or marginal driver circuit or an inadequate heat sink. Perhaps a device with different characteristics was inadvertently substituted in place of the intended device. Or perhaps you chose the wrong transistor for the application.

A possible scenario goes something like this. You build 10 prototypes, and they seem to work okay. You build 100 more, and half of them don't work. You ask me for advice. I ask, "Did they ever work right?" And you reply, "Yes." But wait a minute. There were 10 prototypes that worked, but the circuit design may have been a marginal one. Maybe the prototypes didn't really work all that well. If they're still around, it would be useful to go back and see if they had any margin to spare. If the 10 prototypes had a gain of 22,000, but the current crop of circuits has gains of 18,000 and fails the minimum spec of 20,000, your new units should not be called *failures*. It's not that the circuit isn't working at all, it's just that your expectations were unrealistic.

After all, every engineer has seen circuits that had no right to work, but they did work— for a while. And then when they began to fail, it was obviously just a hopeless case. So, which will burn you quickest, a marginal design or marginal components? That's impossible to say. If you build in some safety margin, you may survive some of each. But you can't design with big margins to cover every possibility, or your design will become a monster. That's where experience and judgment must be invoked. . .

An old friend wrote to me from Japan, "Why do you talk about having to troubleshoot 40% of the units in a batch of switching regulators? In Japan that would be considered a bad design. . ." I replied that I agreed that it sounds like a problem, but until you see what is the cause of the problems, it is unfair to throw any blame around. What if it was a bad workmanship problem? Then that does not sound like a bad design—unless the design was so difficult to execute that the assembly instructions could not be followed. Or maybe a bad part was put in the circuit. Or maybe it was a marginally bad design and part of the circuit does need to be changed—perhaps an extra test or screening of some components—before the circuit can run in production. But you cannot just say that if there is ever trouble, it is the design engineer's fault. What if the design engineer designed a switching regulator that never had any problems in production—never ever—*but* it only puts out 1W per 8 cubic inches, and all the parts are very expensive, and then there is a lot of expensive testing on each component before assembly, to prove that there is a good safety margin. Is that a good design? I doubt it. Because if you tried to build a plane with too big a safety factor, it might be bigger than a 747, but able to carry only 10 passengers. Every circuit should be built with an *appropriate* safety factor. If you only use a transistor that is always SURE to work well, that may be an uneconomic safety factor. Judgment is required to get the right safety factor.

6.7 MOSFETs Avoid Secondary Breakdown

When it comes to power transistors, MOSFETs have certain advantages. For many years, MOSFETs have been available that switch faster than bipolar transistors, with smaller drive requirements. And MOSFETs are inherently stable against secondary breakdown and current hogging because the temperature coefficient of I_{DS} vs. V_{GS} is inherently stable at high current densities. If one area of the power device gets too hot, it tends to carry less current and thus has an inherent mechanism to avoid running away. This self-ballasting characteristic is a major reason for the popularity of MOSFETs over bipolar transistors. However, recent criticism points out that when you run a MOSFET at high-enough voltages and low current, the current density gets very small, the temperature coefficient of I_{DS} vs. V_{GS} reverses, and the device's inherent freedom from current hogging may be lost (Ref. 6.5). So at high voltages and low current densities, watch out for this possibility. When the V_{DS} gets high enough, MOSFETs can exhibit current hogging and "secondary breakdown" similar to that of bipolars!!

The newer power MOSFETs are considerably more reliable and less expensive than the older devices. Even though you may need a lot of transient milliamps to turn the gate ON or OFF quickly, you don't need a lot of amps to hold it ON like you do with a bipolar transistor. You can turn the newer devices OFF quicker, too, if you have enough transient gate drive current available.

However, MOSFETs are not without their problem areas. If you persist in dissipating too many watts into a MOSFET, you can melt it just as you can melt a bipolar device. If you don't overheat a MOSFET, the easiest way to cause a problem is to forget to insert a few dozen or hundred ohms of resistance (or a ferrite bead) right at the gate lead of the device. Otherwise, these devices have such high bandwidths that they can oscillate at much higher frequencies than bipolar transistors.

For example, the first high-fidelity, all-MOSFET audio amplifier I ever saw blew up. It worked okay in the lab, but some misguided engineer decided that if a bandwidth of 5 Hz to 50 kHz was good, then 0.5 Hz to 500 kHz was better. Consequently, when the speaker cables were extended from 10 feet to 20 feet for a demonstration, the amplifier broke into a megahertz-region scream and promptly went up in smoke because of the lack of damping at the sources. I was told that after a minor redesign the amplifier was perfectly reliable. The redesign involved cutting the bandwidth down to a reasonable level, adding some ballasting in the sources, and tying antisnivet resistors directly to the gate pins. (Note: A snivet is a nasty, high-frequency oscillation originally found in vacuum-tube TV sets—an oscillation similar to the oscillation of a MOSFET with no resistance in series with its gate.)

As with bipolar transistors, MOSFETs are very reliable if you don't exceed their voltage, current, or temperature ratings. Dissatisfaction with a device's reliability or performance usually stems from the drivers or the related circuitry. Most MOSFETs have a maximum V_{GS} rating of just 20 or 25V. A MOSFET may temporarily survive operation with 30 or 50V on the gate, but it's not safe to run it up there forever. If you apply excessive gate voltage, gradual gain or threshold degradation may occur. So—please—don't. Also, power MOSFETs are not quite as rugged as bipolars when it comes to surviving ESD transients. A common precaution is to add a little decoupling, clamping, or current-limiting circuitry, so that terminals accessible to the outside world can withstand ESD.

DMOS FETs are so easy to apply that we usually forget about the parasitic bipolar transistor that lurks in parallel with them. If dV/dt is too large at the drain, if the drain junction is avalanched at too high a current and voltage, or if the transistor gets too hot, the bipolar device turns ON and dies an instant death due to current hogging or an excursion from its safe operating area.

But I'm spoiled rotten. I'm accustomed to linear ICs, which have protection transistors built right in, so the user rarely has a problem. (But most of the transistor troubles are left to the IC designer!) Discrete designs are appropriate and cost-effective for many applications, but the availability of linear ICs—especially op-amps—can simplify your design task considerably, at the same time as it improves reliability.

References

[6.1] Leonard, Charles, "Is reliability prediction methodology for the birds?" *Power Conversion and Intelligent Motion,* November 1988, p. 4.

[6.2] Pease, Robert A., "Picoammeter/calibrator system eases low-current measurements," *EDN,* March 31,1982, p. 143.

[6.3] "A 150W IC Op Amp Simplifies Design of Power Circuits," R. J. Widlar and M. Yamatake, AN-446, National Semiconductor Corp, Santa Clara, CA.

[6.4] Applications Engineering Staff, PowerTech Inc., "Speed-up inductor increases switching speed of high current power transistors," *Power Electronics,* May 1989, p. 78.

[6.5] Passafiume, Samuel J., and William J Nicholas, "Determining a MOSFET's real FBSOA," *Power-technics Magazine,* June 1989, p. 48.

CHAPTER 7

Digital Circuit Fundamentals

Ian Grout

Early electronic circuits were analog, and before the advent of digital logic, signal processing was undertaken using analog electronic circuits. The invention of the semiconductor transistor in 1947 at Bell Laboratories [Ref. 7.1] and the improvements in transistor characteristics and fabrication during the 1950s led to the introduction of linear (analog) ICs and the first transistor-transistor logic (TTL) digital logic IC in the early 1960s, closely followed by complementary metal-oxide semiconductor (CMOS) ICs. The early devices incorporated a small number of logic gates. However, rapid growth in the ability to fabricate an increasing number of logic gates in a single IC led to the microprocessor in the early 1970s. This, with the ability to create memory ICs with ever increasing capacities, laid the foundation for the rapid expansion in the computer industry and the types of complex digital systems based on the computer architecture that we have available today. The last fifty years have seen a revolution in the electronics industry.

Fundamentally, a digital circuit will be categorized into one of three general types, each of which is created and fabricated within an integrated circuit:

- *Combinational logic*, in which the response of the circuit is based on a Boolean logic expression of the input only and the circuit responds immediately to a change in the input.

- *Sequential logic*, in which the response of the circuit is based on the current state of the circuit and the sometimes the current input. This may be *asynchronous* or *synchronous*. In *synchronous sequential logic*, the logic changes state whenever an external *clock* control signal is applied. In *asynchronous sequential logic*, the logic changes state on changes of the input data (the circuit does not utilize a *clock* control signal).

- *Memory*, in which digital values can be stored and retrieved some time later. For a user, memory can be either *read-only* (ROM) or *random-access* (RAM). In ROM, the data stored in the memory are initially placed in the memory and can only be read by the user. Data cannot normally be altered in the circuit application. In RAM (also referred to as *read-write memory*, RWM), the user can write data to the memory and read the data back from the memory.

The digital IC consists of a number of logic gates, which are combinational or sequential circuit elements. The logic gates may be implemented using different fabrication processes and different circuit architectures:

- *TTL*, transistor-transistor logic (bipolar)

- *ECL*, emitter-coupled logic (bipolar)

- *CMOS*, complementary metal-oxide semiconductor

- *BiCMOS*, bipolar and CMOS

The material predominantly used to fabricate the digital logic circuits is silicon. However, silicon-based circuits are complemented with the digital logic capabilities of circuits fabricated using gallium arsenide (GaAs) and silicon germanium (SiGe) technologies. Today, silicon-based CMOS is by far the dominant process used for digital logic.

The digital logic gate is actually an abstraction of what is happening within the underlying circuit. All digital logic gates are made up of transistors. The logic gates may take one of a number of different circuit architectures (the way in which the transistors are interconnected) at the transistor level:

- static CMOS

- dynamic CMOS

- pass-transistor logic CMOS

Today, static CMOS logic is by far the dominant logic cell design structure used. The number of logic gates within a digital logic IC will range from a few to hundreds of thousands and ultimately millions for the more complex processors and PLDs.

In previous times, when the potential for higher levels of integration was far less than is now possible, the digital IC was classified by the level of integration—that is, the number of logic gate equivalents per IC (see Table 7.1). With increasing levels of integration, the following levels were identified as follow-on descriptions from VLSI, but these are not in common usage:

- *ULSI*, ultra-large-scale integration

- *WSI*, wafer scale integration

Table 7.1: Levels of integration

Level of integration	Acronym	Number of gate equivalents per IC
Small-scale integration	SSI	<10
Medium-scale integration	MSI	10–100
Large-scale integration	LSI	100–10,000
Very large-scale integration	VLSI	>10,000

The equivalent logic gate consists of four transistors. In static CMOS logic, the 2-input NAND and 2-input NOR are four transistor logic gate structures (2 nMOS +2 pMOS transistors). Figure 7.1 shows the 2-input NAND and NOR gate in static CMOS with both the digital logic gate symbol and the underlying transistor level circuit. At the transistor level, the circuit is connected to a power supply (V_{DD} = positive power supply voltage and V_{SS} = negative power supply voltage). The nMOS transistors are connected toward V_{SS} and the pMOS transistors toward V_{DD}.

7.1   Digital Technology

In the digital domain, the choice of implementation technology is essentially whether to use dedicated- (and fixed-) functionality digital logic, to use a software-programmed processor based system (microprocessor, μP; microcontroller, μC; or digital signal processor, DSP), or to use a hardware-configured programmable logic device (PLD) such as the simple programmable logic device (SPLD), complex programmable logic device (CPLD), or the field-programmable gate array (FPGA). Memory—random access memory (RAM) or read-only memory (ROM)—is also widely used in many digital electronic circuits and systems. The choices are shown in Figure 7.2.

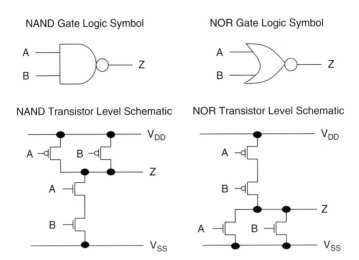

Figure 7.1: Two-input NAND and NOR gates

The initial choice for implementing the digital circuit is between a standard product IC (integrated circuit) and an ASIC (application-specific integrated circuit) [Ref. 7.2]:

- *Standard product IC*, an off-the-shelf electronic component that has been designed and manufactured for a given purpose, or range of use, and that is commercially available. It is purchased either from a component supplier or directly from the designer or manufacturer.

- *ASIC*, an integrated circuit that has been specifically designed and manufactured for a particular application.

For many applications, developing an electronic system based on standard product ICs is more cost effective than ASIC design. Undertaking an ASIC design project also requires access to IC design experience, IC computer-aided design (CAD) tools, and a suitable manufacturing and test capability. Whether a standard product IC or ASIC design approach is taken, the type of IC used or developed will be one of four types:

1. *Fixed-functionality*: These ICs have been designed to implement a specific functionality and cannot be changed. The designer uses a set of these ICs to implement a given overall circuit functionality. Changes to the circuit require a complete redesign of the circuit and the use of different fixed functionality ICs.

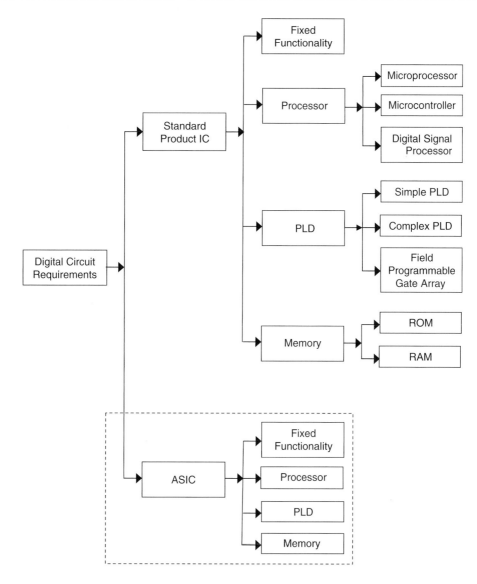

Figure 7.2: Technology choices for digital circuit design

2. *Processor*: Most people are familiar with processors in everyday use; the heart of the PC is a microprocessor. This component runs a software program to implement a required functionality. By changing the software program, the processor will operate a different function. The three types of processor are microprocessor (µP), microcontroller (µC), and digital signal processor (DSP).

3. *Memory*: Memory is used to store, provide access to, and allow modification of data and program code for use within a processor-based electronic circuit/ system. The two types of memory are ROM (read-only memory) and RAM (random access memory). ROM is used for holding program code that must be retained when the memory power is removed; this is *nonvolatile storage*. The code can either be fixed when the memory is fabricated (mask programmable ROM), electrically programmed once (PROM, programmable ROM) or electronically programmed multiple times. Multiple programming capacity requires the ability to erase prior programming, which is available with EPROM (electrically programmable ROM, erased using ultraviolet [UV] light), EEPROM or E^{2}PROM (electrically erasable PROM), or Flash (also electrically erased). PROM is sometimes considered to be in the same category of circuit as simple programmable logic device (SPLD), although in this text, PROM is considered in the memory category only. RAM is used for holding data and program code that require fast access and the ability to modify the contents during normal operation. RAM differs from read-only memory (ROM) in that it can be both read from and written to in the normal circuit application. However, *Flash* memory can also be referred to as nonvolatile RAM (NVRAM). RAM is considered to provide a *volatile storage* since, unlike ROM, the contents of RAM are lost when the power is removed. There are two main types of RAM: static RAM (SRAM) and dynamic RAM (DRAM).

4. *PLD*: The programmable logic device is an IC that contains digital logic cells and programmable interconnect [Refs. 7.3, 7.4] to enable the designer to configure the logic cells and interconnect within the IC itself to form a digital electronic circuit within a single packaged IC. In this, the hardware resources (the available hardware for use) are configured to implement the required functionality. By changing the hardware configuration, the PLD performs a different function. Three types of PLD are available: the simple programmable logic device (SPLD), the complex programmable logic device (CPLD), and the field-programmable gate array (FPGA).

Both the processor and PLD enable the designer to implement and change the functionality of the IC by either changing the software program or the hardware configuration. To avoid potential confusion, the following terms are used to differentiate the PLD from the processor:

*   The PLD will be *configured* using a hardware configuration.

*   The processor will be *programmed* using a software program.

An ASIC can be designed to create any one of the four standard product IC forms (fixed-functionality, processor, memory, or PLD). A standard product IC is designed in the same manner as an ASIC, so anyone who has access to an ASIC design, fabrication, and test facility can create an equivalent to a standard product IC (given that patents and legal issues of IP [intellectual property] for existing designs and devices are taken into account).

No matter how complex the digital circuit design, and the types of operations it is required to undertake, the operation is based on a small number of basic combinational and sequential logic circuit elements that are connected to form the required circuit operation:

- *Combinational logic*: A combinational logic circuit is defined by a Boolean expression, and the output from the circuit (in logic terms) is a function of the logic input values, the logic gates used (AND, OR, etc.), and the way in which the logic gates are connected [Refs. 7.5, 7.6]. The output becomes a final value when the inputs change after a finite time, which is the time required for the logic values to propagate through the circuit given signal propagation delays in each of the logic gates and any delays in the interconnections between the logic gates. The basic combinational logic circuit elements (gates) are:

  ○ AND gate

  ○ NAND gate

  ○ OR gate

  ○ NOR gate

  ○ exclusive-OR (EX-OR) gate

  ○ exclusive-NOR (EX-NOR) gate

  ○ inverter

  ○ buffer

- *Sequential logic*: In a sequential logic circuit, the output from the circuit becomes a value based on the logic input values, the logic gates used, the way in which the logic gates are connected, and on the current state of the circuit [Refs. 7.5, 7.6]. In a synchronous sequential logic circuit, the output change occurs either on the edge of a clock signal change (from 0 to 1 or from 1 to 0) or on a clock signal level (logic 0 or 1). However, an asynchronous sequential logic circuit does not use a clock input. In the sequential logic circuit, the circuit will hold or remember its current value (state) and will change state only on clock or data changes. A sequential logic circuit might also contain additional control signals to reset or set the circuit into a known state either when the circuit is initially turned on or during normal circuit operation. The basic sequential logic circuit elements (gates) are:

○   S-R flip-flop

○   J-K flip-flop

○   toggle flip-flop

○   D-latch

○   D-type flip-flop

References

[7.1] Bell Laboratories (Bell Labs), http:www.bell-labs.com/

[7.2] Smith, M., *Application Specific Integrated Circuits,* Addison-Wesley, 1999, ISBN 0-201-50022-1.

[7.3] Skahill, K., VHDL for Programmable Logic, Addison-Wesley, 1996, ISBN 0-201-89573-0.

[7.4] Maxfield, C., *The Design Warrior's Guide to FPGAs,* Newnes, 2004, ISBN 0-7506-7604-3.

[7.5] Stonham, T. J., *Digital Logic Techniques: Principles and practice,* Second Edition, Van Nostrand Reinhold, UK, 1988, ISBN 0-278-00011-8.

[7.6] Tocci, R. J., Widmer, N. S., and Moss, G. L. K., *Digital Systems,* Ninth Edition, Pearson Education International, USA 2004, ISBN 0-13-121931-6.

CHAPTER 8

Number Systems

Ian Grout

8.1 Introduction

In everyday life, we use the decimal number system (base, or radix, 10), which allows the creation of numbers with digits in the set: 0, 1, 2, 3, 4, 5, 6, 7, 8, 9. The ten possible digits are combined to create integer and real numbers. However, base 10 is not the only number system. Digital circuits and systems use the binary (base, or radix, 2) number system, which allows for the creation of numbers with digits in the set: 0, 1.

The 0 and 1 numbers are logic levels (0 = logic 0, 1 = logic 1), which are created by voltages in a circuit:

- In *positive logic*, 0 is formed by a low voltage level, and 1 is formed by a high voltage level.

- In *negative logic*, 0 is formed by a high voltage level, and 1 is formed by a low voltage level.

In this text, only positive logic will be used and will use the voltage levels shown in Table 8.1.

Table 8.1: Typical voltage levels representing positive logic

Logic level	+5V logic	+3.3V logic
0	+5.0V	+3.3V
1	0V	0V

Decimal and binary number systems are only two of four number systems used in digital circuits and systems:

1. decimal (base 10)

2. binary (base 2)

3. octal (base 8)

4. hexadecimal (base 16)

At some point in the design and analysis of a digital circuit, it will be necessary to convert between the different number systems to view and manipulate values propagating through the design. Such conversion is typically undertaken to aid the interpretation and understanding of the design operation.

In addition, a binary number can have different meanings as different binary coding can be chosen for different design requirement. The main binary coding schemes used are:

1. unsigned (or straight) binary

2. signed binary (1s complement or 2's complement)

3. Gray code

4. binary-coded decimal (BCD)

Unsigned binary numbers are used to represent positive numbers only. Signed binary numbers are used to represent positive and negative numbers that are coded to allow arithmetic using either 1s complement or 2's complement notation. Twos complement notation is more commonly used and will be considered in this text. Gray code allows for a one-bit change when moving from one value to the next (or previous) value. BCD provides a simple conversion between binary and decimal numbers.

All four binary coding schemes are fully discussed in the following sections.

8.2   Decimal–Unsigned Binary Conversion

The conversion between decimal and binary involves identifying the particular decimal value for the particular binary code (or vice versa). Both decimal-to-binary and binary-to-decimal conversion is common and a binary number will be needed to represent each decimal number. If both the decimal and binary numbers are unrestricted in size, then an exact conversion is possible.

In unsigned (or straight) binary, the numbers represented by the binary code will be positive numbers only. Each digit in the binary number will contribute to the magnitude of the value. For example, consider the decimal value 8_{10}. In unsigned binary, this is represented by 1000_2. Each digit in the decimal number has a value in the set of (0, 1, 2, 3, 4, 5, 6, 7, 8, 9). Each digit in the binary number is in the set of (0, 1). A binary digit is referred to as a bit (*binary* digit).

The magnitude of the decimal number is the sum of the product of the value of each digit in the number (d) and its position (n). The position immediately to the left of the decimal point is position zero (0). The value of the digit has a weight of 2^n where n is the position number. Moving left from position 0 (in the integer part of the number), the position increments by 1. Moving right from position zero (into the fractional part of the number), the position decrements by 1. Therefore, the magnitude of the number is given by:

$$\text{Magnitude} = (d_n.10^n) + (d_{n-1}.10^{n-1}) + (d_{n-2}.10^{n-2}) + \ldots + (d_0.10^0) + (d_{-1}.10^{-10}) + \ldots$$
$$+ (d_{-n}.10^{-n})$$

Here, the decimal number is written as:

$$d_n d_{n-1} d_{n-2} \ldots d_0 d_{-1} \ldots d_{-n}$$

Some example decimal numbers are:

> $\textbf{8}_{10}$ is $[(8 \times 10^0)]_{10}$
>
> $\textbf{18}_{10}$ is $[1 \times 10^1 + (8 \times 10^0)]_{10}$
>
> $\textbf{218}_{10}$ is $[(2 \times 10^2) + (1 \times 10^1) + (8 \times 10^0)]_{10}$
>
> $\textbf{218.3}_{10}$ is $[(2 \times 10^2) + (1 \times 10^1) + (8 \times 10^0) + (3 \times 10^{-1})]_{10}$
>
> $\textbf{218.37}_{10}$ is $[(2 \times 10^2) + (1 \times 10^1) + (8 \times 10^0) + (3 \times 10^{-1}) + (7 \times 10^{-2})]_{10.}$

The binary number is a base 2 number whose magnitude is the sum of the product of the value of each digit in the number (b) and its position (n). Moving left from position 0 (in the integer part of the number), the position increments by 1. The value of the digit has a weight of 2^n where n is the position number. Moving right from position zero (into the fractional part of the number), the position decrements by 1. This allows the creation of numbers with digits in the set: 0, 1. Therefore, in general the magnitude of the number (as a decimal number) is given by:

$$\text{Magnitude} = (b_n.2^n) + (b_{n-1}.2^{n-1}) + (b_{n-2}.2^{n-2}) + \ldots + (b_0.2^0) + (b_{-1}.2^{-1})$$
$$+ \ldots + (b_{-n}.2^{-n})$$

Here, the binary number is written as $b_n b_{n-1} \, b_{n-2} \ldots b_0.b_{-1} \ldots b_{-n}$. Some example binary numbers are:

$$1_2$$
$$10_2$$
$$101_2$$
$$101.1_2$$
$$101.1_2$$

The decimal number equivalent for a binary number can be created by taking the binary number and calculating its magnitude (as a decimal number):

$$\text{Magnitude} = (b_n.2^n) + (b_{n-1}.2^{n-1}) + (b_{n-2}.2^{n-2}) + \ldots + (b_0.2^0) + (b_{-1}.2^{-1})$$
$$+ \ldots + (b_{-n}.2^{-n})$$

Some example binary numbers are:

$\mathbf{1_{10}}$ is $[(1 \times 2^0)]_{10} = 1_{10}$

$\mathbf{10_{10}}$ is $[(1 \times 2^1) + (0 \times 2^0)]_{10} = 2_{10}$

$\mathbf{101_{10}}$ is $[(1 \times 2^2) + (0 \times 2^1) + (1 \times 2^0)]_{10} = 5_{10}$

$\mathbf{101.1_{10}}$ is $[(1 \times 2^2) + (0 \times 2^1) + (1 \times 2^0) + (1 \times 2^{-1})]_{10} = 5.5_{10}$

$\mathbf{101.01_{10}}$ is $[(1 \times 2^2) + (0 \times 2^1) + (1 \times 2^0) + (0 \times 2^{-1}) + (1 \times 2^{-2})]_{10} = 5.25_{10}.$

The binary number equivalent of a decimal number is created by dividing the decimal number by 2 until the result of the division is 0. The remainder of the total division forms the binary number digits, the remainder from the first division forms the least significant bit (LSB) of the binary number, and the remainder from the last division forms the most significant bit (MSB) of the binary number.

Consider the number 8_{10}. The conversion procedure is shown in Table 8.2.

Table 8.2: Conversion procedure for number 8_{10}

Action	Division	Remainder	Binary number digit
Start with the decimal number (d = 8)			
Divide by 2	d/2 = 8/2 = 4	0	$b_0 = 0$
Divide by 2	d/2 = 4/2 = 2	0	$b_1 = 0$
Divide by 2	d/2 = 2/2 = 1	0	$b_2 = 0$
Divide by 2	d/2 = 1/2 = 0	1	$b_3 = 1$

The binary number can be read as: $8_{10} = (b_3 b_2 b_1 b_0)_2 = 1000_2$.

Consider now the number 218_{10}. The conversion procedure is shown in Table 8.3.

Table 8.3: Conversion procedure for number 218_{10}

Action	Division	Remainder	Binary number digit
Start with the decimal number (d = 218)			
Divide by 2	d/2 = 218/2 = 109	0	$b_0 = 0$
Divide by 2	d/2 = 109/2 = 54	1	$b_1 = 1$
Divide by 2	d/2 = 54/2 = 27	0	$b_2 = 0$
Divide by 2	d/2 = 27/2 = 13	1	$b_3 = 1$
Divide by 2	d/2 = 13/2 = 6	1	$b_4 = 1$
Divide by 2	d/2 = 6/2 = 3	0	$b_5 = 0$
Divide by 2	d/2 = 3/2 = 1	1	$b_6 = 1$
Divide by 2	d/2 = 1/2 = 0	1	$b_7 = 1$

8.3   Signed Binary Numbers

Unsigned (or straight) binary numbers are used when the operations use only positive numbers and the result of any operations is a positive number. However, in many cases, both the number and the result can be either positive or negative, and the unsigned binary number system cannot be used. The two coding schemes used to achieve this are the 1's complement and 2's complement.

The 1's complement of a number is obtained by changing (or inverting) each of the bits in the binary number (0 becomes a 1 and a 1 becomes a 0):

Original binary number: 10001100

1's complement: 01110011

The 2's complement is formed by adding 1 to the 1's complement:

Original binary number: 10001100

1's complement: 01110011

2's complement: 01110100

The MSB of the binary number is used to represent the sign (0 = positive, 1 = negative) of the number, and the remainder of the number represents the magnitude. It is therefore essential that the number of bits used is sufficient to represent the required range, as shown in Table 8.4. Here, only integer numbers are considered.

Table 8.4: Number range

Number of bits	Unsigned binary range	2's complement number range
4	0_{10} to $+15_{10}$	-8_{10} to $+7_{10}$
8	0_{10} to $+255_{10}$	-128_{10} to $+127_{10}$
16	0_{10} to $+65,535_{10}$	$-32,768_{10}$ to $+32,767_{10}$

Two's complement number manipulation is as follows:

- To create a positive binary number from a positive decimal number, create the positive binary number for the magnitude of the decimal number where the MSB is set to 0 (indicating a positive number).

- To create a negative binary number from a negative decimal number, create the positive binary number for the magnitude of the decimal number where the MSB is set to 0 (indicating a positive number), then invert all bits and add 1 to the LSB. Ignore any overflow bit from the binary addition.

- To create a negative binary number from a positive binary number, where the MSB is set to 0 (indicating a positive number), invert all bits and add 1 to the LSB. Ignore any overflow bit from the binary addition.

- To create a positive binary number from a negative binary number, where the MSB is set to 1 (indicating a negative number), invert all bits and add 1 to the LSB. Ignore any overflow bit from the binary addition.

The 2's complement number coding scheme is widely used in digital circuits and system design and so will be explained further. Table 8.5 shows the binary representations of decimal numbers for a four-bit binary number. In the unsigned binary number coding scheme, the binary number represents a positive decimal number from 0_{10} to $+15_{10}$. In the 2's complement number coding scheme, the decimal number range is -8_{10} to $+7_{10}$.

In this, the most negative 2's complement number is 110 greater in magnitude than the most positive 2's complement number. The number range for an n-bit number is: -2^N to $+(2^N - 1)$.

Addition and subtraction are undertaken by addition and if necessary inversion (creating a negative number from a positive number and vice versa). Table 8.6 shows the cases for addition and subtraction of two numbers (A and B). It is essential to ensure that the two numbers have the same number of bits, the MSB represents the sign of the binary number, and the number of bits used is sufficient to represent the range of possible inputs and the range of possible outputs.

Figure 8.1 shows an arrangement where two inputs are either added or subtracted, depending on the logic level of a control input. This arrangement requires an adder, a complement (a logical inversion of the inputs bits and add 1, disregarding any overflow), and a digital switch (multiplexer).

Input numbers in the range -8_{10} to $+7_{10}$ are represented by four bits in binary. However, the range for the result of an addition is -16_{10} to $+14_{10}$, and the range for the result of a subtraction is -15_{10} to $+15_{10}$. The result requires five bits in binary to represent the number range (one bit more than the number of bits required to represent the inputs), so the number of bits to represent the inputs will be increased by one bit before the addition or subtraction:

Table 8.5: Decimal to binary conversion

Decimal number	4-bit unsigned binary number	4-bit 2's complement signed binary number
+15	1111	—
+14	1110	—
+13	1101	—
+12	1100	—
+11	1011	—
+10	1010	—
+9	1001	—
+8	1000	—
+7	0111	0111
+6	0110	0110
+5	0101	0101
+4	0100	0100
+3	0011	0011
+2	0010	0010
+1	0001	0001
0	0000	0000
−1	—	1111
−2	—	1110
−3	—	1101
−4	—	1100
−5	—	1011
−6	—	1010
−7	—	1001
−8	—	1000

Table 8.6: 2's complement addition and subtraction

Arithmetic operation	Polarity of input A	Polarity of input B	Action
	Augend	Addend	
Addition(A + B)	Positive	Positive	Add the augend to the addend and disregard any overflow.
	Positive	Negative	
	Negative	Positive	
	Negative	Negative	
	Minuend	Subtrahend	
Subtraction(A − B)	Positive	Positive	Negate (invert) the subtrahend, add this to the minuend, and disregard any overflow.
	Positive	Negative	
	Negative	Positive	
	Negative	Negative	

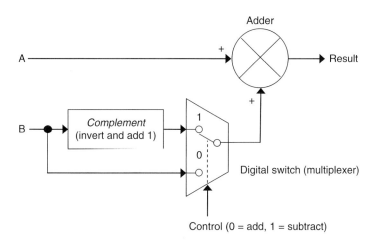

Figure 8.1: Addition and subtraction (2's complement arithmetic)

- In an unsigned binary number, to increase the wordlength (number of bits) by one bit, append a 0 to the number as the new MSB:

$$0010_2 = 00010_2$$

$$1010_2 = 01010_2$$

- In a *2's complement number*, to increase the word length by one bit, then append a bit with the same value as the original MSB to the number as the new MSB:

$$0010_2 = 00010_2$$

$$1010_2 = 11010_2$$

Consider the addition of $+2_{10}$ to $+3_{10}$ using 2's complement numbers. The result should be $+5_{10}$. The two input numbers can be represented by three bits, but if 3-bit addition is undertaken, the result will be in error:

```
 0 1 0            +2₁₀
 0 1 1 +          +3₁₀
 ─────            ─────
 1 0 1            −3₁₀    INCORRECT RESULT
```

If, however, the input word length is increased by one bit, then the addition is undertaken, the result becomes:

```
 0 0 1 0          +2₁₀
 0 0 1 1 +        +3₁₀
 ───────          ─────
 0 1 0 1          +5₁₀    CORRECT RESULT
```

Consider the subtraction of $+3_{10}$ from -2_{10}. The result should be -5_{10}. The two input numbers can be represented by three bits, but if 3-bit addition is undertaken, then the result will be in error:

```
 1 1 0            −2₁₀
 1 0 1 +          −3₁₀       (Subtrahend is complemented)
 ─────            ─────
 1 0 1 1          +3₁₀       INCORRECT RESULT
 ↑
 Overflow is ignored
```

If, however, the input word length is increased by one bit, then the addition is undertaken, the result becomes:

```
  1110          −2₁₀
  1101 +        −3₁₀   (Subtrahend is complemented)
 ─────          ────
 11011          −5₁₀      CORRECT RESULT
 ↑
Overflow is ignored
```

8.4   Gray Code

The Gray code provides a binary code that changes by one bit only when it changes from one value to the next.

The Gray code and the decimal number equivalent of the binary number (in unsigned binary) are shown in Table 8.7. This is no longer a straight binary count sequence.

Table 8.7: Gray code

Decimal number	4-bit Gray code ($d_3d_2d_1d_0$)
0	0000
1	0001
3	0011
2	0010
6	0110
7	0111
5	0101
4	0100
12	1100
13	1101
15	1111
14	1110
10	1010
11	1011
9	1001
8	1000

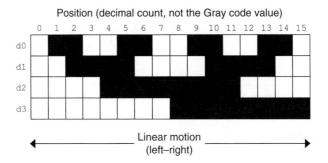

Figure 8.2: Gray code position sensing example

The Gray code is often used in position control systems which represent either a rotary position as in the output shaft of an electric motor or a linear position as in the position of a conveyor belt. Figure 8.2 shows the Gray code used on a sensor to identify the position of an object that can move left and right. Each code represents a point of position or span of distance in length. The Gray code removes the potential for errors when changing from sensing one position to the next position that could occur in a binary code when more than one bit could change. If there is a time delay in the circuitry that senses the individual bits, and the delay for sensing each bit is different, the result will be a short but finite time during which the position code would be wrong. If the circuitry that uses this position signal detects this wrong position code, it will react to a wrong position, and the result would be an erroneous operation of the circuit.

8.5   Binary-Coded Decimal

Binary-coded decimal (BCD) provides a simple conversion between a binary number and the decimal number. For a decimal number, each digit is represented by four bits. For example, the number 12_{10} is represented by 00010010_2.

$$00010010_2 = 0001_2/0010_2$$
$$= (12)_{10}$$

If the MSBs are 0, they might also be left out, so the BCD number could also be represented as 100102. This particular BCD code is referred to as 8421 BCD (or straight binary coding) because the binary number is a direct representation of the decimal value for decimal values 0_{10} to 9_{10}. Decimal values 10_{10} to 15_{10} are not represented in the four bits. Other BCD codes can also be implemented.

It is important to understand that a BCD is not the same as a straight binary (unsigned binary) count. For example, consider the number 12_{10}:

$$12_{10} = 10010_2, \text{BCD}$$
$$12_{10} = 1100_2, \text{straight binary}$$

8.6   Octal-Binary Conversion

The octal number is a number to the base (or radix) 8, and the magnitude of the number is the sum of the product of the value of each digit in the number (o) and its position (n). This allows the creation of numbers with digits in the set: 0, 1, 2, 3, 4, 5, 6, 7.

The position immediately to the left of the decimal point is zero (0). Moving left from position 0 (in the integer part of the number), the position increments by 1. The value of the digit has a weight of 8^n where n is the position number. Moving right from position 0 (into the fractional part of the number), the position decrements by 1. The eight possible digits are combined to create integers and real numbers. Table 8.8 shows the conversion table.

The magnitude of the number (as a decimal number) is given by:

$$\text{Magnitude} = (0_n.8^n) + (o_{n-1}.8^{n-1}) + (o_{n-2}.8^{n-2}) + \ldots + (o_0.8^0) + (0_{-1}.8^1)$$
$$+ \ldots + (o_{-n}.8^{-n})$$

Here, the octal number is written as $o_n o_{n-1} o_{n-2} \ldots o_0 o_{-1} \ldots o_{-n}$ (using the decimal equivalent of the octal number).

Some example octal numbers are:

7_8 is $[(7 \times 8^0)]_{10}$

17_8 is $[(1 \times 8^1) + (7 \times 8^0)]_{10}$

267_8 is $[(2 \times 8^2) + (6 \times 8^1) + (7 \times 16^0)]_{10}$

217.5_8 is $[(2 \times 8^2) + (1 \times 8^1) + (7 \times 8^0) + (5 \times 8^{-1})]_{10}$

217.57_8 is $[(2 \times 8^2) + (1 \times 8^1) + (7 \times 8^0) + (5 \times 8^{-1}) + (7 \times 8^{-2})]_{10}.$

For binary numbers, each octal number represents three bits. Therefore a 6-bit binary number is represented by two octal numbers, an 8-bit binary number is represented by three octal numbers, a 9-bit binary number is also represented by three octal

Table 8.8: Octal-decimal-unsigned 4-bit binary conversion

Octal number	Decimal number	4-bit unsigned binary number
0	0	0000
1	1	0001
2	2	0010
3	3	0011
4	4	0100
5	5	0101
6	6	0110
7	7	0111
10	8	1000
11	9	1001
12	10	1010
13	11	1011
14	12	1100
15	13	1101
16	14	1110
17	15	1111

numbers, a 16-bit binary is represented by six octal numbers, and so on. For example, 7_8 is 111_2 and 17_8 is 001111_2:

$$\underbrace{0 \quad 0 \quad 1}_{1} \quad \underbrace{1 \quad 1 \quad 1_2}_{7_8}$$

Some example octal numbers are:

7_8 is 111_2

17_8 is 001111_2

267_8 is 010110111_2

217.5$_8$ is 010001111.101_2

217.57$_8$ is 010001111.101111_2.

The decimal number equivalent for an octal number is created by calculating the magnitude of the octal number as a decimal number:

$$\text{Magnitude} = (o_n.8^n) + (o_{n-1}.8^{n-1}) + (o_{n-2}.8^{n-2}) + \ldots + (o_0.8^0) + (o_{-1}.8^{-1})$$
$$+ \ldots + (o_{-n}.8^{-n})$$

Converting from decimal to octal is accomplished in a similar manner as converting from decimal to binary, except now dividing by 8 rather than 2. Consider the number 7_{10}. The conversion procedure is shown in Table 8.9.

Table 8.9: Conversion procedure for number 7$_{10}$

Action	Division	Remainder	Octal number digit
Start with the decimal number (d = 7)			
Divide by 8	d/2 = 7/8 = 0	7	o_0= 7

The octal number can be read as: $7_{10} = (o_0)_8 = 7_8$.

Consider the number 100_{10}. The conversion procedure is shown in Table 8.10.

Table 8.10: Conversion procedure for number 100$_{10}$

Action	Division	Remainder	Octal number digit
Start with the decimal number (d = 100)			
Divide by 8	d/2 = 100/8 = 12	4	$o_0 = 4$
Divide by 8	d/2 = 12/8 = 1	4	$o_1 = 4$
Divide by 8	d/2 = 1/8 = 0	1	$o_2 = 1$

The octal number can be read as: $100_{10} = (o_2 o_1 o_0)_8 = 144_8$.

8.7   Hexadecimal-Binary Conversion

The hexadecimal number is a number to the base (or radix) 16, and its magnitude is the sum of the product of the value of each digit in the number (h) and its position (n). This allows the creation of numbers with digits in the set: 0, 1, 2, 3, 4, 5, 6, 7, 8, 9, A, B, C, D, E, F.

The position immediately to the left of the decimal point is zero (0). Moving left from position 0 (in the integer part of the number), the position increments by 1. The value of the digit has a weight of 16^n where n is the position number. Moving right from position zero (into the fractional part of the number), the position decrements by 1. The sixteen possible digits are combined to create integers and real numbers. In a decimal equivalent number, the hexadecimal digits A16 to F16 are the numbers 10_{10} to 15_{10}. Table 8.11 shows the conversion table.

Table 8.11: Hexadecimal-decimal-unsigned four-bit binary conversion

Hexadecimal number	Decimal number	4-bit unsigned binary number
0	0	0000
1	1	0001
2	2	0010
3	3	0011
4	4	0100
5	5	0101
6	6	0110
7	7	0111
8	8	1000
9	9	1001
A	10	1010
B	11	1011
C	12	1100
D	13	1101
E	14	1110
F	15	1111

The magnitude of the number (as a decimal number) is given by:

$$\text{Magnitude} = (h_n.16^n) + (h_{n-1}.16^{n-1}) + (h_{n-2}.16^{n-2}) + \ldots + (h_0.16^0) + (h_{-1}.16^{-1})$$
$$+ \ldots + (h_{-n}.16^{-n})$$

Here, the hexadecimal number is written as $h_n h_{n-1} h_{n-2} \ldots h_0.h_{-1} \ldots h_{-n}$ (using the decimal equivalent of the hexadecimal number).

Some example hexadecimal numbers are:

$\quad$ **8$_{16}$** is $[(8 \times 16^0)]_{10}$

$\quad$ **A8$_{16}$** is $[(10 \times 16^1) + (8 \times 16^0)]_{10}$

$\quad$ **2A8$_{16}$** is $[(2 \times 16^2) + (10 \times 16^1) + (8 \times 16^0)]_{10}$

$\quad$ **218.F$_{16}$** is $[(2 \times 16^2) + (1 \times 16^1) + (8 \times 16^0) + (15 \times 16^{-1})]_{10}$

$\quad$ **218.F7$_{16}$** is $[(2 \times 16^2) + (1 \times 16^1) + (8 \times 16^0) + (15 \times 16^{-1}) + (7 \times 16^{-2})]_{10}$.

For binary numbers, each hexadecimal number represents four bits. Therefore, an 8-bit binary number is represented by two hexadecimal numbers, a 16-bit binary is represented by four hexadecimal numbers, and so on. For example, 8_{16} is 1000_2 and $A8_{16}$ is 10101000_2.

$$\underbrace{1 \quad 0 \quad 1 \quad 1}_{1} \quad \underbrace{1 \quad 0 \quad 0 \quad 0_2}_{8_{16}}$$

Some example hexadecimal numbers are:

$\qquad\qquad$ **8$_{16}$** is 1000_2

$\qquad\qquad$ **A8$_{16}$** is 10101000_2

$\qquad\qquad$ **2A8$_{16}$** is 001010101000_2

$\qquad\qquad$ **218.F$_{16}$** is 001000011000.1111_2

$\qquad\qquad$ **218.F7$_{16}$** is 001000011000.11110111_2.

The decimal number equivalent for a hexadecimal number is created by calculating the magnitude of the hexadecimal number, using the decimal equivalent for hexadecimal numbers A to F, as a decimal number:

$$\text{Magnitude} = (h_n.16^n) + (h_{n-1}.16^{n-1}) + (h_{n-2}.16^{n-2}) + \ldots + (h_0.16^0)$$
$$+ (h_{-1}.16^{-1}) + \ldots + (h_{-n}.16^{-n})$$

Converting from decimal to hexadecimal is accomplished in a similar manner to converting from decimal to binary, except now dividing by 16 rather than 2, and using the letters A to F for decimal remainder values of 10 to 15. Consider the number 7_{10}. The conversion procedure is shown in Table 8.12.

Table 8.12: Conversion procedure for number 7_{10}

Start with the number (d)	Division	Remainder	Hexadecimal number digit
Start with the decimal number (d = 7)			
Divide by 16	d/16 = 7/16 = 0	7	$h_0 = 7$

The hexadecimal number can be read as: $7_{10} = (h_0)_{16} = 7_{16}$.

Consider the number 100_{10}. The conversion procedure is shown in Table 8.13.

Table 8.13: Conversion procedure for number 100_{10}

Action	Division	Remainder	Hexadecimal number digit
Start with the decimal number (d = 100)			
Divide by 16	d/16 = 100/16 = 6	4	$h_0 = 4$
Divide by 16	d/16 = 6/16 = 0	6	$h_1 = 6$

The hexadecimal number can be read as: $100_{10} = (h_1h_0)_{16} = 64_{16}$.

Consider the number 255_{10}. The conversion procedure is shown in Table 8.14.

Table 8.14: Conversion procedure for number 255$_{10}$

Start with the number (d)	Division	Remainder	Hexadecimal number digit
Start with the decimal number (d = 255)			
Divide by 16	d/16 = 255/16 = 15	15	h_0 = F
Divide by 16	d/16 = 15/16 = 0	15	h_1 = F

The hexadecimal number can be read as: $255_{10} = (h_1 h_0)_{16} = FF_{16}$.

Converting from hexadecimal to octal, or vice-versa, is accomplished by converting the number to either a binary or decimal equivalent and from that to the octal to hexadecimal number.

A summary table for the number systems is shown in Table 8.15. Here, unsigned decimal numbers from 0_{10} to 15_{10} are considered.

Both binary and decimal numbers can only be integers or real numbers. Table 8.16 shows the binary and decimal numbers for a real number represented by 40 bits in binary, with 24 bits representing the integer part of the number and 16 bits representing the fractional part of the number.

Table 8.15: Number systems summary

Decimal	Unsigned binary	Octal	Hexadecimal	BCD
0	0000	0	0	0000
1	0001	1	1	0001
2	0010	2	2	0010
3	0011	3	3	0011
4	0100	4	4	0100
5	0101	5	5	0101
6	0110	6	6	0110
7	0111	7	7	0111
8	1000	10	8	1000
9	1001	11	9	1001
10	1010	12	A	00010000
11	1011	13	B	00010001
12	1100	14	C	00010010
13	1101	15	D	00010011
14	1110	16	E	00010100
15	1111	17	F	00010101

Table 8.16: Decimal-binary conversion table, with the positive position to the left of the decimal point and the negative position to the right of the decimal point and the negative

Binary location	Unsigned binary number	Binary weighting	Decimal value
23	10000000000000000000000.0000000000000000	2^{23}	8,388,608
22	01000000000000000000000.0000000000000000	2^{22}	4,194,304
21	00100000000000000000000.0000000000000000	2^{21}	2,097,152
20	00010000000000000000000.0000000000000000	2^{20}	1,048,576
19	00001000000000000000000.0000000000000000	2^{19}	524,288
18	00000100000000000000000.0000000000000000	2^{18}	262,144
17	00000010000000000000000.0000000000000000	2^{17}	131,072
16	00000001000000000000000.0000000000000000	2^{16}	65,536
15	00000000100000000000000.0000000000000000	2^{15}	32,768
14	00000000010000000000000.0000000000000000	2^{14}	16,384
13	00000000001000000000000.0000000000000000	2^{13}	8,192
12	00000000000100000000000.0000000000000000	2^{12}	4,096
11	00000000000010000000000.0000000000000000	2^{11}	2,048
10	00000000000001000000000.0000000000000000	2^{10}	1,024
9	00000000000000100000000.0000000000000000	2^{9}	512
8	00000000000000010000000.0000000000000000	2^{8}	256
7	00000000000000001000000.0000000000000000	2^{7}	128
6	00000000000000000100000.0000000000000000	2^{6}	64
5	00000000000000000010000.0000000000000000	2^{5}	32
4	00000000000000000001000.0000000000000000	2^{4}	16
3	00000000000000000000100.0000000000000000	2^{3}	8
2	00000000000000000000010.0000000000000000	2^{2}	4
1	00000000000000000000001.0000000000000000	2^{1}	2

Decimal point (.)

1	0000000000000000000000.1000000000000000	2^{-1}	0.5
−2	0000000000000000000000.0100000000000000	2^{-2}	0.25
−3	0000000000000000000000.0010000000000000	2^{-3}	0.125
−4	0000000000000000000000.0001000000000000	2^{-4}	0.0625
−5	0000000000000000000000.0000100000000000	2^{-5}	0.03125
−6	0000000000000000000000.0000010000000000	2^{-6}	0.015625
−7	0000000000000000000000.0000001000000000	2^{-7}	0.0078125
−8	0000000000000000000000.0000000100000000	2^{-8}	0.00390625
−9	0000000000000000000000.0000000010000000	2^{-9}	0.001953125
−10	0000000000000000000000.0000000001000000	2^{-10}	0.0009765625
−11	0000000000000000000000.0000000000100000	2^{-11}	0.00048828125
−12	0000000000000000000000.0000000000010000	2^{-12}	0.00024414063
−13	0000000000000000000000.0000000000001000	2^{-13}	0.00012207031
−14	0000000000000000000000.0000000000000100	2^{-14}	0.000061035156
−15	0000000000000000000000.0000000000000010	2^{-15}	0.000030517578
−16	0000000000000000000000.0000000000000001	2^{-16}	0.000015258789

CHAPTER 9

Binary Data Manipulation

Ian Grout

9.1 Introduction

A digital circuit or system utilizes and manipulates binary data to perform a required operation. Essentially, groups of bits of data are converted from one value to another at a particular point in time. Software-programmed processors typically manipulate groups of 8, 16, 32, 64, or 128 bits of data, although a custom design could manipulate as many bits as required.

Binary data is manipulated using the following:

- *Boolean logic* provides a means to display the operations on input signals and produce a result in mathematical terms using AND, NAND, OR, NOR, EXOR, EX-NOR, and NOT logical operations.

- *Truth tables* provide a means to display the operations on input signals and produce a result in table format.

- *Karnaugh maps* provide a means to display the operations on input signals and produce a result on a K-map, which allows logic values to be grouped together with loops.

- *Circuit schematics* provide a graphical representation of the Boolean logic expression using logic gate symbols for the logical operations and the connections between the terminals.

Boolean logic, truth tables, Karnaugh maps, and circuit schematics are used in the design and analysis of digital circuits and systems, and the designer must move between these different representations of circuit and system operation many times during the design process. However, these tools are really only suited for design by hand (as it were) for small circuits; for more complex circuits and systems, hardware description languages (HDL) are more commonly used. Understanding Boolean logic, truth tables, and Karnaugh maps, however, will provide the designer with the necessary skills to design, develop, and debug circuit and system designs of any size and complexity.

9.2   Logical Operations

A digital circuit or system will consist of a number of operations on logic values. The basic logical operators are the:

- AND

- NAND

- OR

- NOR

- exclusive-OR (EX-OR)

- exclusive-NOR (EX-NOR)

- NOT

Considering two inputs (here called A and B) to a logical operator, the AND, OR, and EX-OR operators provide different results:

- The AND operator provides an output when *both* A and B are at the required values.

- The NAND operator provides an output that is the inverse of the AND operator.

- The OR operator provides an output when *either or both* A and B are at the required values.

- The NOR operator provides an output that is the inverse of the OR operator

- The EX-OR operator provides an output when *either but not both* A and B are at the required values.

- The EX-NOR (or equivalence) operator provides an output that is the inverse of the EX-OR operator.

The NOT operator provides an output that is the logical inverse of the input.

In addition, the BUFFER will provide an output that is the same logic level value as the input. The BUFFER is essentially two NOT gates in series.

These logical operators function in electronic hardware as logic gates. Two inputs (A and B) to the logic gate were considered above, but more inputs are possible to certain logic gates.

9.3 Boolean Algebra

Boolean algebra (developed by George Boole and Augustus De Morgan) forms the basic set of rules that regulate the relationship between true-false statements in logic. Applied to digital logic circuits and systems, the true-false statements regulate the relationship between the logic levels (logic 0 and 1) in digital logic circuits and systems. The relationships are based on variables and constants:

- The identifier for the AND logical operator is . (the dot).

- The identifier for the OR logical operator is + (the mathematical addition symbol).

- The identifier for the NOT logical operator is $^{-}$ (a bar across the expression).

- The identifier for the EX-OR logical operator is $\oplus$ (an encircled addition symbol).

Figure 9.1 shows the Boolean logic expression for each of these operators.

Each of the operators can be combined to create more complex Boolean logic expressions. For example, if a circuit has four inputs (A, B, C, and D) and one output (Z), then if Z is a logic 1 when (A *and* B) is a logic 1 *or* when (C *and* D) is a logic 1, the Boolean expression is:

$$Z = (A.B) + (C.D)$$

Here, parentheses are used to group the ANDed variables and to indicate precedence among various operations—similar to their use in other mathematical expressions. The AND logical operator has a higher precedence than the OR logical operator and so would be naturally grouped together in this way.

Boolean expression	Meaning
Z = A . B	Z is A AND B
Z = $\overline{A . B}$	Z is A NAND B
Z = A + B	Z is A OR B
Z = $\overline{A + B}$	Z is A NOR B
Z = A ⊕ B	Z is A XOR B
Z = $\overline{A ⊕ B}$	Z is A XNOR B
Z = $\overline{A}$	Z is NOT A

Figure 9.1: Boolean expressions for the basic logic operators

A Boolean expression written using Boolean algebra can be manipulated according to a number of theorems to modify it into a form that uses the right logic operators (and therefore the right type of logic gate) and to minimize the number of logic gates. The theorems of Boolean algebra fall into three main categories:

1. Logical operations on constants.

2. Logic operations on one variable.

3. Logic operations on two or more variables.

Table 9.1 summarizes the logical operations on constants. Each constant value can be either a logic 0 or 1. The result is either a logic 0 or 1 according to the logic operator. A bar above the constant indicates a logical *inversion* of the constant.

Table 9.1: Logical operations on constants

NOT	AND	OR
$\overline{0} = 1$	0.0 = 0	0 + 0 = 0
$\overline{1} = 0$	0.1 = 0	0 + 1 = 1
	1.0 = 0	1 + 0 = 1
	1.1 = 1	1 + 1 = 1

Table 9.2 summarizes the logical operations on one variable (*A*). The operation is performed on the variable alone or on a variable and a constant value. Each variable

and constant value can be either a logic 0 or 1. The result is either a logic 0 or 1 according to the logic operator.

Table 9.2: Logical operations on one variable

NOT	AND	OR
$\overline{\overline{A}} = A$	$A.0 = 0$	$A + 0 = A$
	$A.1 = A$	$A + 1 = 1$
	$A.A = A$	$A + A = A$
	$A.\overline{A} = 0$	$A + \overline{A} = 1$

A bar above the variable indicates a logical inversion of the variable. A double bar indicates a logical inversion followed by another logical inversion. Using the circuit symbol for the NOT gate (the symbol is a triangle with a circle at the end—see Figure 9.4), this effect is shown in Figure 9.2. Logically, a double inversion of a signal has no logical effect.

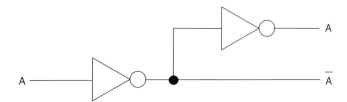

Figure 9.2: Inverting a variable

In practice, the logic gates used to create each of the inversions would create a propagation delay of the value of the variable as it passes through each logic gate. However, a double inversion produces a logic buffer, as shown in Figure 9.3.

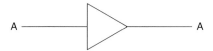

Figure 9.3: Logic buffer schematic symbol

The buffer can be used to allow for a signal to drive a large electrical load.

Table 9.3: Logical operations on two or three variables

Commutation Rule	$A+B = B+A$ $A \cdot B = B \cdot A$
Absorption Rule	$A+A.B = A$ $A \cdot (A+B) = A$
Association Rule	$A + (B+C) = (A+B) + C = (A+C) + B = A+B+C$ $A \cdot (B \cdot C) = (A \cdot B) \cdot C = (A \cdot C) \cdot B = A \cdot B \cdot C$
De Morgan's Theorems	$\overline{A+B} = \overline{A} \cdot \overline{B}$ $\overline{A \cdot B} = \overline{A} + \overline{B}$
Distributive Laws	$A \cdot (B+C) = A \cdot B + A \cdot C$ $A + (B \cdot C) = (A+B) \cdot (A+C)$
Minimization Theorems	$A \cdot B + A \cdot \overline{B} = A$ $(A+B) \cdot (A+\overline{B}) = A$ $A + \overline{A} \cdot B = A+B$ $A \cdot (\overline{A}+B) = A \cdot B$

Table 9.3 summarizes the logical operations on two or more variables. Here, two (*A* and *B*) or three variables (*A, B,* and *C*) are considered. Each variable value can be either a logic 0 or 1. The result is either a logic 0 or 1 according to the logic operator.

The commutation rule states that there is no significance in the order of placement of the variables in the expression. The absorption rule is useful for simplifying Boolean expressions, and the association rule allows variables to be grouped together in any order. De Morgan's theorems are widely used in digital logic design as they allow for AND logical operators to be related to NOR logical operators and OR logical operators to be related to NAND logical operators, which allows Boolean expressions to take different forms and thereby be implemented using different logic gates. The distributive laws allow a process similar to factorization in arithmetic, and the minimization theorems allow Boolean expressions to be reduced to a simpler form.

9.4   Combinational Logic Gates

Each logic gate that implements the logical operators is represented by a circuit symbol. The commonly used symbols are shown in Figure 9.4. Here, for each logic gate, the inputs are A or A and B, and the output is Z.

An alternative set of logic symbols, IEEE/ANSI standard 91-1984 (*Graphics Symbols for Logic Functions* [Refs. 9.1, 9.2]), is shown in Figure 9.5.

Figures 9.4 and 9.5 use only two-input logic gates for the AND, NAND, OR, and NOR gates, but it is common to use these logic gates with more than two inputs. For example, up to six inputs are available for use in many PLD and ASIC design libraries.

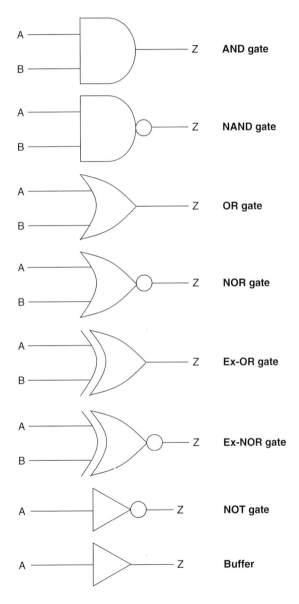

Figure 9.4: Logical operator circuit symbols

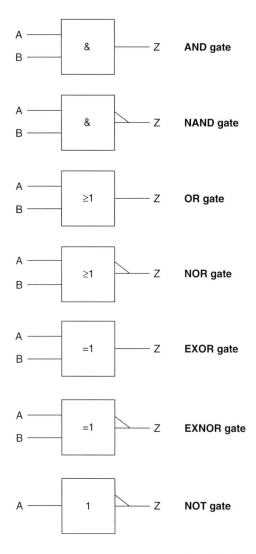

Figure 9.5: Sample IEEE/ANSI standard logic symbols

9.5   Truth Tables

The truth table displays the logical operations on input signals in a table format. Every Boolean expression can be viewed as a truth table. The truth table identifies all possible input combinations and the output for each. It is common to create the table so that the input combinations produce an unsigned binary up-count.

The truth table for the AND gate is shown in Table 9.4. Here, the output Z is a logic 1 only when *both* inputs A and B are logic 1.

Table 9.4: AND gate truth table

A	B	Z
0	0	0
0	1	0
1	0	0
1	1	1

The truth table for the NAND gate is shown in Table 9.5. Here, the output Z is a logic 0 only when *both* inputs A and B are logic 1. This is the logical inverse of the AND gate.

Table 9.5: NAND gate truth table

A	B	Z
0	0	1
0	1	1
1	0	1
1	1	0

The truth table for the OR gate is shown in Table 9.6. Here, the output Z is a logic 1 when *either or both* inputs A and B are logic 1.

Table 9.6: OR gate truth table

A	B	Z
0	0	0
0	1	1
1	0	1
1	1	1

The truth table for the NOR gate is shown in Table 9.7. Here, the output Z is a logic 0 when *either or both* inputs A and B are logic 1. This is the logical inverse of the OR gate.

Table 9.7: NOR gate truth table

A	B	Z
0	0	1
0	1	0
1	0	0
1	1	0

The truth table for the EX-OR gate is shown in Table 9.8. Here, the output Z is a logic 1 when *either but not both* inputs A and B are logic 1.

Table 9.8: EX-OR gate truth table

A	B	Z
0	0	0
0	1	1
1	0	1
1	1	0

The truth table for the EX-NOR gate is shown in Table 9.9. Here, the output Z is a logic 0 when *either but not both inputs* A and B are logic 1. This is the logical inverse of the EX-OR gate.

Table 9.9: EX-NOR gate truth table

A	B	Z
0	0	1
0	1	0
1	0	0
1	1	1

The truth table for the NOT gate (inverter) is shown in Table 9.10. This gate has one input only. The output Z is the logical inverse of the input A.

Table 9.10: NOT gate truth table

A	Z
0	1
1	0

The truth table for the BUFFER is shown in Table 9.11. This gate has one input only. The output Z is the same logical value as that of the input A.

Table 9.11: BUFFER truth table

A	Z
0	0
1	1

Another way to describe a digital circuit or system is by using a suitable HDL such as VHDL [Refs. 9.3, 9.4]. This describes the operation of the circuit or system at different levels of design abstraction. An example VHDL description for each of the basic logic gates using the built-in logical operators in VHDL is shown in Figure 9.6. It is sufficient at this point to note that HDLs exist and for VHDL the basic structure of a VHDL text based description is of the form shown in Figure 9.6.

The EX-OR gate has the Boolean expression:

$$Z = A \oplus B$$

From the truth table for the EX-OR gate, then, a Boolean expression in the *first canonical form* (the *first canonical from* is a set of minterms that are AND logical operators on the variables within the expression with the output of the AND logical operators being logically ORed together) can be written as:

$$Z = \left(\overline{A}.B\right) + \left(A.\overline{B}\right)$$

| ```
LIBRARY IEEE;
USE IEEE.STD_LOGIC_1164.ALL;
ENTITY And_Gate IS
    PORT( A : IN   STD_LOGIC;
          B : IN   STD_LOGIC;
          Z : OUT  STD_LOGIC);
END ENTITY And_Gate;
ARCHITECTURE Dataflow OF And_Gate IS
BEGIN

      Z <= A AND B;

END ARCHITECTURE Dataflow;
``` | **AND gate**<br><br>**Z = (A.B)** |
| ```
LIBRARY IEEE;
USE IEEE.STD_LOGIC_1164.ALL;
ENTITY Nand_Gate IS
    PORT( A : IN   STD_LOGIC;
          B : IN   STD_LOGIC;
          Z : OUT  STD_LOGIC);
END ENTITY Nand_Gate;
ARCHITECTURE Dataflow OF Nand_Gate IS
BEGIN

      Z <= A NAND B;

END ARCHITECTURE Dataflow;
``` | **NAND gate**<br><br>**Z = /(A.B)** |
| ```
LIBRARY IEEE;
USE IEEE.STD_LOGIC_1164.ALL;
ENTITY Or_Gate IS
    PORT( A : IN   STD_LOGIC;
          B : IN   STD_LOGIC;
          Z : OUT  STD_LOGIC);
END ENTITY Or_Gate;
ARCHITECTURE Dataflow OF Or_Gate IS
BEGIN

      Z <= A OR B;

END ARCHITECTURE Dataflow;
``` | **OR gate**<br><br>**Z = (A+B)** |
| ```
LIBRARY IEEE;

USE IEEE.STD_LOGIC_1164.ALL;

ENTITY Nor_Gate IS
        PORT( A : IN   STD_LOGIC;
              B : IN   STD_LOGIC;
              Z : OUT  STD_LOGIC);
END ENTITY Nor_Gate;
ARCHITECTURE Dataflow OF Nor_Gate IS
BEGIN

      Z <= A NOR B;

END ARCHITECTURE Dataflow;
``` | **NOR gate**<br><br>**Z = /(A+B)** |

Figure 9.6: VHDL code examples for the logic gates in Figure 9.4

| ```
LIBRARY IEEE;
USE IEEE.STD_LOGIC_1164.ALL;

ENTITY Xor_Gate IS
    PORT(  A : IN   STD_LOGIC;
           B : IN   STD_LOGIC;
           Z : OUT  STD_LOGIC);
END ENTITY Xor_Gate;

ARCHITECTURE Dataflow OF Xor_Gate IS

BEGIN

    Z <= A XOR B;

END ARCHITECTURE Dataflow;
``` | **EX-OR gate**<br><br>**Z = (A ⊕ B)** |
| ```
LIBRARY IEEE;
USE IEEE.STD_LOGIC_1164.ALL;

ENTITY Xnor_Gate IS
    PORT(  A : IN   STD_LOGIC;
           B : IN   STD_LOGIC;
           Z : OUT  STD_LOGIC);
END ENTITY Xnor_Gate;

ARCHITECTURE Dataflow OF Xnor_Gate IS

BEGIN
    Z <= A XNOR B;
END ARCHITECTURE Dataflow;
``` | **EX-NOR gate**<br><br>**Z = /(A ⊕ B)** |
| ```
LIBRARY IEEE;
USE IEEE.STD_LOGIC_1164.ALL;

ENTITY Not_Gate IS
    PORT(  A : IN   STD_LOGIC;
           Z : OUT  STD_LOGIC);
END ENTITY Not_Gate;

ARCHITECTURE Dataflow OF Not_Gate IS

BEGIN
    Z <= NOT A;
END ARCHITECTURE Dataflow;
``` | **NOT gate**<br><br>**Z = /A** |
| ```
LIBRARY IEEE;
USE IEEE.STD_LOGIC_1164.ALL;

ENTITY Buffer_Gate IS
    PORT(  A : IN   STD_LOGIC;
           Z : OUT  STD_LOGIC);
END ENTITY Buffer_Gate;

ARCHITECTURE Dataflow OF Buffer_Gate IS

BEGIN
    Z <= A;
END ARCHITECTURE Dataflow;
``` | **Buffer**<br><br>**Z = A** |

Figure 9.6: (Cont'd)

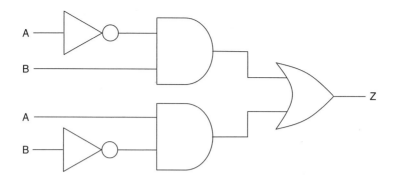

Figure 9.7: EX-OR gate using discrete logic gates

Therefore, the EX-OR gate can be made from AND, OR, and NOT gates as shown in Figure 9.7.

The truth table can be created to identify the input-output relationship for any logic circuit that consists of combinational logic gates and that can be expressed by Boolean logic. It is therefore possible to move between Boolean logic expressions and truth tables. Consider a three-input logic circuit (A, B, and C) with one output (Z), as shown in the truth table in Table 9.12. The inputs are written as a binary count starting at 0_{10} and incrementing to 7_{10}. The output Z is only a logic 1 when inputs A, B, and C are logic 1. This can be written as a Boolean expression:

$$Z = A.B.C$$

Table 9.12: Three-input logic circuit truth table: Z = A.B.C

A	B	C	Z
0	0	0	0
0	0	1	0
0	1	0	0
0	1	1	0
1	0	0	0
1	0	1	0
1	1	0	0
1	1	1	1

Here, where the output Z is a logic 1, the values of inputs A, B, and C are ANDed together. Where a variable is a logic 1, then the variable is used. When the variable is a logic 0, then the inverse (NOT) of the variable is used.

Consider now another three-input logic circuit (inputs A, B, and C) with one output (Z), shown in Table 9.13. The inputs are written as a binary count starting at 0_{10} and incrementing up to 7_{10}. The output Z is only a logic 1 when inputs A, B, and C are logic 0. This can be written as a Boolean expression:

$$Z = \overline{A}.\overline{B}.\overline{C}$$

Here, where the output Z is a logic 1, the values of inputs A, B, and C are ANDed together. Where a variable is a logic 1, then the variable is used. When the variable is a logic 0, then the inverse (NOT) of the variable is used. The expression identified for the truth table in Table 9.13 can be modified using rules and laws identified in Table 9.3:

$$Z = \overline{A}.\overline{B}.\overline{C}$$
$$Z = \overline{\overline{\overline{A}.\overline{B}.\overline{C}}}$$
$$Z = \overline{\overline{\overline{A}} + \overline{\overline{B}} + \overline{\overline{C}}}$$
$$Z = \overline{A + B + C}$$

Table 9.13: Three-input logic circuit truth table: Z = NOT (A + B + C)

A	B	C	Z
0	0	0	1
0	0	1	0
0	1	0	0
0	1	1	0
1	0	0	0
1	0	1	0
1	1	0	0
1	1	1	0

The original expression was manipulated by first double-inverting the expression (which logically makes no change), then breaking one of the inversions (the inversion closest in space to the variables) and changing the AND operator to an OR operator (the second De Morgan theorem). This leaves a NOR expression with double-inverted variables. The double-inversion on each input is then dropped.

Now, combining the operations in Table 9.12 and Table 9.13 produces a more complex operation as shown in Table 9.14.

Table 9.14: Three-input logic circuit truth table: complex logic gate

A	B	C	Z
0	0	0	1
0	0	1	0
0	1	0	0
0	1	1	0
1	0	0	0
1	0	1	0
1	1	0	0
1	1	1	1

The Boolean expression for this is:

$$Z = (A.B.C) + (A+B+C)$$

Each of the ANDed expressions is ORed together. Parentheses group each expression to aid readability of the expression. In this form of expression, the first canonical form, a set of minterms (minimum terms) that are AND logical operators are created (one for each line of the truth table where the output is a logic 1). The outputs for each of the AND logical operators are ORed together. This is also referred to as a sum of products. A circuit schematic for this circuit is shown in Figure 9.8.

The second canonical form is an alternative to the first canonical form. In the second, a set of maxterms that are OR logical operators on the variables within the expression

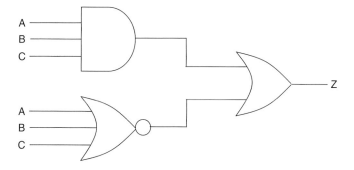

Figure 9.8: Circuit schematic for Boolean expression in Table 9.14

are created (one for each line of the truth table where the output is a logic 0). The outputs for each of the OR logical operators are ANDed together. This is also referred to as a product of sums.

Using these approaches, any Boolean logic expression can be described, analyzed, and possibly minimized.

References

[9.1] The Institute of Electrical and Electronics Engineers, IEEE Standard, 91-1984, *Graphics Symbols for Logic Functions*, IEEE, USA.

[9.2] Overview of IEEE Standard 91-1984. *Explanation of Logic Symbols,* 1996, Texas Instruments, USA.

[9.3] The Institute of Electrical and Electronics Engineers, IEEE Standard, 1076-2002, *IEEE Standard VHDL Language Reference Manual*, IEEE, USA.

[9.4] Zwolinski, M., *Digital System Design with VHDL,* Pearson Education Limited, 2000, England, ISBN 0-201-36063-2.

CHAPTER 10

Combinational Logic Design

Ian Grout

10.1   Introduction

Using the previous ideas, combinational logic circuits can be combined using either the first canonical form (sum of products) or the second canonical form (product of sums). However, in this text only the first canonical form will be considered, taking into account logic level 0 or 1 and propagation (time) delays in the cells.

Within a logic gate is an analog circuit consisting of transistors—either bipolar, using NPN and PNP bipolar junction transistors, or CMOS (complementary metal-oxide semiconductor), using N-channel MOS and P-channel MOS transistors. Logic gates in CMOS are of three different circuit architectures at the transistor level [Ref. 10.1]: static CMOS, dynamic CMOS, and pass transistor logic CMOS. Today, static CMOS logic is by far the dominant type used. It is built on a network of pMOS and nMOS transistors connected between the power supplies, as shown in Figure 10.1.

The input signals are connected to the gates of the transistors, and the output is taken from the common connection between the transistor networks. The transistors will act as switches, with the switch connections between the drain and source of the transistor. Switch control is via a gate voltage:

- An *nMOS* transistor will be switched ON when high voltage (logic 1) is applied to the transistor gate. Low voltage (logic 0) will turn the switch OFF.

- A *pMOS* transistor will be switched ON when low voltage (logic 0) is applied to the transistor gate. High voltage (logic 1) will turn the switch OFF.

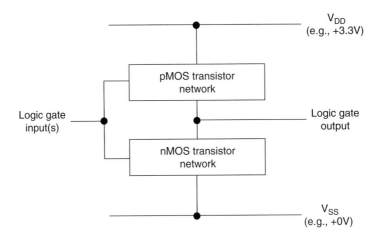

Figure 10.1: Static CMOS logic gate architecture

In the transistor network, a series connection of nMOS transistors will produce an AND effect (i.e., both transistors must be switched ON for the combined effect to be ON). A parallel connection of nMOS transistors will produce an OR effect (i.e., any single transistor must be switched ON for the combined effect to be ON). For the pMOS transistor network, a series connection of nMOS transistors requires a parallel connection of pMOS transistors. A parallel connection of nMOS transistors requires a series connection of pMOS transistors.

The inverter is the most basic logic gate and, in static CMOS, consists of one nMOS and one pMOS transistor. The basic arrangement is shown in Figure 10.2.

The logic gate has both static (DC) and dynamic (time-related) characteristics. Both the voltage (at the different points in the circuit with reference to the common, 0V, node) and the currents (in particular the power supply current) must be considered.

The static characteristics of the inverter are shown in Figure 10.3; in this case, the static (DC) voltages are not time related. Two graphs are shown. The top graph plots the input voltage (V_{IN}) against the output voltage (V_{OUT}). This shows the operating regions (off, saturation, linear) that each transistor will go through during the input and output voltage changes. A logic 0 is a voltage level of V_{SS}, and a logic 1 is a voltage level of V_{DD}. The bottom graph plots the input voltage (V_{IN}) against the current drawn from the power supply (I_{DD}), showing that the current drawn from the power supply peaks during changes in the input and output voltages.

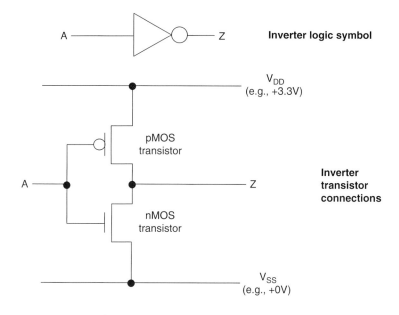

Figure 10.2: Static CMOS inverter

The dynamic characteristics of the inverter are shown in Figure 10.4. These show the operation of the inverter to changes of the inputs and outputs in time. The top graph shows the input test signal, which in this case is a step change for a 0-1-0 logic level change with an instantaneous change in logic value in time. The two voltage levels are V_{OL} and V_{OH}:

- V_{OL} defines the maximum output voltage from the logic gate that would produce a logic 0 output.

- V_{OH} defines the minimum output voltage from the logic gate that would produce a logic 1 output.

- The middle graph shows the output, which changes from a 1 to a 0 and a 0 to a 1 in a finite time. Two values for the propagation time delay are defined, t_{PHL} and t_{PLH}:

  - t_{PHL} defines a propagation time delay from a high level (1) to a low level (0) between the start of the input signal change and the 50% change in output.

  - t_{PLH} defines a propagation time delay from a low level (0) to a high level (1) between the start of the input signal change and the 50% change in output.

- The bottom graph shows the output, which changes from a 1 to a 0 and a 0 to a 1 in a finite time. Two values for the rise and fall times are defined, t_{FALL} and t_{RISE}:

  ○ t_{FALL} defines a fall time from a high level (1) to a low level (0) between the 90% and 10% levels between the high and low levels.

  ○ t_{RISE} defines a rise time from a low level (0) to a high level (1) between the 10% and 90% levels between the low and high levels.

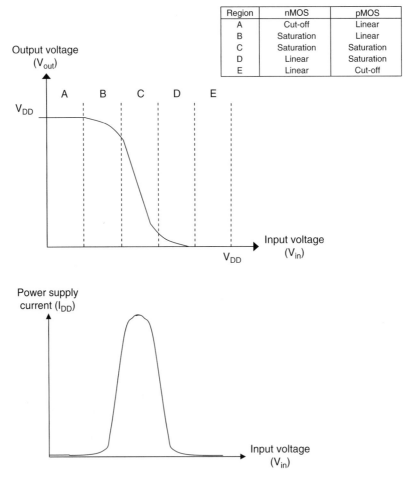

Region	nMOS	pMOS
A	Cut-off	Linear
B	Saturation	Linear
C	Saturation	Saturation
D	Linear	Saturation
E	Linear	Cut-off

Figure 10.3: Static CMOS inverter—static characteristics

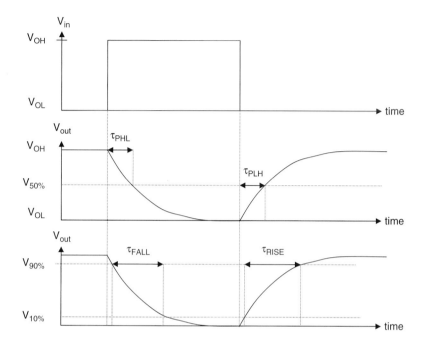

Figure 10.4: Static CMOS inverter—dynamic characteristics

Having considered the static CMOS inverter operation, the logical operation of more complex logic gates will be considered through the following four examples:

1. Two-input multiplexer

2. One-bit half-adder

3. One-bit full-adder

4. Partial odd/even number detector

Example 10.1   Two-Input Multiplexer

Consider a circuit that has two data inputs (A and B) and one data output (Z). An additional control input, Select, is used to select which input appears at the output, such that:

- when Select = 0, A → Z

- when Select = 1, B → Z

This circuit is the multiplexer, and the circuit symbol is shown in Figure 10.5.

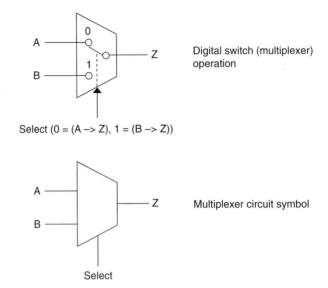

Digital switch (multiplexer) operation

Multiplexer circuit symbol

Figure 10.5: Two-input multiplexer

In general, the multiplexer can have as many data inputs as required, and the number of control signals required will reflect the number of data inputs. For the two-input multiplexer, the truth table has three inputs (for eight possible combinations) and the output as shown in Table 10.1. The Boolean expression can be created for this and a reduced form would be:

$$Z = \left(\overline{\text{Select}}.A\right) + (\text{Select}.B)$$

Table 10.1: Two-input multiplexer truth table

Select	A	B	Z
0	0	0	0
0	0	1	0
0	1	0	1
0	1	1	1
1	0	0	0
1	0	1	0
1	1	0	1
1	1	1	1

Although the multiplexer is normally available as a single logic gate in an ASIC or PLD reference library, the circuit could be created using discrete logic gates. The circuit schematic for this is shown in Figure 10.6. The inverse operation of the *multiplexer* (mux) is the *demultiplexer* (demux). This has one data input and multiple data outputs. The additional control inputs will select one output to be active and will pass the input data logic value to the particular output. As there are multiple outputs in the *demultiplexer*, the remaining outputs (those outputs which have not been selected) will output a logic "0" value.

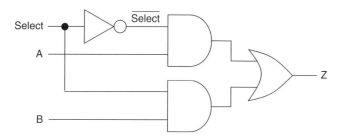

Figure 10.6: Two-input multiplexer using discrete logic gates

Example 10.2 One-Bit Half-Adder

The half-adder is an important logic design created from basic logic gates, as shown in Figure 10.7. This is a design with two inputs (A and B) and two outputs (Sum and Carry-out, Cout). This cell adds the two binary input numbers to produce sum and carry-out terms.

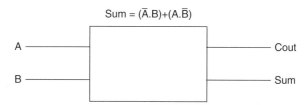

Figure 10.7: One-bit half-adder cell

The truth table for this design is shown in Table 10.2.

From viewing the truth table, the Sum output is only a logic 1 when *either but not both* inputs are logic 1:

$$\text{Sum} = \left(\overline{\text{A}}.\text{B}\right) + \left(\text{A}.\overline{\text{B}}\right)$$

Table 10.2: One-bit half-adder cell truth table

A	B	Sum	Count
0	0	0	0
0	1	1	0
1	0	1	0
1	1	0	1

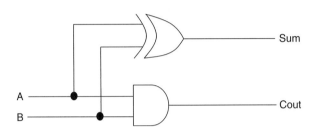

Figure 10.8: One-bit half-adder circuit schematic

This is actually the EX-OR function, so:

$$\text{Sum} = (A \oplus B)$$

From viewing the Cout output in the truth table, the output is logic 1 only when *both* inputs are logic 1 (i.e., A AND B):

$$\text{Cout} = (A.B)$$

This can be drawn as a circuit schematic as shown in Figure 10.8.

Example 10.3 One-Bit Full-Adder

The full-adder extends the concept of the half-adder by providing an additional carry-in (Cin) input, as shown in Figure 10.9. This is a design with three inputs (A, B, and Cin)

Figure 10.9: One-bit full-adder cell

and two outputs (Sum and Cout). This cell adds the three binary input numbers to produce sum and carry-out terms.

The truth table for this design is shown in Table 10.3.

Table 10.3: One-bit full-adder cell truth table

A	B	Cin	Sum	Count
0	0	0	0	0
0	0	1	1	0
0	1	0	1	0
0	1	1	0	1
1	0	0	1	0
1	0	1	0	1
1	1	0	0	1
1	1	1	1	1

From viewing the truth table, the Sum output is only a logic 1 when *one or three* (but not two) of the inputs is logic 1. The Boolean expression for this is (in reduced form):

$$\text{Sum} = \text{Cin} \oplus (\text{A} \oplus \text{B})$$

From viewing the truth table, the Cout output is only a logic 1 when *two or three* of the inputs is logic 1. The Boolean expression for this is (in reduced form):

$$\text{Cout} = (\text{A.B}) + (\text{Cin.(A} \oplus \text{B}))$$

This can be drawn as a circuit schematic as shown in Figure 10.10.

Any number of half- and full-adder cells can be connected together to form an n-bit addition. Figure 10.11 shows the connections for a four-bit binary adder. In this design, there is no Cin input. Inputs A and B are four bits wide, and bit 0 (A(0) and B(0)) are the LSBs.

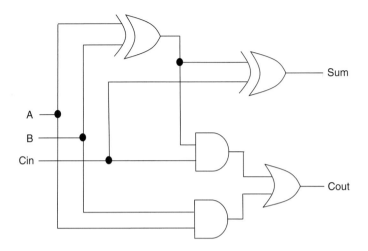

Figure 10.10: One-bit full-adder circuit schematic

Example 10.4   Partial Odd/Even Number Detector

Consider a circuit that receives a three-bit unsigned binary number (A, B, and C where A is the MSB and C is the LSB) and is to detect when the number is ODD or EVEN. The circuit will have two outputs (Odd and Even), as shown in Figure 10.12. The Odd output is a logic 1 when the input number (in decimal) is 1, 3, or 5 *but not* 7. The input 7 is to be considered a forbidden input in this circuit. The Even output is a logic 1 when the input number (in decimal) is 0, 2, 4, 6.

The truth table for this circuit is shown in Table 10.4.

A Boolean expression for each of the outputs can be created. However, because the Odd and Even outputs are inversions of each other (except in the forbidden state), a circuit can be created whereby the Boolean expression for one output is created and the second output is the inverse (NOT) of this output. Considering the Odd output (with three 1s, compared to four in the Even output, making it a smaller Boolean expression), then the Boolean expression for the Odd and Even outputs would be:

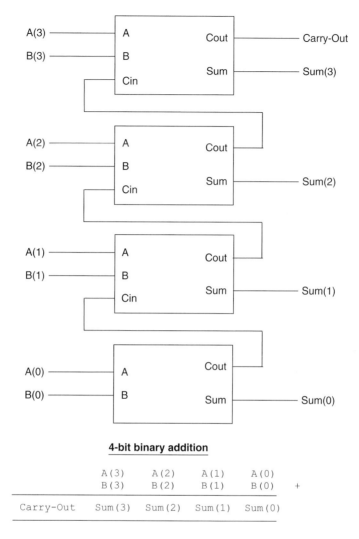

4-bit binary addition

	A(3)	A(2)	A(1)	A(0)	
	B(3)	B(2)	B(1)	B(0)	+
Carry-Out	Sum(3)	Sum(2)	Sum(1)	Sum(0)	

Figure 10.11: Four-bit binary adder

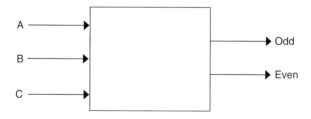

Figure 10.12: Number detector circuit block diagram

Table 10.4: Three-input logic circuit truth table

A	B	C		Odd	Even	
0	0	0		0	1	
0	0	1		1	0	
0	1	0		0	1	
0	1	1		1	0	
1	0	0		0	1	
1	0	1		1	0	
1	1	0		0	1	
1	1	1		0	1	*Forbidden input.*

$$Odd = (\overline{A}.\overline{B}.C) + (\overline{A}.B.C)$$

$$Odd = (\overline{A}.\overline{B}.C) + C.((\overline{A}.B) + (A.\overline{B}))$$

$$Odd = (\overline{A}.\overline{B}.C) + C.(A \oplus B)$$

$$Odd = C.((\overline{A}.\overline{B}) + (A \oplus B))$$

$$Even = \overline{Odd}$$

The circuit schematic for this design is shown in Figure 10.13.

A problem with this circuit is that when the odd number input 7 is applied, the circuit produces a logic 0 on Odd and a logic 1 on Even, which is incorrect. If this circuit is to be used, then the input 7 must be taken into account and the circuit redesigned, or the input 7 must never be applied by design.

If the input 7 is considered in the creation of the Boolean logic expression for the Odd output, then the logic for the Odd output simply becomes the value for the C input.

The basic arrangement is shown in Figure 10.14. Here, the Coolrunner[TM]-II CPLD on the CPLD development board is configured with the digital logic circuit, and the digital I/O board is interfaced to external test and measurement equipment. The CPLD is configured using the pins identified in Table 10.5.

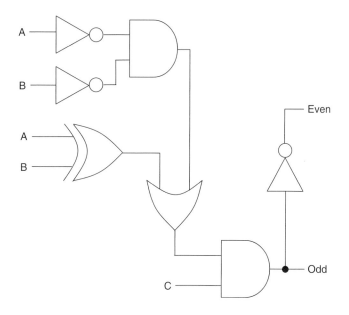

Figure 10.13: Circuit schematic for odd/even number detector

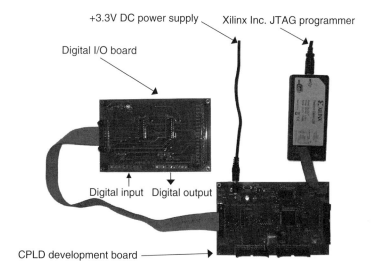

Figure 10.14: Odd/even number detector implementation using the CPLD
development board

Table 10.5: Odd/even number detector CPLD pin assignment

Signal name	CPLD pin number	Digital I/O board identifier	Comment
A	13	B0 (input bit 0)	CPLD input, design A input
B	14	B1 (input bit 1)	CPLD input, design B input
C	15	B2 (input bit 2)	CPLD input, design C input
Odd	3	A0 (output bit 0)	CPLD output, design Odd output
Even	4	A1 (output bit 1)	CPLD output, design Even output
Input buffer enable	12	OE2 (input enable)	CPLD output, tie to logic 0 in CPLD design
Output buffer enable	2	OE1(output enable)	CPLD output, tie to logic 0 in CPLD design

Here, the five design I/Os are defined and are connected to the relevant CPLD pins to connect to Header A for the digital I/O board. In addition, the CPLD design must also incorporate two additional outputs to enable the tristate buffers used on the digital I/O board. Here, the enable (OE, output enable) pins on the tristate buffers must be tied to logic 0 to enable the buffers.

External circuitry is connected to the digital I/O board to provide the logic levels for inputs A, B, and C and to monitor the outputs Odd and Even where:

- logic 0 = 0V

- logic 1 = +3.3V

The CPLD is programmed from using an appropriate JTAG (Joint Test Action Group) programmer.

10.2　NAND and NOR Logic

Logical operations using AND, OR, and NOT logic gates can also be undertaken using either NAND or NOR logic gates. A Boolean expression using AND, OR, and NOT logic can be manipulated to produce NAND and NOR logic. For example, the Boolean expression:

$$Z = (A.B)$$

can also be expressed as:

$$z = \overline{(\overline{A} + \overline{B})}$$

Figure 10.15 shows the two logic gate implementations for these Boolean expressions. Similarly, the Boolean expression:

$$Z = (A + B)$$

can also be expressed as:

$$z = \overline{\overline{A}.\overline{B}}$$

Figure 10.16 shows the two logic gate implementations for these Boolean expressions.

If only NAND and NOR gates are available, any Boolean logic expression can be implemented through such manipulation.

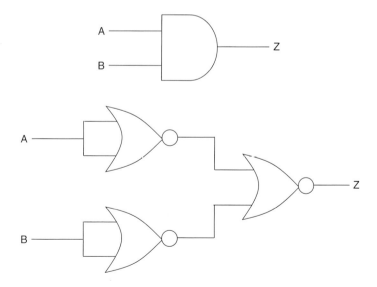

Figure 10.15: NOR implementation for the AND gate

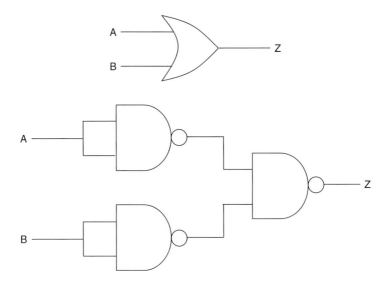

Figure 10.16: NAND implementation for the OR gate

10.3   Karnaugh Maps

The Karnaugh map (or K-map) provides a means to display logical operations on input signals as a map showing the output values for each of the input values. This allows groups of logic values to be looped together with suitably sized loops to minimize the resulting Boolean logic expression. The size of the Karnaugh map depends on the number of inputs to the combinational logic circuit. Karnaugh maps for two-, three-, and four-input circuits are shown in Figure 10.17:

- A *two-input Karnaugh map* contains four cells, one cell for each possible input combination (2^n where n is the number of inputs). Here, the inputs are named A and B.

- A *three-input Karnaugh map* contains eight cells, one cell for each possible input combination (2^n where n is the number of inputs). Here, the inputs are named A, B, and C.

- A *four-input Karnaugh map* contains sixteen cells, one cell for each possible input combination (2^n where n is the number of inputs). Here, the inputs are named A, B, C, and D.

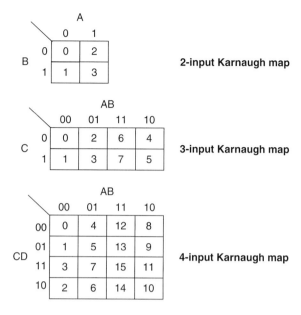

Figure 10.17: Two-, three-, and four-input Karnaugh maps

The Karnaugh map has a direct correspondence with the truth table for a Boolean logic expression. Each K-map cell is filled with the logic value of the output (0 or 1) for the corresponding input combination. In Figure 10.17, the cells are filled with (for reference purposes) the decimal number equivalent for the unsigned binary value of the input combination (A is the MSB of the binary input value). Note the values and locations of the values within the cells.

The Karnaugh maps for the two-input logic gates in Figure 9.4 are shown in Figure 10.18.

The Karnaugh map can then be analyzed, and loops of output logic levels within the cells can be created. In the first canonical form, logic 1s are grouped together. In the second canonical form, logic 0s are grouped together. In this text, the first canonical form will be considered, so in the Karnaugh map, logic 1s are grouped together.

The larger the loop, the smaller the resulting Boolean logic expression (with fewer variables to be considered). The variables in the loop will be ANDed together, and each group will be ORed together.

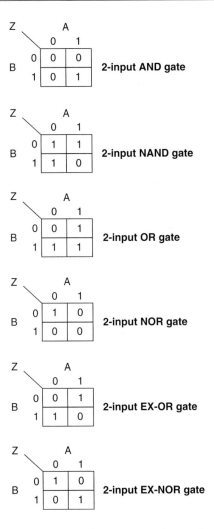

Figure 10.18: Truth table for two-input logic gates

For a *two-input* Karnaugh map, then:

- A group of one logic 1 will result in the ANDing of two variables.

- A group of two logic 1s will result in one variable.

- A group of four logic 1s will result in a constant logic 1.

For a *three-input* Karnaugh map, then:

- A group of one logic 1 will result in the ANDing of three variables.

- A group of two logic 1s will result in the ANDing of two variables.

- A group of four logic 1s will result in one variable.

- A group of eight logic 1s will result in a constant logic 1.

For a *four-input* Karnaugh map, then:

- A group of one logic 1 will result in the ANDing of four variables.

- A group of two logic 1s will result in the ANDing of three variables.

- A group of four logic 1s will result in the ANDing of two variables.

- A group of eight logic 1s will result in one variable.

- A group of sixteen logic 1s will result in a constant logic 1.

Consider the two-input AND gate: it has only one logic 1, so only a loop of 1 can be created, as shown in Figure 10.19. Where the input variable is a logic 1, the variable is used. When the input variable is a logic 0, the inverse (NOT) of the variable is used.

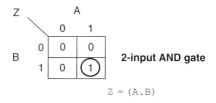

Figure 10.19: Two-input AND gate

Consider now the two-input OR gate: it has three logic 1s, two loops of two can be created, as shown in Figure 10.20. Where the input variable is a logic 1, the variable is used. When the input variable is a logic 0, the inverse (NOT) of the variable is used.

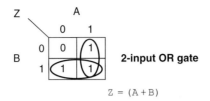

Figure 10.20: Two-input OR gate

When a group of two is created, one of the variables can be a logic 0 or a logic 1 and so can be dropped from the resulting Boolean logic expression. The vertical group of two retains the variable A but drops the variable B. The horizontal group of two retains the variable B but drops the variable A.

The grouping of logic 1s follows the following rules:

1. Loops of 2^n adjacent cells can be made where n is an integer number starting at 0.

2. All cells containing a 1 (first canonical form; or 0 in the second canonical form) must be covered.

3. Loops can overlap provided they contain at least one unlooped cell.

4. Loops must be square or rectangular (diagonal or L-shaped loops are not permitted).

5. Any loop that has all of its cells included in other loops is redundant.

6. The edges of a map are considered to be adjacent—a loop can leave the side of the Karnaugh map and re-enter at the other side, or leave from the top of the Karnaugh map and return at the bottom, as shown in Figure 10.21.

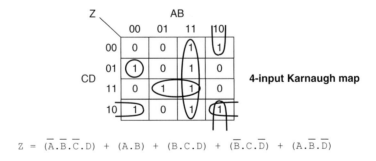

$$Z = (\overline{A}.\overline{B}.\overline{C}.D) + (A.B) + (B.C.D) + (\overline{B}.C.\overline{D}) + (A.\overline{B}.\overline{D})$$

Figure 10.21: Adjacent cells in a Karnaugh map

One potential problem with combinational logic arises from hazards. Here, because of the finite time for a signal change to propagate through the combinational logic (due to any logic gate delays and interconnect delays), there is potential for erroneous output during the time that the change occurs. This results from different time delays in different paths within the combinational logic. Although the final output would be correct, an erroneous output (i.e., wrong logic level) can occur during the change, which would cause problems if detected and used.

If the digital circuit or system can be designed so that the output from the combinational logic with a hazard is only used after it is guaranteed correct, then the hazard, although not eliminated, will not cause a problem in the design.

A way to eliminate hazards using the Karnaugh map is to ensure that all loops are joined together. Although this will introduce a redundant term (see Figure 10.22), the hazard will be removed. However, this is at the expense of using additional logic and introducing potential problems with testing the design [Ref. 10.2].

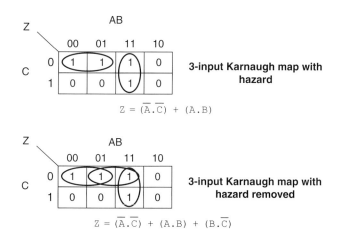

Figure 10.22: Eliminating hazards

Two important points to note with logic gates are:

1. No input to a logic gate may be left unconnected. If an input to a logic gate is not required, then it must be tied to logic level (0 or 1). This is usually achieved by

connecting a high-resistance value resistor (typically 10 to 100 kΩ) between the unused input and one of the power supply connections (V_{DD} for logic 1, V_{SS} or GND for logic 0). In some ICs, specific inputs might be designed for use only under specific circumstances and with a pull-up (to logic 1) or pull-down (to logic 0) component integrated into the IC circuitry. Such integrated pull-up or pull-down components alleviate the need for the designer to place resistors on the PCB and so reduce the PCB design requirements.

2. Where a logic gate only produces a logic 0 or 1 output, then no two or more logic gate outputs are to be connected unless the implementation technology (circuitry within the logic gate) allows it. Certain logic gate outputs can be put into a high-impedance state, which stops the output from producing a logic output and instead turns the output into a high-impedance electrical load. Circuits with a high-impedance output are used where multiple devices are to be connected to a common set of signals (a bus) such as a microprocessor data bus.

Whereas the previous logic gates considered in the design of digital circuits using Boolean logic expressions, truth tables, and Karnaugh maps provided only a logic 0 or 1 output, in many computer architectures, multiple devices share a common set of signals—control signals, address lines, and data lines. In a computer architecture where multiple devices share a common set of data lines, these devices can either receive or provide logic levels when the device is enabled (and all other devices are disabled). However, multiple devices could, when enabled, provide logic levels at the same time; these logic levels typically conflict with the logic levels provided by the other devices. To prevent this, rather than producing a logic level when disabled, a device would be put in a high-impedance state (denoted by the character Z). The tristate buffer, when enabled, passes the logic input level to the output; when disabled, it blocks the input, and the output is seen by the circuit that it is connected to as a high-impedance electrical load. This operation is shown in Figure 10.23, in which the enable signal may be active high (top, 1 to enable the buffer) or active low (bottom, 0 to enable the buffer).

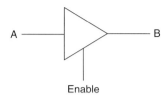

Enable	A	B
0	0	Z
0	1	Z
1	0	0
1	1	1

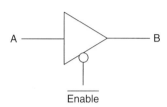

$\overline{\text{Enable}}$	A	B
0	0	0
0	1	1
1	0	Z
1	1	Z

The bar above the *Enable* input name
indicates that the input is active low

Figure 10.23: Tristate buffer symbol

10.4 *Don't Care* Conditions

In some situations, certain combinations of input might not occur, so the designer could consider that these conditions are not important. They are referred to as *Don't care* conditions. As such, the output in these conditions could be either a logic 0 or a logic 1, so the designer is free to choose the output value that results in the simpler output logic (i.e., using fewer logic gates).

References

[10.1] Kang, S., and Leblebici, Y., CMOS *Digital Integrated Circuits Analysis and Design,* McGraw-Hill International Editions, Singapore, 1996, ISBN 0-07-114423-4.
[10.2] Grout, I. A., *Integrated Circuit Test Engineering Modern Techniques,* Springer, 2006, ISBN 1-84628-023-0.

CHAPTER 11

Sequential Logic Design

Ian Grout

11.1 Introduction

Sequential logic circuits are based on combinational logic circuit elements (AND, OR, etc.) working alongside sequential circuit elements (latches and flip-flops). A generic sequential logic circuit is shown in Figure 11.1. Here, the circuit inputs are applied to and the circuit outputs are derived from a combinational logic block. The sequential logic circuit elements store an output from the combinational logic that is fed back to the combinational logic input to constitute the present state of the circuit. The output from the combinational logic that forms the inputs to the sequential logic circuit elements constitutes the next state of the circuit. These sequential logic circuit elements are grouped together to form registers. The circuit changes state from the present state to the next state on a clock control input (as happens in a synchronous sequential logic circuit). Commonly, the D-latch and D-type flip-flops are used (rather than other forms of latch and flip-flop such as the S-R, toggle, and J-K flip-flops), and they will be discussed in this text. The output from the circuit is taken from the output of the combinational logic circuit block.

In general, sequential logic circuits may be asynchronous or synchronous:

1. *Asynchronous sequential logic*. This form of sequential logic does not use a clock input signal to control the timing of the circuit. It allows very fast operation of the sequential logic, but its operation is prone to timing problems where unequal delays in the logic gates can cause the circuit to operate incorrectly.

2. *Synchronous sequential logic*. This form of sequential logic uses a clock input signal to control the timing of the circuit. The timing of changes in states in the

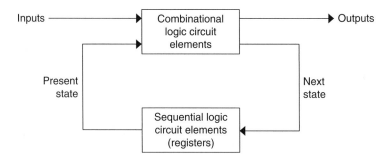

Figure 11.1: Generic sequential logic circuit (counter or state machine)

sequential logic is designed to occur either on the edge of the clock input when flip-flops are used, or at a particular logic level, as when latches are used. State changes that occur on the edge of the clock input, as when flip-flops are used, occur either on a 0 to 1 rise, referred to as positive edge triggered, or on a 1 to 0 fall, referred to as negative edge triggered.

In this text, only synchronous sequential logic will be considered.

An alternative view for the generic sequential logic circuit in Figure 11.1, is shown in Figure 11.2. Here, the combinational logic is separated into input and output logic. Both views are commonly used in the description of sequential logic circuits.

In designing the synchronous sequential logic circuit (from now on simply referred to as the sequential logic circuit), the designer must consider both the type of sequential

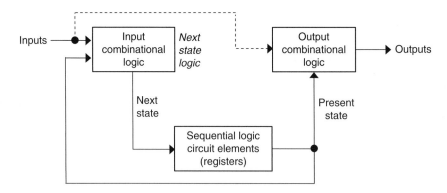

Figure 11.2: Alternative view for the generic sequential logic circuit

logic circuit elements (latch or flip-flop) and the combinational logic gates. The design uses the techniques previously discussed—Boolean logic expressions, truth tables, schematics, and Karnaugh maps—to determine the required input combinational logic (the next state logic) and determine the required output combinational logic.

The sequential logic circuit will form one of two types of machines:

1. In the *Moore machine*, the outputs are a function only of the present state only.

2. In the *Mealy machine*, the outputs are a function of the present state and the current inputs.

In addition, the sequential logic circuit will be designed either to react to an input or to be autonomous. In an autonomous sequential logic circuit, there are no inputs (apart from the clock and reset/set) to control the operation of the circuit, so the circuit moves through states under the control of only the clock input. An example of an autonomous sequential logic circuit is a straight binary up-counter that moves through a binary count sequence taking the outputs directly from the sequential logic circuit element outputs. A sequential logic circuit can also be designed to react to an input: a sequential logic circuit that reacts to an input is called a *state machine* in this text.

Sequential logic circuit design follows a set design sequence aided by:

• state transition diagram, which provides a graphical means to view the states and the transitions between states; and

• state transition table, similar in appearance to a combinational logic truth table, which identifies the current state outputs and the possible next state inputs to the sequential logic circuit elements.

As an example, consider a circuit that is to detect the sequence 1001 on a serial bitstream data input and produce a logic 1 output when the sequence has been detected, as shown in Figure 11.3. The state machine will have three inputs—one Data_In that is to be monitored for the sequence and two control inputs, Clock and Reset—and one output, Detected. Such a state machine could be used in a digital combinational lock circuit.

An example *state transition diagram* for this design is shown in Figure 11.4. The circuit is to be designed to start in *State 0* and has five possible states. With these five states, if D-type flip-flops are to be used, then there will need to be a need for three flip-flops (producing eight possible states although only five will be used when each state is to be

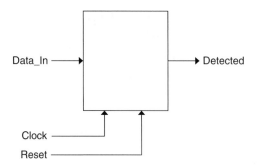

Figure 11.3: 1001 sequence detector

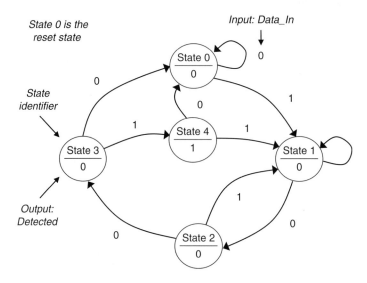

Figure 11.4: 1001 sequence detector state transition diagram (Moore machine)

represented by one value of a straight binary count sequence 0, 1, 2, 3, 4, 0, etc.). The arrangement for the *state transition diagram* is:

1. The *circles* identify the states. The name of the state (the *state identifier*) and the outputs for each state are placed within the circle. Each state is referred to as a *node*.

2. The transition between states uses a *line* with the arrow end identifying the direction of movement. Each line starts and ends at a node.

3. Each line is accompanied by an *identifier* that identifies the logical value of the input (here *Data_In*) that controls the state machine to go to the next particular state.

This form of the state transition diagram is for a *Moore machine* and in this form the outputs for each state are identified within the circles. The alternative to the *Moore machine* is the *Mealy machine*. In the *Mealy machine*, the outputs for a particular state are identified on the lines connecting the states along with the *identifier*.

The state transition table (also referred to as a present state/next state table) for the 1001 sequence detector state diagram is shown in Table 11.1. Each possible input condition has its own column, and each row contains the present state and the next state for each possible input condition. The Detected output is defined in the truth table shown in Table 11.2.

Using the circuit architecture shown in Figure 11.2, the input and output combinational logic blocks are created. Each state is created using the outputs from the sequential logic circuit element block. Flip-flops form a register whose outputs produce a binary value that defines one of the states. It is common to create the states as a straight

Table 11.1: State transition table for the 1001 sequence detector

	Data_In = 0	Data_In = 1
Present state	**Next state**	**Next state**
State 0	State 0	State 1
State 1	State 1	State 2
State 2	State 3	State 1
State 3	State 0	State 4
State 4	State 0	State 1

Table 11.2: Detected output for the 1001 sequence detector

State	**Detected**
State 0	0
State 1	0
State 2	0
State 3	0
State 4	1

binary count. Using n-flip-flops, 2^n states are possible in the register output. However, any count sequence could be used. For example, one-hot encoding uses n-flip-flops to represent n states. In the one-hot encoding scheme, to change from one state to the next, only two flip-flop outputs will change (the first from a 1 to a 0, and the second from a 0 to a 1). The advantage of this scheme is less combinational logic to create the next state values.

11.2   Level-Sensitive Latches and Edge-Triggered Flip-Flops

The two sequential logic circuit elements are the latch and the flip-flop. These elements store a logic value (0 or 1). The basic latches and flip-flops are:

Latches:
- D-latch

- S-R latch (set-reset latch)

Flip-Flops:
- S-R flip-flop (set-reset flip-flop)

- J-K flip-flop

- T-flip-flop (toggle flip-flop)

- D-type flip-flop

Each latch and flip-flop has its own particular characteristics and operation requirements. In this text, only the D-latch and the D-type flip-flop will be considered.

11.3   The D-Latch and D-Type Flip-Flop

The basic D-latch circuit symbol, shown in Figure 11.5, includes two inputs, the data input (D, value to store) and the control input (C). There is one output (Q).

In the D-latch, when the C input is at a logic 1, the Q output is assigned the value of the D input. When the C input is a logic 0, the Q output holds its current value even when the D input changes. In addition, many D-latches also include a logical inversion of the Q output (the NOT-Q output) as an additional output.

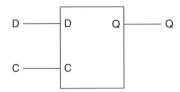

Figure 11.5: D-latch

Latches are normally designed as part of normal circuit operation. However, a problem can occur when writing HDL code in that a badly written design will create unintentional latches. When a design description is synthesized, the synthesis tool will *infer* latches. In VHDL, two common coding mistakes that result in inferred latches are:

- an *If* statement without an *Else* clause;

- a register description without a clock rising or falling edge construct.

An example of an *If* statement without an *Else* clause in VHDL is shown in Figure 11.6. Here, a circuit has two input signals (Data_In and Enable) and one output signal (Data_Out). The output is the logical value of the Data_In input when the Enable input is a logic 1. The operation of the circuit is defined on lines 20 to 30 of the code.

The RTL schematic for this design, as synthesized and viewed as a schematic within the Xilinx® ISE™ tools, is shown in Figure 11.7. The latch is the LD symbol in the middle of the schematic view.

This unintentional latch can be removed by including the *Else* clause, as shown in Figure 11.8. Here, when the Enable input is a 0, then the Data_Out output is a logic 0 also. The schematic for this design, as synthesized and viewed as a RTL schematic within the Xilinx® ISE™ tools, is shown in Figure 11.9. This forms a circuit with a single two-input AND gate. The operation of the circuit is defined on lines 20 to 34 of the code.

The basic D-type flip-flop circuit symbol is shown in Figure 11.10, with two inputs— the data input (D, value to store) and the clock input (CLK)—and one output (Q).

In the D-type flip-flop, when the CLK input changes from a 0 to a 1 (positive edge triggered) or from a 1 to a 0 (negative edge triggered), the Q output is assigned the value of the D input. When the CLK input is steady at a logic 0 or a 1, the Q output holds its current value even when the D input changes.

```
1      ----------------------------------------------
2
3      LIBRARY IEEE;
4      USE IEEE.STD_LOGIC_1164.ALL;
5
6      ----------------------------------------------
7
8      ENTITY Inferred_Latch is
9          PORT ( Enable   : IN    STD_LOGIC;
10                 Data_In  : IN    STD_LOGIC;
11                 Data_Out : OUT   STD_LOGIC);
12     END Inferred_Latch;
13
14     ----------------------------------------------
15
16     ARCHITECTURE Behavioral OF Inferred_Latch IS
17
18     BEGIN
19
20     Enable_Process: PROCESS (Enable, Data_In)
21
22     BEGIN
23
24         IF (Enable = '1') THEN
25
26             Data_Out <= Data_In;
27
28         END IF;
29
30     END PROCESS Enable_Process;
31
32     END Behavioral;
33
34     ----------------------------------------------
```

Figure 11.6: *If* statement without an *Else* clause

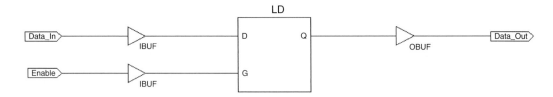

Figure 11.7: Schematic of inferred latch design

```
 1    --------------------------------------------
 2
 3    LIBRARY IEEE;
 4    USE IEEE.STD_LOGIC_1164.ALL;
 5
 6    --------------------------------------------
 7
 8    ENTITY Inferred_Latch is
 9        PORT ( Enable  : IN   STD_LOGIC;
10               Data_In : IN   STD_LOGIC;
11               Data_Out : OUT  STD_LOGIC);
12    END Inferred_Latch;
13
14    --------------------------------------------
15
16    ARCHITECTURE Behavior OF Inferred_Latch IS
17
18    BEGIN
19
20    Enable_Process: PROCESS(Enable, Data_In)
21
22    BEGIN
23
24         IF (Enable = '1') THEN
25
26                Data_Out <= Data_In;
27
28         ELSE
29
30                Data_Out <= '0';
31
32         END IF;
33
34    END PROCESS Enable_Process;
35
36    END Behavior;
37
38    --------------------------------------------
```

Figure 11.8: *If* statement with an *Else* clause

It is common, however, for the flip-flop to have a reset or set input to initialize the output Q to either logic 0 (reset) or logic 1 (set). This reset/set input can either be asynchronous (independent of the clock input) or synchronous (occurs in a clock edge) and active high (reset/set occurs when the signal is a logic 1) or active low (reset/set occurs when the signal is a logic 0). The circuit symbol for the D-type flip-flop with active low reset is shown in Figure 11.11.

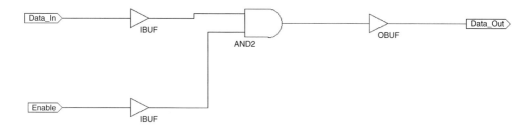

Figure 11.9: Schematic of the design with an *Else* clause

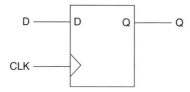

Figure 11.10: D-type flip-flop

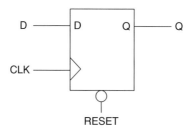

Figure 11.11: D-type flip-flop with active low reset

The circle on the reset input indicates an active low reset: if no circle is used, then the flip-flop is active high reset. Many D-type flip-flops also include a logical inversion of the Q output (the NOT-Q output) as an additional output. The circuit symbol for the D-type flip-flop with a NOT-Q output is shown in Figure 11.12.

When a D-input change is to be stored in the flip-flop, specific timing requirements must be considered for the inputs of the flip-flop in both set-up (how long *before* the clock input must the D input be static?) and hold (how long *after* the clock input must

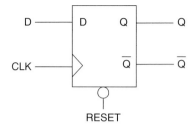

Figure 11.12: D-type flip-flop with active low reset and not-Q output

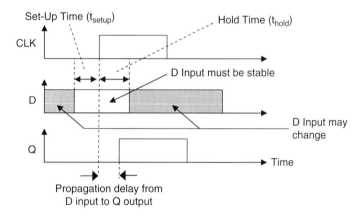

Figure 11.13: D-type flip-flop set-up and hold times

the D input be static?). This is shown in Figure 11.13. If these times are violated, then problems with the flip-flop operation will occur.

A potential problem known as *metastability* can occur when flip-flop set-up and hold times are violated. The flip-flop enters a metastable state in which the output is unpredictable until, after some time, the output becomes a logic 0 or 1. In the metastable state, the flip-flop output oscillates between 0 and 1. A simple way to design a circuit that avoids this problem is to ensure that the clock period is long enough to allow the metastable state to resolve itself and to account for signal delays (resulting from logic gates and interconnect) in the path of the next flip-flop in the circuit.

11.4   Counter Design

The counter is a sequential logic circuit that acts autonomously to perform the functions of a number counter changing its count state (value) on a clock edge. In the following discussions, then:

- positive edge triggered, asynchronous active low or high D-type flip-flops will be used;

- all flip-flops will have a common reset input; and

- All flip-flops will have a common clock input.

In addition, the output from the counter either can be taken directly from the Q outputs of each flip-flop, or can be decoded using output combinational logic to form specific outputs for specific states of the counter.

Because the counter uses flip-flops, for n-flip-flops, there will be 2^n possible combinations of output for the flip-flops. A counter might use all possible states or might use only a subset of the possible states. When a subset is used, the counter should be designed so that it will not enter unused states during normal operation. In addition, it is good practice to design the circuit so that if it does enter one of the unused states, it will have a known operation. For example, if an unused state is entered, the next state would always be the reset state for the counter.

Example 11.1   Three-Bit Straight Binary Up-Counter

Consider a three-bit straight binary up-counter as shown in Figure 11.14, using two inputs, clock and reset, and three outputs. The counter outputs are taken directly from the Q outputs of each flip-flop (Q2, Q1, and Q0), where Q2 is the MSB and Q0 is the LSB.

The design process begins by creating the state transition diagram (Figure 11.15) and the state transition table. The counter is designed to reset (i.e., when reset is a logic 0)

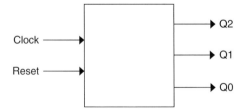

Figure 11.14: Three-bit straight binary up-counter

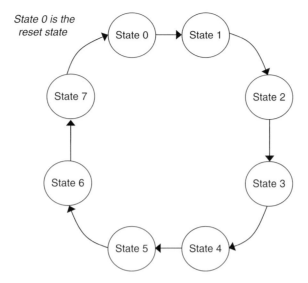

Figure 11.15: Three-bit straight binary up-counter state transition diagram

to a count of 000_2 (0_{10}), which will be state 0. When the reset is removed (i.e., when reset becomes a logic 1), then the counter will count through the sequence 0, 1, 2, 3, 4, 5, 6, 7, 0, etc. This means that when the counter output reaches 111_2 (state 7), it will automatically wrap around back to 000_2.

Each state in the counter will be encoded by the Q outputs of the D-type flip-flops, as shown in Table 11.3, so that it produces the required straight binary count sequence.

The state transition table for the counter can then be created (Table 11.4). For the next state logic, the Q output for each flip-flop in the next state is actually the D input for each flip-flop in the current state. In this view of the state transition table, the current Q outputs and the current D inputs (next state Q outputs) are defined.

The Boolean logic expression can be created for each of the D inputs so that the counter of the form shown in Figure 11.16 is created. Here, the next state logic for each flip-flop (Dff$n$) uses a combination of the Q and NOT-Q outputs from each flip-flop. Manipulation of the Boolean logic expression, the use of truth tables, and Karnaugh maps allow the designer to create a Boolean logic expression of a required form.

Table 11.3: Three-bit straight binary up-counter state encoding

State	Q2	Q1	Q0
State 0	0	0	0
State 1	0	0	1
State 2	0	1	0
State 3	0	1	1
State 4	1	0	0
State 5	1	0	1
State 6	1	1	0
State 7	1	1	1

Table 11.4: Three-bit straight binary up-counter state transition table

Present state				Next state		
	Current Q outputs			Current D inputs		
State name	Q2	Q1	Q0	D2	D1	D0
State 0	0	0	0	0	0	1
State 1	0	0	1	0	1	0
State 2	0	1	0	0	1	1
State 3	0	1	1	1	0	0
State 4	1	0	0	1	0	1
State 5	1	0	1	1	1	0
State 6	1	1	0	1	1	1
State 7	1	1	1	0	0	0

An example of the logic for each flip-flop D input developed using the Karnaugh map is shown in Figure 11.17. Figure 11.18 shows the schematic developed for the counter in which each D-type flip-flop only has a Q output, and the NOT-Q output is created using a discrete inverter. Additionally, each D-type flip-flop has an asynchronous active high reset that must be initially inverted so that the design reset input sees an asynchronous active low reset circuit.

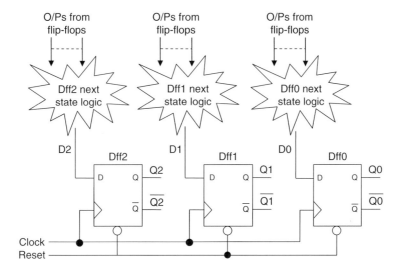

Figure 11.16: Three-bit counter structure

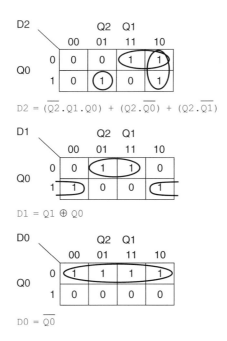

D2 = $(\overline{Q2}.Q1.Q0)$ + $(Q2.\overline{Q0})$ + $(Q2.\overline{Q1})$

D1 = Q1 $\oplus$ Q0

D0 = $\overline{Q0}$

Figure 11.17: Three-bit up-counter D-input Boolean expressions

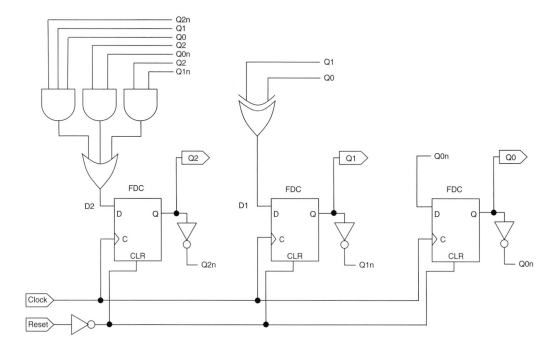

Figure 11.18: Circuit schematic for three-bit up-counter

Example 11.2   Three-Bit Straight Binary Down-Counter

Consider a three-bit straight binary down-counter, as shown in Figure 11.19 with two inputs, clock and reset. The counter outputs are taken directly from the Q outputs of each flip-flop (Q2, Q1, and Q0), where Q2 is the MSB and Q0 is the LSB. This is similar to the up-counter, except now the binary count is downward.

The design process begins by creating the state transition diagram (Figure 11.20) and the state transition table. The counter is designed to reset to a count of 000_2 (0_{10}). When

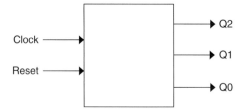

Figure 11.19: Three-bit straight binary down-counter

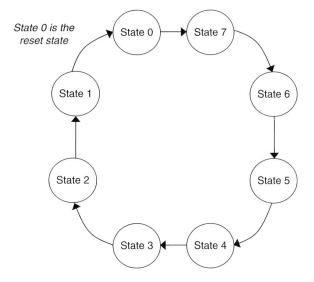

Figure 11.20: Three-bit straight binary down-counter state transition diagram

the reset is removed (i.e., becomes a logic 1), the counter will count through the sequence 0, 7, 6, 5, 4, 3, 2, 1, 0, etc. This means that when the counter output reaches 000_2, it will automatically wrap around back to 111_2.

Each state in the counter is encoded by the Q outputs of the D-type flip-flops, as shown in Table 11.5.

Table 11.5: Three-bit straight binary down-counter state encoding

State	Q2	Q1	Q0
State 0	0	0	0
State 1	0	0	1
State 2	0	1	0
State 3	0	1	1
State 4	1	0	0
State 5	1	0	1
State 6	1	1	0
State 7	1	1	1

Table 11.6: Three-bit straight binary down-counter state transition table

Present state				Next state		
	Current Q outputs			Current D inputs		
State name	Q2	Q1	Q0	D2	D1	D0
State 0	0	0	0	1	1	1
State 1	0	0	1	0	0	0
State 2	0	1	0	0	0	1
State 3	0	1	1	0	1	0
State 4	1	0	0	0	1	1
State 5	1	0	1	1	0	0
State 6	1	1	0	1	0	1
State 7	1	1	1	1	1	0

The state transition table for the counter can then be created (Table 11.6). For the next state logic, the Q output for each flip-flop is actually the D input for each flip-flop in the current state. In this view of the state transition table, the current Q outputs and the current D inputs (next state Q outputs) are defined.

The Boolean logic expression can be created for each of the D inputs so that the counter of the form shown in Figure 11.16 is created. The next state logic for each flip-flop uses a combination of the Q and NOT-Q outputs from each flip-flop.

Manipulation of the Boolean logic expression, use of truth tables, and Karnaugh maps allows the designer to create a Boolean logic expression of a required form.

An example of the logic for each flip-flop D input developed using the Karnaugh map is shown in Figure 11.21; Figure 11.22 shows a schematic for the counter. Each D-type flip-flop has only a Q output, and the NOT-Q output is created using a discrete inverter. Additionally, each D-type flip-flop has an asynchronous active high reset that must be initially inverted so that the design reset input sees an asynchronous active low reset circuit.

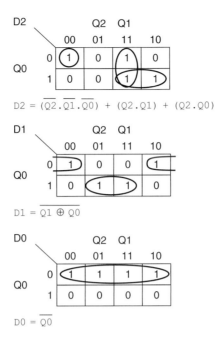

$$D2 = (\overline{Q2}.\overline{Q1}.\overline{Q0}) + (Q2.Q1) + (Q2.Q0)$$

$$D1 = \overline{Q1 \oplus Q0}$$

$$D0 = \overline{Q0}$$

Figure 11.21: Three-bit down-counter D input Boolean expressions

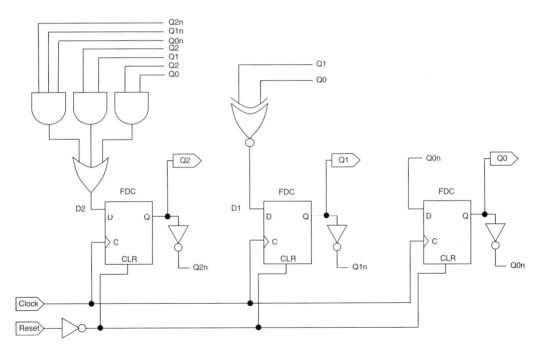

Figure 11.22: Xilinx ISE™ schematic for three-bit down-counter

Example 11.3   Divide-by-5 Circuit

Consider a circuit that receives a clock signal and produces a single output pulse on every fifth clock input pulse. This simple divide-by-5 circuit, shown in Figure 11.23, can be used in a clock divider circuit.

To create this output signal (Divided_Clock), a counter with five count states (0, 1, 2, 3, 4) is created and the output decoded using combinational logic so that on state 4 of the count, the output is a logic 1 only. This ensures that when the counter is reset (either at power on or by an external circuit), the output will be a logic 0. This arrangement is shown in Figure 11.24.

The design process begins by creating the state transition diagram (Figure 11.25) and the state transition table. The counter is designed to reset (into state 0) to a count value of 000_2 (0_{10}). When the reset is removed (i.e., becomes a logic 1), then the counter will count through the sequence 0, 1, 2, 3, 4, 0, etc. When the counter output reaches 100_2, it will automatically wrap around back to 000_2.

Each state in the counter is encoded by the Q outputs of the D-type flip-flops, as shown in Table 11.7.

The state transition table for the counter can then be created (Table 11.8). For the next state logic, the Q output for each flip-flop in the next state is actually the D input for

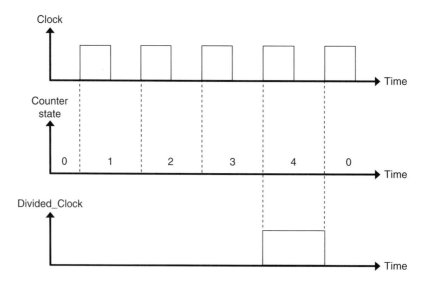

Figure 11.23: Divide-by-5 circuit I/O

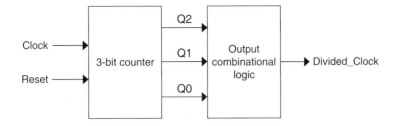

Figure 11.24: Divide-by-5 circuit

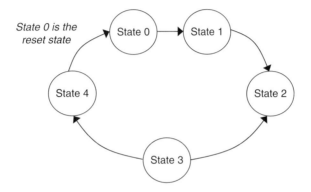

Figure 11.25: Divide-by-5 circuit state transition diagram

each flip-flop in the current state. In this view of the state transition table, the current Q outputs and the current D inputs (next state Q outputs) are defined. The unused states are also shown and are set so that if they are entered, the next state will be state 0.

A Boolean logic expression can be created for each of the D inputs so that a counter of the form shown in Figure 11.16 is created. Here, the next state logic for each flip-flop uses a combination of the Q and NOT-Q outputs from each flip-flop. Manipulation of the Boolean logic expression, the use of truth tables, and Karnaugh maps allow the designer to create a Boolean logic expression of a required form.

An example of the logic for each flip-flop D input developed using the Karnaugh map is shown in Figure 11.26. Figure 11.27 shows the schematic developed for the counter, in which each D-type flip-flop only has a Q output, and the NOT-Q output is created using a discrete inverter. Additionally, each D-type flip-flop has an asynchronous active high reset that must be initially inverted so that the design reset input sees an asynchronous active low reset circuit.

Table 11.7: Divide-by-5 circuit state encoding

State	Q2	Q1	Q0
State 0	0	0	0
State 1	0	0	1
State 2	0	1	0
State 3	0	1	1
State 4	1	0	0
Unused states			
State 5	1	0	1
State 6	1	1	0
State 7	1	1	1

Table 11.8: Divide-by-5 circuit state transition table

Present state				Next state		
	Current Q outputs			Current D inputs		
State name	Q2	Q1	Q0	D2	D1	D0
State 0	0	0	0	0	0	1
State 1	0	0	1	0	1	0
State 2	0	1	0	0	1	1
State 3	0	1	1	1	0	0
State 4	1	0	0	0	0	0
Unused states						
State 5	1	0	1	0	0	0
State 6	1	1	0	0	0	0
State 7	1	1	1	0	0	0

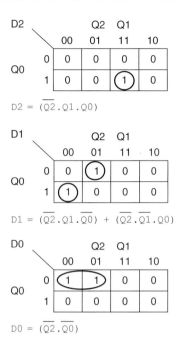

D2 = $(\overline{Q2.Q1.Q0})$

D1 = $(\overline{Q2}.Q1.\overline{Q0}) + (\overline{Q2}.\overline{Q1}.Q0)$

D0 = $(\overline{Q2}.\overline{Q0})$

Figure 11.26: Divide-by-5 D input Boolean expressions

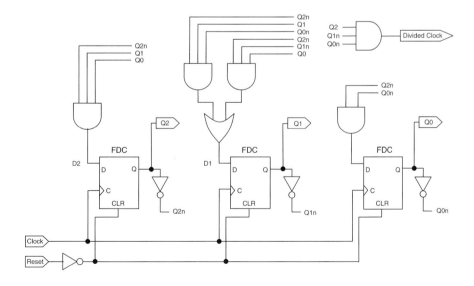

Figure 11.27: Xilinx ISE™ schematic for the divide-by-5 circuit

Table 11.9: Three divide-by-5 circuit output logic decodings

State	Q2	Q1	Q0	Divided_Clock
State 0	0	0	0	0
State 1	0	0	1	0
State 2	0	1	0	0
State 3	0	1	1	0
State 4	1	0	0	1
Unused states				
State 5	1	0	1	0
State 6	1	1	0	0
State 7	1	1	1	0

The output combinational logic is provided in the truth table shown in Table 11.9. The output is a logic 1 only when the counter is in state 4.

The Boolean logic expression for the Divided_Clock output is given as:

$$Divided\_Clock = \left(Q2.\overline{Q1}.\overline{Q0} \right)$$

11.5   State Machine Design

The sequential logic circuit is designed either to react to an input, called a *state machine* in this text, or to be autonomous, in which no inputs control circuit operation. Figure 11.28 shows the basic structure of the state machine.

In the following discussions:

- positive edge triggered, asynchronous active low or high D-type flip-flops will be used;

- all flip-flops will have a common reset input; and

- all flip-flops will have a common clock input.

In addition, the output from the state machine either can be taken directly from the Q outputs of each flip-flop or can be decoded using output combinational logic to form specific outputs for specific states of the counter.

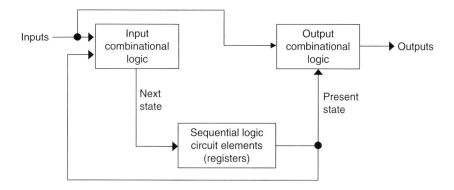

Figure 11.28: State machine structure

The state machine uses flip-flops, so for *n*-flip-flops, there are 2^n possible combinations of output. A state machine might use all possible states or might use only a subset of the possible states. When a sub-set is used, the state machine should be designed so that in normal operation, it will not enter the unused states. However, it is good practice to design the circuit so that if it did enter one of the unused states, it will have a known operation. For example, if an unused state is entered, the next state would always be the reset state for the state machine.

The state machine will be based on either a *Moore* machine or *Mealy* machine. In the Moore machine, the outputs will be a function of the present state only. As such, the outputs will be valid while the state machine is within this state and will not be valid during state transitions. In the Mealy machine, the output is a function of the present state and current inputs. As such, the output of the Mealy machine will change immediately whenever there is a change on the input while the output of the Moore machine would be synchronized to the clock.

Example 11.4 Traffic Light Sequencer

Consider a state machine design to control a set of traffic lights that moves from green to amber to red and back to green whenever a person pushes a button. This is shown in Figure 11.29. There are three inputs—clock, reset, and change—and three outputs—red, amber, and green.

The light begins on green (when the circuit is reset) and stays in the green state when change is a logic 0. When change becomes a 1 (for a duration of 1 clock cycle), the lights will change according to the following sequence:

$$\text{Green} \rightarrow \text{Amber} \rightarrow \text{Red} \rightarrow \text{Red\_Amber} \rightarrow \text{Green}$$

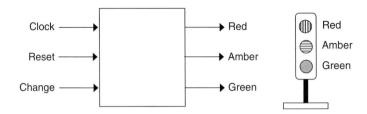

Figure 11.29: Three-bit straight binary up-counter

Table 11.10: Traffic light sequence

State	Green	Amber	Red
Green	ON	OFF	OFF
Amber	OFF	ON	OFF
Red	OFF	OFF	ON
Red_Amber	ON	ON	OFF

The four states and their corresponding outputs (ON = logic 1, OFF = logic 0) are defined in Table 11.10.

The state machine is designed so that when a change input is detected, the lights will change from green to red and back to green. It will then wait for another change input to be detected. During the light changes, the value of change is considered a *Don't care* condition (i.e., it could be a logic 0 or 1).

The design process begins by creating the state transition diagram (Figure 11.30) and the state transition table. There are four distinct states, so two D-type flip-flops are used (where n = 2, giving $2^n = 4$ possible states). The state machine is designed to reset (i.e., when reset is a logic 0) to a count of 00_2 (0_{10}), which is the green state. When the reset is removed (i.e., when reset becomes a logic 1), the state will count through the sequence Green, Amber, Red, Red_Amber, Green, etc. when the Change button is pushed, and when the state machine output reaches 11_2 (state Red_Amber), it will automatically wrap around back to Green. State machine changes are summarized below: ate will count through the sequence Green, Amber, Red, Red_Amber, Green, etc, when the *Change* button is pushed, and when the state machine output reaches 11_2

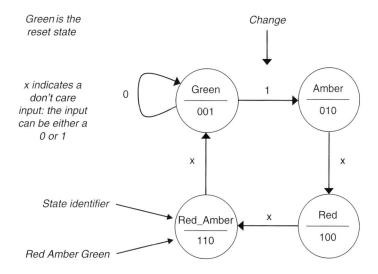

Figure 11.30: Traffic light controller circuit state transition diagram (Moore machine)

(state Red_Amber), it will automatically wrap around back to Green. State machine changes are summarized below:

- State is green and input (change) is a logic 0: state remains green.

- State is green and input (change) is a logic 1: state changes to amber.

- State is amber and input (change) is a logic 0 or 1: state changes to red.

- State is red and input (change) is a logic 0 or 1: state changes to red_amber.

- State is red_amber and input (change) is a logic 0 or 1: state changes to green.

Each state in the state machine will be encoded by the Q outputs of the D-type flip-flops, as shown in Table 11.11.

The state transition table for the counter can then be created, as shown in Table 11.12. For the next state logic, the Q output for each flip-flop in the next state is actually the D input for each flip-flop in the current state. In this view of the state transition table, the current Q outputs and the current D inputs (next state Q outputs) are defined. The change input is included in the state transition table, and the state machine moves into one of two possible next states.

Table 11.11: Divide-by-5 circuit state encoding

State	Q1	Q0
Green	0	0
Amber	0	1
Red	1	0
Red_Amber	1	1

Table 11.12: Traffic light controller state transit

			Change = 0		Change = 1	
Present state			Next state		Next state	
State name	Current Q outputs		Current D inputs		Current D inputs	
	Q1	Q0	D1	D0	D1	D0
Green	0	0	0	0	0	1
Amber	0	1	1	0	1	0
Red	1	0	1	1	1	1
Red_Amber	1	1	0	0	0	0

The Boolean logic expression can be created for each D input so that a state machine like that shown in Figure 11.28 is created. The next state logic for each flip-flop uses a combination of the Q and NOT-Q outputs from each flip-flop along with the change input. Manipulation of the Boolean logic expression, the use of truth tables, and Karnaugh maps allow the designer to create the required Boolean logic expression.

The logic for each flip-flop D input can be developed using the truth table. As there is only one D input to each flip-flop, but two possible input conditions (depending on the value of change), Boolean logic expressions for each possible input are created and the results ORed together to determine the D input of each flip-flop. This idea is shown in Figure 11.31. The resulting Boolean logic expression should then be minimized.

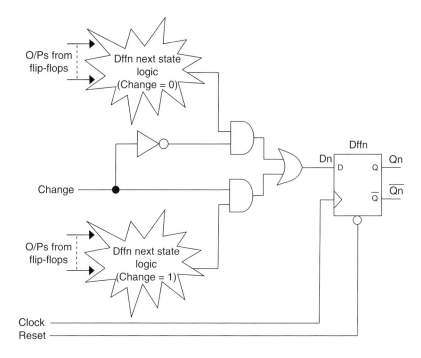

Figure 11.31: ORing the logic expressions to form the flip-flop D input

An example for the Boolean logic expressions for each of the flip-flops is as follows:

$$D1 = (Q1 \oplus Q0)$$
$$D0 = \overline{\text{Change}}.(Q1.\overline{Q0}) + (\text{Change}.\overline{Q0})$$

This shows that the D1 input is actually independent of the change input logic value. The output combinational logic is provided in the truth table shown in Table 11.13.

Table 11.13: Traffic light controller output logic decoding

State	Q1	Q0	Red	Amber	Green
Green	0	0	0	0	1
Amber	0	1	0	1	0
Red	1	0	1	0	0
Red_Amber	1	1	1	1	0

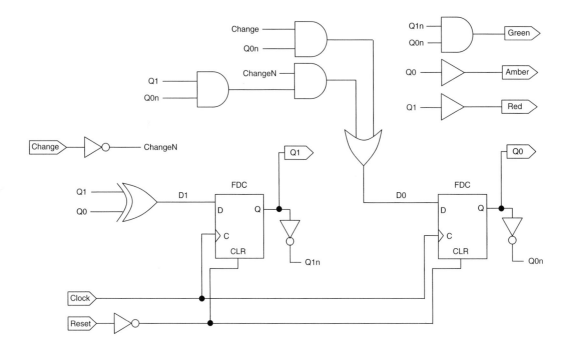

Figure 11.32: Circuit schematic for the traffic light controller

The Boolean logic expressions for the outputs are given as:

$$Green = (\overline{Q1.Q0})$$
$$Amber = (Q0)$$
$$Red = (Q1)$$

Figure 11.32 shows a schematic developed for the counter in which each D-type flip-flop has only a Q output and the NOT-Q output is created using a discrete inverter. Additionally, each D-type flip-flop has an asynchronous active high reset that must be initially inverted so that the design reset input sees an asynchronous active low reset circuit.

Example 11.5   1001 Sequence Detector

Consider the circuit that is to detect the sequence 1001 on a serial bit-stream data input and produce a logic 1 output when the sequence has been detected, as shown in Figure 11.33. The state machine will have three inputs—one Data_In to be monitored

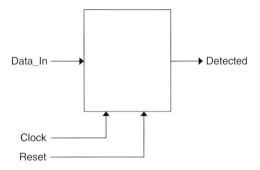

Figure 11.33: 1001 sequence detector

for the sequence and two control inputs, Clock and Reset—and one output, Detected. Such a state machine could be used in a digital combinational lock circuit.

The design process begins by creating the state transition diagram (Figure 11.34) and the state transition table. There are five distinct states, so three D-type flip-flops are used (where n = 3, giving $2^n = 8$ possible states, although only five states are used). The state machine is designed to reset (i.e., when Reset is a logic 0) to a count of 000_2 (0_{10}), which will be the state 0 state. When the reset is removed (i.e., when reset becomes a logic 1), then the state machine becomes active. State machine changes are summarized below:

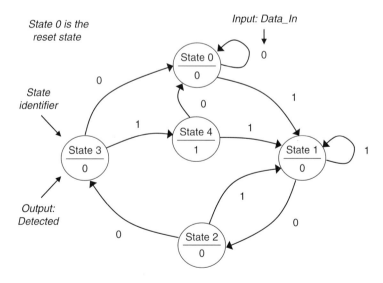

Figure 11.34: 1001 sequence detector state transition diagram (Moore machine)

- At state 0 and input (Data_In) is a logic 0: state remains in state 0.

- At state 0 and input (Data_In) is a logic 1: state changes to state 1.

- At state 1 and input (Data_In) is a logic 0: state changes to state 2.

- At state 1 and input (Data_In) is a logic 1: state remains in state 1.

- At state 2 and input (Data_In) is a logic 0: state changes to state 3.

- At state 2 and input (Data_In) is a logic 1: state changes back to state 1.

- At state 3 and input (Data_In) is a logic 0: state changes back to state 0.

- At state 3 and input (Data_In) is a logic 1: state changes to state 4.

- At state 4 and input (Data_In) is a logic 0: state changes back to state 0.

- At state 4 and input (Data_In) is a logic 1: state changes back to state 1.

Whenever an unused state is encountered, the state machine is designed to enter state 0 on the next clock rising edge.

Each state in the counter is encoded by the Q outputs of the D-type flip-flops, as shown in Table 11.14.

Table 11.14: 1001 sequence detector state encoding

State	Q2	Q1	Q0
State 0	0	0	0
State 1	0	0	1
State 2	0	1	0
State 3	0	1	1
State 4	1	0	0
Unused states			
State 5	1	0	1
State 6	1	1	0
State 7	1	1	1

Table 11.15: 1001 sequence detector state transition table

Present state				Data_In = 0			Data_In = 1		
				Next state			Next state		
	Current Q outputs			Current D inputs			Current D inputs		
State name	Q2	Q1	Q0	D2	D1	D0	D2	D1	D0
State 0	0	0	0	0	0	0	0	0	1
State 1	0	0	1	0	1	0	0	0	1
State 2	0	1	0	0	1	1	0	0	1
State 3	0	1	1	0	0	0	1	0	0
State 4	1	0	0	0	0	0	0	0	1
Unused states									
State 5	1	0	1	0	0	0	0	0	0
State 6	1	1	0	0	0	0	0	0	0
State 7	1	1	1	0	0	0	0	0	0

The state transition table for the counter can then be created, as shown in Table 11.15. For the next state logic, the Q output for each flip-flop in the next state is actually the D input for each flip-flop in the current state. In this view of the state transition table, the current Q outputs and the current D inputs (next state Q outputs) are defined. The change input is included in the state transition table, and the state machine can move into one of two possible next states.

An example for the Boolean logic expressions for each of the flip-flops is as follows:

$$D2 = \text{Data}_{\text{In}}.(\overline{Q2}.Q1.Q0)$$
$$D1 = \overline{\text{Data}_{\text{In}}}.\overline{Q2}.(Q1 \oplus Q0)$$
$$D0 = \overline{\text{Data}_{\text{In}}}.(\overline{Q2}.Q1.Q0) + \text{Data}_{\text{In}}.\left((\overline{Q2}.\overline{Q1}) + (Q1.\overline{Q0})\right)$$

The output combinational logic is provided in the truth table shown in Table 11.16.

The Boolean logic expressions for the output is given as:

$$\text{Detected} = (Q2.\overline{Q1}.\overline{Q0})$$

Table 11.16: 1001 sequence detector output logic decoding

State	Q2	Q1	Q0	Detected
State 0	0	0	0	0
State 1	0	0	1	0
State 2	0	1	0	0
State 3	0	1	1	0
State 4	1	0	0	1
Unused states				
State 5	1	0	1	0
State 6	1	1	0	0
State 7	1	1	1	0

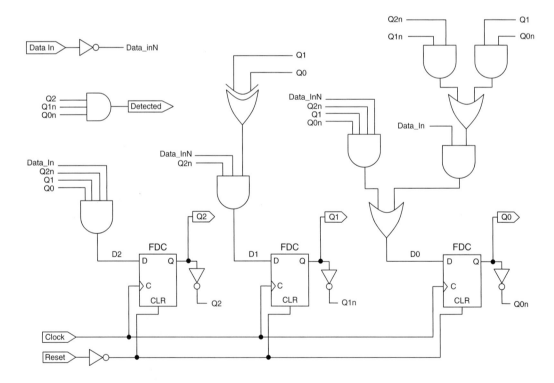

Figure 11.35: Circuit schematic for the 1001 sequence detector

Figure 11.35 shows a schematic developed for the counter in which each D-type flip-flop has only a Q output and the NOT-Q output is created using a discrete inverter. Additionally, each D-type flip-flop has an asynchronous active high reset that must be initially inverted so that the design reset input sees an asynchronous active low reset circuit.

11.6   Moore Versus Mealy State Machines

Sequential logic circuit designs create counters and state machines. The state machines are based on either the Moore machine or the Mealy machine, shown in Figure 11.36.

The diagrams shown in Figure 11.36 are a modification of the basic structure identified in Figure 11.1 by separating the combinational logic block into two blocks, one to create the next state logic (inputs to the state register that store the state of the circuit) and the output logic:

- In the *Moore machine*, the outputs are a function only of the present state only.

- In the *Mealy machine*, the outputs are a function of the present state and the current inputs.

The types of circuits considered here will be synchronous circuits in that activity occurs under the control of a clock control input, these are synchronous circuits. A number of possible circuits can be formed to produce the required circuit functionality.

11.7   Shift Registers

The D-type flip-flops can be connected so that the Q output of one flip-flop is connected to the D input of the next flip-flop, as shown in Figure 11.37. With a single input (Data_In), a serial bit-stream can be applied to the circuit. Whenever a clock edge occurs, the D input of a flip-flop is stored and presented as the Q output of that flip-flop.

If there are *n*-flip-flops in the circuit, the serial bitstream applied at the input appears at the output (as Data_Out) after *n* clock cycles. The serial bitstream input is available as a

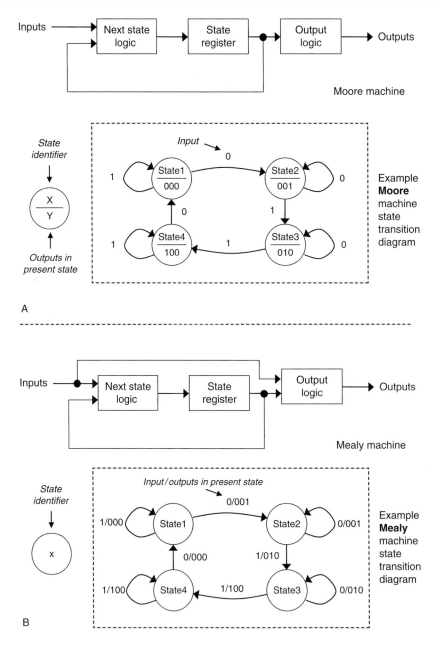

Figure 11.36: Moore and Mealy state machines

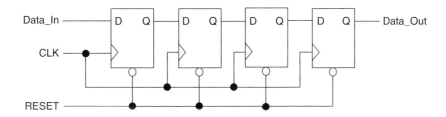

Figure 11.37: Four-bit shift register (serial in, serial out)

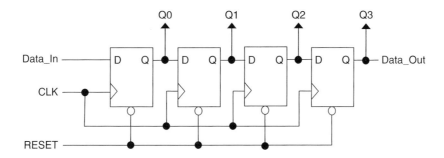

Figure 11.38: Four-bit shift register (serial in, parallel and serial out)

serial bitstream output, which is referred to as a serial-in, serial-out shift register because the input is shifted by the clock signal to become the output.

Modifications to this circuit allow parallel input to the shift register (a parallel data load, rather than a serial data load) and parallel output. A shift register that provides for serial input along with serial and parallel output is shown in Figure 11.38.

11.8   Digital Scan Path

The shift register is used to support circuit and system testing. This arrangement forms a scan path [Ref. 11.1]. Scan path testing is the main method to provide access for internal node controllability and observability of digital sequential logic circuits, where:

- *controllability* is the ability to control specific parts of a design to set particular logic values at specific points.

- *observability* is the ability to observe the response of a circuit to a particular circuit stimulus.

In scan path, the circuit is designed for two modes of operation:

- *normal operating mode*, in which the circuit is running according to its required end-user function; and

- *scan test mode*, in which logic values are serially clocked into circuit flip-flop elements from an external signal source, and the results are serially clocked out for external monitoring.

The incorporation of a scan path into a design requires additional inputs and outputs specifically used for the test procedure. These inputs and outputs, and the scan test circuitry, are not used by the end user.

Scan Test Inputs:

- *Scan data input* (SDI) scans the data to clock serially into the circuit.

- *Scan enable* (SE) enables the scan path mode.

Scan Test Output:

- *Scan data out* (SDO) scans the data (results) that are serially clocked out of the scan path for external monitoring.

Using the basic circuit arrangement shown in Figure 11.38, the D-type flip-flops within the sequential logic circuit are put into a serial-in, serial-out shift register as shown in Figure 11.39, showing SDI and SDO. The parallel outputs (Q0, Q1, Q2, and Q3) form inputs to the combinational logic within the design.

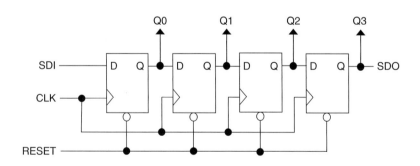

Figure 11.39: Scan test shift register

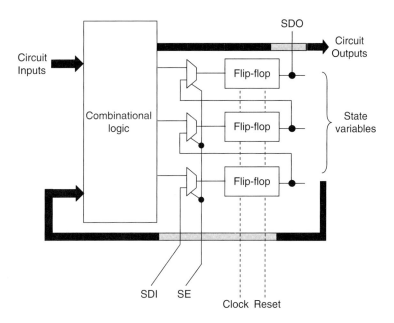

Figure 11.40: Scan path insertion using D-type flip-flops and multiplexers

A typical scan path test arrangement is shown in Figure 11.40, including the combinational logic block and D-type flip-flops. Each flip-flop has a common clock and reset input. Between the flip-flop D input and the combinational logic (the next state logic), a two-input multiplexer is inserted. The first data input to the multiplexer is the output from the next state logic. The second data input comes from the Q output of a flip-flop. This allows either of these signals to be applied to the D input of the flip-flop using the select input on the multiplexer (connected to SE).

In normal operating mode, the next state logic is connected to the flip-flop D input. In scan test mode, the Q output from a previous flip-flop is connected to the flip-flop D input. This isolates the flip-flop from the next state logic, and the flip-flops form a shift register of the form shown in Figure 11.39. Test data can therefore be scanned in (using the SDI input), and test results can be scanned out (using the SDO output). An example operation of this scan path follows:

1. The circuit is put into scan test mode (by control of the SE). The test data is serially scanned into the design to set the flip-flop Q outputs to known values (i.e., to put the circuit into a known, initial state) by applying the test data to the SDI pin.

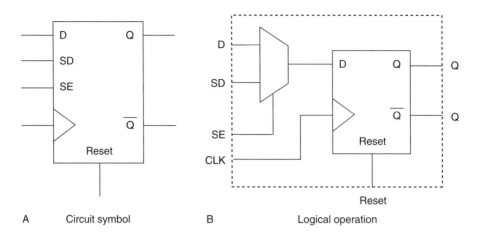

Figure 11.41: Scan D-type flip-flop

2. The circuit is put back into its normal operating mode and operated for a set number of clock cycles.

3. The circuit is again put into scan test mode. The results of the test are stored on the Q outputs of the flip-flops and serially scanned out and monitored on the SDO pin.

4. The monitored values are compared with the expected values, and the circuit is then checked to see if it has passed (expected values received) or failed (the circuit output is not as expected) the test.

The arrangement shown in Figure 11.40 uses a discrete multiplexer and D-type flip-flop to create the scan path. In many circuits, these functions are combined into a single scan D-type flip-flop circuit element, as shown in Figure 11.41. This has the same logic functionality as a discrete flip-flop and multiplexer arrangement, but is optimized for size and speed of operation. It has two data inputs (D, normal data, and SD, scan data input) and a scan enable (SE) control input to select between normal and scan test modes, in addition to the clock, reset (and/or set) inputs and Q/NOT-Q outputs.

Reference

[11.1] Grout I. A., *Integrated Circuit Test Engineering Modern Techniques,* Springer, 2006, ISBN 1-84628-023-0.

CHAPTER 12

Memory

Ian Grout

12.1 Introduction

Memory is used to store, provide access to, and allow modification of data and program code for use within a processor-based electronic circuit or system. The two basic types of memory are ROM (read-only memory, and RAM (random access memory). Memory can be considered for use for one of the following three data or program storage purposes:

1. *Permanent storage* for values that are normally only read within the application and can be changed (if at all) only by removing the memory from the application and reprogramming or replacing it.

2. *Semi-permanent storage* for values that can be read only within the application (as with permanent storage). However, stored values can be modified by reprogramming while the memory remains in the circuit.

3. *Temporary storage* for values needed only for temporary use and requiring fast access or modification (such as data and program code within a computer system that can be removed when no longer needed).

These memories are typically used within a computer architecture of the form shown in Figure 12.1. Here, the ROM and RAM are connected to the other computer functional blocks:

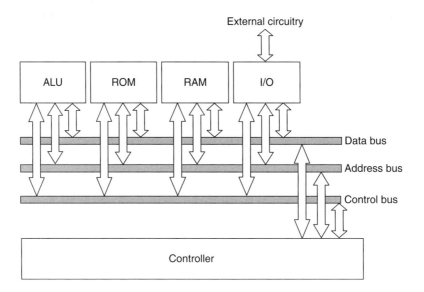

Figure 12.1: Basic computer architecture

- *ALU*, arithmetic and logic unit;

- *I/O*, input/output to external circuitry; and

- *controller* to provide the necessary timing for the circuitry.

Each of the functional blocks is connected to a common set of data, address, and control lines required to access and manipulate the digital data at specific points in time. Also needed is a power supply for each circuit to implement the functional blocks within the computer.

The key drivers for memory development are driven by the end-user, who is constantly demanding more functionality at a lower cost. Hence, the key drivers for memory development are:

1. increased capacity—the amount of data that can be stored within a single memory circuit;

2. increased operating speed—to reduce time to write data to and read data from the memory; and

3. lower cost.

Memory bandwidth, the amount of information that can be transferred to and from memory per unit time, is an increasingly important aspect to memory design and choice for use. This is driven by the increase in processor performance and demanding applications such as multimedia and communications.

12.2 Random Access Memory

RAM (also referred to as read-write memory, RWM) is considered volatile storage because its contents are lost when the power is removed. There are two main types of RAM, static RAM (SRAM) and dynamic RAM (DRAM). In addition, ferromagnetic RAM (FRAM) is also available.

A view of SRAM connections where the SRAM is provided in a dual in-line package (DIP) is shown in Figure 12.2. Here, the SRAM consists of the following connections:

- *Address lines* define the memory location to be selected for reading or writing.

- *Input/output data lines* define the data to write to or read from memory.

- *Write enable* (WE) is a control input that selects between the memory read and write operations (usually active low).

- *Output enable* (OE) is a control input that enables the output buffer for reading data from the memory (usually active low).

- *Chip select* (CS) selects the memory (usually active low).

- *Power supply* provides the necessary power to operate the circuit.

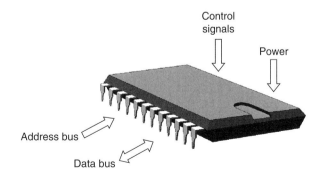

Figure 12.2: SRAM in a DIL package

Where the SRAM is provided as a discrete packaged device, a suitable power supply (V_{DD}/V_{SS}) along with power supply decoupling (capacitors) on the PCB will be needed. Increasingly, SRAMs are provided as macro cells within ICs (such as in the Xilinx® Spartan™-3 FPGA), in which the power supply has already been routed and the SRAM is ready for use.

In some RAM designs, the two *write enable* (WE) and *output enable* (OE) control signals identified above are combined into a single *read/write* (R/W) control signal. This reduces the pin count by one and the logic level of the R/W input will determine if the RAM is written to, or read from.

12.3 Read-Only Memory

ROM is used for holding program code that must be retained when the memory power is removed, so it is considered nonvolatile storage. The code can take one of three forms:

1. Fixed when the memory is fabricated—mask-programmable ROM.

2. Electrically programmed once—PROM, programmable ROM.

3. Electrically programmed multiple times—EPROM (electrically programmable ROM) erased using ultraviolet (UV) light; EEPROM or E^{2}PROM (electrically erasable PROM); and Flash (also electrically erased).

PROM is sometimes considered in the same category of circuit as programmable logic, although in this text, PROM is discussed only in the memory category.

RAM is used for holding data and program code that must be accessed quickly and modified during normal operation. RAM differs from read-only memory (ROM) in that it can be both read from and written to in the normal circuit application. However, flash memory is also referred to as nonvolatile RAM (NVRAM).

A basic ROM design in which ROM is provided in a dual in-line package is shown in Figure 12.3. Here, the ROM consists of the following connections:

-   *Address lines* define the memory location to be selected for reading or writing.

-   *Output data lines* access the data from memory.

-   *Output enable* (OE) is a control input that enables the output buffer for reading data from the memory (usually active low).

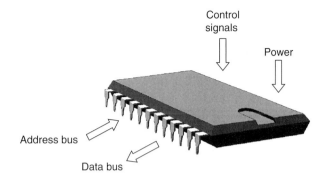

Figure 12.3: Basic ROM in a DIL package

- *Chip select* (CS) selects the memory (usually active low).

- *Power supply provides* the necessary power to operate the circuit.

In this view, the data bus is considered to be unidirectional (i.e., output only). Where the ROM may be electrically programmed, then the data and control line arrangement will be more complex.

CHAPTER 13

Selecting a Design Route

J. Crowe
Barrie Hayes-Gill

13.1 Introduction

This chapter describes the various design routes that can be used to implement a circuit design. The decision regarding which of these design routes to use depends upon the following issues:

- When should the first prototype be ready?

- How many units are needed?

- What are the power requirements?

- What is the budget for the product?

- What are the physical size limitations?

- How complex is the design (gate count, if known)?

- What is the maximum frequency for the design?

- What loads will the system be driving?

- What other components are needed to complete your design?

- What experience have you or your group had to date in the design of digital systems?

These are the questions that must be asked before starting any design. The aim of this chapter is to provide background to the various design routes that are available. Armed with this knowledge, the answers (where possible) to the above questions should allow the reader to decide which route to select or recommend.

13.1.1 Brief Overview of Design Routes

The various design options are illustrated in Figure 13.1. As can be seen, the choice is either to use *standard products* or to enter the world of *application-specific integrated circuits* (ASICs). The "standard product" route is to choose one, or a mixture, of the logic families such as 74HCT, 74LS, 4000 series, etc. On the other hand, an ASIC is simply an IC customized by the designer for a specific application. Various ASIC options exist that can be subdivided into either *field-programmable* or *mask programmable* devices. Field-programmable devices (i.e., ROM, PAL, PLA, GAL, EPLD, and FPGA) are all programmed in the laboratory. However, mask programmable devices must be sent to a manufacturer for at least one mask layer to be implemented. These mask programmable devices may be exclusively digital or analog, or alternatively what is known as a *mixed* ASIC, which will contain both.

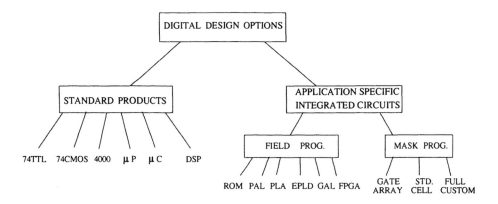

Figure 13.1: Design options

The mask programmable devices can be further subdivided into *full custom, standard cell* and *gate array*. With full custom design, the designer has the option of designing the whole chip, down to the transistor level, exactly as required. Standard cell design again presents the designer with a clean slice of silicon but provides standard cells (e.g., gates, flip-flops, counters, op-amps, etc.) in a software library. These can be automatically positioned and connected on the chip as required (known as *place and route*). Both of these levels of design complexity are used for digital and analog design, and are characterized by long development times and high prototyping costs. The third and lowest level in terms of complexity is the gate array. With the gate array, the designer is presented with a "sea" of universal logic gates and is required only to indicate how these gates are to be connected, which then defines the circuit function. This approach offers a less complex, and hence cheaper, design route than standard cell and full custom.

Until the late 1980s, the cheapest route to a digital ASIC was via the use of a mask programmable gate array. These devices are still widely used, but since the late 1980s, have had to face strong competition from field-programmable gate arrays (FPGAs) where the interconnection and functionality are dictated by electrically programmable links, and hence appear in the field-programmable devices section.

With regard to the previous ten questions, the overriding issue is usually when the first prototype should be ready. ASICs require computer-aided design (CAD) tools of differing complexities. Designs that use such tools provide elegant solutions, but can be very time consuming especially if your team has no experience in this field. However, designs that use "standard products" are quick to realize but can be bulky and expensive when high volumes are required.

With the exception of microcontrollers/processors and DSPs, this chapter will describe the design options in Figure 13.1 in more detail. However, it should be noted that as... you move from left to right across this diagram, each option becomes more complex to implement resulting in a longer design time and greater expenditure.

13.2   Discrete Implementation

The 74 series offers a whole range of devices at various levels of integration. These levels of integration are defined as:

- SSI – Small-scale integration (less than 100 transistors per chip);

- MSI – Medium-scale integration (100–1,000 transistors per chip);

- LSI – Large-scale integration (1,000–10,000 transistors per chip);

- VLSI – Very large-scale integration (greater than 10,000 transistors per chip).

The VLSI devices are mainly microcontrollers and microprocessors which are outside the scope of this book.

Designs using these standard parts are quick to realize and relatively easy to debug. However, they are bulky and expensive when high volumes are required. The various functions available allow all sorts of digital systems to be implemented with minimal overheads and tooling. For expediency these designs can be *ad hoc* and incorporate poor digital design techniques. We shall look at some of these pitfalls and suggest alternative safe design practices.

One such standard product is the 74HCT139 which consists of two 2-to-4 decoders in a single IC package. A logic diagram for this IC is shown in Figure 13.2. A decoder can be used in memories for addressing purposes where only one output goes high for each address applied. Such a device has many other uses. However, one must be careful with this type of circuit since any of the decoder outputs can produce spurious signals called *static hazards*. These static hazards are called *spikes* and *glitches*.

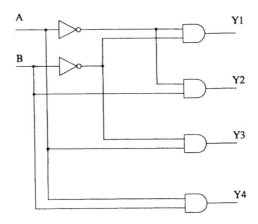

Figure 13.2: 74HCT139: two-to-four decoder

13.2.1   *Spikes and Glitches*

Consider the case of output Y3 in Figure 13.2. A timing diagram is shown in Figure 13.3 for this output for various combinations of *A* and *B*. At first $AB = 00$ and so $Y3 = 0$. Next, $AB = 10$ and Y3 goes high. With *AB* returning to 00 the output goes low again. All seems satisfactory so far but if $AB = 11$, then due to the propagation delay of the inverter the output will go high for a short time equal to the inverter propagation delay. As we shall see, although this spike is only a few nanoseconds in duration, it is sufficiently long enough to create havoc when driving clock lines and may inadvertently clock a flip-flop. This phenomenon is not limited to decoders. All combinational circuits will produce these spikes or glitches.

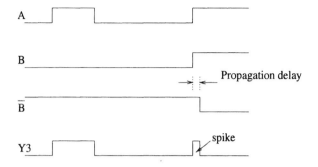

Figure 13.3: Spike generation on output Y3 of the 2-to-4 decoder

To appreciate the problem when driving clock lines, consider a circuit counting the number of times a four-bit counter produces the state 1001. A possible design using 74 series logic is shown in Figure 13.4(A). This consists of a 4-to-16 decoder (74HC154)[1] being used to detect the state 1001 from a four-bit counter (74HC161). (For clarity the four-bit counter output connected to the four inputs of the decoder is represented as a *data bus* having more than one line. The number of signals in the line is indicated alongside the bus.) The 10th output line of the decoder is used to clock a 12-bit counter (74HC4040). However, although this will detect the state 1001 at the required time, it will also detect it at other times due to the differing propagation delays in the 4-to-16 decoder. These spikes and glitches will trigger the larger counter and result in a false count. There are two solutions to this: an elegant one and one that some undergraduates fall mercy to! The latter method, illustrated in Figure 13.4(B), is to use an *RC* network (connected as an integrator or a low-pass filter) and a Schmitt trigger which together remove the spike or glitch. The values of *R* and *C* are chosen so as to filter out this fast transient—usually *RC* is set to be 10 times the glitch or spike pulse width. Due to this long time constant, the signal presented at the input to the Schmitt is now only a fraction of the magnitude of the original spike. To remove this signal completely it is passed through a Schmitt. This device has a voltage transfer characteristic which has two switching points. When the input is rising (from 0V) the Schmitt switches at typically $0.66\,V_{dd}$. However, when the input is falling (from V_{dd}) the Schmitt now switches at $0.33\,V_{dd}$. Hence any signal that does not deviate by more than two-thirds of the supply will be removed. This circuit, although successful, cannot be used in any of the other design options in this chapter since large values of *R* and *C* are not provided on chip. In addition, the provision of extra inputs and outputs

---

[1] This decoder has outputs that are active low; however, for this application we shall assume that the outputs are active high.

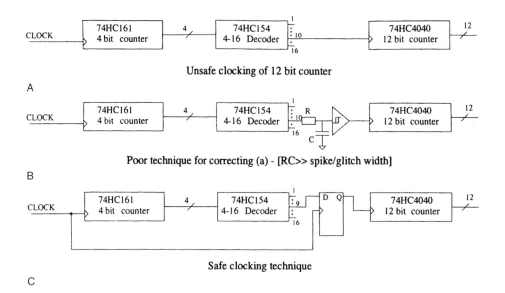

Figure 13.4: Using a decoder as a state detector. (A) Unsafe clocking of 12-bit counter; (B) Poor technique for correcting (A) – [RC>>spike/glitch width]; (C) Safe clocking technique

for these passive components will produce an unnecessarily large chip. The elegant solution, shown in Figure 13.4(C), is to detect the *previous* state with the decoder and present this to the D input of a D-type. The *clean* output of the flip-flop is then used to drive the 12-bit counter.

To summarize, an important rule for all digital designers is that clock inputs must not be driven from *any* combinational circuit, even a single two-input logic gate. This can be stated quite succinctly as *no gated clocks*. In fact the same is true for reset and set lines since these will also respond to spikes and glitches, thus causing spurious resetting of the circuit.

13.2.2   Monostables

Another tempting circuit much frowned upon by the purist is the monostable. The monostable or "*one shot*" produces a pulse of variable width in response to either a *high-to-low* or a *low-to-high* transition at the input. The output pulse width is set via an external resistor and capacitor.

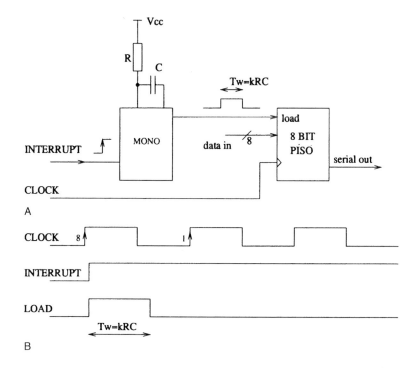

Figure 13.5: Use of a monostable to produce a short pulse

One application of the use of a monostable is shown in Figure 13.5(A). Suppose that we require an eight-bit parallel-in, serial-out (PISO) shift register to be loaded with an eight-bit data word when a line called *interrupt* goes high. An active high *load* signal must be produced that will load the eight-bit data. This *load* signal must be returned low before the *next* rising clock edge so that serial data can continue to be clocked out. It should be noted that in this case the *interrupt* line is assumed to be synchronized with the clock. By adjusting the value of R and C the required parallel *load* pulse width (kRC, where k is a constant) is set to be no longer than the clock pulse width less the *load* to *clock* set-up time. The corresponding timing diagram is shown in Figure 13.5(B).

However, circuits that use monostables exhibit several limitations. The first is that it is necessary to use an external R and C, which will require a redesign when migrating to an ASIC. Other problems related to the analog nature of the device are: the pulse width varies with temperature, V_{cc} and from device to device; poor noise margin thus generating spurious pulses; oscillatory signal edges are generated for narrow pulse widths (less than approximately 30 ns); and long pulses require large capacitors, which are bulky.

An alternative to the circuit in Figure 13.5(A) is to use the circuit in Figure 13.6(A), which uses the *reset technique* with a purely digital synchronous approach. The resulting timing diagram for this circuit is shown in Figure 13.6(B). The circuit operates by using a clock frequency of twice the PISO register clock (*2-clock*). When *interrupt* goes from low to high, Q 1 (i.e., *load*) will go high. This will load in the parallel data. At the next rising *2-clock* edge Q2 goes high (as its input, Q1, is now high) and clears or resets the *load* line. Because of the higher clock frequency used, this all occurs within half a *clock* cycle. A divide-by-two counter is used to divide *2-clock* down to *clock* so that the new data loaded into the PISO can

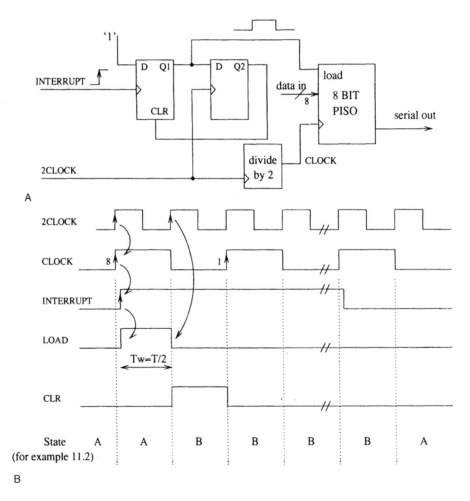

Figure 13.6: Alternative circuit to the monostable circuit in Figure 13.5

be serially shifted out on the immediately following rising edge of *clock*. *Load* will not go high again until another low to high transition on *interrupt* occurs.

The following example shows how pulses of a longer time duration can be produced.

Example 13.1

Consider the circuit in Figure 13.7. What pulse width is produced at the Q output of the D-type (74HC74) device? Assume that both "CLR" and "RESET" are active high.

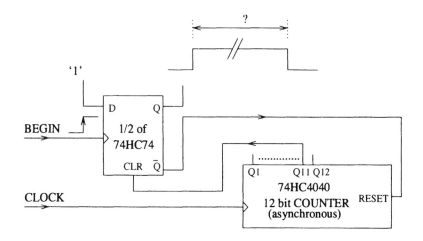

Figure 13.7: Circuit to produce a controlled long pulse width

Solution

When a *BEGIN* low-to-high transition occurs, the Q output goes high, which releases the counter from its reset position. The counter proceeds to count until the Q11 output goes high, at which point the D-type flip-flop is cleared and the Q output goes low again awaiting the arrival of the next *BEGIN* rising edge. The Q output is thus high for 2^{10} clock pulses.

Taking the clear input from any of the other outputs of the counter will produce pulses of varying width. The higher the input clock frequency, the better the resolution of the pulse width.

It should be noted that if *BEGIN* is synchronized with the clock then the rising edge of the output pulse will also be synchronized (albeit delayed by one D-type flip-flop delay). However, the falling edge of the Q output pulse is delayed with respect to the clock. This is because the counter used is an asynchronous or ripple counter. The Q11 output will only go high after the clock signal has passed through 11 flip-flop delays—this could be typically 100–400 ns. This may not cause a problem but is something to be aware of. The solution is to use either a synchronous counter or detect the state before with a 10-input decoder and a D-type as described earlier.

Example 13.2

The circuit of Figure 13.6(A) was designed in an *ad hoc* manner with the reset technique. Using state diagram techniques as discussed earlier produces a circuit that will implement the same timing diagram of Figure 13.6(B).

Solution

The first task is to use the timing diagram of Figure 13.6(B) to produce a state diagram. At the bottom of Figure 13.6(B) are the states *A* and *B* at each rising *2-clock* edge. Remember, that the interrupt (*I*) line is generated by *2-clock* (i.e., synchronized) and thus changes *after* the *2-clock* rising edge. Hence the state diagram, shown in Figure 13.8(A), can be drawn. The corresponding state transition table is shown in Figure 13.8(B) and, since there are only two states, then only one flip-flop is needed. Assigning $A = 0$ and $B = 1$ results in Figure 13.8(C). From this we produce the K-maps for the next state Q^+ and present output *LOAD(L)*. These produce the functions $Q^+ = I$ and $L = \bar{Q}$. The resulting circuit diagram is shown in Figure 13.8(E). It should be remembered that the clock input is the higher frequency *2-clock*.

13.2.3   CR Pulse Generator

The practice of using monostables has already been frowned upon and safe alternative circuit techniques have been suggested. However, monostables are tempting, quick to use and can still be found in many designs. Another design technique that is also simple and tempting to use but should be avoided is the CR pulse generator or differentiator circuit shown in Figure 13.9(A). The circuit is the opposite of the integrator shown in Figure 13.4. This circuit is used to "massage" a long pulse into a shorter one and so gives the appearance of a *one-shot* reacting at either rising or falling edges. If a 5V pulse is applied to the circuit in Figure 13.9(A) two short pulses are produced, one at the rising edge and one at the falling edge. At the rising edge when the input goes instantaneously from 0V to 5V the output momentarily produces 5V. As the capacitor charges the voltage across the resistor starts to fall as the charging current falls, hence the corresponding rising edge waveform. When the

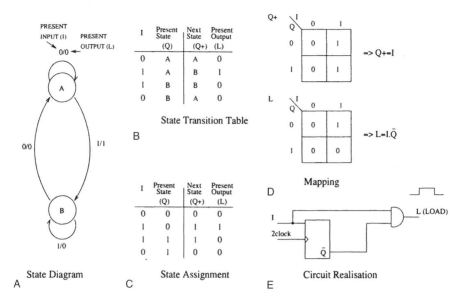

Figure 13.8: Using a state diagram to implement the timing diagram of Figure 13.6 (B); (A) State diagram; (B) State transition table; (C) State assignment; (D) Mapping; (E) Circuit realization

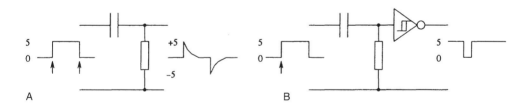

Figure 13.9: Using a CR network to produce a narrow pulse

input changes from 5V to 0V the capacitor cannot change its state instantly and so both plates of the capacitor drop by 5V. Hence, the output momentarily produces −5V. The capacitor then discharges, resulting in the falling edge waveform.

To convert this signal into a digital form the output is fed into a Schmitt trigger and thus produces a short pulse from 5V to 0V whose duration is determined by the value of R and C and the Schmitt switching point. This pulse is only present on the rising edge of the input since the falling edge produces a negative voltage, which the Schmitt does not respond to. However, this circuit should again be avoided as the migration to an ASIC would require a redesign while in addition the negative voltage may in time damage the Schmitt component. Consequently, it is therefore recommended that the pulse shortening techniques described earlier, which use a higher clock frequency, be employed.

13.3 Mask Programmable ASICs

The use of standard products (74 series, etc.) to implement a design becomes inefficient when large volumes are required. Hence, the facility for the independent customer to design their own integrated circuits was provided by IC manufacturers. This required the designer to use either a *gate array, standard cell* or *full custom* approach. In each case the manufacturer uses photomasks (or electron-beam lithography) to fabricate the devices according to the customer's requirements. Therefore, these devices are... collectively named *mask programmable* ASICs.

13.3.1 Safe Design for Mask Programmable ASICs

A limitation of mask programmable ASICs is that, since the layers are etched using these masks, any design errors require a completely new set of masks. This is very expensive and time consuming and hence *safe* design techniques which work first time must be employed. A designer must avoid monostables and *CR/RC* type circuits and be aware that a manufacturing process can vary from run to run and sometimes across a wafer. Consequently, propagation delays vary quite considerably from chip to chip or even across a chip. Hence, the use of gates to provide a delay (see Figure 13.10(A)) is a poor design technique since the value of this delay cannot be guaranteed. Three *poor* ASIC circuit techniques where these delay chains are used are shown in Figs. 13.10 (B)–(D). Essentially the designer must use synchronized signals and a higher clock frequency to generate short predictable pulses.

The use of synchronous techniques is not a panacea for all timing problems. Take for example the master clock in a synchronous system driving several different circuits. The total capacitance being driven by the master clock can be extremely large, thus delaying the clock quite considerably. In order to isolate this large capacitance from the master clock, *buffers* are used leading to each circuit. These are quite simply two CMOS inverters in series. This reduces the capacitance seen directly by the master clock circuit and hence reduces the clock delay to each circuit. However, the input capacitance between the smallest and largest of these circuits may differ by an order of magnitude. Hence, the clock will arrive at different times to each of these circuits and the whole

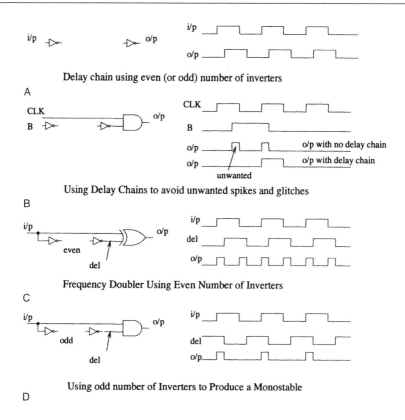

A

Delay chain using even (or odd) number of inverters

B

Using Delay Chains to avoid unwanted spikes and glitches

C

Frequency Doubler Using Even Number of Inverters

D

Using odd number of Inverters to Produce a Monostable

Figure 13.10: Examples of poor ASIC circuit techniques

system will appear asynchronous in nature, A better buffering technique is therefore required. Two improved buffering techniques are shown in Figs 13.11(A) and (B). The first is to use an even number of inverters driving the large load. At first it just looks like our poor delay line shown in Figure 13.10(A). However, each inverter is larger than the previous one by a factor f (i.e., the W/L ratios of the MOS transistors are increased by f at each stage). The load capacitance gradually increases at each stage but the drive strength also increases. The optimum value of f is in fact e or 2.718, but the number of stages required for this case would be quite large. A compromise is to use an increased value of f and a reduced number of stages. Another technique is to use *tree* buffering which consists of several small inverters arranged in a tree structure. This is illustrated in Figure 13.11(B). In this case each inverting buffer is arranged such that it drives the same load. Hence, the relative clock signal delay will be kept to a minimum.

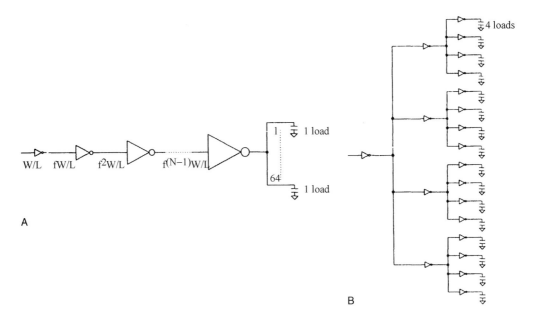

Figure 13.11: Two techniques for buffering the ASIC clock driver from a large capacitance

Example 13.3

One of the small inverters in Figure 13.11(B) is used to drive 64 loads each of 1 pF. Determine the delay of this inverter when driving this load directly and what the delay would be if the tree buffering of Figure 13.11(B) is used. Assume that the inherent delay of a single inverter is 1 ns, its output drive capability is 20 ns/pF and it has an input capacitance of 0.01 pF.

Solution

Unbuffered

$$\text{Delay} = 1 + 20 \times 64 = 1281 \text{ ns}$$

Buffered

$$\text{Delay} = (1 + 20 \times 0.01 \times 4) + (1 + 20 \times 0.01 \times 4) + (1 + 20 \times 4)$$

$$\text{Delay} = 1.8 \text{ ns} + 1.8 \text{ ns} + 81 \text{ ns} = 84.6 \text{ ns}$$

A great saving in delay is achieved at the expense of more gates.

Therefore, these *safe* mask programmable ASIC design techniques can be summarized as follows:

1. no gated clocks or resets;

2. no monostables;

3. no *RC* or *CR* type circuits;

4. use synchronous techniques wherever possible;

5. use a high-frequency clock subdivided down for control;

6. no delay chains; and

7. use clock tree buffering.

In the early days, ASIC designs were breadboarded (i.e., a hardware prototype was produced) using 74 series devices in order to confirm that the design functioned correctly. However, nowadays the designer has available very accurate computer simulators that can be run in conjunction with drawing packages and chip layout. Together these computer programs are called *computer-aided design (CAD)* tools. Since a mask programmable ASIC cannot be modified once fabricated without incurring additional charges, the design cycle relies very heavily upon these CAD tools. The process of fabricating a chip and then finding a design fault is an unforgivable and costly error. We shall look at the various CAD tools employed to guarantee a "right first time" design.

13.3.2   Mask Programmable Gate Arrays

The first mask programmable ASIC that we shall look at is the mask programmable gate array. This device consists of a large array of *unconnected* blocks of transistors called *gates*. All the layers required to form these gates are prefabricated except for the metal interconnect. The IC manufacturer therefore has a "stock-pile" of uncommitted wafers awaiting a metal mask. The user or designer only needs to specify to the manufacturer how these gates are to be connected with the metal layer (i.e., customized).

The basic building block or gate in a CMOS gate array is a versatile cell consisting of four transistors. These blocks of four transistors are repeated many times across the array. Mask programmable gate arrays are characterized in terms of the number of four transistor blocks or *gates* in the array. The gate is called a *versatile cell* since it contains two NMOS and two PMOS transistors which can form simple logic gates such as NOR and NAND.

Two types of arrays exist—*channelled* and *sea of gates*. These are illustrated in Figures 13.12(A) and (B). The channelled array has a routing channel between each row of gates. These routing channels allow metal tracks on a fixed pitch to be used for interconnection across the array. Each channel can contain typically 20 wiring routes. The sea of gates on the other hand does not contain any dedicated routing channels and as a result contains more gates. The routing is implemented *across* each gate at points where no other metal exists. However, with the sea of gates the routing over long distances is more difficult and hence places a limit on the number of gates that can be accessed. This raises the important issue of *utilization*. This is the percentage of gates which the designer can access. As more gates on the array are utilized the routing ability for both array types is reduced. There comes a point where there are not enough routes available to complete the design and because of this manufacturers quote a utilization figure. As you can imagine, the channelled array has a better utilization than the sea of gates. A simple single layer metal channelled array has a utilization of 80% while a double layer metal has a utilization of 95%. Many mask programmable gate array manufacturers use three and four layer metal processes in order to fully utilize the array.

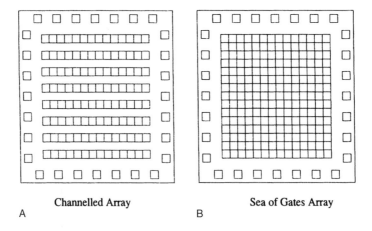

Channelled Array   Sea of Gates Array

A   B

Figure 13.12: Channelled and sea of gates mask programmable gate arrays; (A) Channelled array; (B) Sea of gates array

For any design it is the gate count that is the most important issue. It is therefore useful to know how many gates typical functions consume in CMOS technology. For example a two-input NOR or NAND uses one gate, while a D-type and a JK consume five and eight gates, respectively. Hence, if a design schematic exists, then a quick gate count is always useful to specify what gate array size to use. The selection of an optimum array size is crucial in gate array design since array sizes can vary from 1,000 to 500,000 gates!

The cost of a mask programmable gate array depends upon:

1. number of gates required (or the number of I/Os);

2. number of parts required per year;

3. maximum frequency of operation;

4. number of metal layers.

All mask programmable gate array, manufacturers charge a tooling cost for production of the metal mask(s). This charge is called a *nonrecurring expenditure* or *NRE*. Quotes from three reputable ASIC suppliers for a 2000-gate design, commercialized by the authors, revealed the following prices on a small volume of 1000 parts per year:

- Firm X (2 micron) NRE of £10000 at £4.00 unit cost;

- Firm Y (1.2 micron) NRE of £12000 at £6.80 unit cost;

- Firm Z (3 micron) NRE of £5000 at £2.80 unit cost.

The numbers in brackets indicate the minimum feature size on the chip, which is inversely proportional to the maximum operating frequency. Although the products are not fully comparable, one can see that the costs of mask programmable gate arrays involves the user in large initial charges, hence the importance of accurate CAD simulator tools prior to mask manufacture.

Because the gate array wafers *before* metallization are customer independent, the costs up to this stage are divided among *all* customers. It is only the metallization masks that are customer dependent and so these costs make up the bulk of the NRE. These NRE charges can be greatly reduced by sharing the prototyping costs even further by using a technique called a *multiproject wafer* (MPW). This is a metal mask that contains many different customer designs. The NREs are thus reduced approximately by a factor of N where N is the number of designers sharing that mask. Hence, prototyping costs with mask programmable gate arrays are less of a financial risk when a manufacturer offers an MPW service. The typical prototyping costs for a 2000 gate design, with MPW, are now as low as £1000 for 10 devices.

Of all the mask programmable ASICs the gate array has the fastest fabrication route, since a reduced mask set is required depending upon the number of metal layers used for the interconnect. The typical time to manufacture such a device (referred as the turnaround time) is four weeks.

Example 13.4

How many masks are needed for a double layer metal, mask programmable gate array?

Solution

The answer is not two since it is necessary to insulate one metal layer from the next and provide vias (holes etched in the insulating layers deposited between the first and second layer metal) where connections are needed between layers. Hence, the number is three, i.e., two metal masks and one via mask.

Example 13.5

A schematic for a control circuit consists of four 16 bit D-type based synchronous counters, 20 two-input NAND gates and 24 two-input NOR gates. Estimate the total number of gates required for this design.

Solution

Gate count for each part:

A 16-bit synchronous counter contains 16 D-type bistables plus combinational logic to generate the next state. This logic is typically comparable to the total gate count of the bistable part of the counter. Hence, the total gate count for the counter will be approximately 160 gates (i.e., $16 \times 5 \times 2$). A two-input NAND gate will require four transistors and hence one gate. Thus, 20 will consume 20 gates of the array. Finally a single two-input NOR gate can be made from four transistors. Hence 24 will consume 24 gates of the array.

The total gate count required for this control circuit is $160 + 20 + 24 = 204$ gates.

13.3.2.1   CAD Tools for Mask Programmable Gate Arrays

A mask programmable gate array cannot be modified once it has been fabricated without incurring a second NRE. Consequently a large reliance is placed upon the CAD tools, in particular the simulator, before releasing a design for fabrication. The generic CAD stages involved in the design of both mask- and field-programmable ASICs are illustrated in Figure 13.13. For mask programmable gate arrays this design flow is discussed below:

1. *System description*. The most common way of entering the circuit description is via a drawing package, called *schematic capture*. The user has a library of components to call upon, varying in complexity from a two-input NAND gate through to counters/decoders, PISO/SIPOs and arithmetic logic units. At no stage does the designer see the individual transistors that make up the logic gates. For large circuits (greater than approximately 10,000 gates) the description

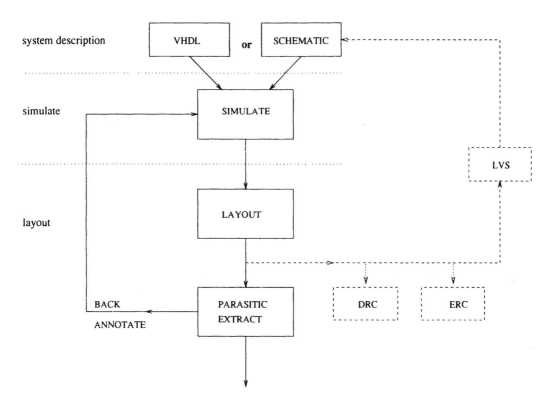

Figure 13.13: Generic CAD stages involved in the design of ASICs

of the circuit using schematic capture becomes rather tedious and error prone. Consequently high-level, textual, programming languages have been developed to describe the system in terms of its behavior. The one language adopted as a standard is that recommended by the USA Department of Defense called *VHDL*. A brief introduction to VHDL is presented later in this chapter.

If the system is described in schematic form it is then converted into a *netlist*. This is a textual description of how the circuit is interconnected and is needed for the simulator. If the system is described in VHDL form, then for the sake of brevity this can be considered as a netlist description already.

2. *Prelayout simulation.* Once the system has been described, the next stage is to simulate the system prior to layout. The components used in the schematic or VHDL are represented as digital (or behavioral) models. A digital simulator, called an *event-driven* simulator, is used to simulate the system by applying input vectors to the system, i.e., a stream of 1s and 0s. This simulator obtains its name since only the gates whose inputs are changing (i.e., an event is occurring) are updated. The outputs then drive other gates and hence a new event is scheduled some time later. In some cases, to simplify the simulation, all gates are assumed to have a 1-ns delay or a *unit delay* and wire delays are set at zero. This is because the chip has not been laid out and therefore no information is available yet about wire delays. This type of simulation is called in some CAD manuals *functional simulation*. It is, however, advisable to simulate with the gate propagation delays which include fan-out loading, thus allowing the simulator to perform more realistic flip-flop timing checks such as: set-up and hold times; minimum clock and reset pulse widths, etc. This will identify, early in the design cycle, poor design techniques such as asynchronous events which violate set-up and hold time, or gated clocks which are revealed as spikes and glitches on clock lines.

3. *Layout.* Next, the chip is *laid out* and this consists of a two-stage process of *place* and *route*. First the gates used to describe the system are placed on to the array and implemented using the versatile four-transistor cell. Optimum placement algorithms are run which aim to reduce the total wire length. The cells are then connected together by using the available routing channels. The I/O positions may be left to the software to decide on the best position so as to assist the place and routing software, or may be specified by the user at the placement stage.

4. *Back annotation of routing delays.* The metal used for the interconnect contains resistance and capacitance and will introduce delays. Hence, these delays need to be added to the original system description, i.e., the schematic or VHDL file. This step is called *back annotation* and these extra delays are referred to as *wiring parasitics.*

5. *Postlayout simulation.* The performance of the original prelayout system will now have changed, which in some cases may result in the delays increasing from 1 ns to 100 ns. The system therefore needs to be resimulated with the parasitic delays included. This final simulation is called *postlayout simulation* and includes the timing delays of both the wiring and logic gates. The simulation is now called a *full timing simulation* since the true delays of the chip are included.[2] Any errors appearing in the simulation at this stage must be corrected by modifying the original schematic or VHDL file and rerunning the layout. This iterative process is characteristic of all ASIC CAD design tools.

An example of a layout-induced timing error is demonstrated with a two-stage shift register in Figure 13.14. The delay element indicated by the dotted box represents additional wire delay on the clock line. If this delay is greater than the propagation delay of the flip-flop, then data is lost. This is because when a shift register shifts data it is assumed that all clocks arrive at the same time at each flip-flop. However, if a clock arrives at the first flip-flop before the second by at least one flip-flop delay, then the data

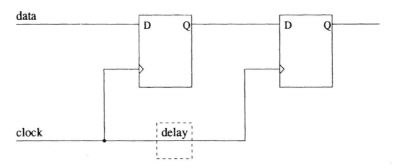

Figure 13.14: Layout delays on clock lines can cause a shift register to malfunction

---

[2] In some environments a separate static timing analyzer is available. This checks all timing delays around bistables with regard to clock and data and confirms that no set-up and hold time violations are present. This removes the time-consuming process of writing a stimulus file for the timing simulator that covers all possible combinations of inputs around all bistables.

at the input to the second flip-flop will change before the arrival of its clock pulse. This data has been overwritten and therefore lost. To avoid this problem occurring, the place and route software allows the designer to influence the layout in several ways. Firstly, the clock line can be given priority (called *seeding*) and it is routed first before all the other routes. It will therefore have the shortest and hence the fastest path. Another technique is to label groups such as shift registers so that they are not broken up during placement. All flip-flops are consequently placed close to each other and hence clock delays are reduced.

When the postlayout simulation has been successfully completed, the designer has to pass an intensive *sign-off* procedure, which needs to be countersigned by the project manager and an engineer at the ASIC manufacturer. The final file that is passed to the manufacturer is in a syntax which is applicable for mask manufacturing machines and allows the metal interconnection layer(s) to be added to the base wafers in order to customize the array.

The CAD tools described here are either supplied by the IC manufacturer or by generic CAD software houses such as Mentor and Cadence. These tools take a design from schematic through to layout. Alternative tools, such as *Viewlogic*, are used for just the prelayout stage. These so-called *front end* tools are popular PC-based commodities and are used extensively in FPGA design.

13.3.3   Standard Cell

The advantages of fast turnaround time and relatively low cost offered by gate arrays are counterbalanced by several problems. The first is that silicon is wasted because a design does not use all the available gates on the array. Also, it is not known by the manufacturer which pad on the array is to be an input or an output and so silicon is further wasted by the inclusion of both input and output circuits at every pad. As the chip price is proportional to die size, then this can be uneconomical when large volumes are required. In addition, because all the transistors in a gate array are the same size, when transistors are placed in series long delays occur. This happens on the PMOS chain for NOR and the NMOS chain for NAND. Consequently the gates cannot be optimally designed and the delays τ_{plh} and τ_{phl} are asymmetrical. If the W/L's of the transistors were individually adjusted for each gate type, the delays would be shorter.

The standard cell approach gets around these problems. Here, the designer again has available a library of logic gates but the design starts with a clean piece of silicon.

Hence, only those gates selected for a design appear on the final chip and no silicon is wasted. It is also known which pads are to be input and output thus further saving silicon. The standard cell chip is therefore smaller than the gate array. This device is also faster partly because it is smaller and the routing is shorter (hence smaller wire delays) and partly because the library of logic gates is optimally designed by the manufacturer. This is achieved by adjusting the W/L's of the transistors in each gate so as to achieve optimum delay.

Since the standard cell only uses those gates that are needed for a design, then each chip is of different size and is unique. Hence, all masks are required, which can be of the order of 8–16 masks where each mask costs £1000–£2000! The NRE costs are therefore considerably higher and the production times longer compared to a mask programmable gate array. This approach is therefore only economical when relatively large volumes are involved. However, reduced prototyping costs are again available by using multiproject wafers.

Libraries for standard cell (and gate arrays) have become quite sophisticated. Not only are the basic and complex gates provided but also counters and UARTs (serial interface) exist. Incredibly some manufacturers are even offering complete processor cores such as the Z180 by VLSI Technology, TMS320C50 by TI and the 80486 by SGS Thomson.

Example 13.6

Compare the transistor count of a complex combinational gate that is offered in the manufacturer's library that produces the function $f = \overline{AB + CD}$ implemented in a mask programmable gate array with a standard cell approach.

Solution

The gate array approach would require De Morgan's theorem to implement this function using the blocks of four transistors (i.e., using either two-input NAND or two-input NOR gates). Choosing NAND gates results in:

$$f = \overline{AB + CD} = \overline{AB} \cdot \overline{CD}$$

The function using NAND gates is shown in Figure 13.15(A). Note that it is not possible to directly produce an AND gate with CMOS. This must be produced by using a NAND with an inverter. Thus the total number of gates required is 3.5 or 14 transistors.

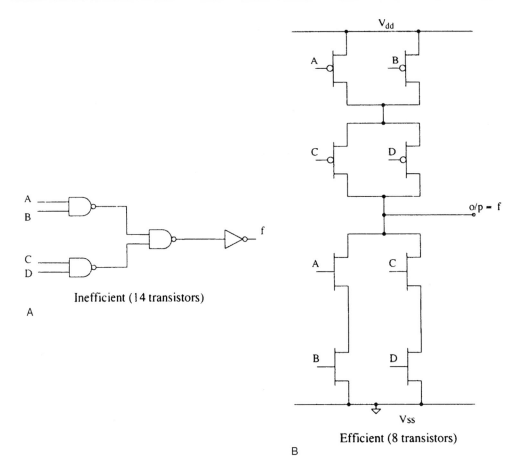

Inefficient (14 transistors)

A

B

Efficient (8 transistors)

Figure 13.15: Inefficient and efficient implementation of the function $\overline{AB + CD}$

Consider now the standard cell. To implement the above function the library designer uses the following technique:

1. Concentrate on the NMOS network first: those terms that are AND'd are placed in series while those OR'd are placed in parallel.

2. The PMOS network is just a reverse of the NMOS network.

The final circuit diagram is shown in Figure 13.15(B). Notice that the number of transistors used is now only eight, a great saving on silicon. In addition the gate array approach uses a three-level logic while the standard cell uses only a single level, giving the gate a much smaller propagation delay.

13.3.3.1   CAD Tools for Standard Cell

The CAD tools for a standard cell follow those for mask programmable gate arrays with a slight exception at the layout stage. Here the designer can interconnect each cell without the restriction of a fixed number of routing channels. This results in a chip that is much easier to route but may cause errors in the layout due to incorrect connectivity caused by designer intervention. To avoid this problem the designer has available layout verification tools which perform various checks on the layout. These are shown dotted in Fig 13.13 and consist of: design rule check (DRC), where the spacing of the metal interconnect is checked; electrical rule check (ERC), where the electrical correctness of the circuit is confirmed, i.e., outputs not shorted to supply, no outputs tied together etc.; and finally layout versus schematic (LVS), where a netlist is extracted from the layout and is compared with the original schematic. Since the NRE costs are high (especially for non-MPW processes) these verification tools are an essential component in standard cell design. Both Mentor and Cadence offer such tools and so are suitable for standard cell design.

13.3.4   Full Custom

This is the traditional method of designing integrated circuits. With a standard cell and gate array the lowest level that the design is performed at is the logic gate level, i.e., NAND, NOR, D-Type, etc. No individual transistors are seen. However, full custom design involves working down at this transistor level where each transistor is handcrafted depending upon what it is driving. Thus a much longer development time occurs and consequently the development costs are larger. The production costs are also large since all masks are required and each design presents new production problems.

Full custom integrated circuits are not so common nowadays unless it is for an analog or a high-speed digital design. A mixed approach tends to be used, which combines full custom and standard cells. In this way a designer can use previously designed cells and for those parts of the circuit that require a higher performance, then a full custom part can be made.

13.3.4.1   CAD Tools for Full Custom

The CAD tools follow the general form described for a standard cell. However, since the design of full custom parts involves more manual human involvement then the chances of error are increased. The designer thus relies very heavily on simulation and verification tools. In addition, since cells are designed from individually handcrafted transistors, then they must be simulated with an analog circuit simulator such as SPICE before being released as a digital part. Needless to say, the choice of a design route that incorporates full custom design is one that should not be taken lightly.

13.4   Field-Programmable Logic

So far we have seen two extremes in the design options available to a digital designer—namely standard products and mask programmable ASICs. Although mask programmable ASICs offer extremely high performance they carry a large risk in terms of time and expenditure. To provide the designer with the flexibility of both, the industry has gradually developed a class of logic that can be programmed with a personal computer in the laboratory. These devices are called *field-programmable logic* and can be either one-time programmable (utilizing small fuses) or many times programmable (using either ultraviolet erasable connections or an SRAM/MUX). Because these devices contain the extra circuitry to control interconnect and functionality this overhead results in a family which is less complex and slower than the mask programmable ASICs. However, the attraction of a much lower risk can outweigh the performance problems especially for prototyping purposes.

These field-programmable logic devices are divided into two groups:

- AND-OR programmable architectures;

- field-programmable gate arrays or FPGAs.

13.4.1   AND-OR Programmable Architectures

The AND-OR programmable architecture devices were the first programmable logic chips available on the market and still exist today. The reason for the interest in such structures is because *all* combinational logic circuits can be expressed in this AND-OR form.

Three types of programmable AND-OR arrays are available:

- fixed AND—programmable OR (ROM);

- programmable AND—fixed OR (PAL);

- programmable AND—programmable OR (PLA).

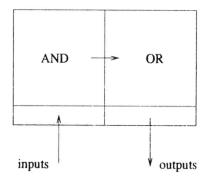

Figure 13.16: Schematic for an AND-OR array

A block schematic of an AND-OR array is shown in Figure 13.16. Inputs are passed to the AND array whose outputs are fed into the OR array which provide the outputs of the chip. Each of these AND-OR array types will now be discussed in more detail.

13.4.2   *ROM: Fixed AND-Programmable OR*

As was seen earlier, a ROM is a read only memory device. It consists of a decoder with n inputs (or addresses) whose 2^n outputs drive a memory array. As seen in Figure 13.2 a decoder can be implemented with AND gates and hence this is called the *AND array*. Since all possible input and output combinations exist then this is classed as a *fixed* array, i.e., an n input decoder requires 2^n n-input AND gates to generate *all* product terms. As we have also seen (see Figure 10.3), the memory array is in fact a NOR array. However, the inclusion of an inverter on each column line will turn this into an OR array. Hence if the decoder has 2^n outputs then the OR array must contain m OR gates with each gate having *up* to 2^n inputs, where in this case m is the number of bits in a word. Notice that we have said "up to" 2^n inputs. This is because the OR array contains the data which is programmable. The ROM architecture is thus a *fixed AND-programmable OR array*.

The complete circuit for a 4×3 bit ROM is shown in Figure 13.17(A). Note that it consists of a fixed AND structure (i.e., a 2-to-4 decoder) and a programmable OR array (i.e., a 4-to-3 encoder).The three-bit words stored in the four addresses are programmed

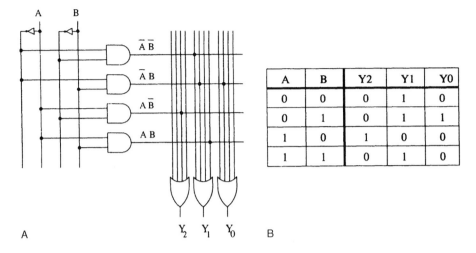

Figure 13.17: A 4 × 3 bit ROM shown storing the data in the
truth table

by simply connecting each decoder output to the appropriate input of an OR gate when a logic "1" is to be stored.

This circuit shows the ROM storing the data in the truth table of Figure 13.17(B). The Boolean equations, in fundamental sum of products form, are:

$$Y_2 = A\overline{B}$$
$$Y_1 = \overline{A}\,\overline{B} + \overline{A}B + AB$$
$$Y_0 = \overline{A}B$$

Note that rather than thinking of the ROM storing four three-bit words, an alternative view is that it is implementing a two-input, three-output truth table.

The same circuit is shown again in Figure 13.18 but this time the 2^n inputs to each OR gate are shown, for simplicity, as a single input. A cross indicates a connection from the address line to the gate. The same data as in Figure 13.17 are shown stored.

As seen earlier, the physical implementation of the programmable OR array is achieved via the presence or absence of a transistor connection. This is achieved either by omitting the source or drain connections of MOS transistors or blowing fuses which are connected to the transistor terminals. Apart from using ROMs to store data or programs they can also be used to perform many digital operations, some of which are described below.

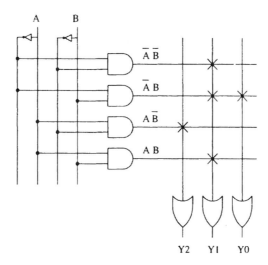

Figure 13.18: A 4 × 3 bit ROM using an abbreviated notation for the OR array

13.4.2.1 Universal Combinational Logic Function

As we have seen a ROM has all fundamental product terms available for summing and can implement an *m*-output, *n*-input truth table. This is simply achieved by connecting the address lines to the *n* input variables, and each output line programmed to give the appropriate output values. The advantages of such a ROM based design are: it is particularly applicable if *n* is large; no minimization is needed; it is cheap if mass produced; and only one IC is needed.

Example 13.7

How would the truth table shown in Figure 13.19(A) be implemented using a ROM?

Solution

A ROM of at least size 16 × 3 would be needed. The four address lines would be connected to the input variables *A, B, C* and *D* with the three outputs providing *X, Y* and *Z*. The required outputs (three-bit word) for each of the 16 possible input combinations would be programmed into the ROM, straight from the truth table. This is shown in Figure 13.19(B) for the first four addresses, where A_n and O_n, are the *n*th address line and output respectively of the ROM.

A	B	C	D	X	Y	Z
0	0	0	0	0	0	0
0	0	0	1	1	0	1
0	0	1	0	0	0	0
0	0	1	1	0	1	1
0	1	0	0	0	1	0
0	1	0	1	1	0	0
0	1	1	0	0	0	0
0	1	1	1	1	0	1
1	0	0	0	0	1	1
1	0	0	1	0	0	0
1	0	1	0	1	1	1
1	0	1	1	0	0	0
1	1	0	0	0	1	0
1	1	0	1	0	0	0
1	1	1	0	1	1	1
1	1	1	1	0	0	0

A

	A0	A1	A2	A3	O0	O1	O2
WORD 0	0	0	0	0	0	0	0
WORD 1	0	0	0	1	1	0	1
WORD 2	0	0	1	0	0	0	0
WORD 3	0	0	1	1	0	1	1

B

Figure 13.19: Truth table used in Example 13.7 for implementation in ROM

Note that because all the fundamental product terms are produced by the fixed AND array of the ROM then no minimization can take place.

13.4.2.2 *Code Converter and Look-Up Table*

A ROM can be used to convert an *n*-bit binary code (presented to the address lines) into an *m*-bit code (which appears at the outputs). The desired *m*-bit code is simply stored at the appropriate address location. Considered in this way it is a general *n*-to-*m* encoder or *code converter*.

Another ROM application similar to the code converter is the *look-up table*. Here, a ROM could be used to look up the values of, for example, a trigonometric function (e.g., sin *x*), by storing the values of the function in ROM. By addressing the appropriate location with a digitized version of *x* the value for the function stored would be output.

13.4.2.3 Sequence Generator and Waveform Generator

A ROM can be used as a *sequence generator* in that if the data from an $n \times m$ ROM are output, address by address, then this will generate n binary data sequences. Also, if the ROM output is passed to an m bit digital-to-analog converter (DAC) then an analog representation of the stored function will be produced. Hence a ROM with a DAC can be used as a *waveform generator*.

13.4.3 PAL: Programmable AND-Fixed OR

ROM provides a fixed AND-programmable OR array in which all fundamental product terms are available, thus providing a universal combinational logic solution. However, ROM is only available in limited sizes and with a restricted number of inputs. Adding an extra input means doubling the size of the ROM. Clearly a means of retaining the flexibility of the AND-OR structure while also overcoming this problem would produce a useful structure.

Virtually all combinational logic functions can be minimized to some degree, therefore allowing nonfundamental product terms to be used. Therefore, a programmable AND array would allow only the necessary product terms, after minimization, to be produced. Followed by a fixed OR array, this would allow a fixed number of product terms to be summed and so a minimized sum of products expression implemented. This type of structure is called a *programmable array logic* or PAL.

The structure of a hypothetical PAL is shown in Figure 13.20. This circuit has two input variables and three outputs, each of which can be composed of two product terms. The product terms are programmable via the AND array. For the connections shown the outputs are:

$$Y_2 = \overline{A}\,\overline{B} + AB$$
$$Y_1 = A + B$$
$$Y_0 = AB$$

(Note that Y_0 only has one product term so only one of the two available AND gates is used.)

Commercially available PAL part numbers are coded according to the number of inputs and outputs. For example the hypothetical PAL shown in Figure 13.20 would be coded PAL2H3, i.e., it is a PAL having two inputs and three outputs. The H indicates

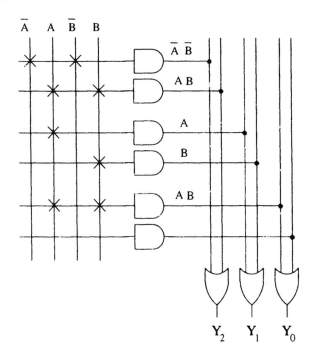

Figure 13.20: A programmable AND-fixed OR logic structure (i.e., PAL) with two inputs, six programmable product terms and three outputs (each summing two of the six product terms)

that the outputs are active high. One of the smallest PALs on the market is a PAL16L8 offered by Texas Instruments, AMD and several other manufacturers. This has 16 input terms and eight outputs. The *L* indicates that the outputs are active low. This device actually shares some of its inputs with its outputs, i.e., it has feedback. Hence, if all eight outputs are required then only eight inputs are available. The other piece of information that is required about a PAL is how many product terms each OR gate can support. This is supplied on the data sheet, and for the PAL16L8, for example, it is seven.

By adding flip-flops at the output, the designer is able to use PALs as sequential elements. The nomenclature for the device would now be PAL16R8 for example where R stands for registered output. The early PALs were fuse programmable. However, companies such as Altera, Intel and Texas Instruments added EPROM technology to these registered output PALs so that the devices could be programmed many times. These devices are called *erasable programmable logic devices* or EPLDs.

Very large PALs exist having gate equivalents of over 2000 gates quoted (remember a *gate* is defined as a two-input NAND gate). The inflexibility of only having the flip-flops at the outputs and not buried within the array (as in mask programmable ASICs) resulted in the GAL. The GAL (generic array logic) is an ultraviolet-erasable PAL with a programmable cell at each output, called an *output logic macro cell* (OLMC). Each OLMC contains a register and multiplexers to allow connections to and from adjacent OLMCs and from the AND/OR array. The GAL (trademark of Lattice Semiconductors) has a similar nomenclature to PALs. For example the GAL16V8 has 16 inputs and eight outputs using a versatile cell (i. e., V in the device name). However, because it uses OLMCs then it can emulate many different PAL devices in one package, having a range of inputs (up to 16) and outputs (up to eight).

Example 13.8

How could the truth table in Figure 13.19(A) be implemented using a (hypothetical) PAL with four inputs, three outputs and a total of 12 programmable product terms (i.e., four to each output)?

Solution

First, we use Karnaugh maps (Figure 13.21) to minimize the functions *X*, *Y* and *Z*.

Z	$\bar{A}\bar{B}$	$\bar{A}B$	AB	$A\bar{B}$
$\bar{C}\bar{D}$	0	0	0	1
$\bar{C}D$	1	0	0	0
CD	1	1	0	0
$C\bar{D}$	0	0	1	1

Y	$\bar{A}\bar{B}$	$\bar{A}B$	AB	$A\bar{B}$
$\bar{C}\bar{D}$	0	1	1	1
$\bar{C}D$	0	0	0	0
CD	1	0	0	0
$C\bar{D}$	0	0	1	1

X	$\bar{A}\bar{B}$	$\bar{A}B$	AB	$A\bar{B}$
$\bar{C}\bar{D}$	0	0	0	0
$\bar{C}D$	1	1	0	0
CD	0	1	0	0
$C\bar{D}$	0	0	1	1

Figure 13.21: Karnaugh maps for Example 13.8

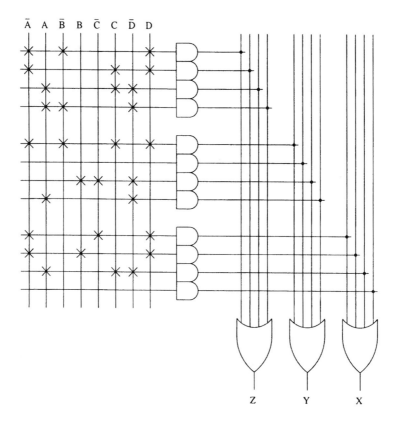

Figure 13.22: Using a PAL to implement the truth table in Figure 13.19

From these Karnaugh maps:

$$Z = \overline{A}\,\overline{B}D + \overline{A}CD + AC\overline{D} + A\overline{B}\,\overline{D}$$
$$Y = \overline{A}\,\overline{B}CD + B\overline{C}\,\overline{D} + A\overline{D}$$
$$X = \overline{A}\,\overline{C}D + \overline{A}BD + AC\overline{D}$$

The PAL, a PAL4H3, would therefore be programmed as shown in Figure 13.22.

13.4.4   PLA: Programmable AND-Programmable OR

The final variant of the AND-OR architectures is the programmable AND-programmable OR array or *programmable logic array* (PLA). With this the desired product terms can be programmed using the AND array and then as many of these terms summed together as required, via a programmable OR array, to give the desired function.

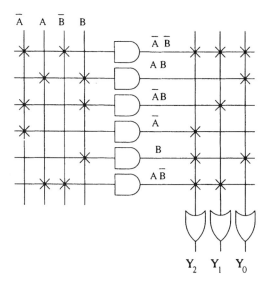

Figure 13.23: A programmable AND-programmable OR logic array (i.e., PLA) with two inputs, six programmable product terms and three programmable outputs

The structure of such an array with two inputs, three outputs and six programmable product terms available is shown in Figure 13.23. For the connections shown the outputs are:

$$Y_0 = \overline{A}\,\overline{B} + AB + B$$
$$Y_1 = \overline{A}\,\overline{B} + \overline{A}B + A\overline{B}$$
$$Y_2 = \overline{A}\,\overline{B} + \overline{A} + B + A\overline{B}$$

Note that any product term can be formed by the AND gates, and that any number of these product terms can be summed by the OR gates.

Example 13.9

How would the truth table shown in Figure 13.19(A) be implemented using a (hypothetical) four-input, three-output PLA with eight product terms?

Solution

From the minimization performed to implement this truth table using the PAL in Example 13.8 it can be seen that the three Boolean expressions for *X, Y* and *Z* contain a total of nine different product terms ($AC\overline{D}$ is common to both *X* and *Z*).

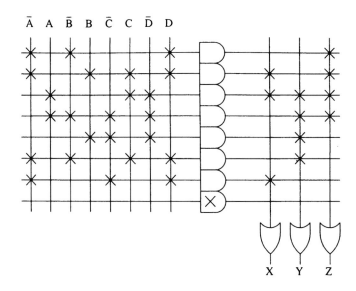

Figure 13.24: A programmable AND-programmable OR logic array (PLA) with four inputs, eight programmable product terms and three programmable outputs

This PLA can only produce eight which means that product terms common to the three expressions must be found, effectively de-minimizing them to some degree.

This can be achieved by reconsidering the Karnaugh maps and *not* fully minimizing them, but rather looking for common implicants in the three expressions:

$$Z = A\overline{B}D + \overline{A}BCD + AC\overline{D} + A\overline{B}\,\overline{C}\,\overline{D}$$
$$Y = B\overline{C}\,\overline{D} + \overline{A}\,\overline{B}CD + AC\overline{D} + A\overline{B}\,\overline{C}\,\overline{D}$$
$$X = \overline{A}\,\overline{C}D + \overline{A}BCD + AC\overline{D}$$

In this form only seven different product terms are required to implement all three functions and so the given PLA can be used as shown in Figure 13.24.

13.4.5 Field-Programmable Gate Arrays

The advancement in on-chip field-programmable techniques combined with ever increasing packing densities has led to the introduction of *field-programmable gate arrays* or FPGAs. These devices can be considered as being the same as mask programmable gate arrays except the functionality and interconnect is programmed in the laboratory at a greatly reduced financial risk. The popularity of FPGAs is indicated by the large number of companies who currently manufacture such devices. These include Actel, Altera, AMD, Atmel, Crosspoint, Lattice, Plessey, Quicklogic, Texas Instruments, and Xilinx, to name but a few. Of these, the three that are perhaps the best known are Altera, Xilinx and Actel. In order to introduce FPGAs, some of the devices provided by these three companies will therefore be discussed. Essentially they differ in terms of: *granularity; programming technique; volatility; and reprogrammability*. All FPGAs consist of a *versatile cell* that is repeated across the chip with its size and hence cell complexity referred to as the *granularity*. These cells are multifunctional such that they can produce many different logic gates from a single cell. The larger the cell, the greater the complexity of gate each cell can produce. Those arrays that use a small simple cell, duplicated many times, are referred to as having fine granularity, while arrays with few, but large, complex cells are defined as coarse grain. These versatile cells have been given different names by the manufacturers, for example: modules; macrocells; and combinatorial logic blocks. The programming of the function of each cell and how each cell is interconnected is achieved via either: small fuses; onboard RAM elements that control multiplexers; or erasable programmable read only memory (EPROM) type transistors. Consequently some devices are *volatile* and lose their functionality when the power is removed, while others retain their functionality even with no supply connected. Finally these devices can be divided into those that can be *reprogrammed* many times and those that are *one-time programmable*.

Let us now look more closely at the FPGA types, which will be divided into: EPROM type; SRAM/MUX type; and fuse type.

13.4.5.1 EPROM Type FPGAs

The most common EPROM type FPGA device is that supplied by Altera. The range of devices available from Altera are the MAX 5000, 7000 and 9000 series (part numbers: EPM5XXX, EPM7XXX and EPM9XXX). These devices are the furthest from the true FPGAs and can be considered really as large PAL structures. They offer coarse granularity and are more an extension to Altera's own range of electrically programmable, ultraviolet-erasable logic devices (EPLD). The versatile cell of these devices is called a *macrocell*. This cell is basically a PAL with a registered output.

Between 16 and 256 macrocells are grouped together into an array inside another block called a *logic array block* (LAB) of which an FPGA can contain between 1 and 16. In addition to the macrocell array each LAB contains an I/O block and an expander which allows a larger number of product terms to be summed. Routing between the LABs is achieved via a programmable interconnect array (PIA) which has a fixed delay (3 ns worst case) that reduces the routing dependence of a design's timing characteristics.

Since these devices are derived from EPLD technology the programming is achieved in a similar manner to an EPROM via an Altera logic programmer card in a PC connected to a master programming unit. The MAX 7000 is similar to the 5000 series except that the logic block has two more input variables. The MAX 9000 is similar to the 7000 device except that it has two levels of PIA. One is a PIA local to each LAB while the other PIA connects all LABs together. Both the 7000 and 9000 series are E^2PROM devices and hence do not need an ultraviolet source to be erased.

It should be noted though that these devices are not true FPGAs and have a limited number of flip-flops available (one per macrocell). Hence, the Altera Max 5000/7000/ 9000 series is more suited to combinatorially intensive circuits. For more register intensive designs Altera offers the Flex 8000 and 10 K series of FPGAs which uses an SRAM memory cell based programming technique (as used by Xilinx—see next section); although currently rather expensive it will in time become an attractive economical option. The Flex 8000 series (part number: EPF8XXX) has gate counts from 2000 to 16000 gates. The 10 K series (part number: EPF10XXX) however, has gate counts from 10000 to 100000 gates!

13.4.5.2   SRAM/MUX Type FPGAs

The most common FPGA that uses the SRAM/MUX programming environment is that supplied by Xilinx. The range of devices provided by Xilinx consists of the XC2000, XC3000 and XC4000. The versatile cell of these devices is the "configurable logic block" (CLB) with each FPGA consisting of an array of these surrounded by a periphery of I/O blocks. Each CLB contains combinational logic, registers and multiplexers and so, like the Altera devices, has a relatively coarse granularity. The Xilinx devices are programmed via the contents of an on-board static RAM array which gives these FPGAs the capability of being reprogrammed (even while in operation). However, the volatility of SRAM memory cells requires the circuit configuration to be held in an EPROM alongside the Xilinx FPGA.

A recent addition to the Xilinx family is the XC6000 range. This family has the same reprogrammability nature as the other devices except it is possible to partially reconfigure these devices. This opens up the potential for fast in-circuit reprogramming of small parts of the device for learning applications such as neural networks.

13.4.5.3 Fuse Type FPGAs

The most common fuse type FPGA is that supplied by Actel. These devices are divided into the Act1, Act2 and Act3 families. The Act1 FPGAs (part numbers: AI0XX) contain two programmable cells: "Actmod" and "IOmod." The versatile core cell is the Actmod which is simply based around a 4-to-1 multiplexer for Act 1. This versatile cell is shown in Figure 13.25. Since this cell is relatively small the array is classed as *fine grain*. By tying the inputs to either a logic "0" or logic "1" this versatile cell can perform 722 different digital functions. The programmable IOmod cell is used to connect the logic created from the Actmods to the outside world. This cell can be configured as various types of inputs and/or outputs (bidirectional, tristate, CMOS, TTL, etc.). Unlike the Xilinx and Altera devices, the Actel range are programmed using fuse technology with *desired* connections simply blown (strictly called an *antifuse*). These devices are therefore "one time

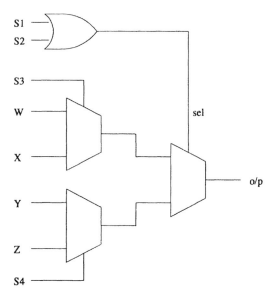

Figure 13.25: Versatile cell used for Actel Act1 range of fused FPGAs

programmable" (OTP) and cannot be reprogrammed. The arrays have an architecture similar to a channelled gate array with the repeating cell (Actmod) arranged in rows with routing between each row. The routing contains horizontal and vertical metal wires with antifuses at the intersection points.

Other devices in the Actel range are the Act2 (part numbers: A12XX) and the Act3 (part numbers: A14XX) devices. These use two repeating cells in the array. The first is a more complex Actmod cell called *Cmod*, used for combinational purposes, while the other cell is a Cmod with a flip-flop.

Table 13.1 shows a comparison of some of the FPGA devices offered by Altera, Xilinx, and Actel.

Example 13.10

Consider the versatile Actel cell shown in Figure 13.25. What functions are produced if the following signals are applied to its inputs:

(a) $S_1S_2S_3S_4 = 000A$ and $WXYZ = 0001$

(b) $S_1S_2S_3S_4 = 0BAA$ and $WXYZ = 1110$

(c) $S_1S_2S_3S_4 = 0CA1$ and $WXYZ = 0B11$.

Assume that the signals A and B are inputs having values of "0" or "1" and that for each multiplexer when the select line is low the lower input is selected.

Solution

(a) In this case the OR gate output is always zero and so the lower line is selected. This is derived from the lower multiplexer whose select line is controlled by the only input "A". With the inputs to this multiplexer at "0" for the upper line and "1" for the lower line, the output follows that of an inverter.

(b) Here it is helpful to construct a truth table and include in this table the output of the OR gate, called *sel* (See Table 13.2). We can thus see that the function is a two-input OR gate.

(c) Again a truth table (including *sel*) is useful to work out the function implemented (See Table 13.3).

A Karnaugh map for the output is shown in Table 13.4, which generates the function:

$$o/p = \overline{C} + \overline{A}B$$

Table 13.1: Comparison of some FPGA types

Manufacturer	Part	Gates(k)	D-types	Cost(£)	Programming technique	Speed (MHz)	I/Os
Altera	EPM5X	8	21–252	14–55	EPROM	60	100
	EPM7X	10	40–400	12–100	EPROM	70	288
	EPM9X	20	400–700	50–140	EPROM	65	100
	EPF8X	2.5–16	78–208	15–90	SRAM	100	208
	EPF10KX	10–100	148–420	25–550	SRAM	120	300
Xilinx	XC2X	0.6–1.5	64–100	10–15	SRAM	60	74
	XC3X	1–7.5	256–1320	10–60	SRAM	70	176
	XC4X	2–25	256–2560	15–190	SRAM	60	256
Actel	A10X (Act1)	1.2–2	147–273	12–20	Antifuse	37	69
	A12X (Act2)	2–8	565–998	17–55	Antifuse	41	140
	A14X (Act3)	1.5–10	264–1153	15–250	Antifuse	125	228

**Table 13.2: Truth table for
Example 13.10(b)**

A	B	Sel	o/p
0	0	0	0
0	1	1	1
1	0	0	1
1 1	1	1	1

**Table 13.3: Truth table for
Example 13.10(c)**

A	B	C	Sel	o/p
0	0	0	0	1
0	0	1	1	0
0	1	0	0	1
0	1	1	1	1
1	0	0	0	1
1	0	1	1	0
1	1	0	0	1
1	1	1	1	0

**Table 13.4: Karnaugh map for
Example 13.10(c) output**

	AB	$\overline{AB}$	$\overline{A}B$	$A\overline{B}$
$\overline{C}$	1	1	1	1
C	0	1	0	0

13.4.6   CAD Tools for Field-Programmable Logic

The programming of field-programmable logic devices is implemented directly via a computer. The software needed for PALs and PLAs is usually a simple matter of producing a programming file called a fuse or an EPROM bit file. This file has a standard format (called *JEDEC*) and contains a list of 1's and 0's. This file is automatically generated from either Boolean equations, truth tables or state diagrams using programs such as ABEL (DataIO Corp.), PALASM (AMD Inc.) and CUPL (Logical Devices Inc.). In other words the minimization is done for you and it is not necessary to draw out any Karnaugh maps. Software programs that can directly convert a schematic representation into a JEDEC file are also available. Since these devices have only an MSI complexity level, the software tools are relatively simple to use and also inexpensive.

The FPGAs, on the other hand, have capacities of LSI and VLSI level and are much more complex. Since FPGAs are similar in nature to mask programmable gate arrays, the associated CAD tools have been derived from mask programmable ASICs and follow that of Figure 13.13; that is: schematic capture (or VHDL), prelayout simulation, layout, back annotation and postlayout simulation.

It should be noted that FPGA simulation philosophy is somewhat different from mask programmable gate arrays. With mask programmable devices, 100% simulation is absolutely essential since these circuits cannot be rectified after fabrication without incurring large financial and time penalties. These penalties are virtually eliminated with FPGA technology due to the fast programming time in the laboratory and the low cost of devices. For one-time programmable devices (such as Actel) the penalty is the price of one chip while for erasable devices (such as Xilinx) the devices can simply be reprogrammed. Hence, the pressure to simulate 100% is not as great.

For those devices that are reprogrammable, this results in an inexpensive iterative procedure whereby a device is programmed and then tested in the final system. If the device fails it can be reprogrammed with the fault corrected. For OTP type FPGAs then a new device will have to be blown at each iteration; although it will incur a small charge the cost is considerably less than mask programmable arrays. It is not uncommon for FPGA designs (both reprogrammable and OTP) to experience four iterations before a working device is obtained. This is totally unthinkable for mask programmable designs where a "right first time approach" has to be employed—hence, the reliance on the simulator.

Since fuses, SRAM/MUX cells, etc., are used to control the connectivity, the delays caused by these elements must be added to the wire delays for postlayout simulation. Hence it is for this reason that FPGAs operate at a lower frequency than mask programmable gate arrays. The large delays in the routing path also mean that timing characteristics are routing dependent. Hence, changing the placement positions of core cells (by altering the pin out for example) will result in a different timing performance. If the design is synchronous then this should not be a problem with the exception of the shift register problem referred to in Figure. 13.14. It should also be noted that the prelayout simulation of FPGAs on some occasions is only a unit delay (i.e., 1 ns for all gates) or functional simulation. It does not take into account fan-out, individual gate delays, set-up and hold time, minimum clock pulse widths (i.e., spike and glitch detector), etc., and does not make any estimate of the wire delay. Hence the simulation at this stage is not reflective of how the final design will perform. To obtain the true delays, the FPGA must be laid out and the delays *back annotated* for a postlayout simulation. This will provide an accurate simulation and hence reveal any design errors. Unfortunately, if a mistake is found then the designer must return all the way back to the original schematic. The design must again be prelayout simulated, laid out and delays back annotated before the postlayout simulation can be repeated. This tedious iterative procedure is another reason why FPGAs are usually programmed prematurely with a limited simulation. It should be mentioned that an FPGA is sometimes used as a prototyping route prior to migrating to a mask programmable ASIC. Hence the practice of postlayout simulation using back annotated delays is an important discipline for an engineer to learn in preparation for moving to mask programmable ASICs.

When all the CAD stages are completed the FPGA netlist file is converted into a programming file to program the device. This is either a standard EPROM bit file for the Xilinx and Altera arrays or a fuse file for the Actel devices. Once a device is programmed, debug and diagnostic facilities are available. These allow the logic state of any node in the circuit to be investigated after a series of signals has been passed to the chip via the PC serial or parallel port. This feature is unique to FPGAs since each node is addressable, unlike mask programmable devices.

FPGA CAD tools are usually divided into two parts. The first is the prelayout stage or front-end software, i.e., schematic and prelayout simulation. The CAD tools here are generic (suitable for any FPGA) and are provided by proprietary packages such as Mentor Graphics, Cadence, Viewlogic, Orcad, etc. However, to access the FPGAs the corresponding libraries are required for schematic symbols and models.

The second part is called the *back-end* software incorporating: layout; back annotation of routing delays; programming file generation and debug. The software for this part is usually tied to a particular type of FPGA and is supplied by the FPGA manufacturer.

For example consider a typical CAD route with Actel on a PC. The prelayout (or front end) tools supplied by Viewlogic can be used to draw the schematic using a package called Viewdraw and the prelayout functional simulation is performed with Viewsim. In both cases library files are needed for the desired FPGA. Once the design is correct it can be converted into an Actel net-list using a *netlist translator*. This new file is then passed into the CAD tools supplied by Actel (called Actel Logic System – ALS) ready for place and routing. The parasitic delays can be extracted and back annotated out of ALS back into Viewlogic so that a post-layout simulation can be performed again with Viewsim. If the simulation is not correct then the circuit schematic must be modified and the array is placed and routed again. Actel provides a static timer to check set-up and hold time and calculate the delays down all wires, indicating which wire is the heaviest loaded. A useful facility is the *net criticality* assignment which allows nets to be tagged depending on how speed critical they are. This facility controls the placing and routing of the logic in order to minimize wiring delays wherever possible. The device is finally programmed by first creating a fuse file and then blowing the fuses via a piece of hardware called an *activator*. This connects to an Actel programming card inside the PC. As an example of the length of time the place and route software can take to complete, the authors ran a design for a 68-pin Actel 1020 device. The layout process took approximately 10 minutes using a 486, 66 MHz PC and utilized 514 (approximately 1200 gates) of the 547 modules available (i.e., a utilization of 94%). In addition, on the same computer the fuse programming via the activator took around 1 minute to complete its program. With mask programmable ASICs, however, the programming step can take at least four weeks to complete! This is one of the great advantages that FPGAs have over mask programmable ASICs. Note, however, that as with mask programmable arrays the FPGA manufacturers only provide a limited range of array sizes. The final design thus never ever uses all of the gates available and hence silicon is wasted. Also, as the gates are used up on the array, the ability for the router to access the remaining gates decreases and hence, although a manufacturer may quote a maximum gate count for the array, the important figure is the percentage utilization.

Actel FPGAs also have comprehensive post-programming test facilities available under the option "Debug." These consist of: the functional debug option; and the in-circuit diagnostic tool. The *functional debug* test involves sending test vectors from the PC to the activator, which houses the FPGA during programming, and simple tests can be carried out. The *in-circuit diagnostic* tool is used to check the real time operation of the device when in the final PCB. This test is 100% observable in that any node within the chip can be monitored in real time with an oscilloscope via two dedicated pins on the FPGA.

The Xilinx FPGA devices are programmed in a similar way by using two pieces of software. Again, typical front-end software for these devices is Viewlogic utilizing Viewdraw and Viewsim for circuit entry and functional simulation, respectively. The netlist for the schematic is this time converted into a Xilinx netlist and the design can now move into the Xilinx development software supplied by Xilinx (called XACT). Although individual programs exist for place and route, parasitic extract, programming file generation, etc., Xilinx provides a simple to use compilation utility called XMAKE. This runs all of these steps in one process. Parasitic delays can again be back annotated to Viewsim for a timing simulation with parasitics included. A static timing analyzer is again available so that the effects of delays can be observed on set-up and hold time without having to apply input stimuli. Bit stream configuration data, used in conjunction with a Xilinx provided cable, allow the data to be downloaded to the chip for configuration. As with Actel, both debug and diagnostic software exist such that the device can be tested and any node in the circuit monitored in real time. The bit stream data can be converted into either Intel (MCS-86), Motorola (EXORMAX) or Tektronix (TEKHEX) PROM file formats for subsequent PROM or EPROM programming. The one disadvantage of these devices as compared to the Actel devices is that when in final use the device needs to have an associated PROM or EPROM, which increases the component count.

13.5   VHDL

As systems become more complex, the use of schematic capture programs to specify the design becomes unmanageable. For designs above 10,000 gates, an alternative design entry technique of *behavioral specification* is invariably employed. This is a high-level programming language that is textual in nature, describes behavior and maps to hardware. The most commonly accepted behavioral language is that standardized by the IEEE (standard 1076) in 1987 called *VHDL*. VHDL is an acronym for *VHSIC Hardware Description Language* where VHSIC (Very High Scale Integrated Circuits) is the next level of integration above VLSI. This language was developed by the USA Department of Defense and is now a worldwide standard for describing general digital hardware. The language allows a system to be described at many different levels from the lowest level of logic gates (called *structural*) through to behavioral level. At behavioral level the design is represented in terms of programming statements which makes no use of logic gates. This behavior can use a digital (i.e., Boolean), integer or real representation of the circuit operation. A system designer can specify the design at a high level (i.e., in integer behavioral) and then pass this source code to another member of the group to break the design down into individual smaller blocks (partitioning). A block in behavioral form requires only a few lines of code and hence is not as complex as a structural logic gate description and hence has a shorter simulation time. Since the code supports mixed levels (i.e., gate and behavior) then the system can be represented with some blocks represented at the gate level and the rest at behavioral. Thus the complete system can be simulated in a much shorter time.

One of the biggest problems of designing an ASIC is the interpretation of the specification required by the customer. Because VHDL has a high-level description capability it can be used also as a formal specification language and establishes a common communication between contractors or within a group. Another problem of ASIC design is that you have to choose a foundry *before* a design is started, thus committing you to that manufacturer. Hence, it is usual to insist on second sourcing outlets to avoid production hold-ups. However, VHDL at the high level is technology and process independent and is therefore transportable into other processes and CAD tools. It is not surprising that many companies are now insisting on a VHDL description for their system as an extra deliverable as well as the chip itself.

A simple example of a VHDL behavioral code for a 2-to-1 multiplexer is shown in Table 13.5. This source code is divided into two parts: *entity* and *architecture*. The *entity lists* the input and output pins and what form they are—bit or binary in this case—while the *architecture* describes the behavior of the multiplexer. The process labeled *f*1 is only run if any of the inputs *d0, d1* or *sel* change, i.e., it is an event driven simulator. If one of these events occurs, the IF statement is processed and the output *q* is set depending upon the value of *sel*.

Notice that since this is a behavioral description then no logic gates are used in the architecture. The next stage would be to convert this design into logic gates. This can be performed in two ways: automatically or manually. With the automatic approach an additional CAD software package is required called a *synthesizer*. These

**Table 13.5: VHDL behavioral code for a
2-to-1 multiplexer**

```
ENTITY mux IS
    PORT (d0, dl, sel:IN bit; q:OUT bit);
END mux;
    ARCHITECTURE behavior OF mux IS
    BEGIN
    f1:
    PROCESS (d0,d1,sel)—sensitivity list
    BEGIN
    IF sel = 0 THEN
        q < = d1;
        ELSE
        q <= d0;
    END IF;
END behavior;
```

are available at an extra charge and will generate the logic gates required to implement the desired behavior. Alternatively this step can be performed manually. A typical VHDL structural description of the above multiplexer implemented with logic gates is shown in Table 13.6.

Since this is only a trivial example, a manual synthesis is possible. It is also apparent that a behavioral code is more succinct than a structural one hence the simulation is faster. As the design becomes more complex then the use of an automatic synthesizer is essential.

13.6   Choosing a Design Route

So we now know all the options available to a digital circuit designer. The decision is now to choose the appropriate route. It is wise at this point to revisit the ten questions that were raised at the beginning of this chapter and to consider them in the light of the summarized information given in Table 13.7.

A standard part (called *Std. Part* in the table) design route (i.e., 74HCT, etc.) is certainly the quickest to get started and can handle large and complex designs. However, it may well be limited when the design needs to be miniaturized or put

**Table 13.6: VHDL structure code for a
2-to-1 multiplexer**

```
ENTITY mux IS
    PORT (d0, d1, sel:IN bit; q:OUT bit);
END mux;
ARCHITECTURE structure OF mux IS
    COMPONENT and2
        PORT (in1, in2:IN bit; f:OUT bit);
    END COMPONENT;
    COMPONENT or2
        PORT (in1, in2:IN bit; f:OUT bit);
    END COMPONENT;
    COMPONENT inv
        PORT (inl, in2:IN bit; f:OUT bit);
    END COMPONENT;
    SIGNAL x, y, nsel: bit;
    FOR U1: inv USE ENTITY work.inv;
    FOR U2: and2 USE ENTITY work.and2;
    FOR U3: or2 USE ENTITY work.or2;
    BEGIN
        U1: inv PORT MAP(sel, nsel)
        U2: and2 PORT MAP(nsel, d1, y)
        U3: and2 PORT MAP(d0, sel, x)
        U4: or2 PORT MAP(x,y,q)
END structure;
```

into production. Also design techniques with standard products often tend to be *ad hoc* and in some instances not synchronous. The design may use *RC* components, 555 timers and gated clocks. If the design is only a one-off and it functions correctly then this will be perfectly satisfactory if size and power are not an issue. However, if the design requires miniaturization or transfer to an ASIC for power consumption reasons, then the circuit will have to be redesigned for a totally synchronous approach.

Of the AND-OR array devices a ROM device can be used to efficiently perform a number of digital tasks, while PALs and PLAs are also widely available, providing much of the capability of ROM but with smaller circuits. However, it is necessary to minimize the Boolean functions before they can be implemented in a PAL or PLA. A PAL allows a fixed number of minimized product terms to be summed, while a PLA

Table 13.7: Comparison of digital design routes

	FC	SC	MPGA	FPGA	PAL/PLA	Std. Part
Design time (months)	6–12	2–6	1–6	1–30 days	1–14 days	1–30 days
Fab. time (months)	2–4	1–3	2–6 wks	1–10 mins	1–5 mins	14 days
Time to mkt. (months)	8–16	3–9	1.5–7.5	30 days	14 days	6 wks
Prototype cost	hi	hi	Med.	lo	V.lo	V.lo
Production cost	med./lo	lo	lo	hi	med./lo	hi
Speed	V.hi	hi	hi/med.	med./slow	med./hi	med./hi
Complexity	V.hi	hi	hi/med.	med./lo	lo	V.hi
Redesign time (months)	3–5	2–4	3–6 wks	5 days	2 days	1–14 days
CAD complexity	V.hi	hi/med.	med.	med./lo	lo	lo
Risk	V.hi	hi	med.	lo	V.lo	V.lo

enables any number of the product terms formed to be summed. However, although PALs and PLAs are quite adequate for gate counts of the order of 500, they are limited due to having no buried registers.

Obviously, no one route will satisfy all options but FPGAs are becoming a strong prototyping contender. At present, FPGA performance is still below that of mask programmable gate arrays (MPGA) which still have the edge in terms of high performance (high speed, low power consumption), high gate density or large volumes. However, for small volumes, FPGAs offer a virtually immediate turn around time and relatively low cost (particularly in terms of the nonrecurring engineering (NRE) costs, which can be very large for many ASIC designs). In addition, since a single FPGA is relatively inexpensive the risk factor is significantly less and hence, the emphasis on the simulation stage is reduced. An FPGA can thus be programmed and tested rapidly. However, if a design fault exists then the fault can be quickly corrected and the device reprogrammed. The use of an FPGA is a very powerful vehicle for testing out an idea before going to volume production. Translators are available which can convert FPGA designs into mask programmable ASICs once the design is confirmed at the prototype stage. FPGAs also/support VHDL description and since VHDL is transportable it can be transferred into standard cell (SC) options if sufficient volume is expected.

CHAPTER 14

Designing with Logic ICs

Tim Williams

14.1 Logic ICs

The interfaces between logic integrated circuits including signal, clock and power supply lines must be considered to achieve a reliable digital design. This applies whether the devices concerned are microprocessors, their support chips, application specific ICs (ASICs), programmable logic arrays (PLAs) or standard "glue" logic.

14.1.1 Noise Immunity and Thresholds

A logic input can take any value of voltage, nominally from one supply rail to the other, although due to transmission line effects the actual voltage can exceed either supply rail on transitions. Each input is designed so that any voltage below one level, conventionally V_{IL}, is regarded as a logic "0" and any voltage above another level, V_{IH}, is regarded as logic "1" (Figure 14.1).

These levels are characterized for each logic or microprocessor family, and worst case values of V_{IL} and V_{IH} can be found on any data sheet. Note that, as with any hardware-determined parameter, they may vary with temperature and you should ensure that the values you use are guaranteed across the device's temperature range. They are also a function of supply voltage. If all ICs are fed from the same supply this is not a problem, but it becomes more significant if you are interfacing logic circuits which may be fed from different supply rails.

The significance of the band between V_{IL} and V_{IH} is that the input logic state (and therefore the output state) is undefined while the voltage is in this band. Therefore,

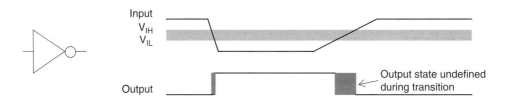

Figure 14.1: Transitions through logic thresholds

transitions between logic states must happen as rapidly as possible and no decisions must be taken while the input is in transit, or for a given period (the "settling time") thereafter. This is why clocked, or synchronous, circuits are normally more reliable than unclocked or asynchronous ones for complex logic operations: the state of the clock determines when logic decisions are taken, and it is arranged that all data transitions occur when the clock is inactive.

14.1.1.1 Susceptibility to Noise

Provided that all signals to logic inputs, whether from other logic outputs or from interfaces to other circuits, lie outside the $V_{IL} - V_{IH}$ band when they are active, then in theory no misinterpretation of the input should occur. The difference between a "low" output logic level (V_{OL}) and V_{IL}, or that between a "high" output level (V_{OH}) and V_{IH}, is the noise immunity (expressed in volts) of the logic interface (Figure 14.2). Notice that noise immunity is not a property of any particular device, but of the interface

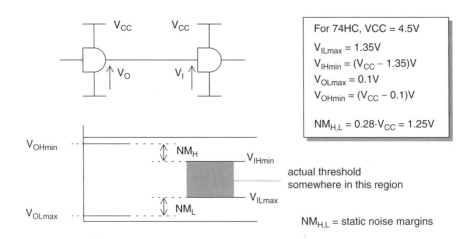

Figure 14.2: Noise immunity of a logic interface

between devices. The noise immunity of a family of devices (such as LVT or HCMOS) only refers to interfaces between devices of the same family.

14.1.1.2 Current Immunity

The noise immunity value gives meaning to the ability of the interface to withstand externally coupled noise without corruption of the perceived logic level. So for instance the HCMOS-LSTTL interface can tolerate a variation of 2.4V in the high state, or 0.47V in the low state. These are worst-case values and the actual circuit could tolerate somewhat more before a change of state occurred. But the voltage difference is only part of the story. When noise is coupled into an interface, the impedance of the interface is just as important, since this determines what voltage will be developed by a given induced interference current. The impedance is normally defined by the output driver (as long as transmission line effects are neglected) and the effective noise current threshold of the interface, given by the noise immunity voltage divided by the driver output impedance, gives a truer picture of the actual noise immunity of a given combination.

The metal-gate 4000B CMOS logic family has a high output impedance at 5V compared with the other families, so that its current immunity is significantly worse. However, as the supply voltage increases so its output impedance goes down, and the combined effect means that its immunity at 15V V_{CC} is about ten times better than at 5V. It is inherently insensitive to low voltage inductively coupled noise, but shows poor rejection of capacitively coupled noise. For general-purpose 5V applications the 74HC family is preferred. It is also true that a microcontroller's high output resistance means that it does not compare favorably with standard logic.

14.1.1.3 Use of a Pull-Up

Note that the figures for high-state and low-state immunities are often different, because of the differing drive impedances and voltage thresholds in the two states. A negative immunity value indicates that, if nothing further is done, this particular interface combination will be unreliable *by design*. For instance, the 2.7V minimum high output of the venerable LS-TTL family is less than the required 3.15V minimum VIH for HCMOS so LS-TTL driving directly into HCMOS is in danger of incorrectly transmitting the logic high level. The standard remedy for this particular situation (if you are still using LS-TTL!) is a pull-up resistor to VCC to ensure a higher output from the LS-TTL (Figure 14.3). The minimum resistor value is a function of the driver

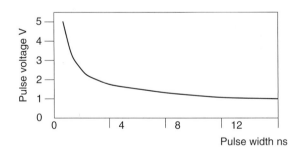

$$R_{min} = [V_{CC} - V_{OL}]/I_{OL}$$

where I_{OL} is the LS-TTL output sink current for an output voltage of V_{OL}, lower than the HCMOS low input threshold

$$R_{max} = t/C_n$$

where t is the maximum edge rise time and C_n is the node capacitance

Figure 14.3: Logic interface pull-up resistor

output capability, and the maximum value depends on permissible timing constraints. Alternatively, use the HCTMOS family, whose inputs are characterized especially for driving from LS-TTL levels.

14.1.1.4 Dynamic Noise Immunity

The static noise margins as discussed above apply until the interference approaches the operating speed of the devices. When very fast interference is present, higher amplitude is necessary to induce upset. The dynamic noise margin is measured by applying an interference pulse of known magnitude and increasing its width until the device just begins to switch. This yields a plot of noise margin versus pulse width such as shown in Figure 14.4. The high level and low level dynamic noise margins may be different.

You may often be forced to interface different logic families. Typically, a 3.3V microprocessor may need to drive 5V buffers or vice versa, or you may not be able to source a particular part in the same family as the rest of the system, or you may need to change families to optimize speed/power product. You can normally expect logic interfaces of the same family to be compatible, but whenever different families or a

Figure 14.4: Dynamic noise immunity of 74HC series devices

custom interface are used you have to check the logic threshold aspects of each one. The voltage level conversion issue is very common, to the extent that there are families of devices such as the 74LVT series which are characterized for an input range of 2.0 V_{IH} and 0.8 V_{IL}, but can still operate from a 5V rail; or vice versa, can accept 5V-swing inputs while being operated from a 3.3V rail.

14.1.2   Fan-Out and Loading

The output voltage levels that are used to fix noise immunity thresholds are not absolute. They depend as usual on temperature, but more importantly on the output current that the driver is required to source or sink. This in turn depends on the type of loading that each output sees (Figure 14.5).

Any driver has an output voltage versus current characteristic which saturates at some level of loading (Figure 14.6). The characteristic is tailored so that at a given load current, the output voltage V_{OH} or V_{OL} is equivalent to the input threshold voltage (V_{IH} or V_{IL}) plus the noise immunity for that particular logic family. This load current

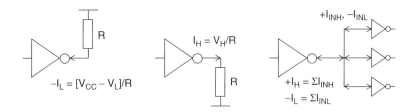

Figure 14.5: Logic output loading

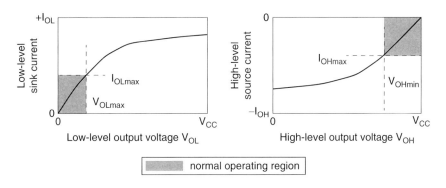

Figure 14.6: Logic driver output characteristics

then corresponds to the sum of the input currents for a given number N of standard gates of that family, and N is called the *fan-out*: that number of standard gates that the output can drive and still keep the interface within the noise threshold limits. The fan-out is normally specified against each output of a device for other devices of the same family, but it can be calculated for other logic family interfaces or indeed for any DC load current simply by comparing the output voltage and current capability for each logic state with the current and voltage requirements of each input. As before, fan-out figures for logic high and low may differ.

For CMOS families there is rarely a limit on the number of inputs that can be driven at DC by one output. When low-power parts are required to drive high-power types, then fan-out may be insufficient and you have to use intermediate buffer devices. The low drive capability of many microprocessor bus outputs severely restricts the number of other components that may be placed on the bus without interposing additional bus buffers—which accounts for the phenomenal popularity of the 74XX244 type of octal bus driver!

14.1.2.1 *Dynamic Loading*

The DC load current taken by the input side of the interface is only part of the total load. Indeed, for CMOS-input logic ICs it is negligible and has no significant effect on fanout calculations. But every input has an associated capacitance to ground and the charging or discharging of this capacitance limits the speed at which the node can operate. Typical logic IC input capacitances are 5–10 pF and these must be summed for all connected inputs, together with an allowance for interconnection capacitance which is layout-dependent but is again typically 5 pF, to reach the total load capacitance facing the driver.

Driver dynamic output current ability is rarely specified on data sheets but some manufacturers give application guidance. For instance, the 74HC range can offer typically $\pm$ 40 mA for standard devices and $\pm$ 60 mA for buffers at 4.5V supply. This current slews the interface node capacitance C_n (Figure 14.7) that you have just calculated and you have to ensure that slewing from logic 0 to the logic 1 threshold, or vice versa, is accomplished before the input data level is required to be valid. As an example, a 100 pF capacitance slewing from 0 to 3V with 40 mA drive will take 7.5 ns, and this time (plus a safety factor) has to be added onto the other specified propagation delays to ensure adequate timing margin. If the figures don't add up, you will need to add extra buffer devices (which add their own propagation delays), reduce the load, reduce the operating frequency or go to a faster logic family.

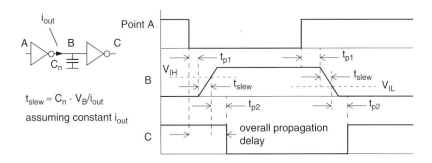

Figure 14.7: Propagation and slewing delays

If you choose to run CMOS devices with a high load capacitance and accept that the edges will be slower, then be aware that this also reduces the reliability of the device because of the higher transient currents that the output drivers are handling.

14.1.3   Induced Switching Noise

This phenomenon is more colloquially known as *ground bounce*. We are not talking here about external noise signals, but about noise which is induced on the supply rails by the switching action of each logic gate in the circuit.

As each gate changes state, a current pulse is taken from the supply pins because of the different device currents required in each state, the external loading, the transient caused by charging or discharging the node capacitance, and the conduction overlap in the totem-pole output stage. All these effects are present in all logic families to some extent, although CMOS types suffer little from the first two. In most cases, the node capacitance charging current dominates, more so with higher-speed circuits. The capacitance C_n must be charged with a current of

$$I = C_n \cdot dV/dt$$

Thus a 74AC-series gate with a dV/dt of around 1.6V/ns will require a 50 mA current pulse when charging a 30 pF node capacitance. Figure 14.8 shows the current paths. The significance of the supply current spike is that it causes a disturbance in the supply voltage and also in the ground line, because of the inductive impedance of the lines. A pulse with a di/dt of 50 mA/ns through a track inductance of 20 nH (one inch of track) will generate a voltage pulse of 1V peak, which is approaching the noise margin of fast logic. Supply voltage spikes are not too much of a problem as the logic high level noise immunity is usually good and they can be attenuated by proper decoupling,

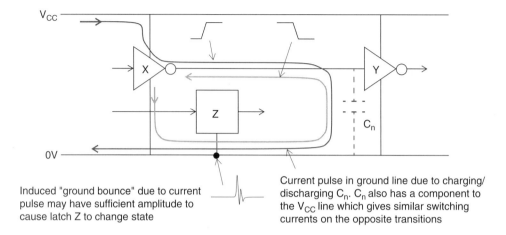

Induced "ground bounce" due to current pulse may have sufficient amplitude to cause latch Z to change state

Current pulse in ground line due to charging/discharging C_n. C_n also has a component to the V_{CC} line which gives similar switching currents on the opposite transitions

Figure 14.8: Induced ground noise due to switching currents

as the next section shows. Ground line disturbances are more threatening. Pulses on a high-impedance ground line can easily exceed the noise threshold and cause spurious switching of innocent gates. Only if a good, low-inductance ground plane system is maintained can this problem be minimized.

14.1.3.1   Synchronous Switching

The supply pin pulse current is magnified in synchronous systems when several gates switch simultaneously. A typical example is an octal bus buffer or latch whose data changes from #FF_H to #00_H. If all outputs are heavily loaded, as may be the case when the device is driving a large data bus, a formidable current pulse—exceeding an amp in fast systems—will pass through the ground pin. Worse, if seven bits of an octal latch change simultaneously, the induced ground bounce may corrupt the state of the eighth bit. You need to ensure that such devices are grounded to their loads with a very low-inductance ground system, preferably a true ground plane.

Ground noise on a microprocessor board can easily be observed by hooking a wide bandwidth oscilloscope to the ground line—you can connect the scope probe tip and its ground together and still see the noise, since the magnetic fields due to the ground currents will induce a signal in the loop formed by the probe leads. What you see is a regular series of narrow, ringing pulses spaced at the clock period. The amplitude of each pulse varies because the sum of the data transitions is random, but the timing does not. Such noise (Figure 14.9) is impossible to remove entirely.

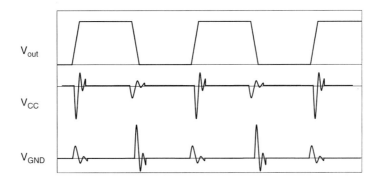

Figure 14.9: Switching noise on power and ground rails

14.1.4   Decoupling

No matter how good the V_{CC} and ground connections are, you cannot eliminate all line inductance. Except on the smallest boards, track distance will introduce an impedance which will create switching noise from the transient currents discussed in the last section. This is the reason for decoupling (Figure 14.10).

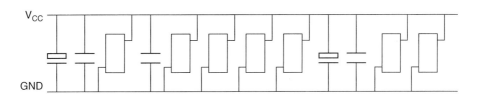

Figure 14.10: Logic decoupling scheme

14.1.4.1   Distance

The purpose of a decoupling capacitor is to maintain a low dynamic impedance from the individual IC supply voltage to ground. This minimizes the local supply voltage droop when a fast current pulse is taken from it. The word "decoupling" means isolating the local circuit from the supply impedance. Bearing in mind the speed of the current pulses just discussed, it should be clear that the capacitor must be located close to the circuit it is decoupling. "Close" in this context means less than half an inch for fast logic such as 74AC or ECL, especially when high current devices such as bus drivers are involved, extending to several inches for low-current, slow devices such as 4000B series CMOS.

If the decoupling current path between IC and capacitor is too long, the track inductance in conjunction with the capacitor forms a high-Q LC tuned circuit, and the ringing it generates will have a worse effect than no decoupling at all.

14.1.4.2 Capacitor Type and Value

The crucial factor for high-speed logic decoupling is lead inductance rather than absolute value. Minimum lead inductance offers a low impedance to fast pulses. Small chip capacitors are preferred, and the smaller the better, since this minimizes the package inductance. 0805, 0603 or even 0402 sizes are acceptable.

You can calculate the value if you want to by matching the transient current demand to the acceptable power rail voltage droop. Take for example a 74HC octal buffer, each of whose outputs when switching takes a transient current of 50 mA for 6 ns (calculated from $I = C_n \cdot dV/dt$). The total peak current demand is then 0.4A.

The acceptable voltage droop is perhaps 0.4V (equivalent to the worst system noise margin). Assume that the local decoupling capacitor supplies all of the current to hold the droop to this level, which is reasonable if other decoupling capacitors on the board are isolated by track inductance. Then the minimum capacitor value is

$$C = I \cdot t/V = 0.4 \cdot 6 \cdot 10^{-9}/0.4 = \underline{\textbf{6 nF}}$$

On the other hand, the actual value is noncritical, especially as the variables in the above calculation tend to be somewhat vague, and you will prefer to use the same component in all decoupling positions for ease of production. Values between 10 and 100 nF are recommended, a good compromise being 22 nF, which has both low self-inductance and respectable reservoir capacitance. It also tends to be cheaper than the higher values, particularly in the low-performance Z5U or Y5V ceramic grades, which are adequate for this purpose.

14.1.4.3 Capacitors Under the IC Package

Very high-speed and high-current logic ICs push the location requirement of the decoupling capacitor to its limits: it has to be right next to the supply pins. In fact, the inductance of the lead-out wires within the chip package becomes significant, and this has meant that locating power and ground pins in the middle of, or all around, the package rather than at opposite corners is necessary for high-performance large scale ICs.

For such devices it is necessary to locate the decoupling capacitors underneath the chip, on the opposite side of the board. The leads to the capacitors are then limited to the vias between the device pads, the planes and the capacitor pads. This is easily achievable with surface mount construction but of course not if you are restricted to through hole.

In fact the power and ground planes themselves are more effective at reducing the high frequency noise than are the decoupling capacitors, because their associated capacitance has no significant inductance. The closer together the planes are, the higher is this capacitance. You still need discrete decoupling capacitors for mid-frequency decoupling but their positioning becomes less important, as long as they are still located close to the relevant IC pins.

14.1.4.4 Low-Frequency Decoupling

You also need to decouple the supply rail against lower-frequency ripple due to varying logic load currents, as distinct from transient switching edges. The frequency of these ripple components is in the megahertz range and lower, so that widely distributed capacitance and low self-inductance are less important. Typically, they can be dealt with by a few tantalum electrolytics of 1–2 μF placed around the board, particularly where there are several devices that can turn on together and produce a significant drain from the power supply, such as burst refresh in dynamic RAM. Additionally, a single large capacitor of 10–47 μF at the power entry to the board is recommended to cope with frequency components in the kHz range.

Under normal circumstances, logic circuits are inherently insensitive to ripple on the supply lines. The exception is when they are faced with slow edges; if the ripple is at a substantially higher frequency than the edge and is modulating the signal, then as the signal passes through the transition region the logic element may undergo spurious switching (Figure 14.11).

Figure 14.11: Spurious switching on slow edges due to ripple

The safest way to deal with slow edges is to apply hysteresis with a Schmitt-trigger logic input.

14.1.4.5   Guidelines

The minimum requirements for good decoupling are:

- one 22 μF bulk capacitor per board;

- one 1 μF tantalum capacitor per 10 packages of SSI/MSI logic or memory;

- one 1 μF tantalum capacitor per 2–3 LSI packages;

- one 10–100 nF ceramic multilayer capacitor for each supply pin of an LSI package with multiple supply pins;

- one 10–100 nF ceramic multilayer capacitor for each octal IC or for each MSI package; and

- one 10–100 nF ceramic multilayer capacitor per 4 packages of SSI logic.

When in doubt, calculate the requirements for individual power/speed-hungry devices to make sure you have enough capacitors, and that they are in the right places.

14.1.5   Unused Gate Inputs

Frequently you will have spare gates or latches in a package left over, or will not be using all the inputs of a multi-input gate or latch. All such unused logic inputs must be tied to a fixed voltage, either high or low, and should never be left floating. Noise immunity of floating inputs is poor, so you should not float spare inputs of used gates, and especially not preset/clear inputs of latches or flip-flops, which are very sensitive to spikes. Figure 14.12 illustrates the options.

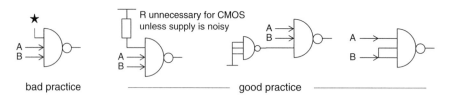

Figure 14.12: Connecting unused inputs

You must connect *all* unused CMOS inputs either to V_{CC} or ground. Floating any input is inadmissible, whether its gate is used or not. This is because the CMOS input has a very high impedance and consequently can float to any voltage if unconnected, and this voltage could be within the threshold switchover region of the gate. At this point both the P-channel and N-channel input transistors are conducting, which results in excessive current drain through the package. Due to the high gain of buffered gates, it is possible for a gate to oscillate, resulting in even higher current drain.

CMOS inputs can be connected directly to either rail; a protection resistor is unnecessary, as long as the supply is not expected to carry noise spikes that would exceed the maximum input voltage.

CHAPTER 15

Interfacing

Tim Williams

15.1 Mixing Analog and Digital

The two main problems that face designers who have to integrate analog and digital circuits on the same PCB are:

- preventing digital switching noise from contaminating the analog signal, and

- interfacing the wide range of analog input voltages to the digital circuit.

Generating analog outputs from digital signals is not usually a problem. Generating digital inputs from analog signals is.

15.1.1 Ground Noise

The high-frequency switching noise that was discussed earlier must be kept out of analog circuits at all costs. An analog-to-digital interface quantizes a variable analog signal into a digital word, and the number of bits in the word determines the resolution that can be achieved of the signal. Assuming a full-scale voltage range of 0 to 10V, which is typical of many analog-digital converters (ADCs), Table 15.1 shows the voltage levels that correspond to one bit change in the digital word.

You can see that the more resolution is demanded of the interface, the smaller the voltage change that will cause one bit change. 8 bits is regarded as commonplace in ADC circuits, 12 bits as reasonably high resolution (0.025%) and 16 bits as precision.

**Table 15.1: ADC resolution voltage for
different word lengths, 10V full-scale**

Word length	Resolution voltage
8 bit	39 mV
10 bit	10 mV
12 bit	2.4 mV
14 bit	0.6 mV
16 bit	0.15 mV

The significance of these diminishing voltage levels is that any noise that is coupled into the analog input will cause unwanted fluctuation of the digital value. For a 12-bit converter, a one-bit uncertainty will be given by noise of 2.4 mV at the converter input; for a 16-bit, this reduces to 150 microvolts. By contrast, the switching noise on the digital ground line is normally tens of millivolts and frequently hundreds of millivolts peak amplitude. If this noise were coupled into the converter input—and it is hard to keep ground noise out of the input—you would be unable to use a converter of greater precision than 8-10 bits.

15.1.2   Filtering

One partial solution to this problem is to filter the bandwidth of the analog signal to well below that of the noise so that the effective noise signal is reduced. For slowly varying analog signals this works reasonably well, especially if the noise injection occurs at the input of the signal-processing amplifier so that bandwidth limitation has maximum effect. Filtering is in any case good practice to minimize susceptibility to external noise.

Filtering the input amplifier is no use if the noise is injected into the ADC itself. For fast ADCs and wide-bandwidth analog signals you cannot take this approach anyway and the only available solution is to prevent the injection of digital noise at its source.

15.1.3   Segregation

The basic rule to follow when designing an analog-to-digital interface is to segregate the circuits, including grounds, completely. This means that:

*   separate analog and digital grounds should be established, connected only at one point, and

- the analog and digital sections of the circuit should be physically separated, with no digital tracks traversing the analog section, or vice versa. This will minimize crosstalk between the circuits.

It should be appreciated that no grounding scheme that establishes a multiplicity of different grounds can ever be optimum, because there will always be circuits that need to communicate signals across different ground areas. These signals are then particularly exposed to the nuances of both internal and external interference, or indeed may be the source of it. You should always strive to make such circuits low-risk in terms of their bandwidth and sensitivity, or else keep a single ground system for all circuits (both digital and analog) and take extreme care in its layout so that ground noise from one noisy part of the system does not circulate in another sensitive part.

15.1.4 Single-Board Systems

The appropriate grounding schemes for single-board and multi-board systems are shown in Figure 15.1. If your system has a single analog-to-digital converter, perhaps

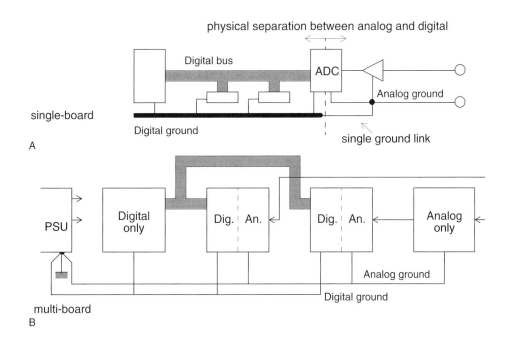

Figure 15.1: Layout for separate analog and digital grounds

with a multiplexer to select from several analog inputs, then the connection between analog and digital grounds can be made at this ADC as in Figure 15.1(A). This scheme requires that the analog and digital power supply returns are not linked together anywhere else, so that two separate power supply circuits are needed. The analog and digital grounds must be treated as entirely separate tracks, despite being nominally at the same potential; unavoidable noise currents circulating in the digital ground will then not couple into the "clean" analog ground. The digital ground should be of gridded or ground plane construction, whereas the analog section may benefit from a single-point grounding system, or may have a separate ground plane of its own. On no account should you extend the digital ground plane over the analog section of the board, since there will then be capacitive coupling from one ground plane to another.

15.1.5 Multi-Board Systems

When your system consists of several boards, some entirely digital, some entirely analog and some a mixture of the two, with an external power supply, then you cannot make the connection between digital and analog grounds at the ADC. There may be several ADCs in the one system. Instead, make the link at the power supply (Figure 15.1 (B)) and run separate analog and digital grounds to each board that requires them. Digital-only boards should be located physically closer to the power supply to minimize the radiating loop area or length.

15.2 Generating Digital Levels from Analog Inputs

The first rule when you want to use a varying analog voltage to generate an on/off digital signal—as distinct from an analog-to-digital conversion—is: always use either a comparator or a Schmitt-trigger gate. *Never* feed an analog signal straight into an ordinary TTL or CMOS gate input.

The reason is that ordinary gates do not have well-defined input voltage switching thresholds. Not only that, but they are also very critical of slow rise-time inputs. Very few analog input signals have the slew rate, typically faster than 5 V/µs, required to produce a clean output from an ordinary logic gate. The result of applying a slow analog voltage to a logic gate is shown in Figure 15.2.

A Schmitt trigger gate, or a comparator with hysteresis, will get over the slow rise time problem. A Schmitt trigger gate has the same output characteristics as an ordinary gate but it includes input hysteresis to ensure a fast transition. The threshold levels

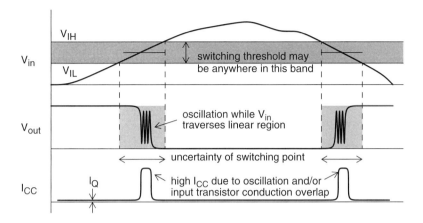

Figure 15.2: The effect of a slow input to a logic gate

of typical Schmitt devices, such as the 74HC14, are specified within wide tolerances and so do not overcome the variability of the actual switching point. When the analog levels corresponding to high and low states can be kept above V_{IH} and below V_{IL}, respectively, a Schmitt is adequate. For more precision you will need to use a comparator with an accurately specified reference voltage.

Second, if the analog supply rail range is greater than the logic supply, interfacing the analog signal straight to the logic input will threaten the gate with damage. This is possible even if the normal signal range is within the logic supply range; abnormal conditions such as turn-on or turn-off may exceed the rails. This, of course, is also a problem with Schmitt trigger gates. Normally, the inputs are protected by clamp diodes to the supply and ground rails, but the current through these must be limited to a safe level so a resistor in series with the input is essential. More positive steps to limit the input voltage, such as running the analog section from the same supply voltage as the logic (heeding the earlier advice about separate digital and analog ground rails), are to be preferred.

15.2.1 DebBouncing Switch Inputs

On the face of it, switch inputs to digital circuitry must be the easiest of interfaces. All you should need are an input port or gate, a pull-up resistor and a single pole switch (Figure 15.3). This circuit, though it undoubtedly works, is prone to a serious problem because of the electromechanical nature of the switch and the speed of logic devices.

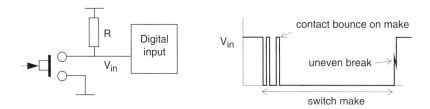

Figure 15.3: Contact bounce

When a switch contact operates, the current flow is not cleanly initiated or interrupted. As the contacts come together or part, the instantaneous contact resistance varies due to contamination, and the mating surfaces may "bounce" apart a few times due to the springiness of the material. As a result the switching edge is irregular and may easily consist of several discrete edges, extending over a period of typically 1 ms. You can verify this behavior simply by observing the input waveform of Figure 15.3 on a storage scope.

Of course, the digital input responds very fast to each crossing of the switching threshold, and consequently the port or gate sees several transitions each time the switch is operated, before it settles to a steady-state 1 or 0. This may not be a problem for level-sensitive inputs, but it undoubtedly is for edge-sensitive ones such as counter or latch clock inputs. Mistriggering of counter circuits that are fed from a switch input is commonly caused by this phenomenon.

The simple solution to contact bounce is to filter the logic input with an RC network (Figure 15.4(A)). The RC time constant must be significantly longer than the bounce period to effectively attenuate the contact noise. This has the extra advantage of protecting against induced impulsive or RF interference, but it requires additional discrete components and demands that the logic input must be a Schmitt-trigger type, since the input rise time has been deliberately slowed.

If the switch input may change state quickly, an RC time constant that is sufficiently long to cure the bounce will slow the response to the switch unacceptably. This can be overcome in two ways: the RS latch, Figure 15.4(B), which requires a changeover rather than single-throw switch, or a software- or hardware-implemented delay. Figure 15.4(C) shows the hardware delay, which uses a continuously clocked shift register and OR gate to effectively "window out" the bounce. The delay can be adjusted to suit the bounce period. These two solutions are most suited to realization with semicustom logic arrays or ASICs, where the overhead of the extra logic is low.

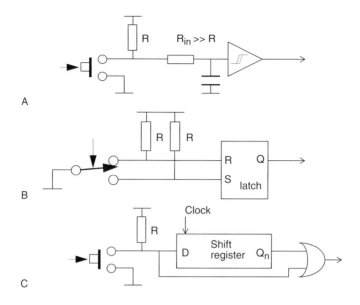

Figure 15.4: Switch debouncing circuits

15.3 Protection Against Externally Applied Overvoltages

Logic inputs and outputs which are taken off-board will be subject at some time in the life of the system to an overvoltage. Your philosophy in this respect should be: if it can happen, it will. Overvoltages can be applied by misconnection of the board or of external equipment, or can be due to static build-up. The latter is a particular threat to CMOS inputs with their high impedance, but the effect of a large static discharge can also be disastrous for other logic families.

There are three major consequences of an overvoltage on a logic signal line:

- immediate damage of the device due to rupture of the track metallization or destruction of the silicon;

- progressive degradation of device characteristics when the overvoltage does not have enough energy to destroy it immediately;

- latch-up, where damage may be caused by excessive power supply current following a transient overvoltage.

Modern logic families include some protection both at their inputs and outputs in the form of clamping diodes to the supply lines, but these diodes are limited in their

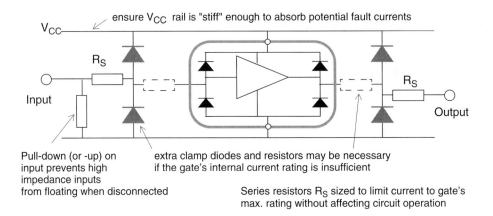

ensure V_{CC} rail is "stiff" enough to absorb potential fault currents

Pull-down (or -up) on
input prevents high
impedance inputs
from floating when disconnected

extra clamp diodes and resistors may be necessary
if the gate's internal current rating is insufficient

Series resistors R_S sized to limit current to gate's
max. rating without affecting circuit operation

Figure 15.5: Logic gate I/O protection

current-handling ability and therefore the potential fault current that can be applied
due to an overvoltage must be limited. This is best achieved by the methods of
Figure 15.5.

The external clamp diodes are used to take the lion's share of the incoming overload
current and divert it to the V_{CC} or 0V rails; the resistors shown dotted are needed if the
IC's internal diodes would otherwise still take too much current because of the ratio of
their forward voltage to the external diodes' forward voltages. The power rail takes
the excess incoming current and must therefore be of a low enough impedance for its
voltage to be substantially unaffected by this current being dumped into it. This may
call for a review of the regulator philosophy, or for extra clamps to be applied on the
power rail local to the interfaces. Series resistors R_S may be adequate in themselves
without external clamp diodes, especially on inputs, where they can be used to limit
the current to what can be handled by the IC's internal diodes.

15.4 Isolation

Even if you take precautions against input/output abuse, it is not good practice to take
logic signals directly into or out of equipment. As well as facing the threat of overload
on individual lines, you also have to extend the ground and/or supply rails outside the
equipment to provide a signal return path. These then act as antennas both to radiate
ground noise out of the equipment and to conduct external interference back into it.
It is very much safer to keep power rails within the bounds of the equipment case.

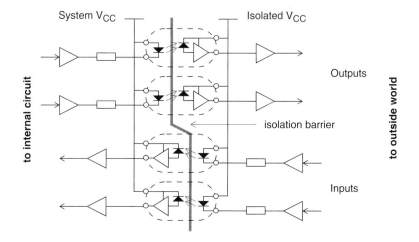

Figure 15.6: Interface isolation using opto-couplers

A common technique to achieve this is to electrically isolate all signal lines entering or leaving the equipment. As well as guarding against interference, this eliminates problems from ground loops and ground differentials. Digital signals lend themselves to the use of opto-couplers. An opto-coupler is basically an LED chip integrated in the same package as a light-sensitive device such as a photodiode or phototransistor, the two components being electrically separate but optically coupled. A typical isolation scheme using such devices is shown in Figure 15.6.

One opto-coupler is needed per digital channel. Opto-couplers can be sourced as single, dual or quad packages, and the price for commercial grade units can vary depending mainly on the required speed and level of integration from 25p to £5 per channel. Clearly, in cost- or space-sensitive applications the number of isolated channels should be minimized. This tends to mean that isolation is more common in industrial products than consumer ones.

15.4.1 Opto-Coupler Trade-Offs

There are a number of quite complex trade-offs to make when you use opto-isolation. Factors to consider are:

- *Speed of the interface*: cheap couplers with standard transistor outputs have switching times of 2–5 μs so are limited to data rates of around 100 kbits/s maximum. High speed devices with data rates of 10 Mbits/s are available but cost over £5 per channel.

- *Power consumption*: standard transistor output types offer a current transfer ratio (CTR) typically between 10 and 80%. This is the ratio of LED input current to transistor output current in the on-state. Thus for a required output current of 1 mA with a CTR of 20%, 5 mA would be needed through the LED. Also, CTR degrades with time and you should include an extra safety margin of between 20 and 50%, depending on expected lifetime and operating current, to ensure end-of-life circuit reliability. Reducing the operating current reduces the speed of the interface. Darlington-output optocouplers are available with CTRs of 200–500%, but these unfortunately have turn-off times of around 100 μs so are only useful for low-speed purposes. Opto-coupler drive current can be a significant fraction of the overall power requirement, especially on the isolated side.

- *Support circuitry*: a simple photo-transistor or photo-Darlington output needs several passive components plus a buffer gate to interface it correctly to logic levels. Alternatively, you can get opto-couplers which have logic-compatible inputs and outputs, especially the faster ones, but at a significantly higher cost. Low-current LED drive requirements can be met directly by a logic gate and series limiting resistor, whereas if you are using a cheaper opto with higher LED current you will need an extra buffer.

15.4.2 Coupling Capacitance

Although an opto-coupler breaks the electrical connection at DC, with an isolation voltage measured in kV, there is still some residual coupling capacitance which reduces the isolation at high frequencies. The specification figures of 0.5–2 pF are increased somewhat by stray wiring capacitance which is layout dependent. Input and output pins are invariably on opposite sides of the package. There is no point in designing in an opto-coupler for isolation if you then run the output tracks back alongside the input tracks!

The coupling capacitance of individual channels, multiplied by the number of channels in the system, means that a significant level of high frequency ground noise may still be coupled out of an isolated system, or fast rise time transients or RFI may still be coupled in. (This is another argument for minimizing the number of channels.) Also, high common mode dV/dt signals can be coupled directly into the photodiode or transistor input through this capacitance and cause false switching. This effect is reduced by incorporating an electrostatic screen across the optical path and connecting it to the output ground pin, and some opto-couplers are available with this screen included. Common mode transient immunity can vary from worse than 100 V/μs to better than 5 kV/μs (for the expensive devices).

15.4.3 Alternatives to Opto-Couplers

Two alternatives to opto-couplers for isolating digital signals are relays and pulse transformers. The relay is a well-established device and is a good choice if its restrictions of size, weight, speed, power consumption and electromechanical nature are acceptable.

Pulse transformers are most useful for passing wide bandwidth, high speed digital data for which opto-couplers are too slow or too expensive. They can also be designed for good immunity to high dV/dt interference. The data must be coded or modulated to remove any DC component. This requires an overhead of a few gates and a latch per channel, but this overhead may be acceptable, especially if you are already using semicustom silicon, and may easily be outweighed by the attractions of high speed and low power consumption.

15.5 Classic Data Interface Standards

When you want to connect logic signals from one piece of equipment to another, it is not sufficient to use standard logic devices and make direct gate-to-gate connections, even if they are isolated from the main system. Standard logic is not suited to driving long lines; line terminations are unspecified and noise immunity is low, so that reflections and interference would give unacceptably high data corruption. External logic interfaces must be specially designed for the purpose.

At the same time, it is essential that there is some commonality of interface between different manufacturers' equipment. This allows the user to connect, say, a computer from manufacturer A to a printer from manufacturer B without worrying about electrical compatibility. There is therefore a need for a standard definition for electrical interface signals. This need has been recognized for many years, and there are a wide variety of data interchange standards available. The logic of the marketplace has dictated that only a small number of these are dominant. This section will consider the two main commercial ones: EIA-232F and EIA-422. EIA-232F is an update of the popular RS-232C standard published in 1969, to bring it into line with the international CCITT V.24 and V.28 and ISO IS2110 standards. EIA-422 is the same as the earlier

RS-422 standard. The prefix changes are cosmetic, purely to identify the source of the standards as the EIA.

15.5.1   EIA-232F

The boom in data communications has led to many products which make interface conformity claims by quoting "RS-232" in their specifications. Some of these claims are in fact quite spurious, and discerning users will regard interface conformity as an indicator of product quality, and test it early on in their evaluation. The major characteristics of the specification are given in Table 15.2. As well as specifying the electrical parameters, EIA-232F also defines the mechanical connections and pin configuration, and the functional description of each data circuit.

By modern standards the performance of EIA-232F is primitive. It was originally designed to link data terminal equipment (DTE) to modems, known as data communications equipment (DCE). It was also used for data terminal-to-mainframe interfaces. These early applications were relatively low speed, less than 20 kbaud, and used cables shorter than 50 feet. Applications which call for such limited capability are now abundant, hence the standard's great popularity. Its new revision recognizes this by replacing the phrase "data communication equipment" with "data circuit-terminating equipment," also abbreviated to DCE. It does not clarify exactly what is a DTE and what is a DCE, and since many applications are simple DTE (computer) to DTE (terminal or printer) connections, it is often open to debate as to what is at which end of the interface. Although a point-to-point connection provides the correct pin terminations for DTE-to-DCE, a useful extra gadget is a cable known as a *null modem* (Figure 15.7), which creates a DTE-to-DTE connection. The common sight of an installation technician crouched over a 9-way connector swapping pins 2 and 3, to make one end's receiver listen to the other end's driver, has yet to disappear.

EIA-232F's transmission distance is limited by its unbalanced design and restricted drive current. The unbalanced design is very susceptible to external noise pick-up and to ground shifts between the driver and receiver. The limited drive current means that the slew rate must be kept slow enough to prevent the cable becoming a transmission line, and this puts a limit on the fastest data rate that can be accommodated. Maximum cable length, originally fixed at 50 feet, is now restricted by a requirement for maximum load capacitance (including receiver input) for each circuit of 2500 pF. As the line length increases so does its capacitance, requiring more current to maintain the same transition time. The graph of Figure 15.8 shows the drive current versus load capacitance required

Table 15.2: Major electrical characteristics of EIA

Interface	EIA-232F	EIA-422	EIA-485
Line type	Unbalanced, point-to-point	Balanced, differential, multidrop (one driver per bus)	Balanced, differential, multiple drivers per bus (half duplex)
Line impedance	Not applicable	100Ω	120Ω
Max. line length	Load dependent, typically 15m depending on capacitance	L ≈105/B meters B = bit rate, kB/s	Max. recommended 1200m, depending on attenuation
Max data rate	20 kB/s	10 MB/s	10 MB/s
Driver			
Output voltage	±5 to ±15V loaded with 3–7 kΩ + V = logic 0, –V = logic 1	±10V max differential unloaded, ±2V min loaded with 100Ω	±6V max differential unloaded, ±1.5V min loaded with 54Ω
Short circuit current	500 mA max	150 mA max	150 mA to gnd, 250 mA to –7 or +12V
Rise time	4% of unit interval (1 ms max) 30 V/µs max. slew rate	10% of unit interval (min 20 ns)	30% of unit interval
Output with power off	>300Ω output resistance	±100 µA max leakage	>12 kΩ output resistance
Receiver			
Sensitivity	±3V max thresholds	±200 mV	±200 mV
Input impedance	3 kΩ –7 kΩ, <2500 pF	4 kΩ min	12 kΩ
Common mode range	Not applicable	±7V	+12 to –7V

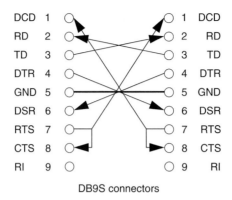

DB9S connectors

Figure 15.7: The null modem

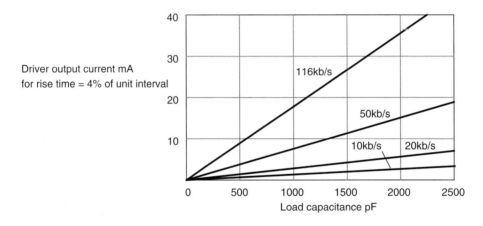

Figure 15.8: EIA-232F transmit driver output current versus C$_L$

to maintain the 4% transition time relationship at different data rates. In practice, the line length is limited to 3 meters or less for data rates more than 20 kb/s. Most drivers can handle the higher transmission rates over such a short length without drawing excessive supply current.

Note that there are several common "enhancements" that are not permitted by strict adherence to the standard. EIA-232F makes no provision for tristating the driver output, so multiple driver access to one line is not possible. Similarly, paralleling receivers is not allowed unless the combined input impedance is held between 3 kΩ and 7 kΩ. It does not consider electrically isolated interfaces: no specification is offered for isolation requirements, despite their desirability. It does not specify the communication data format. The usual "one start bit, eight data bits, two stop bits" format is not part of the standard, just its most common application. It is not directly compatible with another common single-ended standard, EIA-423, although such connections will usually work. Also, you cannot legitimately run EIA-232F off a ±5V supply rail—the minimum driver *output* voltage is specified as ±5V, loaded with 3–7 kΩ and with an output impedance of 300Ω.

The standard calls for slew-rate limiting to 30 V/µs maximum. Although you can do this with an output capacitor, which operates in conjunction with the output transistor's current limit while it is slewing, this will increase the dissipation, and reduces the maximum possible cable length. It is preferable to use a driver that has on-chip slew rate limiting, requiring no external capacitors and making the slew rate independent of cable length.

15.5.2   EIA-422

Many data communications applications now require data rates in the megabaud region, for which EIA-232F is inadequate. This need is fulfilled by the EIA-422 standard, which is an electrical specification for drivers and receivers for use in a balanced or differential, point-to-point or multi-drop high speed interface using twisted pair cable. Table 15.2 summarizes the EIA-422 specification in comparison with EIA- 232F. One driver and up to ten receivers are allowed. The maximum data rate is specified as 10 Mbaud, with a trade-off against cable length; maximum cable length at 100 kbaud is 4000 feet. Note that unlike EIA-232F, EIA-422 does not specify functional or mechanical parameters of the interface. These are included in other standards which incorporate it, notably EIA-449 and EIA-530.

EIA-422 achieves its high-speed and long-distance capabilities by specifying a balanced and terminated design. The balanced design reduces sensitivity to external common mode noise and allows a ground differential of up to a few volts to exist between the driver and one or more of the receivers without affecting the receiver's thresholds. A cable termination, together with increased driver current, allows fast slew rates which in turn allows high data rates. If the cable is not terminated, serious ringing on the edges occurs which may cause spurious switching in the receiver. The specified termination of 100Ω is closely matched to the characteristic impedance of typical twisted pair cables. Only one termination is used, at the receiver at the far end of the cable.

15.5.3   Interface Design

By far the easiest way to realize either EIA-232F or EIA-422 interfaces is to use one of the many specially tailored driver and receiver chip sets that are available. The more common ones, such as the 1488 driver/1489 receiver for EIA-232F or the 26LS31 driver/26LS32 receiver for EIA-422, are available competitively from many sources and in low-power CMOS versions. You can also obtain combined driver/receiver parts so that a small interface can be handled with one IC. Because the 9-pin implementation of EIA-232F is so common, a single package 3-transmitter plus 5-receiver part is also widely sourced. The high-voltage requirement of EIA-232F, typically ±12V supplies, is addressed by some suppliers who offer on-chip DC-to-DC converters from the =5V rail.

Figure 15.9 suggests typical interface circuits for the two standards. Note the inclusion of power supply isolating diodes, to protect the rest of the circuit against

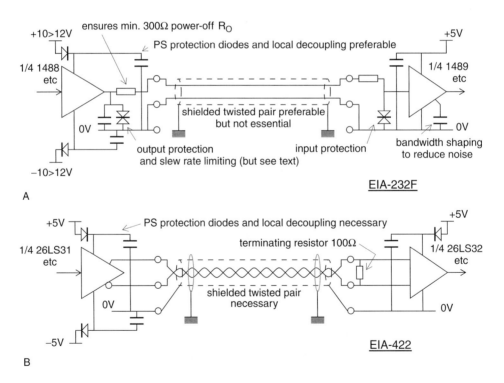

Figure 15.9: Typical EIA-232F and EIA-422 interface circuits

the inevitable overvoltages that will come its way. You can also construct an interface, particularly the simpler EIA-232F, using standard components such as op-amps, comparators, CMOS buffer devices or discrete components if you are prepared to spend some time characterizing the circuit against the requirements of the standard and against expected overload conditions. This may turn out to be marginally cheaper in component cost, but its overall worth is somewhat questionable.

15.6   High Performance Data Interface Standards

This section briefly reviews some of the newer data interface standards that have grown up for high-speed purposes around particular applications and have subsequently become more widely entrenched.

15.6.1   EIA-485

EIA-485 shares many similarities with EIA-422, and is widely used as the basis for inhouse and industrial datacomm systems. For instance, one variant of the SCSI interface (HVD-SCSI: high voltage differential—small computer systems interface) uses 485 as the basis for its electrical specification. 485-compliant devices can be used in 422 systems, though the reverse is not necessarily true. The principal difference is that 485 allows multiple transmitters on the same line, driving up to 32 unit loads, with halfduplex (bidirectional) communication. One unit load is defined as a steady-state load allowing 1 mA of current under a maximum common mode voltage of 12V or 0.8 mA at –7V. ULs may consist of drivers or receivers and failsafe resistors (see below), but do not include the termination resistors. The bidirectional communication means that 485 drivers must allow for line contention and for driving a line that is terminated at each end with 120Ω. The two specifications are compared in Table 15.2.

One further problem that arises in a half-duplex system is that there will be periods when no transmitters are driving the line, so that it becomes high impedance, and it is desirable for the receivers to remain in a fixed state in this situation. This means that a differential voltage of more than 200 mV should be provided by a suitable passive circuit that complies with both the termination impedance requirements and the unit load constraints. A network designed to do this is called a "failsafe" network.

15.6.2 CAN

The Controller Area Network (CAN) standard was originally developed within the automotive industry to replace the complex electrical wiring harness with a two-wire data bus. It has since been standardized in ISO 11898. The specification allows signaling rates up to 1 MB/s, high immunity from electrical interference, and an ability to self-diagnose and repair errors. It is now widespread in many sectors, including factory automation, medical, marine, aerospace and of course automotive. It is particularly suited to applications requiring many short messages in a short period of time with high reliability in noisy operating environments.

The ISO 11898 architecture defines the lowest two layers of the OSI/ISO seven layer model, that is, the data-link layer and the physical layer. The communication protocol is carrier-sense multiple access, with collision detection and arbitration on message priority (CSMA/CD=AMP). The first version of CAN was defined in ISO 11519 and allowed applications up to 125 kB/s with an 11-bit message identifier. The 1 MB/s ISO 11898:1993 version is standard CAN 2.0A, also with an 11-bit identifier, while Extended CAN 2.0B is provided in a 1995 amendment to the standard and provides a 29-bit identifier.

The physical CAN bus is a single twisted pair, shielded or unshielded, terminated at each end with 120Ω. Balanced differential signaling is used. Nodes may be added or removed at any time, even while the network is operating. Un-powered nodes should not disturb the bus, so transceivers should be configured so that their pins are in a high impedance state with the power off. The standard specification allows a maximum cable length of 40m with up to 30 nodes, and a maximum stub length (from the bus to the node) of 0.3m. Longer stub and line lengths can be implemented, with a trade-off in signaling rates. The recessive (quiescent) state is for both bus lines to be biased equally to approximately 2.5V relative to ground; in the dominant state, one line (CANH) is taken positive by 1V while the other (CANL) is taken negative by the same amount, giving a 2V differential signal. The required common mode voltage range is from –2V to =7V, i.e., ±4.5V about the quiescent state.

15.6.3 USB

The Universal Serial Bus (USB) is a cable bus that supports data exchange between a host computer and a wide range of simultaneously accessible peripherals. The attached peripherals share USB bandwidth through a host scheduled, token-based protocol.

The bus allows peripherals to be attached, configured, used, and detached while the host and other peripherals are in operation. There is only one host in any USB system. The USB interface to the host computer system is referred to as the Host Controller, which may be implemented in a combination of hardware, firmware, or software.

USB devices are either hubs, which act as wiring concentrators and provide additional attachment points to the bus, or system functions such as mice, storage devices or data sources or outputs. A root hub is integrated within the host system to provide one or more attachment points.

The USB transfers signal and power over a four-wire point-to-point cable. A differential input receiver must be used to accept the USB data signal. The receiver has an input sensitivity of at least 200 mV when both differential data inputs are within the common mode range of 0.8V to 2.5V. A differential output driver drives the USB data signal with a static output swing in its low state of <0.3V with a 1.5 kΩ load to 3.6V and in its high state of >2.8V with a 15 kΩ load to ground. A full-speed USB connection is made through a shielded, twisted pair cable with a characteristic impedance (Z_0) of 90Ω 15% and a maximum one-way delay of 26 ns. The impedance of each of the drivers must be between 28 and 44Ω. The detailed specification controls the rise and fall times of the output drivers for a range of load capacitances.

In version 1.1, there are two data rates:

- the full-speed signaling bit rate is 12 Mb/s;

- a limited capability low-speed signaling mode is also defined at 1.5 Mb/s.

Both modes can be supported in the same USB bus by automatic dynamic mode switching between transfers. The low-speed mode is defined to support a limited number of low-bandwidth devices, such as mice. In order to provide guaranteed input voltage levels and proper termination impedance, biased terminations are used at each end of the cable. The terminations also allow detection of attachment at each port and differentiate between full-speed and low-speed devices. The USB 2.0 specification adds a high-speed data rate of 480 MB/s between compliant devices using the same cable as 1.1, with both source and load terminations of 45Ω.

The cable also carries supply wires, nominally $=5$V, on each segment to deliver power to devices. Cable segments of variable lengths, up to several meters, are possible. The specification defines connectors, and the cable has four conductors: a twisted signal pair of standard gauge and a power pair in a range of permitted gauges.

The clock is transmitted, encoded along with the differential data. The clock encoding scheme is nonreturn-to-zero with bit stuffing to ensure adequate transitions. A SYNC field precedes each packet to allow the receiver(s) to synchronize their bit recovery clocks.

15.6.4   Ethernet

Ethernet is a well-established specification for serial data transmission. It was first published in 1980 by a multivendor consortium that created the DEC-Intel-Xerox (DIX) standard. In 1985, Ethernet was standardized in IEEE 802.3; since that time it has been extended a number of times. "Classic" Ethernet operates at a data transmission rate of 10 Mbit/s. Since the 1990s, Ethernet has developed in the following areas:

- Transmission media
- Data transmission rates
  - Fast Ethernet at 100 Mbit/s (1995)
  - Gigabit Ethernet at 1 Gbit/s (1999)
- Network topologies.

Nowadays, Ethernet is the most widespread networking technology in the world in commercial information technology systems, and is also gaining importance in industrial automation. All network users have the same rights under Ethernet. Any user can exchange data of any size with another user at any time, and any network device that is transmitting is heard by all other users. Each Ethernet user filters the data packets that are intended for it out from the stream, ignoring all the others.

In the standard Ethernet, all the network users share one collision domain. Network access is controlled by the CSMA/CD procedure (carrier sense multiple access with collision detection). Before transmitting data, a network user first checks whether the network is free (carrier sense). If so, it starts to transmit data. At the same time it checks whether other users have also begun to transmit (collision detection). If that is the case, a collision occurs. All the network users concerned now stop their transmission, wait for a period of time determined according to a randomizing principle, and then start transmission again. The result of this is that the time required to transmit data packets depends heavily on the network loading, and cannot be determined in advance. The more collisions occur, the slower the entire network becomes.

This lack of determinism can be overcome by a variant of the basic approach known as *switched* Ethernet. This refers to a network in which each Ethernet user is assigned a port in a switch, which analyzes all the data packets as they arrive, directing them on to the appropriate port. Switches separate former collision domains into individual point-to-point connections between the network components and the relevant user equipment. Preventing collisions makes the full network bandwidth available to each point-to-point connection. The second pair of conductors in the four-wire Ethernet cable, which otherwise is needed for collision detection, can now be used for transmission, so providing a significant increase in data transfer rate.

The Ethernet interface at each user is defined according to Figure 15.10. It is usual to find structured twisted pair local area network wiring already integrated within a building, and the cabling characteristics are given in IEC 11801 and related standards; hence the 10Base-T and 100Base-T variants are the most popular of the Ethernet implementations, and the appropriate MAU/MDI using the RJ45 connector are included in most types of computer. The maximum lengths are set by signal timing limitations in the Fast Ethernet implementation, and an Ethernet system implementation relies on correct integration of cable lengths, types and terminations.

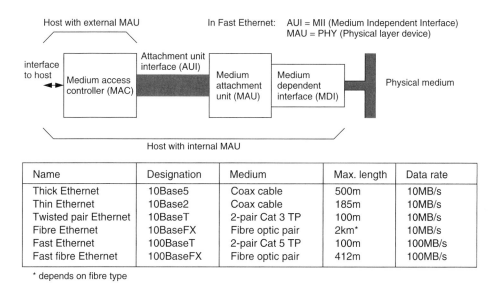

Name	Designation	Medium	Max. length	Data rate
Thick Ethernet	10Base5	Coax cable	500m	10MB/s
Thin Ethernet	10Base2	Coax cable	185m	10MB/s
Twisted pair Ethernet	10BaseT	2-pair Cat 3 TP	100m	10MB/s
Fibre Ethernet	10BaseFX	Fibre optic pair	2km*	10MB/s
Fast Ethernet	100BaseT	2-pair Cat 5 TP	100m	100MB/s
Fast fibre Ethernet	100BaseFX	Fibre optic pair	412m	100MB/s

* depends on fibre type

Figure 15.10: Ethernet interface and media

In contrast to the coaxial versions of Ethernet, which may be connected in multidrop, each segment of twisted pair or fiber route is a point-to-point connection between hosts; this means that a network system that is more than simply two hosts requires a number of hubs or switches, which integrate the connections to each user. A hub will simply pass through the Ethernet traffic between its ports without controlling it in any way, but a switch does control the traffic, separating packets to their destination ports.

The 100Base-T electrical characteristics are a peak differential output signal of 1V into a 100Ω characteristic impedance twisted pair; the 10Base-T level is 2.5V. The rise and fall time and amplitude symmetries are also defined to achieve a high level of balance and hence common mode performance. It is normal to use a transformer and common mode choke to isolate the network connection from the driver electronics.

CHAPTER 16

DSP and Digital Filters

Walt Kester

16.1 Origins of Real-World Signals and Their Units of Measurement

Let's first look at a few key concepts and definitions required to lay the groundwork for things to come.

Webster's *New Collegiate Dictionary* defines a signal as "a detectable (or measurable) physical quantity or impulse (as voltage, current, or magnetic field strength) by which messages or information can be transmitted." Key to this definition are the words: *detectable, physical quantity*, and *information*.

By their very nature, signals are analog, whether DC, AC, digital levels, or pulses. It is customary, however, to differentiate between *analog* and *digital* signals in the following

- Signal Characteristics
  - ◆ Signals Are Physical Quantities
  - ◆ Signals Are Measurable
  - ◆ Signals Contain Information
  - ◆ All Signals Are Analog

- Units of Measurement
  - ◆ Temperature: °C
  - ◆ Pressure: Newtons/m^2
  - ◆ Mass: kg
  - ◆ Voltage: Volts
  - ◆ Current: Amps
  - ◆ Power: Watts

Figure 16.1: Signal characteristics

manner: Analog (or real-world) variables in nature include all measurable physical quantities. In this book, *analog* signals are generally limited to electrical variables, their rates of change, and their associated energy or power levels. Sensors are used to convert other physical quantities such as temperature or pressure to electrical signals. The entire subject of signal conditioning deals with preparing real-world signals for processing, and includes such topics as sensors (temperature and pressure, for example), isolation amplifiers, and instrumentation amplifiers (Ref. 16.1).

Some signals result in response to other signals. A good example is the returned signal from a radar or ultrasound imaging system, both of which result from a known transmitted signal.

On the other hand, there is another classification of signals, called *digital*, where the actual signal has been conditioned and formatted into a digit. These digital signals may or may not be related to real-world analog variables. Examples include the data transmitted over local area networks (LANs) or other high speed networks.

In the specific case of digital signal processing (DSP), the analog signal is converted into binary form by a device known as an analog-to-digital converter (ADC). The output of the ADC is a binary representation of the analog signal and is manipulated arithmetically by the digital signal processor. After processing, the information obtained from the signal may be converted back into analog form using a digital-to-analog converter (DAC).

Another key concept embodied in the definition of *signal* is that there is some kind of *information* contained in the signal. This leads us to the key reason for processing real-world analog signals: the *extraction of information*.

16.2   Reasons for Processing Real-World Signals

The primary reason for processing real-world signals is to extract information from them. This information normally exists in the form of signal amplitude (absolute or relative), frequency or spectral content, phase, or timing relationships with respect to other signals. Once the desired information is extracted from the signal, it may be used in a number of ways.

In some cases, it may be desirable to reformat the information contained in a signal. This would be the case in the transmission of a voice signal over a frequency-division multiple access (FDMA) telephone system. In this case, analog techniques are used to "stack" voice channels in the frequency spectrum for transmission via microwave relay, coaxial cable, or fiber. In the case of a digital transmission link, the analog voice information is first converted into digital using an ADC. The digital information representing the individual voice channels is multiplexed in time (time division multiple access, or TDMA) and transmitted over a serial digital transmission link (as in the T-carrier system).

Another requirement for signal processing is to compress the frequency content of the signal (without losing significant information), then format and transmit the information at lower data rates, thereby achieving a reduction in required channel bandwidth. High speed modems and adaptive pulse code modulation systems (ADPCM) make extensive use of data reduction algorithms, as do digital mobile radio systems, MPEG recording and playback, and high definition television (HDTV).

Industrial data acquisition and control systems make use of information extracted from sensors to develop appropriate feedback signals which in turn control the process itself. Note that these systems require both ADCs and DACs as well as sensors, signal conditioners, and the DSP (or microcontroller). Analog Devices offers a family of MicroConverters™ that includes precision analog conditioning circuitry, ADCs, DACs, microcontroller, and FLASH memory all on a single chip.

In some cases, the signal containing the information is buried in noise, and the primary objective is signal recovery. Techniques such as filtering, autocorrelation, and convolution are often used to accomplish this task in both the analog and digital domains.

- Extract Information about the Signal (Amplitude, Phase, Frequency, Spectral Content, Timing Relationships)
- Reformat the Signal (FDMA, TDMA, CDMA Telephony)
- Compress Data (Modems, Cellular Telephone, HDTV, MPEG)
- Generate Feedback Control Signal (Industrial Process Control)
- Extract Signal from Noise (Filtering, Autocorrelation, Convolution)
- Capture and Store Signal in Digital Format for Analysis (FFT Techniques)

Figure 16.2: Reasons for signal processing

16.3 Generation of Real-World Signals

In most of the previous examples (the ones requiring DSP techniques), both ADCs and DACs are required. In some cases, however, only DACs are required where real-world analog signals may be generated directly using DSP and DACs. Video raster scan display systems are a good example. The digitally generated signal drives a video or RAMDAC. Another example is artificially synthesized music and speech. In reality, however, the real-world analog signals generated using purely digital techniques do rely on information previously derived from the real-world equivalent analog signals. In display systems, the data from the display must convey the appropriate information to the operator. In synthesized audio systems, the statistical properties of the sounds being generated have been previously derived using extensive DSP analysis of the entire signal chain, including sound source, microphone, preamp, and ADC.

16.4 Methods and Technologies Available for Processing Real-World Signals

Signals may be processed using analog techniques (analog signal processing, or ASP), digital techniques (digital signal processing, or DSP), or a combination of analog and digital techniques (mixed-signal processing, or MSP). In some cases, the choice of techniques is clear; in others, there is no clear-cut choice, and second-order considerations may be used to make the final decision.

With respect to DSP, the factor that distinguishes it from traditional computer analysis of data is its speed and efficiency in performing sophisticated digital processing functions such as filtering, FFT analysis, and data compression in real time.

The term *mixed-signal processing* implies that *both* analog and digital processing is done as part of the system. The system may be implemented in the form of a printed circuit board, hybrid microcircuit, or a single integrated circuit chip. In the context of this broad definition, ADCs and DACs are considered to be mixed-signal processors, since both analog and digital functions are implemented in each. Recent advances in very large scale integration (VLSI) processing technology allow complex digital processing as well as analog processing to be performed on the same chip. The very nature of DSP itself implies that these functions can be performed in *real time*.

16.5   Analog Versus Digital Signal Processing

Today's engineer faces a challenge in selecting the proper mix of analog and digital techniques to solve the signal processing task at hand. It is impossible to process real-world analog signals using purely digital techniques, since all sensors, including microphones, thermocouples, strain gages, piezoelectric crystals, and disk drive heads are analog sensors. Therefore, some sort of signal conditioning circuitry is required in order to prepare the sensor output for further signal processing, whether it be analog or digital. Signal conditioning circuits are, in reality, analog signal processors, performing such functions as multiplication (gain), isolation (instrumentation amplifiers and isolation amplifiers), detection in the presence of noise (high common-mode instrumentation amplifiers, line drivers, and line receivers), dynamic range compression (log amps, LOGDACs, and programmable gain amplifiers), and filtering (both passive and active).

Several methods of accomplishing signal processing are shown in Figure 16.3. The top portion of the figure shows the purely analog approach. The latter parts of the figure

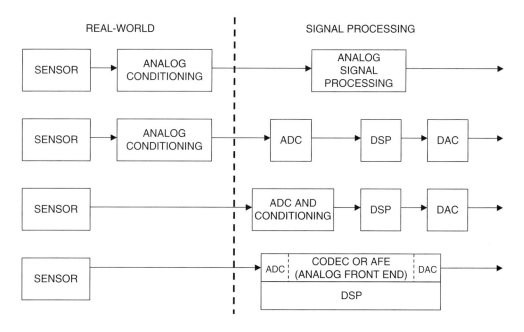

Figure 16.3: Analog and digital signal processing options

show the DSP approach. Note that once the decision has been made to use DSP techniques, the next decision must be where to place the ADC in the signal path.

In general, as the ADC is moved closer to the actual sensor, more of the analog signal conditioning burden is now placed on the ADC. The added ADC complexity may take the form of increased sampling rate, wider dynamic range, higher resolution, input noise rejection, input filtering, programmable gain amplifiers (PGAs), and on-chip voltage references, all of which add functionality and simplify the system. With today's high resolution/high sampling rate data converter technology, significant progress has been made in integrating more and more of the conditioning circuitry within the ADC/DAC itself. In the measurement area, for instance, 24-bit ADCs are available with built-in programmable gain amplifiers (PGAs) that allow full-scale bridge signals of 10 mV to be digitized directly with no further conditioning (e.g., AD773x series). At voice-band and audio frequencies, complete coder/decoders (codecs or analog front ends) are available with sufficient on-chip analog circuitry to minimize the requirements for external conditioning components (AD1819B and AD73322). At video speeds, analog front ends are also available for such applications as CCD image processing and others (e.g., AD9814, AD9816, and the AD984x series).

16.6   A Practical Example

As a practical example of the power of DSP, consider the comparison between an analog and a digital low-pass filter, each with a cut-off frequency of 1 kHz. The digital filter is implemented in a typical sampled data system shown in Figure 16.4. Note that there are several implicit requirements in the diagram. First, it is assumed that an ADC/DAC combination is available with sufficient sampling frequency, resolution, and dynamic range to accurately process the signal. Second, the DSP must be fast enough to complete all its calculations within the sampling interval, $1/f_s$. Third, analog filters are still required at the ADC input and DAC output for antialiasing and anti-imaging, but the performance demands are not as great. Assuming these conditions have been met, the following offers a comparison between the digital and analog filters.

The required cut-off frequency of both filters is 1 kHz. The analog filter is realized as a 6-pole Chebyshev Type 1 filter (ripple in pass-band, no ripple in stop-band), and the response is shown in Figure 16.5. In practice, this filter would probably be realized using three 2-pole stages, each of which requires an op-amp, and several resistors and capacitors. Modern filter design CAD packages make the 6-pole design relatively

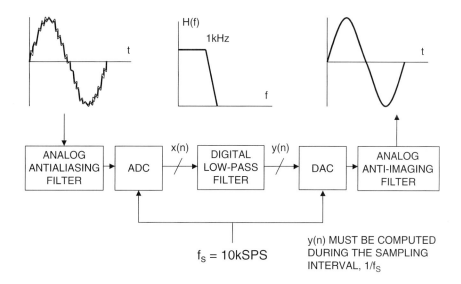

Figure 16.4: Digital filter

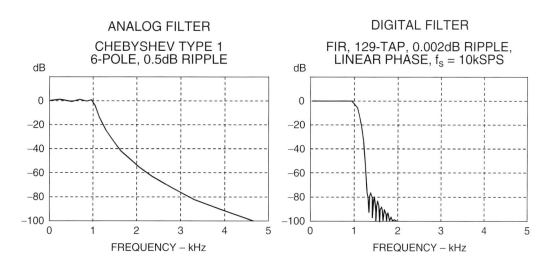

Figure 16.5: Analog vs. digital filter frequency response comparison

straightforward, but maintaining the 0.5 dB ripple specification requires accurate component selection and matching.

On the other hand, the 129-tap digital FIR filter shown has only 0.002 dB pass-band ripple, linear phase, and a much sharper roll-off. In fact, it could not be realized using analog techniques. Another obvious advantage is that the digital filter requires no component matching, and it is not sensitive to drift since the clock frequencies are crystal controlled. The 129-tap filter requires 129 multiply-accumulates (MAC) in order to compute an output sample. This processing must be completed within the sampling interval, $1/f_s$, in order to maintain real-time operation. In this example, the sampling frequency is 10 kSPS; therefore 100 μs is available for processing, assuming no significant additional overhead requirement. The ADSP-21xx family of DSPs can complete the entire multiply-accumulate process (and other functions necessary for the filter) in a single instruction cycle. Therefore, a 129-tap filter requires that the instruction rate be greater than 129/100 μs = 1.3 million instructions per second (MIPS). DSPs are available with instruction rates much greater than this, so the DSP certainly is not the limiting factor in this application. The ADSP-218x 16-bit fixed-point series offers instruction rates up to 75 MIPS.

The assembly language code to implement the filter on the ADSP-21xx family of DSPs is shown in Figure 16.6. Note that the actual lines of operating code have been marked with arrows; the rest are comments.

In a practical application, there are certainly many other factors to consider when evaluating analog versus digital filters, or analog versus digital signal processing in general. Most modern signal processing systems use a combination of analog and digital techniques in order to accomplish the desired function and take advantage of the best of both the analog and the digital worlds.

Digital filtering is one of the most powerful tools of DSP. Apart from the obvious advantages of virtually eliminating errors in the filter associated with passive component fluctuations over time and temperature, op-amp drift (active filters), and other effects, digital filters are capable of performance specifications that would, at best, be extremely difficult, if not impossible, to achieve with an analog implementation. In addition, the characteristics of a digital filter can easily be changed under software control. Therefore, they are widely used in adaptive filtering applications in communications such as echo cancellation in modems, noise cancellation, and speech recognition.

```
        .MODULE             fir_sub;
        {                   FIR Filter Subroutine
                            Calling Parameters
                                    I0 --> Oldest input data value in delay line
                                    I4 --> Beginning of filter coefficient table
                                    L0 = Filter length (N)
                                    L4 = Filter length (N)
                                    M1,M5 = 1
                                    CNTR = Filter length - 1 (N-1)
                            Return Values
                                    MR1 = Sum of products (rounded and saturated)
                                    I0 --> Oldest input data value in delay line
                                    I4 --> Beginning of filter coefficient table
                            Altered Registers
                                    MX0,MY0,MR
                            Computation Time
                                    (N - 1) + 6 cycles = N + 5 cycles
                            All coefficients are assumed to be in 1.15 format. }
        .ENTRY              fir;
→       fir:                MR=0, MX0=DM(I0,M1), MY0=PM(I4,M5)
→                           CNTR = N-1;
→                           DO convolution UNTIL CE;
→       convolution:          MR=MR+MX0*MY0(SS), MX0=DM(I0,M1), MY0=PM(I4,M5);
→                           MR=MR+MX0*MY0(RND);
→                           IF MV SAT MR;
→                           RTS;
        .ENDMOD;
```

Figure 16.6: ADSP-21xx FIR filter assembly code (single precision)

The actual procedure for designing digital filters has the same fundamental elements as that for analog filters. First, the desired filter responses are characterized, and the filter parameters are then calculated. Characteristics such as amplitude and phase response are derived in the same way. The key difference between analog and digital filters is that instead of calculating resistor, capacitor, and inductor values for an analog filter, coefficient values are calculated for a digital filter. So for the digital filter, numbers replace the physical resistor and capacitor components of the analog filter. These numbers reside in a memory as filter coefficients and are used with the sampled data values from the ADC to perform the filter calculations.

The real-time digital filter, because it is a discrete time function, works with digitized data as opposed to a continuous waveform, and a new data point is acquired each sampling period. Because of this discrete nature, data samples are referenced as numbers such as sample 1, sample 2, and sample 3. Figure 16.7 shows a low frequency signal containing higher frequency noise which must be filtered out. This waveform must be digitized with an ADC to produce samples x(n). These data values are fed to the digital filter, which in this case is a low-pass filter. The output data samples, y(n), are used to reconstruct an analog waveform using a low glitch DAC.

Digital filters, however, are not the answer to all signal processing filtering requirements. In order to maintain real-time operation, the DSP processor must be able to execute all the steps in the filter routine within one sampling clock period, $1/f_s$. A fast general-purpose fixed-point DSP (such as the ADSP-2189M at 75 MIPS) can execute a complete filter tap multiply-accumulate instruction in 13.3 ns. The ADSP-2189M requires N + 5 instructions for an N-tap filter. For a 100-tap filter, the total execution time is approximately 1.4 μs. This corresponds to a maximum possible sampling frequency of 714 kHz, thereby limiting the upper signal bandwidth to a few hundred kHz.

However, it is possible to replace a general-purpose DSP chip and design special hardware digital filters that will operate at video-speed sampling rates. In other cases, the speed limitations can be overcome by first storing the high speed ADC data in a buffer memory. The buffer memory is then read at a rate that is compatible with the speed of the DSP-based digital filter. In this manner, pseudo-real-time operation can be maintained as in a radar system, where signal processing is typically done on bursts of data collected after each transmitted pulse.

Another option is to use a third-party dedicated DSP filter engine like the Systolix PulseDSP filter core. The AD7725 16-bit sigma-delta ADC has an on-chip PulseDSP filter that can do 125 million multiply-accumulates per second.

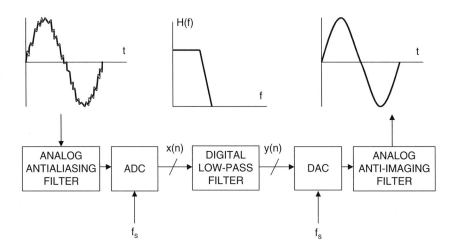

Figure 16.7: Digital filtering

Even in highly oversampled sampled data systems, an analog antialiasing filter is still required ahead of the ADC and a reconstruction (anti-imaging) filter after the DAC. Finally, as signal frequencies increase sufficiently, they surpass the capabilities of available ADCs, and digital filtering then becomes impossible. Active analog filtering is not possible at extremely high frequencies because of op-amp bandwidth and distortion limitations, and filtering requirements must then be met using purely passive components. The primary focus of the following discussions will be on filters that can run in real-time under DSP program control.

As an example, consider the comparison between an analog and a digital filter shown in Figure 16.9. The cut-off frequency of both filters is 1 kHz. The analog filter is realized as a 6-pole Chebyshev Type 1 filter (ripple in pass-band, no ripple in stop-band). In practice, this filter would probably be realized using three 2-pole stages, each of which requires an op-amp, and several resistors and capacitors. The 6-pole design is certainly not trivial, and maintaining the 0.5 dB ripple specification requires accurate component selection and matching.

On the other hand, the digital FIR filter shown has only 0.002 dB pass-band ripple, linear phase, and a much sharper roll-off. In fact, it could not be realized using analog techniques. In a practical application, there are many other factors to consider when evaluating analog versus digital filters. Most modern signal processing systems use a

DIGITAL FILTERS	ANALOG FILTERS
High Accuracy	Less Accuracy – Component Tolerances
Linear Phase (FIR Filters)	Nonlinear Phase
No Drift Due to Component Variations	Drift Due to Component Variations
Flexible, Adaptive Filtering Possible	Adaptive Filters Difficult
Easy to Simulate and Design	Difficult to Simulate and Design
Computation Must be Completed in Sampling Period – Limits Real-Time Operation	Analog Filters Required at High Frequencies and for Antialiasing Filters
Requires High Performance ADC, DAC, and DSP	No ADC, DAC, or DSP Required

Figure 16.8: Digital vs. analog filtering

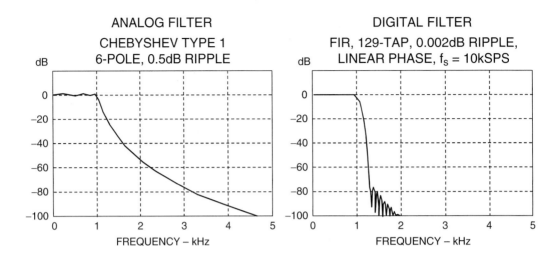

Figure 16.9: Analog vs. digital filter frequency response comparison

combination of analog and digital techniques in order to accomplish the desired function and take advantage of the best of both the analog and the digital world.

There are many applications where digital filters must operate in real-time. This places specific requirements on the DSP, depending upon the sampling frequency and the filter complexity. The key point is *that the* DSP *must finish all computations during the sampling period so it will be ready to process the next data sample.* Assume that the analog signal bandwidth to be processed is f_a. This requires the ADC sampling frequency f_s to be at least $2f_a$. The sampling period is $1/f_s$. All DSP filter computations

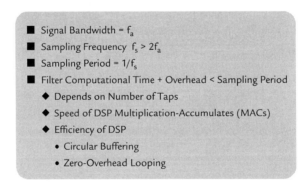

Figure 16.10: Processing requirements for real-time digital filtering

(including overhead) must be completed during this interval. The computation time depends on the number of taps in the filter and the speed and efficiency of the DSP. Each tap on the filter requires one multiplication and one addition (multiply-accumulate). DSPs are generally optimized to perform fast multiply-accumulates, and many DSPs have additional features such as circular buffering and zero-overhead looping to minimize the "overhead" instructions that otherwise would be needed.

16.7   Finite Impulse Response (FIR) Filters

There are two fundamental types of digital filters: finite impulse response (FIR) and infinite impulse response (IIR). As the terminology suggests, these classifications refer to the filter's impulse response. By varying the weight of the coefficients and the number of filter taps, virtually any frequency response characteristic can be realized with a FIR filter. As has been shown, FIR filters can achieve performance levels that are not possible with analog filter techniques (such as perfect linear phase response). However, high performance FIR filters generally require a large number of multiply-accumulates and therefore require fast and efficient DSPs. On the other hand, IIR filters tend to mimic the performance of traditional analog filters and make use of feedback, so their impulse response extends over an infinite period of time. Because of feedback, IIR filters can be implemented with fewer coefficients than for a FIR filter. Lattice filters are simply another way to implement either FIR or IIR filters and are often used in speech processing applications. Finally, digital filters lend themselves to adaptive filtering applications simply because of the speed and ease with which the filter characteristics can be changed by varying the filter coefficients.

- Moving Average
- Finite Impulse Response (FIR)
    - Linear Phase
    - Easy to Design
    - Computationally Intensive
- Infinite Impulse Response (IIR)
    - Based on Classical Analog Filters
    - Computationally Efficient
- Lattice Filters (Can be FIR or IIR)
- Adaptive Filters

Figure 16.11: Types of digital filters

The most elementary form of a FIR filter is a *moving average* filter as shown in Figure 16.12. Moving average filters are popular for smoothing data, such as in the analysis of stock prices. The input samples, x(n) are passed through a series of buffer registers (labeled z^{-1}, corresponding to the z-transform representation of a delay element). In the example shown, there are four taps corresponding to a 4-point moving average. Each sample is multiplied by 0.25, and these results are added to yield the final moving average output y(n). The figure also shows the general equation of a moving average filter with N taps. Note again that N refers to the number of filter taps, and not the ADC or DAC resolution as in previous sections.

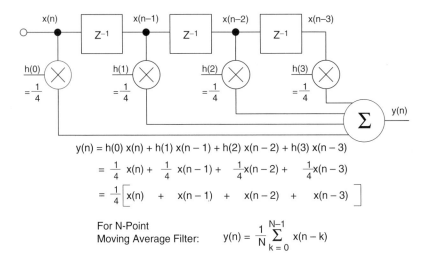

$$y(n) = h(0)\,x(n) + h(1)\,x(n-1) + h(2)\,x(n-2) + h(3)\,x(n-3)$$

$$= \frac{1}{4}\,x(n) + \frac{1}{4}\,x(n-1) + \frac{1}{4}x(n-2) + \frac{1}{4}x(n-3)$$

$$= \frac{1}{4}\Big[x(n) + x(n-1) + x(n-2) + x(n-3) \Big]$$

For N-Point
Moving Average Filter: $y(n) = \dfrac{1}{N}\displaystyle\sum_{k=0}^{N-1} x(n-k)$

Figure 16.12: 4-point moving average filter

Since the coefficients are equal, an easier way to perform a moving average filter is shown in Figure 16.13. Note that the first step is to store the first four samples, x(0), x(1), x(2), x(3) in a register. These quantities are added and then multiplied by 0.25 to yield the first output, y(3). Note that the initial outputs y(0), y(1), and y(2) are not valid because all registers are not full until sample x(3) is received.

When sample x(4) is received, it is added to the result, and sample x(0) is subtracted from the result. The new result must then be multiplied by 0.25. Therefore, the calculations required to produce a new output consist of one addition, one subtraction, and one multiplication, regardless of the length of the moving average filter.

$$y(3) = 0.25 \left[\qquad\qquad\qquad\qquad x(3) + x(2) + x(1) + x(0) \right]$$

$$y(4) = 0.25 \left[\qquad\qquad\qquad x(4) + x(3) + x(2) + x(1) \right]$$

$$y(5) = 0.25 \left[\qquad\qquad x(5) + x(4) + x(3) + x(2) \right]$$

$$y(6) = 0.25 \left[\qquad x(6) + x(5) + x(4) + x(3) \right]$$

$$y(7) = 0.25 \left[x(7) + x(6) + x(5) + x(4) \right]$$

•
•   Each Output Requires:
•   1 Multiplication, 1 Addition, 1 Subtraction
•

Figure 16.13: Calculating output of 4-point moving average filter

The step function response of a 4-point moving average filter is shown in Figure 16.14. Notice that the moving average filter has no overshoot. This makes it useful in signal processing applications where random white noise must be filtered but pulse response preserved. Of all the possible linear filters that could be used, the moving average produces the lowest noise for a given edge sharpness. This is illustrated in Figure 16.15, where the noise level becomes lower as the number of taps are increased. Notice that

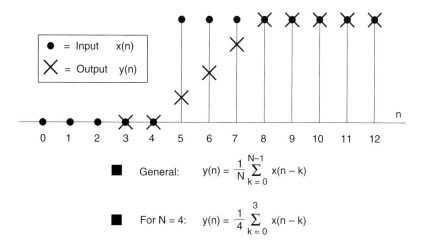

• = Input   x(n)

✕ = Output   y(n)

■ General: $y(n) = \dfrac{1}{N} \displaystyle\sum_{k=0}^{N-1} x(n-k)$

■ For N = 4: $y(n) = \dfrac{1}{4} \displaystyle\sum_{k=0}^{3} x(n-k)$

Figure 16.14: 4-tap moving average filter step response

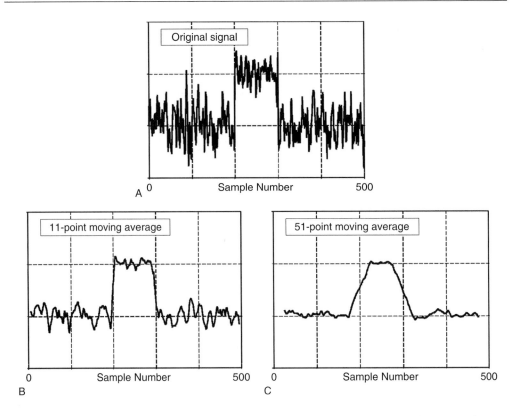

Figure 16.15: Moving average filter response to noise superimposed on step input

the 0% to 100% rise time of the pulse response is equal to the total number of taps in the filter multiplied by the sampling period.

The frequency response of the simple moving average filter is sin(x)/x and is shown on a linear amplitude scale in Figure 16.16. Adding more taps to the filter sharpens the roll-off, but does not significantly reduce the amplitude of the sidelobes which are approximately 14 dB down for the 11- and 31-tap filter. These filters are definitely not suitable where high stop-band attenuation is required.

It is possible to dramatically improve the performance of the simple FIR moving average filter by properly selecting the individual weights or coefficients rather than giving them equal weight. The sharpness of the roll-off can be improved by adding more stages (taps), and the stop-band attenuation characteristics can be improved by properly selecting the filter coefficients. Note that unlike the moving average filter, one

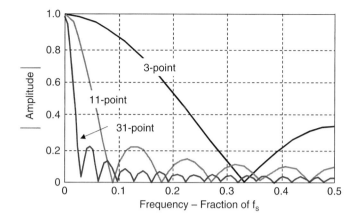

Figure 16.16: Moving average filter frequency response

multiply-accumulate cycle is now required per tap for the generalized FIR filter. The essence of FIR filter design is the appropriate selection of the filter coefficients and the number of taps to realize the desired transfer function H(f). Various algorithms are available to translate the frequency response H(f) into a set of FIR coefficients. Most of this software is commercially available and can be run on PCs. *The key theorem of FIR filter design is that the coefficients h(n) of the FIR filter are simply the quantized values of the impulse response of the frequency transfer function H(f). Conversely, the impulse response is the discrete Fourier transform of H(f).*

The generalized form of an N-tap FIR filter is shown in Figure 16.17. As has been discussed, an FIR filter must perform the following convolution equation:

$$y(n) = h(k) * x(n) = \sum_{k=0}^{N-1} h(k)x(n-k).$$

where h(k) is the filter coefficient array and x(n–k) is the input data array to the filter. The number N, in the equation, represents the number of taps of the filter and relates to the filter performance as has been discussed above. An N-tap FIR filter requires N multiply-accumulate cycles.

FIR filter diagrams are often simplified as shown in Figure 16.18. The summations are represented by arrows pointing into the dots, and the multiplications are indicated by placing the h(k) coefficients next to the arrows on the lines. The z^{-1} delay element is often shown by placing the label above or next to the appropriate line.

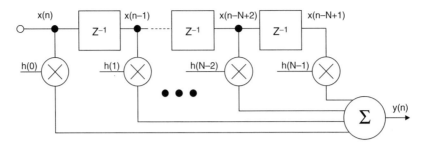

$$\blacksquare \quad y(n) \ = \ h(n) \ * \ x(n) \ = \ \sum_{k=0}^{N-1} h(k)\,x(n-k)$$

$\blacksquare$ $*$ = Symbol for Convolution

$\blacksquare$ Requires N multiply-accumulates for each output

Figure 16.17: N-tap finite impulse response (FIR) filter

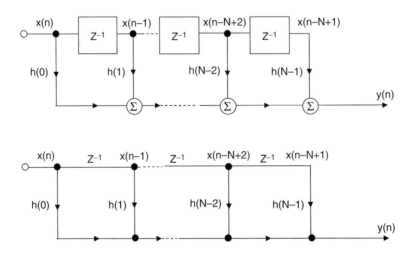

Figure 16.18: Simplified filter notations

16.8   FIR Filter Implementation in DSP Hardware Using Circular Buffering

In the series of FIR filter equations, the N coefficient locations are always accessed sequentially from h(0) to h(N–1). The associated data points circulate through the memory; new samples are added, replacing the oldest each time a filter output is computed. A fixed-boundary RAM can be used to achieve this circulating buffer effect as shown in Figure 16.19 for a four-tap FIR filter. The oldest data sample is replaced by the newest after each convolution. A "time history" of the four most recent data samples is always stored in RAM.

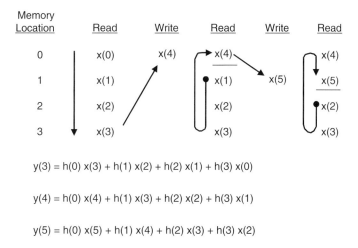

$$y(3) = h(0)\,x(3) + h(1)\,x(2) + h(2)\,x(1) + h(3)\,x(0)$$

$$y(4) = h(0)\,x(4) + h(1)\,x(3) + h(2)\,x(2) + h(3)\,x(1)$$

$$y(5) = h(0)\,x(5) + h(1)\,x(4) + h(2)\,x(3) + h(3)\,x(2)$$

Figure 16.19: Calculating outputs of 4-tap FIR filter using a circular buffer

To facilitate memory addressing, old data values are read from memory starting with the value one location after the value that was just written. For example, x(4) is written into memory location 0, and data values are then read from locations 1, 2, 3, and 0. This example can be expanded to accommodate any number of taps. By addressing data memory locations in this manner, the address generator need only supply sequential addresses, regardless of whether the operation is a memory read or write. This data memory buffer is called *circular* because when the last location is reached, the memory pointer is reset to the beginning of the buffer.

The coefficients are fetched simultaneously with the data. Due to the addressing scheme chosen, the oldest data sample is fetched first. Therefore, the last coefficient must be fetched first. The coefficients can be stored backward in memory: h(N–1) is the first location, and h(0) is the last, with the address generator providing incremental addresses. Alternatively, coefficients can be stored in a normal manner with the accessing of coefficients starting at the end of the buffer, and the address generator being decremented. In the example shown in Figure 16.19, the coefficients are stored in a reverse manner.

1. Obtain sample from ADC (typically interrupt-driven)
2. Move sample into input signal's circular buffer
3. Update the pointer for the input signal's circular buffer
4. Zero the accumulator
5. Implement filter (control the loop through each of the coefficients)
    6. Fetch the coefficient from the coefficient's circular buffer
    7. Update the pointer for the coefficient's circular buffer
    8. Fetch the sample from the input signal's circular buffer
    9. Update the pointer for the input signal's circular buffer
    10. Multiply the coefficient by the sample
    11. Add the product to the accumulator
12. Move the filtered sample to the DAC

```
ADSP-21xx Example code:

CNTR = N-1;
DO convolution UNTIL CE;
convolution:
   MR = MR + MX0 * MY0(SS), MX0 = DM(I0,M1), MY0 = PM(I4,M5);
```

**Figure 16.20: Pseudocode for FIR filter program using a DSP
with circular buffering**

A simple summary flowchart for these operations is shown in Figure 16.20. For Analog Devices DSPs, *all operations within the filter loop are completed in one instruction cycle*, thereby greatly increasing efficiency. This is referred to as *zero-overhead looping*. The actual FIR filter assembly code for the ADSP-21xx family of fixed-point DSPs is shown in Figure 16.21. The arrows in the diagram point to the actual executable instructions, and the rest of the code are simply comments added for clarification.

The first instruction (labeled *fir:*) sets up the computation by clearing the MR register and loading the MX0 and MY0 registers with the first data and coefficient values from data and program memory. The multiply-accumulate with dual data fetch in the *convolution* loop is then executed N–1 times in N cycles to compute the sum of the first N–1 products. The final multiply-accumulate instruction is performed with the rounding mode enabled to round the result to the upper 24 bits of the MR register. The MR1 register is then conditionally saturated to its most positive or negative value, based on the status of the overflow flag contained in the MV register. In this manner, results are accumulated to the full 40-bit precision of the MR register, with saturation of the output only if the final result overflowed beyond the least significant 32 bits of the MR register.

```
        .MODULE           fir_sub;
        {                 FIR Filter Subroutine
                          Calling Parameters
                                  I0 --> Oldest input data value in delay line
                                  I4 --> Beginning of filter coefficient table
                                  L0 = Filter length (N)
                                  L4 = Filter length (N)
                                  M1,M5 = 1
                                  CNTR = Filter length - 1 (N-1)
                          Return Values
                                  MR1 = Sum of products (rounded and saturated)
                                  I0 --> Oldest input data value in delay line
                                  I4 --> Beginning of filter coefficient table
                          Altered Registers
                                  MX0,MY0,MR
                          Computation Time
                                  (N - 1) + 6 cycles = N + 5 cycles
                          All coefficients are assumed to be in 1.15 format. }
        .ENTRY            fir;
 ───▶   fir:              MR=0, MX0=DM(I0,M1), MY0=PM(I4,M5)
 ───▶                     CNTR = N-1;
 ───▶                     DO convolution UNTIL CE;
 ───▶   convolution:        MR=MR+MX0*MY0(SS), MX0=DM(I0,M1), MY0=PM(I4,M5);
 ───▶                     MR=MR+MX0*MY0(RND);
 ───▶                     IF MV SAT MR;
 ───▶                     RTS;
        .ENDMOD;
```

Figure 16.21: ADSP-21xx FIR filter assembly code (single precision)

The limit on the number of filter taps attainable for a real-time implementation of the FIR filter subroutine is primarily determined by the processor cycle time, the sampling rate, and the number of other computations required. The FIR subroutine presented here requires a total of N + 5 cycles for a filter of length N. For the ADSP-2189M 75 MIPS DSP, one instruction cycle is 13.3 ns, so a 100-tap filter would require 13.3 ns $\times$ 100 + 5 $\times$ 13.3 ns = 1330 ns + 66.5 ns = 1396.5 ns = 1.4 μs.

16.9 Designing FIR Filters

FIR filters are relatively easy to design using modern CAD tools. Figure 16.22 summarizes the characteristics of FIR filters as well as the most popular design techniques. *The fundamental concept of FIR filter design is that the filter frequency response is determined by the impulse response, and the quantized impulse response and the filter coefficients are identical.* This can be understood by examining

- ■ Impulse Response has a Finite Duration (N Cycles)
- ■ Linear Phase, Constant Group Delay (N Must be Odd)
- ■ No Analog Equivalent
- ■ Unconditionally Stable
- ■ Can be Adaptive
- ■ Computational Advantages when Decimating Output
- ■ Easy to Understand and Design
  - ◆ Windowed-Sinc Method
  - ◆ Fourier Series Expansion with Windowing
  - ◆ Frequency Sampling Using Inverse FFT – Arbitrary Frequency Response
  - ◆ Parks-McClellan Program with Remez Exchange Algorithm

Figure 16.22: Characteristics of FIR filters

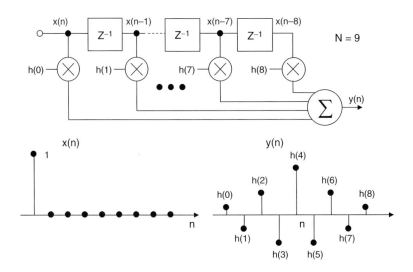

Figure 16.23: FIR filter impulse response determines the filter coefficients

Figure 16.23. The input to the FIR filter is an impulse, and as the impulse propagates through the delay elements, the filter output is identical to the filter coefficients. The FIR filter design process therefore consists of determining the impulse response from the desired frequency response, and then quantizing the impulse response to generate the filter coefficients.

It is useful to digress for a moment and examine the relationship between the time domain and the frequency domain to better understand the principles behind digital filters such as the FIR filter. In a sampled data system, a convolution operation can be carried out by performing a series of multiply-accumulates. The convolution operation in the time or frequency domain is equivalent to point-by-point multiplication in the opposite domain. For example, convolution in the time domain is equivalent to multiplication in the frequency domain. This is shown graphically in Figure 16.24. It can be seen that filtering in the frequency domain can be accomplished by multiplying all frequency components in the pass-band by a 1 and all frequencies in the stop-band by 0. Conversely, convolution in the frequency domain is equivalent to point-by-point multiplication in the time domain.

The transfer function in the frequency domain (either a 1 or a 0) can be translated to the time domain by the discrete Fourier transform (in practice, the fast Fourier transform is used). This transformation produces an impulse response in the time domain. Since the multiplication in the frequency domain (signal spectrum times the transfer function) is equivalent to convolution in the time domain (signal convolved with impulse response), the signal can be filtered by convolving it with the impulse response. The FIR filter is exactly this process. Since it is a sampled data system, the signal and the impulse response are

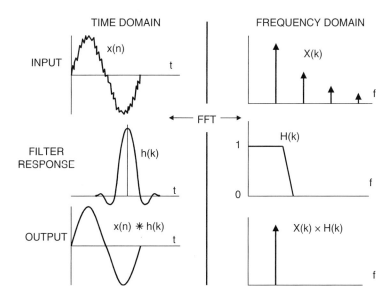

Figure 16.24: Duality of time and frequency

quantized in time and amplitude, yielding discrete samples. The discrete samples comprising the desired impulse response are the FIR filter coefficients.

The mathematics involved in filter design (analog or digital) generally make use of transforms. In continuous-time systems, the Laplace transform can be considered to be a generalization of the Fourier transform. In a similar manner, it is possible to generalize the Fourier transform for discrete-time sampled data systems, resulting in what is commonly referred to as the z-transform. Details describing the use of the z-transform in digital filter design are given in References 1 through 6, but the theory is not necessary for the rest of this discussion.

16.9.1   FIR Filter Design Using the Windowed-Sinc Method

An ideal low-pass filter frequency response is shown in Figure 16.25(A). The corresponding impulse response in the time domain is shown in Figure 16.25(B), and follows the sin(x)/x (sinc) function. If a FIR filter is used to implement this frequency response, an infinite number of taps are required. The windowed-sinc method is used to implement the filter as follows. First, the impulse response is truncated to a reasonable number of N taps as in Figure 16.25(C). The frequency response

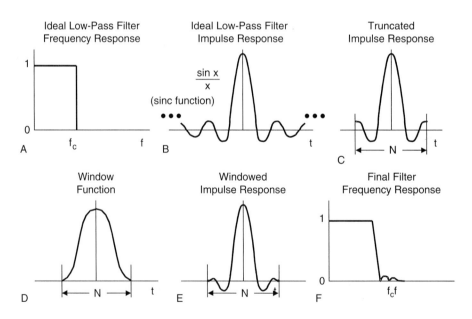

Figure 16.25: FIR filter design using the windowed-sinc method

corresponding to Figure 16.25(C) has relatively poor sidelobe performance because of the end-point discontinuities in the truncated impulse response. The next step in the design process is to apply an appropriate window function as shown in Figure 16.25(D) to the truncated impulse. This forces the endpoints to zero. The particular window function chosen determines the roll-off and sidelobe performance of the filter. There are several good choices of window function, depending upon the desired frequency response. The frequency response of the truncated and windowed-sinc impulse response of Figure 16.25(E) is shown in Figure 16.25(F).

16.9.2   *FIR Filter Design Using the Fourier Series Method with Windowing*

The Fourier series with windowing method (Figure 16.26) starts by defining the transfer function H(f) mathematically and expanding it in a Fourier series. The Fourier series coefficients define the impulse response and therefore the coefficients of the FIR filter. However, the impulse response must be truncated and windowed as in the previous method. After truncation and windowing, an FFT is used to generate the corresponding frequency response. The frequency response can be modified by choosing different window functions, although precise control of the stop-band characteristics is difficult in any method that uses windowing.

- Specify H(f)
- Expand H(f) in a Fourier Series: The Fourier Series Coefficients are the Coefficients of the FIR Filter, h(k), and its Impulse Response
- Truncate the Impulse Response to N Points (Taps)
- Apply a Suitable Window Function to h(k) to Smooth the Effects of Truncation
- Lacks Precise Control of Cutoff Frequency; Highly Dependent on Window Function

Figure 16.26: FIR Filter design using Fourier series method with windowing

16.9.3   *FIR Filter Design Using the Frequency Sampling Method*

This method is extremely useful in generating an FIR filter with an arbitrary frequency response. H(f) is specified as a series of amplitude and phase points in the frequency

- ■ Specify H(k) as a Finite Number of Spectral Points Spread Uniformly between 0 and $0.5f_s$ (512 Usually Sufficient)
- ■ Specify Phase Points (Can Make Equal to Zero)
- ■ Convert Rectangular Form (Real + Imaginary)
- ■ Take the Complex Inverse FFT of H(f) Array to Obtain the Impulse Response
- ■ Truncate the Impulse Response to N Points
- ■ Apply a Suitable Window Function to h(k) to Smooth the Effects of Truncation
- ■ Test Filter Design and Modify if Necessary
- ■ CAD Design Techniques More Suitable for Low-Pass, High-Pass, Band-Pass, or Band-Stop Filters

Figure 16.27: Frequency sampling method for FIR filters with arbitrary frequency response

domain. The points are then converted into real and imaginary components. Next, the impulse response is obtained by taking the complex inverse FFT of the frequency response. The impulse response is then truncated to N points, and a window function is applied to minimize the effects of truncation. The filter design should then be tested by taking its FFT and evaluating the frequency response. Several iterations may be required to achieve the desired response.

16.9.4   FIR Filter Design Using the Parks-McClellan Program

Historically, the design method based on the use of windows to truncate the impulse response and to obtain the desired frequency response was the first method used for designing FIR filters. The frequency-sampling method was developed in the 1970s and is still popular where the frequency response is an arbitrary function.

Modern CAD programs are available today that greatly simplify the design of lowpass, high-pass, band-pass, or band-stop FIR filters. A popular one was developed by Parks and McClellan and uses the Remez exchange algorithm. The filter design begins by specifying the parameters shown in Figure 16.28: pass-band ripple, stop-band ripple (same as attenuation), and the transition region. For this design example, the QED1000 program from Momentum Data Systems was used (a demo version is free and downloadable from http://www.mds.com).

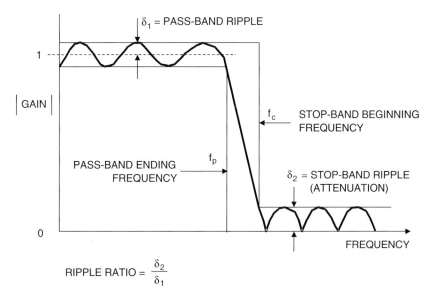

Figure 16.28: FIR CAD Techniques: Parks-McClellan program with Remez exchange algorithm

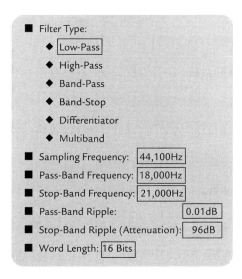

Figure 16.29: Parks-McClellan equiripple FIR filter design: program inputs

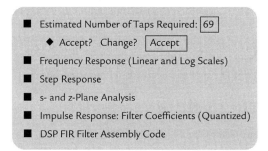

■ Estimated Number of Taps Required: 69
   ◆ Accept?   Change?   Accept
■ Frequency Response (Linear and Log Scales)
■ Step Response
■ s- and z-Plane Analysis
■ Impulse Response: Filter Coefficients (Quantized)
■ DSP FIR Filter Assembly Code

Figure 16.30: FIR filter program outputs

For this example, we will design an audio low-pass filter that operates at a sampling rate of 44.1 kHz. The filter is specified as shown in Figure 16.28: 18 kHz pass-band frequency, 21 kHz stop-band frequency, 0.01 dB pass-band ripple, 96 dB stop-band ripple (attenuation). We must also specify the word length of the coefficients, which in this case is 16 bits, assuming a 16-bit fixed-point DSP is to be used.

The program allows us to choose between a window-based design or the equiripple Parks-McClellan program. We will choose the latter. The program now estimates the number of taps required to implement the filter based on the above specifications. In this case, it is 69 taps. At this point, we can accept this and proceed with the design or decrease the number of taps and see what degradation in specifications occur.

We will accept this number and let the program complete the calculations. The program outputs the frequency response (Figure 16.31), step function response (Figure 16.32), s- and z-plane analysis data, and the impulse response (Figure 16.33). The QED1000 program then outputs the quantized filter coefficients to a program that generates the actual DSP assembly code for a number of popular DSPs, including Analog Devices. The program is quite flexible and allows the user to perform a number of scenarios to optimize the filter design.

The 69-tap FIR filter requires $69 + 5 = 74$ instruction cycles using the ADSP-2189M 75 MIPS processor, which yields a total computation time per sample of 74×13.3 ns $= 984$ ns. The sampling interval is 1/44.1 kHz, or 22.7 μs. This allows 22.7 μs $-$ 0.984 μs $= 21.7$ μs for overhead and other operations.

Other options are to use a slower processor (3.3 MIPS) for this application, a more complex filter that takes more computation time (up to N $= 1700$), or increase the sampling frequency to about 1 MSPS.

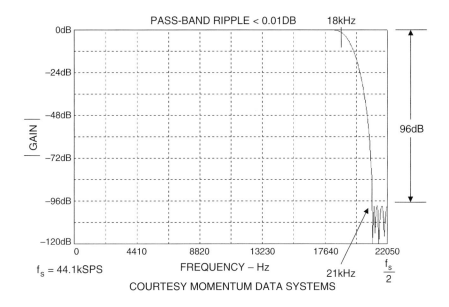

Figure 16.31: FIR design example: frequency response

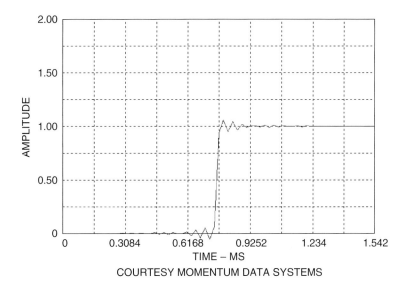

Figure 16.32: FIR filter design example: step response

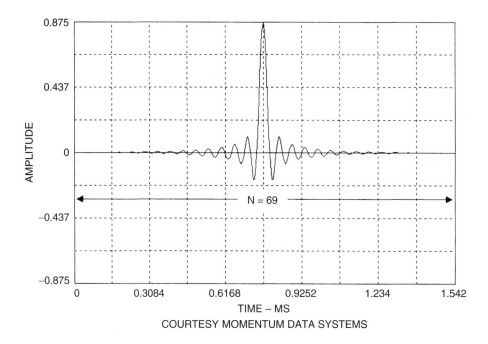

COURTESY MOMENTUM DATA SYSTEMS

Figure 16.33: FIR design example: impulse response (filter coefficients)

- Sampling Frequency f_s = 44.1 kSPS
- Sampling Interval = $1/f_s$ = 22.7 µs
- Number of Filter Taps, N = 69
- Number of Required Instructions = N + 5 = 74
- Processing Time/Instruction = 13.3 ns (75 MIPS)
  (ADSP-2189M)
- Total Processing Time = 74 × 13.3 ns = 984 ns
- Total Processing Time < Sampling Interval with
  22.7 µs – 0.984 µs = 21.7 µs for Other Operations
  - ◆ Increase Sampling Frequency to 1 MHz
  - ◆ Use Slower DSP (3.3 MIPS)
  - ◆ Add More Filter Taps (Up to N = 1700)

**Figure 16.34: Design example using ADSP-2189M: processor time
for 69-tap FIR filter**

16.9.5 Designing High-Pass, Band-Pass, and Band-Stop Filters Based on Low-Pass Filter Design

Converting a low-pass filter design impulse response into a high-pass filter impulse response can be accomplished in one of two ways. In the *spectral inversion method*, the sign of each filter coefficient in the low-pass filter impulse response is changed. Next, 1 is added to the center coefficient. In the *spectral reversal method*, the sign of every other coefficient is changed. This reverses the frequency domain plot. In other words, if the cut-off of the low-pass filter is 0.2 f_s, the resulting high-pass filter will have a cut-off frequency of 0.5 f_s − 0.2 f_s = 0.3 f_s. This must be considered when doing the original low-pass filter design.

Band-pass and band-stop filters can be designed by combining individual low-pass and high-pass filters in the proper manner. Band-pass filters are designed by placing the low-pass and high-pass filters in cascade. The equivalent impulse response of the cascaded filters is then obtained by *convolving* the two individual impulse responses.

A band-stop filter is designed by connecting the low-pass and high-pass filters in parallel and adding their outputs. The equivalent impulse response is then obtained by *adding* the two individual impulse responses.

■ Spectral Inversion Technique:
  ◆ Design Low-Pass Filter (Linear Phase, N Odd)
  ◆ Change the Sign of Each Coefficient in the Impulse Response, h(n)
  ◆ Add 1 to the Coefficient at the Center of Symmetry
■ Spectral Reversal Technique:
  ◆ Design Low-Pass Filter
  ◆ Change the Sign of Every Other Coefficient in the Impulse Response, h(n)
  ◆ This Reverses the Frequency Domain Left-for-Right:
    0 Becomes 0.5, and 0.5 Becomes 0;
    i.e., if the Cut-Off Frequency of the Low-Pass Filter is 0.2, the Cut-Off of the Resulting High-Pass Filter is 0.3

Figure 16.35: Designing high-pass filters using low-pass filter impulse response

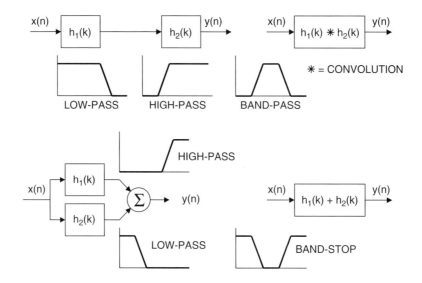

Figure 16.36: Band-pass and band-stop filters designed from low-pass and high-pass filters

16.10   Infinite Impulse Response (IIR) Filters

As was mentioned previously, FIR filters have no real analog counterparts, the closest analogy being the weighted moving average. In addition, FIR filters have only zeros and no poles. On the other hand, IIR filters have traditional analog counterparts (Butterworth, Chebyshev, Elliptic, and Bessel) and can be analyzed and synthesized using more familiar traditional filter design techniques.

Infinite impulse response filters get their name because their impulse response extends for an infinite period of time. This is because they are recursive, i.e., they utilize feedback. Although they can be implemented with fewer computations than FIR filters, IIR filters do not match the performance achievable with FIR filters, and do not have linear phase. Also, there is no computational advantage achieved when the output of an IIR filter is decimated, because each output value must always be calculated.

IIR filters are generally implemented in 2-pole sections called biquads because they are described with a biquadratic equation in the z-domain. Higher order filters are designed using cascaded biquad sections, e.g., a 6-pole filter requires three biquad sections.

- Uses Feedback (Recursion)
- Impulse Response has an Infinite Duration
- Potentially Unstable
- Nonlinear Phase
- More Efficient than FIR Filters
- No Computational Advantage when Decimating Output
- Usually Designed to Duplicate Analog Filter Response
- Usually Implemented as Cascaded Second-Order Sections (Biquads)

Figure 16.37: Infinite impulse response (IIR) filters

The basic IIR biquad is shown in Figure 16.38. The zeros are formed by the feed-forward coefficients b_0, b_1, and b_2; the poles are formed by the feedback coefficients a_1, and a_2.

The general digital filter equation is shown in Figure 16.38, which gives rise to the general transfer function H(z), which contains polynomials in both the numerator and the denominator. The roots of the denominator determine the pole locations of the filter, and the roots of the numerator determine the zero locations. Although it is

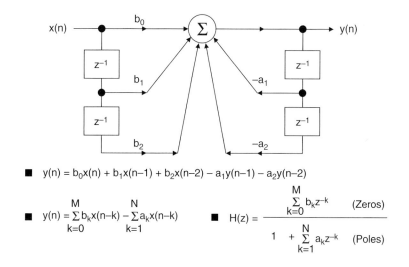

- $y(n) = b_0 x(n) + b_1 x(n-1) + b_2 x(n-2) - a_1 y(n-1) - a_2 y(n-2)$

- $y(n) = \sum_{k=0}^{M} b_k x(n-k) - \sum_{k=1}^{N} a_k x(n-k)$

- $H(z) = \dfrac{\sum\limits_{k=0}^{M} b_k z^{-k}}{1 + \sum\limits_{k=1}^{N} a_k z^{-k}}$ (Zeros) (Poles)

Figure 16.38: Hardware implementation of second-order IIR filter (biquad)
Direct Form 1

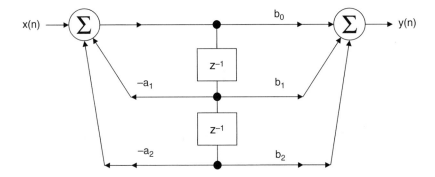

- REDUCES TO THE SAME EQUATION AS DIRECT FORM 1:

$$y(n) = b_0x(n) + b_1x(n-1) + b_2x(n-2) - a_1y(n-1) - a_2y(n-2)$$

- REQUIRES ONLY TWO DELAY ELEMENTS (REGISTERS)

Figure 16.39: IIR biquad filter Direct Form 2

possible to construct a high order IIR filter directly from this equation (called the *direct form* implementation), accumulation errors due to quantization errors (finite word-length arithmetic) may give rise to instability and large errors. For this reason, it is common to cascade several biquad sections with appropriate coefficients rather than use the direct form implementation. The biquads can be scaled separately and then cascaded in order to minimize the coefficient quantization and the recursive accumulation errors. Cascaded biquads execute more slowly than their direct form counterparts, but are more stable and minimize the effects of errors due to finite arithmetic errors.

The Direct Form 1 biquad section shown in Figure 16.38 requires four registers. This configuration can be changed into an equivalent circuit shown in Figure 16.39 that is called the Direct Form 2 and requires only two registers. It can be shown that the equations describing the Direct Form 2 IIR biquad filter are the same as those for Direct Form 1. As in the case of FIR filters, the notation for an IIR filter is often simplified as shown in Figure 16.40.

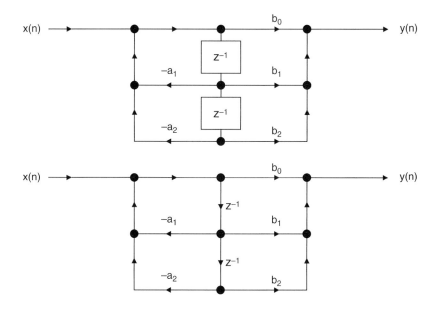

Figure 16.40: IIR biquad filter simplified notations

16.11 IIR Filter Design Techniques

A popular method for IIR filter design is to first design the analog-equivalent filter and then mathematically transform the transfer function H(s) into the z-domain, H(z). Multiple pole designs are implemented using cascaded biquad sections. The most popular analog filters are the Butterworth, Chebyshev, Elliptical, and Bessel (see Figure 16.41). There are many CAD programs available to generate the Laplace transform, H(s), for these filters.

The all-pole Butterworth (also called *maximally flat*) has no ripple in the pass-band or stop-band and has monotonic response in both regions. The all-pole Type 1 Chebyshev filter has a faster roll-off than the Butterworth (for the same number of poles) and has ripple in the pass-band. The Type 2 Chebyschev filter is rarely used, but has ripple in the stop-band rather than the pass-band.

The Elliptical (Cauer) filter has poles and zeros and ripple in both the pass-band and stop-band. This filter has even faster roll-off than the Chebyshev for the same number of poles. The Elliptical filter is often used where degraded phase response can be tolerated.

- Butterworth
  - ◆ All Pole, No Ripples in Pass Band or Stop Band
  - ◆ Maximally Flat Response (Fastest Roll-Off with No Ripple)
- Chebyshev (Type 1)
  - ◆ All Pole, Ripple in Pass Band, No Ripple in Stop Band
  - ◆ Shorter Transition Region than Butterworth for Given Number of Poles
  - ◆ Type 2 has Ripple in Stop Band, No Ripple in Pass Band
- Elliptical (Cauer)
  - ◆ Has Poles and Zeros, Ripple in Both Pass Band and Stop Band
  - ◆ Shorter Transition Region than Chebyshev for Given Number of Poles
  - ◆ Degraded Phase Response
- Bessel (Thompson)
  - ◆ All Pole, No Ripples in Pass Band or Stop Band
  - ◆ Optimized for Linear Phase and Pulse Response
  - ◆ Longest Transition Region of All for Given Number of Poles

Figure 16.41: Review of popular analog filters

Finally, the Bessel (Thompson) filter is an all-pole filter optimized for pulse response and linear phase but has the poorest roll-off of any of the types discussed for the same number of poles.

All of the above types of analog filters are covered in the literature, and their Laplace transforms, H(s), are readily available, either from tables or CAD programs. There are three methods used to convert the Laplace transform into the z-transform: *impulse invariant transformation, bilinear transformation, and the matched z-transform*. The resulting z-transforms can be converted into the coefficients of the IIR biquad. These techniques are highly mathematically intensive and will not be discussed further.

A CAD approach for IIR filter design is similar to the Parks-McClellan program used for FIR filters. This technique uses the Fletcher-Powell algorithm.

In calculating the throughput time of a particular DSP IIR filter, one should examine the benchmark performance specification for a biquad filter section. For the ADSP21xx family, seven instruction cycles are required to execute a biquad filter output sample. For the ADSP-2189M, 75 MIPS DSP, this corresponds to 7×13.3 ns $= 93$ ns, corresponding to a maximum possible sampling frequency of 10 MSPS (neglecting overhead).

- ■ Impulse Invariant Transformation Method
  - ◆ Start with H(s) for Analog Filter
  - ◆ Take Inverse Laplace Transform to Get Impulse Response
  - ◆ Obtain z-Transform H(z) from Sampled Impulse Response
  - ◆ z-Transform Yields Filter Coefficients
  - ◆ Aliasing Effects Must Be Considered
- ■ Bilinear Transformation Method
  - ◆ Another Method for Transforming H(s) into H(z)
  - ◆ Performance Determined by the Analog System's Differential Equation
  - ◆ Aliasing Effects Do Not Occur
- ■ Matched z-Transform Method
  - ◆ Maps H(s) into H(z) for Filters with Both Poles and Zeros
- ■ CAD Methods
  - ◆ Fletcher-Powell Algorithm
  - ◆ Implements Cascaded Biquad Sections

Figure 16.42: IIR filter design techniques

- ■ Determine How Many Biquad Sections are Required to Realize the Desired Frequency Response
- ■ Multiply This by the Execution Time per Biquad for the DSP (7 Instruction Cycles × 13.3 ns = 93 ns for the 75 MIPS ADSP-2189M, for example)
- ■ The Result (Plus Overhead) is the Minimum Allowable Sampling Period ($1/f_s$) for Real-Time Operation

Figure 16.43: Throughput considerations for IIR filters

16.11.1 Summary: FIR Versus IIR Filters

Choosing between FIR and IIR filter designs can be somewhat of a challenge, but a few basic guidelines can be given. Typically, IIR filters are more efficient than FIR filters because they require less memory and fewer multiply-accumulates are needed. IIR filters can be designed based upon previous experience with analog filter designs. IIR filters may exhibit instability problems, but this is much less likely to occur if higher order filters are designed by cascading second-order systems.

IIR FILTERS	FIR FILTERS
More Efficient	Less Efficient
Analog Equivalent	No Analog Equivalent
May Be Unstable	Always Stable
Nonlinear Phase Response	Linear Phase Response
More Ringing on Glitches	Less Ringing on Glitches
CAD Design Packages Available	CAD Design Packages Available
No Efficiency Gained by Decimation	Decimation Increases Efficiency

Figure 16.44: Comparison between FIR and IIR filters

On the other hand, FIR filters require more taps and multiply-accumulates for a given cut-off frequency response, but have linear phase characteristics. Since FIR filters operate on a finite history of data, if some data is corrupted (ADC sparkle codes, for example) the FIR filter will ring for only N–1 samples. Because of the feedback, however, an IIR filter will ring for a considerably longer period of time.

If sharp cut-off filters are needed, and processing time is at a premium, IIR elliptic filters are a good choice. If the number of multiply/accumulates is not prohibitive, and linear phase is a requirement, the FIR should be chosen.

16.12 Multirate Filters

There are many applications in which it is desirable to change the effective sampling rate in a sampled data system. In many cases, this can be accomplished simply by changing the sampling frequency to the ADC or DAC. However, it is often desirable to accomplish the sample rate conversion after the signal has been digitized. The most common techniques used are *decimation* (reducing the sampling rate by a factor of M), and *interpolation* (increasing the sampling rate by a factor of L). The decimation and interpolation factors (M and L) are normally integer numbers. In a generalized sample-rate converter, it may be desirable to change the sampling frequency by a noninteger number. In the case of converting the CD sampling frequency of 44.1 kHz to the digital audio tape (DAT) sampling rate of 48 kHz, interpolating by L = 160 followed by decimation by M = 147 accomplishes the desired result.

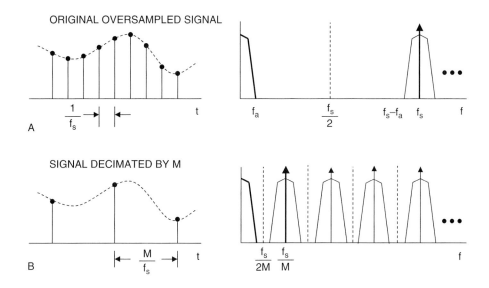

Figure 16.45: Decimation of a sampled signal by a factor of M

The concept of decimation is illustrated in Figure 16.45. The top diagram shows the original signal, f_a, which is sampled at a frequency f_s. The corresponding frequency spectrum shows that the sampling frequency is much higher than required to preserve information contained in f_a, i.e., f_a is oversampled. Notice that there is no information contained between the frequencies f_a and $f_s - f_a$. The bottom diagram shows the same signal where the sampling frequency has been reduced (decimated) by a factor of M. Notice that even though the sampling rate has been reduced, there is no aliasing and loss of information. Decimation by a larger factor than shown in Figure 16.45 will cause aliasing.

Figure 16.46(A) shows how to decimate the output of an FIR filter. The filtered data y(n) is stored in a data register that is clocked at the decimated frequency f_s/M. This does not change the number of computations required of the digital filter; i.e., it still must calculate each output sample y(n).

Figure 16.46(B) shows a method for increasing the computational efficiency of the FIR filter by a factor of M. The data from the delay registers are simply stored in N data registers that are clocked at the decimated frequency f_s/M. The FIR multiply-accumulates now only have to be done every Mth clock cycle. This increase in efficiency could be utilized by adding more taps to the FIR filter, doing other computations in the extra time, or using a slower DSP.

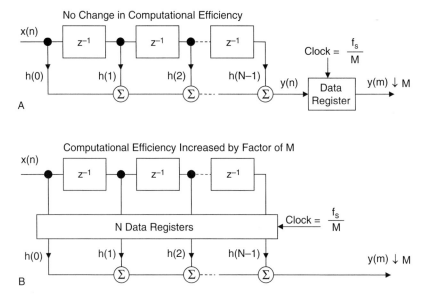

Figure 16.46: Decimation combined with FIR filtering

Figure 16.47 shows the concept of interpolation. The original signal in 16.47(A) is sampled at a frequency f_s. In 16.47(B), the sampling frequency has been increased by a factor of L, and zeros have been added to fill in the extra samples. The signal with added zeros is passed through an interpolation filter, which provides the extra data values.

The frequency domain effects of interpolation are shown in Figure 16.48. The original signal is sampled at a frequency f_s and is shown in 16.48(A). The interpolated signal in 16.48(B) is sampled at a frequency Lf_s. An example of interpolation is a CD player DAC, where the CD data is generated at a frequency of 44.1 kHz. If this data is passed directly to a DAC, the frequency spectrum shown in Figure 16.48(A) results, and the requirements on the anti-imaging filter that precedes the DAC are extremely stringent to overcome this. An oversampling interpolating DAC is normally used, and the spectrum shown in Figure 16.48(B) results. Notice that the requirements on the analog anti-imaging filter are now easier to realize. This is important in maintaining relatively linear phase and also reducing the cost of the filter.

The digital implementation of interpolation is shown in Figure 16.49. The original signal x(n) is first passed through a rate expander that increases the sampling frequency by a factor of L and inserts the extra zeros. The data then passes through an

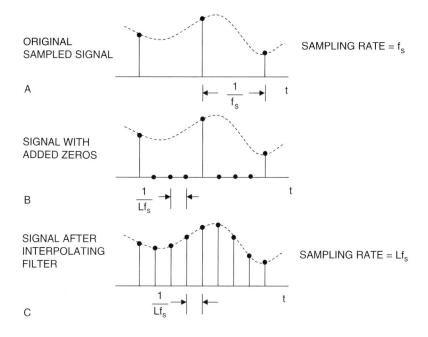

Figure 16.47: Interpolation by a factor of L

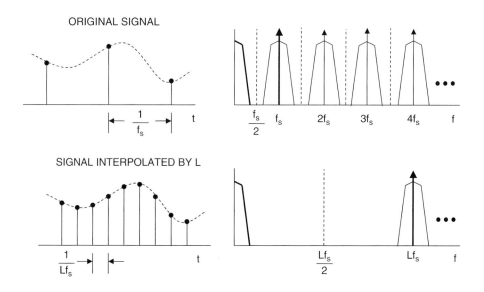

Figure 16.48: Effects of interpolation on frequency spectrum

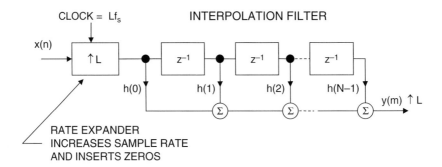

Efficient DSP algorithms take advantage of:

- ■ Multiplications by Zero
- ■ Circular Buffers
- ■ Zero-Overhead Looping

Figure 16.49: Typical interpolation implementation

interpolation filter that smoothes the data and interpolates between the original data points. The efficiency of this filter can be improved by using a filter algorithm that takes advantage of the fact that the zero-value input samples do not require multiply-accumulates. Using a DSP that allows circular buffering and zero-overhead looping also improves efficiency.

Interpolators and decimators can be combined to perform fractional sample rate conversion as shown in Figure 16.50. The input signal $x(n)$ is first interpolated by a factor of L and then decimated by a factor of M. The resulting output sample rate is Lf_s/M. To maintain the maximum possible bandwidth in the intermediate signal, the interpolation must come before the decimation; otherwise, some of the desired frequency content in the original signal would be filtered out by the decimator.

An example is converting from the CD sampling rate of 44.1 kHz to the digital audio tape (DAT) sampling rate of 48.0 kHz. The interpolation factor is 160, and the decimation factor, 147. In practice, the interpolating filter $h'(k)$ and the decimating' filter $h''(k)$ are combined into a single filter, $h(k)$.

The entire sample rate conversion function is integrated into the AD1890, AD1891, AD1892, and AD1893 family, which operates at frequencies between 8 kHz and 56 kHz (48 kHz for the AD1892). The AD1895 and AD1896 operate at up to 192 kHz.

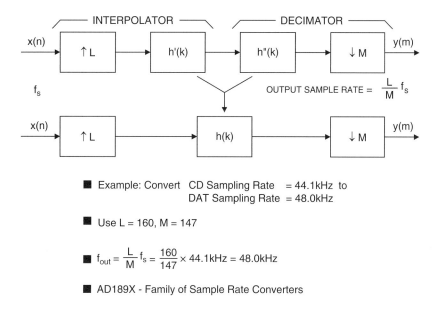

Figure 16.50: Sample rate converters

16.13   Adaptive Filters

Unlike analog filters, the characteristics of digital filters can easily be changed by modifying the filter coefficients. This makes digital filters attractive in communications applications such as adaptive equalization, echo cancellation, noise reduction, speech analysis, and speech synthesis. The basic concept of an adaptive filter is shown in Figure 16.51. The objective is to filter the input signal, x(n), with an adaptive filter in such a manner that it matches the desired signal, d(n). The desired signal, d(n), is subtracted from the filtered signal, y(n), to generate an error signal. The error signal drives an adaptive algorithm that generates the filter coefficients in a manner that minimizes the error signal. The least mean square (LMS) or recursive least square (RLS) algorithms are two of the most popular.

Adaptive filters are widely used in communications to perform such functions as equalization, echo cancellation, noise cancellation, and speech compression. Figure 16.52 shows an application of an adaptive filter used to compensate for the effects of amplitude and phase distortion in the transmission channel. The filter coefficients are determined during a training sequence where a known data pattern is transmitted. The adaptive algorithm adjusts the filter coefficients to force the receive data to match the

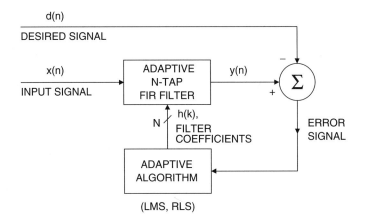

Figure 16.51: Adaptive filter

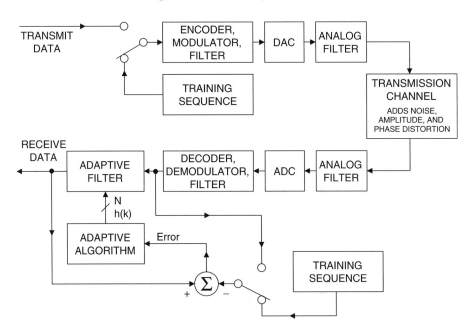

Figure 16.52: Digital transmission using adaptive equalization

training sequence data. In a modem application, the training sequence occurs after the initial connection is made. After the training sequence is completed, the switches are put in the other position, and the actual data is transmitted. During this time, the error signal is generated by subtracting the input from the output of the adaptive filter.

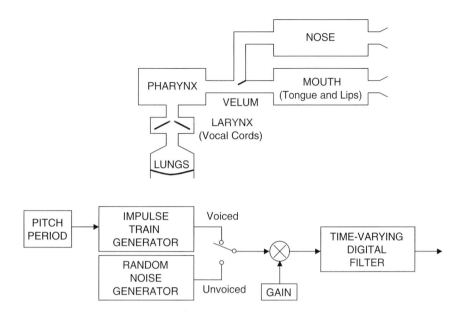

Figure 16.53: Linear predictive coding (LPC) model of speech production

Speech compression and synthesis also makes extensive use of adaptive filtering to reduce data rates. The linear predictive coding (LPC) model shown in Figure 16.53 models the vocal tract as a variable frequency impulse generator (for voiced portions of speech) and a random noise generator (for unvoiced portions of speech such as consonant sounds). These two generators drive a digital filter that in turn generates the actual voice signal.

The application of LPC in a communication system such as GSM is shown in Figure 16.54. The speech input is first digitized by a 16-bit ADC at a sampling frequency of 8 kSPS. This produces output data at 128 Kbps, which is much too high to be transmitted directly. The transmitting DSP uses the LPC algorithm to break the speech signal into digital filter coefficients and pitch. This is done in 20 ms windows, which have been found to be optimum for most speech applications. The actual transmitted data is only 2.4 Kbps, which represents a 53.3 compression factor. The receiving DSP uses the LPC model to reconstruct the speech from the coefficients and the excitation data. The final output data rate of 128 Kbps then drives a 16-bit DAC for final reconstruction of the speech data.

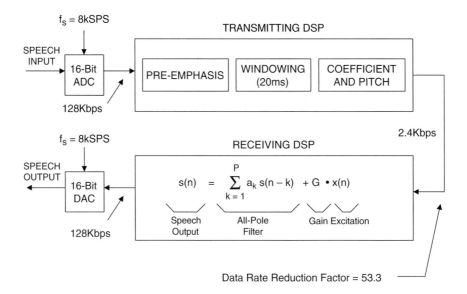

Figure 16.54: LPC Speech companding system

The digital filters used in LPC speech applications can either be FIR or IIR, although all-pole IIR filters are the most widely used. Both FIR and IIR filters can be implemented in a lattice structure as shown in Figure 16.55 for a recursive all-pole filter. This structure can be derived from the IIR structure, but the advantage of the lattice filter is that the coefficients are more directly related to the outputs of algorithms that use the vocal tract model shown in Figure 16.53 than the coefficients of the equivalent IIR filter.

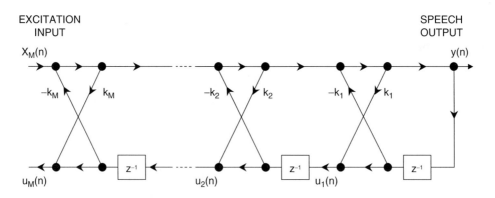

Figure 16.55: All-pole lattice filter

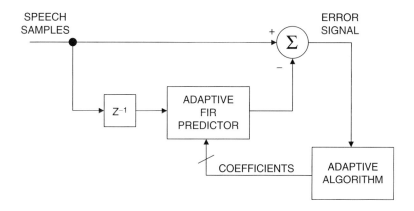

Figure 16.56: Estimation of lattice filter coefficients in transmitting DSP

The parameters of the all-pole lattice filter model are determined from the speech samples by means of linear prediction as shown in Figure 16.56. Due to the nonstationary nature of speech signals, this model is applied to short segments (typically 20 ms) of the speech signal. A new set of parameters is usually determined for each time segment unless there are sharp discontinuities, in which case the data may be smoothed between segments.

References

[16.1] *Practical Design Techniques for Sensor Signal Conditioning,* Analog Devices, 1998.

[16.2] Daniel H. Sheingold, Editor, *Transducer Interfacing Handbook,* Analog Devices, Inc., 1972.

[16.3] Richard J. Higgins, *Digital Signal Processing in VLSI,* Prentice-Hall, 1990.

[16.4] Steven W. Smith, *Digital Signal Processing: A Guide for Engineers and Scientists,* Newnes, 2002.

[16.5] C. Britton Rorabaugh, *DSP Primer,* McGraw-Hill, 1999.

[16.6] Richard J. Higgins, *Digital Signal Processing in VLSI,* Prentice-Hall, 1990.

[16.7] A.V. Oppenheim and R.W. Schafer, *Digital Signal Processing,* Prentice-Hall, 1975.

[16.8] L.R. Rabiner and B. Gold, *Theory and Application of Digital Signal Processing,* Prentice-Hall, 1975.

[16.9] John G. Proakis and Dimitris G. Manolakis, *Introduction to Digital Signal Processing,* MacMillian, 1988.

[16.10] J.H. McClellan, T.W. Parks, and L.R. Rabiner, *A Computer Program for Designing Optimum FIR Linear Phase Digital Filters*, *IEEE Trasactions on Audio and Electroacoustics,* Vol. AU-21, No. 6, December, 1973.

[16.11] Fredrick J. Harris, *On the Use of Windows for Harmonic Analysis with the Discrete Fourier Transform.* Proc. IEEE, Vol. 66, No. 1, 1978, pp. 51–83.

[16.12] Momentum Data Systems, Inc., 17330 Brookhurst St., Suite 140, Fountain Valley, CA 92708, http://www.mds.com.

[16.13] *Digital Signal Processing Applications Using the ADSP-2100* Family, Vol. 1 and Vol. 2, Analog Devices, Free Download at: http://www.analog.com.

[16.14] *ADSP-21000 Family Application Handbook*, Vol. 1, Analog Devices, Free Download at: http://www.analog.com.

[16.15] B. Widrow and S.D. Stearns, *Adaptive Signal Processing*, Prentice-Hall, 1985.

[16.16] S. Haykin, Adaptive Filter Theory, 3rd Edition, Prentice-Hall, 1996.

[16.17] Michael L. Honig and David G. Messerschmitt, *Adaptive Filters – Structures, Algorithms, and Applications*, Kluwer Academic Publishers, Hingham, MA 1984.

[16.18] J.D. Markel and A.H. Gray, Jr., *Linear Prediction of Speech*, Springer-Verlag, New York, NY, 1976.

[16.19] L.R. Rabiner and R.W. Schafer, *Digital Processing of Speech Signals*, Prentice-Hall, 1978.

CHAPTER 17

Dealing with High-Speed Logic

Walt Kester

Much has been written about terminating printed circuit board traces in their characteristic impedance to avoid reflections. A good rule of thumb to determine when this is necessary is: *Terminate the line in its characteristic impedance when the one-way propagation delay of the PCB track is equal to or greater than one-half the applied signal rise/fall time (whichever edge is faster).* A conservative approach is to use a 2-inch (PCB track length)/nanosecond (rise/fall time) criterion. For example, PCB tracks for high speed logic with rise/fall time of 1 ns should be terminated in their characteristic impedance if the track length is equal to or greater than 2 inches (including any meanders). Figure 17.1 shows the typical rise/fall times of several logic families including the SHARC DSPs operating on 3.3V supplies. As would be expected, the rise/fall times are a function of load capacitance.

- GaAs: 0.1 ns,

- ECL: 0.75 ns,

- ADI SHARC DSPs: 0.5 ns to 1 ns (operating on 3.3V supply).

This same 2-inch/nanosecond rule of thumb should be used with analog circuits in determining the need for transmission line techniques. For instance, if an amplifier must output a maximum frequency of f_{max}, then the equivalent rise time, t_r, can be calculated using the equation $t_r = 0.35/f_{max}$. The maximum PCB track length is then calculated by multiplying the rise time by 2 inch/nanosecond. For example, a maximum output frequency of 100 MHz corresponds to a rise time of 3.5 ns, and a track carrying this signal greater than 7 inches should be treated as a transmission line.

■ GaAs: 0.1 ns

■ ECL: 0.75 ns

■ ADI SHARC DSPs: 0.5 ns to 1 ns (Operating on 3.3 V Supply)

ADSP-21060L
SHARC:

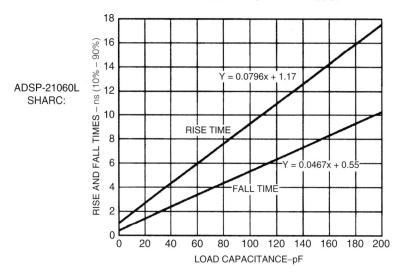

Figure 17.1: Typical DSP output rise times and fall times

Equation 17.1 can be used to determine the characteristic impedance of a PCB track separated from a power/ground plane by the board's dielectric (microstrip transmission line):

$$Z_O(\Omega) = \frac{87}{\sqrt{\varepsilon_r + 1.41}} \ln \left[\frac{5.98d}{0.89w + t} \right] \qquad (17.1)$$

where ε_r = dielectric constant of printed circuit board material:

d = thickness of the board between metal layers, in mils;

w = width of metal trace, in mils;

t = thickness of metal trace, in mils.

The one-way transit time for a single metal trace over a power/ground plane can be determined from Equation 17.2:

$$t_{pd}(ns/ft) = 1.017 \sqrt{0.475\varepsilon_r + 0.67} \qquad (17.2)$$

For example, a standard four-layer PCB board might use 8-mil-wide, 1 oz. (1.4 mils) copper traces separated by 0.021″ FR-4 ($\varepsilon_r = 4.7$) dielectric material. The characteristic impedance and one-way transit time of such a signal trace would be 88W and 1.7 ns/ft (7″/ ns), respectively.

The best ways to keep sensitive analog circuits from being affected by fast logic are to physically separate the two and use no faster logic family than dictated by system requirements. In some cases, this may require the use of several logic families in a system. An alternative is to use series resistance or ferrite beads to slow down the logic transitions where the speed is not required. Figure 17.2 shows two methods. In the first, the series resistance and the input capacitance of the gate form a low-pass filter. Typical CMOS input capacitance is 5 pF to 10 pF. Locate the series resistor close to the driving gate. The resistor minimizes transient currents and may eliminate the necessity of using transmission line techniques. The value of the resistor should be chosen such that the rise and fall times at the receiving gate are fast enough to meet system requirement, but no faster. Also, make sure that the resistor is not so large that the logic levels at the receiver are out of specification because of the voltage drop caused by the source and sink current that flow through the resistor. The second method is suitable for longer distances (>2 inches), where additional capacitance is added to slow down the edge speed. Notice that either one of these techniques increases

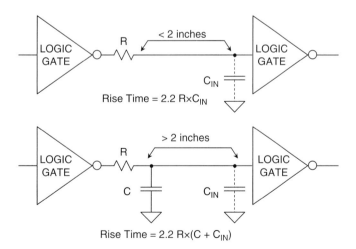

Figure 17.2: Damping resistors slow down fast logic edges to minimize EMI/RFI problems

delay and increases the rise/fall time of the original signal. This must be considered with respect to the overall timing budget, and the additional delay may not be acceptable.

Figure 17.3 shows a situation where several DSPs must connect to a single point, as would be the case when using read or write strobes bidirectionally connected from several DSPs. Small damping resistors shown in Figure 17.3(A) can minimize ringing provided the length of separation is less than about 2 inches. This method will also increase rise/fall times and propagation delay. If two groups of processors must be connected, a single resistor between the pairs of processors as shown in Figure 17.3(B) can serve to damp out ringing.

The only way to preserve 1 ns or less rise/fall times over distances greater than about 2 inches without ringing is to use transmission line techniques. Figure 17.4 shows two popular methods of termination: end termination, and source termination. The end termination method (Figure 17.4(A)) terminates the cable at its terminating point in the characteristic impedance of the microstrip transmission line. Although higher

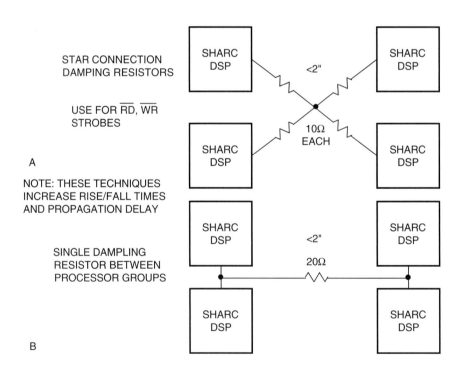

Figure 17.3: Series damping resistors for SHARC DSP interconnections

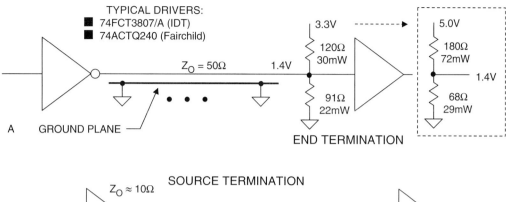

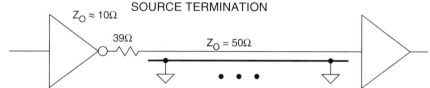

Figure 17.4: Termination techniques for controlled impedance microstrip transmission lines

impedances can be used, 50Ω is popular because it minimizes the effects of the termination impedance mismatch due to the input capacitance of the terminating gate (usually 5 pF to 10 pF). In Figure 17.4(A), the cable is terminated in a Thévenin impedance of 50Ω terminated to 1.4V (the midpoint of the input logic threshold of 0.8V and 2.0V). This requires two resistors (90Ω and 120Ω), which adds about 50 mW to the total quiescent power dissipation to the circuit. Figure 17.4(A) also shows the resistor values for terminating with a 5V supply (68Ω and 180Ω). Note that 3.3V logic is much more desirable in line driver applications because of its symmetrical voltage swing, faster speed, and lower power. Drivers are available with less than 0.5 ns time skew, source and sink current capability greater than 25 mA, and rise/fall times of about 1 ns. Switching noise generated by 3.3V logic is generally less than 5V logic because of the reduced signal swings and lower transient currents.

The source termination method, shown in Figure 17.4(B) absorbs the reflected waveform with an impedance equal to that of the transmission line. This requires about 39Ω in series with the internal output impedance of the driver, which is generally about 10Ω.

This technique requires that the end of the transmission line be terminated in an open circuit; therefore no additional fanout is allowed.

The source termination method adds no additional quiescent power dissipation to the circuit.

Figure 17.5 shows a method for distributing a high speed clock to several devices. The problem with this approach is that there is a small amount of time skew between the clocks because of the propagation delay of the microstrip line (approximately 1 ns/7″). This time skew may be critical in some applications. It is important to keep the stub length to each device less than 0.5″ in order to prevent mismatches along the transmission line.

The clock distribution method shown in Figure 17.6 minimizes the clock skew to the receiving devices by using source terminations and making certain the length of each microstrip line is equal. There is no extra quiescent power dissipation, as would be the case using end termination resistors.

Figure 17.7 shows how source terminations can be used in bidirectional link port transmissions between SHARC DSPs. The output impedance of the SHARC driver is approximately 17Ω, and therefore a 33Ω series is required on each end of the transmission line for proper source termination.

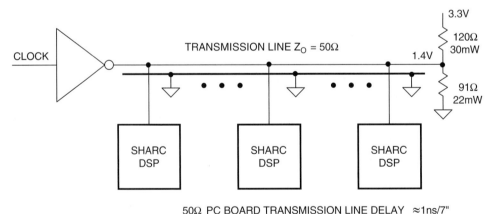

Figure 17.5: Clock distribution using end-of-line termination

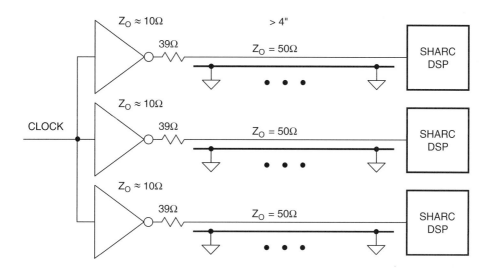

Figure 17.6: Preferred method of clock distribution using source terminated transmission lines

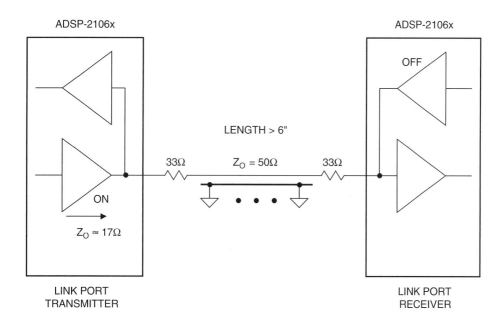

Figure 17.7: Source termination for bidirectional transmission between SHARC DSPs

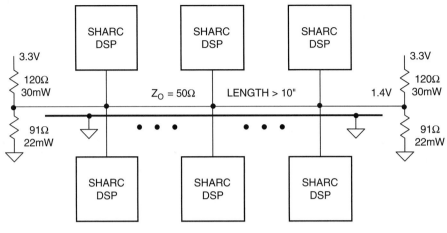

NOTE: KEEP STUB LENGTH < 0.5"

Figure 17.8: Single transmission line terminated at both ends

The method shown in Figure 17.8 can be used for bidirectional transmission of signals from several sources over a relatively long transmission line. In this case, the line is terminated at both ends, resulting in a DC load impedance of 25Ω. SHARC drivers are capable of driving this load to valid logic levels.

References

Howard W. Johnson and Martin Graham, *High-Speed Digital Design,* PTR Prentice Hall, 1993.
"EDN's Designer's Guide to Electromagnetic Compatibility," *EDN*, January, 20, 1994, material reprinted by permission of Cahners Publishing Company, 1995.
Designing for EMC (Workshop Notes), Kimmel Gerke Associates, Ltd., 1994.
Mark Montrose, *EMC and the Printed Circuit Board,* IEEE Press, 1999 (IEEE Order Number PC5756).

CHAPTER 18

Bridging the Gap between Analog and Digital

Bonnie Baker

A few years ago, I was approached by a new graduate engineering applicant at the Embedded Systems Conference (ESC), 2001 in San Francisco. When he found out that I was a manager, he explained that he was looking for a job. He said he knew of Microchip Technology, Inc. and wanted to work for them if he could. He immediately produced his resume. I gave him a few more details about my role at Microchip. At the time, I managed the mixed signal/linear applications group. My department's roles were product definition, technical writing, customer training, and traveling all over the world visiting customers. At the conclusion of this "sales" pitch, he proudly told me that it sounded like a great job. I reemphasized that I was in the Analog arm at Microchip. He obviously thought that he did his homework because he told me that analog is dying and digital will eventually take over. Anyone who knew anything about Microchip would agree! Wow, I had a live one.

I was there, doing my obligatory Microchip booth duty for the afternoon. There was a lot of action on the floor, and the room was full of exhibits. The lights were on, the sound of conversations were projecting across the room. The heating and cooling system was doing a splendid job of keeping us comfortable. Exhibitors in the booths were (believe it or not) demonstrating the operation of sensors, power devices, passive devices, RF products, and so forth. There must have been several hundred booths, all of which were trying to promote their engineering merchandise.

Some of the vendor exhibits had analog signal conditioning demonstrations. As a matter of fact, right in front of us, Microchip had a temperature sensor connected to a computer

Figure 18.1: The Embedded Systems Conference exhibit hall in 2001 had hundreds of booths, many of which were already showing signs of interest in analog systems. This was done even though the emphasis of the conference was digital.

through the parallel port. The temperature sensor board would self-heat, and the sensor would measure this change and show the results on the PC screen. Once the temperature reached a threshold of 85°C, the heating element was turned off. You could then watch the temperature go down on the PC until it reached 40°C, at which point the heating element would be turned on again.

At a second counter, we also had a computer running the new FilterLab® analog filter design program. With this tool, you can specify an analog filter in terms of the number of poles, cut-off frequency and approximation type (Butterworth, Bessel and Chebyshev). Once you type in your information, the software spits out a filter circuit diagram, such as the filter circuit shown in Figure 18.2. You can theoretically build the circuit and take it to the lab for testing and verification. There was a customer at the counter, playing around with the filter software.

At exhibit counter number three, there was a CANbus demonstration with temperature sensing, pressure sensing and DC motor nodes. CANbus networks have been around for over 15 years. Initially, this bus was used in automotive applications requiring

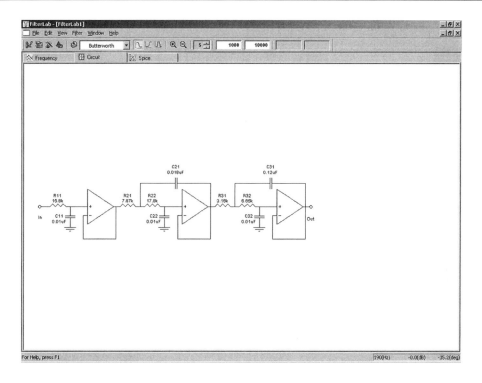

Figure 18.2: One of the views of the FilterLab program from Microchip provided analog filter circuit diagrams. This particular circuit is a 5th order, low-pass Butterworth filter with a cut-off frequency of 1 kHz. The FilterLab program from Microchip is just one example of a filter program from a semiconductor supplier. Texas Instruments, Linear Technology, and Analog Devices have similar programs available on the World Wide Web.

predictable, error-free communications. Recent falling prices of controller area network (CAN) system technologies have made it a commodity item. The CANbus network has expanded past automotive applications. It is now migrating into systems like industrial networks, medical equipment, railway signaling and controlling building services (to name a few). These applications are using the CANbus network, not only because of the lower cost, but because the communication with this network is robust, at a bit rate of up to 1 Mbits/sec.

A CANbus network features a multimaster system that broadcasts transmissions to all of the nodes in the system. In this type of network, each node filters out unwanted messages. An advantage from this topology is that nodes can easily be added or removed with minimal software impact. The CAN network requires intelligence on each node, but the level of intelligence can be tailored to the task at that node. As a result, these individual controllers are usually simpler, with lower pin counts. The CAN network also has higher reliability by using distributed intelligence and fewer wires.

You might say, "What does this have to do with analog circuits?" And the answer is *everything*. The communication channel is important only because you are shipping digitized analog information from one node to another. With this ESC exhibit, three CANbus nodes communicated through the bus to each other. One node measured temperature. The temperature value was used to calibrate the pressure sensor on the second node. You could apply pressure to the pressure-sensing node by manually squeezing a balloon. (This type of demonstration was put together to get the observer more involved.) The sensor circuitry digitized the level pressure applied and sent that data through the CANbus network to a DC motor. The DC motor was configured so that increased pressure would increase the revolutions per minute (RPM) of the motor. Figure 18.3 shows a basic block diagram containing the pressure-sensing node.

Then to finish out the Microchip displays in the booth, there were three counters that had microcontroller demos.

I asked the engineering applicant, giving him a chance to redeem himself, "Out of curiosity, do you see anything analog-ish like in this room?" He looked around the convention room thoughtfully. I was amused when he sympathetically looked at me and answered, "No, not really." I think that he thought I was a bit old-fashioned, behind the times. No regrets from him. He was confident that he gave me an insightful, informed answer.

You guessed it. His resume went into the circular file.

18.1   Try to Measure Temperature Digitally

No, this is not a chapter about interview techniques. This chapter is neither about how to win points and climb the corporate ladder. This is a chapter about the analog design opportunities that surround us every day, all day long, and how we can solve them in a single-supply environment. Reflecting on the applicant's answer, I think that he was partially right. Digital solutions are encroaching into the analog hardware in a majority of applications.

So let's try to measure temperature digitally. The simple, low resolution analog-to-digital (A/D) converter can easily be replaced with a resistor/capacitor (R/C) pair

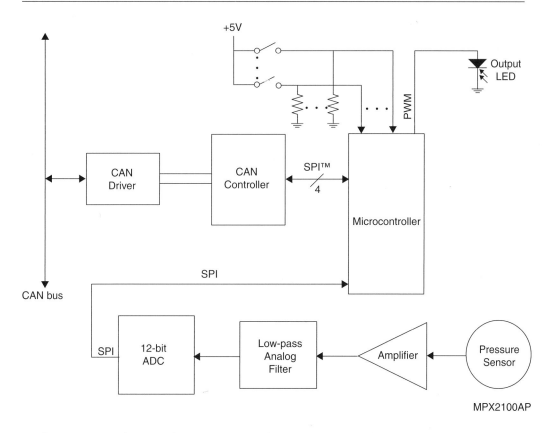

Figure 18.3: The CANbus system at the 2001 Embedded Systems Conference has three different analog function nodes. The node illustrated in this figure measured the pressure applied to a balloon and sent the data across the CANbus network to a DC motor (not illustrated here).

connected to a microcontroller I/O pin. The R/C pair would supply a signal that operates with a single-pole, rise-time function. The controller counts milliseconds, and with its oscillator/timer combination measures the input signal. Why would you want to do this? Maybe you are measuring temperature with a sensor that changes its resistance value with changes in temperature.

The temperature-sensing circuit in Figure 18.4 is implemented by setting GP1 and GP2 of the microcontroller as inputs. Additionally, GP0 is set low to discharge the capacitor, C_{INT}. As the voltage on C_{INT} discharges, the configuration of GP0 is changed to an input and GP1 is set to a high output. An internal timer counts the amount of time (t_1 in

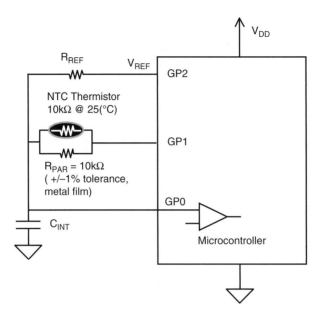

Figure 18.4: This circuit switches the voltage reference on and off at GP1 and GP2. In this manner, the time constant of the NTC thermistor in parallel with a standard resistor ($R_{NTC} \parallel R_{PAR}$) and integrating capacitor (C_{INT}) is compared to the time constant of the reference resistor (R_{REF}) and integrating capacitor.

Figure 18.5) before the voltage at GP0 reaches the threshold (V_{TH}), which changes the recognized input from 0 to 1. In this case, R_{NTC} (a negative temperature coefficient thermistor) is placed in parallel with R_{PAR} or $R_{NTC} \parallel R_{PAR}$. This parallel combination interacts with C_{INT}. After this happens, GP1 and GP2 are again set as inputs and GP0 as an output low. Once the integrating capacitor C_{INT} has time to discharge, GP2 is set to a high output and GP0 as an input. A timer counts the amount of time before GP0 changes to 1 again, but this provides the timed amount of t_2, per Figure 18.5. In this case, R_{REF} is the component interacting with C_{INT}.

The integration time of this circuit can be calculated using:

$$V_{OUT} = V_{REF}(1 - e^{-t/RC}) \text{ or}$$
$$t = RC \ln(1 - V_{TH}/V_{REF})$$

where V_{OUT} is the voltage at the I/O pin, GP0,

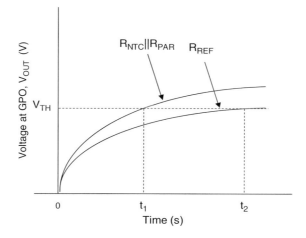

Figure 18.5: The R/C time response of the circuit shown in Figure 18.4 allows for the microcontroller counter to be used to determine the relative resistance of the negative temperature coefficient (R_{NTC}) thermistor element.

V_{REF} is the output, logic-high voltage of the I/O pin, GP1 or GP2;

V_{TH} is the input voltage to GP0 that causes a logic 1 to trigger in the microcontroller. If the ratio of $V_{TH}:V_{REF}$ is kept constant, the unknown resistance of the $R_{NTC} \parallel R_{PAR}$ can be determined with:

$$R_{NTC} \parallel R_{PAR} = R_{REF}(t_2/t_1)$$

Notice that in this configuration, the resistance calculation of the parallel combination of $R_{NTC} \parallel R_{PAR}$ is independent of C_{INT}, but the absolute accuracy of the measurement is dependent on the accuracy of your resistors.

Oops, did I say you can use a linear resistor and a charging device like a capacitor to replace an A/D converter in a temperature measurement system? I guess my applicant at the ESC show was also wrong. Analog will never disappear and the digital engineer will continue to be challenged to delve into these types of issues. The analog solution is many times more efficient and usually more accurate. For instance, the previous R/C example is only as accurate as the number of bits in the timer, the speed of the oscillator, and how accurately you know the value of your resistors.

18.2   Road Blocks Abound

I have worked with a wide spectrum of analog and digital designers. Each one of them has their own quirks and reasons why they can't do everything, but here are some statements that I have received from my digital clientele concerning their analog challenges.

18.2.1   Not My Job!

This statement came about with surprising frankness. "People in my department are avoiding analog circuitry in their design as much as possible, no matter how important it is. Many of them have had experiences where analog circuit performance was hard to predict. Therefore, almost every engineer will find an existing analog circuit and use that as a point of reference. If they have the misfortune of being asked to design part or all of the analog circuit from scratch, they will try to use facts that they remember from their school days. And in their school days they studied mostly digital."

Good luck. It seems from this statement that the dyed-in-the-wool digital designer has no interest in how to get from A to B, but more interest in what the cookbook suggests.

It turns out that the designer who operates in this mode is like a carpenter with a hammer looking for a nail. The designer has a circuit solution and tries to make it fit their application. A good example of applying the cookbook solution to a place where it won't fit is to try to use a standard 12-bit successive approximation register (SAR) in a power-sensing application. This type of application actually requires a delta-sigma converter. The delta-sigma (Δ-Σ) converter can reach a resolution level in the sub-nano volt region. This is an advantage because you not only eliminate the input analog-gain stage, but you reduce the noise in the bandpass region of your signal. Figure 18.6 shows this power meter solution.

In this circuit, the current through the power line is sensed using an inductor on the low side of the load. As a result, the voltage drop across this sensing element must be low. As a result, the voltage drop across this sensing element must be low.

18.2.2   Show Me the Beef

One day, a digital engineer said to me, "Thank God, I have finally found the key to working with analog and now I can go back about my digital business. Thank you for that one insightful tip."

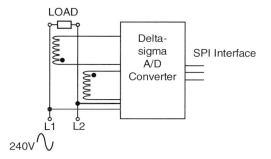

Figure 18.6: A power meter application requires <12-bit resolution in the system. This may imply that a simple 12-bit SAR converter can do the job, but the required LSB size is much smaller than the SAR converter can provide. Consequently, a delta-sigma-sigma converter is often used.

The tip I gave him was not that earthshaking. As a matter of fact, it provided the two primary operational amplifier specifications that an engineer uses when designing an analog low-pass filter.

The *gain-bandwidth product* (GBWP) is a multiplication factor that helps you predict the closed-loop bandwidth of an operational amplifier. You can easily find this parameter by looking at the specification table of the amplifier. You can quickly find this specification out of the 30 or 40 items in the table by looking at the "units" column. That column is usually on the right side of the table. When you are trying to find the gain-bandwidth product specification, look for frequency units in Hz, kHz or MHz. Once you find these abbreviated units, verify that you have found the right item by looking to the left for the specification definition. Now, double-check and ensure you understand the test conditions for this specification by reading the conditions column and the general conditions that are summarized at the top of the table. All of these areas on a typical data sheet are pointed out in Figure 18.7.

A second place where this specification can be found is in the characteristic performance graphs later on in the data sheet (see Figure 18.8). *Open-loop gain versus frequency* is the usual label for this curve. Sometimes the open-loop phase is included in this graph. You will find that the 0dB crossing of the gain curve will usually match the gain-bandwidth product in the specification table.

The gain-bandwidth product (GBWP) will tell you the highest small-signal frequency ($\sim \pm 100$ mV) that you can send through your amplifier circuit without distortion.

AC ELECTRICAL SPECIFICATIONS

Electrical Characteristics: Unless otherwise indicated, T_A = +25°C, V_{DD} = +1.8 to 5.5V, V_{SS} = GND, V_{CM} = V_{DD}/2, V_{OUT} = V_{DD}/2, R_L = 10 kΩ to V_{DD}/2, and C_L = 60 pF.

Parameters	Sym	Min	Typ	Max	Units	Conditions
AC Response						
Gain Bandwidth Product	GBWP	—	1.0	—	MHz	
Phase Margin	PM	—	90	—	°	G = +1
Slew Rate	SR	—	0.6	—	V/μs	
Noise						
Input Noise Voltage	E_{ni}	—	6.1	—	μVp-p	f = 0.1 Hz to 10 Hz
Input Noise Voltage Density	e_{ni}	—	28	—	nV/√Hz	f = 1 kHz
Input Noise Current Density	i_{ni}	—	0.6	—	fA/√Hz	f = 1 kHz

Figure 18.7: A typical electrical specification table for an operational amplifier has seven columns. When searching for a particular specification, the units column is the easiest one to start with.

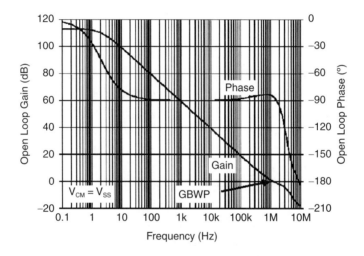

Figure 18.8: These typical performance curves show many of the parameters in the specification table of the data sheet of an amplifier. This graph illustrates the typical open-loop gain, phase vs. frequency response. The arrow in this figure points to the gain-bandwidth product for this unity-gain-stable amplifier.

It also tells you the frequency where a pole is introduced to your closed-loop amplifier circuit. This is particularly critical to know when you design low-pass filters. In this type of circuit, you deliberately put poles in the transfer function by putting resistors and capacitors around the amplifier, as shown in Figure 18.2. If the amplifier adds a pole, your circuit could oscillate. Consequently, the closed-loop bandwidth of the amplifier must be at least 100 times higher than the cut-off frequency ($f_{CUT-OFF}$) of the filter. Another way of stating this is that the gain-bandwidth product of your amplifier should be equal to or greater than $100 \times f_{CUT-OFF}$ (this assumes the filter has a gain of +1 V/V). If you don't take these precautions, you might erroneously be inclined to investigate your filter equations only to find out that the amplifier is not well-suited for your design.

You might ask, "How important is this specification in other amplifier application circuits?" Generally, you will need an amplifier with good bandwidth performance for your signal, but probably won't see instability because of your amplifier selection. Or in another application, you may be more concerned about the quiescent current of the amplifier or power supply capability instead of the bandwidth because you are designing a battery-powered circuit that operates at DC.

The second specification that I mentioned previously is *slew rate*. The slew rate of an amplifier is determined by putting a square wave signal at the input of the amplifier and looking to see how fast the signal changes on the output. The units of this specification are generally V/sec, V/msec, or V/μsec. You can find this specification in the table in the same way we found the gain-bandwidth product. There is also a characteristic curve in most amplifier data sheets that gives a good look at how a typical part will perform. You'll find that the label of this graph is usually "Large signal, noninverting pulse response" (Figure 18.9).

With the filter circuit, this specification will tell you the maximum frequency of the large signals going through your circuit. If you don't pay attention to this specification, you may find that the amplifier distorts your larger, higher frequency signals. A good rule of thumb for this design parameter is: slew rate $\geq$ ($2\pi V_{OUT\ P-P} \times f_{CUT-OFF}$) where $V_{OUT\ P-P}$ is the expected peak-to-peak output voltage swing below $f_{CUT-OFF}$ of your filter.

18.2.3 Don't Bother Me With the Small Stuff—Just Give Me the Data

One of the more common statements as said to me by the ambitious digital engineer is, "Just give me the data. I will fix it in my processor. I know we can design a digital filter with the classical FIR or IIR filters, or better yet implement an FFT response. I can also calibrate the signal if need be. I'm confident that I will be able to get rid of those undesirable, messy analog signals."

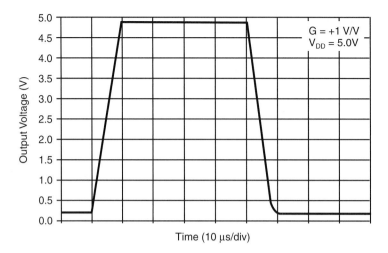

Figure 18.9: This graph illustrates the typical time domain response of the output voltage vs. time of an amplifier.

This comment always brings a smile to my face. See the case in point with the circuit in Figure 18.10.

The analog portion of this circuit has a load cell, a dual-operational amplifier configured as an instrumentation amplifier, a SAR A/D converter, a microcontroller and voltage references for the IA and A/D converter. The sensor is a 1.2Ω, 2 mV/V load cell with a full-scale load range of ±32 ounces. In this 5V system, the electrical full-scale output range of the load cell is ±10 mV. The instrumentation amplifier, consisting of two operational amplifiers (A1 and A2) and five resistors, is configured with a gain of 153 V/V. This gain matches the full-scale output swing of the instrumentation amplifier block to the full-scale input range of the A/D converter. The SAR A/D converter has an internal input sampling mechanism. With this function, a single sample is taken for each conversion. The microcontroller acquires the data from the SAR A/D converter. The controller can also execute calibration and translate the data into a usable format for tasks such as displays or actuator feedback signals.

The transfer function, from sensor to the output of the A/D converter is:

$$D_{OUT} = \left((LC_P - LC_N)(Gain) + V_{REFI} \right)\left(2^{12}/V_{REF2} \right)$$

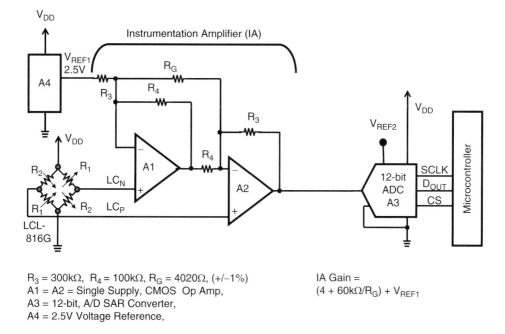

R$_3$ = 300kΩ, R$_4$ = 100kΩ, R$_G$ = 4020Ω, (+/−1%)
A1 = A2 = Single Supply, CMOS Op Amp,
A3 = 12-bit, A/D SAR Converter,
A4 = 2.5V Voltage Reference,

IA Gain =
(4 + 60kΩ/R$_G$) + V$_{REF1}$

Figure 18.10: The circuit in this diagram uses a 12-bit A/D converter in combination with an instrumentation amplifier to convert the low-signal output of a Wheatstone bridge sensor to usable digital codes.

with $LC_P = V_{DD}(R_2/(R_1 + R_2))$

with $LC_N = V_{DD}(R_1/(R_1 + R_2))$

with $GAIN = (1 + R_3/R_4 + 2R_3/R_G)$

where LC_P and LC_N are the positive and negative sensor outputs,

GAIN is the gain of the instrumentation amplifier circuit. The instrumentation amplifier is configured using A1 and A2. The gain is adjusted with R_G,

V_{REF1} is a 2.5V reference which level shifts the instrumentation amplifier output;

V_{REF2} is a 4.096V reference, which determines the A/D converter input range and LSB size;

V_{DD} is the power supply voltage and sensor excitation voltage;

D_{OUT} is a decimal representation of the 12-bit digital output code of the A/D converter (rounded to the nearest integer).

If the design of this system is poorly implemented, it could be an excellent candidate for noise problems. The symptom of a poor implementation is an intolerable level of uncertainty with the digital output results from the A/D converter. It is easy to assume that this type of symptom indicates that the last device in the signal chain generates the noise problem. But, in fact, the root cause of poor conversion results could originate with other active devices or passive components in the signal chain, the PCB layout or even extraneous sources.

In this circuit, noise can be reduced within the analog channel hardware. But, with the first prototype of this circuit, these low noise precautions were not used. Therefore, the data output of the A/D converter illustrated in Figure 18.11 indicates that this was a noisy system. It is fine to design a proto with this level of noise. In addition, it is truly divine to understand the noise and remove it in hardware wherever possible.

But, let's assume that you take the digital route to perform the filtering. On a perfect day, you will need to collect at least 2,048 12-bit data points and calculate the average. I said on *a perfect day* because when I look at this data there seems to be more going on than just white noise. There are small occurrences in the lower 20 codes of

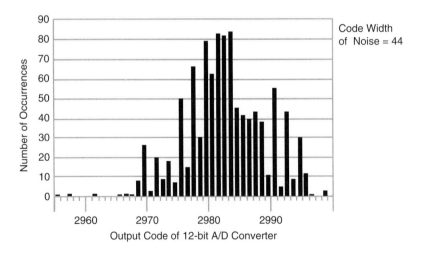

Figure 18.11: A poor implementation of the 12-bit data acquisition system shown in Figure 18.10 could easily have an output range of 44 different codes with a 2,048 sample size.

the data, and the major portion of the data does not form a "normal distribution" type of curve. It seems to have troughs and there is nothing normal about this data at all.

A common, bad scenario is that the problem is never solved through the lifetime of your application circuit. These unknown noise problems are fixed with digital tricks. That overly confident statement ignores the trade-offs inherent in taking the all-digital route. One of the major consequences is time. A digital filter needs to collect several hundreds of samples in order to compete with the analog solution. On top of that, the already digitized signal has been contaminated by aliased high-frequency signals, and you will never be able to tell your original signal from the contaminants. These tricks may or may not work over time.

On the other hand, the analog solution is simple and final. The data loses its erratic behavior and you can get the same converted number every time! What do you do?

1. Put bypass capacitors across the power supply pin to ground with every active device.

2. Use a ground plane. This will usually require at least a two-layer board.

3. Reduce the resistor values in the instrumentation amplifier. When you reduce these resistors (without changing the throughput gain), the noise in the signal chain will also reduce.

4. Use low noise amplifiers.

5. Insert a low-pass filter before the A/D converter. This filter will remove higher frequency noise as well as eliminate aliasing problems.

6. Choke the power supply switching noise to the analog portion of the board with an inductor.

These are all simple solutions to a seemingly impossible, noisy circuit problem. Figure 18.12 shows the results of these actions.

Calibration can be another sticky point when you go to the digital environment. Once you lose your dynamic range in the analog domain, it is impossible to recover it digitally. For instance, if you use amplifiers in this circuit that do not give you good rail-to-rail performance, the outer limits of the signal are lost forever. Another situation, not related to Figure 18.10, could occur if your signal is logarithmic instead of linear. If this is the case, digital manipulation will not take you very far. This type of data can only be fixed in the analog domain.

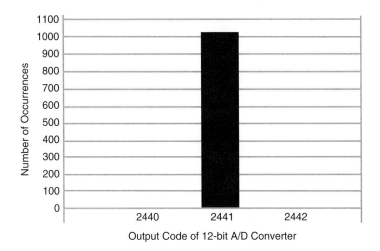

Figure 18.12: When noise reduction techniques are used in the implementation of the circuit in Figure 18.10, it is possible to get a 12-bit system.

18.3   The Ultimate Key to Analog Success

News alert! The ultimate analog key does not exist. And I don't mind turning this around to tell analog engineers that digital engineering is a little more than ones and zeros. The analog mountains that can be climbed are analogous to your digital challenges. Following are three examples.

For the first example, in the spirit of designing a robust design, the digital designer architects the software to identify unforeseen, catastrophic errors. The watchdog timer (WDT) can be used for this purpose. The function of a watchdog timer is easy enough. It counts down using the system clock from an initial value to zero. During implementation, if your firmware does not reset this timer soon enough, the watchdog timer resets or interrupts the system without human intervention when the counter reaches zero. Alternatively, the analog domain protection circuitry is used to minimize the effects of unforeseen errors or transients. In analog disciplines this can be implemented with over-range notifications or protection devices at sensitive nodes, such as zener diodes, metal oxide varistors (MOV), transzorbes, or Schottky diodes. With these types of additions to the hardware, "bad" signals are identified and eliminated before they become part of the signal path.

The second example would be to work on your digital design low-power strategies by effectively using clocking algorithms. Low power should be thought of as a "state of mind." With a low-power mindset, you can throttle down your controller to near inactivity if you really want to save battery power. The hardware approach would be to reduce clock source rate or power supply voltage. An equally effective approach is to operate with a partial or complete controller/processor shutdown mode. Combining these techniques with execution time and a little intelligence, you can easily tackle your most challenging power conservation problems. In your analog design, you will choose the lower power devices and utilize device shutdown features. In this environment, the designer needs to research the market for the best solution, whether it is a similar lower power device, or an alternative silicon topology that runs more efficiently.

A third example would be where you savor and protect your programming tricks from your competition. You can do this by making the code unreadable in the finished product in the same way the analog engineer buries traces inside boards, blacks out device part numbers, or asks vendors to give him proprietary part numbers.

The list goes on. But the thing to remember and understand is that each of us, in our own disciplines, tries to take technology to its limit. So what do you do when your manager says in his one-sided conversational way, "You're an engineer (aren't you)? Good. Since we are understaffed, I need you to do the entire (hardware/software) design. What? You don't know anything about analog. Hmmm, maybe I need to find someone else? I knew you would rise to the occasion. Have your development schedule on my desk by the end of the day so I can set up a deadline schedule."

18.4   How Analog and Digital Design Differ

The basic difference between the analog mindset and digital mindset is embedded in the definitions of precision (calculated risk versus right every time), hardware versus software, and time (or the inverse of). The basic concepts behind analog and digital disciplines are easy to find. In terms of this chapter, I will describe analog design from a practical standpoint. You will find that the in-depth lists and details about product specifications will be a little thin, but there is a detailed discussion about key specifications as they relate to basic analog systems.

18.4.1   Precision

What is "precise enough" in an analog circuit? There are three ways to answer this question. A first aspect of accuracy is "as precise as it needs to be." You will find that some of your circuits will only require accuracy to one or two millivolts. Others will require accuracy to the submicrovolts.

This difference in system requirements will encourage you to settle for "close enough" in some systems, and "What else can I squeeze out of this circuit?" in other systems.

A second aspect of accuracy involves really understanding the components and devices you are working with. In terms of the components, you should know that a 1 kΩ resistor or a 20 pF capacitor is not equal to those absolute values all the time. For instance, temperature can have a dramatic effect on these components. Also, there are variations from device-to-device out of your bin in the lab. The combination of these two major issues can change the performance of your circuit dramatically if you don't take them into consideration.

In terms of devices, you will find product data sheets have maximum guaranteed values and typical values. The maximum guaranteed values are self-explanatory in that you should expect that your devices will not over-range the specifications as stated provided the devices are not overstressed with higher voltages or temperatures. The typical values are another manner. There are a variety of ways to determine what these typical values should be, and you will find that each manufacturer will have their own way to calculate these values along with their justification. Some manufacturers take the average of a large sample of devices prior to the initial product release. Other manufacturers define their typical values as being equal to one standard deviation plus the average. I have also heard of manufacturers using their Spice simulation as a guide for these numbers. Sometimes the Spice simulation is justified because it is impossible to test a particular specification.

The third aspect of accuracy is noise. When you take this issue into consideration, you need to have some understanding of statistical calculations with large samples.

18.4.2   Hardware versus Software

This discussion seems to simplify the problem a bit, but I have a solution for those embarking on the ownership of analog. Think of it in terms of learning the fundamentals about your components, knowing the general behavior of basic building-block devices, and running through a high-level evaluation of your circuits first.

1. *Learn the fundamentals of your components.*

For instance, the fundamentals at the very bottom of the barrel include resistors, capacitors and inductors. You were probably exposed to the devices early in your career, but what do you really need to know as an analog design engineer?

Resistors are simple devices. There are several perspectives that you have to consider when you use this type of component in your design. The first and easiest way of thinking about a resistor is that it influences voltage and current in your design. This is defined through the infamous Thévenin equation:

$$V/RI = 1$$

where V is voltage,

R is resistance in ohms,

I is current in amperes.

I always remember this formula from my elementary school geography lessons. $\underline{V}$ermont is always over $\underline{R}$hode $\underline{I}$sland.

But this is only part of the resistor description in your circuit. For practical purposes, this is a DC equation, not AC. Moving past this formula, you need to be concerned about the parasitic characteristics. Namely, there is a parasitic capacitor in parallel with the resistive element and a parasitic inductor in series. These components are artifacts of the physical device. There is a diagram of the resistor with these parasitics in Figure 18.13.

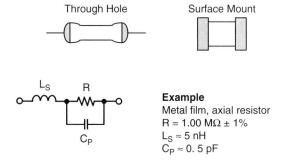

Figure 18.13: This illustrates a typical resistor model. The parasitic elements of a standard resistor are parallel capacitance (C_P) and series inductance (L_S)

The fact is, I never worried about the parasitic capacitance until I started designing transimpedance, optical, photodiode-sensing circuits. An example of this type of circuit is shown in Figure 18.14. If blindly built (without concern for the parasitic capacitance), this photosensing circuit can mysteriously sing like a bird (oscillate) without too much effort. This oscillation is usually caused by an inappropriate choice of C_F, but it can also be caused by that phantom capacitor, C_P. These capacitors, in combination with the photodiode parasitic capacitance and the amplifier's input capacitance interact to establish stability, or not. This is one example, but you can extrapolate this to other circuits if you are using small value discrete capacitors in parallel or series with discrete resistors.

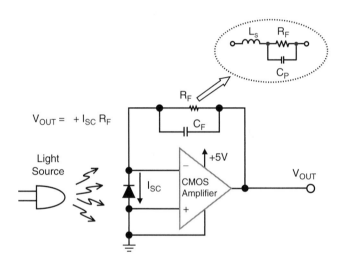

Figure 18.14: If you don't consider the parasitic capacitance of the feedback resistor, a transimpedance photosensing circuit can be unstable.

The parasitic inductance of the resistor (also see Figure 18.15) can affect higher speed systems where lower value resistors are the norm. This inductance can affect the behavior of the current sensing resistor used in switched-mode power supplies.

Generally speaking, the impedance of higher value resistors is more affected by the parasitic capacitance, and that of low value resistors is affected by the parasitic inductance. Figure 18.15 illustrates this point.

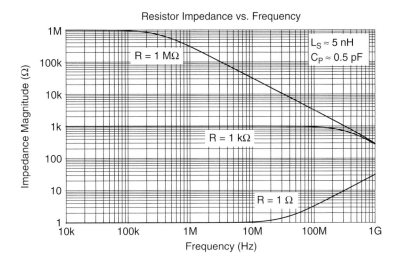

Figure 18.15: The impedance of a resistor changes from the defined DC resistance value to other values over frequency. The parasitic capacitance and impedance influence these changes.

Capacitors, on the other hand, should be considered in the frequency domain when you are designing. There is one formula for the capacitor that I used frequently in my design. This formula is:

$$I = C * \delta V / \delta t$$

where C is capacitance in farads,

δV is change in voltage in volts,

δt is change in time in seconds.

Capacitors are very useful for power supplies, stability, loading low dropout regulators and loading voltage references. But, in all cases, you use capacitors to modify frequencies, not DC signals.

2. *Know the general behavior of basic building blocks.* Consider these basic circuit cells as instruction codes. Start by using them in their most common circuit configurations or the classical approach. In analog, your basic building blocks are:

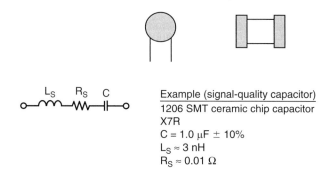

Example (signal-quality capacitor)
1206 SMT ceramic chip capacitor
X7R
$C = 1.0\ \mu F \pm 10\%$
$L_S \approx 3\ nH$
$R_S \approx 0.01\ \Omega$

Figure 18.16: This illustrates a typical ceramic capacitor model. The parasitic elements of a standard capacitor are series resistance (R_s), also known as effective series resistance (ESR), and series inductance (L_s), also known as effective series impedance (ESL).

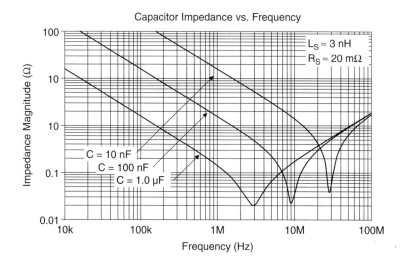

Figure 18.17: The frequency response of a capacitor varies at lower frequencies due to the series resistance and higher frequencies due to the series inductor.

  – Analog-to-digital converters

  – Operational amplifiers

3. *Higher level thinking.* Are you afraid of math? Don't dwell on it at first. Concentrate on the practical side of analog applications. Learn the rules of thumb for analog. For instance, many of us, being indoctrinated in the school

system background, sharpen our pencils, pull out the old calculator and grind through the trees before we have a thought about what the forest looks like. Once you step back and think about it, you will find that your detailed analysis can be way off. If your analysis is correct, it probably is only part of the picture. Here is a perfect example of what I mean.

Problem:

What is the corner frequency of the single-pole, low-pass RC filter shown in Figure 18.18?

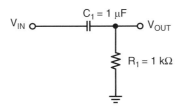

Figure 18.18: Circuit example

Answer

"Hand wave" solution: Wait a minute. This isn't a low-pass filter. This is a high-pass filter. (You probably knew this right away, but you would be amazed at how many would overlook this simple conclusion!) But if I assumed that the author made a mistake and reversed the placement of the resistor and capacitor, the corner frequency would be about $1/(2 \pi R \times C)$ or 160 Hz. How did I get there? Isn't $1/2 \pi$ equal to about 0.16? As a first pass, I think I can accept that error because the capacitor device-to-device error is probably ± 10 or 20% accurate.

Calculated solution (with blinders on):

$$(V_{OUT} - V_{IN})/(1/sC_1) = V_{OUT}/R_1$$
$$V_{OUT}(sC_1 + 1/R_1) = V_{IN}/(1/sC_1)$$
$$V_{OUT}/V_{IN} = (sR_1 \times C_1 + 1)/(sR_1 \times C_1)$$

From this calculation, there is a pole at DC and a zero at 159.1549 Hz.

These two solutions don't agree! And I bet a SPICE simulation would match your calculated solution. The moral to this story is "hand wave," or think yourself through the problem first. SPICE does not mean "don't think," it means "verification of your analysis." With this type of analysis, you should keep in mind the accuracy (or lack thereof) of the various components and devices in your system. After, and only after, you know generally how the circuit works and how the system responds, give your mathematical and SPICE skills a try.

4. *This could be a good career decision.* The universities are graduating fewer and fewer engineers knowledgeable in analog, but as we all know, analog will not be going away any time soon.

18.5   Time and its Inversion

In the digital domain, particularly with real-time operating systems (RTOS), you will find that you are counting minutes, seconds, milliseconds, and nanoseconds. This is also done with analog circuits, but more importantly, the inverse of seconds is counted. Taking the inverse of seconds helps you think in terms of frequency instead of time. Frequency information is much more critical here.

18.6   Organizing Your Toolbox

You need to decide what is important and what is not for your future analog design work. An effective way to do this is to arm yourself with basic, key tools of the trade. You should concentrate as you collect your ammunition on six topics.

First, know how to get data in and out of the digital domain. When this is mastered you will know the different topologies, important specifications, and the art of matching the converter to the application.

Then, sit back and ask yourself, "Where does my data in the controller really come from?" You will usually find some sort of sensor at the origin of the signal path. Further back from the A/D converter is the amplification system. In this system, the signal can either be enhanced through amplification or corrupted because of noise or linearity errors. The key player in the amplification system is the operational amplifier. Volumes of books have been written on this seemingly simple part, but not enough written about the single-supply operational amplifier applied in a simple manner.

Now go back to your strength. Revisit the digital with analog in mind. Can you exploit your digital engine easily with a few analog tricks?

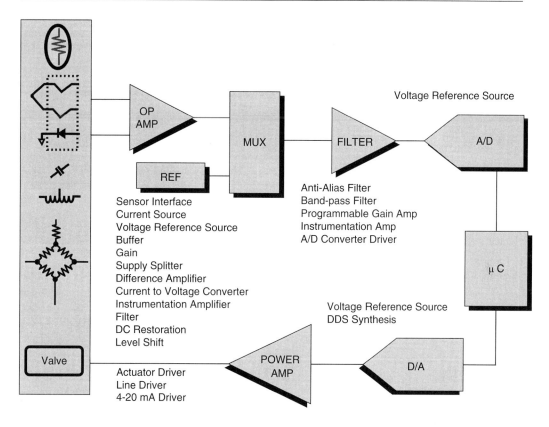

Figure 18.19: This signal chain is somewhat universal in that it deals with the analog signal coming in, conditions it through the amplification system and digitizes it in preparation for the microcontroller or processor.

Go out on a limb. Bring the "art" of some of the essential analog disciplines into your toolbox. In particular, learn about noise sources and noise filters. Think about your layout and how it affects your circuit solution. Then go to the lab with confidence.

18.7   Set Your Foundation and Move On, Out of The Box

Drop your inhibition. Have fun. Work outside your box. Learning a new craft takes persistence, time and a learning attitude. Analog design is a matter of sitting down and doing it, whether it is right or wrong. Then on the next day tweak it, and the next day, and the next day, until the circuit is finally refined. No magic formulas here, just some common sense, and problem solving techniques. First, define the problem. Second identify tools and strategies that can be used to work the problem. Third, work the problem to a solution. Finally, reread your definition of the problem and determine if your solution seems reasonable. Analog only demands good, honest, consistent and persistent work. Sound familiar?

References

"FilterPro™.MFB and Sallen-Key Low-Pass Filter Design Program," Bishop, Trump, Stitt, SBFA001A, Texas Instruments.

FilterLab®2.0 *User's Guide*, DS51419A, Microchip Technology.

"CANbus Networks Break Into Mainstream Use," Marsh, David, *EDN,* Aug. 22, 2002.

"Making the CANbus a "can-do" Bus," Warner, Will, *EDN,* Aug. 21, 2003.

"Implementing Ohmmeter/Temperature Sensor," Cox, Doug, AN512, Microchip Technology.

"Resistance and Capacitance Meter Using a PIC16C622," Richey, Rodger, AN611, Microchip Technology.

CHAPTER 19

Op-Amps

Darren Ashby
Bruce Carter
Ron Mancini
Tim Williams
Bonnie Baker

19.1 The Magical Mysterious Op-Amp

An operational amplifier, often called an *op-amp*, in my opinion, are probably the misunderstood, yet potentially useful IC at the engineer's disposal. It makes sense that if you can understand this device you can put it to use, giving you a great advantage in designing successful products.

19.1.1 What is an Op-Amp Really?

Do you understand how an op-amp works? Would you believe that op-amps were designed to make it *easier* to create a circuit? You probably didn't think that the last time you were puzzling over a misbehaving breadboard in the lab.

In today's digital world, it seems to be common practice to breeze over the topic of op-amps giving the student a dusting of commonly used formulas without really explaining the purpose or theory behind them. Then, the first time an engineer designs an op-amp circuit, the result is utter confusion when the circuit doesn't work as expected. This discussion is intended to give some insight into the guts of an operational amplifier, and to give the reader an intuitive understanding of op-amps.

One last point—make sure you read this section first! It is my opinion that one of the causes of op-fusion (op-amp confusion) as I like to call it, is that the theory is taught out of order. There is a very specific order to this, so please understand each section before moving on.

First, let's take the symbol of an op-amp (Figure 19.1).

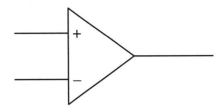

Figure 19.1: Your basic op-amp

There are two inputs, one positive and one negative, identified by the + and − signs.

There is one output.

The inputs are high impedance. I repeat. The inputs are high impedance. Let me say that one more time. THE INPUTS ARE HIGH IMPEDANCE! This means they have (virtually) no effect on the circuit to which they are attached. Write this down, as it is very important. We will talk about this in more detail later. This important fact is commonly forgotten and contributes to the confusion I mentioned earlier.

The output is low impedance. For most analysis it is best to consider it a voltage source.

Now let's represent the op-amp with two separate symbols (Figure 19.2).

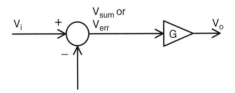

Figure 19.2: What is really inside an op-amp

You see here a summing block and an amplification block. You may remember similar symbols from your control theory class. Actually they are not just similar—they are exactly the same. Control theory works for op-amps. More on that later, too.

First, let's discuss the summing block. You will notice that there is a positive input and a negative input on the summing block, just as on the op-amp. Recognize that the negative input is as if the voltage at that point is multiplied by a −1. Thus, if you have 1V at the positive input and 2V at the negative input, the output of this block is −1. The output of this block is the sum of the two inputs, where one of the inputs is multiplied by –1. It can also be thought of as the difference of the two inputs and represented by this equation: Vs = (V+) – (V–).

Now we come to the amplification block. The variable G inside this block represents the amount of amplification that the op-amp applies to the sum of the input voltages. This is also known as the open-loop gain of the op-amp. In this case, we will use a value of 50,000. You say, how can that be? The amplification circuit I just built with an op-amp doesn't go that high! We will get to the amplification applications in a moment. Go find the open-loop gain in the manufacturer's data sheet. This level of gain or even higher is typical of most op-amps.

Now for some analysis. What will happen at the output if you put 2V on the positive input and 3V on the negative input. I recommend that you actually try this on a breadboard. I want you to see that an op-amp can and will operate with different voltages at the inputs. However, a little math and some common sense will also show us what will happen. For example:

$$V_{out} = 50,000 * (2 - 3), \text{ or } -50,000V \qquad (19.1)$$

Unless you have a 50,000V op-amp hooked up to a 50,000V bipolar supply, you won't see −50,000V at the output. What will you see? Think about it a minute before you read on. The output will go to the minimum rail. In other words, it will try to go as negative as possible. This makes a lot of sense if you think about it like this. The output wants to go to –50,000V and obey the mathematics above. It can't get there, so it will go as close as possible. The rails of an op-amp are like the rails of a train track—a train will stay within its rails if at all possible. Similarly, if an op-amp is forced outside its rails, disaster occurs and the proverbial magic smoke will be let out of the chip. The rail is the maximum and minimum voltage the op-amp can output. As you can intuit, this depends on the power supply and the output specifics of the op-amp.

OK, reverse the inputs. Now the following is true:

$$V_{out} = 50,000 * (3 - 2), \text{ or } +50,000V \qquad (19.2)$$

What will happen now? The output will go to the maximum rail. How do you know where the output rails of the op-amp are? That depends on the power supply you are using and the specific op-amp. You will need to check the manufacturer's data sheet for that information. Let's assume we are using an LM324, with a +5V single-sided supply. In this case the output would get very close to 0V when trying to go negative and around 4V when trying to go positive.

At this time I would like to point something out. The inputs of the op-amp are NOT equal to each other. Many times I have seen engineers expect these inputs to be the same value. During the analysis stage, the designer comes up with currents going into the inputs of the device to make this happen (remember, high impedance inputs, virtually zero current flow). Then when he tries it out, he is confused by the fact that he can measure different voltages at the inputs.

In a special case, you can make the assumption that these inputs are equal. It is NOT the general case. This is a common misconception. You must not fall into this trap or you will not understand op-amps at all.

The examples above indicate a very neat application of op-amps: the comparator circuit. This is a great little circuit to convert from the analog world to the digital one. Using this circuit you can determine if one input signal is higher or lower than another. In fact, many microcontrollers use a comparator circuit in analog-to-digital conversion processes. Comparator circuits are in use all around us. How do you think the streetlight knows when it is dark enough to turn on? It uses a comparator circuit hooked up to a light sensor. How does a traffic light know when there is enough weight on the sensors to trigger a cycle to green? You can bet there is a comparator circuit in there.

Thumb Rules

- The inputs are high impedance; they have negligible effects on the circuit they are hooked to.
- The inputs can have different voltages applied to them; they do NOT have to be equal.
- The open-loop gain of an op-amp is VERY high.
- Due to the high open-loop gain and the output limitations of the op-amp, if one input is higher than the other the output will "rail" to its maximum or minimum value (this application is often called a comparator circuit).

19.1.2   Negative Feedback

If you didn't just finish reading it, go back and read the thumb rules from the last section. They are very important to develop the correct understanding of what an op-amp does. Why are these points important? Let's go over a little history. Up until the invention of op-amps, engineers were limited to the use of transistors in amplification circuits. The problem with transistors is that, being "current-driven" devices, they always affect the signal of the circuit that the designer wants to amplify by loading the circuit. Due to manufacturing tolerances of transistors, the gain of the circuits would vary significantly. All in all, designing an amplifier circuit was a tedious process that required much trial and error. What engineers wanted was a simple device that they could attach to a signal that could multiply the value by any desired amount. The device should be easy to use and require very few external components. To paraphrase, *operation* of this *amplifier* should be a "piece of cake." At least that is the way I remember it. The other way the name *operational amplifier* or op-amp came into being was to describe the fact that these amplifiers were used to create circuits in analog computers, performing such *operations* as multiplication, among others.

To begin with, let's take a look at the special case I mentioned in the previous discussion. First, return to the previous block diagram and add a feedback loop (Figure 19.3).

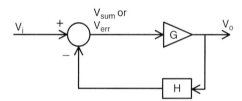

Figure 19.3: Adding negative feedback to the op-amp

You will see that I have represented the forward or open-loop gain with the value G, and the feedback gain with the value H. (This diagram should look very familiar to those of you who have had training in control theory.) First, you see that the output is tied to the negative input. This is called *negative feedback*. What good is negative feedback? Let's try an experiment. Hold your hand an inch over your desk, and keep it there. You are experiencing negative feedback right now. You are observing via sight and feel the distance from your hand to the desk. If your hand moves, you respond with a movement in the opposite direction. That is negative feedback. You invert the signal you receive via your senses and send it back to your arm. The same thing occurs when negative feedback is applied to an op-amp. The output signal is sent back to the negative input. A signal change in one direction at the output causes a V_{sum} to change in the opposite direction.

You should get an intuitive grasp of this negative feedback configuration. Look at the previous diagram and assume a value of 50,000 for *G* and a value of 1 for *H*. Now start by applying a 1 to the positive input. Assume the negative input is at 0 to begin with. That puts a value of 1 at the input of the gain block *G* and the output will start heading for the positive rail. But what happens as the output approaches 1? The negative input also approaches 1. The output of the summing block is getting smaller and smaller. If the negative input goes higher than 1, the input to the gain block G will go negative as well, forcing the output to go in the negative direction. Of course, that will cause a positive error to appear at the input of the gain block *G*, starting the whole process over again. Where will this all stop? It will stop when the negative input is equal to the positive input. In this case since *H* is 1, the output will be 1 also.

You have learned this in control theory. Look at the basic control equation in reference to the previous diagram:

$$V_o = V_i * \frac{G}{1 + G * H} \qquad (19.3)$$

What happens when *G* is very large? The 1 in the denominator becomes insignificant and the equation becomes $V_o =$ approximately $V_i * (1 / H)$. *H* in this case is 1 so it follows that $V_o =$ approximately $V_i * (1 / 1)$ or,

$$V_o = V_i \qquad (19.4)$$

This is the special case where you can assume that the inputs of the op-amp are equal. Apply it ONLY when there is negative feedback. When feedback gain is one, this also demonstrates another neat op-amp circuit, the voltage follower. Whatever voltage is put on the positive input will appear at the output.

Take a look at the following figure. This is an op-amp in the negative feedback configuration. When you look at this, you should see a summer and an amplifier just as in the previous drawing. In this configuration, you can make the assumption that the positive and negative inputs are equal.

Negative feedback is the case that is drilled into you in school, and is the one that often causes confusion. It is a special case, a very widely used special case. Nonetheless, if you do not have negative feedback and the inputs and output are within operational limits, you must NOT assume the inputs of the op-amp are equal.

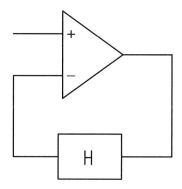

Figure 19.4: Original op-amp symbol with negative feedback

Why is this negative feedback configuration used so much? Remember the reason op-amps were invented? Amplifiers were tough to make. There had to be an easier way. Take a look at the control equation again:

$$V_o = V_i * \frac{G}{1 + G * H} \qquad (19.5)$$

I have already shown that for large values of G, the equation approximates:

$$V_o = V_i * \frac{1}{H} \qquad (19.6)$$

You will see that the amplification of V_i depends on the value of H. For example, if we can make H equal 1/10, then $V_o = V_i * (1 / (1 / 10))$ or,

$$V_o = V_i * 10 \qquad (19.7)$$

How do we go about doing that? Do you remember the voltage divider circuit? That would be very useful here, as we would like H to be the equivalent of dividing by 10. Lets insert the voltage divider circuit in place of H. (Note, V_i will be connected to $V+$.)

Notice that the input to the voltage divider comes from the output of the op-amp, V_o. The output of the voltage divider goes to the negative input of the op-amp $V-$. Now, will the op-amp input $V-$ affect the voltage divider circuit? NO! It has high impedance. It will not affect the divider. (If you didn't get that, go back and read "what's in an op-amp really" till you do!) Since the input to the divider is hooked to a voltage source,

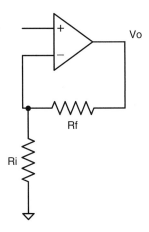

Figure 19.5: Negative feedback is a voltage divider

and the output is not affected by the circuit, we can calculate the gain from V_o to $V-$ very easily with the voltage divider rule.

$$\frac{V-}{V_o} = \frac{R_i}{R_i + R_f} = H \tag{19.8}$$

Thus, it follows that:

$$\frac{1}{H} = \frac{R_i + R_f}{R_i}$$

or with a little algebra,

$$\frac{1}{H} = \frac{R_i}{R_i} + \frac{R_f}{R_i} = \frac{R_f}{R_i} + 1 \text{ or } \frac{1}{H} = \frac{R_f}{R_i} + 1 \tag{19.9}$$

There you have it—the gain of this op-amp circuit. Let's look at it another way. Go back to the previous equation:

$$\frac{V-}{V_o} = \frac{R_i}{R_i + R_f} \tag{19.10}$$

We learned that in this special case of negative feedback we can assume that $V+ = V-$. This is because the negative feedback loop is pushing the output around, trying to reach this state. So let's assume that $V_i = V+$ which is where the input to our amplifier will be hooked up. Now we can replace $V-$ with V_i, and the equation looks like this:

$$\frac{V_i}{V_o} = \frac{R_i}{R_i + R_f} \tag{19.11}$$

What we really want to know is what does the circuit do to V_i to get V_o? Let's do a little math to come up with this equation:

$$V_o = V_i * \frac{R_i + R_f}{R_i} = V_i * \frac{R_f}{R_i + 1} \text{ or } \frac{V_o}{V_i} = \frac{R_f}{V_i} + 1 \tag{19.12}$$

Please note that this is equal to $1/H$. You see, the gain of this circuit is controlled by two simple resistors. Believe me, that is a whole lot easier to understand and calculate than a transistor amplification circuit. As you can see, the operation of this amplifier is pretty easy to understand.

Thumb Rules

- The negative feedback configuration is the only time you can assume that $V- = V+$.
- The high impedance inputs and the low impedance output make it easy to calculate the effects simple resistor networks can have in a feedback loop.
- The high open-loop gain of the op-amp is what makes the output gain of this special case equal to approximately $1/H$.
- Op-amps were meant to make amplification easy, so don't make it hard!

19.1.3 Positive Feedback

What is positive feedback? Let's take a look at a real-world example. You are working one day, and your boss stops by and says, "Hey, you should know that you've handled your project very well, and that new op-amp circuit you built is awesome!" After you bask in his praise for a while, you find yourself working even harder than before. That is positive feedback. The output is sent back to the positive input, which in turn causes the output to move further in the same direction. Let's look at the diagram of an op-amp again (Figure 19.6).

Now we will do a little intuitive analysis. Don't forget the thumb rules we learned in the last two sections. Review them now if you need to.

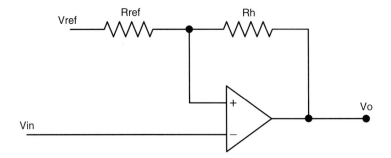

Figure 19.6: Positive feedback on an op-amp

First, apply 0V to V_{in}. In this case the input is connected to $V-$. You also see that the output is connected via a resistor to a reference voltage, V_{ref}. What is the voltage at $V+$? Does the voltage at $V+$ equal the voltage at $V-$? NO! (Don't believe me? Check the thumb rules!)

What is the voltage at $V+$? That depends on two things: the voltage at V_{ref} and the output voltage of the amplifier, V_o. Does the $V+$ input load the circuit at all? No, it does not. To begin the analysis, let $V_{ref} = 2.5V$, and assume the output is equal to 0V. Now what is the voltage at $V+$? What do you know, since V_o is equal to 0, we have a basic voltage divider again. Assume $R_{ref} = 10K$ and $R_h = 100K$:

$$V+ \ = V_{ref} * \frac{R_h}{R_h + R_{ref}} = 2.5 * \frac{100K}{110K} = 2.275V \qquad (19.13)$$

So now there is 2.275V at $V+$ and 0V at $V-$. What will the op-amp do? Let's refer to the block diagram of the op-amp we learned earlier (Figure 19.7).

What do we have? V_{sum} is equal to $V+ - V-$ or, in this case, $V_{sum} = 2.275V$. V_o is equal to $V_{sum} * G$. The output will obviously go to the positive rail (if this is not obvious

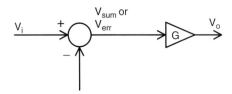

Figure.19.7: Start with what is really inside!

to you, you need to review, "What is an op-amp really?" again). Now we have V_o at the positive rail. Let's assume that it is 4V for this particular op-amp. (Remember, the output rails depend on the op-amp used, and you should always refer to the datasheets for that information. 4V used in this case is typical for an LM324 with a 0 to 5V supply.)

The output is at 4V and $V-$ is at 0V, but what about $V+$? It has changed. We must go back and analyze it again. (Do you feel like you are going in circles? You should. That is what feedback is all about; outputs affect inputs which affect the outputs, and so on, and so on.) The analysis this time has changed slightly. It is no longer possible to use just the voltage divider rule to calculate $V+$. We must also use superposition.

In superposition, you set one voltage source to 0 and analyze the results, and then you set the other source to 0 and analyze the results. Then you add the two results together to get the complete equation. Let's do that now. We already know the result due to V_{ref} from above. Here is the positive feedback diagram again for reference (Figure 19.8).

Here is the result due to V_{ref} using the voltage divider rule:

$$V+ \text{ due to } V_{ref} \ = \ \frac{V_{ref} * R_h}{R_h + R_{ref}} \tag{19.14}$$

Here is the result due to V_o using the voltage divider rule:

$$V+ \text{ due to } V_o \ = \ \frac{V_o * R_{ref}}{R_{ref} + R_h} \tag{19.15}$$

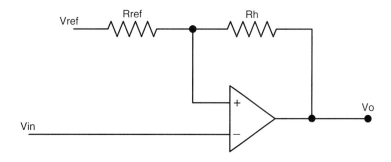

Figure 19.8: Positive feedback on an op-amp

The result due to both is thus:

$$V+ = \left(V+ \text{ due to } V_{ref}\right) + \left(V+ \text{ due to } V_o\right) \text{ or}$$

$$V+ = \frac{V_{ref} * R_h}{R_h + R_{ref}} + \frac{V_o * R_{ref}}{R_h + R_{ref}} \tag{19.16}$$

Now insert all the current values and we have:

$$V+ = \frac{2.5 * 100K}{110K} + \frac{4 * 10K}{110K} = 2.64V \tag{19.17}$$

Is this circuit stable now? Yes, it is. We have 0V at $V-$, and 2.64V at $V+$. This results in a positive error which, when amplified by the open-loop gain of the op-amp, causes the output to go to the positive rail. This is 4V, which is the state that we just analyzed.

Let's change something. Let's start slowly ramping up the voltage at $V-$. At what point will the op-amp output change? Right after the voltage at $V-$ exceeds the voltage at $V+$. This results in a negative error, causing the output to swing to the negative rail. And what happens to $V+$? It changes back to 2.275V as we calculated above. So how do we get the output to go positive again? We adjust the input to less than 2.275V. The positive feedback reinforces the change in the output, making it necessary to move the input farther in the opposite direction to affect another change in the output.

The effect that I have just described is called *hysteresis*. It is an effect very commonly created using a positive feedback loop with an op-amp. What is hysteresis good for, you ask. Well, heating your house for one thing. It is hysteresis that keeps your furnace from clicking on and off every few seconds. Your oven and refrigerator use this principle as well. In fact, the disk drive on the computer I used to write this uses hysteresis to store information.

One important item to note. The size of the hysteresis window depends on the ratio of the two resistors R_{ref} and R_h. In most typical applications R_h is much larger that R_{ref}. If the signal at V_i is smaller than the window, it is possible to create a circuit that latches high or low and never changes. This is usually not desired and can be avoided be performing the analysis above and comparing the calculated limits to the input signal range.

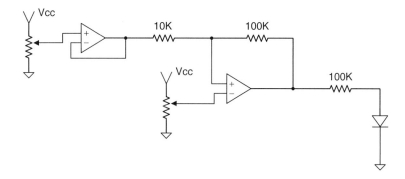

Figure 19.9: Simple op-amp circuit for your bench to understand both positive and negative feedback

Now that we have covered the three basic configurations of an op-amp, let's put together a simple circuit that uses them. Here we have a voltage follower, hooked to a comparator using hysteresis, with an LED as an indicator (Figure 19.9).

You should build this in your lab to gain an intuitive understanding of what has been discussed. Experiment with feedback changes in all parts of the circuit. Note that you can change the input potentiometers from 5K to 100K without affecting the voltage at which the comparator switches.

19.1.4   All About Op-Amps

There you have it, the basics of op-amp circuits. With this information, you can analyze most op-amp circuits you come across and build some really neat circuits yourself. What about filters, you say! Well, a filter is nothing more than an amplifier that changes gain depending on the frequency. Simply replace the resistors with an impedance and thus add a frequency component to the circuit. What about oscillators, you say? These are feedback circuits where timing of the signals is important. They still follow the rules above. I believe that grasping the basics of any discipline is the most important thing you can do. If you understand the basics, you can always build on that foundation to obtain higher knowledge, but if you do not "get the basics" you will flounder in your chosen field.

Thumb Rules

- Op-amp inputs are high impedance (that means no current flows into the inputs); this can't be said too much so forgive me for repeating it.
- Op-amp outputs are low impedance.
- $V+ = V-$ only if negative feedback is present, they don't have to equal if feedback is positive.
- Positive feedback creates hysteresis when properly set up.
- Positive feedback can make an output latch to a state and stay there.
- Positive feedback with a delay can cause an oscillation.
- Op-amps were designed to make it easy, so don't make it hard!

19.2   Understanding Op-Amp Parameters

This section is about op-amp data sheet parameters. The designer must have a clear understanding of what op-amp parameters mean and their impact on circuit design. The section is arranged for speedy access to parameter information. Their definitions, typical abbreviations, and units appear in Section 19.2.1. Section 19.2.2 digs deeper into important parameters for the designer needing more in-depth information.

While these parameters are the ones most commonly used at Texas Instruments, the same parameter may go by different names and abbreviations at other manufacturers. Not every parameter listed here may appear in the data sheet for a given op-amp. An op-amp that is intended only for AC applications may omit DC offset information. The omission of information is not an attempt to "hide" anything. It is merely an attempt to highlight the parameters of most interest to the designer who is using the part the way it was intended. There is no such thing as an ideal op-amp—or one that is universally applicable. The selection of any op-amp must be based on an understanding of what particular parameters are most important to the application.

If a particular parameter cannot be found in the data sheet, a review of the application may well be in order and another part, whose data sheet contains the pertinent information, might be more suitable. Texas Instruments manufactures a broad line of op-amps that can implement almost any application. The inexperienced designer could easily select an op-amp that is totally wrong for the application. Trying to use an audio op-amp with low total harmonic distortion in a high-speed video circuit, for example, will not work—no matter how superlative the audio performance might be.

Some parameters have a statistically normal distribution. The typical value published in the data sheet is the mean or average value of the distribution. The typical value listed is the 1σ value. This means that in 68% of the devices tested, the parameter is found to be ± the typical value or better. Texas Instruments currently uses 6σ to define minimum and maximum values. Usually, typical values are set when the part is characterized and never change.

19.2.1 Operational Amplifier Parameter Glossary

There are usually three main sections of electrical tables in op-amp data sheets. The absolute *maximum ratings* table and the *recommended operating conditions* table list constraints placed upon the circuit in which the part will be installed. *Electrical characteristics* tables detail device performance.

Absolute maximum ratings are those limits beyond which the life of individual devices may be impaired and are never to be exceeded in service or testing. Limits, by definition, are maximum ratings, so if double-ended limits are specified, the term will be defined as a range (e.g., operating temperature range).

Recommended operating conditions have a similarity to maximum ratings in that operation outside the stated limits could cause unsatisfactory performance. Recommended operating conditions, however, do not carry the implication of device damage if they are exceeded.

Electrical characteristics are measurable electrical properties of a device inherent in its design. They are used to predict the performance of the device as an element of an electrical circuit. The measurements that appear in the electrical characteristics tables are based on the device being operated within the recommended operating conditions.

Table 19.1 is a list of parameters and operating conditions that are commonly used in TI op-amp data sheets. The glossary is arranged alphabetically by parameter name. An abbreviation cross-reference is provided after the glossary in Table 19.2 to help the designer find information when only an abbreviation is given. More detail is given about important parameters in Section 19.2.2.

19.2.2 Additional Parameter Information

Depending on the application, some op-amp parameters are more important than others. This section contains additional information for parameters that impact a broad range of designs.

Table 19.1: Op-amp parameter glossary

Parameter	ABBV	Units	Definition	Info
Bandwidth for 0.1 dB flatness		MHz	The range of frequencies within which the gain is ± 0.1 dB of the nominal value.	
Case temperature for 60 seconds		°C	Usually specified as an absolute maximum—It is meant to be used as guide for automated soldering processes.	
Common-mode input capacitance	C_{ic}	pF	Input capacitance a common-mode source would see to ground.	11.3.7.1
Common-mode input impedance	Z_{ic}	Ω	The parallel sum of the small-signal impedance between each input terminal and ground.	
Common-mode input voltage	V_{IC}	V	The average voltage at the input pins.	11.3.3
Common-mode rejection ratio	CMRR or k_{CMR}	dB	The ratio of differential voltage amplification to common-mode voltage amplification. Note: This is measured by determining the ratio of a change in input common-mode voltage to the resulting change in input offset voltage.	11.3.9
Continuous total dissipation		mW	Usually specified as an absolute maximum. It is the power that can be dissipated by the op-amp package, including the load power. This parameter may be broken down by ambient temperature and package style in a table.	
Crosstalk	X_T	dBc	The ratio of the change in output voltage of a driven channel to the resulting change in output voltage from another channel that is not driven.	
Differential gain error	A_D	%	The change in AC gain with change in DC level. The AC signal is 40 IRE (0.28 V_{PK}) and the DC level change is ± 100 IRE ($\pm 0.7V$). Typically tested at 3.58 MHz (NTSC) or 4.43 MHz (PAL) carrier frequencies.	

(Continued)

Table 19.1: Op-amp parameter glossary (Cont'd)

Parameter	ABBV	Units	Definition	Info
Differential input capacitance	C_{ic}	pF	(See common mode input capacitance.)	11.3.7.1
Differential input resistance	r_{id}	Ω	The small-signal resistance between two ungrounded input terminals.	
Differential input voltage	V_{ID}	V	The voltage at the noninverting input with respect to the inverting input.	
Differential phase error	Φ_D	°	The change in AC phase with change in DC level. The AC signal is 40 IRE (0.28 V_{PK}) and the DC level change is ±100 IRE (±0.7 V). Typically tested at 3.58 MHz (NTSC) or 4.43 MHz (PAL) carrier frequencies.	
Differential voltage amplification	A_{VD}	dB	(See open-loop voltage gain.)	11.3.6
Fall time	t_f	ns	The time required for an output voltage step to change from 90% to 10% of its final value.	
Duration of short-circuit current			Amount of time that the output can be shorted to network ground—usually specified as an absolute maximum.	
Input common-mode voltage range	V_{ICR}	V	The range of common-mode input voltage that, if exceeded, may cause the operational amplifier to cease functioning properly. This is sometimes is taken as the voltage range over which the input offset voltage remains within a set limit.	11.3.3
Input current	I_I	mA	The amount of current that can be sourced or sinked by the op-amp input—usually specified as an absolute maximum rating.	
Input noise current	I_n	$\dfrac{pA}{\sqrt{Hz}}$	The internal noise current reflected back to an ideal current source in parallel with the input pins.	11.3.13

(Continued)

Table 19.1: Op-amp parameter glossary (Cont'd)

Parameter	ABBV	Units	Definition	Info
Input noise voltage	V_n	$\dfrac{nV}{\sqrt{Hz}}$	The internal noise voltage reflected back to an ideal voltage source in parallel with the input pins.	11.3.13
Gain-bandwidth product	GBWP	MHz	The product of the open-loop voltage gain and the frequency at which it is measured.	11.3.13
Gain margin	Am	dB	The reciprocal of the open-loop voltage gain at the frequency where the open-loop phase shift first reaches $-180°$.	
High-level output voltage	V_{OH}	V	The highest positive op-amp output voltage for the bias conditions applied to the power pins.	11.3.5
Input bias current	I_{IB}	μA	The average of the currents into the two input terminals with the output at a specified level.	11.3.2
Input capacitance	c_i	pF	The capacitance between the input terminals with either input grounded.	11.3.7.1
Input offset current	I_{IO}	μA	The difference between the currents into the two input terminals with the output at the specified level.	11.3.2
Input offset voltage	V_{IO}, V_{OS}	mV	The DC voltage that must be applied between the input terminals to cancel DC offsets within the op-amp.	11.3.1
Input offset voltage long-term drift		$\dfrac{\mu V}{month}$	The ratio of the change in input offset voltage to the change time. It is the average value for the month.	11.3.1
Input resistance	r_i	$M\Omega$	The DC resistance between the input terminals with either input grounded.	11.3.7.1
Input voltage range	V_I	V	The range of input voltages that may be applied to either the IN+ or IN– inputs	11.3.15
Large-signal voltage amplification	A_V	dB	(See open-loop voltage gain.)	

(Continued)

Table 19.1: Op-amp parameter glossary (Cont'd)

Parameter	ABBV	Units	Definition	Info
Lead temperature for 10 or 60 seconds		°C	Usually specified as an absolute maximum. It is meant to be used as guide for automated and hand soldering processes.	
Low-level output current	I_{OL}	mA	The current into an output with input conditions applied that according to the product parameter will establish a low level at the output.	
Low-level output voltage	V_{OL}	V	The smallest positive op-amp output voltage for the bias conditions applied to the power pins.	11.3.5
Maximum peak output volt-age swing	$V_{OM\pm}$	V	The maximum peak-to-peak output voltage that can be obtained without clipping when the op-amp is operated from a bipolar supply.	11.3.5
Maximum peak-to-peak out-put voltage swing	$V_{O(PP)}$	V	The maximum peak-to-peak voltage that can be obtained without waveform clipping when the DC output voltage is zero.	
Maximum-output-swing bandwidth	B_{OM}	MHz	The range of frequencies within which the maximum output voltage swing is above a specified value or the maximum frequency of an amplifier in which the output amplitude is at the extents of it's linear range. Also called full power bandwidth.	11.3.15
Noise figure	NF	dB	The ratio of the total noise power at the output of an amplifier, referred to the input, to the noise power of the signal source.	
Open-loop transimpedance	Z_t	MΩ	In a trans-impedance or current feedback amplifier, it is the frequency dependent ratio of change in output voltage to the frequency dependent change in current at the inverting input.	

(Continued)

Table 19.1: Op-amp parameter glossary (Cont'd)

Parameter	ABBV	Units	Definition	Info
Open-loop transresistance	R_t	MΩ	In a trans-impedance or current feedback amplifier, it is the ratio of change in DC output voltage to the change in DC current at the inverting input.	
Open-loop voltage gain	A_{OL}	dB	The ratio of change in output voltage to the change in voltage across the input terminals. Usually the DC value and a graph showing the frequency dependence are shown in the data sheet.	
Operating temperature	T_A	°C	Temperature over which the op-amp may be operated. Some of the other parameters may change with temperature, leading to degraded operation at temperature extremes.	
Output current	I_O	mA	The amount of current that is drawn from the op-amp output. Usually specified as an absolute maximum rating—not for long term operation at the specified level.	
Output impedance	Z_o	Ω	The frequency dependent small-signal impedance that is placed in series with an ideal amplifier and the output terminal.	11.3.8
Output resistance	r_o	Ω	The DC resistance that is placed in series with an ideal amplifier and the output terminal.	
Overshoot factor	–	–	The ratio of the largest deviation of the output voltage from its final steady-state value to the absolute value of the step after a step change at the output.	
Phase margin	Φ_m	°	The absolute value of the open-loop phase shift at the frequency where the open-loop-amplification first equals one.	11.3.15

(Continued)

Table 19.1: Op-amp parameter glossary (Cont'd)

Parameter	ABBV	Units	Definition	Info
Power supply rejection ratio	PSRR	dB	The absolute value of the ratio of the change in supply voltages to the change in input offset voltage. Typically, both supply voltages are varied symmetrically. Unless otherwise noted, both supply voltages are varied symmetrically.	11.3.10
Rise time	t_r	nS	The time required for an output voltage step to change from 10% to 90% of its final value.	
Settling time	t_s	nS	With a step change at the input, the time required for the output voltage to settle within the specified error band of the final value. Also known as total response time, t_{tot}.	
Short-circuit output current	I_{OS}	mA	The maximum continuous output current available from the amplifier with the output shorted to ground, to either supply, or to a specified point. Sometimes a low value series resistor is specified.	
Slew rate	SR	V/µs	The rate of change in the output voltage with respect to time for a step change at the input.	11.3.12
Storage temperature	T_S	°C	Temperature over which the op-amp may be stored for long periods of time without damage.	
Supply current	I_{CC}/I_{DD}	mA	The current into the V_{CC+}/V_{DD+} or V_{CC-}/V_{DD-} terminal of the op-amp while it is operating.	
Supply current (shutdown)	I_{CC-}/I_{DD-} SHDN	mA	The current into the V_{CC+}/V_{DD+} or V_{CC-}/V_{DD-} terminal of the amplifier while it is turned off.	
Supply rejection ratio	k_{SVR}	dB	(See power supply rejection ratio.)	11.3.10

(Continued)

Table 19.1: Op-amp parameter glossary (Cont'd)

Parameter	ABBV	Units	Definition	Info
Supply voltage sensitivity	k_{SVS}, $\Delta V_{CC\pm}$, $\Delta V_{DD\pm}$, or ΔV_{IO}	dB	The absolute value of the ratio of the change in input offset voltage to the change in supply voltages.	11.3.10
Supply voltage	V_{CC}/V_{DD}	V	Bias voltage applied to the op-amp power supply pin(s). Usually specified as a $\pm$ value, referenced to network ground.	
Temperature coefficient of input offset current	αI_{IO}	$\mu A/^{\circ}C$	The ratio of the change in input offset current to the change in free-air temperature. This is an average value for the specified temperature range.	11.3.2
Temperature coefficient of input offset voltage	αV_{IO}	$\mu V/^{\circ}C$	The ratio of the change in input offset voltage to the change in free-air temperature. This is an average value for the specified temperature range.	
Total current into VCC+/ VDD+		mA	Maximum current that can be supplied to the positive power terminal of the op-amp—usually specified as an absolute maximum.	
Total current out of VDD–		mA	Maximum current that can be drawn from the negative power terminal of the op-amp—usually specified as an absolute maximum.	
Total harmonic distortion	THD	dB	The ratio of the RMS voltage of the first nine harmonics of the fundamental signal to the total RMS voltage at the output.	
Total harmonic distortion plus noise	THD+N	dB	The ratio of the RMS noise voltage and RMS harmonic voltage of the fundamental signal to the total RMS voltage at the output.	11.3.14
Total power dissipation	P_D	mW	The total DC power supplied to the device less any power delivered from the device to a load. Note: At no load: $P_D = V_{CC+} \times I$ or $P_D = V_{DD+} \times I$.	

(Continued)

Table 19.1: Op-amp parameter glossary (Cont'd)

Parameter	ABBV	Units	Definition	Info
Turn-on voltage (shutdown)	$V_{IH-SHDN}$	V	The voltage required on the shutdown pin to turn the device on.	
Turn-off voltage (shutdown)	$V_{IL-SHDN}$	V	The voltage required on the shutdown pin to turn the device off.	
Turn-on time (shutdown)	t_{EN}	µs	The time from when the turn-on voltage is applied to the shutdown pin to when the supply current has reached half of its final value.	
Turn-off time (shutdown)	t_{DIS}	µs	The time from when the turn-off voltage is applied to the shutdown pin to when the supply current has reached half of its final value.	
Unity gain-bandwidth	B_1	MHz	The range of frequencies within which the open-loop voltage amplification is greater that unity.	11.3.15

Table 19.2: Cross-reference of op-amp parameters

ABBV	Parameter
αI_{IO}	Temperature coefficient of input offset current
$\Delta V_{CC\pm}/ \Delta V_{DD\pm}$	Supply voltage sensitivity
αV_{IO}	Temperature coefficient of input offset voltage
ΔV_{IO}	Supply voltage sensitivity
A_D	Differential gain error
A_m	Gain margin
A_{OL}	Open-loop voltage gain
A_V	Large-signal voltage amplification
A_{VD}	Differential voltage amplification
B_1	Unity gain-bandwidth
B_{OM}	Maximum-output-swing bandwidth

(Continued)

Table 19.2: Cross-reference of op-amp parameters (Cont'd)

ABBV	Parameter
c_i	Input capacitance
C_{ic}	Common-mode input capacitance
CMRR	Common-mode rejection ratio
GBWP	Gain-bandwidth product
$I_{CC-SHDN}/I_{DD-SHDN}$	Supply current (shutdown)
I_{CC}/I_{DD}	Supply current
I_I	Input current
I_{IB}	Input bias current
I_{IO}	Input offset current
I_n	Input noise current
I_O	Output current
I_{OL}	Low-level output current
I_{OS}	Short-circuit output current
k_{CMR}	Common-mode rejection ratio
k_{SVR}	Supply rejection ratio
k_{SVS}	Supply voltage sensitivity
NF	Noise figure
P_D	Total power dissipation
PSRR	Power supply rejection ratio
r_i	Input resistance
R_{id}	Differential input resistance
R_o	Output resistance
R_t	Open-loop transresistance
SR	Slew rate
T_A	Operating temperature
t_{DIS}	Turn-off time (shutdown)

(Continued)

Table 19.2: Cross-reference of op-amp parameters (Cont'd)

ABBV	Parameter
t_{EN}	Turn-on time (shutdown)
THD	Total harmonic distortion
t_f	Fall time
THD+N	Total harmonic distortion plus noise
t_r	Rise time
t_s	Settling time
T_S	Storage temperature
V_{CC}/V_{DD}	Supply voltage
V_I	Input voltage range
V_{IC}	Common-mode input voltage
V_{ICR}	Input common-mode voltage range
V_{ID}	Differential input voltage
V_{IHSHDN}	Turn-on voltage (shutdown)
$V_{IL-SHDN}$	Turn-off voltage (shutdown)
V_{IO}, V_{OS}	Input offset voltage
V_n	Input noise voltage
$V_{O(PP)}$	Maximum peak-to-peak output voltage swing
V_{OH}	High-level output voltage
V_{OL}	Low-level output voltage
$V_{OM\pm}$	Maximum peak output voltage swing
X_T	Crosstalk
Z_{ic}	Common-mode input impedance
Z_o	Output impedance
Z_t	Open-loop transimpedance
Φ_D	Differential phase error
Φ_m	Phase margin

19.2.2.1 Input Offset Voltage

All op-amps require a small voltage between their inverting and noninverting inputs to balance mismatches due to unavoidable process variations. The required voltage is known as the input offset voltage and is abbreviated V_{IO}. V_{IO} is normally modeled as a voltage source driving the noninverting input.

Figure 19.10 shows two typical methods for measuring input offset voltage—DUT stands for device under test. Test circuit (a) is simple, but since V_{out} is not at zero volts, it does not really meet the definition of the parameter. Test circuit (b) is referred to as a *servo loop*. The action of the loop is to maintain the output of the DUT at zero volts.

Bipolar input op-amps typically offer better offset parameters than JFET or CMOS input op-amps.

TI data sheets show two other parameters related to V_{IO}; the average temperature coefficient of input offset voltage, and the input offset voltage long-term drift.

The average temperature coefficient of input offset voltage, αV_{IO}, specifies the expected input offset drift over temperature. Its units are $\mu V/°C$. V_{IO} is measured at the temperature extremes of the part, and αV_{IO} is computed as $\Delta V_{IO}/\Delta °C$.

Normal aging in semiconductors causes changes in the characteristics of devices. The input offset voltage long-term drift specifies how V_{IO} is expected to change with time. Its units are $\mu V/month$.

V_{IO} is normally attributed to the input differential pair in a voltage feedback amplifier. Different processes provide certain advantages. Bipolar input stages tend to have lower offset voltages than CMOS or JFET input stages.

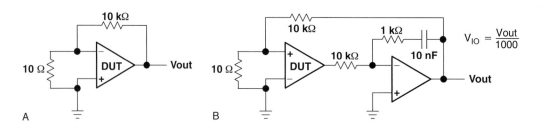

Figure 19.10: Test circuits for input offset voltage

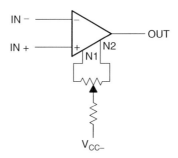

Figure 19.11: Offset voltage adjust

Input offset voltage is of concern anytime that DC accuracy is required of the circuit. One way to null the offset is to use external null inputs on a single op-amp package (Figure 19.11). A potentiometer is connected between the null inputs with the adjustable terminal connected to the negative supply through a series resistor. The input offset voltage is nulled by shorting the inputs and adjusting the potentiometer until the output is zero.

19.2.2.2 *Input Current*

The input circuitry of all op-amps requires a certain amount of bias current for proper operation. The input bias current, I_{IB}, is computed as the average of the two inputs:

$$I_{IB} = \frac{(I_N + I_P)}{2} \tag{19.18}$$

CMOS and JFET inputs offer much lower input current than standard bipolar inputs. Figure 19.12 shows a typical test circuit for measuring input bias currents.

The difference between the bias currents at the inverting and noninverting inputs is called the input offset current, $I_{IO} = I_N - I_P$. Offset current is typically an order of magnitude less than bias current.

Input bias current is of concern when the source impedance is high. If the op-amp has high input bias current, it will load the source and a lower than expected voltage is seen. The best solution is to use an op-amp with either CMOS or JFET input. The source impedance can also be lowered by using a buffer stage to drive the op-amp that has high input bias current.

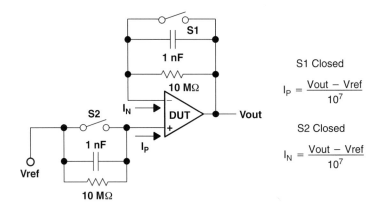

Figure 19.12: Test Circuit $-I_{IB}$

In the case of bipolar inputs, offset current can be nullified by matching the impedance seen at the inputs. In the case of CMOS or JFET inputs, the offset current is usually not an issue and matching the impedance is not necessary.

The average temperature coefficient of input offset current, αI_{IO}, specifies the expected input offset drift over temperature. Its units are $\mu A/°C$. I_{IO} is measured at the temperature extremes of the part, and αI_{IO} is computed as $\Delta I_{IO}/\Delta °C$.

19.2.2.3   *Input Common Mode Voltage Range*

The input common voltage is defined as the average voltage at the inverting and noninverting input pins. If the common mode voltage gets too high or too low, the inputs will shut down and proper operation ceases. The common mode input voltage range, VICR, specifies the range over which normal operation is guaranteed.

Different input structures allow for different input common-mode voltage ranges:

The LM324 and LM358 use bipolar PNP inputs that have their collectors connected to the negative power rail. This allows the common-mode input voltage range to include the negative power rail.

The TL07X and TLE207X type BiFET op-amps use P-channel JFET inputs with the sources tied to the positive power rail via a bipolar current source. This allows the common-mode input voltage range to include the positive power rail.

TI LinCMOS op-amps use P-channel CMOS inputs with the substrate tied to the positive power rail. This allows the common-mode input voltage range to include the negative power rail.

Rail-to-rail input op-amps use complementary n- and p-type devices in the differential inputs. When the common-mode input voltage nears either rail, at least one of the differential inputs is still active, and the common-mode input voltage range includes both power rails.

The trends toward lower, and single-supply voltages make V_{ICR} of increasing concern.

Rail-to-rail input is required when a noninverting unity-gain amplifier is used and the input signal ranges between both power rails. An example of this is the input of an analog-to digital-converter in a low-voltage, single-supply system.

High-side sensing circuits require operation at the positive input rail.

19.2.2.4   *Differential Input Voltage Range*

Differential input voltage range is normally specified as an absolute maximum. Exceeding the differential input voltage range can lead to breakdown and part failure.

Some devices have protection built into them, and the current into the input needs to be limited. Normally, differential input mode voltage limit is not a design issue.

19.2.2.5   *Maximum Output Voltage Swing*

The maximum output voltage, $V_{OM\pm}$, is defined as *the maximum positive or negative peak output voltage that can be obtained without wave form clipping, when quiescent DC output voltage is zero*. $V_{OM\pm}$ is limited by the output impedance of the amplifier, the saturation voltage of the output transistors, and the power supply voltages. This is shown pictorially in Figure 19.13.

This emitter follower structure cannot drive the output voltage to either rail. Rail-to-rail output op-amps use a common emitter (bipolar) or common source (CMOS) output stage. With these structures, the output voltage swing is only limited by the saturation voltage (bipolar) or the on resistance (CMOS) of the output transistors, and the load being driven.

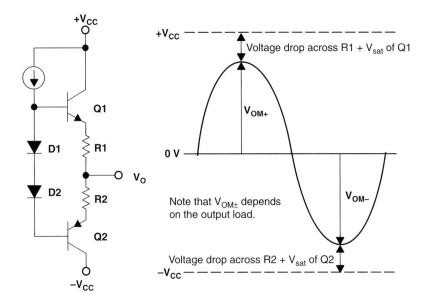

Figure 19.13: $V_{OM\pm}$

Because newer products are focused on single-supply operation, more recent data sheets from Texas Instruments use the terminology V_{OH} and V_{OL} to specify the maximum and minimum output voltage.

Maximum and minimum output voltage is usually a design issue when dynamic range is lost if the op-amp cannot drive to the rails. This is the case in single-supply systems where the op-amp is used to drive the input of an A-to-D converter, which is configured for full scale input voltage between ground and the positive rail.

19.2.2.6   Large Signal Differential Voltage Amplification

Large signal differential voltage amplification, A_{VD}, is similar to the open-loop gain of the amplifier except open-loop is usually measured without any load. This parameter is usually measured with an output load. Figure 19.20 shows a typical graph of A_{VD} vs. frequency. A_{VD} is a design issue when precise gain is required. The gain equation of a noninverting amplifier:

$$\text{Gain} = \frac{1}{\beta} \times \frac{1}{1 + \dfrac{1}{A_{VD}\beta}} \tag{19.19}$$

β is a feedback factor, determined by the feedback resistors. The term $1/A_{VD}\beta$ in the equation is an error term. As long as A_{VD} is large in comparison with $1/\beta$ it will not greatly affect the gain of the circuit.

19.2.2.7 *Input Parasitic Elements*

Both inputs have parasitic impedance associated with them. Figure 19.14 shows a model of the resistance and capacitance between each input terminal and ground and between the two terminals. There is also parasitic inductance, but the effects are negligible at low frequency.

Input impedance is a design issue when the source impedance is high. The input loads the source.

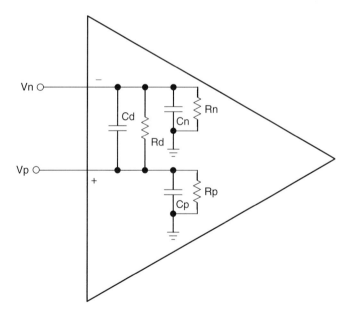

Figure 19.14: Input parasitic elements

Input Capacitance

Input capacitance, C_i, is measured between the input terminals with either input grounded. C_i is usually a few pF. In Figure 19.14, if V_p is grounded, then $C_i = C_d \parallel C_n$.

Sometimes common-mode input capacitance, C_{ic} is specified. In Figure 19.14, if V_p is shorted to V_n, then $C_{ic} = C_p \parallel C_n \cdot C_{ic}$ is the input capacitance a common mode source would see referenced to ground.

Input Resistance

Input resistance, r_i is the resistance between the input terminals with either input grounded.

In Figure 19.14, if V_p is grounded, then $r_i = R_d \parallel R_n$. r_i ranges from $10^7\Omega$ to $10^{12}\Omega$, depending on the type of input.

Sometimes common-mode input resistance, r_{ic}, is specified. In Figure 19.14, if V_p is shorted to V_n, then $r_{ic} = R_p \parallel R_n \cdot r_{ic}$ is the input resistance a common mode source would see referenced to ground.

19.2.2.8   Output Impedance

Different data sheets list the output impedance under two different conditions. Some data sheets list *closed-loop* output impedance while others list open-loop output impedance, both designated by Z_o.

Z_o is defined as the small signal impedance between the output terminal and ground. Data sheet values run from 50Ω to 200Ω.

Common emitter (bipolar) and common source (CMOS) output stages used in rail-to-rail output op-amps have higher output impedance than emitter follower output stages.

Output impedance is a design issue when using rail-to-rail output op-amps to drive heavy loads. If the load is mainly resistive, the output impedance will limit how close to the rails the output can go. If the load is capacitive, the extra phase shift will erode phase margin.

Figure 19.15 shows how output impedance affects the output signal assuming Z_o is mostly resistive.

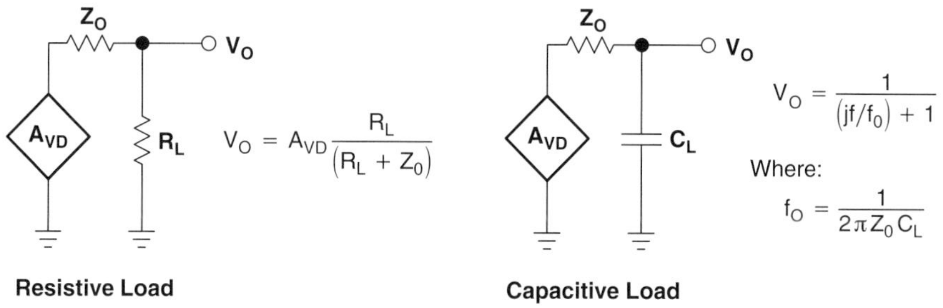

Figure 19.15: Effect of output impedance

Some new audio op-amps are designed to drive the load of a speaker or headphone directly. They can be an economical method of obtaining very low output impedance.

19.2.2.9 Common-Mode Rejection Ratio

Common-mode rejection ratio, CMRR, is defined as the ratio of the differential voltage amplification to the common-mode voltage amplification, A_{DIF}/A_{COM}. Ideally this ratio would be infinite with common mode voltages being totally rejected.

The common-mode input voltage affects the bias point of the input differential pair. Because of the inherent mismatches in the input circuitry, changing the bias point changes the offset voltage, which, in turn, changes the output voltage. The real mechanism at work is $\Delta V_{OS}/\Delta V_{COM}$.

In a Texas Instruments data sheet, CMRR $= \Delta V_{COM}/\Delta V_{OS}$, which gives a positive number in dB. CMRR, as published in the data sheet, is a DC parameter. CMRR, when graphed vs. frequency, falls off as the frequency increases.

A common source of common-mode interference voltage is 50-Hz or 60-Hz AC noise. Care must be used to ensure that the CMRR of the op-amp is not degraded by other circuit components. High values of resistance make the circuit vulnerable to common mode (and other) noise pick up. It is usually possible to scale resistors down and capacitors up to preserve circuit response.

19.2.2.10 Supply Voltage Rejection Ratio

Supply voltage rejection ratio, k_{SVR} (a.k.a. power supply rejection ratio, PSRR), is the ratio of power supply voltage change to output voltage change.

The power voltage affects the bias point of the input differential pair. Because of the inherent mismatches in the input circuitry, changing the bias point changes the offset voltage, which, in turn, changes the output voltage.

For a dual supply op-amp, $K_{SVR} = \Delta V_{CC\pm}/\Delta V_{OS}$ or $K_{SVR} = \Delta V_{DD\pm}/\Delta V_{OS}$. The term $\Delta V_{CC\pm}$ means that the plus and minus power supplies are changed symmetrically. For a single-supply op-amp, $K_{SVR} = \Delta V_{CC}/\Delta V_{OS}$ or $K_{SVR} = \Delta V_{DD}/\Delta V_{OS}$

Also note that the mechanism that produces k_{SVR} is the same as for CMRR. Therefore k_{SVR} as published in the data sheet is a DC parameter like CMRR. When k_{SVR} is graphed vs. frequency, it falls off as the frequency increases.

Switching power supplies produce noise frequencies from 50 kHz to 500 kHz and higher. k_{SVR} is almost zero at these frequencies so that noise on the power supply results in noise on the output of the op-amp. Proper bypassing techniques must be used to control high-frequency noise on the power lines.

19.2.2.11 Supply Current

Supply current, I_{DD}, is the quiescent current draw of the op-amp(s) with no load. In a Texas Instruments data sheet, this parameter is usually the total quiescent current draw for the whole package. There are exceptions, however, such as data sheets that cover single and multiple packaged op-amps of the same type. In these cases, I_{DD} is the quiescent current draw for each amplifier.

In op-amps, power consumption is traded for noise and speed.

19.2.2.12 Slew Rate at Unity Gain

Slew rate, SR, is the rate of change in the output voltage caused by a step input. Its units are V/µs or V/ms. Figure 19.16 shows slew rate graphically. The primary factor controlling slew rate in most amps is an internal compensation capacitor C_C, which is added to make the op-amp unity-gain stable. Referring to Figure 19.17, voltage change in the second stage is limited by the charging and discharging of the compensation capacitor C_C. The maximum rate of change is when either side of the differential pair is conducting $2I_E$. Essentially $SR = 2I_E/C_C$. Remember, however, that not all op-amps have compensation capacitors. In op-amps without internal compensation capacitors, the slew rate is determined by internal op-amp parasitic

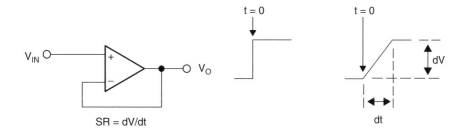

Figure 19.16: Slew rate

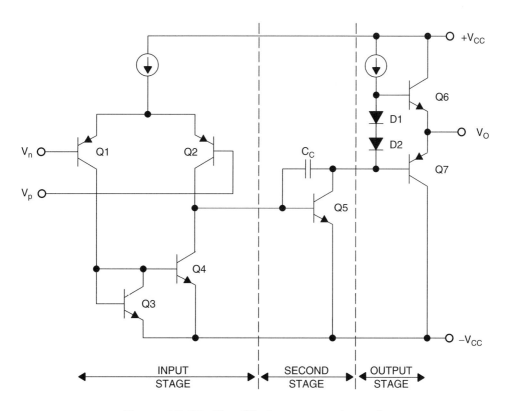

Figure 19.17: Simplified op-amp schematic

capacitances. Noncompensated op-amps have greater bandwidth and slew rate, but the designer must ensure the stability of the circuit by other means.

In op-amps, power consumption is traded for noise and speed. In order to increase slew rate, the bias currents within the op-amp are increased.

19.2.2.13   Equivalent Input Noise

All op-amps have parasitic internal noise sources. Noise is measured at the output of an op-amp, and referenced back to the input. Therefore, it is called equivalent input noise.

Equivalent input noise parameters are usually specified as voltage, V_n, (or current, I_n) per root hertz. For audio frequency op-amps, a graph is usually included to show the noise over the audio band.

Spot Noise

The spectral density of noise in op-amps has a pink and a white noise component. Pink noise is inversely proportional to frequency and is usually only significant at low frequencies. White noise is spectrally flat. Figure 19.18 shows a typical graph of op-amp equivalent input noise.

Usually spot noise is specified at two frequencies. The first frequency is usually 10 Hz where the noise exhibits a 1/f spectral density. The second frequency is typically

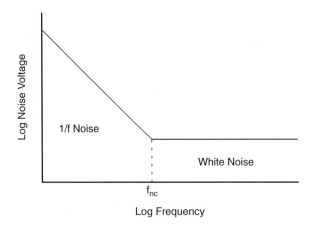

Figure 19.18: Typical op-amp input noise spectrum

1 kHz where the noise is spectrally flat. The units used are normally $nV_{rms}/\sqrt{Hz}$ or $pA_{rms}/\sqrt{Hz}$ for current noise). In Figure 19.18 the transition between 1/f and white is denoted as the corner frequency, f_{nc}.

Broadband Noise

A noise parameter like $V_{N(PP)}$, is the a peak to peak voltage over a specific frequency band, typically 0.1 Hz to 1 Hz, or 0.1 Hz to 10 Hz. The units of measurement are typically nV P-P.

Given the same structure within an op-amp, increasing bias currents lowers noise (and increases SR, GBW, and power dissipation).

Also the resistance seen at the input to an op-amp adds noise. Balancing the input resistance on the noninverting input to that seen at the inverting input, while helping with offsets due to input bias current, adds noise to the circuit.

19.2.2.14 Total Harmonic Distortion Plus Noise

Total harmonic distortion plus noise, THD + N, compares the frequency content of the output signal to the frequency content of the input. Ideally, if the input signal is a pure sine wave, the output signal is a pure sine wave. Due to nonlinearity and noise sources within the op-amp, the output is never pure.

THD + N is the ratio of all other frequency components to the fundamental and is usually specified as a percentage:

$$\text{THD} + \text{N} = \left[\frac{(\sum \text{Harmonic voltages} + \text{Noise Voltages})}{\text{Fundamental}}\right] \times 100\% \qquad (19.20)$$

Figure 19.19 shows a hypothetical graph where THD + N = 1%. The fundamental is the same frequency as the input signal. Nonlinear behavior of the op-amp results in harmonics of the fundamental being produced in the output. The noise in the output is mainly due to the input noise of the op-amp. All the harmonics and noise added together make up 1% of the fundamental.

Two major reasons for distortion in an op-amp are the limit on output voltage swing and slew rate. Typically an op-amp must be operated at or below its recommended operating conditions to realize low THD.

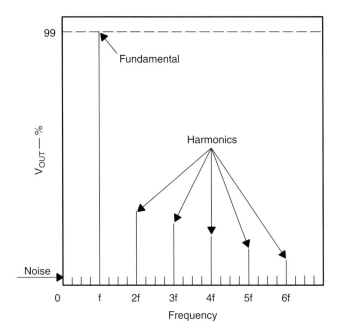

Figure 19.19: Output spectrum with THD + N = 1%

19.2.2.15   Unity gain-bandwidth and phase margin

There are five parameters relating to the frequency characteristics of the op-amp that are likely to be encountered in Texas Instruments data sheets. These are unity-gain bandwidth (B1), gain-bandwidth product (GBWP), phase margin at unity-gain (φ_m), gain margin (A_m), and maximum output-swing bandwidth (B_{OM}).

Unity-gain bandwidth (B_1) and gain-bandwidth product (GBWP) are very similar. B1 specifies the frequency at which A_{VD} of the op-amp is 1:

$$B_1 = (f)A_{VD} = 1 \qquad (19.21)$$

GBW specifies the gain-bandwidth product of the op-amp in an open-loop configuration and the output loaded:

$$GBW = A_{VD} \times f \qquad (19.22)$$

GBW is constant for voltage-feedback amplifiers. It does not have much meaning for current-feedback amplifiers because there is not a linear relationship between gain and bandwidth.

Phase margin at unity-gain (φ_m) is the difference between the amount of phase shift a signal experiences through the op-amp at unity-gain and 180°:

$$\varphi_m = 180 - \varphi \text{ @ B1} \tag{19.23}$$

Gain margin is the difference between unity-gain and the gain at 180° phase shift:

$$\text{Gain margin} = 1 - \text{Gain @ 180° phase shift} \tag{19.24}$$

Maximum output-swing bandwidth (B_{OM}) specifies the bandwidth over which the output is above a specified value:

$$B_{OM} = f_{MAX}, \text{ while } V_O > V_{MIN} \tag{19.25}$$

The limiting factor for B_{OM} is slew rate. As the frequency gets higher and higher the output becomes slew rate limited and can not respond quickly enough to maintain the specified output voltage swing.

In order to make the op-amp stable, capacitor, C_C, is purposely fabricated on chip in the second stage (Figure 19.17). This type of frequency compensation is termed dominant pole compensation. The idea is to cause the open-loop gain of the op-amp to roll off to unity before the output phase shifts by 180°. Remember that Figure 19.17 is very simplified, and there are other frequency shaping elements within a real op-amp.

Figure 19.20 shows a typical gain vs. frequency plot for an internally compensated op-amp as normally presented in a Texas Instruments data sheet.

As noted earlier, A_{VD} falls off with frequency. A_{VD} (and thus, B_1 or GBW) is a design issue when precise gain is required of a specific frequency band.

Phase margin (φ_m) and gain margin (A_m) are different ways of specifying the stability of the circuit. Since rail-to-rail output op-amps have higher output impedance, a significant phase shift is seen when driving capacitive loads. This extra phase shift erodes the phase margin, and for this reason most CMOS op-amps with rail-to-rail outputs have limited ability to drive capacitive loads.

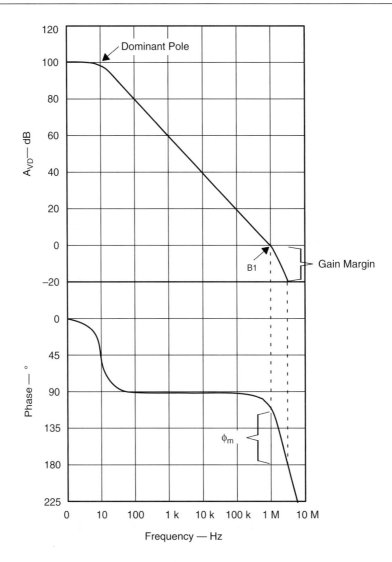

Figure 19.20: Voltage amplification and phase shift vs. frequency

19.2.2.16 Settling Time

It takes a finite time for a signal to propagate through the internal circuitry of an op-amp. Therefore, it takes a period of time for the output to react to a step change in the input. In addition, the output normally overshoots the target value, experiences damped oscillation, and settles to a final value. Settling time, ts, is the time required for the output voltage to settle to within a specified percentage of the final value given a step input.

Figure 19.21 shows this graphically:

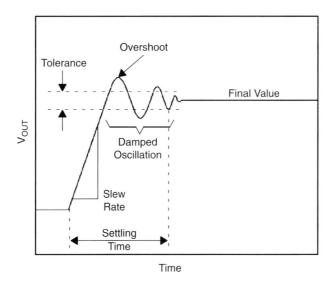

Figure 19.21: Settling time

Settling time is a design issue in data acquisition circuits when signals are changing rapidly. An example is when using an op-amp following a multiplexer to buffer the input to an A-to-D converter. Step changes can occur at the input to the op-amp when the multiplexer changes channels. The output of the op-amp must settle to within a certain tolerance before the A-to-D converter samples the signal.

19.3 Modeling Op-Amps

Virtually every op-amp supplier provides Spice models, which are a very useful approximation of device performance. There are two opposing criteria for such a model. It should use the fewest internal elements to ease computing, but it should also give an accurate representation of the device as a "black box." You can use these models as a necessary (but not entirely sufficient) step in the design process. Models cannot capture a device's every sensitivity to supply variations or temperature and load changes. Dynamic performance such as slew rate and overshoot are especially difficult to model, and peculiarities such as behavior at or beyond the common mode limits will be entirely absent.

The circuit design must be characterized for the entire range of performance characteristics that an off-the-shelf part might show, but generally available Spice models use typical rather than worst-case specs.

Even a perfect model would not capture what is just as critical in high-performance analog design: your physical circuit that surrounds the part. Just a few picofarads of circuit-board capacitance will change the frequency response, for example. Common impedances in the power or ground circuits can affect stability and power supply rejection. Conductive residues on the circuit provide a leakage path between IC pins. No model of itself will detail your circuit layout strays or ground topology.

This doesn't mean you shouldn't use models. Use them for initial assessments of your circuit, to about $\pm20\%$ accuracy. At the same time, recognize that the model itself is neither perfect nor does it include the subtleties of your design. Check with the supplier to understand which modeled parameters are typical, which are worst-case, which are at room temperature, and other similar limitations and simplifications. The typically short development timescale, and the project manager champing at the bit, may constrain your ability to experiment and tempt you to go straight from the model to the final layout. But if there is any critical performance issue which you know is not covered by the model, be prepared for a few design iterations, and don't be afraid to breadboard the design if possible.

Most suppliers offer evaluation boards and suggested circuit-board layout drawings, especially for high-performance or complex parts. An evaluation board shows you what the part can do in a reference design. The layout can serve as a starting point for your own implementation, so you won't waste time discovering mistakes the application engineers have already made and dealt with. The first question an applications support engineer asks when a designer calls with a problem such as oscillation in high-frequency current-feedback circuits is, "Did you use the evaluation board layout?"

19.4 Finding the Perfect Op-Amp

The operational amplifier's operation and circuits are easy to find in the books in your local university library. The amplifier operation and circuit descriptions found in these reference books take you through computational algorithms that theoretically will provide the solutions to your analog amplifier design woes. If there were a perfect amplifier on the market today, the designs found in these books would indeed be easy to implement successfully. But there isn't a perfect amplifier—yet. Throughout the history of analog system design, circuits have required special care in key areas in order to ensure success. As luck would have it, a little common sense and bench sense will pull you out of most of your amplifier design disasters.

In an ideal world, the perfect amplifier would look like the one described in Figure 19.22.

The input stage design of this perfect amplifier would use devices whose inputs (IN+ and IN−) extend all the way to the power supply rails. Some single-supply amplifiers are able to do this with some distortion, but the perfect amplifier would be distortion-free. As a matter of fact, it would be nice if the inputs operated beyond the rails. If this were the case, the common-mode range goes beyond the rails as well.

Additionally, the inputs would not source or sink current—that is, they would have zero-input bias current. This allows source impedances to the amplifier to be infinite.

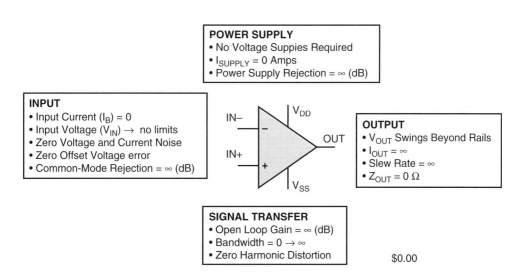

Figure 19.22: A perfect amplifier has an infinite input impedance, open-loop gain, power supply rejection ratio, common-mode rejection ratio, bandwidth, slew rate and output current. It also has zero offset voltage, input noise, output impedance, power dissipation and most importantly, zero cost.

This implies no common-mode or differential-mode input capacitance. Since voltage errors across the two inputs are usually gained by closed-loop circuit configurations around the amplifier, any DC voltage error (offset voltage) or AC error (noise) would be zero. The absence of these errors removes all of your calibration worries!

As for the power supply requirements of this ideal amplifier, there would be none. As you know, industry trends are always working on requests for lower supply voltages, and consequently, lower power consumption from active components. The ideal amplifier wouldn't need a voltage supply across V_{DD} and V_{SS} and would have zero power dissipation in its quiescent state.

The output of this amplifier would be capable of really swinging rail-to-rail, or even beyond. This would eliminate the problem of losing bits on the outer rim in the following A/D conversion. The output impedance would be zero at DC, as well as over frequency ensuring that the device connected to the input of the amplifier is perfectly isolated from the external output device. The op-amp would respond to input signals instantaneously—that is, the slew rate would be infinite and it would be able to drive any load (resistive or capacitive) while maintaining an infinite open-loop gain and rail-to-rail output swing performance. Finally, in the frequency domain, the open-loop gain would be infinite at DC as well as over frequency, and the bandwidth of the amplifier would also be infinite. Oh, did I forget price? We would all love to have this ideal amplifier for $0.00.

Welcome to op-amp 101! This describes the textbook amplifier.

I know that if I'm able to figure out how to design this amplifier, I guarantee you, I will become a multizillionaire. At this point, you are probably saying, "Only in your dreams!" Well, maybe not a multizillionaire, mainly because the profits are $0.00. However, it is certain that I will become a very popular (still poor) person.

It is interesting to note that many of these design imperfections are used to an advantage by most designers. For example, an amplifier circuit design uses a less than infinite bandwidth to limit the noise and high-speed transients in circuits. An infinite slew rate is not as good as it sounds. The amplifier users enjoy slower signals. This reduces the glitches further down the signal path and simplifies the layout.

So, for today, we know that there isn't an ideal amplifier for all circuit situations. The best we can do with the choices available is to pick the best amplifier for our application circuit and then use it properly.

19.4.1 Choose the Technology Wisely

CMOS and bipolar are the two silicon technologies that single-supply operational amplifiers commonly use. Figure 19.23 shows the differences between these two operational amplifier technologies. The most important difference between CMOS and bipolar is in the input stage transistors. These transistors have a profound effect on the overall operation of the amplifier.

Because of the difference between the input transistors of these two types of amplifiers, the CMOS amplifier has lower input current noise and higher input impedance. Because of the high input impedance, the input bias current of the CMOS amplifier is much lower. In fact, the electrostatic discharge (ESD) cells at the input of the CMOS amplifier cause the input bias current errors. As will be shown in circuits later in this chapter, we can use this to an advantage for high impedance sources, such as photosensing transimpedance amplifiers.

The CMOS amplifier typically has a higher open-loop gain than bipolar amplifiers. This can minimize gain error in applications where the closed-loop gain is extremely high (60 dB or greater).

In contrast with the CMOS amplifier, the bipolar amplifier usually has lower input-voltage noise, room temperature offset-voltage and offset-drift. Bipolar amplifiers are

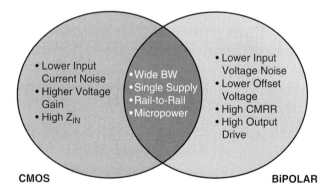

Figure 19.23: The two silicon technologies that single-supply amplifiers are man-ufactured with are CMOS or bipolar processes. By using the CMOS process, you can manufacture bipolar amplifiers. In these designs, the input transistors are bipolar, and the remaining transistors are CMOS.

more likely to provide higher output drive. They also exhibit a higher common-mode rejection capability. This is useful if the amplifier is in a buffer configuration. Although these specifications are typically better than the CMOS amplifier counterpart, the input bias current and input current noise is considerably higher.

Single-supply operating conditions are perfect for both CMOS and bipolar amplifiers. With the proper IC design, they are also capable of input and (near) output rail-to-rail operation.

19.4.2 Fundamental Operational Amplifier Circuits

The op-amp is the analog building block that is analogous to the digital gate. By using the op-amp in the design, circuits can be configured to modify the signal in the same fundamental way that the inverter, AND and OR gates do in digital circuits. This section of the chapter will show the fundamental circuits using this building block. The list of circuits we will discuss include the voltage follower, noninverting gain and inverting gain circuits. This will be followed by more complex circuits, including a difference amplifier, summing-amplifier and current-to-voltage converter.

19.4.2.1 Voltage Follower Amplifier

Starting with the most basic op-amp circuit, the buffer amplifier (shown in Figure 19.24) is used to drive heavy loads, solve impedance matching problems, or isolate high power circuits from sensitive, precise circuitry. Usually, heavy loads require an additional specialized amplifier that is capable of supplying the higher output currents that are greater than 20 mA. You will find that the amplifier data sheet has specifications for the magnitude of the amplifier output current, capable of driving higher currents.

Solving impedance matching problems is also a good reason to use a buffer amplifier. This type of problem exists when the signal path has a high impedance device or resistor that creates an undesirable voltage divider in the circuit. A buffer amplifier breaks up this type of impedance path because of the high impedance input and low impedance output of the amplifier.

Another use for a buffer is to keep high thermal changes away from sensitive circuits. In this scenario the buffer follows the sensitive circuit and serves the purpose of driving high output currents.

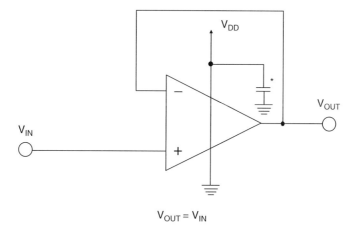

$V_{OUT} = V_{IN}$

* Bypass capacitor, 1μF or 0.1μF

Figure 19.24: A buffer amplifier, also called a voltage follower, is useful when you want to provide a high-current drive stage, match impedances or electrically isolate signals.

The buffer amplifier as shown in Figure 19.24 can be implemented with any single-supply, unity-gain stable amplifier. In this circuit, as with all amplifier circuits, bypassing the op-amp power with a capacitor is a must. For single-supply amplifiers that operate in bandwidths from DC to 1 MHz, a 1 μF capacitor is usually appropriate. Sometimes a smaller bypass capacitor is required for amplifiers that have bandwidths of up to the tens to hundreds of megahertz. In these cases, a 0.1 μF capacitor would be appropriate. If the selection of the value of the bypass capacitor is an inappropriate value or placed too far from the power supply pin and not connected to ground directly on the PCB, the op-amp circuit may oscillate. If you are unsure of what the bypass capacitor value should be, refer to the product data sheet for details.

The analog gain of the circuit in Figure 19.24 is +1 V/V. Notice that this circuit has positive overall gain, but the feedback loop is tied from the output of the amplifier to the inverting input. An all too common error is to assume that an op-amp circuit that has a positive gain requires positive feedback. You can configure this amplifier with positive feedback if you connect the noninverting input to the output. I know this sounds unbelievable, but I have had applicants draw buffers with positive feedback during their interviews. If positive feedback is used, the amplifier will most likely drive to either rail at the output.

This amplifier circuit will give good linear performance across the bandwidth of the amplifier. And, you may be looking at this discussion and saying to yourself, "There is that textbook description, again." You are right; however, here are the land mines in this type of circuit.

The only restrictions on the signal will occur as a result of a violation of the input common-mode voltage and output swing limits. You need to scrutinize these performance characteristics in your amplifier data sheet and your application's demands on this type of circuit. Oh, by the way, ensure that the bandwidth of the amplifier is at least $100\times$ higher than the bandwidth of your signal. However, be aware that you need to look at the input and output of the amplifier.

When using this circuit to drive heavy loads, the specifications of the amplifier must indicate that it is capable of providing the required output currents. Another application where this circuit may be used is to drive capacitive loads. Not every amplifier is capable of driving capacitors without becoming unstable. If an amplifier can drive capacitive loads, the product data sheet will highlight this feature. However, if an amplifier can't drive capacitive loads, the product data sheets will not explicitly say so. This is an instance where features are not in the advertisements or promotions and there is no mention of average performance.

Another use for the buffer amplifier is to solve impedance-matching problems. This would be applicable in a circuit where the analog signal source has relatively high impedance as compared to the impedance of the following circuitry. If this occurs, there will be a voltage loss with the signal because of the voltage divider between the source's impedance and the following circuitry's impedance. The buffer amplifier is a perfect solution to the problem. The input impedance of the noninverting input of an amplifier can be as high as 1013Ω for CMOS amplifiers. In addition, the output impedance of this amplifier configuration is usually less than 100Ω.

Yet another use of this configuration is to separate a heat source from sensitive precision circuitry, as shown in Figure 19.25. Imagine that the input circuitry to this buffer amplifier is amplifying a 100 mV signal. This type of amplification is difficult to do with any level of accuracy in the best of situations. Assigning the output current drive to the device that is doing the precision, amplification work can easily disrupt this measurement. An increase in current drive will cause self-heating of the chip, which will induce an offset change. In this circuit (Figure 19.25), the front-end circuitry makes precision measurements, while an analog buffer performs the function of driving a heavy load.

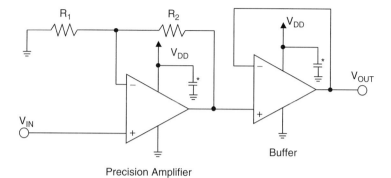

Precision Amplifier

* Bypass capacitor, 1 µF or 0.1 µF

Figure 19.25: A buffer amplifier helps achieve load isolation in this circuit. The buffer separates any high-current output requirements from this input amplifier.

19.4.2.2 Amplifying Analog Signals

The buffer solves many analog signal problems; however, there are instances in circuits where you need to gain a signal. Two fundamental types of amplifier circuits can provide gain. With the first type, the signal gain is positive (or not inverted) as shown in Figure 19.26. This type of circuit is useful in single-supply amplifier applications where negative voltages are usually not present, difficult to produce or just not possible.

The input signal to this circuit is presented to the high impedance, noninverting input of the op-amp. The gain that the amplifier circuit applies to the signal is equal to:

$$V_{OUT} = (1 + R_2/R_1)V_{IN} \qquad (19.26)$$

Typical values for these resistors in single-supply circuits are above 5 kΩ to 25 kΩ for R_2. For the input resistor, R_1, restrictions are dependent on the amount of gain desired versus the amount of amplifier noise and input offset voltage as specified in the product data sheet of the op-amp.

Again, this circuit has some restrictions in terms of the input and output range. The common-mode range of the amplifier restricts the noninverting input. The output swing of the amplifier is also restricted as stated in the product data sheet of the individual amplifier. Most typically, the larger signal at the output of the amplifier causes more signal-clipping errors than the smaller signal at the input. Reducing the gain of this circuit may eliminate undesirable output clipping errors.

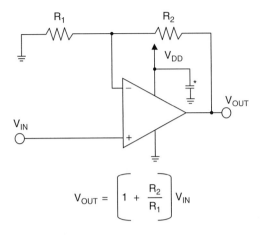

$$V_{OUT} = \left(1 + \frac{R_2}{R_1} \right) V_{IN}$$

* Bypass capacitor, 1 µF or 0.1 µF

Figure 19.26: This is an operational amplifier configured in a noninverting gain circuit. This circuit applies a positive gain to a signal in your circuit. Therefore, you won't need a reference level-shift voltage to keep the output of the amplifier within its operating range.

Figure 19.27 illustrates an inverting amplifier configuration. This circuit gains and inverts the signal present at the input resistor, R_1. The gain equation for this circuit is:

$$V_{OUT} = -(R_2/R_1)V_{IN} + (1 + R_2/R_1)V_{BIAS} \qquad (19.27)$$

The ranges for R_1 and R_2 are the same as in the noninverting circuit shown in Figure 19.26.

This circuit has a minor pitfall in single-supply circuits. In single-supply applications, this circuit is easy to misuse. The problem is rooted in the selection of the voltage at V_{BIAS}. You need to select a value for V_{BIAS} so that the output of the amplifier always remains between the supplies.

For example, let R_2 equal 10 kΩ, R_1 equal 1 kΩ, V_{BIAS} equal 0V, and the voltage at the input resistor, R_1, equal to 100 mV, the output voltage would be −1V. This would violate the output swing range of the operational amplifier. In reality, the output of the amplifier would try to go as near to ground as possible.

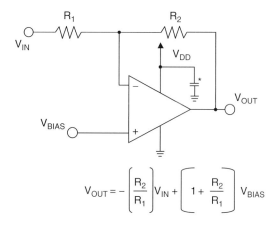

$$V_{OUT} = -\left[\frac{R_2}{R_1}\right]V_{IN} + \left[1 + \frac{R_2}{R_1}\right]V_{BIAS}$$

* Bypass capacitor, 1 μF or 0.1 μF

Figure 19.27: This is an operational amplifier configured in an inverting gain circuit. Single-supply environments usually require V_{BIAS} to ensure the output stays above ground.

The inclusion of a positive DC voltage at V_{BIAS} in this circuit solves this problem. In the previous example, a voltage of 225 mV applied to V_{BIAS} would level shift the output signal up 2.475V. This would make the output signal equal (2.475V − 1V) or 1.475V at the output of the amplifier. Typically, you want to make the target average output voltage of the amplifier equal to $V_{DD}/2$.

19.4.2.3   The Difference Amplifier

The difference amplifier combines the noninverting amplifier and inverting amplifier circuits of Figure 19.26 and Figure 19.27 into a signal block that subtracts two signals. Figure 19.28 illustrates an example of the difference amplifier circuit.

Figure 19.28 illustrates a straightforward implementation of this function. A difference amplifier or op-amp subtractor uses this arrangement of resistors around an amplifier. The DC transfer function of this circuit is equal to:

$$\begin{aligned}V_{OUT} = {} & V_{IN+} \times R_4(R_1 + R_2)/(R_1 \times (R_3 + R_4)) - V_{IN-} \times (R_2/R_1) + \\ & V_{SHIFT} \times R_3(R_1 + R_2)/((R_3 + R_4)R_1)\end{aligned} \qquad (19.28)$$

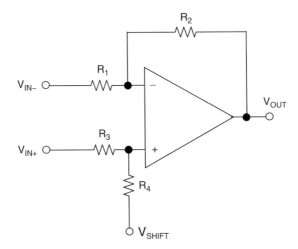

Figure 19.28: This is an operational amplifier circuit configured in a difference amplifier circuit. A difference amplifier implements the subtraction and gain function in one stage.

If R_1/R_2 is equal to R_3/R_4, the closed loop system gain of this circuit equals:

$$V_{OUT} = (V_{IN+} - V_{IN-})(R_2/R_1) + V_{SHIFT} \qquad (19.29)$$

This circuit configuration will reliably take the difference of two signals as long as the signal-source impedances are low. If the signal source impedances are high with respect to R_1, there will be a signal loss due to the voltage divider action between the source and the input resistors to the difference amplifier. Additionally, errors can occur if the two signal source impedances are mismatched. With this circuit, it is possible to have gains equal to or higher than one.

The fact that R_1/R_2 is equal to R_3/R_4 simplifies the mathematics in this system considerably. Since the gain of both signals is equal, the difference amplifier conveniently subtracts the common-mode voltage of the two signals from the system. It is also easy to implement gain by setting the two resistor ratios to be equal or greater than one.

One limitation of this circuit is the lack of flexibility with gain adjustments. If you change the gain dynamically in the application, you must adjust two resistors. In a single-supply environment, a voltage reference centers the output signal between ground and the power supply. Figure 19.28 shows this voltage, "V_{SHIFT}." The purpose

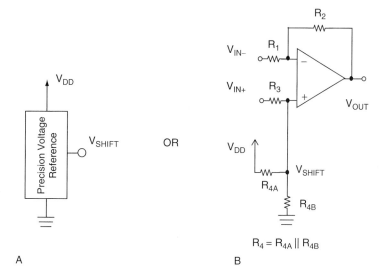

Figure 19.29: A precision voltage reference, (a) or a less expensive solution of replacing R_4 of the voltage divider between the supply, (b) provides the voltage, V_{SHIFT}, of this difference amplifier.

of this reference voltage is to simply shift the output signal into the linear region of the amplifier. A precision, voltage-reference, or a resistive network implements the V_{SHIFT} circuit function as shown in Figure 19.29.

19.4.2.4 Summing Amplifier

You can use summing amplifiers to combine multiple signals by addition or subtraction. Since the difference amplifier can only process two signals, it is a subset of the summing amplifier.

The transfer function of this circuit as shown in Figure 19.30 is:

$$V_{OUT} = (V_1 + V_2 - V_3 - V_4)(R_2/R_1) \qquad (19.30)$$

You can use any number of inputs on either the inverting or noninverting input sides as long as there are an equal number of both with equivalent resistors. All of the inputs to this circuit should be connected to a signal source or (if unused) to the ground.

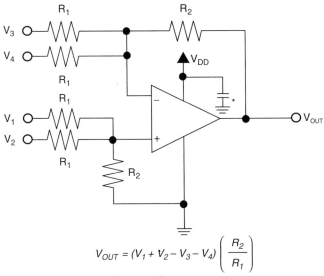

$$V_{OUT} = (V_1 + V_2 - V_3 - V_4)\left(\frac{R_2}{R_1}\right)$$

* Bypass capacitor, 1 µF or 0.1 µF

Figure 19.30: Operational amplifier configured in a summing amplifier circuit.

19.4.2.5   Current-to-Voltage Conversion

If you use a photodetector, feedback resistor and an operational amplifier in your circuit you can sense light. This type of circuit converts the output current of a photodetector into a voltage. The single resistor and an optional capacitor are in the feedback-loop of the amplifier as shown in Figure 19.31.

In the circuits shown in Figure 19.31, light impinging on the photodetector generates a current. This current flows in the reverse bias direction of the diode. If a CMOS op-amp is used (with low input bias current), the current from the detector (I_{D1}) primarily goes through the feedback resistor, R_2. Additionally, the op-amp input bias current error is low because it is CMOS (typically <200 pA). You would ground the noninverting input of the op-amp, which keeps the entire circuit biased to ground. These two circuits will only work if the common mode range of the amplifier includes zero and you are not concerned about a zero level of light. If your light source has zero luminance, the output of the single-supply amplifier is unable to go all the way to ground.

The two circuits in Figure 19.31 provide precision-sensing from the photodetector (top circuit in the figure) and higher speed sensing (bottom circuit in the figure). In the top

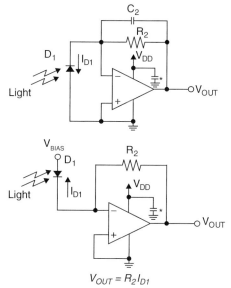

Figure 19.31: These circuits show how to convert current to voltage by using an amplifier and one resistor. The top light-sensing circuit is appropriate for precision applications. The bottom circuit is appropriate for high-speed applications.

circuit, the voltage across the detector is nearly zero and equal to the offset voltage of the amplifier. With this configuration, current that appears across the resistor, R_2, is primarily a result of the light excitation on the photodetector.

The photosensing circuit at the bottom of the figure works best in a high-speed, digital environment. By reverse biasing the photodetector (which reduces the parasitic capacitance of the diode), this sensing circuit can respond very quickly to digital signals. There is more leakage through the photodetector in this bottom circuit, which causes a higher DC error.

19.4.3   Using these Fundamentals

You can use several amplifiers to build instrumentation amplifiers and floating current sources.

19.4.3.1   Instrumentation Amplifier

You will find instrumentation amplifiers in a large variety of applications from medical instrumentation to process control. The instrumentation amplifier is similar to the difference amplifier in that it subtracts one analog signal from another, but it differs in terms of the quality of the input stage. Figure 19.32 illustrates a classic, three op-amp instrumentation amplifier.

In this circuit, the high impedance, noninverting inputs of the input amplifiers (A_1, A_2) acquire the two input signals. This is a distinct advantage over the difference amplifier configuration where source impedances are high or mismatched. The first stage also gains the two incoming signals. One resistor, R_G, adjusts the gain.

Following the first stage of this circuit is a difference amplifier (A_3). The function of this portion of the circuit is to reject the common-mode voltage of the two input signals as well as take the difference between them. The source impedances of the signals into the input of the difference amplifier are low, equivalent and well controlled.

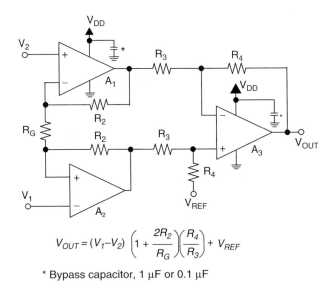

$$V_{OUT} = (V_1 - V_2)\left(1 + \frac{2R_2}{R_G}\right)\left(\frac{R_4}{R_3}\right) + V_{REF}$$

* Bypass capacitor, 1 μF or 0.1 μF

Figure 19.32: You can design an instrumentation amplifier with three amplifiers. The input operational amplifiers (A_1, A_2) provide signal gain. The output operational amplifier converts the signal from the two input amplifiers to a single-ended output with a difference amplifier (A_3).

The reference voltage (V_{REF}) of the difference stage of this instrumentation amplifier is capable of spanning a wide range. Typically, you would connect the voltage reference to half of the supply voltage in a single-supply application. The transfer function of this circuit is:

$$V_{OUT} = (V_1 - V_2)(1 + 2R_2/R_G)(R_4/R_3) + V_{REF}(R_4/R_3) \qquad (19.31)$$

Figure 19.33 shows a second type of instrumentation amplifier. In this circuit, the two amplifiers serve the functions of load isolation and signal gain. The second amplifier also takes the difference between the two input signals (V_1, V_2).

You would connect the circuit reference voltage to the first op-amp in the signal chain. Typically, this voltage is half of the supply voltage in a single-supply environment.

The transfer function of this circuit is:

$$V_{OUT} = (V_1 - V_2)(1 + R_1/R_2 + 2R_1/R_G) + V_{REF} \qquad (19.32)$$

19.4.3.2 *Floating Current Source*

A floating current source (Figure 19.34) can come in handy when driving a variable resistance, like a resistance temperature device (RTD). This particular configuration

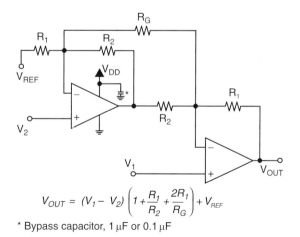

$$V_{OUT} = (V_1 - V_2)\left(1 + \frac{R_1}{R_2} + \frac{2R_1}{R_G}\right) + V_{REF}$$

* Bypass capacitor, 1 μF or 0.1 μF

Figure 19.33: You can design an instrumentation amplifier with two amplifiers. This configuration is best suited for higher gains (gain ≥ 3 V/V).

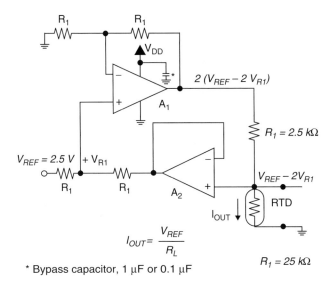

Figure 19.34: A floating current source uses two operational amplifiers and a precision voltage reference.

produces an appropriate 1 mA source for an RTD-type sensor. However, you can change this current reference magnitude to any current.

With this configuration, R_1 reduces the voltage of V_{REF} by the voltage V_{R1}. The voltage applied to the noninverting input of the top op-amp is $V_{REF} - V_{R1}$. This voltage is gained to the amplifier's output by two, to equal $2 \times (V_{REF} - V_{R1})$. Meanwhile, the output for the bottom op-amp (A_2) is presented with the voltage $V_{REF} - 2\,V_{R1}$. Subtracting the voltage at the output of the top-amplifier from the noninverting input of the bottom amplifier gives $2 \times (V_{REF} - V_{R1}) - (V_{REF} - 2\,V_{R1})$, which equals V_{REF}.

The transfer function of the circuit is:

$$I_{OUT} = V_{REF}/R_2 \tag{19.33}$$

19.4.4   Amplifier Design Pitfalls

Theoretically, the circuits within this chapter work. Beyond the theory, however, there are a few tips that will help get the circuit right the first time. This section lists common problems associated with using an op-amp on a PC board. The following discussion has two categories: general suggestions and single-supply pitfalls.

19.4.4.1 In General

- Be careful of the supply pins. Don't make them too high per the amplifier specification sheet and don't make them too low. High supplies will damage the part. In contrast, low supplies won't bias the internal transistors and the amplifier won't work or it may not operate properly.

- Make sure the negative supply is, in fact, tied to a low impedance potential. Additionally, make sure the positive supply is the voltage you expect with respect to the negative supply pin of the op-amp. Placing a voltmeter across the negative and positive supply pins will verify that you have the right relationship between the pins.

- Ground can't be trusted, especially in digital circuits. Plan your grounding scheme carefully. If the circuit has a lot of digital circuitry, consider separate analog and digital grounds and power planes. It is very difficult, if not impossible, to remove digital switching noise from an analog signal.

- Bypass the amplifier power supplies with bypass capacitors as close to the amplifier as possible. Amplifiers usually use a 1 µF or 0.1 µF capacitor. Also, bypass the power supply at the source with a 10 µF capacitor.

- Use short lead lengths to the inputs of the amplifier. If you have a tendency to use the white perf boards for prototyping, be aware that their capacitance and inductance can cause noise and oscillation. There is a good chance that these problems won't be a problem with the PCB implementation of the circuit.

- Amplifiers are static sensitive! If damage has occurred, they may fail immediately or exhibit a soft error (like offset voltage or input bias current changes) that will get worse over time.

19.4.4.2 Single-Supply Rail-to-Rail Amplifiers

- Operational amplifier output drivers are capable of driving a limited amount of current to the load. Check your product data sheet for that number

- Capacitive loading an amplifier is risky business. Make sure the amplifier can handle any loads that you may have.

- It is very rare that a single-supply amplifier will truly swing rail-to-rail. In reality, the output of most of these amplifiers can only come within 50 to 300 mV from each rail. Check the product data sheets of your amplifier.

References

Design with Operational Amplifiers and Analog Integrated Circuits, Sergio Franco, McGraw-Hill, 1998, USA.

Intuitive Operational Amplifiers, Frederiksen, Thomas, McGraw-Hill, 1998, USA.

Analog Circuit Design, Williams, Jim, Butterworth-Heinemann, 1998, USA.

"Operational Amplifiers: 6 Part," Baker, Bonnie C., first published in analogZone (2002, 2003) and reproduced with permission.

CHAPTER 20

Analog-to-Digital Converters

Stuart Ball

Although this chapter is primarily about analog-to-digital converters (ADCs), an understanding of digital-to-analog converters (DACs) is important to understanding how ADCs work. Figure 20.1 shows a simple resistor ladder with three switches. The resistors are arranged in an R/2R configuration. The actual values of the resistors are unimportant; R could be 10K or 100K or almost any other value. Each switch, S0–S2, can switch one end of one 2R resistor between ground and the reference input voltage, VR. The Figure shows what happens when switch S2 is on (connected to VR) and S1 and S2 are OFF (connected to ground). By calculating the resulting series/parallel resistor network, the final output voltage (VO) turns out to be $0.5 \times$ VR. If we similarly calculate VO for all the other switch combinations, we get the results shown in Table 20.1:

If the three switches are treated as a 3-bit digital word, then we can rewrite the table as shown in Table 20.2 (using ON = 1; OFF = 0).

The output voltage is a representation of the switch value. Each additional table entry adds VR/8 to the total voltage. Or, put another way, the output voltage is equal to the binary, numeric value of S0–S2, times VR/8. This 3-switch DAC has 8 possible states and each voltage step is VR/8.

We could add another R/2R pair and another switch to the circuit, making a 4-switch circuit with 16 steps of VR/16 volts each. An 8-switch circuit would have 256 steps of VR/256 volts each. Finally, we can replace the mechanical switches in the schematic with electronic switches to make a true DAC.

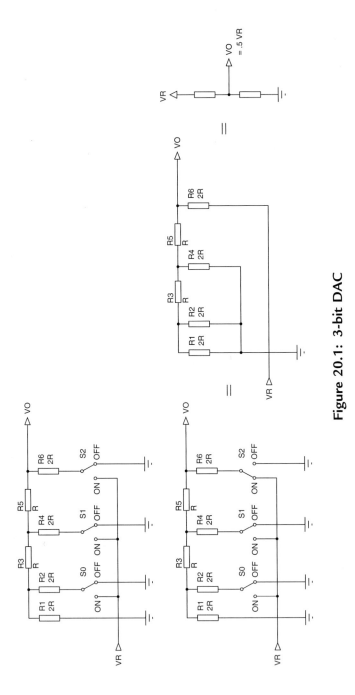

Figure 20.1: 3-bit DAC

Table 20.1: VO for all switch combinations

S2	S1	S0	Vo
OFF	OFF	OFF	0
OFF	OFF	ON	$0.125 \times VR$ ($1/8 \times VR$)
OFF	ON	OFF	$0.25 \times VR$ ($2/8 \times VR$)
OFF	ON	ON	$0.375 \times VR$ ($3/8 \times VR$)
ON	OFF	OFF	$0.5 \times VR$ ($4/8 \times VR$)
ON	OFF	ON	$0.625 \times VR$ ($5/8 \times VR$)
ON	ON	OFF	$0.75 \times VR$ ($6/8 \times VR$)
ON	ON	ON	$0.875 \times VR$ ($7/8 \times VR$)

Table 20.2: Same results with three switches treated as 3-bit digital word

S2	S1	S0	Equivalent Logic State			S0–S2 NUMERIC EQUIVALENT
			S2	S1	S0	
OFF	OFF	OFF	0	0	0	0
OFF	OFF	ON	0	0	1	1
OFF	ON	OFF	0	1	0	2
OFF	ON	ON	0	1	1	3
ON	OFF	OFF	1	0	0	4
ON	OFF	ON	1	0	1	5
ON	ON	OFF	1	1	0	6
ON	ON	ON	1	1	1	7

20.1 ADCs

The usual method of bringing analog inputs into a microprocessor is to use an ADC. An ADC accepts an analog input, a voltage or a current, and converts it to a digital word that can be read by a microprocessor. Figure 20.2 shows a simple ADC. This hypothetical part has two inputs: a reference and the signal to be measured.

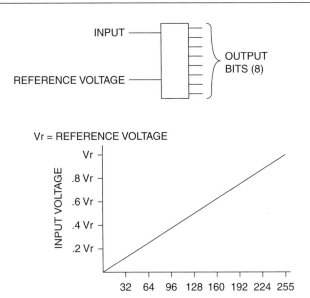

Figure 20.2: Simple ADC

It has one output, an 8-bit digital word that represents, in digital form, the input value. For the moment, ignore the problem of getting this digital word into the microprocessor.

20.1.1   Reference Voltage

The reference voltage is the maximum value that the ADC can convert. Our example 8-bit ADC can convert values from 0V to the reference voltage. This voltage range is divided into 256 values, or steps. The size of the step is given by:

$$\frac{\text{Reference Voltage}}{256} = \frac{5V}{256} = 0.0195V, \text{or } 19.5 \text{ mv}$$

This is the step size of the converter. It also defines the converter's resolution.

20.1.2   Output Word

Our 8-bit converter represents the analog input as a digital word. The most significant bit of this word indicates whether the input voltage is greater than half the reference

(20.5V, with a 5V reference). Each succeeding bit represents half of the previous bit, like this:

Bit	Bit 7	Bit 6	Bit 5	Bit 4	Bit 3	Bit 2	Bit 1	Bit 0
Volts	2.5	1.25	0.625	0.3125	0.156	0.078	0.039	0.0195

So a digital word of 0010 1100 represents this:

Bit	Bit 7	Bit 6	Bit 5	Bit 4	Bit 3	Bit 2	Bit 1	Bit 0
Volts	2.5	1.25	0.625	0.3125	0.156	0.078	0.039	0.0195
Output Value	0	0	1	0	1	1	0	0

Adding the voltages corresponding to each bit, we get:

$$0.625 + 0.156 + 0.078 = 0.859 \text{ volts}$$

20.1.3 Resolution

The resolution of an ADC is determined by the reference input and the word width. The resolution defines the smallest voltage change that can be measured by the ADC. As mentioned earlier, the resolution is the same as the smallest step size, and can be calculated by dividing the reference voltage range by the number of possible conversion values.

For the example we've been using so far, an 8-bit ADC with a 5V reference, the resolution is 0.0195V (19.5 mv). This means that any input voltage below 19.5 mv will result in an output of 0. Input voltages between 19.5 and 39 mv will result in an output of 1. Between 39 mv and 58.6 mv, the output will be 2. Resolution can be improved by reducing the reference input. Changing from 5V to 2.5V gives a resolution of 2.5/256, or 9.7 mV. However, the maximum voltage that can be measured is now 2.5V instead of 5V.

The only way to increase resolution without changing the reference is to use an ADC with more bits. A 10-bit ADC using a 5V reference has 2^{10}, or 1024 possible output codes. So the resolution is 5V/1024, or 4.88 mv.

20.2 Types of ADCs

ADCs come in various speeds, use different interfaces, and provide differing degrees of accuracy. Three types of ADCs are illustrated in Figure 20.3.

20.2.1 Tracking ADC

The tracking ADC has a comparator, a counter, and a digital-to-analog converter. The comparator compares the input voltage to the DAC output voltage. If the input is higher than the DAC voltage, the counter counts up. If the input is lower than the DAC voltage, the counter counts down.

The DAC input is connected to the counter output. Say the reference voltage is 5V. This would mean that the converter could convert voltages between 0V and 5V. If the most significant bit of the DAC input is "1," the output voltage is 2.5V. If the next bit is "1," 1.25V is added, making the result 3.75V. Each successive bit adds half the voltage of the previous bit, so the DAC input bits correspond to the following voltages:

Bit:	Bit 7	Bit 6	Bit 5	Bit 4	Bit 3	Bit 2	Bit 1	Bit 0
Volts:	2.5	1.25	0.625	0.3125	0.156	0.078	0.039	0.0195

Figure 20.3 shows how the tracking ADC resolves an input voltage of 0.37V. The counter starts at zero, so the comparator output will be high. The counter counts up once for every clock pulse, stepping the DAC output voltage up. When the counter passes the binary value that represents the input voltage, the comparator output will switch and the counter will count down. The counter will eventually oscillate around the value that represents the input voltage.

The primary drawback to the tracking ADC is speed—a conversion can take up to 256 clocks for an 8-bit output, 1024 clocks for a 10-bit value, and so on. In addition, the conversion speed varies with the input voltage. If the voltage in this example were 0.18V, the conversion would take only half as many clocks as the 0.37V example.

The maximum clock speed of a tracking ADC depends on the propagation delay of the DAC and the comparator. After every clock, the counter output has to propagate through the DAC and appear at the output. The comparator then takes some amount of time to respond to the change in DAC voltage, producing a new up/down control input to

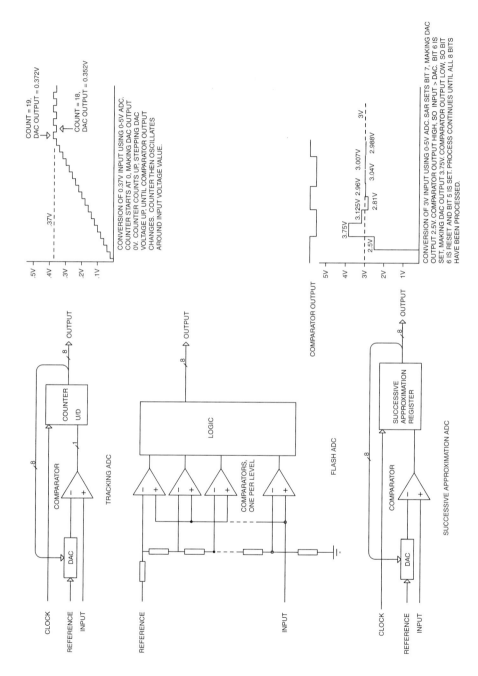

Figure 20.3: ADC types

the counter. Tracking ADCs are not commonly available; in looking at the parts available from Analog Devices, Maxim, and Burr-Brown (all three are manufacturers of ADC components), not one tracking ADC is shown. This only makes sense: a successive approximation ADC with the same number of bits is faster. However, there is one case where a tracking ADC can be useful; if the input signal changes slowly with respect to the sampling clock, a tracking ADC may produce an output in fewer clocks than a successive approximation ADC. However, since there are no commercial tracking ADCs available, a tracking ADC would have to be built from discrete hardware.

20.2.2   Flash ADC

The Flash ADC is the fastest type available. A Flash ADC has one comparator per voltage step. A 4-bit ADC will have 16 comparators, an 8-bit ADC will have 256 comparators. One input of all the comparators is connected to the input to be measured. The other input of each comparator is connected to one point in a string of resistors. As you move up the resistor string, each comparator trips at a higher voltage. All of the comparator outputs connect to a block of logic that determines the output based on which comparators are low and which are high.

The conversion speed of the Flash ADC is the sum of the comparator delays and the logic delay (the logic delay is usually negligible). Flash ADCs are very fast, but take enormous amounts of IC real estate to implement. Because of the number of comparators required, they tend to be power hogs, drawing significant current. A 10-bit Flash ADC IC may use half an amp.

20.2.3   Successive Approximation Converter

The successive approximation converter is similar to the tracking ADC in that a DAC/counter drives one side of a comparator and the input drives the other. The difference is that the successive approximation register performs a binary search instead of just counting up or down by one. As shown in Figure 20.3, say we start with an input of 3V, using a 5V reference. The successive approximation register would perform the conversion as follows:

```
Set MSB of SAR, DAC voltage = 2.5 V

    Comparator output high, so leave MSB set

    Result = 1000 0000
```

Set bit 6 of SAR, DAC voltage = 3.75V (2.5 + 1.25)

Comparator output low, reset bit 6

Result = 1000 0000

Set bit 5 of SAR, DAC voltage = 3.125V (2.5 + 0.625)

Comparator output low, reset bit 5

Result = 1000 0000

Set bit 4 of SAR, DAC voltage = 2.8125V (2.5 + 0.3125)

Comparator output high, leave bit 4 set

Result = 1001 0000

Set bit 3 of SAR, DAC voltage = 2.968V (2.8125 + 0.15625)

Comparator output high, leave bit 3 set

Result = 1001 1000

Set bit 2 of SAR, DAC voltage = 3.04V (2.968 + 0.078125)

Comparator output low, reset bit 2

Result = 1001 1000

Set bit 1 of SAR, DAC voltage = 3.007V (2.8125 + 0.039)

Comparator output low, reset bit 1

Result = 1001 1000

Set bit 0 of SAR, DAC voltage = 2.988V (2.8125 + 0.0195)

Comparator output high, leave bit 0 set

Final result = 1001 1001

Using the 0-to-5V, 8-bit DAC, this corresponds to:

$$2.5 + 0.3125 + 0.15625 + 0.0195 \text{ or } 2.988 \text{ volts}$$

This is not exactly 3V, but it is as close as we can get with an 8-bit converter and a 5V reference.

An 8-bit successive approximation ADC can do a conversion in 8 clocks, regardless of the input voltage. More logic is required than for the tracking ADC, but the conversion speed is consistent and usually faster.

20.2.4 Dual-Slope (Integrating) ADC

A dual-slope converter (Figure 20.4) uses an integrator followed by a comparator, followed by counting logic. The integrator input is first switched to the input signal, and the integrator output charges toward the input voltage. After a specified number of clock cycles, the integrator input is switched to a reference voltage (VREF1 in Figure 20.4) and the integrator charges down toward this value.

When the switch occurs to VREF1, a counter is started, and it counts using the same clock that determined the original integration time. When the integrator output falls past a second reference voltage (VREF2 in Figure 20.4), the comparator output goes high, the counter stops, and the count represents the analog input voltage. Higher input voltages will allow the integrator to charge to a higher voltage during the input time,

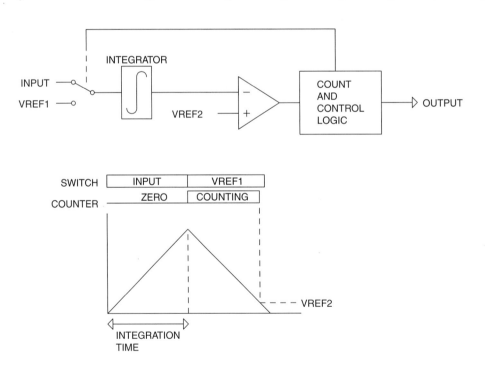

Figure 20.4: Dual-slope ADC

taking longer to charge down to VREF2, and resulting in a higher count at the output. Lower input voltages result in a lower integrator output and a smaller count.

A simpler integrating converter, the single-slope, runs the counter while charging up and stops counting when a reference voltage is reached (instead of charging for a specific time). However, the single-slope converter is affected by clock accuracy. The dual-slope design eliminates clock accuracy problems because the same clock is used for charging and incrementing the counter. Note that clock jitter or drift within a single conversion will affect accuracy. The dual-slope converter takes a relatively long time to perform a conversion, but the inherent filtering action of the integrator eliminates noise.

20.2.4 Sigma-Delta

Before describing the sigma-delta converter, we need to look at how oversampling works, because it is key to understanding the sigma-delta architecture. Figure 20.5 shows a noisy 3V signal, with 0.2V peak-to-peak of noise. As shown in the figure, we can sample this signal at regular intervals. Four samples are shown in the figure; by averaging these we can filter out the noise:

$$(3.05V + 3.1V + 2.9V + 2.95V)/4 = 3V$$

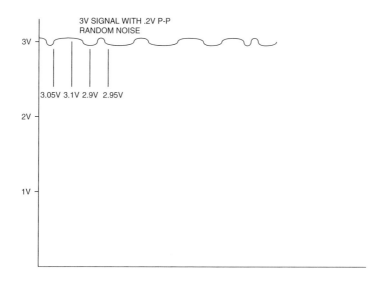

Figure 20.5: Oversampling

Obviously this example is a little contrived, but it illustrates the point. If our system can sample the signal four times faster than data is actually needed, we can average four samples. If we can sample ten times faster, we can average ten samples for an even better result. The more samples we can average, the closer we get to the actual input value. The catch, of course, is that we have to run the ADC faster than we actually need the data, and must have software to do the averaging.

Figure 20.6 shows how a sigma-delta converter works. The input signal passes through one side of a differential amp, through a low-pass filter (integrator), and on to a comparator. The output of the comparator drives a digital filter and a 1-bit DAC. The DAC output can switch between +V and V. In the example shown in Figure 20.6, the +V is 0.5V, and the −V is −0.5V. The output of the DAC drives the other side of the differential amp, so the output of the differential amp is the difference between the input voltage and the DAC output. In the example shown, the input is 0.3V, so the output of the differential amp is either 0.8V (when the DAC output is −0.5V) or −0.2V (when the DAC output is 0.5V).

The output of the low-pass filter drives one side of the comparator, and the other side of the comparator is grounded. So any time the filter output is above ground, the comparator output will be high, and any time the filter output is below ground, the comparator output will be low. The thing to remember is that the circuit tries to keep the filter output at 0V.

As shown in Figure 20.6, the duty cycle of the DAC output represents the input level; with an input of 0.3V (80% of the −0.5 to 0.5V range), the DAC output has a duty cycle of 80%. The digital filter converts this signal to a binary digital value.

The input range of the sigma-delta converter is the plus-and-minus DAC voltage. The example in Figure 20.6 uses 0.5 and −0.5V for the DAC, so the input range is −0.5V to 0.5V, or 1V total. For ±1V DAC outputs, the range would be ±1V, or 2V total.

The primary advantage of the sigma-delta converter is high resolution. Because the duty cycle feedback can be adjusted with a resolution of one clock, the resolution is limited only by the clock rate. Faster clock equals higher resolution.

All of the other types of ADCs use some type of resistor ladder or string. In the Flash ADC the resistor string provides a reference for each comparator. On the tracking and successive approximation ADCs, the ladder is part of the DAC in the feedback path. The problem with the resistor ladder is that the accuracy of the resistors directly affects the accuracy of the conversion result. Although modern ADCs use very precise,

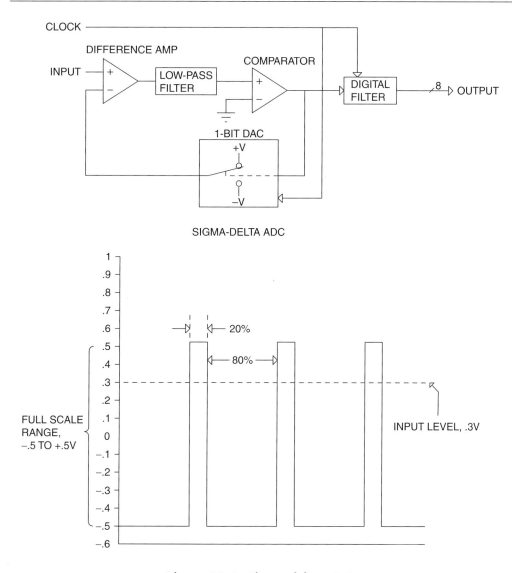

Figure 20.6: Sigma-delta ADC

laser-trimmed resistor networks (or sometimes capacitor networks), there are still some inaccuracies in the resistor ladders. The sigma-delta converter does not have a resistor ladder; the DAC in the feedback path is a single-bit DAC, with the output swinging between the two reference endpoints. This provides a more accurate result.

The primary disadvantage of the sigma-delta converter is speed. Because the converter works by oversampling the input, the conversion takes many clocks. For a given clock rate, the sigma-delta converter is slower than other converter types. Or, to put it another way, for a given conversion rate, the sigma-delta converter requires a faster clock.

Another disadvantage of the sigma-delta converter is the complexity of the digital filter that converts the duty cycle information to a digital output word. Single-IC sigma-delta converters have become more commonly available with the ability to add a digital filter or DSP to the IC die.

20.2.6   Half-Flash

Figure 20.7 shows a block diagram of a half-Flash converter. This example implements an 8-bit ADC with 32 comparators instead of 256. The half-Flash converter has a 4-bit (16 comparators) Flash converter to generate the MSB of the result. The output of this Flash converter then drives a 4-bit DAC to generate the voltage represented by the 4-bit result. The output of the DAC is subtracted from the input signal, leaving a remainder that is converted by another 4-bit Flash to produce the LS 4 bits of the result.

If the converter shown in Figure 20.7 were a 0–5V converter converting a 3.1V input, then the conversion would look like this:

Upper Flash converter output = 9

DAC output = 2.8125V(9 × 16 × 19.53 mv)

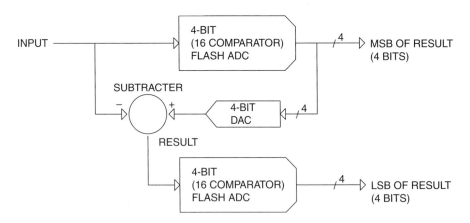

Figure 20.7: Half-Flash converter

Subtracter output $= 3.1\text{V} - 2.8125\text{V} = 0.2875\text{V}$

Lower Flash converter output $= \text{E (hex)}$

Final result $= 9\text{E (hex)}, 158 \text{ (decimal)}$

Half-Flash converters can also use three stages instead of two; a 12-bit converter might have three stages of 4 bits each. The result of the MS 4 bits would be subtracted from the input voltage and applied to the middle 4-bit state. The result of the middle stage would be subtracted from its input and applied to the least significant 4-bit stage. A half-Flash converter is slower than an equivalent Flash converter, but uses fewer comparators, so it draws less current.

20.3   ADC Comparison

Figure 20.8 shows the range of resolutions available for integrating, sigma-delta, successive approximation, and Flash converters. The maximum conversion speed for each type also is shown. As you can see, the speed of available sigma-delta ADCs reaches into the range of the SAR ADCs, but is not as fast as even the slowest Flash ADCs. What these charts do not show is trade-offs between speed and accuracy.

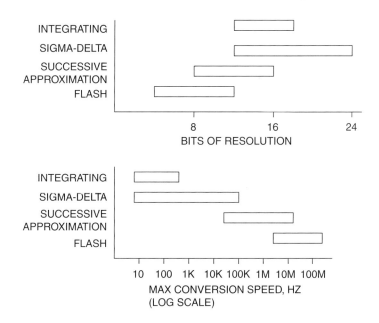

Figure 20.8: ADC comparison

For instance, although you can get SAR ADCs that range from 8 to 16 bits, you won't find the 16-bit version to be the fastest in a given family of parts. The fastest Flash ADC won't be the 12-bit part, it will be a 6-or 8-bit part.

These charts are snapshots of the current state of the technology. As CMOS processes have improved, SAR conversion times have moved from tens of microseconds to microseconds to tens of nanoseconds. Not all technology improvements affect all types of converters; CMOS process improvements speed up all families of converters, but the ability to put increasingly sophisticated DSP functionality on the ADC chip doesn't improve SAR converters. It does improve sigma-delta types.

20.4   Sample and Hold

ADC operation is straightforward when a DC signal is being converted. What happens when the signal is changing? Figure 20.9 shows a successive-approximation ADC attempting to convert a changing input. When the ADC starts the conversion, the input voltage is 2.3V. This should result in an output code of 117 (decimal) or 75 (hex). The SAR register sets the MSB, making the internal DAC voltage 2.5V. Because the signal is below 2.5V, the SAR resets bit 7 and sets bit 6 on the next clock. The ADC "chases" the input signal, ending up with a final result of $127_{10}(7F_{16})$. The actual voltage at the end of the conversion is 2.8V, corresponding to a code of $143_{10}(8F_{16})$.

The final code out of the ADC (127d) corresponds to a voltage of 2.48V. This is neither the starting voltage (2.3V) nor the ending voltage (2.8V). This example used a relatively

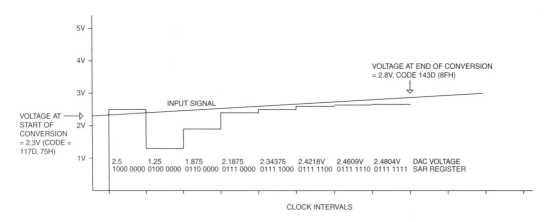

Figure 20.9: ADC inaccuracy caused by a changing input

fast input to show the effect; a slowly changing input has the same effect, but the error will be smaller. One way to reduce these errors is to place a low-pass filter ahead of the ADC. The filter parameters are selected to ensure that the ADC input does not change appreciably within a conversion cycle.

Another way to handle changing inputs is to add a sample-and-hold (S/H) circuit ahead of the ADC. Figure 20.10 shows how a sample-and-hold circuit works. The S/H circuit has an analog (solid state) switch with a control input. When the switch is closed, the input signal is connected to the hold capacitor and the output of the buffer follows the input. When the switch is open, the input is disconnected from the capacitor.

Figure 20.10 shows the waveform for S/H operation. A slowly rising signal is connected to the S/H input. While the control signal is low (sample), the output follows the input. When the control signal goes high (hold), disconnecting the hold capacitor from the input, the output stays at the value the input had when the S/H switched to hold mode. When the switch closes again, the capacitor charges quickly and the output again follows the input. Typically, the S/H will be switched to hold mode just before the ADC conversion starts, and switched back to sample mode after the conversion is complete.

In a perfect world, the hold capacitor would have no leakage and the buffer amplifier would have infinite input impedance, so the output would remain stable forever. In the

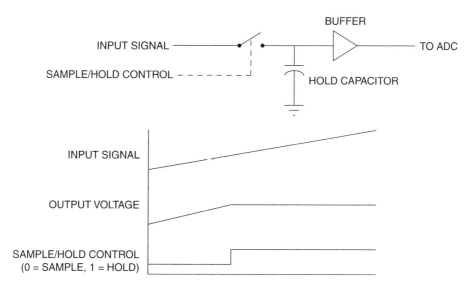

Figure 20.10: Sample-and-hold circuit

real world, the hold capacitor will leak and the buffer amplifier input impedance is finite, so the output level will slowly drift down toward ground as the capacitor discharges.

The ability of an S/H to maintain the output in hold mode is dependent on the quality of the hold capacitor, the characteristics of the buffer amplifier (primarily input impedance), and the quality of the sample-and-hold switch (real electronic switches have some leakage when open). The amount of drift exhibited by the output when in hold mode is called the *droop rate*, and is specified in millivolts per second, microvolts per microsecond, or millivolts per microsecond.

A real S/H also has finite input impedance, because the electronic switch isn't perfect. This means that, in sample mode, the hold capacitor is charged through some resistance. This limits the speed with which the S/H can acquire an input. The time that the S/H must remain in sample mode in order to acquire a full-scale input is called the *acquisition time*, and is specified in nanoseconds or microseconds.

Because there is some impedance in series with the hold capacitor when sampling, the effect is the same as a low-pass RC filter. This limits the maximum frequency the S/H can acquire. This is called the *full power bandwidth*, specified in kHz or MHz.

As mentioned, the electronic switch is imperfect and some of the input signal appears at the output, even in hold mode. This is called *feedthrough*, and is typically specified in dB.

The *output offset* is the voltage difference between the input and the output. S/H datasheets typically show a hold mode offset and sample mode offset, in millivolts.

20.5   Real Parts

Real ADC ICs come with a few real-world limitations and some added features.

20.5.1   *Input Levels*

The examples so far have concentrated on ADCs with a 0–5V input range. This is a common range for real ADCs, but many of them operate over a wider range of voltages. The Analog Devices AD570 has a 10V input range. The part can be configured so that this 10V range is either 0 to 10V or –5V to +5V, using one pin. Of course, having a negative input voltage range implies that the ADC will need a negative voltage supply. Other common input voltage ranges are ±2.5V and ±3V.

With the trend toward lower-powered devices and small consumer equipment, the trend in ADC devices is to lower-voltage, single-supply operation. Traditional single-supply ADCs have operated from +5V and had an input range between 0V and +5V. Newer parts often operate at 3.3 or 2.7V, and have an input range somewhere between 0V and the supply.

20.5.2 Internal Reference

Many ADCs provide an internal reference voltage. The Analog Devices AD872 is a typical device with an internal 2.5V reference. The internal reference voltage is brought out to a pin and the reference input to the device is also connected to a pin. To use the internal reference, the two pins are connected together. To use your own external reference, connect it to the reference input instead of the internal reference.

20.5.3 Reference Bypassing

Although the reference input is usually high impedance with low DC current requirements, many ADCs will draw current from the reference briefly while a conversion is in process. This is especially true of successive approximation ADCs, which draw a momentary spike of current each time the analog switch network is changed. Consequently, most ADCs require that the reference input be bypassed with a capacitor of 0.1 µf or so.

20.5.4 Internal S/H

Many ADCs, such as the Maxim MAX191, include an internal S/H. An ADC with an internal S/H may have a separate pin that controls whether the S/H is in sample or hold mode, or the switch to hold mode may occur automatically when a conversion is started.

20.6 Microprocessor Interfacing

20.6.1 Output Coding

The examples used so far have been based on binary codes, where each bit in the result represents a voltage value and the sum of these voltages in the output word is the analog input voltage value. Some ADCs produce 2's complement outputs, where a negative voltage is represented by a negative 2's complement value. A few ADCs output values in BCD. Obviously this requires more bits for a given range; a 12-bit binary output can represent values from 0 to 4095, but a 12-bit BCD output can only represent values from 0 to 999.

20.6.2 Parallel Interfaces

ADCs come in a variety of interfaces, intended to operate with multiple processors. Some parts include more than one type of interface to make them compatible with as many processor families as possible.

The Maxim MAX151 is a typical 10-bit ADC with an 8-bit "universal" parallel interface. As shown in Figure 20.11, the processor interface on the MAX151 has 8 data bits, a chip select (–CS), a read strobe (–RD), and a –BUSY output. The MAX151 includes an internal S/H. On the falling edge of –RD and –CS, the S/H is placed into hold mode and a conversion is started. If –CS and –RD do not go low at the same time, the last falling edge starts a conversion. In most systems, –CS is connected to an address decode and will go low before –RD. As soon as the conversion starts, the ADC drives –BUSY low (active). –BUSY remains low until the conversion is complete.

In the first mode of operation, which Maxim calls *Slow Memory Mode*, the processor waits, holding –RD and –CS low, until the conversion is complete. In such a system, the –BUSY signal would typically be connected to the processor –RDY or –WAIT signal. This holds the processor in a wait state until the conversion is complete. The maximum conversion time for the MAX151 is 2.5 μs.

The second mode of operation is called *ROM* mode. In this mode the processor performs a read cycle, which places the S/H in hold mode and starts a conversion. During this read, the processor reads the results of the previous conversion. The –BUSY signal is not used to extend the read cycle. Instead, –BUSY is connected to an interrupt, or is polled by the processor to indicate when the conversion is complete. When –BUSY goes high, the processor does another read to get the result and start another conversion. Although the data sheets refer to two different modes of operation, the ADC works the same way in both cases:

- Falling edge of –RD and –CS starts a conversion.

- Current result is available on bus after read access time has elapsed.

- As long as –RD and –CS stay low, current result remains available on bus.

- When conversion completes, new conversion data is latched and available to the processor; if –RD and –CS are still low, this data replaces result of previous conversion on bus.

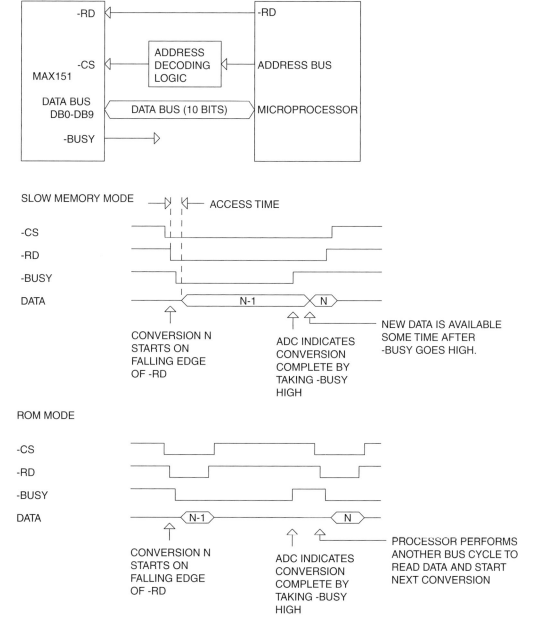

Figure 20.11: Maxim MAX151 interface

The MAX151 is designed to interface to most microprocessors. Actually interfacing to a specific processor requires analysis of the MAX151 timing and how it relates to the microprocessor timing.

20.6.3   Data Access Time

The MAX151 specifies a maximum access time of 180 ns over the full temperature range (see Figure 20.12). This means that the result of a conversion will be available on the bus no more than 180 ns after the falling edge of –RD (assuming –CS is already low when –RD goes low). The processor will need the data to be stable some time before the rising edge of –RD. If there is a data bus buffer between the MAX151 and the processor, the propagation delay through the buffer must be included. This means that the processor bus cycle (the time that –RD is low) must be at least as long as the access time of the MAX151, plus the processor data setup time, plus any bus buffer delays.

20.6.4   –BUSY Output

The –BUSY output of the MAX151 goes low a maximum of 200 ns after the falling edge of –RD. This is too long for the signal to directly drive most microprocessors if you want to use the slow memory mode. Most microprocessors require that the RDY or –WAIT signal be driven low earlier than this in the bus cycle. Some require the wait request signal to be low one clock after –RD goes low. The only solution to this problem is to artificially insert wait states to the bus cycle until the –BUSY signal goes low. Some microprocessors, such as the 80188 family, have internal wait-state generators that can add wait states to a bus cycle. The 80188 wait-state generator can be programmed to add 0, 1, 2, or 3 wait states.

As shown in Figure 20.12, in Slow Memory mode the –BUSY signal goes high just before the new conversion result is available; according to the data sheet, this time is a maximum of 50 ns. For some processors, this means that the wait request must be held active for an additional clock cycle after –BUSY goes high to ensure that the correct data is read at the end of the bus cycle.

20.6.5   Bus Relinquish

The MAX151 has a maximum bus relinquish time of 100 ns. This means that the MAX151 can drive the data bus up to 100 ns after the –RD signal goes high. If the

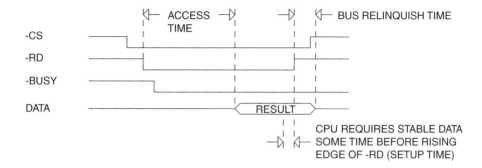

ADDING A BUFFER TO REDUCE
BUS RELINQUISH TIME

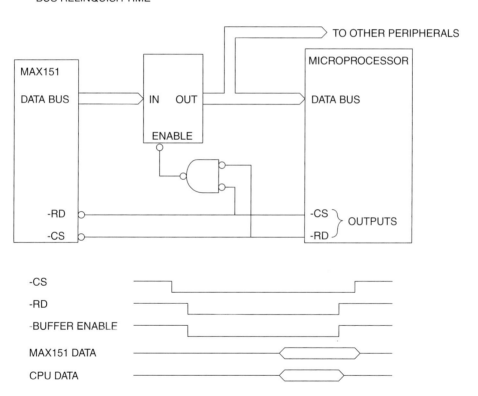

Figure 20.12: MAX151 data access and bus relinquish timing

processor tries to start another cycle immediately after reading the MAX151 result, this may result in bus contention. A typical example would be the 80186 processor, which multiplexes the data bus with the address bus; at the start of a bus cycle the data bus is not tristated, but the processor drives the address onto the data bus. If the MAX151 is still driving the bus, this can result in an incorrect bus address being latched. The solution to this problem is to add a data bus buffer between the MAX151 and the processor. The buffer inputs are connected to the MAX151 data bus outputs, and the buffer outputs are connected to the processor data bus. The buffer is turned on when –RD and –CS are both low, and turned off when either goes high. Although the MAX151 will continue to drive the buffer inputs, the outputs will be tristated and so will not conflict with the processor data bus. A buffer may also be required if you are interfacing to a microprocessor that does not multiplex the data lines but does have a very high clock rate. In this case, the processor may start the next cycle before the MAX151 has relinquished the bus. A typical example would be a fast 80960-family processor, which we will look at later in the chapter.

20.6.6 Coupling

The MAX151 has an additional specification, not found on some ADCs, that involves coupling of the bus control signals into the ADC. Because modern ADCs are built as monolithic ICs, the analog and digital portions share some internal components such as the power supply pins and the substrate on which the IC die is constructed. It is sometimes difficult to keep the noise generated by the microprocessor data bus and control signals from coupling into the ADC and affecting the result of a conversion. To minimize the effect of coupling, the MAX151 has a specification that the –RD signal be no more than 300 ns wide when using ROM mode. This prevents the rising edge of –RD from affecting the conversion.

20.6.7 Delay between Conversions

When the MAX151 S/H is in sampling mode the hold capacitor is connected to the input. This capacitance is about 150 pF. When a conversion starts, this capa-citor is disconnected from the input. When a conversion ends, the capacitor is again connected to the input, and it must charge up to the value of the input pin before another conversion can start. In addition, there is an internal 150 ohm resistor in series with the input capacitor. Consequently, the MAX151 specifies a delay between conversions of at least 500 ns if the source impedance driving the input is less than 50Ω. If the source impedance is more than 1 KΩ, the delay must be at least 1.5 μs. This delay is the time from the rising edge of –BUSY to the falling edge of –RD.

20.6.8 LSB Errors

In theory, of course, an infinite amount of time is required for the capacitor to charge up, because the charging curve is exponential and the capacitor never reaches the input voltage. In practice, the capacitor does stop charging. More important, the capacitor only has to charge to within 1 bit (called 1 *LSB*) of the input voltage; for a 10V converter with a ±4V input range, this is 8 V/1024, or 7.8 mV. To simplify the concept, errors that fall within one bit of resolution have no effect on conversion accuracy. The other side of that coin is that the accumulation of errors (op-amp offsets, gain errors, etc.) cannot exceed one bit of resolution or they *will* affect the result.

20.7 Clocked Interfaces

Interfacing the MAX151 to a clocked bus, such as that implemented on the Intel 80960 family, is shown in Figure 20.13. Processors such as the 960 use a clock-synchronized bus without a –RD strobe. Data is latched by the processor on a clock edge, rather than on the rising edge of a control signal such as –RD. These buses are often implemented on very fast processors and are usually capable of high-speed burst operation.

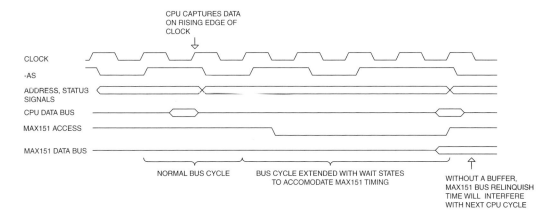

Figure 20.13: Interfacing to a clocked microprocessor bus

Shown in Figure 20.13 is a normal bus cycle without wait states. This bus cycle would be accessing a memory or peripheral able to operate at the full bus speed. The address and status information is provided on one clock, and the CPU reads the data on the next clock.

Following this cycle is an access to the MAX151. As can be seen, the MAX151 is much slower than the CPU, so the bus cycle must be extended with wait states (either internally or externally generated). This diagram is an example; the actual number of wait states that must be added depends on the processor clock rate. The bus relinquish time of the MAX151 will interfere with the next CPU cycle, so a buffer is necessary. Finally, because the CPU does not generate a –RD signal, one must be synthesized by the logic that decodes the address bus and generates timing signals to memory and peripherals. The normal method of interfacing an ADC like this to a fast processor is to use the ROM mode. Slow Memory mode holds the CPU in a wait state for a long time—the 2.5 μs conversion time of the MAX151 would be 82 clocks on a 33-MHz 80960. This is time that could be spent executing code.

20.8   Serial Interfaces

Many ADCs use a serial interface to connect to the microprocessor. This has the advantage of providing a processor-independent interface that does not affect processor wait states, bus hold times, or clock rates. The primary disadvantage is speed, because the data must be transferred one bit at a time.

20.8.1   SPI/Microwire

SPI is a serial interface that uses a clock, chip select, data in, and data out bits. Data is read from a serial ADC a bit at a time (Figure 20.14). Each device on the SPI bus requires a separate –CS signal.

The Maxim MAX1242 is a typical SPI ADC. The MAX1242 is a 10-bit successive approximation ADC with an internal S/H, in an 8-pin package. Figure 20.15 shows the

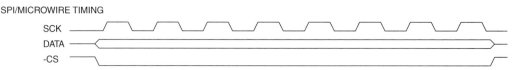

SPI/MICROWIRE TIMING

SCK

DATA

-CS

Figure 20.14: SPI bus

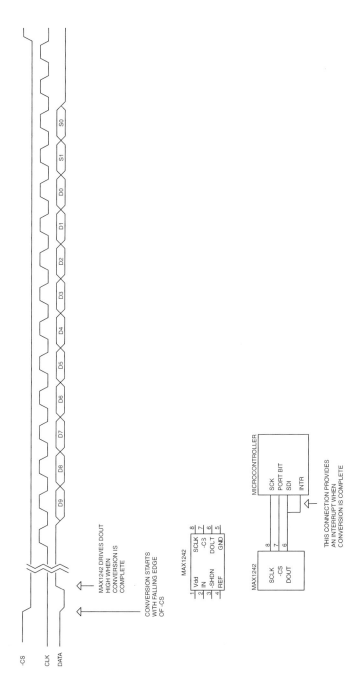

Figure 20.15: Maxim MAX1242 interface

MAX1242 interface timing. The falling edge of –CS starts a conversion, which takes a maximum of 7.5 μs. When –CS goes low, the MAX1242 drives its data output pin low. After the conversion is complete, the MAX1242 drives the data output pin high. The processor can then read the data a bit at a time by toggling the clock line and monitoring the MAX1242 data output pin. After the 10 bits are read, the MAX1242 provides two sub-bits, S1 and S0. If further clock transitions occur after the 13 clocks, the MAX1242 outputs zeros.

Figure 20.15 shows how a MAX1242 would be connected to a microcontroller with an on-chip SPI/Microwire interface. The SCLK signal goes to the SPI SCLK signal on the microcontroller, and the MAX1242 DOUT signal connects to the SPI data input pin on the microcontroller. One of the microcontroller port bits generates the –CS signal to the MAX1242. Note that the –CS signal starts the conversion and must remain low until the conversion is complete. This means that the SPI bus is unavailable for communicating with other peripherals until the conversion is finished and the result has been read. If there are interrupt service routines that communicate with SPI devices in the system, they must be disabled during the conversion. To avoid this problem, the MAX1242 could communicate with the microcontroller over a dedicated SPI bus. This would use three more pins on the microcontroller. Since most microcontrollers that have on-chip SPI have only one, the second port would have to be implemented in software.

Finally, it is possible to generate an interrupt to the microcontroller when the ADC conversion is complete. An extra connection is shown in Figure 20.15, from the MAX1242 DOUT pin to an interrupt on the microcontroller. When –CS is low and the conversion is completed, DOUT will go high, interrupting the microcontroller. To use this method, the firmware must disable or otherwise ignore the interrupt except when a conversion is in process.

Another ADC with an SPI-compatible interface is the Analog Devices AD7823. Like the MAX1242, the AD7823 uses three pins: SCLK, DOUT, and –CONVST. The AD7823 is an 8-bit successive approximation ADC with internal S/H. A conversion is started on the falling edge of –CONVST, and takes 5.5 μs. The rising edge of –CONVST enables the serial interface.

Unlike the MAX1242, the AD7823 does not drive the data pin until the microcontroller reads the result, so the SPI bus can be used to communicate with other devices while the conversion is in process. However, there is no indication to the microprocessor when the conversion is complete—the processor must start the conversion, then wait until the conversion has had time to complete before reading the result. One way to handle this is with a regular timer interrupt; on each interrupt, the result of the previous conversion is read and a new conversion is started.

20.8.2 I^2C Bus

The I^2C bus uses only two pins: SCL (SCLock) and SDA (SDAta). SCL is generated by the processor to clock data into and out of the peripheral device. SDA is a bidirectional line that serially transmits all data into and out of the peripheral. The SDA signal is open-collector, so several peripherals can share the same two-wire bus.

When sending data, the SDA signal is allowed to change only while SCL is in the low state. Transitions on the SDA line while SCL is high are interpreted as start and stop conditions. If SDA goes low while SCL is high, all peripherals on the bus will interpret this as a START condition. SDA going high while SCL is high is a STOP or END condition. Figure 20.16 illustrates a typical data transfer. The processor initiates the START condition and then sends the peripheral address, which is 7 bits long, and tells the devices on the bus which one is to be selected. This is followed by a read/write bit (1 for read, 0 for write).

After the read/write bit, the processor programs the I/O pin connected to the SDA bit to be an input and clocks an acknowledge bit in. The selected peripheral will drive the SDA line low to indicate that it has received the address and read/ write information.

After the acknowledge bit, the processor sends another address, which is the internal address within the peripheral that the processor wants to access. The length of this field varies with the peripheral. After this is another acknowledge; then the data is sent. For a write operation, the processor clocks out 8 data bits, and for a read operation, the processor treats the SDA pin as an input and clocks in 8 bits. After the data comes another acknowledge.

Some peripherals permit multiple bytes to be read or written in one transfer. The processor repeats the data/acknowledge sequence until all the bytes are transferred. The peripheral will increment its internal address after each transfer.

One drawback to the I^2C bus is speed—the clock rate is limited to about 100 kHz. A newer Fast-mode I^2C bus that operates to 400 kbits/sec is also available, and a high-speed mode that goes to 3.4 Mbits/sec is also available. High speed and fast-mode buses both support a 10-bit address field so up to 1024 locations can be addressed.

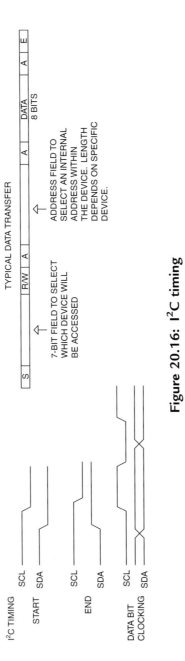

Figure 20.16: I²C timing

High-speed and fast-mode devices are capable of operating in the older system, but older peripherals are not useable in a higher-speed system. The faster interfaces have some limitations, such as the need for active pullups and limits on bus capacitance. Of course, the faster modes of operation require hardware support and are not suitable for a software-controlled implementation.

A typical ADC that uses I^2C is the Philips PCF8591. This part includes both an ADC and a DAC. Like many I^2C devices, the 8591 has three addressing pins: A0, A1, and A2. These can be connected to either "1" or "0" to select which address the device responds to. When the peripheral address is decoded, the PCF8591 will respond to address 1001xxx, where xxx matches the value of the A2, A1, and A0 pins. This allows up to eight PCF8591 devices to share a single I^2C bus.

20.8.3 SMBus

SMBus is a variation on I^2C, defined by Intel in 1995. I^2C is primarily defined by hardware and varies somewhat from one device to the next, but SMBus defines the bus as more of a network interface between a processor and its peripherals. The SMBus specification defines things such as powerdown operation of devices (no bus loading) and operating voltage range (3–5V) that all devices must meet. The primary difference between SMBus and I^2C is that SMBus defines a standard set of read and write protocols, rather than leaving these specifics up to the IC manufacturers.

20.8.4 Proprietary Serial Interfaces

Some ADCs have proprietary interfaces. The Maxim MAX1101 is a typical device. This is an 8-bit ADC that is optimized for interfacing to CCDs. The MAX1101 uses four pins: MODE, LOAD, DATA, and SCLK. The MODE pin determines whether data is being written or read (1 = read, 0 = write). The DATA pin is a bidirectional signal, the SCLK signal clocks data into and out of the device, and the LOAD pin is used after a write to clock the write data into the internal registers. The clocked serial interface of the MAX1101 is similar to SPI, but because there is no chip select signal, multiple devices cannot share the same data/clock bus. Each MAX1101 (or similar device) needs four signals from the processor for the interface.

Many proprietary serial interfaces are intended for use with microcontrollers that have on-chip hardware to implement synchronous serial I/O. The 8031 family, for example, has a serial interface that can be configured as either an asynchronous interface or as a synchronous interface. Many ADCs can connect directly to these types of microprocessors. The problem with any serial interface on an ADC is that it limits conversion speed. In addition, the type of interface limits speed as well. Because every I^2C exchange involves at least 20 bits, an I^2C device will never be as fast as an equivalent SPI or proprietary device. For this reason, there are many more ADCs available with SPI/Microwire than with I^2C interfaces.

The required throughput of the serial interface drives the design. If you need a conversion speed of 100,000 8-bit samples per second and you plan to implement an SPI-type interface in software, then your processor will not be able to spend more than $1 / (100; 000 \times 8)$ or 1.25 μS transferring each bit. This may be impractical if the processor has any other tasks to perform, so you may want to use an ADC with a parallel interface or choose a processor with hardware support for the SPI.

The bandwidth of the bus must be considered as well as the throughput of the processor. If there are multiple devices on the SPI bus, then you have to be sure the bus can support the total throughput required of all the devices. Of course, the processor has to keep up with the overall data rate as well.

20.9   Multichannel ADCs

Many ADCs are available with multiple channels—anywhere from two to eight. The Analog Devices AD7824 is a typical device, with eight channels. The AD7824 contains a single 8-bit ADC and an 8-channel analog multiplexer. The microprocessor interface to the AD7824 is similar to the Maxim MAX151, but with the addition of three address lines (A0–A2) to select which channel is to be converted. Like the MAX151, the AD7824 may be used in a mode in which the microprocessor starts a conversion and is placed into a wait state until the conversion is complete. The microprocessor can also start a conversion on any channel (by reading data from that channel), then wait for the conversion to complete and perform another read to get the result. The AD7824 also provides an interrupt output that indicates when a conversion is complete.

20.10   Internal Microcontroller ADCs

Many microcontrollers contain on-chip ADCs. Typical devices include the Microchip PIC167C7xx family and the Atmel AT90S4434. Most microcontroller ADCs are successive approximation because this gives the best trade-off between speed and IC real estate on the microcontroller die.

The PIC16C7xx microcontrollers contain an 8-bit successive approximation ADC with analog input multiplexers. The microcontrollers in this family have from four to eight channels. Internal registers control which channel is selected, start of conversion, and so on. Once an input is selected, there is a settling time that must elapse to allow the S/H capacitor to charge before the A/D conversion can start. The software must ensure that this delay takes place.

20.10.1 Reference Voltage

The Microchip devices allow you to use one input pin as a reference voltage. This is normally tied to some kind of precision reference. The value read from the A/D converter after a conversion is:

$$\text{Digital word} = (\text{Vin}/\text{Vref}) \times 256$$

The Microchip parts also permit the reference voltage to be internally set to the supply voltage, which permits the reference input pin to be another analog input. In a 5V system, this means that Vref is 5V. So measuring a 3.2V signal would produce the following result:

$$\text{Result} = \frac{\text{Vin} \times 256}{\text{Vref}} = \frac{3.2\text{V} \times 256}{5\text{V}} = 163_{10} = A3_{16}$$

However, the result is dependent on the value of the 5V supply. If the supply voltage is high by 1%, it has a value of 5.05V. Now the value of the A/D conversion will be:

$$\frac{3.2\text{V} \times 256}{5.05\text{V}} = 162_{10} = A2_{16}$$

So a 1% change in the supply voltage causes the conversion result to change by one count. Typical power supplies can vary by 2 or 3%, so power supply variations can have a significant effect on the results. The power supply output can vary with loading, temperature, AC input variations, and from one supply to the next.

This brings up an issue that affects all ADC designs: the accuracy of the reference. The Maxim MAX1242, which we have already looked at, uses an internal reference. The part can convert inputs from 0V to the reference voltage. The reference is nominally

2.5V, but it can vary between 2.47V and 2.53V. Converting a 2V input at the extremes of the reference ranges gives the following result:

$$\text{At Vref} = 2.47\text{V, Result} = \frac{2\text{V} \times 1024}{2.47} = 829_{10}$$

$$\text{At Vref} = 2.53\text{V, Result} = \frac{2\text{V} \times 1024}{2.53} = 809_{10}$$

(Note: Multiplier is 1024 because the MAX1242 is a 10-bit converter.)

So the variation in the reference voltage from part to part can result in an output variation of 20 counts.

20.11   Codecs

The term *codec* has two meanings: it is short for compressor/decompressor, or for coder/decoder. In general, a codec (either type) will have two-way operation; it can turn analog signals into digital and vice versa, or it can convert to and from some compression standard.

The National Semiconductor LM4546 is an audio codec intended to implement the sound system in a personal computer. It contains an internal 18-bit ADC and DAC. It also includes much of the audio-processing circuitry needed for 3D PC sound. The LM4546 uses a serial interface to communicate with its host processor.

The National TP3054 is a telecom-type codec, and includes ADC, DAC, filtering, and companding circuitry. The TP3054 also has a serial interface.

20.12   Interrupt Rates

The MAX151 can perform a conversion every 3.3 μs, or 300,000 conversions per second. Even a 33 MHz processor operating at one instruction per clock cycle can execute only 110 instructions in that time. The interrupt overhead of saving and restoring registers can be a significant portion of those instructions.

In some applications, the processor does not need to process every conversion. An example would be a design in which the processor takes four samples, averages them, and then does something with the average. In cases like this, using a processor with DMA capability can reduce the interrupt overhead. The DMA controller is programmed to read the ADC at regular intervals, based on a timer (the ADC has to be a type that starts a new conversion as soon as the previous result is read). After all the conversions are complete, the DMA controller interrupts the processor. The accumulated ADC data is processed and the DMA controller is programmed to start the sequence over. Processors that include on-chip DMA controllers include the 80186 and the 386EX.

20.13   Dual-Function Pins on Microcontrollers

If you work with microcontrollers, you sometimes find that you need more I/O pins than your microcontroller has. This is most often a problem when working with smaller devices, such as the 8-pin Atmel ATtiny parts, or the 20-and 28-pin Atmel AVR and Microchip PIC devices. In some cases, you can make an analog input double as an output or make it handle two inputs. Figure 20.17(A) shows how an analog input can also control two outputs. In this case, the analog input is connected to a 2.5V reference diode. A typical use for this design would be in an application where you are using the 5V supply as the ADC reference, but you want to correct the readings for the actual supply value. A precise 2.5V reference permits you to do this, because you know that the value of the reference should read as 80 (hex) if the power supply is exactly 5V.

The pin on the microcontroller is also tied to the inputs of two comparators. A voltage divider sets the noninverting input of comparator A at 3V, and the inverting input of comparator B at 2V. By configuring the pin as an analog input, the reference value can be read. If the pin is then configured as a digital output and set low, the output of comparator A will go low. If the pin is configured as a digital output and set high, the output of comparator B will go low. Of course, this scheme works only if the comparator outputs drive signals that never need to both be low at the same time. The resistor values must be large enough that the microcontroller can source enough current to drive the pin high. This technique will also work for a digital-only I/O pin; instead of a 2.5V reference, a pair of resistors is used to hold the pin at 2.5V when it is configured as an input.

Figure 20.17(B) shows how a single analog input can be used to read two switches. When both switches are open, the analog input will read 5V. When switch S1 is closed, the analog input will read 3.9V. When switch S2 is closed, the input will read 3.4V, and when both switches are closed, the input will read 2.9V. Instead of switches, you could also use this technique to read the state of open-collector or open-drain digital signals.

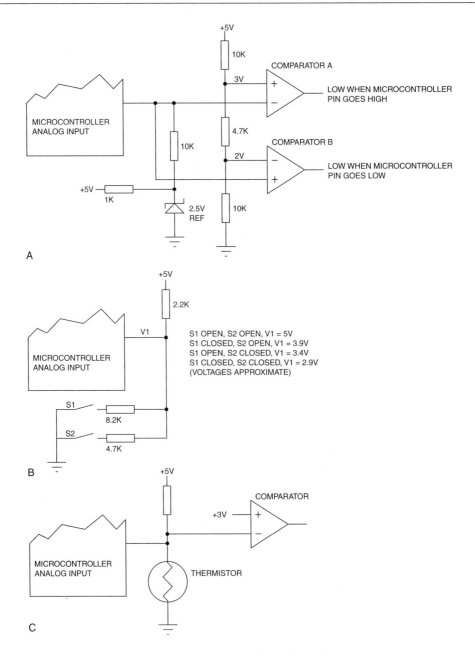

Figure 20.17: Dual-function pins

Figure 20.17(C) shows how a thermistor or other variable-resistance sensor can be combined with an output. The microcontroller pin is programmed as an analog input to read the temperature. When the pin is programmed as an output and driven high, the comparator output will go low. To make this work, the operating temperature range must be such that the voltage divider created by the thermistor and the pull-up resistor never brings the analog input above 3V. Like the example shown in 2.17(A), this circuit works best if the output is something that periodically changes state, so the software has a regular opportunity to read the analog input.

20.14   Design Checklist

- Be sure ADC bus interface is compatible with microprocessor timing. Pay particular attention to bus setup, hold, and min/max pulse width timings.

- If using SPI and an ADC that requires the bus to be inactive during conversion, ensure that the system will work with this limitation or provide a separate SPI bus for the ADC.

- If using an ADC that does not indicate when conversion is complete, ensure that software allows conversion to complete before reading result.

- Be sure reference accuracy meets requirements of the design.

- Bypass reference input as recommended by ADC manufacturer.

- Be sure the processor can keep up with the conversion rate.

CHAPTER 21

Sensors

Mike Tooley

Sensors provide us with a means of generating signals that can be used as inputs to electronic circuits. The things that we might want to sense include physical parameters such as temperature, light level, and pressure. Being able to generate an electrical signal that accurately represents these quantities allows us not only to measure and record these values but also to control them.

Sensors are, in fact, a subset of a larger family of devices known as *transducers* so we will consider these before we look at sensors and how we condition the signals that they produce in greater detail. We begin, however, with a brief introduction to the instrumentation and control systems in which sensors, transducers, and signal conditioning circuits are used

21.1   Instrumentation and Control Systems

Figure 21.1 shows the arrangement of an instrumentation system. The physical quantity to be measured (e.g., temperature) acts upon a sensor that produces an electrical output signal. This signal is an electrical analog of the physical input but note that there may not be a linear relationship between the physical quantity and its electrical equivalent. Because of this and since the output produced by the sensor may be small or may suffer from the presence of noise (i.e., unwanted signals) further signal conditioning will be required before the signal will be at an acceptable level and in an acceptable form

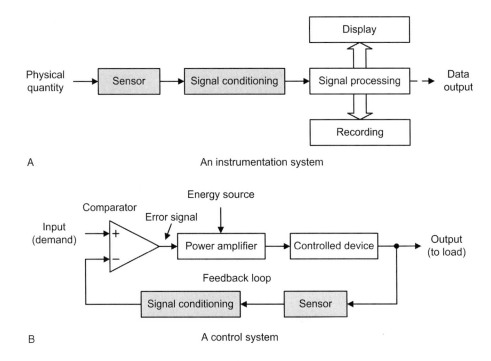

A   An instrumentation system

B   A control system

Figure 21.1: Instrumentation and control system

for signal processing, display and recording. Furthermore, because the signal processing may use digital rather than analog signals an additional stage of analog-to-analog conversion may be required.

Figure 21.1(B) shows the arrangement of a control system. This uses *negative feedback* in order to regulate and stabilize the output. It thus becomes possible to set the input or *demand* (i.e., what we desire the output to be) and leave the system to regulate itself by comparing it with a signal derived from the output (via a sensor and appropriate signal conditioning).

A *comparator* is used to sense the difference in these two signals and where any discrepancy is detected the input to the power amplifier is adjusted accordingly. This signal is referred to as an *error signal* (it should be zero when the output exactly matches the demand). The input (demand) is often derived from a simple potentiometer connected across a stable DC voltage source while the controlled device can take many forms (e.g., a DC motor, linear actuator, heater, etc.).

21.2   Transducers

Transducers are devices that convert energy in the form of sound, light, heat, etc., into an equivalent electrical signal, or vice versa.

Before we go further, let's consider a couple of examples that you will already be familiar with. A loudspeaker is a transducer that converts low-frequency electric current into audible sounds. A microphone, on the other hand, is a transducer that performs the reverse function, i.e., that of converting sound pressure variations into voltage or current. Loudspeakers and microphones can thus be considered as complementary transducers.

Transducers may be used as both inputs to electronic circuits and outputs from them. From the two previous examples, it should be obvious that a loudspeaker is an *output transducer* designed for use in conjunction with an audio system, whereas a microphone is an *input transducer* designed for use with a recording or sound reinforcing system.

There are many different types of transducer and Tables 21.1 and 21.2 provide some examples of transducers that can be used to input and output three important physical quantities; sound, temperature, and angular position.

Table 21.1: Some examples of input transducers

Physical quantity	Input transducer	Notes
Sound (pressure change)	Dynamic microphone (see Figure 21.3)	Diaphragm attached to a coil is suspended in a magnetic field. Movement of the diaphragm causes current to be induced in the coil.
Temperature	Thermocouple (see Figure 21.2)	Small e.m.f. generated at the junction between two dissimilar metals (e.g., copper and constantan). Requires reference junction and compensated cables for accurate measurement
Angular position	Rotary potentiometer	Fine wire resistive element is wound around a circular former. Slider attached to the control shaft makes contact with the resistive element. A stable DC voltage source is connected across the ends of the potentiometer. Voltage appearing at the slider will then be proportional to angular position.

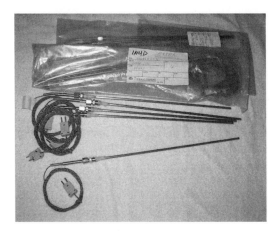

Figure 21.2: A selection of thermocouple probes

Table 21.2: Some examples of output transducers

Physical quantity	Output transducer	Notes
Sound (pressure change)	Loudspeaker (see Figure 21.3)	Diaphragm attached to a coil is suspended in a magnetic field. Current in the coil causes movement of the diaphragm which alternately compresses and rarefies the air mass in front of it.
Temperature	Heating element (resistor)	Metallic conductor is wound onto a ceramic or mica former. Current flowing in the conductor produces heat.
Angular position	Rotary potentiometer	Multi-phase motor provides precise rotation in discrete steps of 15° (24 steps per revolution), 7.5° (48 steps per revolution) and 1.8° (200 steps per revolution)

21.3   Sensors

A *sensor* is a special kind of transducer that is used to generate an input signal to a measurement, instrumentation or control system. The signal produced by a sensor is an *electrical analogy* of a physical quantity, such as distance, velocity, acceleration, temperature, pressure, light level, etc. The signals returned from a sensor, together with control inputs from the user or controller (as appropriate) will subsequently be used to determine the output from the system. The choice of sensor is governed by a number of factors including accuracy, resolution, cost, and physical size.

Figure 21.3: A selection of audible transducers

Figure 21.4: Various switch sensors

Sensors can be categorized as either *active* or *passive*. An active sensor *generates* a current or voltage output. A passive transducer *requires a source of current or voltage* and it modifies this in some way (e.g., by virtue of a change in the sensor's resistance). The result may still be a voltage or current *but it is not generated by the sensor on its own.*

Sensors can also be classed as either *digital* or *analog*. The output of a digital sensor can exist in only two discrete states, either "on" or "off", "low" or "high", "logic 1" or "logic 0", etc. The output of an analog sensor can take any one of an infinite number of voltage or current levels. It is thus said to be *continuously variable.* Table 21.3 provides details of some common types of sensor.

Table 21.3: Some examples of output transducers

Physical quantity	Output transducer	Notes
Angular position	Resistive rotary position sensor (see Figure 21.5)	Rotary track potentiometer with linear law produces analog voltage proportional to angular position.
	Optical shaft encoder	Encoded disk interposed between optical transmitter and receiver (infrared LED and photodiode or photo-transistor).
Angular velocity	Tachogenerator	Small DC generator with linear output characteristic. Analog output voltage proportional to shaft speed.
	Toothed rotor tachometer	Magnetic pick-up responds to the movement of a toothed ferrous disk. The pulse repetition frequency of the output is proportional to the angular velocity.
Flow	Rotating vane flow sensor (see Figure 21.9)	Turbine rotor driven by fluid. Turbine interrupts infrared beam. Pulse repetition frequency of output is proportional to flow rate.
Linear position	Resistive linear position sensor	Linear track potentiometer with linear law produces analog voltage proportional to linear position. Limited linear range.
	Linear variable differential transformer (LVDT)	Miniature transformer with split secondary windings and moving core attached to a plunger. Requires AC excitation and phase-sensitive detector.
	Magnetic linear position sensor	Magnetic pick-up responds to movement of a toothed ferrous track. Pulses are counted as the sensor moves along the track.
Light level	Photocell	Voltage-generating device. The analog output voltage produced is proportional to light level.
	Light dependent resistor (LDR) (see Figure 21.8)	An analog output voltage results from a change of resistance within a cadmium sulphide (CdS) sensing element. Usually connected as part of a potential divider or bridge.

(Continued)

Table 21.3: Some examples of output transducers (Cont'd)

Physical quantity	Output transducer	Notes
	Photodiode (see Fig, 21.8)	Two-terminal device connected as a current source. An analog output voltage is developed across a series resistor of appropriate value.
	Phototransistor (see Figure 21.8)	Three-terminal device connected as a current source. An analog output voltage is developed across a series resistor of appropriate value.
Liquid level	Float switch (see Figure 21.7)	Simple switch element which operates when a particular level is detected.
	Capacitive proximity switch	Switching device which operates when a particular level is detected. Ineffective with some liquids.
	Diffuse scan proximity switch	Switching device which operates when a particular level is detected. Ineffective with some liquids.
Pressure	Microswitch pressure sensor (see Figure 21.4)	Microswitch fitted with actuator mechanism and range setting springs. Suitable for high-pressure applications.
	Differential pressure vacuum switch	Microswitch with actuator driven by a diaphragm. May be used to sense differential pressure. Alternatively, one chamber may be evacuated and the sensed pressure applied to a second input.
	Piezo-resistive pressure sensor	Pressure exerted on diaphragm causes changes of resistance in attached piezo-resistive transducers. Transducers are usually arranged in the form of a four active element bridge which produces an analog output voltage.
Proximity	Reed switch (see Figure 21.4)	Reed switch and permanent magnet actuator. Only effective over short distances.
	Inductive proximity switch	Target object modifies magnetic field generated by the sensor. Only suitable for metals (nonferrous metals with reduced sensitivity).
	Capacitive proximity switch	Target object modifies electric field generated by the sensor. Suitable for metals, plastics, wood, and some liquids and powders.

(Continued)

Table 21.3: Some examples of output transducers (Cont'd)

Physical quantity	Output transducer	Notes
	Optical proximity switch (see Figure 21.4)	Available in diffuse and through scan types. Diffuse scan types require reflective targets. Both types employ optical transmitters and receivers (usually infrared emitting LEDs and photo-diodes or photo-transistors). Digital input port required.
Strain	Resistive strain gauge	Foil type resistive element with polyester backing for attachment to body under stress. Normally connected in full bridge configuration with temperature-compensating gauges to provide an analog output voltage.
	Semiconductor strain gauge	Piezo-resistive elements provide greater outputs than comparable resistive foil types. More prone to temperature changes and also inherently nonlinear.
Temperature	Thermocouple (see Figure 21.2)	Small e.m.f. generated by a junction between two dissimilar metals. For accurate measurement, requires compensated connecting cables and specialized interface.
	Thermistor (see Figure 21.6)	Usually connected as part of a potential divider or bridge. An analog output voltage results from resistance changes within the sensing element.
	Semiconductor temperature sensor (see Figure 21.6)	Two-terminal device connected as a current source. An analog output voltage is developed across a series resistor of appropriate value.
Weight	Load cell	Usually comprises four strain gauges attached to a metal frame. This assembly is then loaded and the analog output voltage produced is proportional to the weight of the load.
Vibration	Electromagnetic vibration sensor	Permanent magnet seismic mass suspended by springs within a cylindrical coil. The frequency and amplitude of the analog output voltage are respectively proportional to the frequency and amplitude of vibration.

Figure 21.5: Resistive linear position sensor

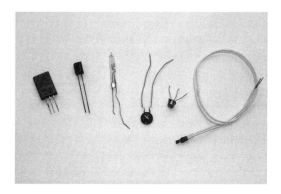

Figure 21.6: Various temperature and gas sensors

Figure 21.7: Liquid level float switch

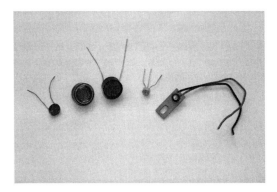

Figure 21.8: Various optical and light sensors

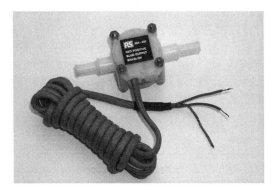

Figure 21.9: Liquid flow sensor (digital output)

Figure 21.10: Contactless joystick

21.4 Switches

Switches can be readily interfaced to electronic circuits in order to provide manual inputs to the system. Simple toggle and push-button switches are generally available with *normally open* (NO), *normally closed* (NC), or *changeover* contacts. In the latter case, the switch may be configured as either an NO or an NC type, depending upon the connections used.

Toggle, lever, rocker, rotary, slide, and push-button types are all commonly available in a variety of styles. Illuminated switches and key switches are also available for special applications. The choice of switch type will obviously depend upon the application and operational environment.

An NO switch or push-button may be interfaced to a logic circuit using nothing more than a single *pull-up resistor* as shown in Figure 21.11.

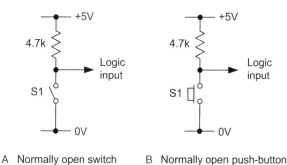

A Normally open switch   B Normally open push-button

Figure 21.11: Interfacing a normally open switch or push-button to a digital input port

The relevant bit of the input will then return 0 when the switch contacts are closed (i.e., when the switch is operated or where the pushbutton is depressed). When the switch is inactive, the logic input will return 1.

Unfortunately, this simple method of interfacing has a limitation when the state of a switch is being sensed regularly (e.g., during program execution). However, a typical application which is unaffected by this problem is that of using one or more PCB mounted switches (e.g., a DIL switch package) to configure a logic system in one of a number of preset modes. In such cases, the switches would be set once only and the software would read the state of the switches and use the values returned to initially configure the system. Thereafter, the state of the switches would then only be changed in order to modify the operational parameters of the system (e.g., when changing input sensors or output transducers). A typical DIL switch input interface to a digital input port is shown in Figure 21.12.

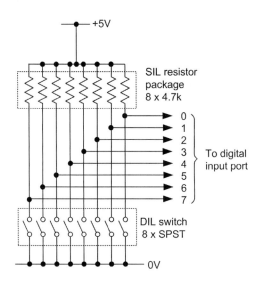

Figure 21.12: Interfacing a DIL switch input to a digital input port

As mentioned earlier, the simple circuit of Figure 21.11 is unsuitable for use when the state of the switch is regularly changing. The reason for this is that the switching action of most switches is far from "clean" (i.e., the switch contacts make and break several times whenever the switch is operated). This may not be a problem when the state of a switch remains static during program execution but it can give rise to serious problems when dealing with, for example, an operator switch bank or keypad.

The contact "bounce" that occurs when a switch is operated results in rapid making and breaking of the switch until it settles into its new state. Figure 21.13 shows the waveform generated by the simple switch input circuit of Figure 21.11 as the contacts close. The spurious states can cause problems if the switch is sensed during the period in which the switch contacts are in motion, and hence steps must be taken to minimize the effects of bounce. This may be achieved by using some extra circuitry in the form of a *debounce circuit* or by including appropriate software delays (of typically 4 to 20 ms) so that spurious switching states are ignored. We shall discuss these two techniques separately.

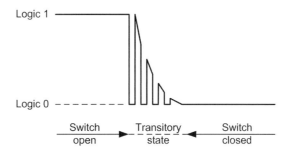

Figure 21.13: Typical waveform produced by a switch closure

Immunity to transient switching states is generally enhanced by the use of active-low inputs (i.e., a logic 0 state at the input is used to assert the condition required). The debounce circuit shown in Figure 21.14 is adequate for most toggle, slide and push-button type switches. The value of the 100Ω resistor takes into account the low-state sink current required by IC1 (normally 1.6 mA for standard TTL and 400 μA for LS-TTL). This resistor should not be allowed to exceed approximately 470Ω in order to maintain a valid logic 0 input state. The values quoted generate an approximate 1 ms delay (during which the switch contacts will have settled into their final state). It should be noted that, on power-up, this circuit generates a logic 1 level for approximately 1 ms before the output reverts to a logic 0 in the inactive state. The circuit obeys the following state table:

Switch condition	Logic output
closed	1
open	0

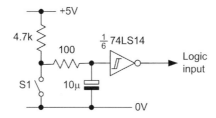

Figure 21.14: Simple debounce circuit

An alternative, but somewhat more complex, switch de-bouncing arrangement is shown in Figure 21.15. Here a single-pole double-throw (SPDT) changeover switch is employed. This arrangement has the advantage of providing complementary outputs and it obeys the following state table:

Switch condition	Q
Q → 1	1
Q → 0	0

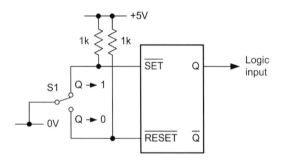

Figure 21.15: Debounce circuit based on an RS bistable

Rather than use an integrated circuit RS bistable in the configuration of Figure 21.15 it is often expedient to make use of "spare" two-input NAND or NOR gates arranged to form bistables using the circuits shown in Figs 21.16(A) and 21.16(B), respectively. Figure 21.17 shows a rather neat extension of this theme in the form of a touch-operated switch. This arrangement is based on a 4011 CMOS quad two-input NAND gate (though only two gates of the package are actually used in this particular configuration).

Finally, it is some times necessary to generate a latching action from a normally-open push-button switch. Figure 21.18 shows an arrangement in which a 74LS73 JK bistable is clocked from the output of a debounced switch.

Pressing the switch causes the bistable to change state. The bistable then remains in that state until the switch is depressed a second time. If desired, the complementary outputs provided by the bistable may be used to good effect by allowing the unused output to drive an LED. This will become illuminated whenever the Q output is high.

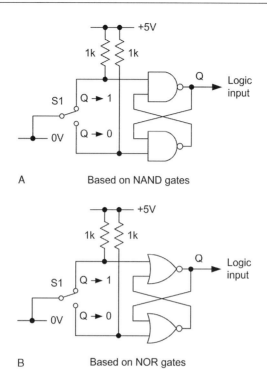

A   Based on NAND gates

Figure 21.16: Alternative switch debounce circuits: (A) based on NAND gates; (B) based on NOR gates

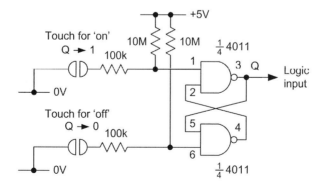

Figure 21.17: Touch-operated switch

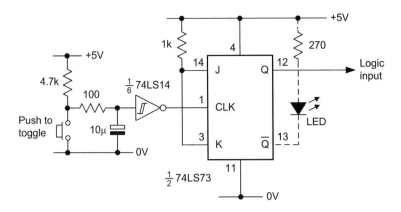

Figure 21.18: Latching action switch

21.5   Semiconductor Temperature Sensors

Semiconductor temperature sensors are ideal for a wide range of temperature-sensing applications. The popular AD590 semiconductor temperature sensor, for example, produces an output current that is proportional to absolute temperature and which increases at the rate of 1 μA/K. The characteristic of the device is illustrated in Figure 21.19. The AD590 is laser trimmed to produce a current of 298.2 μA (±2.5 μA) at a temperature of 298.2°C (i.e., 25°C). A typical interface between the AD590 and an analog input is shown in Figure 21.20.

21.6   Thermocouples

Thermocouples comprise a junction of dissimilar metals which generate an e.m.f. proportional to the temperature differential which exists between the measuring junction and a reference junction. Since the measuring junction is usually at a greater temperature than that of the reference junction, it is sometimes referred to as the *hot junction*. Furthermore, the reference junction (i.e., the *cold junction*) is often omitted in which case the sensing junction is simply terminated at the signal conditioning board. This board is usually maintained at, or near, normal room temperatures.

Thermocouples are suitable for use over a very wide range of temperatures (from −100°C to +1100°C). Industry standard "type K" thermocouples comprise a positive arm (conventionally colored brown) manufactured from nickel/chromium alloy while the negative arm (conventionally colored blue) is manufactured from nickel/aluminum.

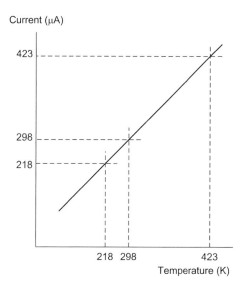

Figure 21.19: AD590 semiconductor temperature sensor characteristic

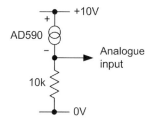

Figure 21.20: Typical input interface for the AD590 temperature sensor (the output voltage will increase at the rate of 10 mV per °C

The characteristic of a *type K thermocouple* is defined in BS 4937 Part 4 of 1973 (International Thermocouple Reference Tables) and this standard gives tables of e.m.f. versus temperature over the range 0°C to +1100°C. In order to minimize errors, it is usually necessary to connect thermocouples to appropriate signal conditioning using compensated cables and matching connectors. Such cables and connectors are available from a variety of suppliers and are usually specified for use with type K thermocouples. A selection of typical thermocouple probes for high temperature measurement was shown earlier in Figure 21.2.

21.7   Threshold Detection

Analog sensors are sometimes used in situations where it is only necessary to respond to a pre-determined threshold value. In effect, a two-state digital output is required. In such cases a simple one-bit analog-to-digital converter based on a comparator can be used. Such an arrangement is, of course, very much simpler and more cost-effective than making use of a conventional analog input port!

Simple threshold detectors for light level and temperature are shown in Figure 21.21, 21.22, and 21.24. These circuits produce TTL-compatible outputs suitable for direct connection to a logic circuit or digital input port.

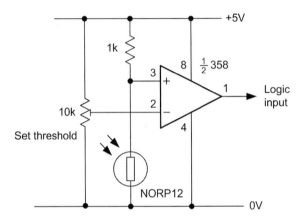

Figure 21.21: Light-level threshold detector based on a light-dependent resistor (LDR)

Figure 21.21 shows a light level threshold detector based on a comparator and light-dependent resistor (LDR). This arrangement generates a logic 0 input whenever the light level exceeds the threshold setting, and vice versa. Figure 21.22 shows how light level can be sensed using a photodiode. This circuit behaves in the same manner as the LDR equivalent but it is important to be aware that circuit achieves peak sensitivity in the near infrared region. Figure 21.23 shows how the spectral response of a typical light-dependent resistor (NORP12) compares with that of a conventional photodiode (BPX48). Note that the BPX48 can also be supplied with an integral daylight filter (BPX48F).

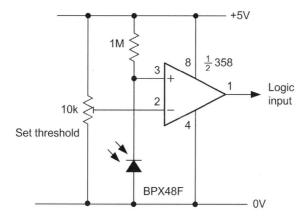

Figure 21.22: Light level threshold detector based on a photodiode

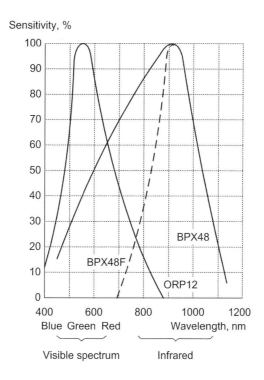

Figure 21.23: Comparison of the spectral response of an LDR and some common photodiodes

Figure 21.24 shows how temperature thresholds can be sensed using the AD590 sensor described earlier. This arrangement generates a logic 0 input whenever the temperature level exceeds the threshold setting, and vice versa.

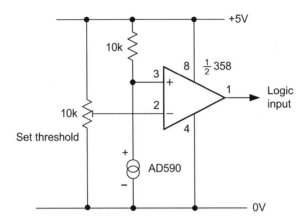

Figure 21.24: Temperature threshold detector based on an AD590 semiconductor temperature sensor

21.8   Outputs

Having dealt at some length with input sensors, we shall now focus our attention on output devices (such as relays, loudspeakers, and LED indicators) and the methods used for interfacing them. Integrated circuit output drivers are available for more complex devices, such as LCD displays and stepper motor. However, many simple applications will only require a handful of components in order to provide an effective interface.

21.9   LED Indicators

Indicators based on light emitting diodes (LEDs) are inherently more reliable than small filament lamps and also consume considerably less power. They are ideal for providing visual status and warning displays. LEDs are available in a variety of styles and colors and "high brightness" types can be employed where high-intensity displays are required.

A typical red LED requires a current of around 10 mA to provide a reasonably bright display and such a device may be directly driven from a buffered digital output port. Different connections are used depending upon whether the LED is to be illuminated for a logic 0 or logic 1 state. Several possibilities are shown in Figure 21.25.

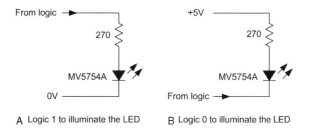

Figure 21.25: Driving an LED from a buffered logic gate or digital I/O port

Where drive current is insufficient to operate an LED, an auxiliary transistor can be used as shown in Figure 21.26. The LED will operate when the output from a logic circuit card is taken to logic 1 and the operating current should be approximately 15 mA (thereby providing a brighter display than the arrangements previously described). The value of LED series resistance will be dependent upon the supply voltage and can be selected from the data shown in Table 21.4.

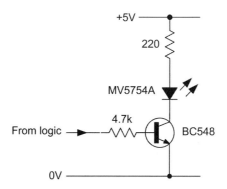

Figure 21.26: Using an auxiliary transistor to drive an LED where current drive is limited

Table 21.4: Typical waveform produced by a switch closure

Voltage	Series resistance (all 0.25W)
3V to 4V	100Ω
4V to 5V	150Ω
5V to 8V	220Ω
8V to 12V	470Ω
12V to 15V	820Ω
15V to 20V	1.2kΩ
20V to 28V	1.5kΩ

21.10   Driving High-Current Loads

Due to the limited output current and voltage capability of most standard logic devices and I/O ports, external circuitry will normally be required to drive anything other than the most modest of loads. Figure 21.27 shows some typical arrangements for operating various types of medium- and high-current load. Figure 21.27(B) shows how an NPN transistor can be used to operate a low-power relay. Where the relay requires an appreciable operating current (say, 150 mA, or more) a plastic encapsulated Darlington power transistor should be used as shown in Figure 21.27(B). Alternatively, a power MOSFET may be preferred, as shown in Figure 21.27(C). Such devices offer very low values of "on" resistance coupled with a very high "off" resistance. Furthermore, unlike conventional bipolar transistors, a power FET will impose a negligible load on an I/O port. Figure 21.27(D) shows a filament lamp driver based on a plastic Darlington power transistor. This circuit will drive lamps rated at up to 24V, 500 mA. Finally, where visual indication of the state of a relay is desirable it is a simple matter to add an LED indicator to the driver stage, as shown in Figure 21.28.

21.11   Audible Outputs

Where simple audible warnings are required, miniature piezoelectric transducers may be used. Such devices operate at low voltages (typically in the range 3V to 15V) and can be interfaced with the aid of a buffer, open-collector logic gate, or transistor. Figure 21.29(A)–(C) show typical interface circuits which produce an audible output when the port output line is at logic 1.

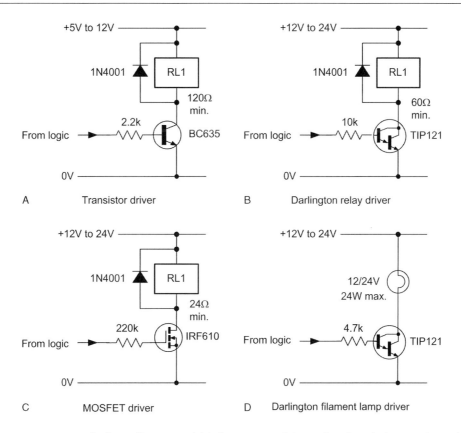

Figure 21.27: Typical medium- and high-current driver circuits: (A) transistor low-current relay driver; (B) Darlington medium/high-current relay driver; (C) MOSFET relay driver; (D) Darlington low-voltage filament lamp driver

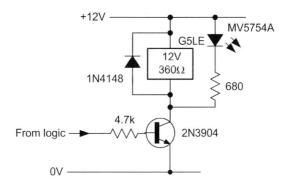

Figure 21.28: Showing how an LED indicator can easily be added to a relay driver

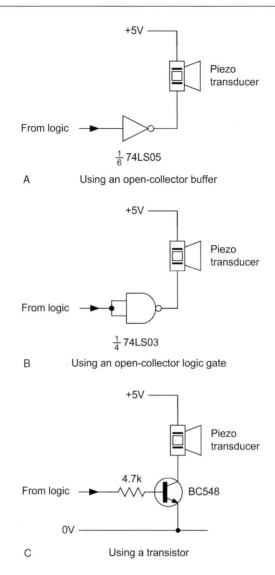

Figure 21.29: Audible output driver circuits

Where a pulsed rather than continuous audible alarm is required, a circuit of the type shown in Figure 21.30 can be employed. This circuit is based on a standard 555 timer operating in astable mode and operates at approximately 1 Hz. A logic 1 from the port output enables the 555 and activates the pulsed audio output.

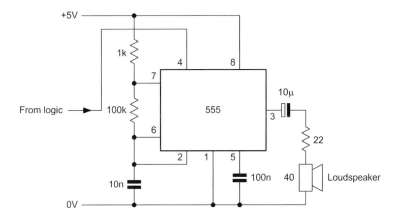

Figure 21.30: Audible alarm circuit based on a 555 astable oscillator and a piezoelectric transducer

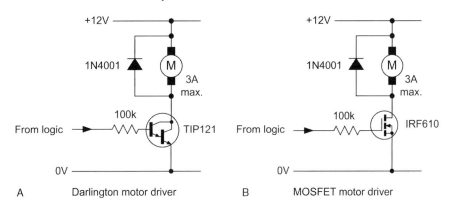

A   Darlington motor driver   B   MOSFET motor driver

Figure 21.31: Audible alarm circuit based on a 555 astable oscillator and a 40Ω loudspeaker

Finally, the circuit shown in Figure 21.31 can be used where a conventional moving-coil loudspeaker is to be used in preference to a piezoelectric transducer. This circuit is again based on the 555 timer and provides a continuous output at approximately 1 kHz whenever the port output is at logic 1.

21.12   Motors

Circuit arrangements used for driving DC motors generally follow the same lines as those described earlier for use with relays. As an example, the circuits shown in Figure 21.32 show how a Darlington driver and a power MOSFET can be used to drive

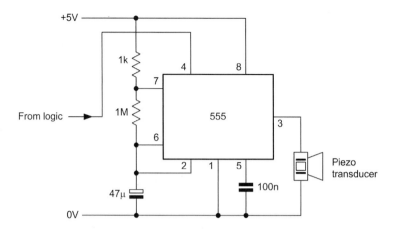

Figure 21.32: Motor driver circuits

a low-voltage DC motor. These circuits are suitable for use with DC motors rated at up to 12V with stalled currents of up to 3A. In both cases, a logic 1 from the output port will operate the motor.

21.13   Driving Mains Connected Loads

Control systems are often used in conjunction with mains connected loads. Modern *solid-state relays* (SSRs) offer superior performance and reliability when compared with conventional relays in such applications. SSRs are available in a variety of encapsulations (including DIL, SIL, flat-pack, and plug-in octal) and may be rated for RMS currents between 1A and 40A.

In order to provide a high degree of isolation between input and output, SSRs are optically coupled. Such devices require minimal input currents (typically 5 mA, or so, when driven from 5V) and they can thus be readily interfaced with an I/O port that offers sufficient drive current. In other cases, it may be necessary to drive the SSR from an unbuffered I/O port using an open-collector logic gate. Typical arrangements are shown in Figure 21.33. Finally, it is important to note that, when an inductive load is to be controlled, a *snubber network* should be fitted, as shown in Figure 21.34.

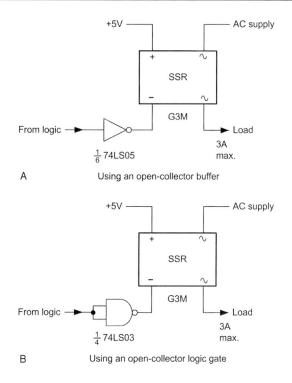

A   Using an open-collector buffer

B   Using an open-collector logic gate

Figure 21.33: Interface circuits for driving solid-state relays

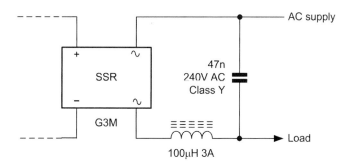

Figure 21.34: Using a snubber circuit with an inductive load

CHAPTER 22

Active Filters

Ron Mancini
Thomas Kugelstadt

22.1   Introduction

What is a filter?

A filter is a device that passes electric signals at certain frequencies or frequency ranges while preventing the passage of others.
—*Webster.*

Filter circuits are used in a wide variety of applications. In the field of telecommunication, band-pass filters are used in the audio frequency range (0 kHz to 20 kHz) for modems and speech processing. High-frequency band-pass filters (several hundred MHz) are used for channel selection in telephone central offices. Data acquisition systems usually require anti-aliasing low-pass filters as well as low-pass noise filters in their preceding signal conditioning stages. System power supplies often use band-rejection filters to suppress the 60-Hz line frequency and high frequency transients.

In addition, there are filters that do not filter any frequencies of a complex input signal, but just add a linear phase shift to each frequency component, thus contributing to a constant time delay. These are called *all-pass* filters.

At high frequencies (> 1 MHz), all of these filters usually consist of passive components such as inductors (L), resistors (R), and capacitors (C). They are then called *LRC* filters.

In the lower frequency range (1 Hz to 1 MHz), however, the inductor value becomes very large and the inductor itself gets quite bulky, making economical production difficult.

In these cases, active filters become important. Active filters are circuits that use an operational amplifier (op-amp) as the active device in combination with some resistors and capacitors to provide an LRC-like filter performance at low frequencies (Figure 22.1).

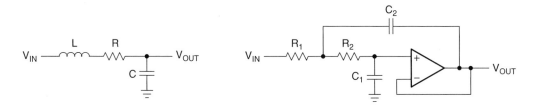

Figure 22.1: Second-order passive low-pass and second-order active low-pass

This chapter covers active filters. It introduces the three main filter optimizations (Butterworth, Tschebyscheff, and Bessel), followed by five sections describing the most common active filter applications: low-pass, high-pass, band-pass, band-rejection, and all-pass filters. Rather than resembling just another filter book, the individual filter sections are written in a cookbook style, thus avoiding tedious mathematical derivations. Each section starts with the general transfer function of a filter, followed by the design equations to calculate the individual circuit components. The chapter closes with a section on practical design hints for single-supply filter designs.

22.2   Fundamentals of Low-Pass Filters

The most simple low-pass filter is the passive RC low-pass network shown in Figure 22.2.

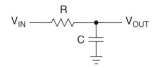

Figure 22.2: First-Order Passive RC Low-Pass

Its transfer function is:

$$A(s) = \frac{\dfrac{1}{RC}}{s + \dfrac{1}{RC}} = \frac{1}{1 + sRC}$$

where the complex frequency variable, $s = j\omega + \sigma$, allows for any time variable signals. For pure sine waves, the damping constant, σ, becomes zero and $s = j\omega$.

For a normalized presentation of the transfer function, s is referred to the filter's corner frequency, or -3 dB frequency, ω_C, and has these relationships:

$$s = \frac{s}{\omega_C} = \frac{j\omega}{\omega_C} = j\frac{f}{f_C} = j\Omega$$

With the corner frequency of the low-pass in Figure 22.2 being $f_C = 1/2\pi RC$, s becomes $s = sRC$ and the transfer function A(s) results in:

$$A(s) = \frac{1}{1+s}$$

The magnitude of the gain response is:

$$|A| = \frac{1}{\sqrt{1+\Omega^2}}$$

For frequencies $\Omega \gg 1$, the roll-off is 20 dB/decade. For a steeper roll-off, **n** filter stages can be connected in series as shown in Figure 22.3. To avoid loading effects, op-amps, operating as impedance converters, separate the individual filter stages.

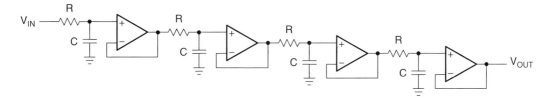

Figure 22.3: Fourth-order passive RC low-pass with decoupling amplifiers

The resulting transfer function is:

$$A(s) = \frac{1}{(1+\alpha_1 s)(1+\alpha_2 s)\ldots(1+\alpha_n s)}$$

In the case that all filters have the same cut-off frequency, f_C, the coefficients become $\alpha_1 = \alpha_2 = \ldots \alpha_n = \alpha = \sqrt{\sqrt[n]{2} - 1}$ and f_C of each partial filter is $1/\alpha$ times higher than f_C the overall filter.

Figure 22.4 shows the results of a fourth-order RC low-pass filter. The roll-off of each partial filter (Curve 1) is –20 dB/decade, increasing the roll-off of the overall filter (Curve 2) to 80 dB/decade.

NOTE

Filter response graphs plot gain versus the normalized frequency axis $\Omega(\Omega = f/f_C)$.

The corner frequency of the overall filter is reduced by a factor of $\alpha \approx 2.3$ times versus the –3 dB frequency of partial filter stages.

In addition, Figure 22.4 shows the transfer function of an ideal fourth-order low-pass function (Curve 3).

In comparison to the ideal low-pass, the RC low-pass lacks in the following characteristics:

- The pass-band gain varies long before the corner frequency, f_C; thus, amplifying the upper pass-band frequencies less than the lower pass-band.

- The transition from the pass-band into the stop-band is not sharp, but happens gradually, moving the actual 80-dB roll off by 1.5 octaves above f_C.

- The phase response is not linear; thus, increasing the amount of signal distortion significantly.

The gain and phase response of a low-pass filter can be optimized to satisfy *one* of the following three criteria:

1) A maximum pass-band flatness,

2) An immediate pass-band to stop-band transition,

3) A linear phase response.

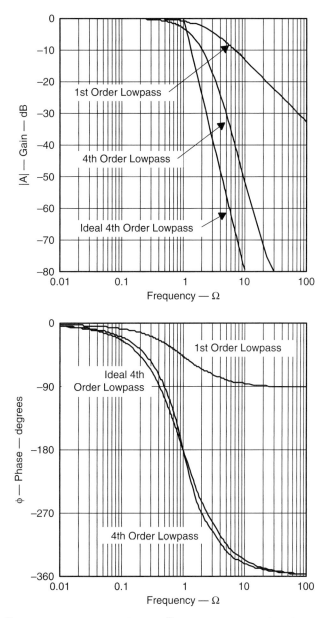

Note: Curve 1: 1st-order partial low-pass filter, Curve 2: 4th-order overall low-pass filter, Curve 3: Ideal 4th-order low-pass filter

Figure 22.4: Frequency and phase responses of a fourth-order passive RC low-pass filter

For that purpose, the transfer function must allow for complex poles and needs to be of the following type:

$$A(s) = \frac{A_0}{(1 + a_1 s + b_1 s^2)(1 + a_2 s + b_2 s^2) \ldots (1 + a_n s + b_n s^2)} = \frac{A_0}{\prod_i (1 + a_i s + b_i s^2)}$$

where A_0 is the pass-band gain at DC, and a_i and b_i are the filter coefficients.

Since the denominator is a product of quadratic terms, the transfer function represents a series of cascaded second-order low-pass stages, with a_i and b_i being positive real coefficients. These coefficients define the complex pole locations for each second-order filter stage; thus, determining the behavior of its transfer function.

The following three types of predetermined filter coefficients are available listed in table format in Section 22.9:

- The Butterworth coefficients, optimizing the pass-band for maximum flatness.

- The Tschebyscheff coefficients, sharpening the transition from pass-band into the stop-band.

- The Bessel coefficients, linearizing the phase response up to f_C.

The transfer function of a passive RC filter does not allow further optimization, due to the lack of complex poles. The only possibility to produce conjugate complex poles using passive components is the application of LRC filters. However, these filters are mainly used at high frequencies. In the lower frequency range (< 10 MHz) the inductor values become very large and the filter becomes uneconomical to manufacture. In these cases active filters are used.

Active filters are RC networks that include an active device, such as an operational amplifier (op-amp).

Section 22.3 shows that the products of the RC values and the corner frequency must yield the predetermined filter coefficients a_i and b_i, to generate the desired transfer function.

The following paragraphs introduce the most commonly used filter optimizations.

22.2.1 *Butterworth Low-Pass Filters*

The Butterworth low-pass filter provides maximum pass-band flatness. Therefore, a Butterworth low-pass is often used as anti-aliasing filter in data converter applications where precise signal levels are required across the entire pass-band.

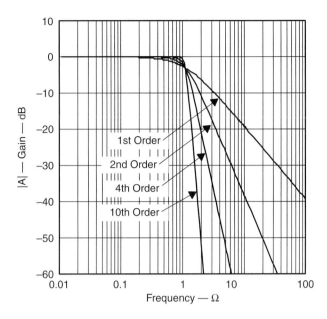

Figure 22.5: Amplitude responses of Butterworth low-pass filters

Figure 22.5 plots the gain response of different orders of Butterworth low-pass filters versus the normalized frequency axis, $\Omega(\Omega = f/f_C)$; the higher the filter order, the longer the pass-band flatness.

22.2.2   Tschebyscheff Low-Pass Filters

The Tschebyscheff low-pass filters provide an even higher gain roll-off above f_C. However, as Figure 22.6 shows, the pass-band gain is not monotone, but contains ripples of constant magnitude instead. For a given filter order, the higher the pass-band ripples, the higher the filter's roll-off.

With increasing filter order, the influence of the ripple magnitude on the filter roll-off diminishes.

Each ripple accounts for one second-order filter stage. Filters with even order numbers generate ripples above the 0-dB line, while filters with odd order numbers create ripples below 0 dB.

Tschebyscheff filters are often used in filter banks, where the frequency content of a signal is of more importance than a constant amplification.

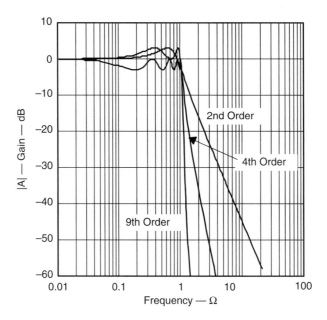

Figure 22.6: Gain responses of Tschebyscheff low-pass filters

22.2.3 Bessel Low-Pass Filters

The Bessel low-pass filters have a linear phase response (Figure 22.7) over a wide frequency range, which results in a constant group delay (Figure 22.8) in that frequency range. Bessel low-pass filters, therefore, provide an optimum square-wave transmission behavior. However, the pass-band gain of a Bessel low-pass filter is not as flat as that of the Butterworth low-pass, and the transition from pass-band to stop-band is by far not as sharp as that of a Tschebyscheff low-pass filter (Figure 22.9).

22.2.4 Quality Factor Q

The quality factor Q is an equivalent design parameter to the filter order n. Instead of designing an nth order Tschebyscheff low-pass, the problem can be expressed as designing a Tschebyscheff low-pass filter with a certain Q.

For band-pass filters, Q is defined as the ratio of the mid frequency, f_m, to the bandwidth at the two −3 dB points:

$$Q = \frac{f_m}{(f_2 - f_1)}$$

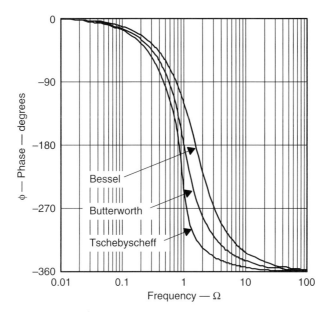

Figure 22.7: Comparison of phase responses of fourth-order low-pass filters

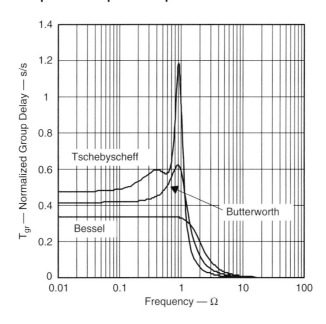

Figure 22.8: Comparison of normalized group delay (Tgr) of fourth-order low-pass filters

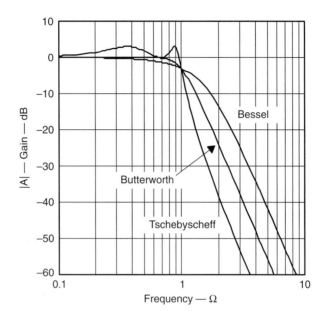

Figure 22.9: Comparison of gain responses of fourth-order low-pass filters

For low-pass and high-pass filters, Q represents the pole quality and is defined as:

$$Q = \frac{\sqrt{b_i}}{a_i}$$

High Qs can be graphically presented as the distance between the 0-dB line and the peak point of the filter's gain response. An example is given in Figure 22.10, which shows a tenth-order Tschebyscheff low-pass filter and its five partial filters with their individual Qs.

The gain response of the fifth-filter stage peaks at 31 dB, which is the logarithmic value of Q_5:

$$Q_5[dB] = 20 \cdot \log Q_5$$

Solving for the numerical value of Q_5 yields:

$$Q_5 = 10^{\frac{31}{20}} = 35.48$$

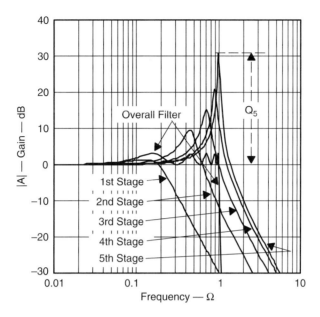

Figure 22.10: Graphical presentation of quality factor Q on a tenth-order Tschebyscheff low-pass Filter with 3-dB pass-band ripple

which is within 1% of the theoretical value of $Q = 35.85$ given in Section 22.9, Table 22.11, last row.

The graphical approximation is good for $Q > 3$. For lower Qs, the graphical values differ from the theoretical value significantly. However, only higher Qs are of concern, since the higher the Q is, the more a filter inclines to instability.

22.2.5 Summary

The general transfer function of a low-pass filter is :

$$A(s) = \frac{A_0}{\prod\limits_{i} (1 + a_i s + b_i s^2)} \qquad (22.1)$$

The filter coefficients a_i and b_i distinguish between Butterworth, Tschebyscheff, and Bessel filters. The coefficients for all three types of filters are tabulated down to the tenth order in Section 22.9 Tables 22.6 through 22.12.

The multiplication of the denominator terms with each other yields an nth order polynomial of S, with n being the filter order.

While n determines the gain roll-off above f_C with $-n \cdot 20$ dB/decade, a_i and b_i determine the gain behavior in the pass-band.

In addition, the ratio $\sqrt{b_i}/ai = Q$ is defined as the pole quality. The higher the Q value, the more a filter inclines to instability.

22.3   Low-Pass Filter Design

Equation 22.1 represents a cascade of second-order low-pass filters. The transfer function of a single stage is:

$$A_i(s) = \frac{A_0}{(1 + a_1 s + b_i s^2)} \qquad (22.2)$$

For a first-order filter, the coefficient b is always zero ($b_1 = 0$), thus yielding:

$$A(s) = \frac{A_0}{1 + a_1 s} \qquad (22.3)$$

The first-order and second-order filter stages are the building blocks for higher-order filters.

Often the filters operate at unity-gain ($A_0 = 1$) to lessen the stringent demands on the op-amp's open-loop gain.

Figure 22.11 shows the cascading of filter stages up to the sixth order. A filter with an even order number consists of second-order stages only, while filters with an odd order number include an additional first-order stage at the beginning.

Figure 22.10 demonstrated that the higher the corner frequency of a partial filter, the higher its Q. Therefore, to avoid the saturation of the individual stages, the filters need to be placed in the order of rising Q values. The Q values for each filter order are listed (in rising order) in Section 22.9, Tables 22.6 through 22.12.

22.3.1   First-Order Low-Pass Filter

Figures 22.12 and 22.13 show a first-order low-pass filter in the inverting and in the noninverting configuration.

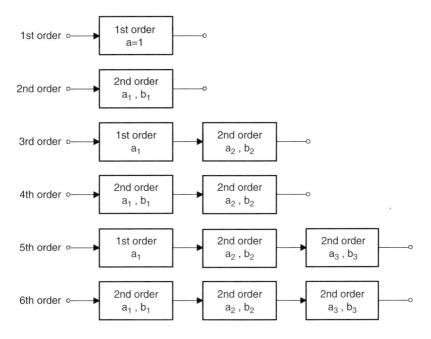

Figure 22.11: Cascading filter stages for higher-order filters

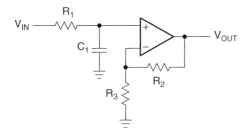

Figure 22.12: First-order noninverting low-pass filter

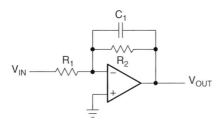

Figure 22.13: First-order inverting low-pass filter

The transfer functions of the circuits are:

$$A(s) = \frac{1 + \dfrac{R_2}{R_3}}{1 + \omega_C R_1 C_1 s} \quad \text{and} \quad A(s) = \frac{-\dfrac{R_2}{R_1}}{1 + \omega_C R_2 C_1 s}$$

The negative sign indicates that the inverting amplifier generates a 180° phase shift from the filter input to the output.

The coefficient comparison between the two transfer functions and Equation 22.3 yields:

$$A_0 = 1 + \frac{R_2}{R_3} \quad \text{and} \quad A_0 = -\frac{R_2}{R_1}$$

$$a_1 = \omega_C R_1 C_1 \quad \text{and} \quad a_1 = \omega_C R_2 C_1$$

To dimension the circuit, specify the corner frequency (f_C), the DC gain (A_0), and capacitor C_1, and then solve for resistors R_1 and R_2:

$$R_1 = \frac{a_1}{2\pi f_C C_1} \quad \text{and} \quad R_2 = \frac{a_1}{2\pi f_C C_1}$$

$$R_2 = R_3(A_0 - 1) \quad \text{and} \quad R_1 = -\frac{R_2}{A_0}$$

The coefficient a1 is taken from one of the coefficient tables, Tables 22.6 through 22.12 in Section 22.9.

Note, that all filter types are identical in their first order and $a_1 = 1$. For higher filter orders, however, $a_1 \neq 1$ because the corner frequency of the first-order stage is different from the corner frequency of the overall filter.

Example 22.1   First-Order Unity-Gain Low-Pass Filter

For a first-order unity-gain low-pass filter with $f_C = 1$ kHz and $C_1 = 47$ nF, R_1 calculates to:

$$R_1 = \frac{a_1}{2\pi f_C C_1} = \frac{1}{2\pi \cdot 10^3 \text{Hz} \cdot 47 \cdot 10^{-9}\text{F}} = 3.38 \text{ k}\Omega$$

However, to design the first stage of a third-order unity-gain Bessel low-pass filter, assuming the same values for f_C and C_1, requires a different value for R_1. In this case,

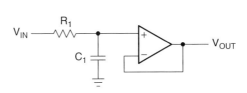

Figure 22.14: First-order noninverting low-pass filter with unity-gain

obtain a1 for a third-order Bessel filter from Table 22.6 in Section 22.9 (Bessel coefficients) to calculate R_1:

$$R_1 = \frac{a_1}{2\pi f_C C_1} = \frac{0.756}{2\pi \cdot 10^3 Hz \cdot 47 \cdot 10^{-9}F} = 2.56\ k\Omega$$

When operating at unity-gain, the noninverting amplifier reduces to a voltage follower (Figure 22.14); thus, inherently providing a superior gain accuracy. In the case of the inverting amplifier, the accuracy of the unity-gain depends on the tolerance of the two resistors, R_1 and R_2.

22.3.2 *Second-Order Low-Pass Filter*

There are two topologies for a second-order low-pass filter, the Sallen-Key and the Multiple Feedback (MFB) topology.

22.3.2.1 *Sallen-Key Topology*

The general Sallen-Key topology in Figure 22.15 allows for separate gain setting via $A_0 = 1 + R_4/R_3$. However, the unity-gain topology in Figure 22.16 is usually applied in filter designs with high-gain accuracy, unity-gain, and low Qs (Q < 3).

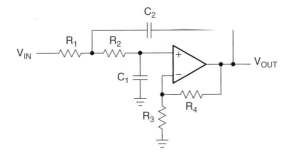

Figure 22.15: General Sallen-Key low-pass filter

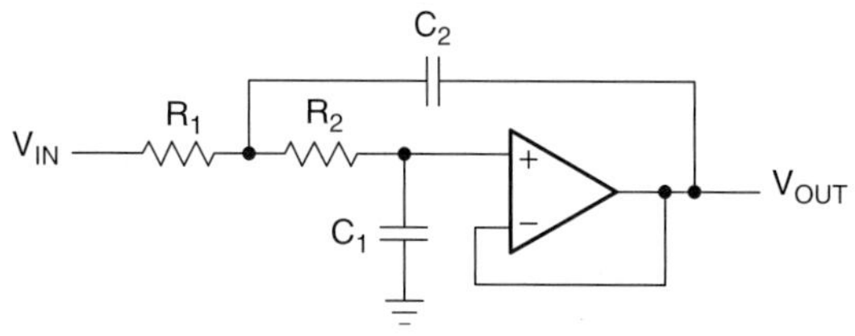

Figure 22.16: Unity-gain Sallen-Key low-pass filter

The transfer function of the circuit in Figure 22.15 is:

$$A(s) = \frac{A_0}{1 + \omega_C[C_1(R_1+R_2)+(1-A_0)\,R_1C_2]s + \omega_C^2\,R_1R_2C_1C_2s^2}$$

For the unity-gain circuit in Figure 22.16 ($A_0 = 1$), the transfer function simplifies to:

$$A(s) = \frac{1}{1 + \omega_CC_1(R_1+R_2)s + \omega_C^2\,R_1R_2C_1C_2s^2}$$

The coefficient comparison between this transfer function and Equation 22.2 yields:

$$A_0 = 1$$
$$a_1 = \omega_CC_1(R_1+R_2)$$
$$b_1 = \omega_C^2R_1R_2C_1C_2$$

Given C_1 and C_2, the resistor values for R_1 and R_2 are calculated through:

$$R_{1,2} = \frac{a_1C_2 \mp \sqrt{a_1^2C_2^2 - 4b_1C_1C_2}}{4\pi f_CC_1C_2}$$

In order to obtain real values under the square root, C_2 must satisfy the following condition:

$$C_2 \geq C_1\frac{4b_1}{a_1^2}$$

Example 22.2   Second-Order Unity-Gain Tschebyscheff Low-Pass Filter

The task is to design a second-order unity-gain Tschebyscheff low-pass filter with a corner frequency of $f_C = 3$ kHz and a 3-dB pass-band ripple.

From Table 22.11 (the Tschebyscheff coefficients for 3-dB ripple), obtain the coefficients a_1 and b_1 for a second-order filter with $a_1 = 1.0650$ and $b_1 = 1.9305$.

Specifying C_1 as 22 nF yields in a C_2 of:

$$C_2 \geq C_1 \frac{4b_1}{a_1^2} = 22 \cdot 10^{-9} \, \text{nF} \cdot \frac{4 \cdot 1.9305}{1.065^2} \cong 150 \, \text{nF}$$

Inserting a_1 and b_1 into the resistor equation for $R_{1,2}$ results in:

$$R_1 = \frac{1.065 \cdot 150 \cdot 10^{-9} - \sqrt{\left(1.065 \cdot 150 \cdot 10^{-9}\right)^2 - 4 \cdot 1.9305 \cdot 22 \cdot 10^{-9} \cdot 150 \cdot 10^{-9}}}{4\pi \cdot 3 \cdot 10^3 \cdot 22 \cdot 10^{-9} \cdot 150 \cdot 10^{-9}} = 1.26 \, \text{k}\Omega$$

and

$$R_2 = \frac{1.065 \cdot 150 \cdot 10^{-9} + \sqrt{\left(1.065 \cdot 150 \cdot 10^{-9}\right)^2 - 4 \cdot 1.9305 \cdot 22 \cdot 10^{-9} \cdot 150 \cdot 10^{-9}}}{4\pi \cdot 3 \cdot 10^3 \cdot 22 \cdot 10^{-9} \cdot 150 \cdot 10^{-9}} = 1.30 \, \text{k}\Omega$$

with the final circuit shown in Figure 22.17.

A special case of the general Sallen-Key topology is the application of equal resistor values and equal capacitor values: $R_1 = R_2 = R$ and $C_1 = C_2 = C$.

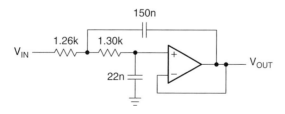

Figure 22.17: Second-order unity-gain Tschebyscheff low-pass with 3-dB ripple

The general transfer function changes to:

$$A(s) = \frac{A_0}{1 + \omega_C RC(3 - A_0)s + (\omega_C RC)^2 s^2} \quad \text{with} \quad A_0 = 1 + \frac{R_4}{R_3}$$

The coefficient comparison with Equation 22.2 yields:

$$a_1 = \omega_C RC(3 - A_0)$$
$$b_1 = (\omega_C RC)^2$$

Given C and solving for R and A_0 results in:

$$R = \frac{\sqrt{b_1}}{2\pi f_C C} \quad \text{and} \quad A_0 = 3 - \frac{a_1}{\sqrt{b_1}} = 3 - \frac{1}{Q}$$

Thus, A_0 depends solely on the pole quality Q and vice versa; Q, and with it the filter type, is determined by the gain setting of A_0:

$$Q = \frac{1}{3 - A_0}$$

The circuit in Figure 22.18 allows the filter type to be changed through the various resistor ratios R_4/R_3.

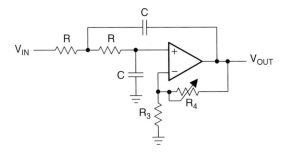

Figure 22.18: Adjustable Second-Order Low-Pass Filter

Table 22.1 lists the coefficients of a second-order filter for each filter type and gives the resistor ratios that adjust the Q.

Table 22.1: Second-order filter coefficients

SECOND-ORDER	BESSEL	BUTTERWORTH	3-dB TSCHEBYSCHEFF
a1	1.3617	1.4142	1.065
b1	0.618	1	1.9305
Q	0.58	0.71	1.3
R4/R3	0.268	0.568	0.234

22.3.2.2 Multiple Feedback Topology

The MFB topology is commonly used in filters that have high Qs and require a high-gain.

The transfer function of the circuit in Figure 22.19 is:

$$A(s) = - \frac{\dfrac{R_2}{R_1}}{1 + \omega_C C_1 \left(R_2 + R_3 + \dfrac{R_2 R_3}{R_1} \right) s + \omega_C^2 C_1 C_2 R_2 R_3 s^2}$$

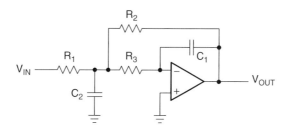

Figure 22.19: Second-order MFB low-pass filter

Through coefficient comparison with Equation 22.2 one obtains the relation:

$$A_0 = - \frac{R_2}{R_1}$$

$$a_1 = \omega_C C_1 \left(R_2 + R_3 + \frac{R_2 R_3}{R_1} \right)$$

$$b_1 = \omega_C^2 C_1 C_2 R_2 R_3$$

Given C_1 and C_2, and solving for the resistors R_1–R_3:

$$R_2 = \frac{a_1 C_2 - \sqrt{a_1^2 C_2^2 - 4 b_1 C_1 C_2 (1 - A_0)}}{4 \pi f_C C_1 C_2}$$

$$R_1 = \frac{R_2}{-A_0}$$

$$R_3 = \frac{b_1}{4 \pi^2 f_C^2 C_1 C_2 R_2}$$

In order to obtain real values for R_2, C_2 must satisfy the following condition:

$$C_2 \geq C_1 \frac{4 b_1 (1 - A_0)}{a_1^2}$$

22.3.3   Higher-Order Low-Pass Filters

Higher-order low-pass filters are required to sharpen a desired filter characteristic. For that purpose, first-order and second-order filter stages are connected in series, so that the product of the individual frequency responses results in the optimized frequency response of the overall filter.

In order to simplify the design of the partial filters, the coefficients ai and bi for each filter type are listed in the coefficient tables (Tables 22.6 through 22.12 in Section 22.9), with each table providing sets of coefficients for the first 10 filter orders.

Example 22.3   Fifth-Order Filter

The task is to design a fifth-order unity-gain Butterworth low-pass filter with the corner frequency $f_C = 50\,\text{kHz}$.

First the coefficients for a fifth-order Butterworth filter are obtained from Table 22.7 Section 22.9:

Table 22.2: Coefficients for fifth-order Butterworth filter

	a_i	b_i
Filter 1	$a_1 = 1$	$b_1 = 0$
Filter 2	$a_2 = 1.6180$	$b_2 = 1$
Filter 3	$a_3 = 0.6180$	$b_3 = 1$

Then dimension each partial filter by specifying the capacitor values and calculating the required resistor values.

First Filter

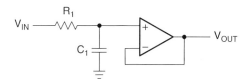

Figure 22.20: First-Order Unity-Gain Low-Pass

With $C_1 = 1$ nF,

$$R_1 = \frac{a_1}{2\pi f_C C_1} = \frac{1}{2\pi \cdot 50 \cdot 10^3 \text{Hz} \cdot 1 \cdot 10^{-9} \text{F}} = 3.18 \text{ k}\Omega$$

The closest 1% value is 3.16 kΩ.

Second Filter

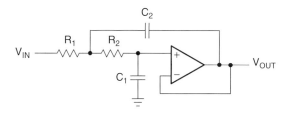

Figure 22.21: Second-order unity-gain Sallen-Key low-pass filter

With $C_1 = 820$ pF,

$$C_2 \geq C_1 \frac{4b_2}{a_2^2} = 820 \cdot 10^{-12} \text{ F} \cdot \frac{4 \cdot 1}{1.618^2} = 1.26 \text{ nF}$$

The closest 5% value is 1.5 nF.

With $C_1 = 820$ pF and $C_2 = 1.5$ nF, calculate the values for R_1 and R_2 through:

$$R_1 = \frac{a_2C_2 - \sqrt{a_2^2C_2^2 - 4b_2C_1C_2}}{4\pi f_C C_1 C_2} \quad \text{and} \quad R_1 = \frac{a_2C_2 + \sqrt{a_2^2C_2^2 - 4b_2C_1C_2}}{4\pi f_C C_1 C_2}$$

and obtain

$$R_1 = \frac{1.618 \cdot 1.5 \cdot 10^{-9} - \sqrt{(1.618 \cdot 1.5 \cdot 10^{-9})^2 - 4 \cdot 1 \cdot 820 \cdot 10^{-12} \cdot 1.5 \cdot 10^{-9}}}{4\pi \cdot 50 \cdot 10^3 \cdot 820 \cdot 10^{-12} \cdot 1.5 \cdot 10^{-9}} = 1.87 \text{ k}\Omega$$

$$R_2 = \frac{1.618 \cdot 1.5 \cdot 10^{-9} + \sqrt{(1.618 \cdot 1.5 \cdot 10^{-9})^2 - 4 \cdot 1 \cdot 820 \cdot 10^{-12} \cdot 1.5 \cdot 10^{-9}}}{4\pi \cdot 50 \cdot 10^3 \cdot 820 \cdot 10^{-12} \cdot 1.5 \cdot 10^{-9}} = 4.42 \text{ k}\Omega$$

R_1 and R_2 are available 1% resistors.

Third Filter

The calculation of the third filter is identical to the calculation of the second filter, except that a_2 and b_2 are replaced by a_3 and b_3, thus resulting in different capacitor and resistor values.

Specify C_1 as 330 pF, and obtain C_2 with:

$$C_2 \geq C_1 \frac{4b_3}{a_3^2} = 330 \cdot 10^{-12}\text{F} \cdot \frac{4 \cdot 1}{0.618^2} = 3.46 \text{ nF}$$

The closest 10% value is 4.7 nF.

With $C_1 = 330$ pF and $C_2 = 4.7$ nF, the values for R1 and R2 are:

- $R_1 = 1.45$ kΩ, with the closest 1% value being 1.47 kΩ

- $R_2 = 4.51$ kΩ, with the closest 1% value being 4.53 kΩ

Figure 22.22 shows the final filter circuit with its partial filter stages.

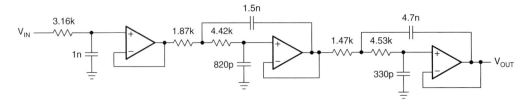

Figure 22.22: Fifth-order unity-gain Butterworth low-pass filter

22.4   High-Pass Filter Design

By replacing the resistors of a low-pass filter with capacitors, and its capacitors with resistors, a high-pass filter is created.

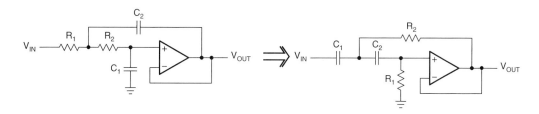

Figure 22.23: Low-pass to high-pass transition through components exchange

To plot the gain response of a high-pass filter, mirror the gain response of a low-pass filter at the corner frequency, $\Omega = 1$, thus replacing Ω with $1/\Omega$ and S with $1/S$ in Equation 22.1.

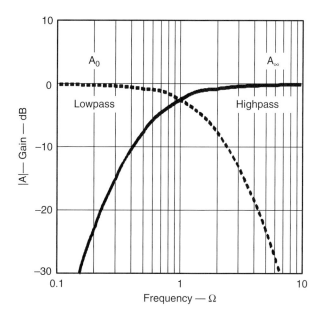

Figure 22.24: Developing the gain response of a high-pass filter

The general transfer function of a high-pass filter is then:

$$A(s) = \frac{A_\infty}{\prod\limits_{i}\left(1 + \dfrac{a_i}{s} + \dfrac{b_i}{s^2}\right)} \tag{22.4}$$

with A_∞ being the pass-band gain.

Since Equation 22.4 represents a cascade of second-order high-pass filters, the transfer function of a single stage is:

$$A_i(s) = \frac{A_\infty}{\left(1 + \dfrac{a_i}{s} + \dfrac{b_i}{s^2}\right)} \tag{22.5}$$

with $b = 0$ for all first-order filters, the transfer function of a first-order filter simplifies to:

$$A(s) = \frac{A_0}{1 + \dfrac{a_i}{s}} \tag{22.6}$$

22.4.1 First-Order High-Pass Filter

Figure 22.25 and 22.26 show a first-order high-pass filter in the noninverting and the inverting configuration.

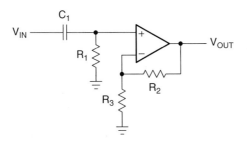

Figure 22.25: First-order noninverting high-pass filter

The transfer functions of the circuits are:

$$A(s) = \frac{1 + \dfrac{R_2}{R_3}}{1 + \dfrac{1}{\omega_C R_1 C_1}\cdot\dfrac{1}{s}} \quad \text{and} \quad A(s) = \frac{1 + \dfrac{R_2}{R_1}}{1 + \dfrac{1}{\omega_C R_1 C_1}\cdot\dfrac{1}{s}}$$

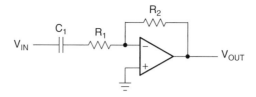

Figure 22.26: First-order inverting high-pass filter

The negative sign indicates that the inverting amplifier generates a 180° phase shift from the filter input to the output.

The coefficient comparison between the two transfer functions and Equation 22.6 provides two different pass-band gain factors:

$$A_\infty = 1 + \frac{R_2}{R_3} \quad \text{and} \quad A_\infty = -\frac{R_2}{R_1}$$

while the term for the coefficient a_1 is the same for both circuits:

$$a_1 = \frac{1}{\omega_C R_1 C_1}$$

To dimension the circuit, specify the corner frequency (f_C), the DC gain (A_∞), and capacitor (C_1), and then solve for R_1 and R_2:

$$R_1 = \frac{1}{2\pi f_C a_1 C_1}$$

$$R_2 = R_3(A_\infty - 1) \quad \text{and} \quad R_2 = -R_1 A_\infty$$

22.4.2 Second-Order High-Pass Filter

High-pass filters use the same two topologies as the low-pass filters: Sallen-Key and Multiple Feedback. The only difference is that the positions of the resistors and the capacitors have changed.

22.4.2.1 Sallen-Key Topology

The general Sallen-Key topology in Figure 22.27 allows for separate gain setting via $A_0 = 1 + R_4/R_3$.

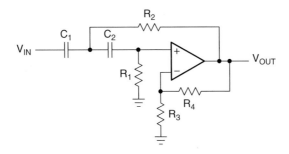

Figure 22.27: General Sallen-Key high-pass filter

The transfer function of the circuit in Figure 22.27 is:

$$A(s) = \frac{\alpha}{1 + \dfrac{R_2(C_1 + C_2) + R_1C_2(1 - \alpha)}{\omega_C R_1 R_2 C_1 C_2} \cdot \dfrac{1}{s} + \dfrac{1}{\omega_C R_1 R_2 C_1 C_2} \cdot \dfrac{1}{s}} \qquad \text{with} \quad \alpha = 1 + \frac{R_4}{R_3}$$

The unity-gain topology in Figure 22.28 is usually applied in low-Q filters with high-gain accuracy.

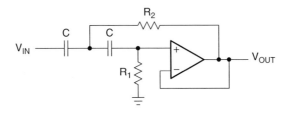

Figure 22.28: Unity-gain Sallen-Key high-pass filter

To simplify the circuit design, it is common to choose unity-gain ($\alpha = 1$), and $C_1 = C_2 = C$. The transfer function of the circuit in Figure 22.28 then simplifies to:

$$A(s) = \frac{1}{1 + \dfrac{2}{\omega_C R_1 C} \cdot \dfrac{1}{s} + \dfrac{1}{\omega_C^2 R_1 R_2 C^2} \cdot \dfrac{1}{s^2}}$$

The coefficient comparison between this transfer function and Equation 22.5 yields:

$$A_\infty = 1$$

$$a_1 = \frac{2}{\omega_C R_1 C}$$

$$b_1 = \frac{1}{\omega_C{}^2 R_1 R_2 C^2}$$

Given C, the resistor values for R_1 and R_2 are calculated through:

$$R_1 = \frac{1}{\pi f_C C a_1}$$

$$R_2 = \frac{1}{4\pi f_C C b_1}$$

22.4.2.2 Multiple Feedback Topology

The MFB topology is commonly used in filters that have high Qs and require a high gain.

To simplify the computation of the circuit, capacitors C_1 and C_3 assume the same value ($C_1 = C_3 = C$) as shown in Figure 22.29.

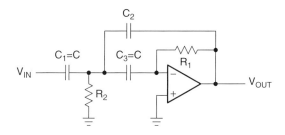

Figure 22.29: Second-order MFB high-pass filter

The transfer function of the circuit in Figure 22.29 is:

$$A(s) = \frac{-\dfrac{C}{C_2}}{1 + \dfrac{2C_2 + C}{\omega_C R_1 C_2 C} \cdot \dfrac{1}{s} + \dfrac{1}{\omega_C^2 R_2 R_1 C_2 C} \cdot \dfrac{1}{s^2}}$$

Through coefficient comparison with Equation 22.5, obtain the following relations:

$$A_\infty = \frac{C}{C_2}$$

$$a_1 = \frac{2C + C_2}{\omega_C R_1 C C_2}$$

$$b_1 = \frac{2C + C_2}{\omega_C R_1 C C_2}$$

Given capacitors C and C_2, and solving for resistors R_1 and R_2:

$$R_1 = \frac{1 - 2A_\infty}{2\pi f_C \cdot C \cdot a_1}$$

$$R_2 = \frac{a_1}{2\pi f_C \cdot b_1 C_2 (1 - 2A_\infty)}$$

The pass-band gain (A_∞) of a MFB high-pass filter can vary significantly due to the wide tolerances of the two capacitors C and C_2. To keep the gain variation at a minimum, it is necessary to use capacitors with tight tolerance values.

22.4.3 Higher-Order High-Pass Filter

Likewise, as with the low-pass filters, higher-order high-pass filters are designed by cascading first-order and second-order filter stages. The filter coefficients are the same ones used for the low-pass filter design, and are listed in the coefficient tables (Tables 22.6 through 22.12 in Section 22.9).

Example 22.4 Third-Order High-Pass Filter with $f_C = 1$ kHz

The task is to design a third-order unity-gain Bessel high-pass filter with the corner frequency $f_C = 1$ kHz. Obtain the coefficients for a third-order Bessel filter from Table 22.6 Section 22.9: and compute each partial filter by specifying the capacitor values and calculating the required resistor values.

Table 22.3: Coefficients for third-order Bessel filter

	a_i	b_i
Filter 1	$a_1 = 0.756$	$b_1 = 0$
Filter 2	$A_2 = 0.9996$	$b_2 = 0.4772$

First Filter
With $C_1 = 100$ nF,

$$R_1 = \frac{1}{2\pi f_C a_1 C_1} = \frac{1}{2\pi \cdot 10^3 \text{Hz} \cdot 0.756 \cdot 100 \cdot 10^{-9}\text{F}} = 2.105 \text{ k}\Omega$$

Closest 1% value is 2.1 kΩ.

Second Filter
With $C = 100$ nF,

$$R_1 = \frac{1}{\pi f_C C a_1} = \frac{1}{\pi \cdot 10^3 \cdot 100 \cdot 10^{-9} \cdot 0.756} = 3.18 \text{ k}\Omega$$

Closest 1% value is 3.16 kΩ.

$$R_2 = \frac{a_1}{4\pi f_C C b_1} = \frac{0.9996}{4\pi \cdot 10^3 \cdot 100 \cdot 10^{-9} \cdot 0.4772} = 1.67 \text{ k}\Omega$$

Closest 1% value is 1.65 kΩ.

Figure 22.30 shows the final filter circuit.

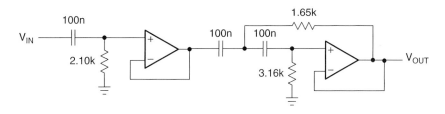

Figure 22.30: *Third-order unity-gain Bessel high-pass*

22.5 Band-pass Filter Design

In Section 22.4, a high-pass response was generated by replacing the term S in the lowpass transfer function with the transformation $1/S$. Likewise, a band-pass characteristic is generated by replacing the S term with the transformation:

$$\frac{1}{\Delta\Omega}\left(s + \frac{1}{s}\right) \tag{22.7}$$

In this case, the pass-band characteristic of a low-pass filter is transformed into the upper pass-band half of a band-pass filter. The upper pass-band is then mirrored at the mid frequency, fm ($\Omega = 1$), into the lower pass-band half.

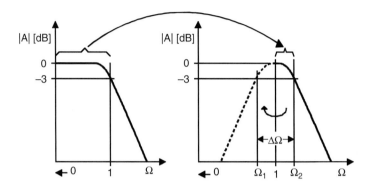

Figure 22.31: Low-pass to bandpass transition

The corner frequency of the low-pass filter transforms to the lower and upper -3 dB frequencies of the band-pass, Ω_1 and Ω_2. The difference between both frequencies is defined as the normalized bandwidth $\Delta\Omega$:

$$\Delta\Omega = \Omega_2 - \Omega_1$$

The normalized mid frequency, where Q = 1, is:

$$\Omega_m = 1 = \Omega_2 \cdot \Omega_1$$

In analogy to the resonant circuits, the quality factor Q is defined as the ratio of the mid frequency (f_m) to the bandwidth (B):

$$Q = \frac{f_m}{B} = \frac{f_m}{f_2 - f_1} = \frac{1}{\Omega_2 - \Omega_1} = \frac{1}{\Delta\Omega} \tag{22.8}$$

The simplest design of a band-pass filter is the connection of a high-pass filter and a lowpass filter in series, which is commonly done in wide-band filter applications. Thus, a first-order high-pass and a first-order low-pass provide a second-order band-pass, while a second-order high-pass and a second-order low-pass result in a fourth-order band-pass response.

In comparison to wide-band filters, narrow-band filters of higher order consist of cascaded second-order band-pass filters that use the Sallen-Key or the Multiple Feedback (MFB) topology.

22.5.1 Second-Order Bandpass Filter

To develop the frequency response of a second-order band-pass filter, apply the transformation in Equation 22.7 to a first-order low-pass transfer function:

$$A(s) = \frac{A_0}{1 + s}$$

Replacing s with $\frac{1}{\Delta\Omega}\left(s + \frac{1}{s}\right)$

yields the general transfer function for a second-order band-pass filter:

$$A(s) = \frac{A_0 \cdot \Delta\Omega \cdot s}{1 + \Delta\Omega \cdot s + s^2} \tag{22.9}$$

When designing band-pass filters, the parameters of interest are the gain at the mid frequency (A_m) and the quality factor (Q), which represents the selectivity of a band-pass filter.

Therefore, replace A_0 with A_m and $\Delta\Omega$ with $1/Q$ (Equation 22.7) and obtain:

$$A(s) = \frac{\dfrac{A_m \cdot s}{Q}}{1 + \dfrac{1}{Q} \cdot s + s^2} \tag{22.10}$$

Figure 22.32 shows the normalized gain response of a second-order band-pass filter for different Qs.

The graph shows that the frequency response of second-order band-pass filters gets steeper with rising Q, thus making the filter more selective.

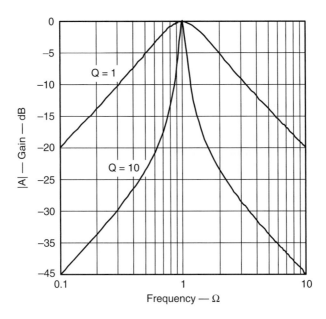

Figure 22.32: Gain response of a second-order bandpass filter

22.5.1.1 Sallen-Key Topology

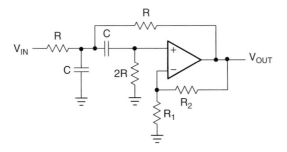

Figure 22.33: Sallen-Key bandpass

The Sallen-Key band-pass circuit in Figure 22.33 has the following transfer function:

$$A(s) = \frac{G \cdot RC\omega_m \cdot s}{1 + RC\omega_m(3 - G) \cdot s + R^2 C^2 \omega_m \cdot s^2}$$

Through coefficient comparison with Equation 22.10, obtain the following equations:

$$\text{mid} - \text{frequency}: \quad f_m = \frac{1}{2\pi RC}$$

$$\text{inner gain}: \quad G = 1 + \frac{R_2}{R_1}$$

$$\text{gain at } f_m: \quad A_m = \frac{G}{3-G}$$

$$\text{filter quality}: \quad Q = \frac{1}{3-G}$$

The Sallen-Key circuit has the advantage that the quality factor (Q) can be varied via the inner gain (G) without modifying the mid frequency (f_m). A drawback is, however, that Q and A_m cannot be adjusted independently.

Care must be taken when G approaches the value of 3, because then A_m becomes infinite and causes the circuit to oscillate.

To set the mid frequency of the band-pass, specify f_m and C and then solve for R:

$$R = \frac{1}{2\pi f_m C}$$

Because of the dependency between Q and A_m, there are two options to solve for R_2: either to set the gain at mid frequency:

$$R_2 = \frac{2A_m - 1}{1 + A_m}$$

or to design for a specified Q:

$$R_2 = \frac{2Q - 1}{Q}$$

22.5.1.2 Multiple Feedback Topology

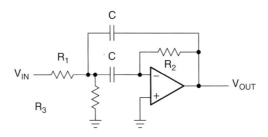

Figure 22.34: MFB Bandpass

The MFB band-pass circuit in Figure 22.34 has the following transfer function:

$$A(s) = \frac{-\dfrac{R_2 R_3}{R_1 + R_3} C\omega_m \cdot s}{1 + \dfrac{2R_1 R_3}{R_1 + R_3} C\omega_m \cdot s + \dfrac{R_1 R_2 R_3}{R_1 + R_3} C^2 \cdot \omega_m{}^2 \cdot s^2}$$

The coefficient comparison with Equation 22.9, yields the following equations:

$$\text{mid} - \text{frequency}: \quad f_m = \frac{1}{2\pi C}\sqrt{\frac{R_1 + R_3}{R_1 R_2 R_3}}$$

$$\text{gain at fm}: \quad -A_m = \frac{R_2}{2R_1}$$

$$\text{filter quality}: \quad Q = \pi f_m R_2 C$$

$$\text{bandwidth}: \quad B = \frac{1}{\pi R_2 C}$$

The MFB band-pass allows to adjust Q, Am, and fm independently. Bandwidth and gain factor do not depend on R_3. Therefore, R_3 can be used to modify the mid frequency without affecting bandwidth, B, or gain, A_m. For low values of Q, the filter can work without R_3, however, Q then depends on A_m via:

$$-A_m = 2Q^2$$

Example 22.5   Second-Order MFB Bandpass Filter with f_m = 1 kHz

To design a second-order MFB band-pass filter with a mid frequency of fm = 1 kHz, a quality factor of Q = 10, and a gain of A_m = –2, assume a capacitor value of C = 100 nF, and solve the previous equations for R_1 through R_3 in the following sequence:

$$R_2 = \frac{Q}{\pi f_m C} = \frac{10}{\pi \cdot 1 \text{ kHz} \cdot 100 \text{ nF}} = 31.8 \text{ k}\Omega$$

$$R_1 = \frac{R_2}{-2A_m} = \frac{31.8 \text{ k}\Omega}{4} = 7.96 \text{ k}\Omega$$

$$R_3 = \frac{-A_m R_1}{2Q^2 + A_m} = \frac{2 \cdot 7.96 \text{ k}\Omega}{200 - 2} = 80.4 \Omega$$

22.5.2   Fourth-Order Bandpass Filter (Staggered Tuning)

Figure 22.32 shows that the frequency response of second-order band-pass filters gets steeper with rising Q. However, there are band-pass applications that require a flat gain response close to the mid frequency as well as a sharp pass-band to stop-band transition. These tasks can be accomplished by higher-order band-pass filters.

Of particular interest is the application of the low-pass to band-pass transformation onto a second-order low-pass filter, since it leads to a fourth-order band-pass filter.

Replacing the *S* term in Equation 22.2 with Equation 22.7 gives the general transfer function of a fourth-order band-pass:

$$A(s) = \frac{\dfrac{s^2 - A_0(\Delta\Omega)^2}{b_1}}{1 + \dfrac{a_1}{b_1}\Delta\Omega \cdot s + \left[2 + \dfrac{(\Delta\Omega)^2}{b_1}\right] \cdot s^2 + \dfrac{a_1}{b_1}\Delta\Omega \cdot s^3 + s^4} \tag{22.11}$$

Similar to the low-pass filters, the fourth-order transfer function is split into two second-or-der band-pass terms. Further mathematical modifications yield:

$$A(s) = \frac{\dfrac{A_{mi}}{Q_i} \cdot \alpha s}{\left[1 + \dfrac{\alpha s}{Q_1} + (\alpha s)^2\right]} \cdot \frac{\dfrac{A_{mi}}{Q_i} \cdot \dfrac{s}{\alpha}}{\left[1 + \dfrac{1}{Q_i}\left(\dfrac{s}{\alpha}\right) + \left(\dfrac{s}{\alpha}\right)^2\right]} \tag{22.12}$$

Equation 22.12 represents the connection of two second-order band-pass filters in series, where

- A_{mi} is the gain at the mid frequency, f_{mi}, of each partial filter.

- Q_i is the pole quality of each filter

- α and $1/\alpha$ are the factors by witch the mid frequencies of the individual filters, f_{m1} and f_{m2}, deriver from the mid frequency, f_m, of the overall bandpass

In a fourth-order band-pass filter with high Q, the mid frequencies of the two partial filters differ only slightly from the overall mid frequency. This method is called *staggered tuning*.

Factor α needs to be determined through successive approximation, using equation 22.13:

$$\alpha^2 + \left[\frac{\alpha \cdot \Delta\Omega \cdot a_1}{b_1(1+\alpha^2)}\right]^2 + \frac{1}{\alpha^2} - 2 - \frac{(\Delta\Omega)^2}{b^1} = 0 \qquad (22.13)$$

with a_1 and b_1 being the second-order low-pass coefficients of the desired filter type.

To simplify the filter design, Table 22.4 lists those coefficients, and provides the α values for three different quality factors, $Q = 1$, $Q = 10$, and $Q = 100$.

Table 22.4: Values of α for different filter types and different Qs

	Bessel				Butterworth				Tschebyscheff		
a_1	1.3617			a_1	1.4142			a_1	1.0650		
b_1	0.6180			b_1	1.0000			b_1	1.9305		
Q	100	10	1	Q	100	10	1	Q	100	10	1
$\Delta\omega$	0.01	0.1	1	$\Delta\Omega$	0.01	0.1	1	$\Delta\Omega$	0.01	0.1	1
α	1.0032	1.0324	1.438	α	1.0035	1.036	1.4426	α	1.0033	1.0338	1.39

After α has been determined, all quantities of the partial filters can be calculated using the following equations

The mid frequency of filter 1 is:

$$f_{m1} = \frac{f_m}{\alpha} \qquad (22.14)$$

the mid frequency of filter 2 is:

$$f_{m2} = f_m \cdot \alpha \qquad (22.15)$$

with f_m being the mid frequency of the overall fourth-order band-pass filter.

The individual pole quality, Q_i, is the same for both filters:

$$Q_i = Q \cdot \frac{(1 + \alpha^2)b_1}{\alpha \cdot a_1} \qquad (22.16)$$

with Q being the quality factor of the overall filter.

The individual gain (A_{mi}) at the partial mid frequencies, f_{m1} and f_{m2}, is the same for both filters:

$$A_{mi} = \frac{Q_i}{Q} \cdot \sqrt{\frac{A_m}{B_1}} \qquad (22.17)$$

with A_m being the gain at mid frequency, f_m, of the overall filter.

Example 22.6 Fourth-Order Butterworth Bandpass Filter

The task is to design a fourth-order Butterworth band-pass with the following parameters:

- mid frequency· $f_m = 10\,kHz$
- bandwidth, $B = 1000\,Hz$
- and gain, $A_m = 1$

From Table 22.4 the following values are obtained:

- a1 = 1.4142
- b1 = 1
- $\alpha = 1.036$

In accordance with Equations 22.14 and 22.15, the mid frequencies for the partial filters are:

$$f_{mi} = \frac{10\,kHz}{1.036} = 9.653\,kHz \quad \text{and} \quad f_{m2} = 10\,kHz \cdot 1.036 = 10.36\,kHz$$

The overall Q is defined as $Q + f_m/B$, and for this example results in $Q = 10$.

Using Equation 22.16, the Q_i of both filters is:

$$Q_i = 10 \cdot \frac{(1 + 1.036^2) \cdot 1}{1.036 \cdot 1.4142} = 14.15$$

With Equation 22.17, the pass-band gain of the partial filters at f_{m1} and f_{m2} calculates to:

$$A_{mi} = \frac{14.15}{10} \cdot \sqrt{\frac{1}{1}} = 1.415$$

The Equations 22.16 and 22.17 show that Q_i and A_{mi} of the partial filters need to be independently adjusted. The only circuit that accomplishes this task is the MFB band-pass filter in section 22.5.1.2.

To design the individual second-order band-pass filters, specify $C = 10$ nF, and insert the previously determined quantities for the partial filters into the resistor equations of the MFB band-pass filter. The resistor values for both partial filters are calculated below.

Filter 1 :

$$R_{21} = \frac{Q_i}{\pi f_{m1} C} = \frac{14.15}{\pi \cdot 9.653 \text{ kHz} \cdot 10 \text{ nF}} = 46.7 \text{ k}\Omega$$

$$R_{11} = \frac{R_{21}}{-2A_{mi}} = \frac{46.7 \text{ k}\Omega}{-2 \cdot -1.415} = 16.5 \text{ k}\Omega$$

$$R_{11} = \frac{-A_{mi}R_{11}}{2Q_i^2 + A_{mi}} = \frac{1.415 \cdot 16.5 \text{ k}\Omega}{2 \cdot 14.15^2 + 1.415} = 58.1 \text{ }\Omega$$

Filter 2 :

$$R_{22} = \frac{14.15}{\pi \cdot 10.36 \text{ kHz} \cdot 10 \text{ nF}} = 43.5 \text{ k}\Omega$$

$$R_{12} = \frac{R_{22}}{-2A_{mi}} = \frac{43.5 \text{ k}\Omega}{-2 \cdot -1.415} = 15.4 \text{ k}\Omega$$

$$R_{32} = \frac{-A_{mi}R_{12}}{2Q_i^2 + A_{mi}} = \frac{1.415 \cdot 15.4 \text{ k}\Omega}{2 \cdot 14.15^2 + 1.415} = 54.2 \text{ }\Omega$$

Figure 22.35 compares the gain response of a fourth-order Butterworth band-pass filter with $Q = 1$ and its partial filters to the fourth-order gain of Example 22.4 with $Q = 10$.

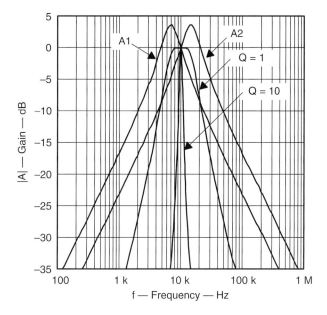

Figure 22.35: Gain Responses of a Fourth-Order Butterworth bandpass and its partial filters

22.6   Band-Rejection Filter Design

A band-rejection filter is used to suppress a certain frequency rather than a range of frequencies.

Two of the most popular band-rejection filters are the active twin-T and the active Wien-Robinson circuit, both of which are second-order filters.

To generate the transfer function of a second-order band-rejection filter, replace the S term of a first-order low-pass response with the transformation in 22.18:

$$\frac{\Delta\Omega}{s + \dfrac{1}{s}} \qquad (22.18)$$

which gives:

$$A(s) = \frac{A_0(1 + s^2)}{1 + \Delta\Omega \cdot s + s^2} \qquad (22.19)$$

Thus the pass-band characteristic of the low-pass filter is transformed into the lower pass-band of the band-rejection filter. The lower pass-band is then mirrored at the mid frequency, f_m ($\Omega = 1$), into the upper pass-band half (Figure 22.36).

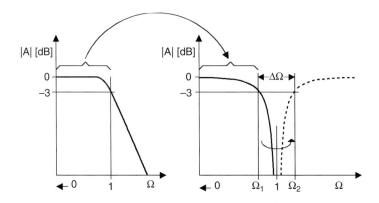

Figure 22.36: Low-pass to band-rejection transition

The corner frequency of the low-pass transforms to the lower and upper -3-dB frequencies of the band-rejection filter Ω_1 and Ω_2. The difference between both frequencies is the normalized bandwidth $\Delta\Omega$:

$$\Delta\Omega = \Omega_{max} - \Omega_{min}$$

Identical to the selectivity of a band-pass filter, the quality of the filter rejection is defined as:

$$Q = \frac{f_m}{B} = \frac{1}{\Delta\Omega}$$

Therefore, replacing Ω in Equation 22.19 with $1/Q$ yields:

$$A(s) = \frac{A_0(1 + s^2)}{1 + \dfrac{1}{Q} \cdot s + s^2} \qquad (22.20)$$

22.6.1 Active Twin-T Filter

The original twin-T filter, shown in Figure 22.37, is a passive RC-network with a quality factor of Q = 0.25. To increase Q, the passive filter is implemented into the feedback loop of an amplifier, thus turning into an active band-rejection filter, shown in Figure 22.38.

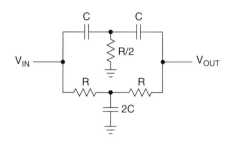

Figure 22.37: Passive Twin-T Filter

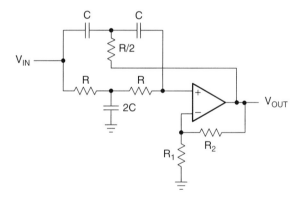

Figure 22.38: Active Twin-T Filter

The transfer function of the active twin-T filter is:

$$A(s) = \frac{k(1 + s^2)}{1 + 2(2 - k) \cdot s + s^2} \qquad (22.21)$$

Comparing the variables of Equation 22.21 with Equation 22.20 provides the equations that determine the filter parameters:

$$\text{mid-frequency:} \quad f_m = \frac{1}{2\pi RC}$$

$$\text{inner gain:} \quad G = 1 + \frac{R_2}{R_1}$$

$$\text{pass-band gain:} \quad A_0 = G$$

$$\text{rejection quality:} \quad Q = \frac{1}{2(2 - G)}$$

The twin-T circuit has the advantage that the quality factor (Q) can be varied via the inner gain (G) without modifying the mid frequency (f_m). However, Q and A_m cannot be adjusted independently.

To set the mid frequency of the band-pass, specify f_m and C, and then solve for R:

$$R = \frac{1}{2\pi f_m C}$$

Because of the dependency between Q and A_m, there are two options to solve for R_2: either to set the gain at mid-frequency:

$$R_2 = (A_0 - 1)R_1$$

or to design for a specific Q:

$$R_2 = R_1 \left(1 - \frac{1}{2Q} \right)$$

22.6.2   Active Wien-Robinson filter

The Wien-Robinson bridge in Figure 22.39 is a passive band-rejection filter with differential output. The output voltage is the difference between the potential of a constant voltage divider and the output of a band-pass filter. Its Q-factor is close to that

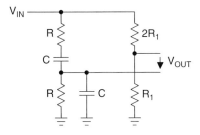

Figure 22.39: Passive Wien-Robinson bridge

of the twin-T circuit. To achieve higher values of Q, the filter is connected into the feedback loop of an amplifier.

The active Wien-Robinson filter in Figure 22.40 has the transfer function:

$$A(s) = -\frac{\dfrac{\beta}{1+\alpha}\left(1+s^2\right)}{1+\dfrac{3}{1+\alpha}\cdot s+s^2} \tag{22.22}$$

$$\text{with } \alpha = \frac{R_2}{R_3} \quad \text{and,} \quad \beta = \frac{R_2}{R_4}$$

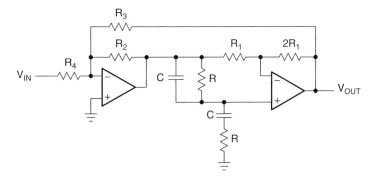

Figure 22.40: Active Wien-Robinson filter

Comparing the variables of Equation 22.22 with Equation 22.20 provides the equations that determine the filter parameters:

$$\text{mid-frequency}: \quad f_m = \frac{1}{2\pi RC}$$

$$\text{pass-band gain}: \quad A_0 = -\frac{\beta}{1+\alpha}$$

$$\text{rejection quality}: \quad Q = \frac{1+\alpha}{3}$$

To calculate the individual component values, establish the following design procedure:

1) Define f_m and C and calculate R with:

$$R = \frac{1}{2\pi f_m C}$$

2) Specify Q and determine α via:

$$\alpha = 3Q - 1$$

3) Specify A_0 and determine β via:

$$\beta = -A_0 \cdot 3Q$$

4) Define R_2 and calculate R_3 and R_4 with:

$$R_3 = \frac{R_2}{\alpha}$$

and,

$$R_4 = \frac{R_2}{\beta}$$

In comparison to the twin-T circuit, the Wien-Robinson filter allows modification of the pass-band gain, A_0, without affecting the quality factor, Q.

If f_m is not completely suppressed due to component tolerances of R and C, a fine-tuning of the resistor $2R_2$ is required.

Figure 22.41 shows a comparison between the filter response of a passive band-rejection filter with $Q = 0.25$, and an active second-order filter with $Q = 1$, and $Q = 10$.

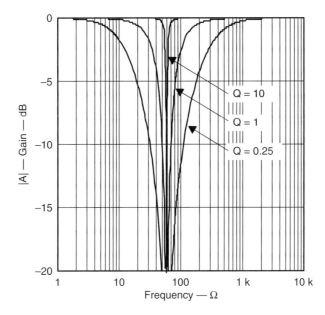

Figure 22.41: Comparison of Q between passive and active band-rejection filters

22.7 All-Pass Filter Design

In comparison to the previously discussed filters, an all-pass filter has a constant gain across the entire frequency range, and a phase response that changes linearly with frequency.

Because of these properties, all-pass filters are used in phase compensation and signal delay circuits.

Similar to the low-pass filters, all-pass circuits of higher order consist of cascaded first-order and second-order all-pass stages. To develop the all-pass transfer function from a low-pass response, replace A_0 with the conjugate complex denominator.

The general transfer function of an allpass is then:

$$A(s) = \frac{\prod_i (1 - a_i s + b_i s^2)}{\prod_i (1 + a_i s + b_i s^2)} \tag{22.23}$$

with a_i and b_i being the coefficients of a partial filter. The all-pass coefficients are listed in Table 22.12 of Section 22.9.

Expressing Equation 22.23 in magnitude and phase yields:

$$A(s) = \frac{\prod\limits_i \sqrt{\left(1 - b_i\Omega^2\right)^2 + a_i^2\ \Omega^2} \cdot e^{-ja}}{\prod\limits_i \sqrt{\left(1 - b_i\Omega^2\right)^2 + a_i^2\ \Omega^2} \cdot e^{+ja}} \tag{22.24}$$

This gives a constant gain of 1, and a phase shift, φ, of:

$$\phi = -2\alpha = -2\sum_i \arctan\frac{a_i\Omega}{1 - b_i\Omega^2} \tag{22.25}$$

To transmit a signal with minimum phase distortion, the all-pass filter must have a constant group delay across the specified frequency band. The group delay is the time by which the all-pass filter delays each frequency within that band.

The frequency at which the group delay drops to $1/\sqrt{2}$ times its initial value is the corner frequency, f_C.
The group delay is defined through:

$$t_{gr} = -\frac{d\phi}{d\omega} \tag{22.26}$$

To present the group delay in normalized form, refer t_{gr} to the period of the corner frequency, T_C, of the all-pass circuit:

$$T_{gr} = \frac{t_{gr}}{T_C} = t_{gr} \cdot f_C = t_{gr} \cdot \frac{\omega_C}{2\pi} \tag{22.27}$$

Substituting t_{gr} through Equation 22.26 gives:

$$T_{gr} = -\frac{1}{2\pi} \cdot \frac{d\phi}{d\Omega} \tag{22.28}$$

Inserting the term in Equation 22.25 into Equation 22.28 and completing the derivation, results in:

$$T_{gr} = \frac{1}{\pi}\sum_i \frac{a_i\left(1 + b_i\Omega^2\right)}{1 + \left(a_i^2 - 2b_1\right) \cdot \Omega^2 + b_i^2\Omega^4} \tag{22.29}$$

Setting $\Omega = 0$ in Equation 22.29 gives the group delay for the low frequencies, $0 < \Omega < 1$, which is:

$$T_{gr0} = \frac{1}{\pi} \sum_i a_i \qquad (22.30)$$

The values for T_{gr0} are listed in Table 22. 12, Section 22.9, from the first to the tenth order.

In addition, Figure 22.42 shows the group delay response versus the frequency for the first ten orders of all-pass filters.

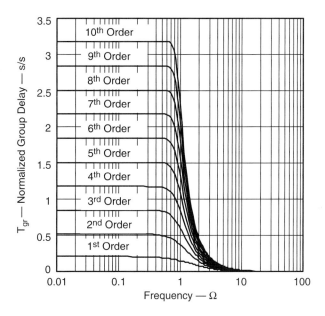

Figure 22.42: Frequency Response of the Group Delay for the First 10 Filter Orders

22.7.1 First-Order All-Pass Filter

Figure 22.43 shows a first-order all-pass filter with a gain of $+1$ at low frequencies and a gain of -1 at high frequencies. Therefore, the magnitude of the gain is 1, while the phase changes from $0°$ to $-180°$.

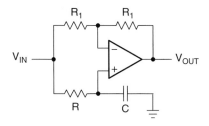

Figure 22.43: First-order all-pass

The transfer function of the circuit above is:

$$A(s) = \frac{1 - RC\omega_C \cdot s}{1 + RC\omega_C \cdot s}$$

The coefficient comparison with Equation 22.23 ($b_1 = 1$), results in:

$$a_i = RC \cdot 2\pi f_C \tag{22.31}$$

To design a first-order all-pass, specify f_C and C and then solve for R:

$$R = \frac{a_i}{2\pi f_C \cdot C} \tag{22.32}$$

Inserting Equation 22.31 into 22.30 and substituting ω_C with Equation 22.27 provides the maximum group delay of a first-order all-pass filter:

$$t_{gr0} = 2RC \tag{22.33}$$

22.7.2 Second-Order All-Pass Filter

Figure 22.44 shows that one possible design for a second-order all-pass filter is to subtract the output voltage of a second-order band-pass filter from its input voltage.

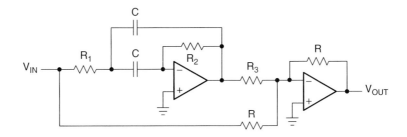

Figure 22.44: Second-order all-pass filter

The transfer function of the circuit in Figure 22.44 is:

$$A(s) = \frac{1 + (2R_1 - \alpha R_2)C\omega_C \cdot s + R_1 R_2 C^2 \omega_C^2 \cdot s^2}{1 + 2R_1 C\omega_C \cdot s + R_1 R_2 C^2 \omega_C^2 \cdot s^2}$$

The coefficient comparison with Equation 22.23 yields:

$$a_1 = 4\pi f_C R_1 C \tag{22.34}$$

$$b_1 = a_1 \pi f_C R_2 C \tag{22.35}$$

$$\alpha = \frac{a_1^2}{b_1} = \frac{R}{R_3} \tag{22.36}$$

To design the circuit, specify f_C, C, and R, and then solve for the resistor values:

$$R_1 = \frac{a_1}{4\pi f_C C} \tag{22.37}$$

$$R_2 = \frac{b_1}{a_1 \pi f_C C} \tag{22.38}$$

$$R_3 = \frac{R}{\alpha} \tag{22.39}$$

Inserting Equation 22.34 into Equation 22.30 and substituting ω_C with Equation 22.27 gives the maximum group delay of a second-order all-pass filter:

$$t_{gr0} = 4R_1 C \tag{22.40}$$

22.7.3   Higher-Order All-Pass Filter

Higher-order all-pass filters consist of cascaded first-order and second-order filter stages.

Example 22.7   2-ms Delay All-Pass Filter

A signal with the frequency spectrum, $0 < f < 1$ kHz, needs to be delayed by 2 ms. To keep the phase distortions at a minimum, the corner frequency of the all-pass filter must be $f_C \geq 1$ kHz.

Equation 22.27 determines the normalized group delay for frequencies below 1 kHz:

$$T_{gr0} = \frac{t_{gr0}}{T_C} = 2 \text{ ms} \cdot 1 \text{ kHz} = 2.0$$

Figure 22.42 confirms that a seventh-order all-pass is needed to accomplish the desired delay. The exact value, however, is $T_{gr0} = 2.1737$. To set the group delay to precisely 2 ms, solve Equation 22.27 for f_C and obtain the corner frequency:

$$f_C = \frac{T_{gr0}}{t_{gr0}} = 1.087 \text{ kHz}$$

To complete the design, look up the filter coefficients for a seventh-order all-pass filter, specify C, and calculate the resistor values for each partial filter.

Cascading the first-order all-pass with the three second-order stages results in the desired seventh-order all-pass filter (Figure 22.45).

22.8   Practical Design Hints

This section introduces DC-biasing techniques for filter designs in single-supply applications, which are usually not required when operating with dual supplies. It also provides recommendations on selecting the type and value range of capacitors and resistors as well as the decision criteria for choosing the correct op-amp.

22.8.1   Filter Circuit Biasing

The filter diagrams in this chapter are drawn for dual supply applications. The op-amp operates from a positive and a negative supply, while the input and the output voltage are referenced to ground (Figure 22.46).

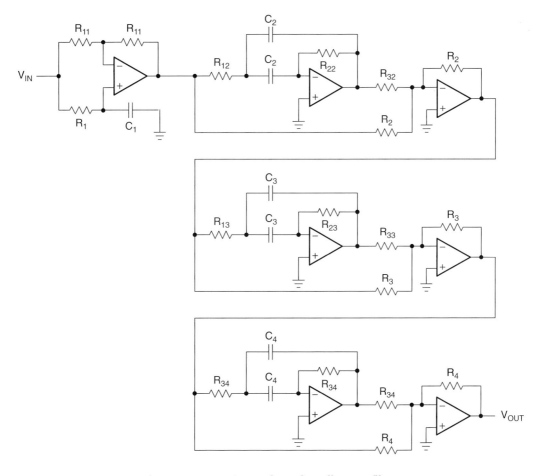

Figure 22.45: Seventh-order all-pass filter

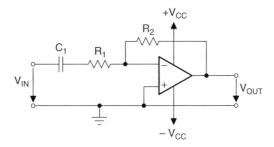

Figure 22.46: Dual-supply filter circuit

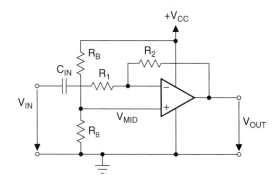

Figure 22.47: Single-supply filter circuit

For the single-supply circuit in Figure 22.47, the lowest supply voltage is ground. For a symmetrical output signal, the potential of the noninverting input is level-shifted to midrail.

The coupling capacitor, CIN in Figure 22.47, AC-couples the filter, blocking any unknown DC level in the signal source. The voltage divider, consisting of the two equal-bias resistors R_B, divides the supply voltage to V_{MID} and applies it to the inverting op-amp input.

For simple filter input structures, passive RC networks often provide a low-cost biasing solution. In the case of more complex input structures, such as the input of a second-order low-pass filter, the RC network can affect the filter characteristic. Then it is necessary to either include the biasing network into the filter calculations, or to insert an input buffer between biasing network and the actual filter circuit, as shown in Figure 22.48.

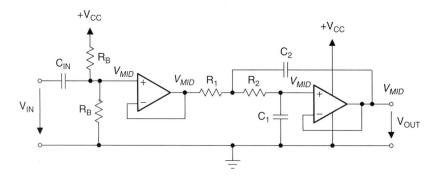

Figure 22.48: Biasing a Sallen-Key low-pass

C_{IN} AC-couples the filter, blocking any DC level in the signal source. V_{MID} is derived from V_{CC} via the voltage divider. The op-amp operates as a voltage follower and as an impedance converter. V_{MID} is applied via the DC path, R_1 and R_2, to the noninverting input of the filter amplifier.

Note that the parallel circuit of the resistors, R_B , together with C_{IN} create a high-pass filter. To avoid any effect on the low-pass characteristic, the corner frequency of the input high-pass must be low versus the corner frequency of the actual low-pass.

The use of an input buffer causes no loading effects on the low-pass filter, thus keeping the filter calculation simple.

In the case of a higher-order filter, all following filter stages receive their bias level from the preceding filter amplifier.

Figure 22.49 shows the biasing of an multiple feedback (MFB) low-pass filter.

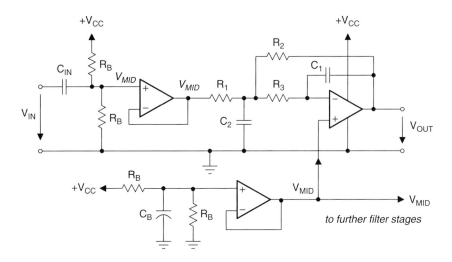

Figure 22.49: Biasing a second-order MFB low-pass filter

The input buffer decouples the filter from the signal source. The filter itself is biased via the noninverting amplifier input. For that purpose, the bias voltage is taken from the output of a V_{MID} generator with low output impedance. The op-amp operates as a difference amplifier and subtracts the bias voltage of the input buffer from the bias voltage of the V_{MID} generator, thus yielding a DC potential of V_{MID} at zero input signal.

A low-cost alternative is to remove the op-amp and to use a passive biasing network instead. However, to keep loading effects at a minimum, the values for R_B must be significantly higher than without the op-amp.

The biasing of a Sallen-Key and an MFB high-pass filter is shown in Figure 22.50.

The input capacitors of high-pass filters already provide the AC-coupling between filter and signal source. Both circuits use the V_{MID} generator from Figure 22.50 for biasing. While the MFB circuit is biased at the noninverting amplifier input, the Sallen-Key high-pass is biased via the only DC path available, which is R_1. In the AC circuit, the input signals travel via the low output impedance of the op-amp to ground.

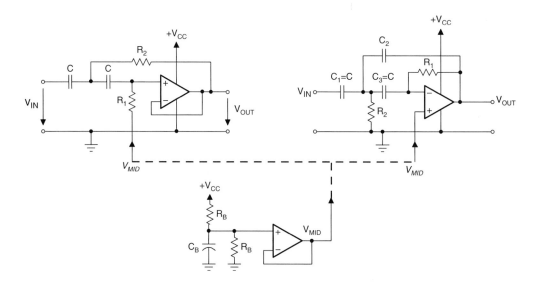

Figure 22.50: Biasing a Sallen-Key and an MFB high-pass filter

22.8.2 Capacitor selection

The tolerance of the selected capacitors and resistors depends on the filter sensitivity and on the filter performance.

Sensitivity is the measure of the vulnerability of a filter's performance to changes in component values. The important filter parameters to consider are the corner frequency, f_C, and Q.

For example, when Q changes by $\pm 2\%$ due to a $\pm 5\%$ change in the capacitance value, then the sensitivity of Q to capacity changes is expressed as: $S \cdot Q/C = 2\%/5\% = 0.4 \%/\%$ The following sensitivity approximations apply to second-order Sallen-Key and MFB filters:

$$ S\frac{Q}{C} \approx S\frac{Q}{R} \approx S\frac{f_C}{C} \approx S\frac{f_C}{R} \approx \pm\, 0.5\frac{\%}{\%} $$

Although 0.5 %/% is a small difference from the ideal parameter, in the case of higherorder filters, the combination of small Q and f_C differences in each partial filter can significantly modify the overall filter response from its intended characteristic.

Figures 22.51 and 22.52 show how an intended eighth-order Butterworth low-pass can turn into a low-pass with Tschebyscheff characteristic mainly due to capacitance changes from the partial filters.

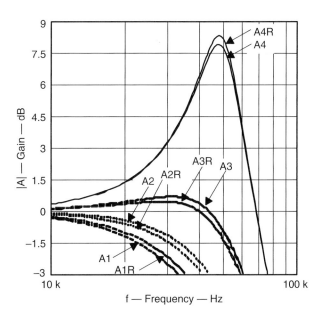

Figure 22.51: Differences in Q and f_C in the partial filters of an eighth-order Butterworth low-pass

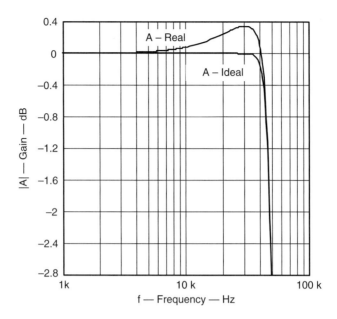

**Figure 22.52: Modification of the intended Butterworth response
to a Tschebyscheff-type characteristic**

Figure 22.51 shows the differences between the ideal and the actual frequency responses of the four partial filters. The overall filter responses are shown in Figure 22.52.

The difference between ideal and real response peaks with 0.35 dB at approximately 30 kHz, which is equivalent to an enormous 4.1% gain error can be seen.

If this filter is intended for a data acquisition application, it could be used at best in a 4-bit system. In comparison, if the maximum full-scale error of a 12-bit system is given with ½ LSB, then maximum pass-band deviation would be – 0.001 dB, or 0.012%.

To minimize the variations of f_C and Q, NPO (COG) ceramic capacitors are recommended for high-performance filters. These capacitors hold their nominal value over a wide temperature and voltage range. The various temperature characteristics of ceramic capacitors are identified by a three-symbol code such as: COG, X7R, Z5U, and Y5V.

COG-type ceramic capacitors are the most precise. Their nominal values range from 0.5 pF to approximately 47 nF with initial tolerances from ±0.25 pF for smaller values and up to ±1% for higher values. Their capacitance drift over temperature is typically 30ppm/°C.

X7R-type ceramic capacitors range from 100 pF to 2.2 μF with an initial tolerance of +1% and a capacitance drift over temperature of ±15%.

For higher values, tantalum electrolytic capacitors should be used.

Other precision capacitors are silver mica, metallized polycarbonate, and for high temperatures, polypropylene or polystyrene.

Since capacitor values are not as finely subdivided as resistor values, the capacitor values should be defined prior to selecting resistors. If precision capacitors are not available to provide an accurate filter response, then it is necessary to measure the individual capacitor values, and to calculate the resistors accordingly.

For high performance filters, 0.1% resistors are recommended.

22.8.3 Component Values

Resistor values should stay within the range of 1 kΩ to 100 kΩ. The lower limit avoids excessive current draw from the op-amp output, which is particularly important for single-supply op-amps in power-sensitive applications. Those amplifiers have typical output currents of between 1 mA and 5 mA. At a supply voltage of 5V, this current translates to a minimum of 1 kΩ.

The upper limit of 100 kΩ is to avoid excessive resistor noise.

Capacitor values can range from 1 nF to several μF. The lower limit avoids coming too close to parasitic capacitances. If the common-mode input capacitance of the op-amp, used in a Sallen-Key filter section, is close to 0.25% of C1, (C1/400), it must be considered for accurate filter response. The MFB topology, in comparison, does not require in-put-capacitance compensation.

22.8.4 Op-Amp Selection

The most important op-amp parameter for proper filter functionality is the unity-gain bandwidth. In general, the open-loop gain (A_{OL}) should be 100 times (40 dB above) the peak gain (Q) of a filter section to allow a maximum gain error of 1%.

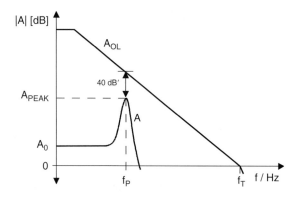

Figure 22.53: Open-loop gain (A_{OL}) and filter response (A)

The following equations are good rules of thumb to determine the necessary unity-gain bandwidth of an op-amp for an individual filter section.

1)   First-order filter:

$$f_T = 100 \cdot \text{Gain} \cdot f_C$$

2)   Second-order filter ($Q < 1$):

$$f_T = 100 \cdot \text{Gain} \cdot f_c \cdot k_i \quad \text{with} \quad k_i = \frac{f_{ci}}{f_c}$$

3)   Second-order filter ($Q > 1$):

$$f_T = 100 \cdot \text{Gain} \cdot \frac{f_c}{a_i} \sqrt{\frac{Q_i^2 - 0.5}{Q_i^2 - 0.25}}$$

For example, a fifth-order, 10-kHz, Tschebyscheff low-pass filter with 3-dB pass-band ripple and a DC gain of $A_0 = 2$ has its worst case Q in the third filter section. With $Q_3 = 8.82$ and $a_3 = 0.1172$, the op-amp needs to have a unity-gain bandwidth of:

$$f_T = 100 \cdot 2 \cdot \frac{10 \text{ kHz}}{0.1172} \sqrt{\frac{8.82^2 - 0.5}{8.82^2 - 0.25}} = 17 \text{ MHz}$$

In comparison, a fifth-order unity-gain, 10-kHz, Butterworth low-pass filter has a worst case Q of $Q_3 = 1.62$; $a_3 = 0.618$. Due to the lower Q value, f_T is also lower and calculates to only:

$$f_T = 100 \cdot \frac{10 \text{ kHz}}{0.618} \sqrt{\frac{1.62^2 - 0.5}{1.62^2 - 0.25}} = 1.5 \text{ MHz}$$

Besides good DC performance, low noise, and low signal distortion, another important parameter that determines the speed of an op-amp is the slew rate (SR). For adequate full-power response, the slew rate must be greater than:

$$SR = \pi \cdot V_{PP} \cdot f_C$$

For example, a single-supply, 100-kHz filter with 5 V_{PP} output requires a slew rate of at least:

$$SR = \pi \cdot 5V \cdot 100 \text{ kHz} = 1.57 \frac{V}{\mu s}$$

Texas Instruments offers a wide range of op-amps for high-performance filters in single supply applications. Table 22.5 provides a selection of single-supply amplifiers sorted in order of rising slew rate.

Table 22.5: Single-Supply Op-Amp Selection Guide *(TA = 25°C, VCC = 5V)*

OP-AMP	BW (MHz)	FPR (kHz)	SR (V/s)	V_{IO} (mv)	Noise (nV/Hz)
TLV2721	0.51	11	0.18	0.6	20
TLC2201A	1.8	159	2.5	0.6	8
TLV2771A	4.8	572	9	1.9	21
TLC071	10	1000	16	1.5	7
TLE2141	5.9	2800	45	0.5	10.5
THS4001	270	127 MHz (1VPP)	400	6	7.5

22.9   Filter coefficient tables

The following tables contain the coefficients for the three filter types, Bessel, Butterworth and Tschebyscheff. The Tschebyscheff tables (Table 22.9) are split into categories for the following pass-band ripples: 0.5 dB, 1 dB, 2 dB, and 3 dB.

The table headers consist of the following quantities:

n   is the filter order.

i   is the number of the partial filter.

a_i, b_i are the filter coefficients.

k_i is the ratio of the corner frequency of a partial filter, f_{Ci}, to the corner frequency of the overall filter, f_C. This ratio is used to determine the unity-gain bandwidth of the op-amp, as well as to simplify the test of a filter design by measuring f_{Ci} and comparing it to f_C.

Q_i is the quality factor of the partial filter.

f_i / f_C this ratio is used for test purposes of the allpass filters, where f_i is the frequency, at which the phase is 180° for a second-order filter, respectively 90° for a first-order all-pass.

T_{gr0} is the normalized group delay of the overall all-pass filter.

Table 22.6: Bessel coefficients

n	i	a_i	b_i	$k_i = fc_i/fc$	Q_i
1	1	1.0000	0.0000	1.000	—
2	1	1.3617	0.6180	1.000	0.58
3	1	0.7560	0.0000	1.323	—
	2	0.9996	0.4772	1.414	0.69
4	1	1.3397	0.4889	0.978	0.52
	2	0.7743	0.3890	1.797	0.81
5	1	0.6656	0.0000	1.502	—
	2	1.1402	0.4128	1.184	0.56
	3	0.6216	0.3245	2.138	0.92
6	1	1.2217	0.3887	1.063	0.51
	2	0.9686	0.3505	1.431	0.61
	3	0.5131	0.2756	2.447	1.02
7	1	0.5937	0.0000	1.648	—
	2	1.0944	0.3395	1.207	0.53
	3	0.8304	0.3011	1.695	0.66
	4	0.4332	0.2381	2.731	1.13
8	1	1.1112	0.3162	1.164	0.51
	2	0.9754	0.2979	1.381	0.56
	3	0.7202	0.2621	1.963	0.71
	4	0.3728	0.2087	2.992	1.23
9	1	0.5386	0.0000	1.857	—
	2	1.0244	0.2834	1.277	0.52
	3	0.8710	0.2636	1.574	0.59
	4	0.6320	0.2311	2.226	0.76
	5	0.3257	0.1854	3.237	1.32
10	1	1.0215	0.2650	1.264	0.50
	2	0.9393	0.2549	1.412	0.54
	3	0.7815	0.2351	1.780	0.62
	4	0.5604	0.2059	2.479	0.81
	5	0.2883	0.1665	3.466	1.42

Table 22.7: Butterworth coefficients

n	i	a_i	b_i	$K_i = c_i/f_c$	Q_i
1	1	1.0000	0.0000	1.000	—
2	1	1.4142	1.0000	1.000	0.71
3	1	1.0000	0.0000	1.000	—
	2	1.0000	1.0000	1.272	1.00
4	1	1.8478	1.0000	0.719	0.54
	2	0.7654	1.0000	1.390	1.31
5	1	1.0000	0.0000	1.000	—
	2	1.6180	1.0000	0.859	0.62
	3	0.6180	1.0000	1.448	1.62
6	1	1.9319	1.0000	0.676	0.52
	2	1.4142	1.0000	1.000	0.71
	3	0.5176	1.0000	1.479	1.93
7	1	1.0000	0.0000	1.000	—
	2	1.8019	1.0000	0.745	0.55
	3	1.2470	1.0000	1.117	0.80
	4	0.4450	1.0000	1.499	2.25
8	1	1.9616	1.0000	0.661	0.51
	2	1.6629	1.0000	0.829	0.60
	3	1.1111	1.0000	1.206	0.90
	4	0.3902	1.0000	1.512	2.56
9	1	1.0000	0.0000	1.000	—
	2	1.8794	1.0000	0.703	0.53
	3	1.5321	1.0000	0.917	0.65
	4	1.0000	1.0000	1.272	1.00
	5	0.3473	1.0000	1.521	2.88
10	1	1.9754	1.0000	0.655	0.51
	2	1.7820	1.0000	0.756	0.56
	3	1.4142	1.0000	1.000	0.71
	4	0.9080	1.0000	1.322	1.10
	5	0.3129	1.0000	1.527	3.20

Table 22.8: Tschebyscheff coefficients for 0.5-dB pass-band ripple

n	i	a_i	b_i	$k_i = f_{c_i}/f_c$	Q_i
1	1	1.0000	0.0000	1.000	—
2	1	1.3614	1.3827	1.000	0.86
3	1	1.8636	0.0000	0.537	—
	2	0.0640	1.1931	1.335	1.71
4	1 2	2.6282	3.4341	0.538	0.71
		0.3648	1.1509	1.419	2.94
5	1	2.9235	0.0000	0.342	—
	2	1.3025	2.3534	0.881	1.18
	3	0.2290	1.0833	1.480	4.54
6	1	3.8645	6.9797	0.366	0.68
	2	0.7528	1.8573	1.078	1.81
	3	0.1589	1.0711	1.495	6.51
7	1	4.0211	0.0000	0.249	—
	2	1.8729	4.1795	0.645	1.09
	3	0.4861	1.5676	1.208	2.58
	4	0.1156	1.0443	1.517	8.84
8	1	5.1117	11.9607	0.276	0.68
	2	1.0639	2.9365	0.844	1.61
	3	0.3439	1.4206	1.284	3.47
	4	0.0885	1.0407	1.521	11.53
9	1	5.1318	0.0000	0.195	—
	2	2.4283	6.6307	0.506	1.06
	3	0.6839	2.2908	0.989	2.21
	4	0.2559	1.3133	1.344	4.48
	5	0.0695	1.0272	1.532	14.58
10	1	6.3648	18.3695	0.222	0.67
	2	1.3582	4.3453	0.689	1.53
	3	0.4822	1.9440	1.091	2.89
	4	0.1994	1.2520	1.381	5.61
	5	0.0563	1.0263	1.533	17.99

Table 22.9: Tschebyscheff coefficients for 1-dB pass-band ripple

n	i	a_i	b_i	$k_i = fc_i/fc$	Q_i
1	1	1.0000	0.0000	1.000	—
2	1	1.3022	1.5515	1.000	0.96
3	1	2.2156	0.0000	0.451	—
	2	0.5442	1.2057	1.353	2.02
4	1	2.5904	4.1301	0.540	0.78
	2	0.3039	1.1697	1.417	3.56
5	1	3.5711	0.0000	0.280	—
	2	1.1280	2.4896	0.894	1.40
	3	0.1872	1.0814	1.486	5.56
6	1	3.8437	8.5529	0.366	0.76
	2	0.6292	1.9124	1.082	2.20
	3	0.1296	1.0766	1.493	8.00
7	1	4.9520	0.0000	0.202	—
	2	1.6338	4.4899	0.655	1.30
	3	0.3987	1.5834	1.213	3.16
	4	0.0937	1.0432	1.520	10.90
8	1	5.1019	14.7608	0.276	0.75
	2	0.8916	3.0426	0.849	1.96
	3	0.2806	1.4334	1.285	4.27
	4	0.0717	1.0432	1.520	14.24
9	1	6.3415	0.0000	0.158	—
	2	2.1252	7.1711	0.514	1.26
	3	0.5624	2.3278	0.994	2.71
	4	0.2076	1.3166	1.346	5.53
	5	0.0562	1.0258	1.533	18.03
10	1	6.3634	22.7468	0.221	0.75
	2	1.1399	4.5167	0.694	1.86
	3	0.3939	1.9665	1.093	3.56
	4	0.1616	1.2569	1.381	6.94
	5	0.0455	1.0277	1.532	22.26

Table 22.10: Tschebyscheff coefficients for 2-dB pass-band ripple

n	i	a_i	b_i	$k_i = fc_i/fc$	Q_i
1	1	1.0000	0.0000	1.000	—
2	1	1.1813	1.7775	1.000	1.13
3	1	2.7994	0.0000	0.357	—
	2	0.4300	1.2036	1.378	2.55
4	1	2.4025	4.9862	0.550	0.93
	2	0.2374	1.1896	1.413	4.59
5	1	4.6345	0.0000	0.216	—
	2	0.9090	2.6036	0.908	1.78
	3	0.1434	1.0750	1.493	7.23
6	1	3.5880	10.4648	0.373	0.90
	2	0.4925	1.9622	1.085	2.84
	3	0.0995	1.0826	1.491	10.46
7	1	6.4760	0.0000	0.154	—
	2	1.3258	4.7649	0.665	1.65
	3	0.3067	1.5927	1.218	4.12
	4	0.0714	1.0384	1.523	14.28
8	1	4.7743	18.1510	0.282	0.89
	2	0.6991	3.1353	0.853	2.53
	3	0.2153	1.4449	1.285	5.58
	4	0.0547	1.0461	1.518	18.39
9	1	8.3198	0.0000	0.120	—
	2	1.7299	7.6580	0.522	1.60
	3	0.4337	2.3549	0.998	3.54
	4	0.1583	1.3174	1.349	7.25
	5	0.0427	1.0232	1.536	23.68
10	1	5.9618	28.0376	0.226	0.89
	2	0.8947	4.6644	0.697	2.41
	3	0.3023	1.9858	1.094	4.66
	4	0.1233	1.2614	1.380	9.11
	5	0.0347	1.0294	1.531	29.27

Table 22.11: Tschebyscheff coefficients for 3-dB pass-band ripple

n	i	a_i	b_i	$k_i = fc_i/fc$	Q_i
1	1	1.0000	0.0000	1.000	—
2	1	1.0650	1.9305	1.000	1.30
3	1	3.3496	0.0000	0.299	—
	2	0.3559	1.1923	1.396	3.07
4	1	2.1853	5.5339	0.557	1.08
	2	0.1964	1.2009	1.410	5.58
5	1	5.6334	0.0000	0.178	—
	2	0.7620	2.6530	0.917	2.14
	3	0.1172	1.0686	1.500	8.82
6	1	3.2721	11.6773	0.379	1.04
	2	0.4077	1.9873	1.086	3.46
	3	0.0815	1.0861	1.489	12.78
7	1	7.9064	0.0000	0.126	—
	2	1.1159	4.8963	0.670	1.98
	3	0.2515	1.5944	1.222	5.02
	4	0.0582	1.0348	1.527	17.46
8	1	4.3583	20.2948	0.286	1.03
	2	0.5791	3.1808	0.855	3.08
	3	0.1765	1.4507	1.285	6.83
	4	0.0448	1.0478	1.517	22.87
9	1	10.1759	0.0000	0.098	—
	2	1.4585	7.8971	0.526	1.93
	3	0.3561	2.3651	1.001	4.32
	4	0.1294	1.3165	1.351	8.87
	5	0.0348	1.0210	1.537	29.00
10	1	5.4449	31.3788	0.230	1.03
	2	0.7414	4.7363	0.699	2.94
	3	0.2479	1.9952	1.094	5.70
	4	0.1008	1.2638	1.380	11.15
	5	0.0283	1.0304	1.530	35.85

Table 22.12: All-pass coefficients

n	i	a_i	b_i	$k_i = f_{ci}/f_c$	Q_i	igr
1	1	0.6436	0.0000	1.554	—	0.2049
2	1	1.6278	0.8832	1.064	0.58	0.5181
3	1	1.1415	0.0000	0.876	—	0.8437
	2	1.5092	1.0877	0.959	0.69	
4	1	2.3370	1.4878	0.820	0.52	1.1738
	2	1.3506	1.1837	0.919	0.81	
5	1	1.2974	0.0000	0.771	—	1.5060
	2	2.2224	1.5685	0.798	0.56	
	3	1.2116	1.2330	0.901	0.92	
6	1	2.6117	1.7763	0.750	0.51	1.8395
	2	2.0706	1.6015	0.790	0.61	
	3	1.0967	1.2596	0.891	1.02	
7	1	1.3735	0.0000	0.728	—	2.1737
	2	2.5320	1.8169	0.742	0.53	
	3	1.9211	1.6116	0.788	0.66	
	4	1.0023	1.2743	0.886	1.13	
8	1	2.7541	1.9420	0.718	0.51	2.5084
	2	2.4174	1.8300	0.739	0.56	
	3	1.7850	1.6101	0.788	0.71	
	4	0.9239	1.2822	0.883	1.23	
9	1	1.4186	0.0000	0.705	—	2.8434
	2	2.6979	1.9659	0.713	0.52	
	3	2.2940	1.8282	0.740	0.59	
	4	1.6644	1.6027	0.790	0.76	
	5	0.8579	1.2862	0.882	1.32	

References

D. Johnson and J. Hilburn, *Rapid Practical Designs of Active Filters,* John Wiley & Sons, 1975.

U. Tietze and Ch. Schenk, *Halbleiterschaltungstechnik,* Springer-Verlag, 1980.

H. Berlin, *Design of Active Filters with Experiments,* Howard W. Sams & Co, 1979.

M. Van Falkenburg, *Analog Filter Design,* Oxford University Press, 1982.

S. Franko, *Design with Operational Amplifiers and Analog Integrated Circuits,* McGrawHill, 1988.

CHAPTER 23

Radio-Frequency (RF) Circuits

Ian Hickman

Radio-frequency equipment is used for a vast range of purposes, including heat treating special steels, medical diathermy treatment for cancer, heat sealing plastic bags, and experiments in atomic physics. Nevertheless, as the name implies, the original use was in connection with the transmission of information by radio waves. The earliest form of this was wireless telegraphy (WT) using Morse code. This was followed by wireless telephony and, much later, broadcasting—radio and television. So, before diving into RF circuits in detail, a word might be in order about the different forms of modulation employed to impress the information to be transmitted onto the radio wave.

23.1 Modulation of Radio Waves

Figure 23.1A shows how information is transmitted by means of an interrupted continuous wave, often called simply *continuous wave* (CW). This type of modulation is frequently employed in the high-frequency (HF) band; that is, from 1.6 to 30 MHz. In a simple transmitter either the oscillator would be "keyed" on and off with a Morse key, or alternatively the drive signal or the power supply to the output stage would be likewise keyed. In the simplest possible transmitter there would be no separate output stage, only a keyed oscillator. Using CW, amateur radio enthusiasts can contact others in any country in the world using only a few watts, but only as and when propagation conditions are favorable.

Broadcasting on medium wave (MW) uses *amplitude modulation*, which is illustrated in Figure 23.lB. Here, the frequency of the radio-frequency or *carrier* wave does not change, but its amplitude is modulated in sympathy with the program material, usually

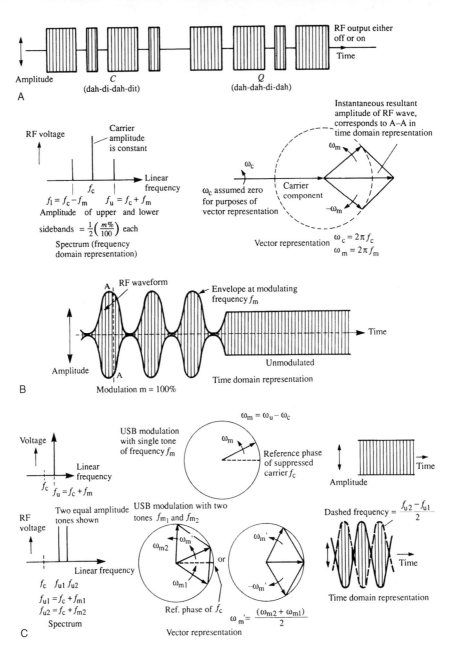

Figure 23.1: Types of modulation of radio waves. (A) CW modulation. The letters CQ in Morse (seek you?) are used by amateurs to invite a response from any other amateur on the band, to set up a QSO (Morse conversation). (B) AM: 100% modulation by a single sinusoidal tone shown. (c) SSB (USB) modulation. Note that with two-tone modulation, the signal is indistinguishable from a double-sideband suppressed carrier signal with a suppressed carrier frequency of = $(f_{u1} + f_{u2})/2$. This can be seen by subtracting the carrier component from the 100% AM signal in (b).

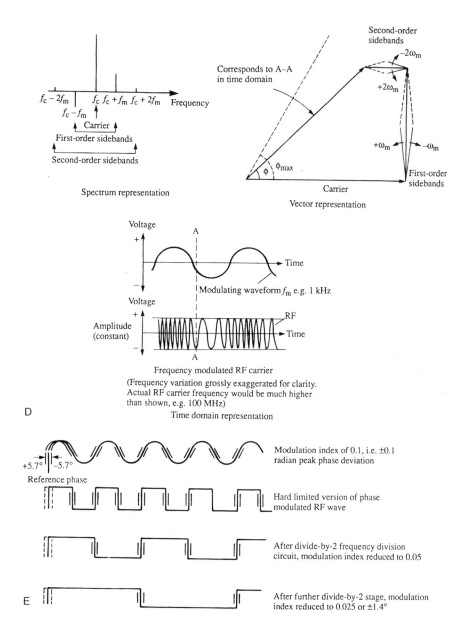

Figure 23.1: (Cont'd) The upper and lower halves of the envelope will then overlap as in (C), with the RF phase alternating between 0° and 180° in successive lobes. (D) FM. For maximum resultant phase deviation φ up to about 60° as shown, third- and higher-order sidebands are insignificant. (E) Reduction of phase deviation when a phase modulated signal passes through a frequency divider chain, showing—for example—how a divide-by-4 (two-stage binary divider) reduced modulation index by a factor of 4.

speech or music. This gives rise to *sidebands*, which are limited to –4.5 kHz about the carrier frequency by limiting the bandwidth of the *baseband* modulating signal to 4.5 kHz maximum. This helps to minimize interference between adjacent stations on the crowded MW band, where frequency allocations are only 9 kHz apart (10 kHz in the USA). With maximum modulation by a single sinusoidal tone, the transmitted power is 50% greater than with no modulation; this is the *100% modulation case*. Note that the power in the carrier is unchanged from the 0% or *unmodulated* case. Thus, at best only one-third of the transmitted power actually conveys the program information, and during average program material the proportion is much lower even than this. For this reason, the *single-sideband* (SSB) mode of modulation has become very popular for voice communication at HF. With this type of modulation, illustrated in Figure 23.1C, only one of the two sidebands is transmitted, the other and the carrier being suppressed. As there is no carrier, all of the transmitted power represents wanted information, and as all of this is concentrated in one sideband, "spectrum occupancy" is halved. At the receiver, the missing carrier must be supplied from a carrier reinsertion oscillator at exactly the appropriate frequency in order to demodulate the signal and recover the original. Although this is a trivial exercise with modern synthesized receivers, historically it was difficult. Amplitude modulation, with its uncritical tuning requirements, continues to be used by broadcasters for both local audiences on MW and international broadcasting on SW. There are a number of bands of frequencies allocated by international agreement to broadcasting in the short-wave band between 1.6 and 30 MHz.

Figure 23.1D illustrates *frequency modulation*. FM was proposed as a modulation method even before the establishment of AM broadcasting, but any enthusiasm for it waned as a result of an analysis which showed that it produced sidebands exceeding greatly the bandwidth of the baseband signal [Ref. 23.1].With the limited bandwidth available in the LW and MW bands, this was obviously an undesirable characteristic. However, following the Second World War the technology had advanced to the point where it was possible to use the considerable bandwidth available in the then largely unused very high-frequency (VHF) band. The lower part of the 30–300 MHz VHF band had already been used before the war for television, and now a high-quality sound broadcasting service was established using FM in the band 88-108 MHz. The standard adopted was a maximum deviation from the center or carrier frequency of –75 kHz, and a baseband frequency response extending from 50 Hz to 15 kHz. This represented real hi-fi compared with the 4.5 kHz limitation on MW, and the much lower level of interference from unwanted stations was a real blessing. The modulation index for an

FM signal is defined in terms of a single sinusoidal modulating tone, as "m", where m $= f_d/f_m$, the peak frequency deviation of the carrier, divided by the modulating frequency. It is shown below that m is also equal to the peak phase deviation of the carrier in radians. With the 75 kHz peak deviation being five times the highest modulating frequency, broadcast FM (also known as WBFM—wide band FM) is a type of spread spectrum signal. This confers a degree of immunity to adjacent- and co-channel interference due to the "capture effect." This is particularly effective on mono reception, the advantage being much less for stereo reception.

Figure 23.1 shows the characteristics of the various modulation methods in three ways: in the frequency domain, in the time domain, and as represented by vector diagrams. Each illustrates one aspect of the signal particularly well, and it is best to be familiar with all the representations. Choosing one and sticking to it is likely to be misleading since they each tell only a part of the story. Note that in Figure 23.1D a very low level of modulation is shown, corresponding to a low amplitude of the modulating sine wave (frequency f_m). Even so, it is clear that if only the sidebands at the modulating frequency are considered, the amplitude of the signal would be greatest at those instants when its phase deviation from the unmodulated position is greatest. It is the presence of the second-order side-bands at $2f_m$ which compensate for this, maintaining the amplitude constant. At wider deviations many more FM sidebands appear, all so related in amplitude and phase as to maintain the amplitude constant. They arise automatically as a result of frequency modulating an oscillator whose output amplitude is constant; their existence is predicted by the maths and confirmed by the spectrum analyser.

Note that the maximum phase deviation of the vector representing the FM signal will occur at the end of a half-cycle of the modulating frequency, since during the whole of this half-cycle the frequency will have been above (or below) the center frequency. Thus the phase deviation is 90° out of phase with the frequency deviation. Note also that, for a given peak frequency deviation, the peak phase deviation is inversely proportional to the modulating frequency, as may be readily shown. Imagine the modulating signal is a 100 Hz square wave and the deviation is 1 kHz. Then during the 10 ms occupied by a single cycle of the modulation, the RF will be first 1000 Hz higher in frequency than the nominal carrier frequency and then, during the second 5 ms, 1000 Hz lower in frequency. So the phase of the RF will first advance steadily by five complete cycles (or 10π radians) and then crank back again by the same amount, i.e., the phase deviation is $\pm 5\pi$ radians relative to the phase of the unmodulated carrier. Now the average value of a half-cycle of a sine wave is $2/\pi$ of that of a half-cycle of square wave of the same peak amplitude; so if the modulating signal had been a sine wave, the peak phase deviation would have been just ± 10 radians. Note that the peak phase deviation in radians (for sine wave modulation) is just f_d/f_m, the peak frequency deviation divided by the modulating frequency: this is known as the *modulation index* of an FM signal. If the modulating sine wave had been 200 Hz, the deviation being 1 kHz as before, the shorter period of the modulating frequency would result in the peak-to-peak phase change being halved to ± 5 radians; that is, for a given peak frequency deviation the peak phase deviation is inversely proportional to the modulating frequency.

For monophonic FM broadcasting the peak deviation at full modulation is 75 kHz, so the peak phase deviation corresponding to full sine wave modulation would be ± 5 radians at 15 kHz and ± 1500 radians at 50 Hz modulating frequency. If the modulation index of an FM signal is much less than unity, the second-order and higher-order FM sidebands are insignificant. If, on the other hand, the modulation index is very large compared with unity, there are a large number of significant sidebands and these occupy a bandwidth virtually identical to $2f_d$; that is, the bandwidth over which the signal sweeps. The usual approximation for the bandwidth of an FM signal is $BW = (2f_d + f_m)$.

You can see in Figure 23.1B that the vectors representing the two sidebands of an AM signal are always symmetrically disposed about the vector representing the carrier. As they rotate at the same rate but in opposite directions, their resultant is always directly adding to or reducing the length of the carrier vector. The second and higher even-order sidebands of an FM signal behave in the same way. But as Figure 23.1D shows, the first-order sidebands (at the modulating frequency) are symmetrical about a line at right angles to the carrier, and the same goes for higher odd-order sidebands. Note that if one of the first-order FM sidebands was reversed, they would look exactly like a pair of AM sidebands: this is why one of the first-order FM sideband signals in the frequency domain representation in Figure 23.1D has been shown as inverted. A spectrum analyzer will show the carrier and sidebands of either an AM or a low-deviation FM signal as identical, as the analyser responds only to the amplitudes of the individual sidebands, not their phases. However, if the first-order sidebands displayed on the analyzer are unequal in amplitude, this indicates that there is both AM and FM present on the modulated wave.

An important principle in connection with phase modulation is illustrated in Figure 23.1E. This shows how dividing the frequency of a phase or frequency modulated wave divides the modulation index in the same proportion. In the figure,

a sinusoidal modulating waveform has been assumed; in this case the peak phase deviation in radians is numerically equal to the modulation index, i.e., to the peak frequency deviation divided by the modulating frequency, as noted above. However, whatever the modulating waveform—and even in the case of a nonrepetitive signal such as noise—passing the modulated carrier through a divide-by-N circuit will reduce the peak deviation by a factor of N, as should be apparent from Figure 23.1E. For the time variations on the edges of the divider output remain unaffected but they now represent a smaller proportion of a complete cycle. Conversely, if a phase or frequency modulated signal is passed through a frequency multiplier (described later in this chapter), any phase noise on the signal is multiplied
pro rata.

23.2   Low-Power RF Amplifiers

Having looked at some typical radio-frequency signals (there are many other sorts, for example frequency-shift keying (FSK), numerous varieties of digital modulation, and of course television), it is time to look at some of the wide range of RF circuits, both passive and active, used to process them. These include amplifiers of all sorts, but only low-power RF amplifiers are discussed, for a very good reason. This is a very exciting time for the high-power RF engineer, with devices of ever higher power becoming available almost daily. There are regular improvements in high-power bipolar RF transistors. RF MOSFETs are improving in terms of both power handling and reduced capacitances, particularly the all-important drain/gate feedback capacitance C_{dg}; they are also available now as matched pairs in a single package, for push-pull applications. Meanwhile other exciting developments are on the horizon, including the static induction transistor (SIT). This device is half-way between a bipolar and an FET, and its notable feature is an unusually high voltage capability. This eases the difficulties associated with the design of high-power RF circuits due to the very low impedance levels at which lower-voltage devices necessarily work. Even more exciting is the prospect of high-power devices using not silicon or gallium arsenide (GaAs), or even indium phosphide (InP), but diamond. The technology is currently being researched in the USA, Japan and the USSR, and already diodes (operating up to 700°C!) have been produced. With a carrier velocity three times that of silicon and a thermal conductivity twenty times that of silicon (four times that of copper, even) the possibilities are immense. So any detailed discussion of RF power devices is fated to be out of date by the time it appears in print. So only low-power amplifiers are discussed below.

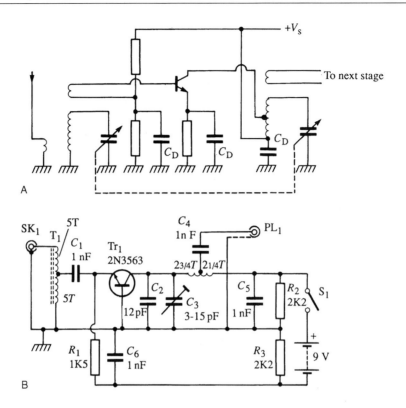

Figure 23.2: RF amplifier stages. (A) Common emitter RF amplifier stage with both input and output circuits tuned. C_D are decoupling capacitors. (B) Common base RF amplifier with aperiodic (broad band) input and tuned output stages (reproduced from "VHF preamplifier for band II," Ian Hickman, _Practical Wireless_, June 1982, p. 68, by courtesy of _Practical Wireless_).

Figure 23.2 shows two class A NPN _bipolar transistor_ amplifier stages. In Figure 23.2A, both the input and output circuits are tuned. This is by no means the invariable practice but, for the input RF stage of a high-quality communications receiver, for example, it enables one to provide more selectivity than could be achieved with only one tuned circuit, while avoiding some of the complications of coupled tuned circuits. The latter can provide a better band-pass shape—in particular a flatter pass-band—but, for a communications receiver covering say 2 to 30 MHz, two single-tuned circuits such as in Figure 23.2A provide an adequate pass-band width in any case. With the continuing heavy usage of the 2 to 30 MHz HF band, which seems to become even more congested yearly rather than dying as the pundits were once predicting, RF stages are coming back into favour again. However, an RF amplifier with both input and output circuits tuned needs very careful design to ensure stability, especially when using the *common emitter* configuration. The potential source of trouble is the collector/base capacitance, which provides a path by which energy from the output tuned circuit can be fed back to the base input circuit. The common emitter amplifier provides inverting gain, so that the output is effectively 180° out of phase with the input. The current fed back through the collector base capacitance will of course lead the collector voltage by 90°. At a frequency somewhat below *resonance* (Figure 23.3) the collector voltage will lead the collector current, and the feedback current via the collector/base capacitance will produce a leading voltage across the input tuned circuit. At the frequency where the lead in each tuned circuit is 45°, there is thus a total of 180° of lead, cancelling out the inherent phase reversal of the stage and the feedback becomes positive. The higher the stage gain and the higher the *Q* of the tuned circuits, the more likely is the feedback to be sufficient to cause oscillation, since when the phase shift in each tuned circuit is 45°, its amplitude response is only 3 dB down (see Figure 23.3). Even if oscillation does not result, the

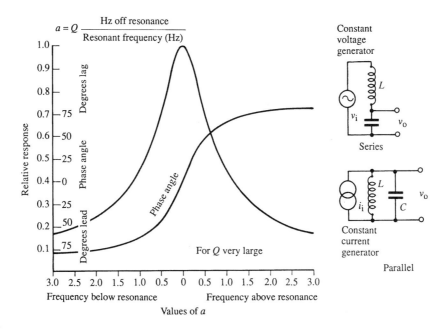

Figure 23.3: Universal resonance curve for series resonant circuit. For a *Q* of greater, the phase and amplitude curves depart by only a very small amount from the previous. Also applies to the response of a parallel tuned circuit, for *Q* > 20. In both cases, curves give v_o/v_{max} in magnitude and phase.

stage is likely to show a much steeper rate of fall of gain with detuning on one side of
the tuned frequency than on the other—a sure sign of significant internal feedback.
The *grounded base* stage of Figure 23.2B may prove a better choice, since some bipolar
transistors exhibit a significantly smaller feedback capacitance in the grounded base
connection; that is, C_{ce} is smaller than their C_{cb}. The N channel *junction depletion
FET* (JFET) is also a useful RF amplifier, and can be used in either the grounded source or
grounded gate configuration, corresponding to the circuits of Figure 23.2. It is particularly
useful in the grounded gate circuit as a VHF amplifier.

23.3   Stability

There are a number of circuit arrangements that are used to ensure the *stability* of
an RF amplifier stage. One of these, the cascode, is shown in Figure 23.4A. The
cascode stage consists of two active devices; bipolar transistors are shown in the
figure, but JFETs or RF MOSFETs are equally applicable. The input transistor is used
in the grounded emitter configuration, which provides much more current gain than
the grounded base configuration. However, there is no significant feedback from the
collector circuit to the base tuned circuit since the collector load of the input transistor
consists of the very low emitter input impedance of the second transistor. This is used in

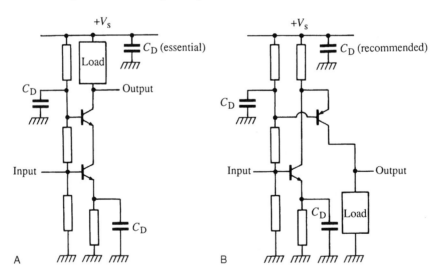

**Figure 23.4: (A) Cascode amplifier. (B) Complementary cascode. The load may be
a resistor, an *RL* combination (peaking circuit), a tuned circuit or a wide band RF
transformer. C_D are decoupling capacitors.**

the grounded base configuration, which again results in very low feedback from its output to its input. With a suitable type of transistor the cascode circuit can provide well over 20 dB of gain at 100 MHz together with a reverse isolation of 70 dB. *Reverse isolation* is an important parameter of any RF amplifier, and is simply determined by measuring the "gain" of the circuit when connected back to front; that is, with the signal input applied to the output port and the "output" taken from the input port. This is easily done in the case of a stand-alone amplifier module, but is not so easy when the amplifier is embedded in a string of circuitry in equipment. In the days of valves, one could easily derive a stage's reverse isolation (knowing its forward gain beforehand) by simply disconnecting one of the heater leads and seeing how much the gain fell! When a valve is cold it provides no amplification, so signals can only pass via the interelectrode capacitances, and these are virtually the same whether the valve is hot or cold. With no gain provided by the valve, the forward and reverse isolation are identical. Much the same dodge can be used with transistors by open-circuiting the emitter to DC but leaving it connected as before at AC. However, the results are not nearly so reliable as in the valve case, as many of the transistor's parasitic reactances will change substantially when the collector current is reduced to zero. For an RF amplifier stage to be stable, clearly its reverse isolation should exceed its forward gain by a reasonable margin, which need not be anything like the 40 to 80 dB obtainable with the cascode mentioned above. A difference of 20 dB is fine and of 10 dB adequate, while some commercially available broad band RF amplifier modules exhibit a reverse isolation which falls to as little as 3 dB in excess of the forward gain at the top end of their frequency range.

An interesting feature of the cascode stage of Figure 23.4A arises from the grounded base connection of the output transistor. In this connection its collector/base breakdown voltage is higher than in the common emitter connection, often by a considerable margin, as transistor data sheets will show. This fact makes the cascode circuit a favourite choice for amplifiers which have to handle a very wide range of frequencies while producing a very large peak-to-peak output voltage swing. Examples include the range from DC to RF in the *Y* deflection amplifier of an oscilloscope, and that from 50 Hz to RF in the video output amplifiers in a TV set. Figure 23.4B shows a *complementary cascode* stage. This has the advantage of not drawing any appreciable RF current from the positive supply rail, easing decoupling requirements.

Figure 23.5A shows what is in effect a cascade circuit, but in the *dual-gate RF MOSFET* the two devices are integrated into one, the drain region of the input device acting as the source of the output section. Thus, the dual-gate MOSFET is a "semiconductor tetrode"

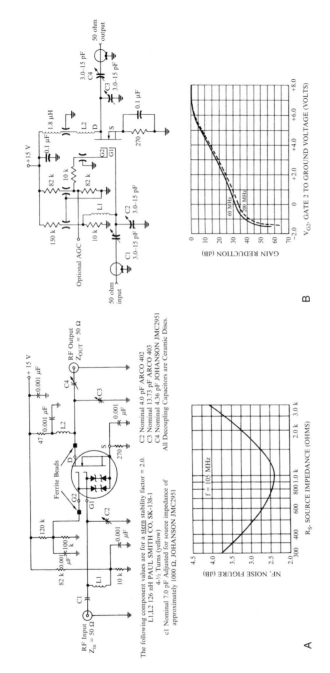

Figure 23.5: Dual-gate MOSFET RF amplifiers. (A) Low-noise dual-gate MOSFET VHF amplifier stage and noise figure curve. The Motorola MFE140 shown incorporates gate protection Zener diodes, to guard against static electricity discharge damage. (B) Dual-gate MOSFET VHF amplifier with AGC, with gain reduction curve. Maximum gain 27 (20) db at 60 (200) MHz with no gain reduction (V_{g2} at +7.5V). The Motorola MPF131 provides an AGC range featuring up to 60 dB of gain reduction. (Reproduced by courtesy of Motorola Inc.)

and, as in the thermionic tetrode and pentode, the feedback capacitance internal to the device is reduced to a very low level (for the Motorola MFE140 the drain/gate1 capacitance amounts to little more than 0.02 pF). The dual-gate MOSFET exhibits a very high output slope resistance, again like its thermionic counterpart, and also an AGC capability. The circuit of Figure 23.5B provides up to 27 dB gain at 60MHz when the AGC voltage V_{gg} is +8V and up to 60 dB of gain reduction as V_{gg} is reduced to below 0V.

A common technique to increase the stability margin of transistor RF amplifiers is *mismatching*. This simply means accepting a stage gain less than the maximum that could be achieved in the absence of feedback. In particular, if the collector (or drain) load impedance is reduced, the stage will have a lower voltage gain, so the voltage available to drive current through the feedback capacitance (C_{cb} in a bipolar transistor, C_{dg} or C_{rss} in an FET) is reduced *pro rata*. Likewise, if the source impedance seen by the base (or gate) is reduced, the current fed back will produce less voltage drop across the input circuit. Both measures reduce gain and increase stability: it may well be cheaper to recover the gain sacrificed by simply adding another amplifier stage than to add circuit complexity to obtain the extra gain from fewer stages by *unilateralization*. This cumbersome term is used to indicate any type of scheme to reduce the effective internal feedback in an amplifier stage, i.e., to make the signal flow in only one direction— forward. Data sheets for RF transistors often quote a figure for the maximum available gain (MAG) and a higher figure for maximum unilateralized gain (MUG).

The traditional term for unilateralization is *neutralization*, and I shall use this term hereafter as it is just a little shorter, even though they are not quite the same thing. Figure 23.6A shows one popular neutralization scheme, sometimes known as *bridge* neutralization. The output tuned circuit is center tapped so that the voltage at one end of the inductor is equal in amplitude to, and in antiphase with, the collector voltage. The neutralizing capacitor C_n has the same value as the typical value of the transistor's C_{cb}, or C_n can be a trimmer capacitance, set to the same value as the C_{cb} of the particular transistor. The criterion for setting the capacitor is that the response of the stage should be symmetrical. This occurs when there is no net feedback, either positive or negative. The series capacitance of C_{cb} and C_n appears across the output tuned circuit and is absorbed into its tuning capacitance, while the parallel capacitance of C_{cb} and C_n appears across the input tuned circuit and is absorbed into its tuning capacitance. Neutralization can be very effective for small-signal amplifiers, but is less so for stages handling a large voltage swing. This is because the feedback capacitance C_{cb}, owing to the capacitance of the reverse biased collector/base junction, is not constant but varies (approximately) inversely as the square of the collector/base voltage.

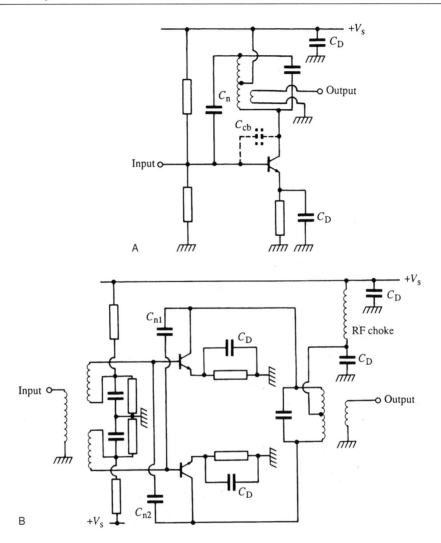

Figure 23.6: Neutralization. (A) Bridge neutralization. The internal feedback path is not an ideal capacitor C_{cb} as shown, but will have an in-phase component also. If the phase angle of the neutralization via C_n is adjusted, e.g., by means of an appropriate series resistance, the neutralization is more exact—at that particular frequency. The stage is then described as "unilateralized" at that frequency. (B) Cross-neutralization, push-pull amplifier.

Neutralization can be applied to a push-pull stage as in Figure 23.6B, but great care is necessary when so doing. The scheme works fine just so long as the voltage at the collectors can be guaranteed to be in antiphase. This will indeed be the case at the resonant frequency where the collector load is a tuned circuit, or over the desired band of output frequencies where the collector load is a wideband RF transformer. However, at some other (usually higher) frequency this may no longer apply, owing to leakage inductance between the two halves of the collector circuit's inductor or transformer. The two collector voltages may then be able to vary in phase with each other, and the circuit simply becomes two identical amplifiers in parallel, each with a total feedback capacitance equal to twice its internal feedback capacitance. If the amplifier devices still have substantial power gain left at the frequency at which this condition exists, then the circuit can oscillate in a parallel single-ended mode.

23.4 Linearity

All of the amplifier circuits discussed so far have operated in class A, that is to say the peak current swing is less than the standing current, so that at no time is the transistor cut-off. Where the collector circuit of an amplifier is a tuned circuit, this will have a "flywheel" effect so that the collector voltage is approximately sinusoidal even though the collector current is not. Thus, a transistor can amplify the signal even though it conducts for more than $180°$ but less than $360°$—i.e., operates in class AB. Likewise, for class B ($180°$ conduction angle) and class C (conduction angle less than $180°$). These modes offer higher efficiency than class A, but whether one or other of them is appropriate in any given situation depends upon the particular application. Consider the earlier low and intermediate power amplifier stages of an FM transmitter, for example. Here, the amplitude of the signal to be transmitted is constant and it is the only signal present; there are no unwanted signals such as one inevitably finds in the earlier stages of a receiver. Consequently, a class B or C stage is entirely appropriate in this application. However, an AM or SSB transmitter requires a linear amplifier, i.e., one that faithfully reproduces the variations in signal amplitude which constitute the *envelope* of the signal.

In the receiver the requirement for *linearity* is even more pressing, at least in the earlier stages where many unwanted signals, some probably very much larger than the wanted signal, are present. Second-order nonlinearity—second-harmonic distortion—results in sum and difference products when more than one signal is present, and third-order nonlinearity in products of the form $2f_1 \pm f_2$. These latter intermodulation products, resulting from two unwanted frequencies f_1 and f_2, are particularly embarrassing in radio reception. Imagine that f_1 is, say, 20 kHz higher than the wanted signal at f_0, and that f_2 is 20 kHz higher still. Then $2f_1 - f_2$ turns out to be exactly at f_0. If the two unwanted frequencies were on the low-frequency side, f_2 being 20 kHz lower than f_0 and f_1 20 kHz lower still, then it would be the intermodulation product $2f_1 - f_2$ that falls on the wanted frequency. The intermediate-frequency (IF) amplifier section of a superheterodyne receiver—or *superhet* for short, shown in block diagram form in Figure 23.7A—is preceded by a highly selective

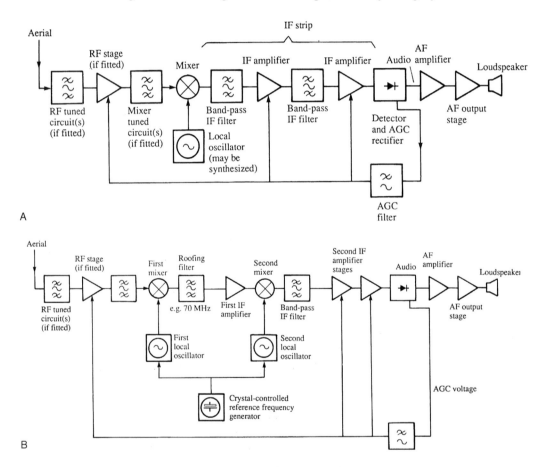

Figure 23.7: Supersonic heterodyne (superhet) receivers. (A) Single-conversion superhet. Several filters may be used throughout the IF strip. (B) Double-conversion superhet, with synthesized first local oscillator and second local oscillator both crystal reference controlled.

filter which, in a good quality communications receiver, will attenuate frequencies 20 kHz or more off tune by at least 80 dB. However, it is not possible to provide that sort of selectivity in a tunable filter; the comparative ease of obtaining high selectivity at a fixed frequency is the whole *raison d'être* of the superhet. So the RF amplifier stage (if any) and the mixer must be exceedingly linear to avoid interference caused by third-order intermodulation products.

In a *double-conversion superhet* such as shown in Figure 23.7B, this requirement applies also to the first IF amplifier and second mixer, although the probability of interference from odd-order intermodulation products is reduced by the *roofing filter* preceding the first IF amplifier. This is always a crystal filter offering 30 or 40 dB of attenuation at frequencies 30 kHz or more off tune. Indeed, recent developments in crystal filter design and manufacture permit the roofing crystal filter to be replaced by a crystal filter, operating at 70 MHz, with the same selectivity as previously obtained in the second IF filter at 1.4 MHz, enabling the design of an "up-converting single superhet." An up-converting superhet removes the *image* problem encountered with a down-converting single superhet such as in Figure 23.7A. With a 1.4 MHz IF and a local oscillator tuning from 3 to 31.4 MHz, it is difficult to provide enough selectivity at the top end of the HF band. For example, when the receiver is tuned to 25 MHz the local oscillator frequency will be 26.4 MHz, and an unwanted signal at 27.8 MHz will also produce an IF output from the mixer at 1.4 MHz. This represents a fractional detuning of only 11.2%, and reference to the universal tuned circuit curves of Figure 23.3 will verify that even with high-Q tuned signal frequency circuits, it is difficult adequately to suppress the image response; hence the popularity of the up-converting double superhet of Figure 23.7B. Here, signal frequency tuned circuits can be replaced by *suboctave filters* (band-pass filters each covering a frequency range of about 1.5 to 1), or simply omitted entirely—although this sacrifices the protection against second-order intermodulation products afforded by suboctave filters.

The linearity of amplifiers, both discrete components and multistage amplifier stages, and of mixers is often quoted in terms of *intercept points*. You may recall that if the input to an amplifier with some second-order curvature in its transfer characteristic is increased by 1 dB, the second-harmonic distortion rises by 2 dB, and the sum and difference terms due to two different input frequencies applied simultaneously behave likewise. Also, with third-order (S-shaped) curvature, both third-harmonic and third-order intermodulation products rise three times as fast as the input, at least for small inputs.

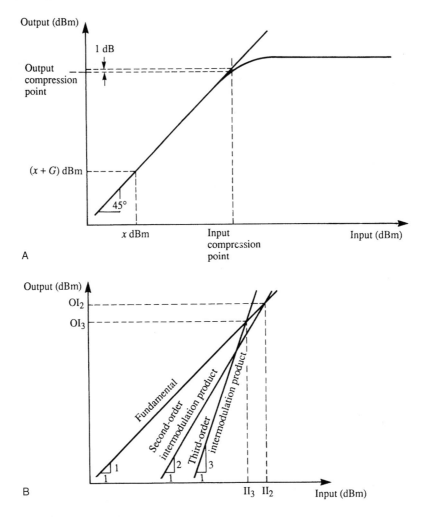

Figure 23.8: Compression and intermodulation. (A) Compression point of an amplifier, mixer or other device with gain _G_ dB. (Single tone input. (B) Second- and third-order input and output intercept points (II and OI); see text. (Two tones of equal amplitude)

Of course, for very large inputs an amplifier will be driven into limiting and the output will eventually cease to rise: the output is said to be _compressed_. Figure 23.8A illustrates this: the point where the gain is 1 dB less than it would have been if overload did not occur is called the _compression level_. It is found that for levels up to about 10 dB below compression, it is a good rule that _N_th-order intermodulation products rise by N dB for every 1 dB by which the two inputs rise. Figure 23.8B shows the behavior of an imaginary but not untypical amplifier. The level of second- and third-order intermodulation products, as well as of the wanted output, have been plotted against input level. All three characteristics have then been produced on past the region of linearity, and it can be seen that eventually they cross. The higher the level at which an amplifier's second- and third-order intercept points occur, the less problem there will be with unwanted responses due to intermodulation products, provided always that it also has a high enough compression point to cope linearly with the largest signals. A mixer (frequency changer) can be characterized in a similar way, except of course that the intermodulation products (coloquially called *intermods*) now appear translated to the intermediate frequency.

23.5   Noise and Dynamic Range

For an amplifier forming part of a receiver, high linearity is only one of several very desirable qualities. The input stage must exhibit a low noise figure, as indeed must all the stages preceding the IF filter defining the final bandwidth. For it makes sense to supply most of the gain after this filter; this way, large unwanted signals are amplified as little as possible before being rejected by the filter. Remember that unwanted signals may be 60, 80 or even 100 dB larger than the wanted signal!

The *noise figure* of an amplifier is related to the amount of noise at its output, in the absence of any intentional input, and its gain. Noise is an unavoidable nuisance, and not only in amplifiers. A current in a metallic conductor consists of a flow of electrons jostling their way through a more or less orderly jungle of atoms, of copper maybe or some other metal; and current is produced by carriers—electrons or holes—flowing in a semiconductor. Since at room temperature, indeed at any temperature above absolute zero, the atoms are in a state of thermal agitation, the flow of current will not be smooth and orderly but noisy, like the boisterous rushing of a mountain stream. Like the noise of a stream, no one frequency predominates. Electrical noise of this sort is called *thermal agitation noise* or just *thermal noise*, and its intensity is independent of frequency (or "white") for most practical purposes. The available noise power associated with a resistor is independent of its resistance and is equal to -174 dBm/Hz, e.g., -139 dB relative to a level of 1 milliwatt in a 3 kHz bandwidth.

This means that the wider the bandwidth of a filter, the more noise it lets through. It would seem that if we have no filter at all to limit the bandwidth, there would be an infinite amount of noise power available from a resistor—free heating for evermore!

This anomaly had theoretical physicists in the late nineteenth century worrying about an ultraviolet catastrophy, but all is well; at room temperature thermal noise begins to tail off beyond 1000 GHz (10% down), the noise density falling to 50% at 7500 GHz. At very low temperatures such as are used with maser amplifiers, say 1 kelvin (−272°C), the noise density is already 10% down by 5 GHz (see Figure 23.9B).

Returning to RF amplifiers then, if one is driven from a 50Ω source there will be noise power fed into its input therefrom (see Figure 23.9A). If the amplifier is matched to

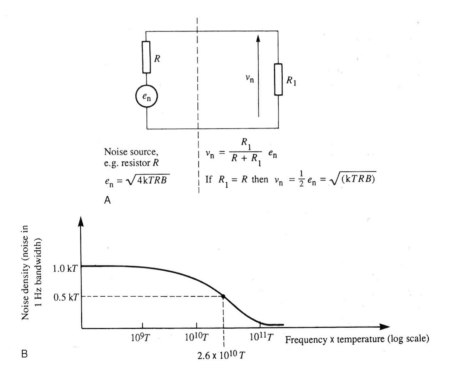

Noise source,
e.g. resistor R

$e_n = \sqrt{4kTRB}$

$v_n = \dfrac{R_1}{R + R_1}\, e_n$

If $R_1 = R$ then $v_n = \frac{1}{2} e_n = \sqrt{(kTRB)}$

A

B

Figure 23.9: Thermal noise (A) A noisy source such as a resistor can be represented by a noise-free resistor R of the same resistance, in series with a noise voltage generator of EMF $e_n = \sqrt{(4kTRB)}$ volts. Available noise power $= v_n^2/R = (e_n/2)^2/R = P_n$ say. At room temperature (290 K) $P_n = -204$ dBW in a 1 Hz bandwidth $= -174$ dBm in a 1 Hz bandwidth. If $B = 3000$ Hz then $P_n = -139$ dBm, and if $R = R_1 = 50\Omega$ then $v_n = 0.246$ μV in 3 kHz bandwidth. (B) Thermal noise is "white" for all practical purposes. The available noise power density falls to 50% at a frequency of $2{:}6 \times 10^{10}T$, i.e., at about 8000 GHz at room temperature, or 26 GHz at $T = 1$K.

the source, i.e., its input impedance is 50Ω resistive, the RMS noise voltage at the amplifier's input v_n is equal to half the source resistor's open-circuit noise voltage, i.e., to $\sqrt{(kTRB)}$, where R is 50Ω, k is Boltzmann's constant = $1:3803 \times 10^{-23}$ joules per kelvin, T is the absolute temperature in kelvin (i.e., degrees centigrade plus 273) and B is the bandwidth of interest. At a temperature of 290K (17°C or roughly room temperature) this works out at 24.6 nV in 50Ω in a 3 kHz bandwidth. If the amplifier were perfectly noise free and had a gain of 20 dB (i.e., a voltage gain of ×10, assuming its output impedance is also 50Ω), we would expect 0.246 μV RMS noise at its output: if the output noise voltage were twice this, 0.492 μV RMS, we would describe the amplifier as having a noise figure of 6 dB. Thus, the noise figure simply expresses the ratio of the actual noise output of an amplifier to the noise output of an ideal noise-free amplifier of the same gain. The amplifier's equivalent input noise is its actual output noise divided by its gain.

The *dynamic range* of an amplifier means the ratio between the smallest input signal which is larger than the equivalent input noise, and the largest input signal which produces an output below the compression level, expressed in decibels.

23.6 Impedances and Gain

The catalogue of desirable features of an amplifier is still not complete; in addition to low noise, high linearity and wide dynamic range, the input and output impedances need to be well defined, and the gain also. Further, steps to define these three parameters should not result in deterioration of any of the others. Figure 23.10A shows a broadband RF amplifier with its gain, input impedance and output impedance determined by negative feedback [Ref. 23.2]. The resistors used in the feedback network necessarily contribute some noise to the circuit. This can be avoided by the scheme known as *lossless feedback,* [Ref. 23.3], shown in Figure 23.10B. Here, the gain and the input and output impedances are determined by the ampere-turn ratios of the windings of the transformer.

While in a high-quality receiver the stages preceding the final bandwidth crystal filter need to be exceedingly linear, this requirement is relaxed in the stages following the filter; a little distortion in these will merely degrade the wanted signal marginally, since by that stage in the circuit all the unwanted signals have been rejected by the filter. It is usual to apply *automatic gain control* so that the level of the wanted signal at the receiver's output does not vary by more than a few decibels for an input level change of 100 dB or more. This is achieved by measuring the level of the signal, for example the

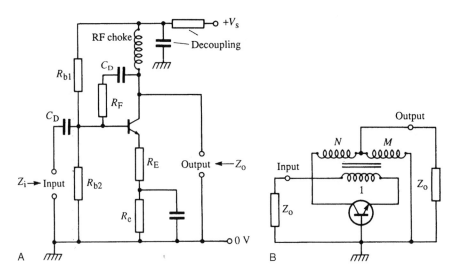

Figure 23.10: Input and output impedance determining arrangements. (A) Gain, input and output impedances determined by resistive feedback. R_{b1}, R_{b2} and R_e determine the stage DC conditions. Assuming the current gain of the transistor is 10 at the required operating frequency, then for input and output impedances in the region of 50Ω, $R_F = 50^2/R_E$. For example, if $R_E = 10\Omega$, $R_F = 250\Omega$, then $Z_i \approx 35\Omega$, $Z_o \approx 65\Omega$ and stage gain $\approx$ 10 dB, while if $R_E = 4.7\Omega$, $R_F = 470\Omega$, then $Z_i \approx 25\Omega$, Zo $\approx 95\Omega$ and gain $\approx$ 15 dB. C_D are blocking capacitors, e.g., 0.1 μF. (B) Gain, input and output impedances determined by lossless (transformer) feedback. The absence of resistive feedback components results in a lower noise figure and higher compression and third-order intercept points. Under certain simplifying assumptions, a two-way match to Z_o results if $N = M^2 - M - 1$. Then power gain = M^2, impedance seen by emitter = $2Z_o$ and by the collector.= $(N + M)Z_o$. This circuit arrangement is used in various broadband RF amplifier modules produced by Anzac Electronics Division of Adams Russel and is protected by US Patent 3 891 934: 1975 (DC biasing arrangements not shown). (Reprinted by permission of *Microwave Journal*.)

level of the carrier in the case of an AM signal, and using this to control the gain of the receiver. Since most of the gain is in the IF amplifier, this is where most of the gain reduction occurs, starting with the penultimate stage and progressing toward the earlier stages the larger the gain reduction required. The final IF stage may also be gain-controlled, but this must be done in such a way that it can still handle the largest received signals. Finally, in the presence of a very large wanted signal it may be necessary to reduce the gain of the RF amplifier. The application of AGC is usually "scheduled" to reduce the gain of successive stages in the order described, as this ensures that the overall noise figure of the receiver is not compromised.

A number of different schemes are used to vary the gain of radio-frequency amplifier stages, one of which, the dual-gate FET, has already been mentioned. The gain of a bipolar transistor can also be reduced, by reducing its collector current, but this also reduces its signal handling capability, so that only a few tens of millivolts of RF signal may be applied to the base. The available output is also reduced when AGC is applied. At one time, bipolar transistors designed specifically for gain-controlled IF amplifier stages were available. These used forward rather than reverse control, i.e., the collector current was increased to reduce gain. This had the advantage that the signal handling capability of the stage was actually increased rather than reduced with large signals. The change of gain was brought about by a spectacular fall in the f_T of the transistor as the collector current increased. At the constant intermediate frequency at which the device was designed to operate, this resulted in a fall in stage gain.

Discrete transistor IF stages are giving way to integrated circuits purpose designed to provide stable gain and wide range AGC capability. A typical example is the Plessey SL600/6000 series of devices, the SL6I0C and 611C being RF amplifiers and the 612C an IF amplifier. The devices provide 20 to 34 dB gain according to type, and a 50 dB AGC range. The range also contains the SL621 AGC generator. When receiving an AM signal, the automatic gain control voltage can be derived from the strength of the carrier component at the detector. With an SSB signal there is no carrier; the signal effectively disappears in pauses between words or sentences. So audio derived AGC is used, with a fast attack capable of reducing the gain to maintain constant output in just a few milliseconds, and at a rate of decay or recovery of gain of typically 20 dB per second. The disadvantage of this scheme is that a stray plop of interference can wind the receiver's gain right down, blanking the wanted signal for several seconds. The SL621 avoids this problem. It provides a "hold" period to maintain the AGC level during pauses in speech, but will nevertheless smoothly follow the fading signals characteristic of HF communication. In addition, interaction between two detector time constants, a level detector and a charge/discharge pulse generator, prevent stray plops and crashes from inappropriately winding the receiver gain down.

In critical applications such as the RF stage of a professional communications receiver, a different approach to gain variation is often employed. As noted previously, with an increasing input signal level the AGC scheduling would reduce the RF stage gain last. But if it is difficult to achieve sufficient linearity in the RF stage in the first place, it is

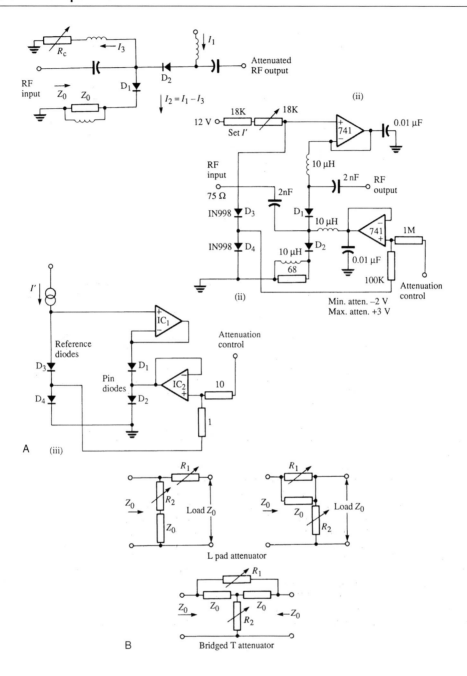

L pad attenuator

Bridged T attenuator

virtually impossible to maintain adequate linearity if the gain is reduced. So instead the gain is left constant and an electronically controlled *attenuator* is introduced ahead of the RF stage. The attenuator uses PIN diodes. PIN diodes can only operate as current-controlled linear variable resistors at frequencies at which the minority carrier lifetime in the intrinsic region is long compared with the period of one cycle of the RF. Even so, PIN diodes are available capable of operation down to 1 MHz or so, and can exhibit an on resistance, when carrying a current of several tens of millamperes, of an ohm or less. When off, the diode looks like a capacitance of 1 pF or less, depending on type. While a single PIN diode can provide control of attenuation when used as a current-controlled variable resistor in series with the signal path, the source and load circuits will be mismatched when attenuation is introduced. Two or more diodes can therefore be used, and the current through each controlled in such a manner as to implement an L pad [Ref. 23.4], which is matched in one direction (see Figure 23.11A), or a T or π pad, which is matched from both sides. In principle, an attenuator matched both ways can be implemented with only two diodes if the bridged-T circuit is used (Figure 23.11B).

It is only when receiving signals where the modulation results in variations of signal amplitude, such as AM and SSB, that AGC is required. With FM, PM and certain other signal types, no information is contained in the signal amplitude—other than an indication as to how strong the signal is. Any variations in amplitude are therefore entirely adventitious and are due to fading or noise or interference. The effect of fading can be suppressed, and that of noise or interference reduced by using a *limiting IF strip*; that is, one in which there is sufficient gain to overload the last IF stage even with the smallest usable signal. With larger signals, more and more of the IF stages operate in over-load; all the stages are designed to overload "cleanly," that is to accept an input as

Figure 23.11: Voltage-controlled RF attenuators using PIN diodes (A) (i) Pair of PIN diodes in L pad configuration, used to attenuate RF signals controlled by DC. Both I_1 and I_2 must be varied appropriately to control attenuation and keep Z_o constant. (ii) Working PIN diode attenuator must provide separation of the DC control current and RF signal paths. (iii) Constant attenuator impedance and temperature compensation are attained when the PIN diodes are matched against reference diodes in this arrangement. Op-amp IC_1 keeps the voltage drive to both sets of diodes equal, and IC_2 acts as a current sink control for the PIN diodes and as a temperature compensator. Control of attenuation is logarithmic (dB law). (B) L pad attenuators can provide a constant characteristic impedance Z_o as the attenuation is varied, but only at the input terminals. A bridged-T configuration can keep Z_o constant at both input and output terminals.

large as their output. Thus stages in limiting provide a gain of unity; in this way the effective gain of the IF strip is always just sufficient to produce a limited output, however small or large the input, without the need for any form of AGC. Here again, ICs have taken over from discrete devices in limiting IF strips, and other stages as well. For example, the Plessey SL6652 is a complete single-chip mixer/oscillator, IF amplifier and detector for FM cellular radio, cordless telephones and low-power radio applications. Its limiting IF strip has a maximum gain to small signals, before limiting sets in, of 90 dB, while the whole chip typically draws a mere 1.5 mA from a supply in the range 2.5 to 7.5V.

In contrast to FM and PM signals, for some signals the amplitude is the only useful information. For example, in a low-cost radar receiver a successive detection *log IF strip* is used to detect the returns from targets. As the strength of a return varies enormously depending upon the range and size of the target, an IF strip with a wide dynamic range is needed. The Plessey SL1613C is an IC wideband log IF stage with 12 dB gain RF input to RF output and a rectified output providing 1mA video current for a 500mV RMS signal input. The video output currents of successive stages are summed to provide an output whose amplitude is proportional to the logarithm of the signal amplitude, with a video rise time of only 70 ns. Six or more stages may be cascaded to provide 60MHz IF strips with up to 108 dB gain with better than 2 dB log linearity.

23.7 Mixers

Most modern receivers are of the superheterodyne type, with most of the amplification provided by the IF stages. This applies to broadcast receivers of all sorts, both sound and television; to professional communications, both civil and military, whether at HF (up to 30 MHz), VHF or UHF; and to receivers of other sorts, such as radar and navigation beacons. A *frequency changer, converter or mixer*—all names for the same thing—is used to translate the incoming signal from whatever frequency it was transmitted at to a fixed frequency, which is more convenient for providing high selectivity. In a single superhet such as Figure 23.7A the RF signal is applied, following amplification by one or more RF stages if fitted, to a mixer. This stage has two input ports and one output port. To the second input port is applied an RF signal generated locally in the receiver; this is called the *local oscillator* (LO). The mixer is a nonlinear device and thus produces sum and difference frequency components. For example, if the receiver of Figure 23.7B were tuned to receive a signal at 10 MHz (it might be the WWV standard time and frequency transmission, broadcast from the USA on that frequency) the local oscillator frequency could be either 23.6 MHz or 11.4 MHz, since in either case the difference frequency is equal to 1.4 MHz, the intermediate frequency. The sum frequency will also appear at the output of the mixer, but the IF filter rejects not only the sum frequency but the original RF and local oscillator signals as well, accepting only the wanted 1.4 MHz IF. In many cases the local oscillator frequency will be higher than the signal frequency ("high side injection," "LO runs high"); for example, the first LO in the 100 kHz to 30 MHz double superhet of Figure 23.7B would run from 70.1 MHz to 100 MHz.

It has already been noted that any device with second-order curvature of its transfer characteristic will produce not only second-harmonic distortion but also second-order intermodulation products, i.e., sum and difference tones. The mixer in an early valve superhet, also called the *first detector*, worked in exactly this manner: a half-wave rectifier circuit would do just as well. However, this type of mixer exhibits a large number of *spurious responses*. At its broadest, a receiver spurious response is any frequency at which a receiver produces an output other than the wanted frequency to which it is tuned. One example, the *image frequency* (formerly called the *second channel*), has already been noted: this is really a special case. Given sufficient front end selectivity, there will be no image response since no energy at that frequency can reach the mixer. In the up-converting superhet of Figure 23.7B, the image frequency will always be higher than 70.2 MHz, so a low-pass filter at the front end can suppress the image response entirely. This same filter will also prevent a response at the IF frequency by preventing any signals at 70 MHz reaching the mixer. However, the image and IF rejection are usually quoted separately in a receiver's specification, the term *spurious response* being reserved for unwanted responses due to much subtler and more insidious causes.

A mixer necessarily works by being nonlinear. It would be nice if the mixer produced only the wanted IF output, usually the difference frequency between the RF signal and local oscillator inputs. In practice the mixer may also produce an output at the intermediate frequency due to signals not at the wanted RF at all. A mixer, being a nonlinear device, will produce harmonics of the frequencies present at its inputs, and these harmonics themselves are in effect inputs to the mixer. So imagine the single superhet of Figure 23.7A tuned to receive a signal at 25 MHz. The LO will be at 26.4 MHz, and the second harmonic of this, at 52.8 MHz, will be lurking in the mixer just waiting to cause trouble. Imagine an unwanted signal at 25.7 MHz, too close to the wanted frequency to be much attenuated by the RF tuned circuits. The second harmonic of this, at 51.4 MHz, is exactly 1.4 MHz away from the second harmonic of the local oscillator and will therefore be translated to IF. This is variously called the *2–2 response* or the "half IF away" response, being removed from the wanted frequency by half the IF frequency. Similarly, the *3–3 response* will occur at a frequency removed from the wanted frequency by 1.4/3 MHz. Clearly these responses will not be a problem in the up-converting superhet of Figure 23.7B, which is one reason for the popularity of this design. However, it is not entirely immune from spurious responses. Imagine that it is tuned to 23 MHz, so that its first LO is at 93 MHz, and that there is a strong unwanted signal at 23.2 MHz. The fifth harmonic of the latter, at 116 MHz, is removed from the second harmonic (186 MHz) of the LO by 70 MHz. Admittedly this is a seventh-order response, and fortunately the magnitude of spurious responses falls off fairly rapidly as the order increases. But it does indicate that the ideal mixer is a very peculiar device: it must be very linear to two or more unwanted signals applied at the RF port (to avoid unwanted responses due to intermodulation), and should ideally only produce an output due to the RF and LO signals themselves, not their harmonics.

The spurious responses just described are termed *external* spurious responses, in that they appear in response to an externally applied signal which bears a particular relation to the LO frequency, and thus to the wanted frequency. *Internal* spurious responses, on the other hand, are totally self-generated in the receiver. Most professional communications receivers nowadays contain a microprocessor to service the front panel, to accept frequency setting data from a remote control input, to display the tuned frequency, and so on. Harmonics of the microprocessor's clock frequency can beat with either the first or the second local oscillator, to produce the same effect as an externally applied CW interfering signal. Needless to say, in a well-designed receiver such responses are usually at, or below, the receiver's noise level. However, there is also the possibility of the odd spurious response due to interaction of the first and the second LO, which makes the up-converting single superhet an attractive proposition now that advances in crystal filter technology make it possible. Most modern communications receivers have the odd internal "spur" in addition to the inevitable external spurious responses or "spurii".

A dual-gate FET can be used as a multiplicative mixer by applying the RF and LO voltages to gate 1 and 2 respectively. If the RF and LO voltages are represented by pure sinusoidal waveforms sin r and sin L, where sin r stands for $\sin(2\pi f_{RF} t)$ and sin (L) for sin ($2\pi f_{LO} t$), then, ignoring a few constants, the mutual conductance can be represented by sin r sin L. So the drain output current can be represented by $[\cos(r - L) - \cos(r + L)]/2$, courtesy of your friendly neighborhood math text-book; that is, it contains the sum and difference frequencies. The constants ignored in such a cavalier fashion are responsible for the presence in the drain current of components at the RF and LO frequencies, so the dual-gate FET mixer is described as *unbalanced*. However, if its operation were ideally multiplicative then these would be the only unwanted outputs, i.e., it would be free of spurious responses.

The presence in a mixer's output of components at the RF and LO frequencies can be a serious embarrassment. Consider the communications receiver of Figure 23.7B, for example. Such a receiver is typically specified to operate right down to an input frequency of 10 kHz. At this tuned frequency the LO will be running at 70.001 MHz, which is uncomfortably close to the IF at 70 MHz, bearing in mind that the LO signal is very large compared with a weak RF signal. So a balanced mixer is used. A *single-balanced* mixer is arranged so that the signal at one of the input ports (usually the LO port) does not appear at the output port; thus, it can effectively "reject" the LO. In a *double-balanced* mixer (DBM) neither of the inputs appears at the output, at least in the ideal case—and in practice this condition is nearly met, with RF and LO rejection figures typically greater than 20 dB.

Figure 23.12 shows three DBMs. The first is the basic *diode ring* mixer, so called because if you follow round the four diodes you will find they are connected head to tail (anode to cathode) like four dogs chasing each other in a circle. On positive-going half-cycles of the LO drive two of the diodes conduct, connecting one phase of the RF input to the IF port. On the other half-cycle the other two diodes conduct, reversing the phase fed to the IF port. A very large LO drive is used, so that for virtually all the time either one pair of diodes or the other is conducting heavily: the diodes (which are selected for close matching, or are monolithic) are in fact used simply as switches. The ring DBM is double balanced, produces the sum and difference frequencies, and exhibits about half as many spurious responses as an unbalanced mixer. The *conversion loss* (ratio of IF output power to RF input power) is about 7 dB; this is attributable to several different causes. Half of the input RF energy will contribute to the sum output and half to the difference: as only one of these is required there is an inherent 3 dB conversion loss, the other 3 or 4 dB being due to resistive losses in the on resistance of the Schottky diodes, and to transformer losses. The IF port is "DC coupled," and thus operates down to 0 Hz. This is the mode of operation when the diode DBM is used as a phase sensitive detector, the RF and LO frequencies then being identical. Where an IF response down to DC is not required, the inputs can be applied differently. For example, the LO can be applied to the DC coupled port and the IF output taken from one of the transformer coupled ports. While this has certain advantages in special cases, it is not usually used in a receiver, since LO radiation via the receiver's input port is then likely to be worse.

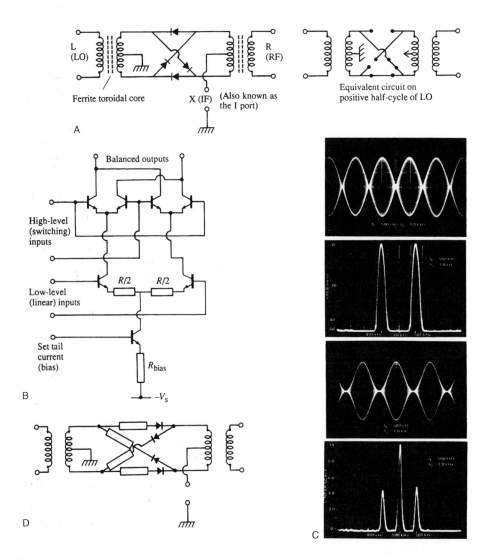

Figure 23.12: Double-balanced mixers (DBMs) (A) The ring modulator. The frequency range at the R and L ports is limited by the transformers, as also is the upper frequency at the X port. However, the low-frequency response of the X port extends down to 0 H$_z$ (DC). **(B)** Basic seven-transistor tree active double-balanced mixer. Emitter-to-emitter resistance *R*, in conjunction with the load impedances at the outputs, sets the conversion gain. **(C)** The transistor tree circuit can be used as a demodulator (see text). It can also, as here, be used as a modulator, producing a double-sideband suppressed carrier output if the carrier is nulled, or AM if the null control is offset. The MC1496 includes twin constant current tails for the linear stage, so that the gain setting resistor does not need to be split as in (B). (Reproduced by courtesy of Motorola Inc.). **(D)** High dynamic range DBM (see text).

Another well-known scheme (not illustrated here) uses MOSFETs instead of diodes as the switches [Ref. 23.5]. It is thus, like the Schottky diode ring DBM, a passive mixer, since the active devices are used solely as voltage-controlled switches and not as amplifiers. Reference 23.6 describes a single-balanced active MOSFET mixer providing 16 dB conversion gain and an output third-order intercept point of +45 dBm.

Figure 23.12B shows a double-balanced active mixer of the *seven-transistor tree* variety; the interconnection arrangement of the upper four transistors is often referred to as a *Gilbert cell*. The emitter-to-emitter resistance R sets the conversion gain of the stage; the lower it is made the higher the gain but the worse the linearty, i.e., the lower the third-order intercept point. This circuit is available in IC form (see Figure 23.12C) from a number of manufacturers under type numbers such as LM1496/1596 (National Semiconductor), MC1496/1596 (Motorola, Mullard/Signetics) and SG1496/1596 (Silicon General), while derivatives with higher dynamic range are also available.

Finally, in this whistle-stop tour of mixers, Figure 23.12D shows one of the simplest of the many ingenious ways in which the performance of the basic Schottky diode ring DBM has been improved—almost invariably, as here, at the expense of a requirement for greater LO power (up to +27 dBm is not uncommon). The resistors in series with the diodes waste LO power and increase the insertion loss, but they have beneficial effects as well. They permit a larger LO drive to be applied, which reduces the fraction of the LO cycle which is taken up by commutation, that is changing from one pair of diodes conducting to the other pair. They stabilize the effective on resistance of the diodes, which would otherwise vary throughout each half-cycle owing to the sinusoidal current waveform. Finally, they cause an additional voltage drop across the on diodes; this increases the reverse bias of the off diodes, thus reducing their reverse capacitance.

23.8 Demodulators

The DBM is also popular as both a modulator and a demodulator. A modern transmitter works rather like a superhet receiver in reverse, that is to say that the signal to be transmitted is modulated onto a carrier at a fixed IF and then translated to the final transmit frequency by a mixer, for amplification in the power output stages. In an SSB transmitter, the voice signal to be transmitted can be applied to the DC coupled port (also known as the X or I port) of a double-balanced mixer, while the LO signal is applied to one of the transformer coupled ports as in Figure 23.13. The output from the other transformer coupled port is a double-sideband suppressed carrier signal as shown,

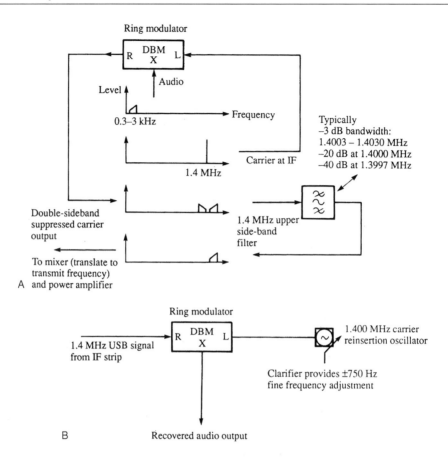

Figure 23.13: DBM used as modulator and demodulator. (A) DBM used as a modulator in an HF SSB transmitter. The carrier rejection of the mixer plus the 20 dB selectivity of the USB filter at 1.4MHz ensure that the residual carrier level is more than 40 dB down on the peak transmitter power. (B) DBM used as an SSB demodulator in an HF SSB receiver.

which can then be filtered to leave the SSB signal, either upper side-band (USB) or lower sideband (LSB) as required. (Amateur radio practice is to use LSB below 10 MHz and USB above, but in commercial and military applications USB is the norm regardless of frequency.) In the receiver, the reverse process can be applied to demodulate an SSB signal, i.e., the output of the IF strip is applied to one of the transformer coupled ports of a diode ring mixer, and a carrier wave at the frequency of the missing suppressed carrier at the other. The beat frequency between the two is simply the original modulating voice signal, but offset by a few cycles if the reinserted carrier is not at exactly the appropriate frequency. This results in reduced intelligibility and has been likened to the sound of Donald Duck talking through a drainpipe.

A control called a *clarifier* is usually provided on an SSB receiver to permit adjustment of the frequency of the reinserted carrier for maximum intelligibility. In practice an IC such as 1496DBM is often used for the demodulator: linearity is not of paramount importance in this application, since any signal in the pass band of the IF is either the wanted signal or unavoidable cochannel interference.

Having touched on the subject of SSB demodulation, it is appropriate to cover here demodulators—often called *detectors*—for other types of signals as well.

Figure 23.14A shows a diode detector as used in an AM broadcast receiver. It recovers the audio modulation riding on a DC level proportional to the strength of the carrier component the signal. This DC level is used as an AGC voltage, being fed back to control the gain of the IF stages, so as to produce an effectively constant signal even though the actual level may change due to fading. The result is usually acceptable, but AGC can give rise to unfortunate effects. For example, on medium wave after dark, signals from distant stations can be received but the nature of the propagation (via reflections from the ionosphere) can give rise to frequency selective fading, resulting in quite sharp notches in the received RF spectrum. If one of these coincides with the carrier component of an AM signal, the AGC will increase the IF gain to compensate. At the same time, as the sidebands have not faded in sympathy, the result is that the signal is effectively modulated by greater than 100%, resulting in gross distortion in the detected audio. It is unfortunate that this coincides with the increased output due to AGC action, resulting in a very loud and unpleasant noise!

Figure 23.14B shows one of the types of demodulator used for FM signals. It depends for its action upon the change of phase of the voltage across a parallel tuned circuit relative to the current as the signal frequency deviates first higher then lower in frequency than the resonant frequency. The reference voltage v_{ref} in the small closely coupled winding at the earthy end of the collector tuned circuit is in phase with the voltage across the latter. The center tapped tuned circuit is very lightly coupled to the collector tuned circuit, so the reference voltage is in quadrature with the voltage across the center tapped tuned circuit. The resulting voltages applied to the detector diodes are as indicated by the vector diagrams. Capacitors C' have a value of around 330 pF, so that they present a very low impedance at the usual FM IF of 10.7 MHz but a very high impedance at audio frequencies. As the detected output voltage from one diode rises, that from the other falls, so that the recovered audio appears in antiphase

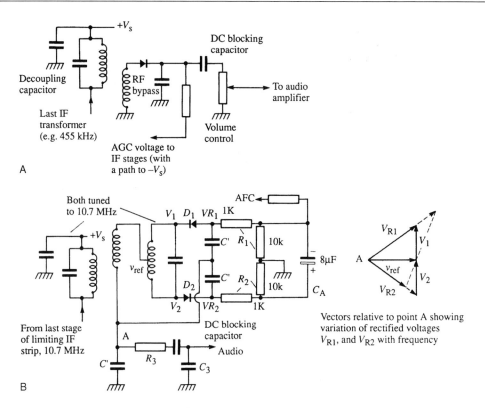

Figure 23.14: AM and FM demodulators (detectors) (A) Diode AM detector. In the "infinite impedance detector," e.g., Tr₃ in Figure 23.21, a transistor base/emitter junction is used in place of the diode. The emitter is bypassed to RF but not to audio, the audio signal being taken from the emitter. Since only a small RF base current is drawn, the arrangement imposes much less damping on the previous stage, e.g., the last IF transformer, while the transistor, acting as an emitter follower, provides a low-impedance audio output. (B) Ratio detector for FM, with de-emphasis. C' = RF bypass capacitor, 330 pF.

across R_1 and R_2, while the voltage across C_A is constant. R_3C_3 provides de-emphasis to remove the treble boost applied at the transmitter for the purpose of improving the signal/noise ratio at high frequencies: the time constant is 50 μs (75 μs is used in the USA). This type of frequency discriminator, known as the *ratio detector*, was popular in the early days of FM broadcasting, since it provided a measure of AM rejection to back up the limiting action of the IF strip. Any rapid increase or decrease in the peak-to-peak IF voltage applied to the diodes would result in an increase or decrease of the damping on the center tapped tuned circuit by the detectors, as C_4 was charged up or discharged again. This tended to stabilize the detected output level, while slow variations in level, due to fading for example, were unaffected. Modern FM receivers use IC IF strips with more than enough gain to provide hard limiting on the smallest usable signal, so an on-chip discriminator based upon quadrature detection by the Gilbert cell is normally used.

23.9   Oscillators

The next major category of circuit considered in this chapter is the RF oscillator. Every transmitter needs (at least) one, and receivers of the superhet variety also need one in the shape of the local oscillator. The frequency of oscillation is determined by a tuned circuit of some description. The hallmarks of a good oscillator are stability (of both output frequency and output level), good wave-form (low harmonic content) and low noise. An oscillator can be considered either as an amplifier whose output is applied via a band-pass filter back to its input, so as to provide positive feedback with a loop gain of just unity at one frequency; or as a circuit in which an active device is arranged to reflect a negative resistance in parallel with a tuned circuit, of value just sufficient to cancel out the losses and raise its Q to infinity. In practice, there is seldom any real difference between these apparently divergent views: Figure 23.15 illustrates the two approaches.

In Figure 23.15A a single tuned circuit with no coupled windings is employed. For the circuit to oscillate, Z_2 and Z_3 must be impedances of the same sign (both positive, i.e., inductances, or both negative, i.e., capacitances), while Z_1 must be of the opposite sign. The funny symbol is a shorthand sign for any three-terminal active device, be it valve, bipolar transistor or FET. Figure 23.16 shows a number of Figure 23.15A type oscillators, with their usual names. Of these, the *Clapp* (or *Gouriet*) is a circuit where the value of the two capacitors of the corresponding Colpitts oscillator has been increased and the original operating frequency restored by connecting another capacitor in series with them. To understand how this improves the stability of the oscillator, remember that any excess phase shift through the active maintaining device, resulting in its phase shift departing from exactly 180°, must be compensated for by a shift of the frequency of oscillation away from the resonant frequency of the tuned circuit, so that the voltage applied to the "grid" lags or leads the "anode" current by the opposite amount. This restores zero net loop phase shift, one of the necessary conditions for oscillation. By just how much the frequency of operation has to change to allow for any

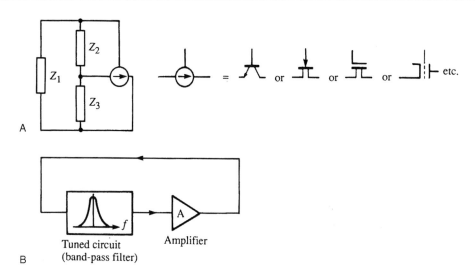

A

B

**Figure 23.15: Oscillator types. (A) Negative resistance oscillator: see text.
(B) Filter/amplifier oscillator.**

nonideal phase shift in the active device depends upon the Q of the tuned circuit. True, the Q is infinity, in the sense that the amplitude of the oscillation is not dying away, but that is only because the active device is making up the losses as they occur. As far as rate of change of phase with frequency in the tuned circuit is concerned, the Q is determined by the dynamic resistance R_d of the tuned circuit itself, in parallel with the loading reflected across it by the presence of the active device. In the case of a valve or FET, the anode or drain slope resistance is often the main factor: in the case of a bipolar transistor, the low base input impedance is equally important.

The additional capacitor C_1 in the Clapp circuit effectively acts in the same way as a step-down transformer, reducing the resistive loading on the tuned circuit, so that its loaded Q approaches more nearly to its unloaded Q. This improves the frequency stability by increasing the isolation of the tuned circuit from the vagaries of the maintaining circuit, but of course does nothing to reduce frequency drift due to variation of the value of the inductance and of the capacitors with time and temperature variations. The improved isolation of the tuned circuit from the active device cuts both ways. There is less drive voltage available at the active device's input and, at the same time, the load resistance reflected into its output circuit is reduced: both of these factors reduce the stage gain. Thus the Clapp circuit needs a device, be it valve, transistor or FET, with a high power gain.

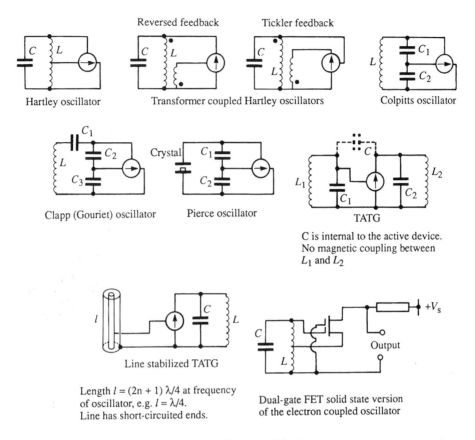

Figure 23.16: Negative resistance oscillators (biasing arrangements not shown).

Clearly the higher is the unloaded Q of the tuned circuit, the lower are the losses to be made up and hence the less gain is demanded of the maintaining circuit. Assuming a high output slope resistance in the active device, the losses will nearly all be in the inductor. If this is replaced by a crystal, which at a frequency slightly below its parallel resonant frequency will look inductive, a very high-Q resonant circuit results, and indeed the Clapp oscillator is a deservedly popular configuration for a high-stability crystal oscillator.

Figure 23.17 shows oscillator circuits of the Figure 23.15B variety. The TATG circuit in Figure 23.16 (named from its valve origins: tuned anode, tuned grid) is like the Meissner oscillator in Figure 23.17, except that the feedback occurs internally in the

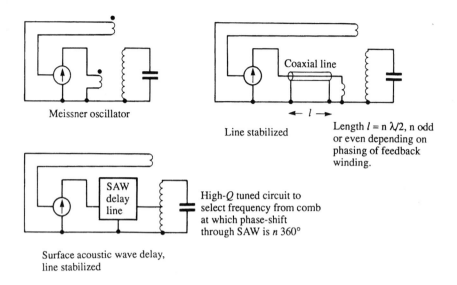

Meissner oscillator

Line stabilized

Coaxial line

$\leftarrow l \rightarrow$

Length $l = n\,\lambda/2$, n odd
or even depending on
phasing of feedback
winding.

SAW
delay
line

High-Q tuned circuit to
select frequency from comb
at which phase-shift
through SAW is $n\,360°$

Surface acoustic wave delay,
line stabilized

Figure 23.17: Filter/amplifier oscillators

device. The *line stabilized* oscillator is an interesting circuit, sometimes used at UHF where a coaxial line one wavelength long becomes a manageable proposition. By increasing the rate of change of loop phase shift with frequency, the line increases the stability of the frequency of oscillation. A *surface acoustic wave* (SAW) device can provide at UHF a delay equal to many wave-lengths. If the SAW device provides N complete cycles of delay, the rate of change of phase shift with frequency will be N times as great as for a single-wavelength delay. The SAW stabilized oscillator can thus oscillate at any one of a "comb" of closely spaced frequencies, a conventional tuned circuit being used to force operation at the desired frequency.

Figure 23.18 shows two oscillator circuits which use two active devices. In principle, two devices can provide a higher gain in the maintaining amplifier and thus permit it to be more lightly coupled to the tuned circuit, improving stability. But on the other hand the tuned circuit has to cope with the vagaries of two active devices instead of just one. The maintaining amplifier need not use discrete devices at all. The maintaining device can be an integrated circuit amplifier, or even an inverting logic gate used as an amplifier, as shown in Figure 23.19.

Where high stability of frequency is required, a crystal oscillator is the usual choice. For the most critical applications, an ovened crystal oscillator can be used. Here, the crystal

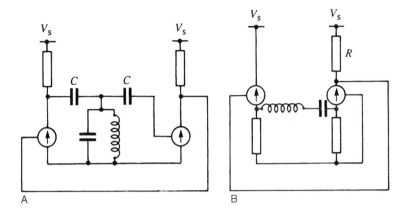

Figure 23.18: Two-device oscillators (A) Franklin oscillator. The two stages provide a very high noninverting gain. Consequently the two capacitors C can be very small and the tuned circuit operates at close to its unloaded value of Q. (B) Emitter coupled oscillator. This circuit is unusual in employing a series tuned resonant circuit. Alternatively it is suitable for a crystal operating at or near series resonance, in which case R can be replaced by a tuned circuit to ensure operation at the fundamental or desired harmonic, as appropriate.

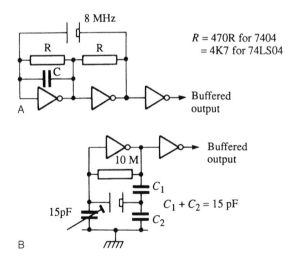

Figure 23.19: Crystal-controlled computer clock oscillators. (A) TTL type with crystal operating at series resonance. (B) CMOS type with crystal operating at parallel resonance.

itself and the maintaining amplifier are housed within a container, the interior of which is maintained at a constant temperature such as $+75°C$. Oven-controlled crystal oscillators (OCXOs) can provide a temperature coefficient of output frequency in the range 10^{-7} to 10^{-9} per °C, but stabilities of much better than one part in 10^{6} per annum are difficult to achieve. The best stability is provided by the glass encapsulated crystal, the worst by the solder seal metal can crystal, with cold weld metal cans providing intermediate performance. Where the time taken for oven warm-up is unacceptable and the heater cannot be left permanently switched on, a temperature-compensated crystal oscillator (TCXO) is used. In this, the ambient temperature is sensed by one or more thermistors and a voltage with an appropriate law is derived for application to a voltage-controlled variable capacitor (varicap). Both OCXOs and TCXOs are provided with adjustment means—a trimmer capacitor or varicap diode controlled by a potentiometer—with sufficient range to cover several years drift, allowing periodic readjustment to the nominal frequency.

In any oscillator circuit, some mechanism is needed to maintain the loop gain at unity at the desired amplitude of oscillation. Thus the gain must fall if the amplitude rises and vice versa. In principle, one could have a detector circuit which measures the amplitude of oscillation, compares it with a reference voltage and adjusts the amplifier's gain accordingly, just like an AGC loop. In this scheme, called an *automatic level control* (ALC) loop, the amplifier operates in a linear manner, for example in class A. However, it requires a detector circuit with a very rapid response, otherwise the level will "hunt" or, worse, the oscillator will "squegg" (operate only in short bursts). Most oscillator circuits therefore forsake class A and allow the collector current to be nonsinusoidal. This does not of itself ensure a stable amplitude of oscillation, but the circuit is arranged so that as the amplitude of oscillation increases, the device biases itself further back into class C. Thus the energy delivered to the tuned circuit at the fundamental frequency decreases, or at least increases less rapidly than the losses, leading to an equilibrium amplitude. In a transistor oscillator, stability is often brought about by the collector voltage bottoming, thus imposing heavy additional damping upon the tuned circuit. This is most undesirable from the point of view of frequency stability, and the current switching circuit of Figure 23.20B is much to be preferred.

Figure 23.20 also shows various ways in which the net loop gain of an oscillator can vary with amplitude. The characteristic of Figure 23.20B is often met though not particularly desirable. That of Figure 23.20C will not commence to oscillate unless kicked into oscillation by a transient such as at switch-on, a most undesirable characteristic. That of Figure 23.20D is representative of the current switching and

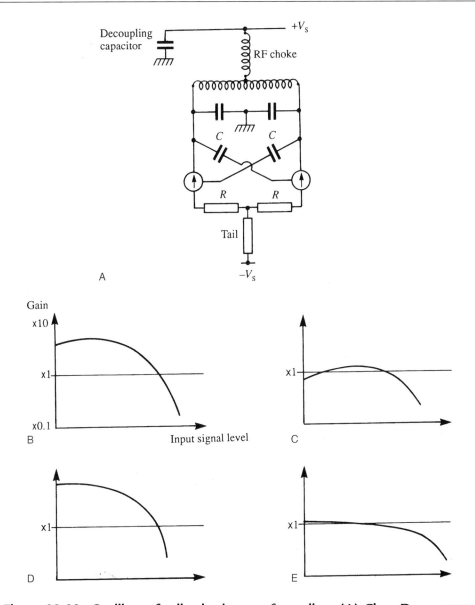

Figure 23.20: Oscillator feedback: degree of coupling. (A) Class D or current switching oscillator, also known as the Vakar oscillator. With *R* zero, the active devices act as switches, passing push-pull square waves of current. Capacitors *C* may be replaced by a feedback winding. *R* may be zero, or raised until circuit only just oscillates. "Tail" resistor approximates a constant current sink. (B–E) Characteristics (see text).

Vakar oscillators and is very suitable for a high-stability oscillator. That of Figure 23.20E results in an amplitude of oscillation which is very prone to amplitude variations due to outside influences. It is therefore excellent for a simple radio receiver designed to achieve most of its sensitivity by means of *reaction*, also known as *regeneration*, such as that shown in Figure 23.21 [Ref. 23.7]. With this circuit, as the reaction is turned up, the effective circuit Q rises toward infinity, providing a surprising degree of sensitivity. The greatest sensitivity occurs when the RF amplifier is actually oscillating very weakly; it is thus able to receive both CW and SSB signals. With AM signals its frequency becomes locked to that of the incoming signal and its amplitude varies in sympathy; anyone who has never played with a "straight" set (that is, not a superhet) with reaction has missed an experience.

It is not always convenient to generate an RF signal using an oscillator running at that frequency: an example is when a crystal-controlled VHF or UHF frequency is required, as crystals are only readily available for frequencies up to around 70 MHz. A common procedure in these cases is to generate the signal at a frequency of a few tens of megahertz and then multiply it in a series of doubler and/or tripler stages.

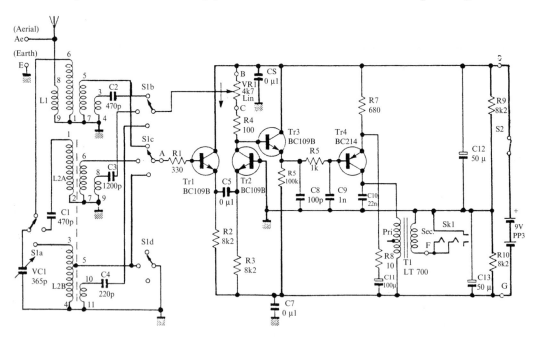

Figure 23.21: A straight receiver with reaction (regeneration). (Reproduced by courtesy of *Practical Wireless*.)

A multiplier stage is simply a class C amplifier with frequency *f* MHz applied to its input and with a tuned circuit resonant at *Nf* MHz as its collector load. The collector current contains harmonics of the input frequency, since for a single-ended amplifier stage only class A operation provides distortion-free amplification. The output tuned circuit selects the desired harmonic. In principle, the bias and drive level can be adjusted to optimize the proportion of the desired harmonic in the collector current; however, while this is worth doing in a one-off circuit, it is difficult to achieve in production.

It is important at any frequency, and particularly in RF circuits, to ensure that the signals to be amplified, multiplied, converted to another frequency or whatever, only proceed by the intended paths and do not sneak into places where they are not wanted, there to cause spurious responses, oscillations or worse. The main means of achieving this are *decoupling*, to prevent RF signals travelling along the DC supply rails, and *screening*, to avoid unintended capacitive or inductive coupling between circuits. At radio frequencies, screens of nonmagnetic metal are equally effective at suppressing unwanted magnetic coupling as well as electrostatic coupling. Supply rail decoupling is achieved by bypassing RF currents to ground with decoupling capacitors whose reactance is very low at the frequency involved, while placing a high series impedance in the supply rail, in the form of an inductance so as not to incur any voltage drop at DC. For a more detailed coverage of radio-frequency technology, see Ref. 23.8.

References

[23.1] Notes on the Theory of Modulation. J. R. Carson. *Proc. I.R.E.* Vol. 10, p. 57. February 1922.

[23.2] *Solid State Design for the Radio Amateur*. Hayward and DeMaw. 2nd printing 1986. p. 189, American Radio Relay League Inc.

[23.3] High Dynamic Range Transistor Amplifiers Using Lossless Feedback, D. E. Norton, p. 53. *Microwave Journal*. May 1976.

[23.4] Need a PIN-Diode Attenuator? R. S. Viles, *Electronic Design* 7, p. 100. March 29, 1977.

[23.5] Symmetric MOSFET Mixers of High Dynamic Range, R. P. Rafuse, p. 122. Session XI, 1968 International Solid State Conference.

[23.6] Single Balanced Active Mixer Using MOSFETs. E. S. Oxner, p. 292, *Power FETs and Their Applications*, 1982, Prentice-Hall.

[23.7] The PW Imp 3-Waveband Receiver. I. Hickman, p. 41. *Practical Wireless*, May 1979. NOTE: 'Plessey' devices are now manufactured by GEC Plessey Semiconductors Ltd.

[23.8] *Practical RF Handbook*, 2nd edition, 1997, Ian Hickman, Bitterworth-Heinemann.

CHAPTER 24

Signal Sources

Ian Hickman

Signal sources play an important role in electronic test and measurements, but their use is far from limited to that. They form an essential part of many common types of equipment. For example, a stabilized power supply needs an accurate DC voltage source as a reference against which to compare its output voltage. Many pieces of electronic equipment incorporate an audio-frequency signal source as an essential part of their operation, from the mellifluous warble of a modern push-button telephone to the ear-shattering squeal of a domestic smoke detector. And RF sources—oscillators— form an essential part of every radio transmitter and of virtually every receiver. So let's start with the DC signal source or voltage reference circuit.

24.1   Voltage References

The traditional voltage reference was the Weston standard cell, and these are still used in calibration laboratories. However, in most electronic instruments nowadays, from power supplies to digital voltmeters (DVMs), an electronic reference is used instead.

A Zener diode exhibits a voltage drop, when conducting in the reverse direction, which is to a first approximation independent of the current flowing through it, i.e., it has a low slope resistance. Thus if a Zener diode is supplied with current via a resistor from say the raw supply of a power supply (Figure 24.1(A), the voltage variations across the Zener—both AC due to supply frequency ripple and DC due to fluctuations of the mains voltage—will be substantially less than on the raw supply, provided that the value of the resistor is much greater than the diode's slope resistance. In practice, this means that about as many volts must be "thrown away" across the resistor as appear across the diode. Even so, the improvement is inadequate for any purposes other than

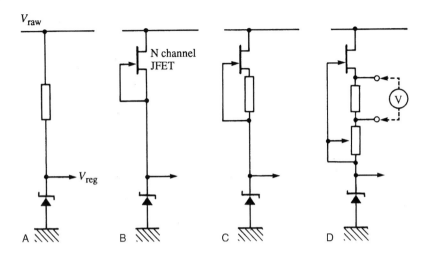

Figure 24.1: Zener DC voltage references, simple and improved (reproduced by courtesy of *New Electronics*)

the cheapest and simplest stabilized power supply. Figure 24.1(B) shows how the performance of the regulator can be notably improved by using the high drain slope resistance of a junction FET in place of the resistor. Unfortunately an FET is a lot dearer than a resistor. Two-lead FETs with the gate and source internally connected as shown are available as "constant current diodes" and work very well; unfortunately they are even more expensive than FETs, which themselves have always commanded a price ratio relative to small-signal bipolar transistors of about five to one. If an FET is used, the problem of the usual 5:1 spread in I_{dss} can be alleviated by including a source bias resistor, as in Figure 24.1(C), or even by adjusting it for a given drain current as in Figure 24.1(D).

Zener diodes have been much improved over the years. Earlier types left one with the difficult choice of going for lowest slope resistance—which was found in devices with a rating of about 8.2V—or for lowest temperature coefficient (TC or "tempco"), then found in 5.1V devices. With modern devices such as the Philips BZX79 series, lowest TC and lowest slope resistance occur for the same voltage rating device; that is, $+0.4\%/$°C and 10Ω at 5 mA respectively in the BZX79 C6V2 with its 6.2V $\pm 5\%$ voltage rating. A point to watch out for is that the measurement of a Zener diode's slope resistance is usually an adiabatic measurement. This means that a small alternating current is superimposed upon the steady DC and the resulting alternating potential is measured. The frequency of the AC is such that the diode's temperature does not have time to change in sympathy with each cycle of the current. If now there is a change in the value of the steady DC component of current through the diode, there will be an accompanying instantaneous small change in voltage δV due to δI , the change in current flowing through the slope resistance Rs, followed by a slower change of voltage due to the TC as the operating temperature of the diode changes. This clearly highlights the benefit of a range of diodes where the minimum slope resistance and TC can be had in one and the same device.

Returning to Figure 24.1(A), this arrangement can provide a stabilization ratio V_{raw}/V_{reg} of about 100:1 or 1%, whereas the FET aided version improves on this by a factor of about 30, depending on the FET's slope resistance. However, a useful if not quite so great improvement can be provided by the arrangement of Figure 24.2 [Ref. 24.1] Here the diode current is stabilized at a value of approximately $0.6/R_2$, since the PNP transistor's V_{be} changes little with change of emitter current. Consequently, if V_{raw} increases, most of the resultant increase in current through R_1 is shunted via the collector to ground rather than through the Zener diode. Where a modest performance, about 10 times better than Figure 24.1(A), is adequate, the circuit of Figure 24.2 offers a very cheap solution. Where substantially better performance is required, a voltage reference IC is nowadays the obvious choice. These are available from most manufacturers of linear ICs and operate upon the *bandgap* principle. A typical example is the micropower two-lead LM385-1.2 from National Semiconductor, which is used in series with a resistor or

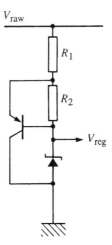

Figure 24.2: Inexpensive improved Zener voltage references (reproduced by courtesy of *New Electronics*)

constant current circuit, just like a Zener diode. This 1.2V reference device is available in 1% or 2% selection tolerance, operates over a current range of 10 μA to 20 mA, and features a dynamic impedance of 1Ω; the suffix X version features a TC at 100 μA of less than 30 PPM/°C. A 2.5V device, the LM385-2.5, is also available. Other commonly available reference voltage ICs come in various output voltages, including 5.0V, 10.0V and 10.24V.

24.2 Nonsinusoidal Waveform Generators

Sources of AC signals can be divided into two main categories: sine wave generators, and generators of nonsinusoidal waveforms. The latter can be subdivided again into pulse generators and other types. *Pulse generators* provide pulses of positive- or negative-going polarity with respect to earth or to a presettable DC offset voltage. The pulse repetition frequency, pulse width, amplitude and polarity are all adjustable; on some pulse generators, so too are the rise and fall times. Commonly also the output may be set to provide "double pulses," that is pulse pairs with variable separation, and a pulse delay with respect to a prepulse, which is available at a separate output for test and synchronization purposes. Pulse generators of this type are used mainly for test purposes in digital systems, so they are not considered further here. So let's press straight on and look at those "other types."

Nonsinusoidal or *astable waveform generators* may be categorized as operating in one of two modes, both of which are varieties of *relaxation oscillator*. As the name implies, the oscillation frequency is determined by the time taken by the circuit to relax or recover from a positive extreme of voltage excursion, toward a switching level at which a transient occurs. The transient carries the output voltage to a negative extreme and the circuit then proceeds to relax toward the switching level again, but from the opposite polarity. On reaching it, the circuit switches rapidly again, finishing up back at the positive extreme.

The two modes are those in which *differentiated* (phase advanced) *positive* feedback is combined with broad band negative feedback on the one hand, and types in which broad band positive feedback is combined with *integrated* (phase retarded) *negative* feedback on the other. Figures 24.3 and 24.4 show both discrete component and IC versions of these two types, respectively. The circuit operation should be clear from the circuit diagrams and waveforms given.

There is no reason why such an oscillator should not use differentiated positive feedback and integrated negative feedback, as in Figure 24.5(A); indeed, there is a

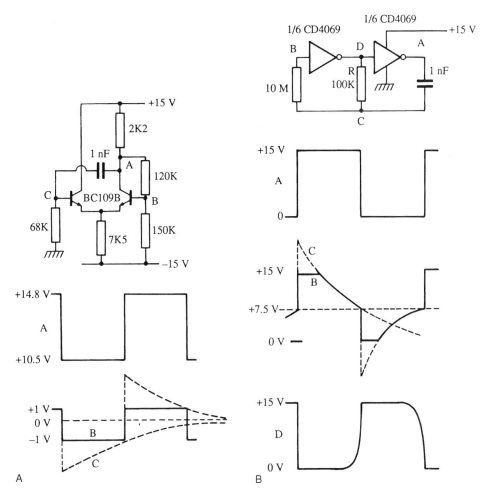

Figure 24.3: Astable (free-running) circuits using differentiated positive feedback and flat (broad band) negative feedback. (A) Cross-coupled astable circuit. The dashed line shows the 0V level at which the discharge at point C is aiming when it reaches the switching level. (B) Astable circuit using CMOS inverters. The waveform at B is similar to that at C except that the excursions outside the 0V and +15V supply rails have been clipped off by the device's internal gate protection diodes.

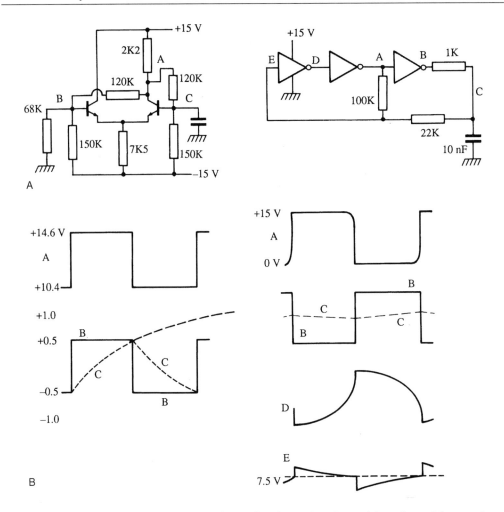

Figure 24.4: Astable (free-running) circuits using broad band positive and integrated (delayed) negative feedback. (A) Cross-coupled astable circuit. (B) Astable circuit using CMOS inverters.

definite advantage in so doing. It results in a greater angle between the two changing voltage levels at the point at which regeneration occurs, and this makes that instant less susceptible to influence by external or internal circuit noise. Thus, the frequency of oscillation is more stable, a worthwhile improvement since the frequency purity of astable oscillators generally is very much poorer than that of sinusoidal oscillators using an *LC* resonant circuit. In the latter the stored energy is much greater than any circuit noise, which consequently has less effect. However, a circuit such as Figure 24.5(A)

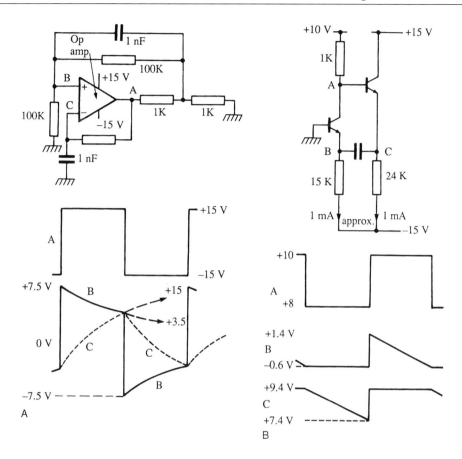

Figure 24.5: Other types of astable circuit. (A) Astable circuit using both differ-entiated positive and integrated negative feedback. Aiming potentials of points B and C prior to switching shown dashed. (B) The Bowes, White or emitter coupled astable does not have separate positive and negative feedback paths, so differing from the oscillators of Figures 24.3, 24.4 and 24.5(A).

contains two time constants, both of which play a part in determining the frequency. The circuit of Figure 24.3(B) will provide a 10:1 variation of frequency for a 10:1 variation of the resistance R forming part of the frequency determining time constant *CR*. The same applies to the circuit of Figure 24.5(A) only if more than one resistor is varied in sympathy. Thus the circuit of Figure 24.5(A) is more attractive in fixed frequency applications or where a tuning range of less than an octave is required. For wide frequency applications, as in a function generator providing sine, triangular and square output waveforms, it is not uncommon to opt for the economy of single resistor control.

Figure 24.5(B) shows a popular and simple astable oscillator circuit. There is only a single path around the circuit, for both the positive and the negative feedback. At any time (except during the switching transients) only one of the two transistors conducts, both tail currents being supplied via the 1K resistor or from the +15V rail.

The circuit of Figure 24.6 works on a slightly more sophisticated principle than the circuits of Figures 24.3 and 24.4, where the feedback voltage relaxes exponentially [Ref. 24.2]. It uses the *Howland current pump* to charge a capacitor, providing a linearly rising ramp. When this reaches the trigger level of half the supply rail voltage (at the noninverting input of the comparator), the trigger level, and the voltage drive to the current pump, both reverse their polarity, setting the voltage on the capacitor charging linearly in the opposite direction. The frequency is directly proportional to the output of the current pump and hence to the setting of the 10K potentiometer, which can

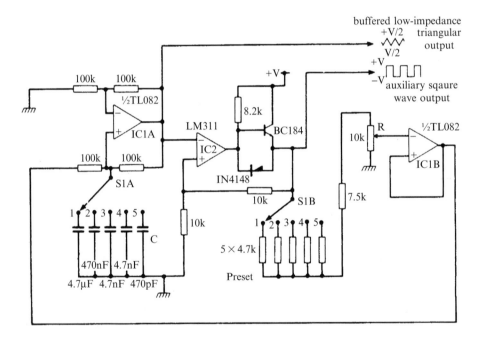

Figure 24.6: Function generator using a Howland current pump. The five 4K7 preset potentiometers enable the maximum frequency of the ranges to be set to 1 Hz to 10 kHz exactly; range capacitors C can thus be inexpensive 10% or even 20% tolerance types. If 10K resistor R is a ten-turn digital dial potentiometer, it will indicate the output frequency directly. +V and –V supplies must be equal, but frequency is independent of the value of V. (Reproduced by courtesy of *New Electronics*.)

be a multiturn type with a ten-turn digital dial. With the values shown the circuit provides five frequency ranges from 0 to 1 Hz up to 0 to 10 kHz, with direct read-out of frequency. Each range determining capacitor has an associated 4K7 preset resistor associated with it, enabling the full-scale frequency to be set up for each range, even though ordinary 10% tolerance capacitors are used. The circuit provides buffered low-impedance triangular and square wave outputs.

Most function generators provide a sine wave output of sorts. The popular 8038 function generator IC includes an on-chip shaping stage to produce a sine wave output by shaping the triangle waveform. This operates purely on a waveform shaping basis and thus works equally well at any frequency. An alternative scheme is to use an integrator: a triangular (linear) waveform is integrated to a parabolic (square law) waveform which forms a passable imitation of a sine wave, the total harmonic distortion being about 3.5%. However, the disadvantage of the integrator approach is that the output amplitude varies inversely with frequency, unless the value of the integrator's input resistor is varied to compensate for this.

An aperiodic (nonfrequency dependent) method of shaping a triangular wave into an approximation to a sine wave is to use an amplifier which runs gently into saturation on each peak of the triangular waveform. Unlike the integrator method, where the sharp point at the peak of the triangle wave becomes a slope discontinuity at the zero crossing point of the pseudo-sine wave, it is difficult with the aperiodic shaping method to avoid some residual trace of the point at the peak of the sine-shaped waveform. A scheme which has been used to avoid this is to slice off the peaks of the triangular wave before feeding it to the shaping circuit [Ref. 24.3]. In the reference cited, by choosing the optimum degree of preclipping and of nonlinearity of the shaping amplifier gain, distortion as low as 0.2% is achieved at low frequencies (a times ten improvement on the results usually achieved by this method). The shaping amplifier is implemented in an IC using a 1-GHz device process, resulting in good conversion of triangular waveforms to sine waves at frequencies up to 100 MHz.

Some function generators are capable of producing other waveforms besides the usual square/triangle/sine waves. A popular waveform is the sawtooth and its close cousin the asymmetrical triangle (see Figure 24.7). This figure also indicates how a stepwise approximation to any arbitrary waveform can be produced by storing the data values corresponding to say 256 successive samples of the waveform over one whole cycle in a read-only memory (ROM), and then reading them out sequentially to a digital-to-analog converter (DAC). In this way it is possible to reproduce natural sounds which have been

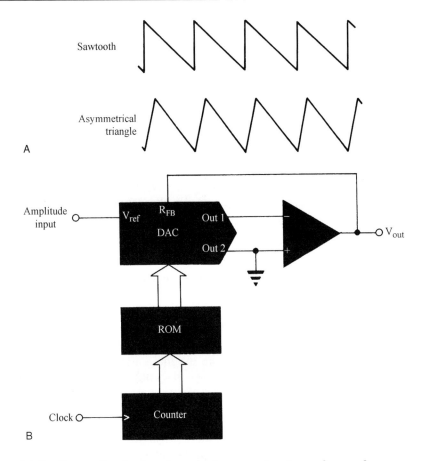

Figure 24.7: Generalized triangle waveform and universal waveform generator. (A) Sawtooth and asymmetrical triangle waveforms; both are generally provided by the more versatile type of function generator. The sawtooth and the triangular wave (Figure 24.6) can both be considered as limiting cases of the asymmetrical triangular wave. (B) Simple ROM waveform generator (reproduced by courtesy of *Electronic Engineering*).

recorded and digitized, for example the sound of a diapason or reed pipe from a real pipe organ, as is done in some electronic organs. The step nature of the output will correspond to very high-frequency harmonics of the fundamental, which in the organ application may well be beyond the range of hearing, but where necessary the steps can be smoothed off with a low-pass filter. This can still have a high enough cut-off frequency to pass all the harmonics of interest in the output waveform.

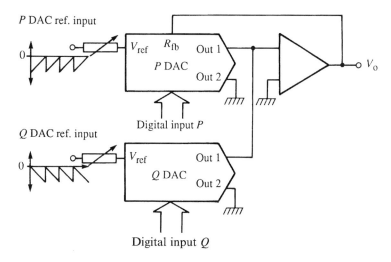

Figure 24.8: Interpolating DACs. (reproduced by courtesy of *Electronic Engineering*)

Another way of achieving a smooth, step-free output waveform is to make use of the multiplying capability of a DAC. The output current from a DAC is equal to the input bit code times the reference voltage input. Figure 24.8 shows two multiplying DACs with reverse sawtooth waveforms applied to their reference inputs so that as the output of the *P* DAC decreases, that of the Q DAC increases. Sample values are fed to the DACs at the same rate as the sawtooth frequency. When the output of the *P* DAC reaches zero, its input code is changed to that currently present at the input of the *Q* DAC. Immediately after this the sawtooth waveforms fly back to their initial values, so that the output from the *Q* DAC is now zero, and its input bit code is promptly changed to that of the next waveform sample. The output currents of the two DACs are summed to give a smoothly changing voltage output from the op-amp. The generation of a sine wave by this means is illustrated in Figure 24.9, but any arbitrary waveform can be produced once the appropriate values are stored in ROM. In practice, both the new DAC values are simply applied at the same instant that the sawtooth waveforms fly back to their starting values: any "glitch" in the output voltage, if appreciable, can be smoothed out with a little integrating capacitor across the summing op-amp's feedback resistor, which in Figure 24.8 is internal to the DAC.

An interesting application of this is for writing data on the screen of a real-time oscilloscope. Such an oscilloscope uses the electron beam to write the traces under control of the *X* and *Y* deflection plates, but it does not produce a raster scan like a TV

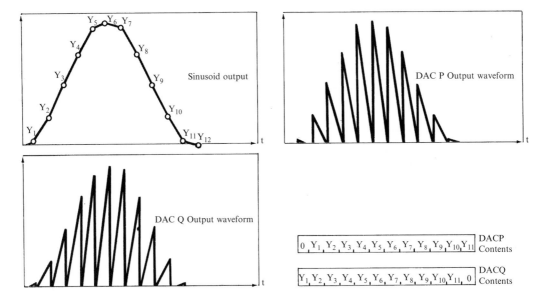

The DACP and DACQ contents tables:

0	Y_1	Y_2	Y_3	Y_4	Y_5	Y_6	Y_7	Y_8	Y_9	Y_{10}	Y_{11}	DACP Contents

Y_1	Y_2	Y_3	Y_4	Y_5	Y_6	Y_7	Y_8	Y_9	Y_{10}	Y_{11}	0	DACQ Contents

Figure 24.9: Waveform synthesis (reproduced by courtesy of *Electronic Engineering*).

display, so some other means is needed if information such as control settings is to be displayed on the screen. Two completely separate but complementary voltage waveform generators such as Figure 24.9 can be used to produce the appropriate X and Y deflection voltages to write alphanumeric data on the screen, the appropriate DAC data being stored in ROM. This scheme is used on many makes of oscilloscope. When the display read-out is on, it is possible under certain conditions to observe short breaks in the trace where the beam goes away temporarily to write the read-out data.

24.3   Sine Wave Generators

Turning now to sine wave generators, let's look first at *audio-frequency generators*. These generally do not use LC tuned circuits to determine the frequency, and therefore have a degree of frequency stability intermediate between that of tuned circuit oscillators and relaxation oscillators, and in some cases not much better than the latter. To measure the distortion of a high-fidelity audio power amplifier, one needs, in addition to a distortion meter, a sine wave source of exceptional purity. Not only must the source's distortion be exceedingly low, but its frequency stability must be of a very high order. This is because the usual sort of distortion meter works by rejecting the fundamental component of the amplifier's output with a narrow notch filter, so that the harmonics, residual noise and hum can be measured. Their level relative to the total output signal, expressed as a percentage, is the *total harmonic distortion* (THD) or, more strictly, the total residual signal if noise and hum are significant. Clearly, if the frequency of the sine wave generator drifts it will be difficult to set and keep it in the notch long enough to take a measurement. However, even if its drift is negligible, it may exhibit very short-term frequency fluctuations. Thus, it will "shuffle about" in the notch, resulting in a higher residual output than if its frequency were perfectly steady, as it tends to peep out first one side of the notch and then the other.

Now this is simply an explanation in the time domain of something which can equally well be explained in the frequency domain. Figure 24.10(A) shows an ideal sinusoidal signal, while Figure 24.10(B) shows, much exaggerated for clarity, a practical sine wave, warts and all. In addition to the ideal sine wave there are *close-in noise sidebands* of two sorts, AM and FM. These represent energy at frequencies very close to that of the sine wave, falling rapidly in amplitude as the frequency difference increases. The FM noise sidebands are the manifestation in the frequency domain of slight phase variations, which were noted as frequency shuffle in the time domain and which are shown as FM sidebands in Figure 24.10(B). There are also AM sidebands corresponding to slight amplitude variations in the sine wave, and these also will contribute to the residual. The residual may be considered as being responsible for it being impossible to say exactly where the tip of the vector in Figure 24.10(B) is at any time; it will be somewhere in the much exaggerated "circle of uncertainty" shown. (Note that noise sidebands, both AM and FM, are also found either side of the output frequency of an *LC* oscillator and even of a crystal oscillator; it is just that in those cases they are restricted by the high Q of the frequency determining components to a very much narrower fractional bandwidth about the centre frequency.) In a well-designed audio oscillator, the energy in the noise sidebands which is not rejected by the notch of the distortion meter is always lower in level than the energy of the harmonics.

Figure 24.11 shows an audio oscillator using the popular *Wien bridge* configuration. In Figure 24.11(A) you can see the principle of the thing. By using the idea of extremums—replacing a capacitor by an open-circuit at 0 Hz and by a short-circuit at infinite frequency—there will clearly be no signal at B at these frequencies. It turns out (the sums are not difficult, have a go) that at the frequency $f = 1/2\pi RC$ the amplitude at B is one-third of that at A and the two waveforms are in phase. At other frequencies the attenuation is greater and the waveforms are out of phase. If the bridge is just out of balance sufficiently to provide the necessary input to the maintaining amplifier, then the latter will drive the bridge at an amplitude adequate to produce the said input. If this

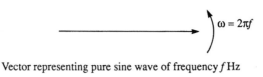

Vector representing pure sine wave of frequency f Hz

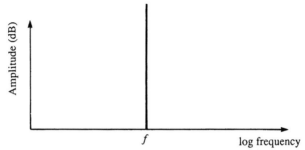

A   Pure sine wave of frequency f Hz represented in the frequency domain

Sine wave with AM and FM noise sidebands (A, F), grossly exaggerated

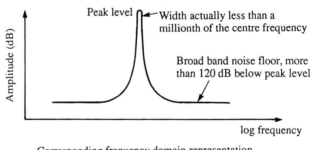

B   Corresponding frequency domain representation

Figure 24.10: Sine waves (A) Ideal pure sine wave. (B) Real-life sine wave.

sounds like a specious circular argument, it is: in the practical circuit of Figure 24.11(B) the necessary degree of bridge imbalance is provided by a thermistor. The usual type is an R53, which has a cold resistance of 5K (or $5 \times 10^3 \Omega$, hence the type number). At switch-on, the bridge is unbalanced by much more than is necessary, so that the positive feedback via the CR network exceeds the negative feedback via the thermistor/ resistor combination. Therefore, the circuit commences to oscillate at the frequency at

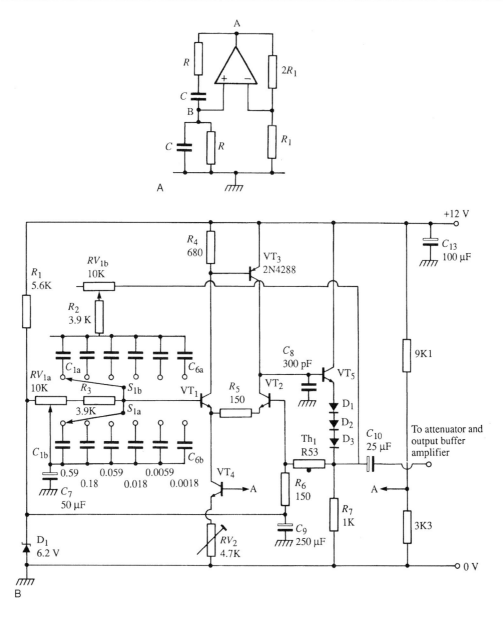

Figure 24.11: Audio-frequency Wien bridge sine wave oscillator. (A) Principle of oscillator using Wien bridge. (B) Low-distortion sine wave oscillator: 20–66 Hz, 66–200 Hz etc. up to 6.6–20 kHz. *RV*1 is semilog; *S*1 frequency range; all transistors BC109 except VT3: D1–D3 IN4148.

which the phase shift and attenuation of the *CR* network is least. As the amplitude of the oscillation builds up, the current through the thermistor heats it up. Now the thermistor consists of a pellet of amorphous semiconductor whose resistance falls rapidly with increasing temperature; the negative feedback via the thermistor/resistor arm therefore increases, and the bridge approaches balance. At an output voltage of about 3V peak to peak, the dissipation in the thermistor, with the circuit values shown, is approaching the rated maximum, corresponding to a temperature of the pellet inside its evacuated glass envelope of 125°C, and the output amplitude is stabilized. Oscillators operating on this principle are commercially available from many manufacturers, such are their popularity. The oscillator can even be made to cover the frequency range 10 Hz to 10 MHz, although it is not then possible to optimize the circuit for the lowest possible distortion in the audio-frequency range.

The main problems with a thermistor stabilized Wien bridge oscillator are amplitude bounce and poorish distortion. The former is due to the thermistor: it is found that on changing frequency, the amplitude of the output oscillates up and down several times before settling to a steady value. Running the thermistor near its maximum permitted dissipation helps to minimize this. The other problem is due to the limited selectivity of the Wien bridge, which does little to reduce any distortion in the maintaining amplifier, and (at frequencies below 100 Hz) to the finite thermal time constant of the thermistor.

The Wien bridge oscillator shown in Figure 24.11(B) uses a two-gang variable resistor to vary both resistors of the frequency determining network simultaneously. This keeps constant the attenuation through the network at the zero phase shift frequency. It can also provide a 10 to 1 frequency tuning range for a 10 to 1 resistance variation, as can be deduced from the formula for the frequency quoted above. There are numerous sine wave oscillator circuits which provide frequency variation using only a single variable resistor, but in these the frequency ratio obtained is only equal to the square root of the resistance variation [Refs. 24.4, 24.5].

An improved audio oscillator can be based on the *state variable filter*. Oscillation is ensured by the addition of fixed positive feedback and variable negative feedback applied to the inverting and noninverting inputs, respectively. The degree of NFB can be controlled by an FET, used as a variable resistor. In turn, the FET's resistance is controlled by a DC voltage proportional to the oscillator's peak-to-peak output voltage, so as to make the positive and negative feedback balance at the desired output amplitude. A variation on this scheme is shown in the SVF-based oscillator of Figure 24.12, which covers the frequency range 200–2000 Hz: here the PFB is variable, while the NFB or

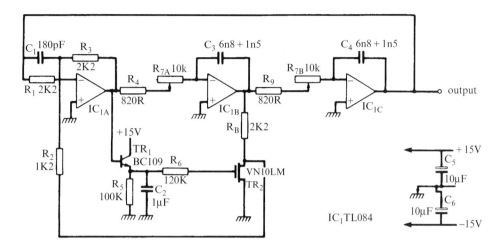

Figure 24.12: SVF-based oscillator with FET stabilizing: the FET costs one-tenth of an R53 thermistor (reproduced by courtesy of *New Electronics*).

damping is provided by the phase advance due to C_1 [Ref. 24.6]. Interestingly, Tr_1 can be replaced by a IN4148 diode with virtually no increase in the harmonic distortion, which is about 0.02%. This is because the resultant slight dent in the positive peak of the sine wave at IC_{1a}'s output, as C_2 is topped up, is composed of high-order even harmonics. These are heavily attenuated in the two following integrators IC_{1b} and IC_{1c}.

The circuit of Figure 24.12 exhibits two undesirable features, which a little lateral circuit design can circumvent. First, if other values of C_3 and C_4 are switched in, to provide 20–200 Hz and 2–20 kHz ranges, the smoothing time constant R_5C_2 is inadequate on the lower range, leading to increased distortion owing to the FET's resistance varying in sympathy with the ripple. Worse still, the time constant is excessive on the top frequency range, leading to amplitude bounce just like a thermistor stabilized oscillator and even complete instability of the level control loop. Second, the frequency is inversely proportional to the integrator time constants, leading to a very nonlinear frequency scale if the two-gang fine frequency control potentiometer (pot) R_7 has a linear resistance law. The scale is excessively open at the low-frequency end and terribly cramped at the top end. A somewhat better, more linear, scale results if a reverse taper log pot is used, but this is a rather specialized component. If the frequency scale is marked on the skirt of the knob rather than on the panel of the instrument, a normal log pot can be used, but there is still a problem due to the wide selection tolerance and poor law repeatability of log pots.

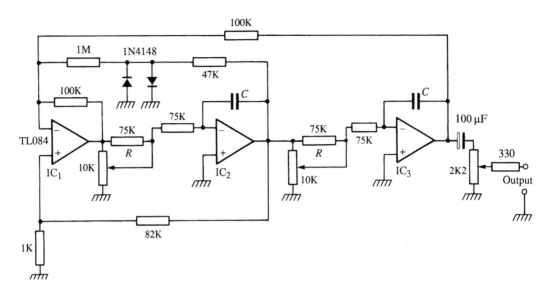

Figure 24.13: Low-distortion audio-frequency oscillator. Additional 75K resistors R result in an almost linear frequency scale. C is 10 nF for 0–200 Hz; 1 nF for 0–2 kHz; 100 pF for 0–20 kHz.

Both of these drawbacks are avoided by the circuit of Figure 24.13. Here, the degree of NFB applied to the noninverting terminal is fixed, while the PFB is applied to the inverting terminal via a diode clipping network. Thus, the oscillator works as a high-Q filter with a small square wave input of approximately fixed amplitude at its corner frequency. Amplitude control does not involve any control loop time constant and there is no rise in distortion below 100 Hz, such as is always found with thermistor-controlled circuits. The NFB or damping is set by the two resistors feeding back to the non-inverting terminal of IC_1. The ratio is one part in 83 and for feedback organized in this way the Q is one-third of the ratio, that is just over 27 in this case. If you assume that the input to the filter is a perfect square wave at the corner frequency, one would expect the third-harmonic component, which in a square wave amounts to one-third of the fundamental amplitude, to be attenuated by a factor of nine at the low-pass output from IC_3. For that is the theoretical attenuation of a frequency three times higher than the corner frequency at 12 dB/octave, relative to the flat response at low frequencies. Meanwhile, the fundamental at the corner frequency is actually accentuated by the factor Q, in this case 27. So for a per unit input, the output from IC_3 should consist of 27 per unit at the fundamental (approximately) and one-ninth of one-third at the third harmonic. So the third-harmonic component in the output should be $1/27^2$ per unit, which works out at about 0.15%. Despite a few approximations (the fundamental of a per unit square wave is itself slightly larger than unity, and the fifth and higher odd harmonics also contribute marginally to the distortion at the low-pass output), that is just about the distortion level measured in the output of the circuit of Figure 24.13 at all frequencies from 20 Hz to 20 kHz. Furthermore, if the integrators' frequency determining resistors are high in value compared with the value of the two-gang fine frequency control pot, a linear law pot will provide a substantially linear scale.

Having covered audio-frequency oscillators in some detail, let's turn our attention next to *radio-frequency oscillators*. The basic oscillator circuits have already been covered in the previous chapter, so here I will just look at a couple of interesting variations before moving on to see how oscillators can be integrated with other circuits to increase their flexibility and accuracy. Figure 24.14(A) shows a two-transistor RF oscillator designed to be free of all time constants other than that of the tuned circuit itself, so that it cannot "squegg," i.e., oscillate in bursts instead of continuously, as sometimes happens [Ref. 24.7]. *Squegging* is a form of relaxation oscillation usually involving a *CR* time constant forming part of the stage's biasing circuit. The two transistors form a DC coupled pair with 100% NFB. They can only oscillate at the frequency at which the tuned circuit provides phase inversion or 180° phase shift. Further, if the total resistance of $R_2 + R_3$ is greater than R1, then the tuned circuit must also provide a voltage step-up. C_1 and C_2 are in series as far as determining the frequency goes, and, by making $C_2 \gg C_1$, a wide tuning range can be achieved with the variable capacitor. If waveform is unimportant, R_2 and R_3 can be replaced by a single 10K resistor.

The question of suppressing unwanted modes of oscillation is particularly important in crystal oscillators, since in most cases the quartz crystal (which is simply a high-Q electromechanical resonator) can vibrate in several different modes, rather like the harmonics of a violin string or the overtones of a bell. Indeed many crystals are designed specifically to operate at a harmonic (often the third) rather than at their fundamental resonance, since for a given frequency the crystal is then larger and has a higher Q. In the highest-quality crystal oscillators, the *strain compensated* (SC) cut is used. This is a *doubly rotated* crystal, where the angle of the cut is offset from two of the three cystallographic axes. It has the advantage of a much lower temperature coefficient than the commoner cuts such as AT or BT, together with less susceptibility to shock and improved ageing characteristics (Figure 24.14(B)). These advantages do not come without an appropriate price tag, in that the SC crystal is more difficult and expensive to produce and has more spurious modes than other crystal types. For example, the 10 MHz SC crystal used in the Hewlett-Packard 10811A/B ovened

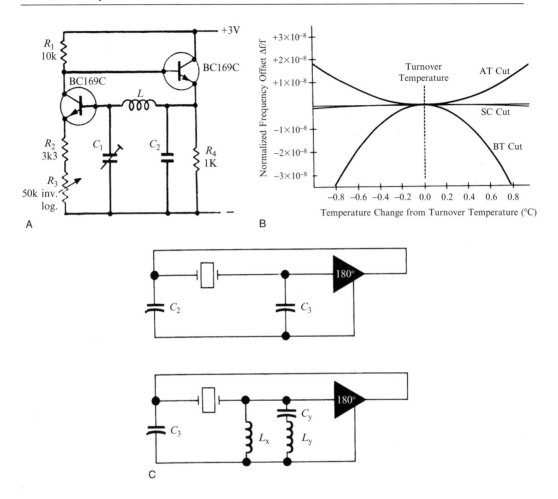

Figure 24.14: Oscillators with unwanted mode suppression. (A) This *LC* oscillator can only oscillate at the frequency at which the tuned circuit supplies 180° phase shift and voltage step-up (reproduced by courtesy of *Electronics and Wireless World*). (B) Temperature performance of SC, AT and BT crystal cuts. (C) Standard Colpitts oscillator (top) and the same oscillator with SC mode suppression (10811A/B oscillator). (Parts (B) and (C) reproduced with permission of Hewlett-Packard Co.)

reference oscillator is designed to run in the third-overtone C mode resonance. The third-overtone B mode resonance is at 10.9 MHz, the fundamental A mode is at 7 MHz, and below that are the strong fundamental B and C modes. The circuit of Figure 24.14(C) shows the SC cut crystal in the 10811A/B oscillator connected in what is basically a Colpitts oscillator, so as to provide the 180° phase inversion at the input of the inverting maintaining amplifier [Ref. 24.8]. With the correct choice of values for L_x, L_y and C_y, they will appear as a capacitive reactance over a narrow band of frequencies centred on the desired mode at 10.0 MHz, but as an inductive reactance at all other frequencies. Thus all of the unwanted modes are suppressed.

24.4 Voltage-Controlled Oscillators and Phase Detectors

Increasingly today receivers, and more particularly transmitters, have digital read-out and control of frequency. Consequently, where the old-time RF designer strove to design a variable frequency oscillator tuned with a mechanical variable capacitor, whose frequency (at any given capacitor setting) was very stable regardless of temperature and supply voltage, the design problem nowadays is usually slightly different. The oscillator is likely to be a voltage-controlled oscillator (VCO) forming part of a *phase-locked loop* synthesizer, where the oscillator's output frequency is locked to a stable reference such as a crystal oscillator. Not that synthesizers are the only application for VCOs. Figure 24.15 shows a VCO covering the frequency range 55–105 MHz, which is suitable for the generation of very wide deviation FM [Ref. 24.9]. Alternatively, if its output is mixed with a fixed 55 MHz oscillator, a very economical 0–50 MHz sweeper results.

Probably the earliest description of a phase-locked loop as such is to be found in the article "La Reception synchrone," by H. de Bellescize, published in *L'Onde Electronique*, vol. 11, pp. 230–40, June 1932. This described the synchronous reception of radio signals, in which a local oscillator operates at the same frequency as the incoming signal. If the latter is an amplitude modulated wave, such as in MW broadcasting, the audio can be recovered by mixing the received signal with the local oscillator, provided the local oscillator has exactly the same frequency as the carrier and approximately the same phase. This can be achieved by locking the frequency and phase of the local oscillator to that of the signal's carrier—in other words, a phase-locked loop. When the local oscillator and the mixer are one and the same stage, as in Figure 23.21, the result is a simple synchrodyne receiver, in which the carrier of the received signal takes control of the frequency and phase of the (just oscillating) detector circuit. In a synthesizer employing a phase-locked loop, however, the oscillator, the phase detector (mixer) and the all-important loop filter are all separate, distinct stages.

The operating principle of a synthesizer incorporating a phase-locked loop is indicated in the block diagram of Figure 24.16(A). A sample of the output of the oscillator is fed by a buffer amplifier to a variable ratio divider; let's call the division ratio N.

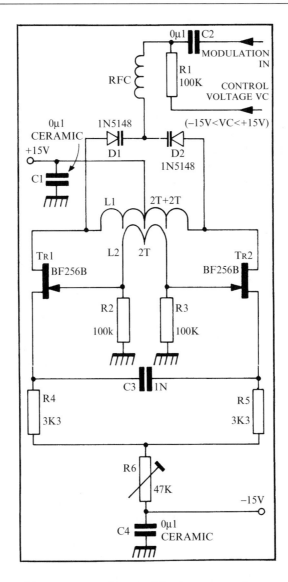

Figure 24.15: A wide range VCO, suitable for generating very wide deviation FM signals (reproduced by courtesy of *New Electronics*).

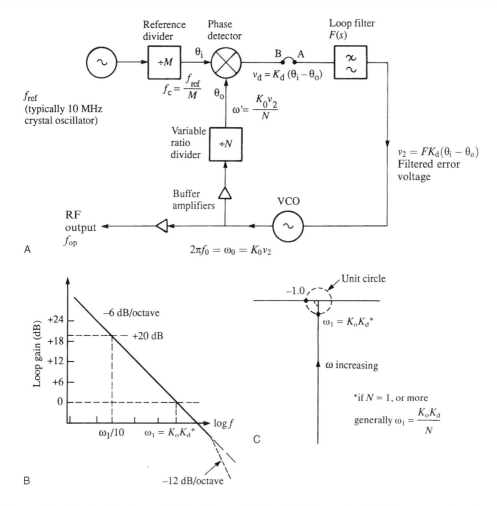

**Figure 24.16: Basic phase-locked loop (PLL). (A) Phase-locked loops synthesizer.
(B) Bode plot, first-order loop. (C) Nyquist diagram, first-order loop.**

The divider output is compared with a comparison frequency, derived from a stable
reference frequency source such as a crystal oscillator. An error voltage is derived
which, after smoothing, is fed to the VCO in such a sense as to reduce the frequency
difference between the variable ratio divider's a output and the comparison frequency.
If the comparison is performed by a frequency discriminator, there will be a standing
frequency error in the synthesizer's output, albeit small if the loop gain is high. Such an
arrangement is called a *frequency lock-loop* (FLL); these are used in some specialized
applications. However, the typical modern synthesizer operates as a phase-locked loop,

where there is only a standing phase difference between the ratio N divider's output and the comparison frequency. The oscillator's output frequency is simply Nf_c, where f_c is the comparison frequency. Thus f_c equals 12.5 kHz gives a simple means of generating any of the transmit frequencies used in the VHF private mobile radio (PMR) band, used by taxi operators, delivery drivers, etc., which is channelized in steps of 12.5 kHz in Europe (15 kHz in North America).

In fact, there is a practical difficulty in that variable ratio divide-by-N counters which work at frequencies up to 150 MHz or more are not available, but this problem is circumvented by the use of a prescaler. If a fixed prescaler ratio, say divide by 10, is used, then the comparison frequency must be reduced to 1.25 kHz to compensate. However, the lower the comparison frequency, the more difficult it is to avoid comparison frequency ripple at the output of the phase comparator passing through the loop filter and reaching the VCO, causing comparison frequency FM sidebands. Of course one could just use a lower cut-off frequency in the loop filter, but this makes the synthesizer slower to settle to the new channel frequency following a change in N. The solution is a 'variable modulus prescaler' such as a divide by 10/11 type. Such prescalers are available in many ratios, through divide by 64/65 up to divide by 512/514, thus a high comparison frequency can still be used: their detailed mode of operation is not covered in this book, as it is a purely digital topic.

A phase-locked loop synthesizer is an NFB loop and, as with any NFB loop, care must be taken to roll off all the loop gain safely before the phase shift exceeds 180°. This is easier if the loop gain itself does not vary wildly over the frequency range covered by the synthesizer. In this respect, a VCO whose output frequency is a linear function of the control voltage is a big help (see Figure 24.17) [Ref. 24.10]. The other elements of the loop equally need to be correctly proportioned to achieve satisfactory operation, so let's analyze the loop in a little more detail. Returning to Figure 24.16(A), the parameters of the various blocks forming the circuit have been marked in, following for the most part the terminology used in what is probably the most widely known treatise on phase-locked loops [Ref. 24.11]. Assuming that the loop is in lock, then both inputs to the phase detector are at the comparison frequency f_c, but with a standing phase difference $\theta_i - \theta_o$. This results in a voltage v_d out of the phase detector equal to $K_d(\theta_I - \theta_o)$.

In fact, as Figure 24.20 shows, the phase detector output will usually include ripple, i.e., alternating frequency components at the comparison frequency or at $2f_c$, although there are types of phase detector which produce very little (ideally zero) ripple. The ripple is suppressed by the low-pass loop filter, which passes v_2, the DC component

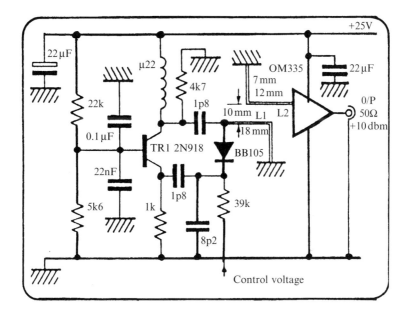

Figure 24.17: A linear VCO (reproduced by courtesy of *Electronic Engineering*).

of v_d, to the VCO. Assuming that the VCO's output radian frequency ω_o is linearly related to v_2 then $\omega_o = K_o v_2 = K_o F K_d (\theta_i - \theta_o)$, where F is the response of the low-pass filter. Because the loop is in lock, ω' (i.e., ω_o/N) is the same radian frequency as ω_c, the reference. If the loop gain $K_o F K_d/N$ is high, then for any frequency in the synthesizer's operating range, $\theta_i - \theta_o$ will be small. The loop gain must at least be high enough to tune the VCO over the frequency range without $\theta_i - \theta_o$ exceeding $\pm 90°$ or $\pm 180°$, whichever is the maximum range of operation of the particular phase detector being used.

Let's check up on the dimensions of the various parameters. K_d is measured in volts per radian phase difference between the two phase detector inputs. F has units simply of volts per volt at any given frequency. K_o is in hertz per volt, i.e., radians per second per volt. Thus while the filtered error voltage v_2 is proportional to the difference in phase between the two phase detector inputs, v_2 directly controls not the VCO's phase but its frequency. Any change in frequency of ω_o/N, however small, away from exact equality with ω_c will result in the phase difference $\theta_i - \theta_o$ increasing indefinitely with time. Thus the phase detector acts as a perfect integrator, whose gain falls at 6 dB per octave from an infinitely large value at DC. It is this infinite gain of the phase detector, considered as a frequency comparator, which is responsible for there being zero net average frequency error between the comparison frequency and f_o/N. (Some other writers alternatively consider the VCO as the integrator, producing an output phase which is proportional to the integral of the error voltage. It comes to the same thing either way; it's just that the explanation I have given seems clearer to me. You pays your money and you takes your choice.)

Consider a first-order loop, that is, one in which the filter F is omitted, or where $F = 1$ at all frequencies, which comes to the same thing. At some frequency ω_1 the loop gain, which is falling at 6 dB/octave due to the phase detector, will be unity (0 dB). This is illustrated in Figure 24.16(B) and (C), which show the critical loop unity-gain frequency ω_1 on both an amplitude (Bode) plot and a vector (Nyquist) diagram. To find ω_1 in terms of the loop parameters K_o and K_d without resort to the higher mathematics, break the loop at B, the output of the phase detector, and insert at A a DC voltage exactly equal to what was there previously. Now superimpose upon this DC level a sinusoidal signal, say 1V peak. The resultant peak FM deviation of ω_o will be K_o radians per second. If the frequency of the superimposed sinusoidal signal were itself K_o radians/second, then the modulation index would be unity, corresponding to a peak VCO phase deviation of ±1 radian. This would result in a phase detector output of K_1 volts (assuming for the moment that $N = 1$). If we increase the frequency of the input at A from K_o to $K_o K_d$, the peak deviation will now be $1/K_d$ and so the voltage at B will unity. So the loop unity-gain frequency ω_1 is $K_o K_d$ radians/second, or, more generally, $K_o K_2/N$, as shown in Figure 24.16(B) and (C). With a first-order loop there is no independent choice of gain and bandwidth; quite simply $\omega_1 = K_o K_d/N_1$. We could reintroduce the filter F as a simple passive CR cutting off at a corner frequency well above ω_1, as indicated by the dashed line in Figure 24.16(B) and by the teacup handle at the origin in Figure 24.16(C), to help suppress any comparison frequency ripple. This technically makes it a low-gain second-order loop, but it still behaves basically as a first-order loop provided that the corner frequency of the filter is well clear of ω_1 as shown.

Synthesizers usually make use of a high-gain second-order loop and I will examine that in a moment, but first a word as to why this type is preferred. Figure 24.18(A) compares the close-in spectrum of a crystal oscillator with that of a mechanically tuned LC oscillator and a VCO. Whereas, the output of an ideal oscillator would consist of energy solely at the wanted output frequency f_o, that of a practical oscillator is accompanied by undesired noise sidebands, representing minute variations in the oscillator's amplitude and frequency. In a crystal oscillator these are very low, so the noise sidebands, at 100 Hz either side, are typically –120 dB relative to the wanted output, falling to a noise floor further out of about –150 dB. The Q of an LC tuned circuit is only about

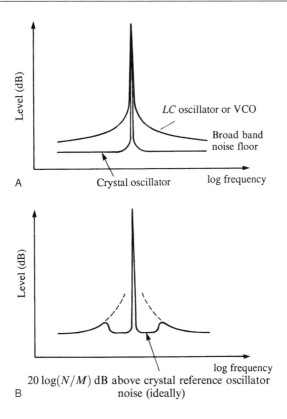

A

B

Figure 24.18: Purity of radio-frequency signal sources. (A) Comparison of spectral purity of a crystal and an LC oscillator. (B) At low-frequency offsets, where the loop gain is still high, the purity of the VCO (a buffered version of which forms the synthesizer's output) can approach that of the crystal derived reference frequency, at least for small values of *N/M*.

one-hundredth of the Q of a crystal, so the noise of a well-designed LC oscillator reaches –120 dB at more like 10 kHz off tune. In principle a VCO using a varicap should not be much worse than a conventional LC oscillator, provided the varicap diode has a high Q over the reverse bias voltage range; however, with the high value of K_o commonly employed (maybe as much as 10 MHz/volt), noise on the control voltage line is a potential source of degradation. Like any NFB loop, a phase-locked loop will reduce distortion in proportion to the loop gain. "Distortion" in this context includes any phase deviation of ω' from the phase of the reference f_c. Thus, over the range of offset from the carrier for which there is a high loop gain, the loop can clean up the VCO output substantially, as illustrated in Figure 24.18(B). In fact, the high loop gain will force θ_o, the instantaneous phase of ω', the output of the divider N, to mirror almost exactly ω_i, the instantaneous phase of the output of the reference divider M. Thus, the purity of ω' will equal that of ω_c, the reference input to the phase detector. It follows that the VCO output phase noise should be reduced to only N times that of the reference input to the phase detector, i.e., N/M times that of the crystal reference oscillator itself, as indicated in Figure 24.18(B). This assumes that the phase noise sidebands of f_c are in fact 20 log M dB below the phase noise of the crystal reference f_{ref}. Unfortunately this is often not so, since dividers, whether ripple dividers or synchronous (clocked) dividers, are not themselves noise free, owing to inevitable jitter on the timing edges. Likewise, the variable ratio divider will not in practice be noise free, so that the close-in VCO phase noise in Figure 24.18(B) will be rather more than 20 log(f_o/f_{ref}) dB higher than the phase noise of the crystal reference.

A second-order loop enables one to maintain a high loop bandwidth up to a higher frequency, by rolling off the loop gain faster. Consider the case where the loop filter is an integrator as in Figure 24.19(C); this is an example of a high-gain second-order loop. With the 90° phase lag of the active loop filter added to that of the phase detector, there is no phase margin whatever at the unity-gain frequency; as Figure 24.19(B) shows,

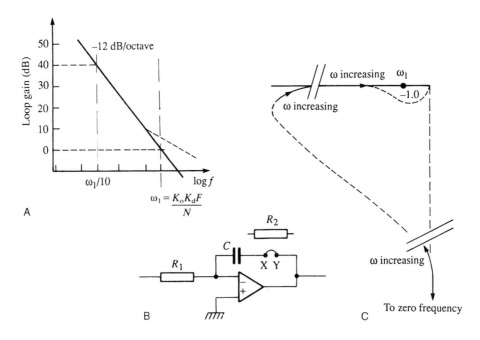

Figure 24.19: PLL with second-order active loop filter.

disaster (or at least instability) looms at ω_1. By reducing the slope of the roll-off in Figure 24.19(A) to 6 dB per octave before the frequency reaches ω_1 (dashed line), a phase margin is restored, as shown dashed in Figure 24.19(B), and the loop is stable. This is achieved simply by inserting a resistor R_2 in series with the integrator capacitor C at X–Y in Figure 24.19(C). This is the active counterpart of a passive transitional lag. If $R_1 = (2)R_2$, then at the corner frequency $\omega_c = 1/CR_2$ the gain of the active filter is unity and its phase shift is 45°, while at higher frequencies it tends to –3 dB and zero phase shift. If ω_c equals $K_o K_d = N$, then ω_1 (the loop unity-gain frequency) is unaffected but there is now a 45° phase margin. As Figure 24.19(b) shows, at frequencies well below ω_1, the loop gain climbs at 12 dB/octave accompanied by a 180° phase shift, until the op-amp runs out of open loop gain. This occurs at the frequency ω, where $1/\omega C$ equals AR_1, where A is the open loop gain of the op-amp; an op-amp integrator only approximates a perfect integrator. Below that frequency the loop gain continues to rise for evermore, but at just 6 dB/octave with an associated 90° lag due to the phase detector which, as noted, is a perfect integrator. This change occurs at a frequency too low to be shown in Figure 24.19(A); it is off the page to the top left. I have only managed to show it in Figure 24.19(B) by omitting chunks of the open loop locus of the tip of the vector.

For a high-gain second-order loop, analysis by the root locus method [Ref. 24.12] shows that the damping (phase margin) increases with increasing loop gain; so provided the loop is stable at that output frequency (usually the top end of the tuning range) where K_o is smallest, then stability is assured. This is also clear from Figure 24.19. For if K_o or K_d increases, then so will ω_1, the unity-gain frequency of the corresponding first-order loop. Thus ω_c (the corner frequency of the loop filter) is now lower than ω_1, so the phase margin will now be greater than 45°.

Having found a generally suitable filter, let's return for another look at phase detectors and VCOs. Figure 24.20 shows several types of phase detector and indicates how they work. The logic types are fine for an application such as a synthesizer, but not so useful when trying to lock onto a noisy signal, e.g., from a distant tumbling spacecraft; here the ex-OR type is more suitable. Both pump-up/pump-down and sample-and-hold types exhibit very little ripple when the standing phase error is very small, as is the case in a high-gain second-order loop. However, the pump-up/pump-down types can cause problems. Ideally, pump-up pulses—albeit very narrow—are produced however small the phase lead of the reference with respect to the variable ratio divider output; like-wise, pump-down pulses are produced for the reverse phase condition. In practice, there may be a very narrow band of relative phase shift around

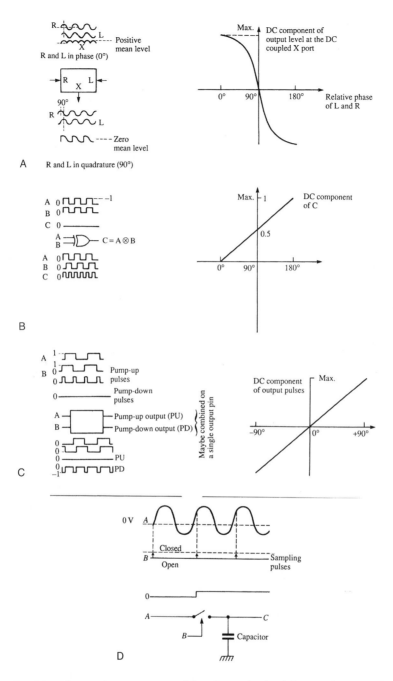

Figure 24.20: Phase detectors used in phase-locked loops (PLLs). (A) The ring DBM used as a phase detector is only approximately linear over say ±45° relative to quadrature. (B) The exclusive-OR gate used as a phase detector. (C) One type of logic phase detector. (D) The sample-and-hold phase detector. In the steady state following a phase change, this detector produces no comparison frequency ripple.

the exactly in-phase point, where neither pump-up nor pump-down pulses are produced. The synthesizer is thus entirely open loop until the phase drifts to one end or other of the "dead space," when a correcting output is produced. Thus the loop acts as a bang-bang servo, bouncing the phase back and forth from one end of the dead space to the other—evidenced by unwanted noise sidebands. Conversely, if both pump-up and pump-down pulses are produced at the in-phase condition, the phase detector is no longer ripple free when in lock and, moreover, the loop gain rises at this point. Ideally the phase detector gain K_d should, like the VCO gain K_o, be constant. Constant gain, and absence of ripple when in lock, are the main attractions of the sample-and-hold phase detector.

In the quest for low noise sidebands in the output of a synthesizer, many ploys have been adopted. One very powerful aid is to minimize the VCO noise due to noise on the tuning voltage, by substantially minimizing K_o, to the point where the error voltage can only tune the VCO over a fraction of the required frequency range. The VCO is pretuned by other means to approximately the right frequency, leaving the phase-locked loop with only a fine tuning role. Figure 24.21 shows an example of this arrangement [Ref. 24.13].

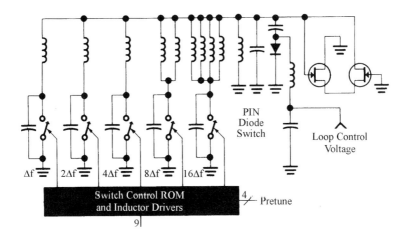

Figure 24.21: This VCO used in the HP8662A synthesized signal generator is pretuned to approximately the required frequency by the microcontroller. The PLL error voltage therefore only has to tune over a small range, resulting in spectral purity only previously attainable with a cavity tuned generator, and an RF settling time of less than 500 μs. (Reproduced with permission of Hewlett-Packard Co.)

References

[24.1] Voltage Reference Circuit, I. March, p. 22, *Electronic Product Design*, April 1981.

[24.2] Triangle Generator Reads Frequency Directly, I. March, p. 24, *New Electronics, 13 August* 1985.

[24.3] New Sine Shaper, p. 37, *Hewlett Packard Journal*, Vol. 34, no. 5, June 1983.

[24.4] RC Oscillators, Single-element Frequency Control, P. Williams, p. 82, *Wireless World*, December 1980.

[24.5] Easily Tuned Sinewave Oscillators, R.C. Dobkin, p. 116, *Electronic Engineering,* December 1971.

[24.6] FET Controlled Oscillator, I. Hickman, p. 19, *New Electronics*, 25 November 1986.

[24.7] Good-tempered LC Oscillator, G.W. Short, p. 84, *Wireless World*, February 1973.

[24.8] SC-Cut Quartz Oscillator Offers Improved Performance, J.R. Burgoon, R.L. Wilson, p. 20, *Hewlett Packard Journal* Vol. 32, no. 3, March 1981.

[24.9] A Wide Range VCO, I. March, p. 19, *New Electronics*, 14 December 1982.

[24.10] A Linear Voltage Controlled Oscillator, J. Dearden, p. 26, *Electronic Engineering*, June 1983.

[24.11] *Phaselock Techniques*, F.M. Gardner, John Wiley and Sons Inc., 1966.

[24.12] *Automatic Feedback Control System Synthesis*, J. G. Truxal, McGraw Hill, New York, 1955.

[24.13] Low-Noise RF Signal Generator Design, p. 12, *Hewlett Packward Journal*, Vol. 32, no. 2, February 1981.

CHAPTER 25

EDA Design Tools for Analog and RF

Bonnie Baker
Peter Wilson

They say a computer-based simulation of your analog circuit is important. This is because the use of your preferred computer Simulation Program with Integrated Circuit Emphasis (SPICE) program can reduce initial errors and development time. If you use your SPICE simulator correctly, you can drum out circuit errors and nuances before you go to your breadboard. In this manner, you will verify your design before you spend the time soldering your circuit. SPICE helps troubleshoot bench problems; it is a great place to try out different hypotheses. It is also great at "what if" scenarios (for example, exploratory design).

You can view the results from these software tools on a PC with user-friendly, graphical user interface (GUI) suites. This tool will fundamentally provide DC operating (quiescent) points, small signal (AC) gain, time domain behavior and DC sweeps. At a more sophisticated level, it will help you analyze harmonic distortion, noise power, gain sensitivity and perform pole-zero searches. This list is not complete, but generally, SPICE software manufacturers have many of these fundamental features available for the user. By finessing the Monte Carlo and worst-case analysis tools in SPICE, you can predict the yields of your final product. If you use your breadboard for this type of investigation, it could be very expensive and time consuming. All of these SPICE simulation advantages will speed up your application circuit time-to-market.

But, beware. You can effectively evaluate analog products if your SPICE models or macromodels are accurate enough for your application. The key words here are "accurate enough." Such models, or macromodels, should reflect the actual performance of the component without carrying the burden of too many circuit details.

Too many details can lead to convergence problems and extremely long simulation times. Not enough details can hide some of the intricacies of your circuit's performance. Worse yet, your simulation, whether you use complete models or just macromodels, may give you a misrepresentation of what your circuit will really do. Remember that a SPICE simulation is simply a pile of mathematical equations that, hopefully, represent what your circuit will do. In essence, a computer product produces imaginary results.

So you might ask, "Why bother?" Is a SPICE simulation worth the time and effort? A pop quiz will help you clarify this question. The circuit in Figure 25.1 shows a fundamental, basic circuit. Is this circuit stable or does it oscillate? Would the output of the amplifier have an unacceptable ring? I would think that you would quickly look at this and say, "That is a silly question. Of course it is stable!" But then again, if you are always looking for the trick question you may be suspicious. So what is the answer?

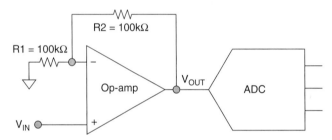

Figure 25.1: A variety of applications throughout the industry have this simple sub-circuit embedded in the system. This circuit simply takes an analog input signal and gains that signal to the output of the amplifier. For instance, an input signal of +1 V_{DC} would become a +2 VDC signal at V_{OUT}. The question is, would this DC signal oscillate? Or, would a 50 kHz sinusoidal signal oscillate or ring? The bandwidth of this amplifier is 2.8 MHz.

This simple amplifier circuit uses an amplifier in a gain of +2 V/V. The amplifier has a 100 kΩ resistor connected to its inverting-input to ground, and 100 kΩ resistor in the feedback loop. It would be easy to assume that this circuit is stable. However, tedious calculations will verify that this amplifier circuit will ring. This is due to the parasitic capacitances around the resistors and the high differential/common-mode capacitance of the amplifier's input stage. For this particular amplifier, the input common-mode capacitance is 6 pF and the differential-mode capacitance is 3 pF. These capacitances interact with the feedback resistor causing a semi-unstable condition. If you bench-test this circuit, you will immediately see this condition on the oscilloscope. Parasitics on the breadboard will aggravate this instability.

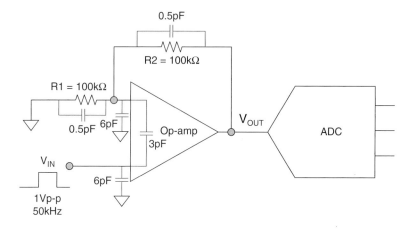

Figure 25.2: By enhancing the circuit diagram in Figure 25.1 with the parasitic capacitances of the resistors and amplifier, a simple of a circuit is not so simple. In the DC domain these capacitors will operate as open circuits. In the AC domain, the capacitors will affect the perfect square wave from the input to output. The perfect square wave will have quite a ring at the V$_{OUT}$ node.

For example, the 100 kΩ resistor in the feedback loop of the amplifier will also have approximately a parallel 0.5 pF parasitic capacitor (see Figure 25.2). This parasitic capacitance on the breadboard to ground could be as high as 2 pF or 3 pF. If you use the amplifier's SPICE macromodel, with input impedances in the model and board parasitics, you will see this problem immediately in your simulation. If you breadboard the circuit, you most certainly will see this ringing.

Changing the values of the two resistors in this circuit solves this problem. Hand calculations will help you find the correct values. A SPICE simulation will facilitate the process. This is a little easier than swapping out resistors on the breadboard until you find the right values. In SPICE, you can also look at the response of the amplifier using various resistors. This will help you find the "corner" of this oscillation. If you go back and change both values to 10 kΩ, you will have great success in SPICE and on the bench.

Figure 25.3. shows the simulation result.

In this chapter, we will discuss how to best determine if your SPICE simulation is telling you the truth. We do this by using three techniques. First, we will go through a short list of a few rules of thumb, which will help you examine the validity of your simulations; second, we will use common sense (at least where your circuit is concerned) when you first

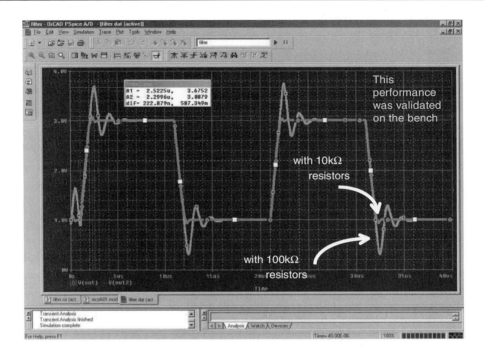

Figure 25.3: You can quickly verify that this simple circuit will ring using a SPICE simulation. If you need to double-check this with a breadboard circuit, that is also a good idea, however, reducing the 100 kΩ resistors down to 10 kΩ resistors solves the problem. You do need to understand where the problem came from before you continue with your circuit design. But this simulation caught a significant stability problem. This ringing problem was an easy one to miss by inspection of the schematic.

examine the results of your simulation; third, we will engage in an overview of what a macromodel can (or can't) do for you. With this arsenal, you will be able to effectively use a SPICE simulation to weed out most of your circuit problems.

I'll center the discussion on signal quality operational amplifiers. This is only because an amplifier is embedded somewhere in most purely analog circuits. I am going to leave it up to you to explore the remainder of analog macromodels, such as instrumentation amplifiers, difference amplifiers, references and so forth.

What won't we cover in this chapter? I don't intend to give you tips on how to use your favorite SPICE simulator tool. I'll leave that topic up to your SPICE vendor and the numerous books on this topic. I also will not attempt to think for you by giving you cookbook answers to your problems. Rather, I am going to ask you to think through things yourself. Basically, you need to ask, "What do you expect from your circuit? Does your SPICE simulation match your expectations? Why or why not?"

The naysayers in the industry will tell you that your computer-based simulation tools will not work and using them will be a waste of time. These people are a bit misguided, and in my opinion have a superficial view of what this tool can really do. Sure, SPICE tools can lead you astray. But, like any tool, it is only as good as the user. Any insight that you gain from your simulations are provided if your SPICE tools are understood and used properly. Better yet, SPICE simulations will point out problems that you had never anticipated. In most cases, they use double-precision calculations. This makes it easier to detect low-level problems that are impossible to find on the bench. But step back and look at what you have. You very likely will have a simulation circuit that is built out of macromodels that are generated by various manufacturers of the products you are interested in using in your circuit.

The questions that bear asking are, "Does my model or macromodel simulate over temperature? Distortion? AC spec? Over process? Am I expecting the model to simulate these parameters? And to what degree of accuracy? What information do the macromodels I am using really provide?" The only way you can answer these questions is to have a feel for what your circuit will do (in real life), and ask challenging questions about your SPICE simulation results.

To give you a little taste of my experiences with SPICE, during one of my many lives, I was an analog IC designer. As a designer, I was using a transistor in an unorthodox manner. Mind you, my transistor models were the best that we had. My models were full-blown transistor-level models, designed to simulation the accurate behavior of my amplifier design. I was not using a macromodel substitution, but the "real" thing. I believed that a certain configuration of a transistor in the output stage would allow me to design a single-supply amplifier that could go nearly (within a few millivolts) to the negative rail. Although this type of circuit operation is not new today, in 1990 it had a degree of innovation. I arrived at this belief by examining and thinking through the circuit operation on paper. It was a very cool idea!

The simulation told me that there was no way my amplifier would even go near the negative rail. Therefore, I questioned my paper calculations and then the SPICE models. This exercise did not produce any answers, so I finally built the circuit on a breadboard. The tests from the breadboard circuit proved that my paper calculations were correct.

With this verification, I went back and created a special model for the culprit transistor in my circuit. I went to the bench to create the new transistor model, using a single transistor in a TO-99 package. This model was required because I was using the transistor in an unorthodox configuration. With those transistor changes, my amplifier circuit model did simulate the circuit accurately. At the "end-of-the-day," I had all three tools in agreement. Modifying the transistor model did this. The problem could have been in my calculations or on the bench, but with these three votes, I was certain that I had a winner. The required three votes were the hand calculations, SPICE simulation and breadboard.

You probably will not dig into your circuits to this level of detail, however, the three steps to a successful, expeditious circuit design still hold true. These steps are: 1) Draw up the concept on paper, 2) Simulate your circuit to your satisfaction, 3) Breadboard critical portions of the circuit. All three of these steps are critical. Your SPICE simulations will not replace any of these steps; it will improve the likelihood of your success.

All of these steps were necessary:

1. Imagine what the circuit will do;

2. Simulate the circuit to match what you imagine it should do;

3. Breadboard the circuit and ensure the pencil design and simulation match. You need all three for a good level of confidence;

4. Adjust the model (or circuit) to match the three conditions (imagined operation, simulated operation and actual operation);

5. Always build a prototype.

You might ask me why I bothered with this level of detail in my amplifier design project. Not only did I double-check the operation of the circuit using a breadboard circuit, but I also learned a lot about the nuances of the transistor through developing the new model. I found that the culprit transistor was quite useful. I was able to get the proper output-swing from my amplifier, and I used the same transistor configuration in other areas of my op-amp circuit. Not a waste of time for me! I gained a comfort level that is still paying me back. The product is still on the market, 14 years later.

The SPICE simulation tool is a good thing. It will help organize your thoughts and priorities. You can look for faults in terms of how you think the circuit or system will work versus reality. The best of all worlds is to have your tools point out where your mistakes are so you can initiate corrective actions to fix the circuit. The best place to find these problems is at the beginning of the design cycle, not the end. If you misunderstand the nuances of your circuit—that is, parasitics, you will find that your tools will tell you that all is well, when in fact it is not well at all.

25.1   The Old Pencil and Paper Design Process

The generation of your "pencil 'n paper" circuit design is critical. Most likely, you will generate this circuit diagram using your SPICE software, but this is the stage where you take a good look at what you are trying to design. This is the time where you will develop an insider's view of your circuit. In this stage of the design, you can labor through the mathematics of your circuit or better yet, hand-wave your way through the system. During this process, you should think about device and layout parasitics. You will also define the various circuit excitation parameters in preparation for your upcoming simulation.

We've all done the mathematical calculations of our designs in our good 'ole school days. In those days, you boiled every part of the circuit down to a fundamental mathematical representation. However, the most important part of this stage is not the mathematics. In this stage, you should develop an intuition of what you think the circuit will do; then run some critical calculations in preparation for your next step, simulation. Hardware solutions (usually analog circuits) or firmware solutions (involving microprocessors, FPGAs or microcontrollers) require this process during the design phase. This technique is shown in Figure 25.4.

Figure 25.4 illustrates a floating-current-source circuit design. What I want to do with this circuit is to determine what the reference current (I_{REF}) will be in relation to the reference voltage, A_5. Conceptually, the current through R_1 and R_2 is equivalent. This assumes that the input current at the noninverting input of A_1 is zero amps. If A_1 is a CMOS-input amplifier, this is a pretty good assumption. The current through R_1 and R_2 can sink into A_5 or source from A_5. The voltage at the output of A2 can be higher or lower than the reference voltage of 2.5V. This is actually a good thing, because we did want a floating current source.

The voltage drop across the inputs of A_1 is zero volts. If you are going to be exact, the voltage drop across these inputs is equal to the offset voltage of the amplifier, but we are going to let that go for now. The output voltage of A_1 will be at least twice as high

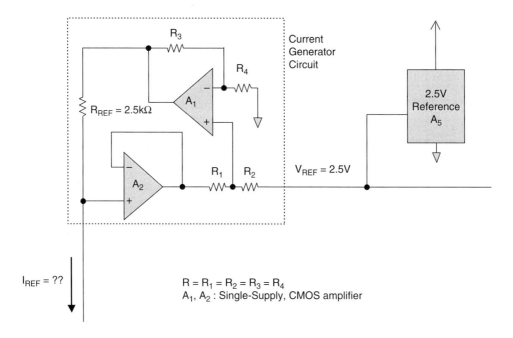

Figure 25.4: Sometimes a constant current source is required to excite a sensor in the system. This constant, floating current source contains a voltage reference, five resistors and two operational amplifiers. The fact that this current source is floating is a little tricky. This requires the pencil and paper evaluation as well as the breadboard. Conceptually, the noninverting input voltage of A_2 is independent of the 2.5V reference (A_5).

(if not higher) as the input voltage of that same amplifier. Since the reference for R_4 is ground, the voltage of the output of A_1 will always be equal to or greater than the voltage reference.

Since this is a floating supply, the best way to get a feel for the circuit is to assign an arbitrary voltage to a node and then work out the rest of the circuit. For example, if we assume the voltage at the noninverting input of A_2 is equal to 0.5V, the voltage at the noninverting input of A_1 is equal to 1.5V. Given this condition, the voltage at the output of A1 is 3.0V. Therefore, the voltage drop across the reference resistor, R_{REF}, is 2.5V. If R_{REF} is equal to 2.5 kΩ, the constant current source will be 1 mA.

That assumption took us a long way. It appears the impedance that I_{REF} flows through determines the voltage at the noninverting input of A1. But, let's not be too hasty. As an exercise for you, assume that the voltage at the noninverting input of

A2 is equal to 3V. You will find that the voltage at the output of A_1 is equal to 5.5V. You may notice that if you have a power supply voltage of 5V, this operation point is not possible. But, that is okay. Now we know most of the basics of this circuit. We also know the limits of the value of the resistor, R_{LOAD} in Figure 25.5.

Figure 25.5 contains a summary of the calculations for this circuit. If you follow the logic in the formulas in this figure, the voltage at the output of A_1 is equal to ½ the voltage at V1. The voltage at the noninverting input of A2 is equal to the voltage at the output of A1. This voltage is also equal to the reference voltage of A5 minus twice the difference between the reference voltage. The output of A2 is also equal to twice V1 minus the 2.5V reference of A5. The resistor value, R, is equal to 25 kΩ. This value ensures amplifier stability and to keep the output currents from the amplifiers relatively low.

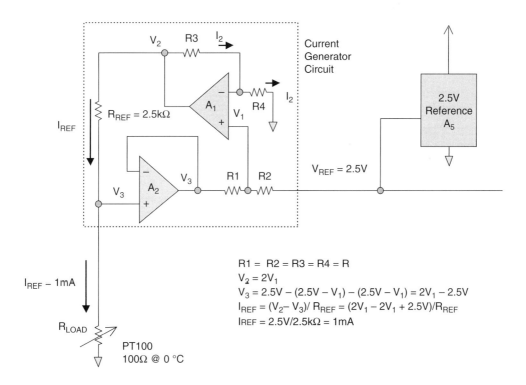

$$R1 = R2 = R3 = R4 = R$$
$$V_2 = 2V_1$$
$$V_3 = 2.5V - (2.5V - V_1) - (2.5V - V_1) = 2V_1 - 2.5V$$
$$I_{REF} = (V_2 - V_3)/R_{REF} = (2V_1 - 2V_1 + 2.5V)/R_{REF}$$
$$I_{REF} = 2.5V/2.5k\Omega = 1mA$$

Figure 25.5: Hand-waving your way through a circuit will give you a good instinct about the circuit operation. As a final step, working through the calculations will validate your initial assumptions.

Knowing this, you can determine the current through the reference resistor, R_{REF}. Thevenin says that this current is equal to the voltage drop across R_{REF} divided by R_{REF}. We can calculate this voltage drop by using the earlier equations to equal 2.5V. As you work the real voltage and resistance values, you will summarize that the current through R_{REF} is equal to 1 mA.

Now it is time to load the circuit. This circuit requires a low impedance load, such as a resistance temperature detector (RTD). If R_{LOAD} is a PT100 RTD, it is equal to 100Ω at 0°C. In this the case, the voltage at the noninverting input of A2 is equal to 100 mV. Consequently, the voltage at the output of A2 is equal to 100 mV. In this application, the resistance range of the PT100 is 100Ω @ 0°C to 254Ω @ 400°C. At higher temperatures the output of A1 is equal to 2.754 mV. If the power supply voltage of the amplifiers is 5V, both amplifiers in this circuit are operating within their linear ranges.

The last steps in this portion of the process is to define the input signals, output representations of the signals, and parasitic resistances, capacitances or inductances that appear as a result of your layout of your circuit. The input signals would include transient signals in the time domain and AC signals in the frequency domain. The input signal definitions will be included at the front-end of your SPICE simulation listing. Further circuit examination will highlight the parasitic elements. For instance, the resistors in Figure 25.5 will have a parasitic capacitance (~0.5 pF) in parallel with the resistor element. Your layout may contribute additional capacitance in the ones of pico-farads because of the traces or wires that you are using. You need to determine if your PCB parasitics are an issue in your circuit. If they are, you need to quantify their values.

The stability of this circuit is another issue. Injecting a current spike with your upcoming SPICE simulator into a high impedance node, such as the inverting input of A2, will cause the circuit to ring if unstable. In the circuit in Figure 25.5, the parasitic capacitances of the resistors do not present a stability issue.

25.2   Is Your Simulation Fundamentally Valid?

Assuming you have worked through the "pencil 'n paper" design of your circuit, you are ready to simulate. The output of this first design phase should be a circuit diagram as well as the operating points throughout the circuit. Defining the operating points of the initial, DC operating points and basic operation of the circuit over time is critical. The initial DC operating point should primarily provide the node voltages, but the current magnitudes of various portions of your circuit may also be important.

Once you finish designing your SPICE model, initiate your first simulation. At the conclusion of the simulation, you should first check the validity all of the operating points in your circuit. If you miss this step, you may be looking at erroneous AC or transient simulation data. The most critical initial DC operating points are the voltages throughout the circuit. For instance, verify that you have correct power supply connections. Then check to see that all of the DC voltages in your circuit are between the power-supply voltages. If any node in the DC operating points exceed your power-supply voltage, you probably have a bad connection in your circuit net list.

Figure 25.6 contains an example of several "red flags" in the DC operating points of the Figure 25.4 circuit. This listing initially shows the simulation circuit connections. Following the OP statement, the simulation listing calls out operational amplifier model ("ideal.mod"). Then there is a listing of the elements of the circuit, their associated device numbers and node assignments.

Everything in this listing in Figure 25.6 looks in order to this point (unless you have already found the error). All of the amplifier nodes and resistor nodes are connected. The indication that something is wrong shows up in the NODE/VOLTAGE table. This is a SPICE generated table of simulation numbers. All of the nodes are present. You won't recognize some of them because they are nodes that are internal to the two amplifier macromodels. You should immediately notice that there are negative voltages assigned to some nodes. It is a red flag that some of the negative nodes are internal in the amplifiers. This is a single-supply circuit. The supply voltages are ground and 5V.

If you return to the top of Figure 25.6 you will notice there is something peculiar with the op-amp, node assignments. The order of nodes versus function is:

inp1 – noninverting input,

inm1 – inverting input,

ps – positive power connect,

1 – negative power connect,

out1 – output.

That seems fine, but there is also a node called *ns*, and it is a ground connect for the rest of the circuit. There is the error. The amplifier-macromodel, negative-supply nodes attaches to ground. The schematic capture tool generated this error.

```
* SHELL FOR floating current source

***************************************************************************
.OP
.lib   ideal.mod
x1    inp1    inm1   ps    1    out1    mcp601
x2    inp2    out2   ps    1    out2    mcp601
* reistorsin circuit
r1            out2  inp1       25k
r2            inp1    2        25k
r3            out1  inm1  25k
r4            inm1  ns         25k
rref   out1  inp2   2.5k
rsens  inp2  ns          100
vref   2             ns        2.5V
vps        ps    0    5
vns        ns    0    0
.END
```

Some simulation results

```
****     SMALL SIGNAL BIAS SOLUTION      TEMPERATURE =  27.000 DEG C
******************************************************************

NODE    VOLTAGE  NODE   VOLTAGE   NODE    VOLTAGE    NODE   VOLTAGE

(   1)    −4.8814   (   2)    2.5000    (  ns)    0.0000    (  ps)    5.0000
( inm1)    1.3000   ( inp1)   1.3000    ( inp2)    .1000    ( out1)   2.6000
( out2)     .1000   ( x1.5)   4.9777    ( x1.6)   4.9777    ( x1.7)    .0753
( x2.5)    4.9777   ( x2.6)   4.9777    ( x2.7)   −1.1247   (x1.23)  −8.3335
(x1.33)     .0593   (x1.34)   3.5113    (x1.43)   2.7000    (x1.44)   2.5000
(x1.45)    4.2235   (x1.46)  45.5970    (x2.23)  −4.0503    (x2.33)    .0593
(x2.34)    −.7718   (x2.43)    .2000    (x2.44)  4.276E-09  (x2.45)   4.2485
(x2.46)   45.5720
```

Figure 25.6: The first portion that you should inspect of any SPICE simulation is the DC operating points. You should check for appropriate voltage and currents in all of the elements of your circuit. This figure shows a portion of complete DC analysis. You will note that some of the nodes are negative values. Since this is a single-supply circuit, this is a warning that something is wrong.

This is just one example of where the SPICE simulation can go wrong. If you continue with any type of analysis, such as AC or time transients, you will always wonder why the results look bad. Even worse, you will do what I did in the beginning of my career and assume that these types of bad results are true. A worse case scenario is to not look at the DC operating points at all. It always pays to question results and challenge the outcome. In a particular instance that I can recall, I chased my tail for most of the week only to find out that one of the nodes was not properly connected. If I had examined the DC analysis results, I would have immediately seen the problem. But, that is what experience is all about, right?

Another place you may want to look for the correctness of your DC analysis (or further on in the simulation) is places where the default values give you erroneous results. The OPTION statement of SPICE contains these default values that affect a variety of conditions. The OPTION statement sets all options, limits, and simulation analysis control parameters. The list of limits includes current accuracy, charge accuracy and the minimum conductance between branches, to name a few. It might be worth your time to look at this list. Usually these OPTION statement defaults won't affect your simulation. However, an attitude that challenges the results of your SPICE simulation may bring errors into focus.

For instance, it may be critical to have the correct input-bias current values with a low-bias CMOS operational amplifier. An error in this parameter appears where your application circuit has high input impedances, such as a transimpedance amplifier or a low-pass filter. The SPICE program will insert a noiseless resistor inside components that have a discontinuity. The gate of the CMOS transistor is essentially floating, or not connected at DC. Although this node does have gate-to-source and gate-to drain capacitors, your SPICE will "view" this as an open circuit in the DC analysis. The SPICE software "fixes" this during the simulation by inserting a minimum conductance between discontinuous nodes.

In the SPICE macromodel in Figure 25.7, the input bias current of the CMOS transistor should be zero. Although, the ESD cells are not modeled here, they generate the input bias current of the amplifier. The "real" amplifier input CMOS transistors Q_1 and Q_2, will not generate gate current.

In SPICE, the input or output current of the gates of these transistors is dependent on the voltage that appears across the gate-to-drain and gate-to-source nodes. This additional current is the SPICE default value, GMIN. The default value of GMIN is 1×10^{-12} S (S = Siemens = $1/\Omega$). If inverted, this is equal to 10^{12} Ω. At first glance, this may not seem to be a problem. However, a voltage across that impedance will cause several picoamperes of error. A transimpedance amplifier (see Figure 25.8) is a circuit where this error will manifest itself as an output voltage error. To solve this problem, you can change the default value of GMIN or insert voltage-dependent-current-sources from the gate to ground of Q_1 and Q_2. As a note, when you change the default values through the OPTIONS statement, your changes will apply to the entire circuit simulation. Use this strategy with care. I prefer to make the changes to these defaults more local, which gives credence to the insertion of the voltage-dependent-current-sources over changing GMIN.

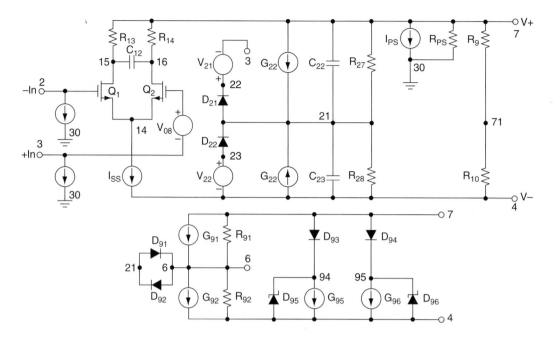

Figure 25.7: This is an example of a SPICE macromodel for a CMOS input amplifier. In this circuit, the SPICE simulation generates an input bias current error as an artifact of the SPICE constraints. Alexander and Bowers were the first to go public with this macromodel. This information appeared in the *Electronic Design* magazine in 1990.

Figure 25.8 highlights this problem. In Figure 25.8, impinging light generates current from the photodiode. The current then flows through RF creating a voltage change at V_{OUT}. The light source to the photodiode generates a low-level, full-scale current of several nano-amperes. If the light is not at full-scale, generating a lower current, the amplifier input bias current errors could cause voltage errors throughout the circuit.

Your DC analysis is the most important part of the validation of your SPICE simulation. If you take the time to meticulously perform this task you will have the confidence that the rest of you simulation has a good chance of being accurate (or as accurate as the simulation can be).

You should always challenge the validity of your SPICE simulation. If you know what to expect from your simulation, you can perform these challenges. If you plan to not evaluate your circuit and just "wait and see" what your SPICE simulation produces,

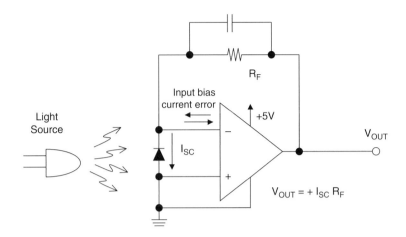

Figure 25.8: An amplifier with a low-input bias current in a transient-impedance amplifier circuit, like this one, is critical if you want to preserve reasonable accuracy. For this reason, CMOS or FET input amplifiers are preferred. If you try to simulate this circuit without the proper input bias current values, you will see an output voltage error at V_{OUT}.

there is a good chance that you will either chase your tail or go back to the pencil and paper evaluation. Either way, you will have wasted valuable time.

At this point, you may have noticed a small, nagging skepticism about SPICE simulations. Well, you are right. The simulation is only as good as your imaginary SPICE circuit. The fact that you place all of the components at the proper location in your circuit diagram and you are using the manufacturers approved macromodels may give you a false sense of security! Your simulations are only as good as your models. So, how are these models defined?

25.3 Macromodels: What Can They Do?

The concept of macromodels first came about in the 1970s. ("Macromodeling of Integrated Circuit Operational Amplifiers," Graeme R. Boyle, Barry M. Cohn, Donald O. Pederson, James E. Solomon, *IEEE Journal of Solid-state Circuits*, Vol. SC-9, No. 6, December 1974, pp. 353–363.) These types of SPICE models provide a tool that reduces the system designer's SPICE simulation time and convergence errors. It allows the designer to focus their efforts at a higher level of simulation. However, system designers, where there are many devices in the circuit, find macromodels very useful. Macromodels allow the SPICE user to simulate results successfully, in a timely fashion.

The engineers that develop IC semiconductors and use SPICE tools only use the macromodel during their circuit development to get proof of concept. They require more transistor-level detail inside their SPICE simulation when they look at the details of their integrated circuits. As the amplifier design process progresses to completion, the macromodel simplifies the transistor-level design too much for the IC designer. Contrary to popular belief, the IC designer's models are also subject to discrepant behavior. There is no such thing as a 100% accurate SPICE model, whether it is a behavioral model, macromodel or transistor-level model.

A macromodel is actually a simple thing. On occasion, I have designed a few macromodels for existing ICs. The macromodel treats the device that you are trying to model like a black box. There are three general classes of simulation models. They are the behavioral model, the macromodel and the transistor-level model. The complexity of each of these types of models increases in the order of this list.

The behavioral model is closest to representing a "black box" with little or no relation to the actual device, except that it tries to emulate the real thing. The macromodel provides a circuit with more complexity. This type of model usually has the actual transistors on the input and output nodes. With this level of complexity, the model emulates the actual device more closely, but not completely.

In the interior of the macromodel, there are variety of dependent sources and independent sources. The most common dependent sources that are used are the linear voltage-controlled voltage source (VCVS), voltage-controlled current source (VCCS), current-controlled voltage source (CCVS), and the current-controlled current source (CCCS). Additionally, inside the macromodel there can be nonlinear dependent sources. The nonlinear dependent sources utilize a polynomial function in their definition. The macromodel's author defines the coefficients of this polynomial. All of these sources attempt to emulate the actual performance of the device.

The transistor-level SPICE model includes all of the transistors, resistors and inductors of the actual device. Each of the elements has a SPICE model definition. Within each of these model definitions, the user can adjust various variables. For instance, the MOSFET SPICE model has 24 variables. With these variables, the SPICE user can adjust parameters such as lateral diffusion length, lateral diffusion width, zero-bias threshold voltage, transconductance coefficient and so forth.

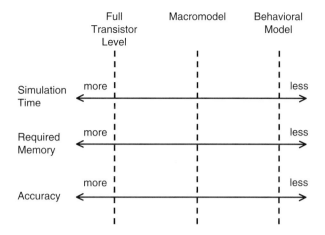

Figure 25.9: Transistor-level models are more complex than the macromodel or the behavioral model. Therefore, the transistor-level model will require more time to simulate and more computer memory. The increase in these require-ments can make it difficult to complete a simulation, particularly if there are several transistor-level models in the simulation circuit. However, if you are able to tolerate longer simulation time, the transistor-level will produce results that are more accurate. In all three cases, the models are quite accurate for their intended application, and work poorly for other applications.

The transistor-level SPICE representation of the circuit is more complex than the behavioral model or the macromodel. Each of these levels has their place in your simulation strategy. The transistor SPICE model has more complexity and provides accuracy particularly for the IC designs. The transistor-level model details how the transistors in the circuit interact. This level of detail is not appropriate for the systems level designs. This type of design requires listings that can simulate several devices at one time. Under these conditions, the transistor-level models will be slower and less accurate. This is due to the increase in the node count of the entire simulation circuit.

Although the elements of a macromodel are complex, there are fewer elements in the total model listing. The overall complexity or sophistication of the macromodel is less than the transistor-level model. The macromodel simulates a list of specific parameters and no more. Many vendors will tell you what those parameters are in their SPICE macromodel listing. An example list of the parameters modeled for an operational amplifier would be: input voltage offset, DC PSRR, DC CMRR, input impedance, input bias current, open-loop gain, voltage ranges and supply current (typical performance at room temperature, 25°C).

A system-level simulation may include hundreds of building blocks. At this level it may be impossible to simulate in SPICE at the transistor or macromodel level. Even if these two levels of models do manage to give results, the system level simulation accuracy is much greater with correctly modeled behavioral building blocks. The transistor-level and macromodel complexity causes reduced accuracy in this environment and greater simulation time.

Consequently, the complete device model will consume more computer memory during simulation and take longer. If you simulate several device models at the same time, such as five or six transistor-level operational amplifiers, it is possible that the SPICE simulation will "crash" and not be able to complete the simulation of the circuit. Another setback that you will find with the transistor-level model is availability. Device vendors are very reluctant to provide the transistor-level models to their customers, or for that fact, anyone. This is because the transistor-level model contains proprietary information about their circuit. You can imagine that if a competitor got their hands on this type of model, the second source design work would be reduced significantly. This would make it easy for the competition to reverse-engineer a part and to quickly start stealing market share. More importantly, transistor-level models are less accurate for board and system level designs.

Figure 25.10 shows an example of an operational amplifier macromodel.

The circuit in Figure 25.10 has some limitations. You will notice the ground connects in several places. Because of these ground connects, and the way they affect the macromodel's behavior, the model will not operate in a single-supply environment. Figure 25.7 shows a model that is used in single-supply or in floating-supply environments. You should notice that in the operational amplifier input stages of the models in Figure 25.7 and Figure 25.10 have transistors. The input stage is the only place in these macromodels that have any resemblance to the actual device. With transistors inserted at this point, the model emulates the unique nonlinearities of the op-amp inputs. If you go beyond that stage, there is no longer a resemblance between the actual transistor-level model and these macromodels. The creator of the final model in Figure 25.7 and Figure 25.10 uses the dependent current and voltage sources to produce the real amplifier behavior. Keep in mind, the objective of the macromodel is not to copy the transistor model, but to imitate the operation of the device. In particular, the amplifier macromodel imitates the amplifier's operation in applications, and not as a stand-alone model.

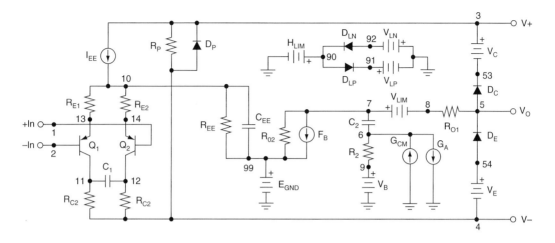

Figure 25.10: This PNP operational amplifier macromodel was originally designed in 1974 by Graeme R. Boyle, Barry M. Cohn, Donald O. Pederson, James E. Solomon. It was the first legitimate model published. Several other operational amplifier macromodel templates have been developed since then. Some SPICE vendors provide tools in their software to generate this type of macromodel.

With all of these issues in mind, it is easy to understand that macromodels do not produce the complete performance of the actual amplifier circuit. The more simplistic models are able to simulate a limited number of the amplifier attributes. For example, in its most basic form, the macromodel illustrated in Figure 25.10 will only provide a small subset of the amplifier's attributes. This macromodel models the input bias current and input impedance. It does not model the input rail-to-rail swing very accurately. The output characteristics that this macromodel can simulate are output current limit, output resistance and output voltage swing.

The output-voltage swing limits are set as if the amplifier is in a comparator configuration. This macromodel will not assist in demonstrating the nonlinear behavior of the output stage as the output gets close the rails. The macromodel's attributes in the AC domain include gain versus frequency, phase versus frequency and a symmetrical slew rate. This macromodel will not reproduce a real amplifier's asymmetrical (low to high, high to low) slew rate. Finally, the macromodel accurately reflects the DC quiescent current in a simulation. If the output of the amplifier is loaded and exercised, the current required will be pulled from ground and not from the power supplies. All of these performance attributes are only good at room temperature, or 25°C.

Enhancements that are added by the vendor to this limited list of attributes are offset voltage at room temperature, input noise and input offset current.

Most vendors design their macromodels to produce typical performance attributes and they don't reflect the minimums and maximums you will find in the data sheet. If you want the macromodel to show you the amplifier's operation with minimum or maximum performance specifications, you will have to proceed with caution and tweak the macromodel yourself. As a final shortcoming of this type of macromodel, these simulation attributes do not change as the real amplifier would if you were to vary the simulation temperature.

The model shown in Figure 25.10 is less flexible than the model shown in Figure 25.7. The model in Figure 25.7 overcomes quite a few limitations found in the model in Figure 25.10. For example, it is possible to use this macromodel for single-supply amplifiers. In addition, the output current in Figure 25.7 flows from the power supplies rather than the ground connect. Such attributes as power supply reduction (PSRR) and common-mode rejection (CMRR) ratios are included in this model. This model also lends itself to easily include over-temperature attributes.

The long and short of this discussion is that the capability of every macromodel is dependent on the whims of the macromodel designer. Figure 25.11 has a short list of a few amplifier macromodels that have various levels of capability.

Computer-based simulations can significantly reduce the development time and therefore speed up the time-to-market of your designs. These facts alone make SPICE simulations attractive to the IC designer as well as the systems designer. With the increasing use of SPICE-based simulations, there is also a rising demand for accurate models. The expectation is that your models, or macromodels, reflect the actual performance of the component. This should be done without carrying the burden of too many circuit details. Companies in industry have responded to this need by providing macromodels for a broad range of products. The selection of SPICE macromodels from these companies ranges from op-amp, difference amps, instrumentation amps, isolation amps, to analog function circuits.

Analog manufacturers will also provide other tools that will facilitate your design process. An analog filter's design tools are the most prevalent. Some of the tools allow the designer to use any manufacturers' devices in the circuit while others require that the user use only their products. Another popular tool can help you with power supply design. In this case, the manufacturer controls the selection of the devices that will work

	INPUT BIAS CURRENT	INPUT OFFSET CURRENT	OFFSET VOLTAGE	INPUT VOLTAGE NOISE	INPUT CURRENT NOISE	INPUT PROTECTION	INPUT IMPEDANCE	INPUT BIAS CURRENT CORRECTION	OUTPUT RESISTANCE	OUTPUT CURRENT LIMIT	OUTPUT FLOWING FROM POWER SUPPLIES	OUTPUT VOLTAGE SWING	QUIESCENT CURRENT	QUIESCENT CURRENT vs POWER SUPPLY	QUIESCENT CURRENT vs TEMPERATURE	GAIN vs FREQUENCY	GAIN vs TEMPERATURE	PHASE RESPONSE	CMRR vs FREQUENCY	PSRR	PSRR vs FREQUENCY	SLEW RATE	PAD PARASITICS	NO GROUND REFERENCE
Op-amp A	X							X	X	X		X	X			X		X				X		
Op-amp B	X	X		X	X		X		X	X	X	X	X	X		X		X		X		X		
Op-amp C	X	X	X	X	X		X		X	X	X	X	X			X		X	X	X	X	X		X
Op-amp D	X	X	X	X	X		X		X	X	X	X	X			X		X	X		X	X	X	X
Op-amp E	X		X	X	X		X		X	X	X	X	X			X		X	X			X	X	X

Figure 25.11: The capability of the amplifier macromodel varies from author to author and vendor to vendor. The best line of defense is to find out what your macromodel can or can't do before you use it in your simulation. You can determine this by asking the vendor for that information, or by running your own tests in SPICE to determine what the macromodel's capabilities are.

in the circuits that their tool creates. They do this by knowing the particulars of their products and assisting design-in for their products.

25.4   VHDL-AMS

With the increasingly high level of system integration it is becoming necessary to model not only electronic behavior of systems, but also interfaces to 'real-world' applications and the detailed physical behavior of elements of the system in question. The emergence of standard languages such as VHDL-AMS has made it possible to now describe a variety of physical systems using a single design approach and simulate a complete system. Application areas where this is becoming increasingly important include mixed-signal electronics, electro-magnetic interfaces, integrated thermal modeling, electro-mechanical and mechanical systems (including micro-electro-mechanical systems, MEMS), fluidics (including hydraulics and microfluidics), power electronics with digital control and sensors of various kinds.

In this section, we will show how the behavioral modeling of multiple energy domains is achieved using VHDL-AMS, demonstrating with the use of examples how the interactions between domains takes place, and provide an insight into design techniques for a variety of these disciplines. The basic framework is described, showing how standard packages can define a coherent basis for a wide range of models, and specific examples used to illustrate the practical details of such an approach. Examples such as integrated simulation of power electronics systems including electrical, magnetic and thermal effects, mixed-domain electronics and mechanical systems are presented to demonstrate the key concepts involved in multiple energy domain behavioral modeling.

The basic approach for modeling devices in VHDL-AMS is to define a model entity and architecture(s). The model entity defines the interface of the model to the system and includes connection points and parameters. A number of architectures can be associated with an entity to describe the model behavior such as a behavioral or a physical level description. A complete model consists of a single entity combined with a single architecture. The domain or technology type of the model is defined by the type of terminal used in the entity declaration of the ports. The IEEE Std 1076.1.1 defines standard types for multiple energy domains including electrical, thermal, magnetic, mechanical and radiant systems. Within the architecture of the model, each energy domain type has a defined set of through and across variables (in the electrical domain these are voltage and current, respectively) that can be used to define the relationship between the model interface pins and the internal behavior of the model.

In the 'conventional' electronics arena, the nature of the VHDL-AMS language is designed to support 'mixed-signal' systems (containing digital elements, analog elements and the boundary between them) with a focus on IC design. Where the strengths of the VHDL-AMS language have really become apparent, however, is in the multi-disciplinary areas of mechatronic and MEMS. In this chapter, I have highlighted several interesting examples that illustrate the strengths of this modeling approach, with emphasis on multiple domain simulations.

25.4.1 *Introduction to VHDL-AMS*

VHDL-AMS is a set of analog extensions to standard digital VHDL to allow mixed-signal modeling of systems. The VHDL-AMS language was approved as IEEE Std 1076.1 in 1999; however, it is important to note that IEEE 1076.1-1999 encompasses the complete digital VHDL 1076 standard and is not a subset.

The standard does not specify any libraries for analog disciplines (e.g., electrical, mechanical, etc.). This is a separate exercise and is covered by a subset working group IEEE 1076.1.1, which was released as an IEEE Std 1076.1.1 in 2004.

In order to put the extensions into context it is useful to show the scope of VHDL, and then VHDL-AMS alongside it and this is shown in Figure 25.12.

The key extensions for VHDL-AMS is the ability to look upward to transfer functions (behavioral and in the Laplace domain) and downward to differential equations at the circuit level.

The extensions to VHDL for VHDL-AMS can be summarized as follows:

1. A new type of ports called TERMINALS—basically analog pins.

2. A new type of TYPE called a NATURE that defines the relationship between analog pins and variables.

3. A new type of variable called a QUANTITY that is an analog variable.

4. A new type of variable assignment that is used to define analog equations that are solved simultaneously.

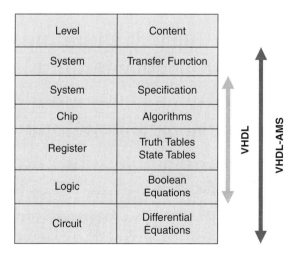

Figure 25.12: Scope of VHDL-AMS

5. Differential equation operators for derivative ('DOT) and integration ('INTEG) with respect to time.

6. IF statements for equations (IF USE).

7. Break statement to initialize the nonlinear solver.

8. STEP LIMIT control for limiting the analog time step in the solver.

25.4.2   Analog Pins: TERMINALS

In order to define analog pins in VHDL-AMS we need to use the TERMINAL keyword in a standard entity PORT declaration. For example, if we have a two pins device that has two analog pins (of type electrical, more on this later), then the entity would have the basic form as shown below:

```
LIBRARY IEEE;
USE IEEE.ELECTRICAL_SYSTEMS.ALL;
ENTITY model IS
GENERIC();
PORT(
     TERMINAL p : electrical;
     TERMINAL m : electrical
   );
END ENTITY;
```

Notice that as the VHDL-AMS extensions are defined as an IEEE standard, then the use of a standard library such as electrical pins requires the use of the electrical_systems.all; packages from the IEEE library.

Notice that the pins do not have a direction assigned as analog pins are part of a conserved energy system and are therefore solved simultaneously.

25.4.3   Mixed-Domain Modeling

In order to use standard models, there has to be a framework for terminals and variables which is where the standard packages are used. There is a complete IEEE Std (1076.1.1) which defines the standard packages in their entirety; however, it is useful to look at a simplified package (electrical systems in this case) to see how the package is put together.

For example, electrical systems models need to be able to handle several key aspects:

- Electrical connection points

- Electrical "through" variables (i.e., current)

- Electrical "across" variables (i.e., voltages)

The electrical systems 'package' needs to encompass these elements.

First, the basic subtypes need to be defined. In ALL the analog systems and types, the basic underlying VHDL type is always REAL, and so the voltage and current must be defined as subtypes of REAL:

```
Subtype voltage is REAL;
Subtype current is REAL;
```

Notice that there is no automatic unit assignment for either, but this is handled separately by the UNIT and SYMBOL attributes in IEEE Std 1076.1.1. For example, for voltage the unit is defined as "Volt" and the symbol is defined as "V".

The remainder of the basic electrical type definition then links these subtypes to the through and across variable of the type, respectively:

```
PACKAGE electrical_system IS
    SUBTYPE voltage IS real;
    SUBTYPE current IS real;
    NATURE electrical IS
        voltage ACROSS
        current THROUGH
        ground REFERENCE;
END PACKAGE electrical_system;
```

25.4.4 Analog Variables: Quantities

Quantities are purely analog variables and can be defined in one of three ways. Free quantities are simply analog variables that do not have a relationship with a conserved energy system. Branch quantities have a direct relationship between one or more analog terminals and finally source quantities are used to define special source functions (such as AC sources or noise sources).

For example to define a simple analog variable called *x*, that is a voltage (but not related directly to an electrical connection (TERMINAL), then the following VHDL could be used:

```
QUANTITY x : voltage;
```

On the other hand, a branch between two electrical pins has a through variable (current) and an across variable (voltage) and this requires a "branch" quantity so that the complete description can be solved simultaneously. For example, the complete quantity declaration for the voltage (v) and current (i) of a component between two pins (p & m) could be defined as:

```
QUANTITY v across i through p to m;
```

25.4.5 Simultaneous Equations in VHDL-AMS

In VHDL-AMS the equations are analog and solved simultaneously, which is in contrast to signals that are solved concurrently using logic techniques and variables which are evaluated sequentially. For example in VHDL-AMS to solve an equation use the '==' operator:

```
Y == x**2;
```

where both Y and X have to be defined as real numbers (quantities or other VHDL variable types).

25.4.6 A VHDL-AMS Example

25.4.6.1 A DC Voltage Source

In order to illustrate some of these basic concepts consider a simple example of a DC voltage source. This has two electrical pins p & m, and a single parameter dc_value that is used to define the output voltage of the source (Figure 25.13).

This can be modeled in VHDL-AMS in two parts, the entity and architecture. First, consider the entity. This has two electrical pins, so we need to use the ieee. electrical_systems.all; package and therefore the ports are to be declared as TERMINALS. Also the generic de_value must be defined as a real number with the default value also defined as a real number (e.g., 1.0):

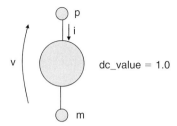

Figure 25.13: Basic voltage source

```
LIBRARY IEEE;
USE IEEE.ELECTRICAL_SYSTEMS.ALL;
ENTITY v_dc IS
GENERIC(
     dc_value : real := 1.0);
PORT(
     TERMINAL p : electrical;
     TERMINAL m : electrical
     );
END ENTITY;
```

The architecture must define the quantities for voltage and current through the source and then link those to the terminal pin names. Also, the output equation of the source must be modeled as an analog equation in VHDL-AMS using the '= =' operator to implement the function v = dc_value:

```
ARCHITECTURE simple OF v_dc IS
     QUANTITY v ACROSS I THROUGH p TO m;
BEGIN
     v == dc_value;
END ARCHITECTURE simple;
```

25.4.6.2 *Resistor*

In the case of the resistor, the basic entity is very similar to the voltage source with two electrical pins p & m with a single generic, this time for the nominal resistance rnom (Figure 25.14).

This can be modeled in VHDL-AMS in two parts, the entity and architecture. First consider the entity. This has two electrical pins, so we need to use the ieee.

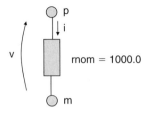

Figure 25.14: VHDL-AMS resistor symbol

electrical_systems.all; package and therefore the ports are to be declared as TERMINALS. Also the generic rnom must be defined as a real number with the default value also defined as a real number (e.g. 1000.0):

```
LIBRARY IEEE;
USE IEEE.ELECTRICAL_SYSTEMS.ALL;
ENTITY resistor IS
GENERIC(
      rnom : real := 1000.0);
PORT(
      TERMINAL p : electrical;
      TERMINAL m : electrical
      );
END ENTITY;
```

The architecture must define the quantities for voltage and current through the resistor and then link those to the terminal pin names.

Also, the output equation of resistor must be modeled as an analog equation in VHDL-AMS using the '==' operator to implement the function v = I * rnom:

```
ARCHITECTURE simple OF resistor IS
      QUANTITY v ACROSS I THROUGH p TO m;
BEGIN
      v == I * rnom;
END ARCHITECTURE simple;
```

25.4.7   Differential Equations in VHDL-AMS

VHDL-AMS also allows the modeling of linear differential equations using the two differential operators:

1. 'DOT (Differentiate the variable with respect to time)

2. 'INTEG (Integrate the variable with respect to time)

We can illustrate this by taking two examples, a capacitor and an inductor. First, consider the basic equation of a capacitor:

$$i = C\frac{dV}{dt}$$

Using a similar model structure as the resistor, we can define a model entity and architecture, but what about the equation? In VHDL-AMS, the 'DOT operator is used on the voltage to represent the differentiation as follows:

```
i == c* v'DOT;
```

Therefore, a complete capacitor model in VHDL-AMS could be implemented as follows:

```
LIBRARY IEEE;
USE IEEE.ELECTRICAL_SYSTEMS.ALL;
ENTITY capacitor IS
GENERIC(
     cap : real := 1.0e-9);
PORT(
     TERMINAL p : electrical;
     TERMINAL m : electrical
     );
END ENTITY;
ARCHITECTURE simple OF capacitor IS
     QUANTITY v ACROSS I THROUGH p TO m;
BEGIN
   I == cap * v'DOT;
END ARCHITECTURE simple;
```

What about an inductor? The basic equation for an inductor is given below:

$$i = \frac{1}{L}\int v \, dt$$

Obviously, the most direct way to implement this equation would be to use the 'INTEG operator, however care should be taken with the integration operator.

Obviously, the initial condition must be considered and in addition different implementations can occur across simulators. However, the resulting implementation in its simplest form could be as follows:

```
LIBRARY IEEE;
USE IEEE.ELECTRICAL_SYSTEMS.ALL;
ENTITY inductor IS
GENERIC(
      ind : real := 1.0e-9);
PORT(
      TERMINAL p : electrical;
      TERMINAL m : electrical
      );
END ENTITY;

ARCHITECTURE simple OF inductor IS
      QUANTITY v ACROSS I THROUGH p TO m;
BEGIN
      I == (1.0/ind) * v'INTEG;
END ARCHITECTURE simple;
```

25.4.8   Mixed-Signal Modeling with VHDL-AMS

Most design engineers are familiar with the concepts of "digital" or "analog" modeling; however, a true understanding of "mixed-signal" modeling is often lacking. In order to explain the term *mixed-signal modeling*, it is necessary to review what we mean by analog and digital modeling first. First, consider digital modeling techniques.

Digital systems can be modeled using digital gates or events. This is a fast way of simulating digital systems structurally and is based on VHDL or Verilog gate level models. Digital simulation with digital computers relies on an event-based approach, so rather than solve differential equations, events are scheduled at certain points in time, with discrete changes in level. The resolution of multiple events and connections is achieved using logic methods. The digital models are usually gates, or logic based and the resulting simulation waveforms are of fixed, predefined levels (such as "0" or "1"). Also, "instantaneous" changes can take place, that is the state can change from '0' to '1' with zero risetime.

In the analog world, in contrast, the lowest level of detail in practical electrical system design is the use of analog equation models in an analog simulator—the benchmark of this approach is historically the SPICE simulator. In many cases the circuit is extracted in the form of a netlist. The netlist is a list of the components in the design, their connection points and any parameters (such as length, width or scaling) that customize the individual devices.

Each device is modeled using nonlinear differential equations that must be solved using a Newton–Raphson type approach. This approach can be very accurate, but is also fraught with problems such as:

- *Convergence*: If the model does not converge, then the simulation will not give any meaningful result or fail altogether.

- *Oscillation*: If there are discontinuities, the solution may be impossible to find.

- *Time*: The simulations can take hours to complete, days for large designs with detailed device models.

In the analog domain the Newton-Raphson approach is generally used to find a solution which relies on calculating the derivatives as well as the function value to obtain the next solution. The basic Newton-Raphson method for nonlinear equations is defined as:

$$x_{n+1} = x_n - \frac{F(x_n)}{F'(x_n)}$$

$F(x_n)$ and $F'(x_n)$ must be explicitly known and coded into the simulator (for SPICE) and this gives an approximate solution to the exact problem. For VHDL-AMS simulators the derivatives must be estimated using a Secant method (or similar) (Figure 25.15).

So given these diametrically opposed methods, how can we put them together? What about mixed-signal systems? In these cases, there is a mixture of continuous analog variables and digital events. The models need to be able to represent the boundaries and transitions between these different domains effectively and efficiently. The basic mechanism to checking if an analog variable crosses a threshold is to use the ABOVE operator in VHDL-AMS.

For example, to check if a voltage "vin" is above 1.0V, the following VHDL-AMS could be used:

```
if ( vin'above(1.0) ) then
    flag <= true;
end if;
```

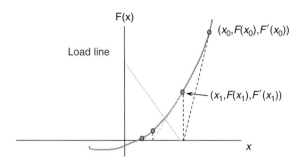

Figure 25.15: Newton-Raphson Method

This can be extended to use parameters in the model—say a threshold voltage parameter (vth)—defined previously as a generic or constant:

```
if ( vin'above(vth) ) then
    flag <= true;
end if;
```

Notice that flag is a signal and is therefore able to be used in the sensitivity list to a process enabling digital behavior to be triggered when the threshold is crossed. If the opposite condition is required, that is BELOW the threshold, then the condition is simply inverted using the NOT operator:

```
if ( NOT vin'above(vth) ) then
    flag <= true;
end if;
```

The digital-to-analog interface is slightly more complex than the analog-to-digital interface inasmuch as the output variable needs to be controlled in the analog domain.

When a digital event changes (this can be easily monitored by a sensitivity list in a process) the analog variable needs to have the correct value and the correct rate of change. To achieve this we use the RAMP attribute in VHDL-AMS.

Consider a simple example of a digital-logic-to-analog-voltage interface:

- When Din = '1' Vout = 5V

- When Din = '0' Vout = 0V

This can be implemented using VHDL-AMS as follows:

```
process (din) :
begin
    if ( din = '1' ) then
        vdin = 5.0;
    else
        vdin = 0.0;
    end if;
end process;
vout == vdin;
```

Clearly, there will be problems with this simplistic interface as the transition of vout will be instantaneous—causing potential convergence problems. The technique to solve this problem is to introduce a RAMP on the definition of the value of vout with a transition time to change continuously from one value to another:

```
vout == dvin'RAMP(tt)
```

where tt (the transition time) is defined as a real number (e.g., tt : real : = 1.0e = − 9;).

An alternative to the specific transition time definition is to limit the slew rate using the SLEW operator. The technique to solve this problem is to introduce a slew rate definition on the definition of the value of vout with a transition time to change continuously from one value to another:

```
vout == dvin'SLEW(max_slew_rate)
```

where max_slew_rate is defined as a real number (e.g., max_slew_rate : real : = 1.0e6;).

25.4.9 A basic switch model

Consider a simple digitally controlled switch that has the following characteristics:

- Digital control input (d)
- Two electrical terminals (p & m)
- On resistance (Ron)
- Off resistance (Roff)
- Turn on time (Ton)
- Turn off time (Toff)

Using this simple outline a basic switch model can be created in VHDL-AMS. The entity is given below:

```
USE ieee.electrical_system.ALL;
USE ieee.std_logic_1164.ALL;
ENTITY switch IS
    GENERIC ( ron : real := 0.1;– On resistance
            roff : real := 1.0e6; – Off resistance
            ton : real := 1.0e-6; – turn on time
            toff : real := 1.0e-6); – turn off time
    PORT (
    d : IN std_logic;
    TERMINAL p,m : electrical);
END ENTITY switch;
```

The basic structure of the architecture requires that the voltage and current across the terminals of the switch be dependent on the effective resistance of the switch (reff):

```
ARCHITECTURE simple OF switch IS
        QUANTITY v ACROSS i THROUGH p TO m;
        QUANTITY reff : real;
        SIGNAL r_eff : real := roff;
BEGIN
        PROCESS (d)
        BEGIN
            ...
        END;

        i = v / reff;
END;
```

The process waits for changes on the input digital signal (d) and schedules a signal r_eff to take the value of the effective resistance (ron or roff) depending on the logic value of the input signal. The VHDL for this functionality is shown below:

```
PROCESS (d)
 BEGIN
        if ( d = '1' ) then
            r_eff <= ron;
        else
            r_eff <= roff;
        end if;
END;
```

When the signal r_eff changes, then this must be linked to the analog quantity reff using the ramp function. Previously we showed how the ramp could define a risetime, but in fact it can also define a falltime. Implementing this in the switch model architecture we get the following VHDL-AMS:

```
reff == r_eff'RAMP ( ton, toff );
i == v / reff;
```

The complete VHDL-AMS model for the switch is given below:

```
ARCHITECTURE simple OF switch IS
      QUANTITY v ACROSS i THROUGH p TO m;
      QUANTITY reff : real;
      SIGNAL r_eff : real := roff;
BEGIN
      PROCESS (d)
        BEGIN
           if ( d = '1' ) then
                 r_eff <= ron;
           else
                 r_eff <= roff;
           end if;
        END PROCESS;

        reff == r_eff'RAMP ( ton, toff );
        i == v / reff;
END;
```

25.4.10   Basic VHDL-AMS Comparator Model

Consider a simple comparator that has two electrical inputs (p & m), an electrical ground (gnd) and a digital output (d). The comparator has a digital output of "1"when p is greater than m and "0" otherwise (Figure 25.16).

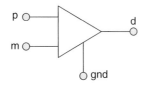

Figure 25.16: Comparator

The entity defines the terminals (p, m, gnd), digital output (d), input hysteresis (hys) and the propagation delay (td):

```
USE ieee.electrical_system.ALL;
USE ieee.std_logic_1164.ALL;
ENTITY comparator IS
    GENERIC (
td : time := 10 ns;
hys : real := 1.0e-6);
    PORT (
d : OUT std_logic := '0' ;
TERMINAL p,m,gnd : electrical);
END ENTITY comparator;
```

The first step in the architecture is to define the input voltage and basic process structure:

```
architecture simple of comparator is
  quantity vin across p to m;
begin
 p1 : process
      constant vh : real := ABS(hys)/2.0;
      constant vl : real := -ABS(hys)/2.0;
  begin
      ...
      wait on vin'above(vh), vin'above(vl);
  end process;
end architecture simple;
```

The quantity vin is defined as the voltage across the input pins p and m:

```
quantity vin across p to m;
```

Notice that no current is defined, assumed to be zero, so there is no input current to the comparator. Also notice that there is no input voltage offset defined—this could be added as a refinement to the model later. The process defines the upper and lower thresholds (vh and vl) based on the hysteresis:

```
constant vh : real  := ABS(hys)/2.0;
constant vl : real  := -ABS(hys)/2.0;
```

The process then defines a wait statement checking vin for crossing either of those threshold values:

```
wait on vin'above(vh), vin'above(vl);
```

The final part of the process is to add the digital output logic state dependent on the threshold status of vin:

```
if vin'above(vh) then
    d <= '1' after td;
elsif not vin'above(vl) then
    d <= '0' after td;
end if;
```

The output state (d) is then scheduled after the delay time defined by td.

The completed architecture is shown below:

```
architecture simple of comparator is
  quantity vin across p to m;
begin
  p1 : process
      constant vh : real := ABS(hys)/2.0;
      constant vl : real := -ABS(hys)/2.0;
begin
      if vin'above(vh) then
          d <= '1' after td;
      elsif not vin'above(vl) then
          d <= '0' after td;
      end if;
      wait on vin'above(vh), vin'above(vl);
  end process;
end architecture simple;
```

25.4.11 Multiple Domain Modeling

A final significant application area for VHDL-AMS has been the modeling of electro-mechanical systems, particularly micromachines (or MEMS). Exactly the same principles are used for these devices, with the mechanical domain models defined as required for the mechanical equations. It is worth noting that the mechanical models are divided into rotational (angular velocity and torque) and translational (force and distance) types. A typical simple example of a mixed-domain system is a motor, in this case a simple DC motor. Taking the standard motor equations as shown below, it can be seen that the parameter ke links the rotor speed to the electrical domain (back emf) and the parameter kt links the current to the torque:

$$V = L\frac{di}{dt} + iR + Ke\omega$$

$$T = Kti - J\frac{d\omega}{dt} - D\omega$$

This is implemented using the VHDL-AMS model shown below:

```
Library ieee;
use ieee.electrical_systems.all;
use ieee.mechanical_systems.all;

entity dc_motor is
      generic (kt : real;
               j : real;
               r : real;
               ke : real;
               d : real;
               l : real);
      port (terminal p, m : electrical;
            terminal rotor : rotational_v);
end entity dc_motor;

architecture behav of dc_motor is
      quantity w across t through rotor
  to rotational_v_ref;
      quantity v across i through p to m;
begin
   v == l*I' DOT + i*r + ke*w;
   t == i*kt - j*w' DOT - d*w;
end architecture behav;
```

25.4.12   Summary

It has become crucial for effective design of integrated systems, whether on a macro- or microscopic scale, to accurately predict the behavior of such systems prior to manufacture. Whether it is ensuring that sensors or actuators operate correctly, or integrated components such as magnetics also operate correctly, or analyzing the effect of parasitics and nonideal effects such as temperature, losses and nonlinearities, the requirement for multiple domain modeling has never been greater.

Now languages such as VHDL-AMS offer an effective and efficient route for engineers to describe these systems and effects, with the added benefit of standardization leading to interoperability and model exchange. The challenge for the EDA industry is to provide adequate simulation and particularly modeling tools to support engineering design.

The opportunity for field-programmable gate array (FPGA) designers is to take advantage of this huge advance in modeling technology and use it to make sure that digital controllers and designs can operate effectively and robustly in real-world applications.

References

Introduction to Pspice Manual for Electronic Circuits Using OrCad Release 9.1, Nilsson, Riedel, Prentice-Hall, 2000.
Inside SPICE: Overcoming the Obstacles of Circuit Simulation, Kielkowski, Ron M., McGraw Hill, 1994.
Macromodeling with SPICE, Choi, Pyung and Connelly, J. Alvin, Prentice-Hall, 1992.
"Spice Models Low-bias Op Amps Correctly," Baker, Bonnie C., *EDN Magazine*, August 20, 1992.
"Macromodeling of Integrated Circuit Operational Amplifiers," Graeme R. Boyle, Barry M. Cohn, Donald O. Pederson, James E. Solomon, IEEE Journal of Solid-State Circuits, Vol. SC-9, No. 6, December 1974, pp. 353–363.
"Designer's Guide to Spice-Compatible Op-amp Macromodels – Part 1," Alexander, Bowers, *Electronic Design News*, Volume 35, No. 4, February 15, 1990.
Semiconductor Device Modeling with Spice, Antognetti, Massobrio, McGraw Hill, 1980.

Useful texts for VHDL

Digital Systems Design

Digital System Design with VHDL by Mark Zwolinski, published by Pearson Education, is a superb introduction to designing with VHDL. It is used in many universities worldwide for teaching VHDL at an undergraduate level and has numerous basic examples to enable a student to get started. I would also recommend this to an engineer getting started with VHDL.

The Designers Guide to VHDL

The Designers Guide to VHDL by Peter Ashenden is perhaps the most comprehensive book on VHDL from a variety of perspectives. It covers the syntax and language rigorously, has plenty of examples, and is a great desk top reference book. For nonbeginners in VHDL, this is the book I would recommend.

VHDL: Analysis and Modeling of Digital Systems

VHDL: Analysis and Modeling of Digital Systems (McGraw-Hill Series in Electrical and Computer Engineering) by Zainalabedin Navabi is a detailed look at not only how VHDL can be used to model digital systems, but many of the detailed issues regarding timing and analysis that are often skipped over by other texts on VHDL. It is perhaps not a beginner's book, but is especially useful for those who require a deeper understanding of issues relating to timing.

VHDL for Logic Synthesis

VHDL for Logic Synthesis by Andrew Rushton, published by Wiley, is a useful background text for those who perhaps need to understand how VHDL can be used for practical synthesis. The book discusses what and what is not synthesizable and also explains how some useful and somewhat arcane VHDL functions operate.

CHAPTER 26

Useful Circuits

Ron Mancini
Darren Ashby
Richard Palmer

26.1   Introduction

Portable and single-supply electronic equipment is becoming more popular each day. The demand for single-supply op-amp circuits increases with the demand for portable electronic equipment because most portable systems have one battery. Split- or dual-supply op-amp circuit design is straightforward because op-amp inputs and outputs are referenced to the normally grounded center tap of the supplies. In the majority of split-supply applications, signal sources driving the op-amp inputs are referenced to ground, thus with one input of the op-amp referenced to ground, as shown in Figure 26.1, there is no need to consider input common-mode voltage problems.

$$V_{OUT} = -V_{IN}\frac{R_F}{R_G} \qquad (26.1)$$

When the signal source is not referenced to ground (see Figure 26.2 and Equation 26.19), the voltage difference between ground and the reference voltage shows up amplified in the output voltage. Sometimes this situation is OK, but other times the difference voltage must be stripped out of the output voltage.

$$V_{OUT} = -(V_{IN} + V_{REF})\frac{R_F}{R_G} \qquad (26.2)$$

An input bias voltage is used to eliminate the difference voltage when it must not appear in the output voltage (see Figure 26.3 and Equation 26.3). The voltage, V_{REF},

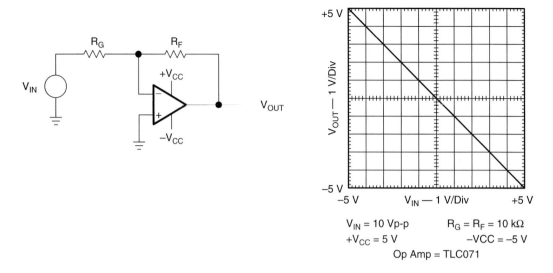

Figure 26.1: Split-supply op-amp circuit

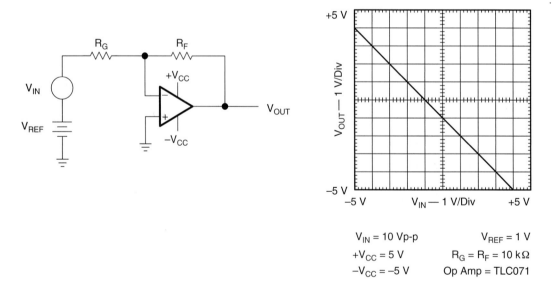

Figure 26.2: Split-supply op-amp circuit with reference voltage input

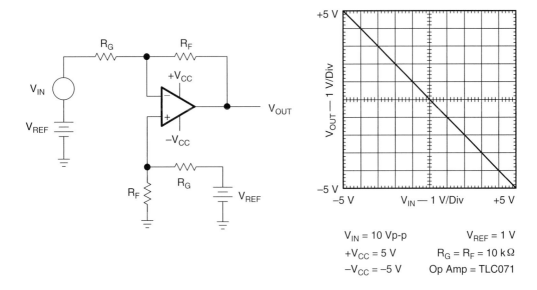

$V_{IN} = 10$ Vp-p $V_{REF} = 1$ V

$+V_{CC} = 5$ V $R_G = R_F = 10$ kΩ

$-V_{CC} = -5$ V   Op Amp = TLC071

Figure 26.3: Split-supply op-amp circuit with common-mode voltage

is in both input circuits, hence it is named a common-mode voltage. Voltage-feedback op-amps, like those used in this document, reject common-mode voltages because their input circuit is constructed with a differential amplifier (chosen because it has natural common-mode voltage rejection capabilities).

$$V_{OUT} = -(V_{IN} + V_{REF})\frac{R_F}{R_G} + V_{REF}\left(\frac{R_F}{R_F + R_G}\right)\left(\frac{R_F + R_G}{R_G}\right) = -V_{IN}\frac{R_F}{R_G} \quad (26.3)$$

When signal sources are referenced to ground, single-supply op-amp circuits always have a large input common-mode voltage. Figure 26.4 shows a single-supply op-amp circuit that has its input voltage referenced to ground. The input voltage is not referenced to the midpoint of the supplies like it would be in a split-supply application, rather it is referenced to the lower power supply rail. This circuit malfunctions when the input voltage is positive because the output voltage would have to go negative—hard to do with a positive supply. It operates marginally with small negative input voltages because most op-amps do not function well when the inputs are connected to the supply rails.

The constant requirement to account for inputs connected to ground or other reference voltages makes it difficult to design single-supply op-amp circuits. This chapter presents a collection of single-supply op-amp circuits, including their description and transfer equation. Those without a good working knowledge of op-amp

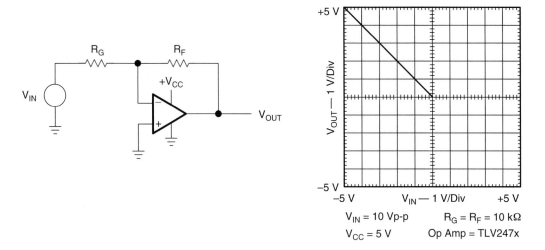

Figure 26.4: Single-supply op-amp circuit

equations should reference the *Understanding Basic Analog...* series of application notes available from Texas Instruments. Application note SLAA068, *Understanding Basic Analog—Ideal Op-Amps* develops the ideal op-amp equations. Circuit equations in this chapter are written with the ideal op-amp assumptions as specified in that document. The assumptions appear in Table 26.1 for easy reference.

Detailed information about designing single-supply op-amp circuits appears in application note SLOA030, *Single-Supply Op-Amp Design Techniques*. Unless otherwise specified, all op-amp circuits shown here are single-supply circuits. The single supply may be wired with the negative or positive lead connected to ground, but as long as the supply polarity is correct, the wiring does not affect circuit operation.

Table 26.1: Ideal op-amp assumptions

Parameter name	Parameter symbol	Value
Input current	I_{IN}	0
Input offset voltage	V_{OS}	0
Input impedance	Z_{IN}	∞
Output impedance	Z_{OUT}	0
Op-amp gain	a	∞

26.2 Boundary Conditions

All op-amps are constrained to output voltage swings less than or equal to their power supply. Use of a single supply limits the output voltage to the range of the supply voltage. For example, when the supply voltage VCC equals +10V, the output voltage is limited to the range $0 \leq V_{OUT} \leq 10$. This limitation precludes negative output voltages when the circuit has a positive supply voltage, but it does not preclude negative input voltages. As long as the voltage on the op-amp input leads does not become negative, the circuit can handle negative voltages applied to the input resistors.

Beware of working with negative (positive) input voltages when the op-amp is powered from a positive (negative) supply because op-amp inputs are highly susceptible to reverse-voltage breakdown. Also, ensure that no *start-up* condition reverse biases the op-amp inputs when the input and supply voltage are opposite polarity. It may be advisable to protect the op-amp inputs with a diode (Schottky or germanium) connected anode to ground and cathode to the op-amp input.

26.3 Amplifiers

Many types of amplifiers can be created using op-amps. This section consists of a selection of some basic, single-supply op-amp circuits that are available to the designer during the concept stage of a design. The circuit configuration and correct single-supply DC biasing techniques are presented for the following cases: inverting, noninverting, differential, T-network, buffer and AC-coupled amplifiers.

26.3.1 Inverting Op-Amp with Noninverting Positive Reference

The ideal transfer equation is given in Equation 26.4.

$$V_{OUT} = -V_{IN}\frac{R_F}{R_G} + V_{REF}\left(\frac{R_F + R_G}{R_G}\right) \qquad (26.4)$$

The transfer equation for this circuit (Figure 26.5) takes the form of $Y = -mX + b$. The transfer function slope is negative, and the DC intercept is positive. R_F and R_G are contained in both halves of the equation, thus it is hard to obtain the desired slope and DC intercept without modifying V_{REF}. This is the minimum component count configuration for this transfer function. When the reference voltage is 0, the input

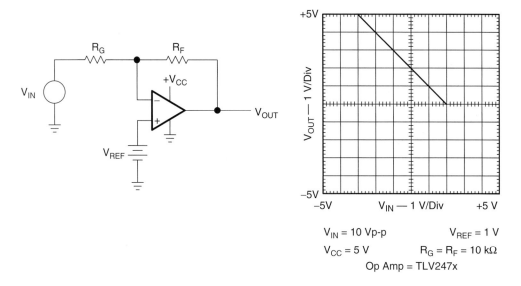

$V_{IN} = 10$ Vp-p $V_{REF} = 1$ V

$V_{CC} = 5$ V $R_G = R_F = 10$ kΩ

Op Amp = TLV247x

Figure 26.5: Inverting op-amp with noninverting positive reference

voltage is constrained to negative voltages because positive input voltages would cause the output voltage to saturate at ground.

26.3.2   Inverting Op-Amp with Inverting Negative Reference

The transfer equation takes the form of Y = −mX + b and is given in Equation 26.5.

$$V_{OUT} = -V_{IN}\frac{R_F}{R_{G1}} + V_{REF}\frac{R_F}{R_{G2}} \tag{26.5}$$

The transfer function slope is negative, and the DC intercept is positive (Figure 26.6). R_{G1} and R_{G2} are contained in the equation, thus it is easy to obtain the desired slope and DC intercept by adjusting the value of both resistors. Because of the virtual ground at the inverting input, R_{G2} is the terminating impedance for V_{REF}. When the reference voltage is 0, the input voltage is constrained to negative voltages because positive input voltages would cause the output voltage to saturate at ground.

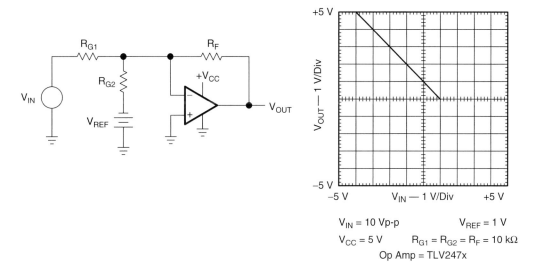

Figure 26.6: Inverting op-amp with inverting negative reference

26.3.3 Inverting Op-Amp with Noninverting Negative Reference

The transfer equation takes the form of Y = −mX − b and is given in Equation 26.6.

$$V_{OUT} = -V_{IN}\frac{R_F}{R_G} - V_{REF}\left(\frac{R_F + R_G}{R_G}\right) \tag{26.6}$$

The transfer function slope is negative, and the DC intercept is negative. R_F and R_G are contained in both halves of the equation, thus it is hard to obtain the desired slope and DC intercept without modifying V_{REF}. This is the minimum component count configuration for this transfer function. The slope and DC intercept terms in Equation 26.6 are both negative, hence, unless the correct input voltage range is selected, the output voltage will saturate at ground. The negative input voltage must be limited to less than −400 mV because op-amp inputs either break down or have protection circuits that forward bias when large negative voltages are applied to the inputs.

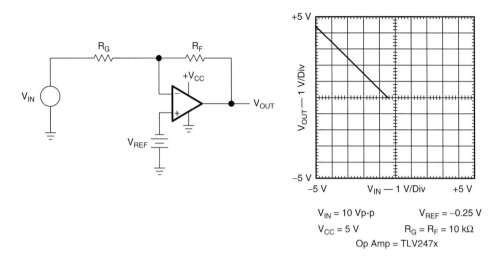

Figure 26.7: Inverting op-amp with noninverting negative reference

26.3.4   Inverting Op-Amp with Inverting Positive Reference

The transfer equation takes the form of $Y = -mX - b$ and is given in Equation 26.7.

$$V_{OUT} = -V_{IN}\frac{R_F}{R_{G1}} - V_{REF}\frac{R_F}{R_{G2}} \tag{26.7}$$

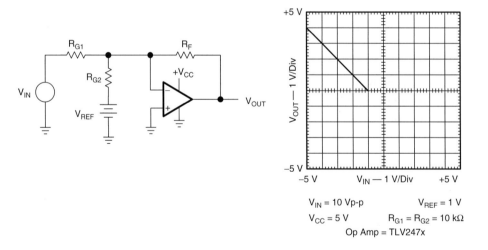

Figure 26.8: Inverting op-amp with inverting positive reference

The transfer function slope is negative, and the DC intercept is negative. R_{G1} and R_{G2} are contained in the equation, thus it is easy to obtain the desired slope and DC intercept by adjusting the value of both resistors. Because of the virtual ground at the inverting input, R_{G2} is the terminating impedance for V_{REF}. The slope and DC intercept terms in Equation 26.7 are both negative, hence, unless the correct input voltage range is selected, the output voltage will saturate.

26.3.5 Noninverting Op-Amp with Inverting Positive Reference

The transfer equation takes the form of $Y = mX - b$ and is given in Equation 26.8.

$$V_{OUT} = V_{IN} \frac{R_F + R_G}{R_G} - V_{REF} \frac{R_F}{R_G} \qquad (26.8)$$

The transfer function slope is positive, and the DC intercept is negative. This is the minimum component count configuration for this transfer function. The reference termination resistor is connected to a virtual ground, so R_G is the load across V_{REF}. R_F and R_G are contained in both halves of the equation, thus it is hard to obtain the desired slope and DC intercept without modifying V_{REF} or placing an attenuator in series with V_{IN}.

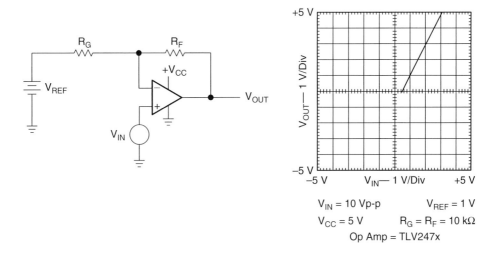

Figure 26.9: Noninverting op-amp with inverting positive reference

26.3.6   Noninverting Op-Amp with Noninverting Negative Reference

The transfer equation takes the form of $Y = mX - b$ and is given in Equation 26.9.

$$V_{OUT} = V_{IN} \left(\frac{R_2}{R_1 + R_2} \right) \frac{R_F + R_G}{R_G} - V_{REF} \left(\frac{R_1}{R_1 + R_2} \right) \frac{R_F + R_G}{R_G} \tag{26.9}$$

The transfer function slope is positive, and the DC intercept is negative. The reference is terminated in R_1 and R_2. R_1 and R_2 can be selected independent of R_F and R_G to obtain the desired slope and DC intercept. The price for the extra degree of freedom is two resistors.

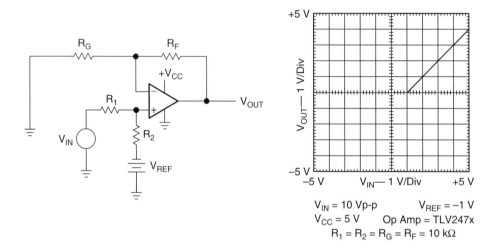

Figure 26.10: Noninverting op-amp with noninverting negative reference

26.3.7   Noninverting Op-Amp with Inverting Negative Reference

The transfer equation takes the form of $Y = mX + b$ and is given in Equation 26.10.

$$V_{OUT} = V_{IN} \frac{R_F + R_G}{R_G} + V_{REF} \frac{R_F}{R_G} \tag{26.10}$$

The transfer function slope is positive, and the DC intercept is positive. This is the minimum component count configuration for this transfer function. The reference termination resistor is connected to a virtual ground, so R_G is the load across V_{REF}. R_F and R_G are contained in both halves of the equation. Thus, it is hard to obtain the desired slope and DC intercept without modifying V_{REF} or placing an attenuator in series with V_{IN}.

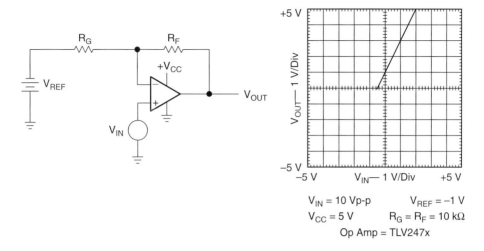

V_{IN} = 10 Vp-p V_{REF} = −1 V
V_{CC} = 5 V R_G = R_F = 10 kΩ
Op Amp = TLV247x

Figure 26.11: Noninverting op-amp with inverting positive reference

26.3.8   *Noninverting Op-Amp with Noninverting Positive Reference*

The transfer equation takes the form of $Y = mX + b$ and is given in Equation 26.11.

$$V_{OUT} = V_{IN}\left(\frac{R_2}{R_1 + R_2}\right)\frac{R_F + R_G}{R_G} + V_{REF}\left(\frac{R_1}{R_1 + R_2}\right)\frac{R_F + R_G}{R_G} \qquad (26.11)$$

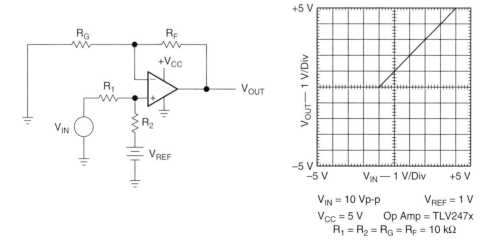

V_{IN} = 10 Vp-p V_{REF} = 1 V
V_{CC} = 5 V   Op Amp = TLV247x
R_1 = R_2 = R_G = R_F = 10 kΩ

Figure 26.12: Noninverting op-amp with noninverting positive reference

The transfer function slope is positive, and the DC intercept is positive. The reference is terminated in R_1 and R_2. R_1 and R_2 can be selected independent of R_F and R_G to obtain the desired slope and DC intercept. The price for the extra degree of freedom is two resistors.

26.3.9 Differential Amplifier

When R_F is set equal to R_2 and R_G is set equal to R_1, Equation 26.12 reduces to Equation 26.13.

$$V_{OUT} = V_{IN2}\frac{R_F + R_G}{R_G}\left(\frac{R_2}{R_1 + R_2}\right) - V_{IN1}\frac{R_F}{R_G} \tag{26.12}$$

$$V_{OUT} = (V_{IN2} - V_{IN2})\frac{R_F}{R_G} \tag{26.13}$$

These resistors must be matched very closely to obtain good differential performance. The mismatch error in these resistors reduces the common-mode performance, and the mismatch shows up in the output as an amplified common-mode voltage.

Consider Equation 26.13. Note that only the difference signal is amplified, thus this configuration is called a *differential amplifier*. The differential amplifier is a popular

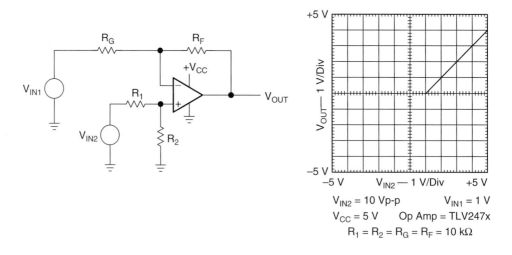

Figure 26.13: Differential amplifier

circuit in precision applications where it is used to amplify sensor outputs while rejecting common-mode noise.

The inverting input impedance is R_G because of the virtual ground at the inverting op-amp input. The noninverting input impedance is $R_F + R_G$ because the noninverting op-amp input impedance approaches infinity. The two input impedances are different, and this leads to two problems with this circuit.

First, mismatched input impedances preclude any attempts to cancel input bias currents through resistor matching. Often R_2 is set equal to $R_F \parallel R_G$ so that the bias currents develop equal common-mode voltages which the op-amp rejects. This is not possible when $R_2 = R_F$ and $R_1 = R_G$ unless the source impedances are matched. Second, high output impedance sensors are often used, and when high output sensors work into mismatched input impedances, errors occur.

26.3.10 Differential Amplifier with Bias Correction

When R_F is set equal to R_2 and R_G is set equal to R_1, Equation 26.14 reduces to Equation 26.15.

$$V_{OUT} = V_{IN2} \frac{R_F + R_G}{R_G} \left(\frac{R_2}{R_1 + R_2} \right) + V_{REF} \frac{R_F + R_G}{R_G} \left(\frac{R_1}{R_1 + R_2} \right) - V_{IN1} \frac{R_F}{R_G} \quad (26.14)$$

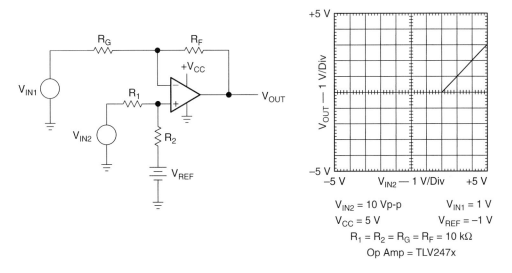

$V_{IN2} = 10$ Vp-p $\qquad V_{IN1} = 1$ V
$V_{CC} = 5$ V $\qquad V_{REF} = -1$ V
$R_1 = R_2 = R_G = R_F = 10$ kΩ
Op Amp = TLV247x

Figure 26.14: Differential amplifier with bias correction

$$V_{OUT} = (V_{IN2} - V_{IN1})\frac{R_F}{R_G} + V_{REF} \qquad (26.15)$$

When an offset voltage must be eliminated from or added to the input signal, this differential amplifier circuit is employed. The reference voltage can be positive or negative depending upon the polarity offset required, but care must be taken to protect the op-amp inputs and not exceed the output range.

26.3.11   High Input Impedance Differential Amplifier

When R_F is set equal to R_1 and R_G is set equal to R_2, Equation 26.16 reduces to Equation 26.17.

$$V_{OUT} = V_{IN1}\frac{R_1 + R_2}{R_1}\left(-\frac{R_F}{R_G}\right) + V_{IN2}\frac{R_F + R_G}{R_G} \qquad (26.16)$$

$$V_{OUT} = (V_{IN2} - V_{IN1})\frac{R_F + R_G}{R_G} \qquad (26.17)$$

Each input signal is connected to an op-amp noninverting input that is very high impedance. The input impedance of the circuit is very high, and it is matched, so this

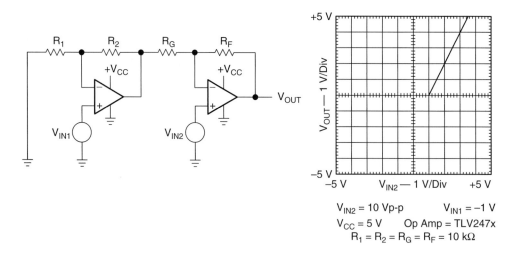

Figure 26.15: High input impedance differential amplifier

circuit is often used to interface to high-impedance sensors. Each op-amp has a signal propagation time, and V_{IN1} experiences two propagation delays versus V_{IN2}'s one propagation delay. At high frequencies, the propagation delay becomes a significant portion of the signal period, and this configuration is not usable at that frequency.

R_F and R_2, and R_G and R_1 should be matched to achieve good common-mode rejection capability. Bias current cancellation resistors equal to $R_F \parallel R_G$ should be connected in series with the input sources for precision applications.

26.3.12 High Common-Mode Range Differential Amplifier

When all resistors are equal, Equation 26.18 reduces to Equation 26.19.

$$V_{OUT} = V_{IN2}\left(-\frac{R_2}{R_1}\right)\left(-\frac{R_F}{R_{G1}}\right) + V_{REF1}\left(\frac{R_1 + R_2}{R_1}\right)\left(-\frac{R_F}{R_{G1}}\right) + V_{IN1}\left(-\frac{R_F}{R_{G2}}\right) +$$

$$V_{REF2}\left(\frac{R_F + R_{G1} \parallel R_{G2}}{R_{G1} \parallel R_{G2}}\right) \tag{26.18}$$

$$V_{OUT} = (V_{IN2} - V_{IN1}) - 2V_{REF1} + 3V_{REF2} \tag{26.19}$$

R_1 and R_{G2} are equal-value resistors terminated into a virtual ground; hence, the input sources are equally terminated. This configuration has high common-mode capability

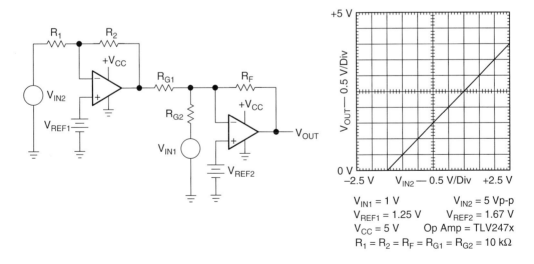

Figure 26.16: High common-mode range differential amplifier

because R_1 and R_{G2} limit the current that can flow into or out of the op-amp. Thus, the input voltage can rise to any value that does not exceed the op-amp's drive capability. The voltage references, V_{REF1} and V_{REF2}, are added for bias purposes. Without bias, the output voltage of the op-amps would saturate at ground, and the bias voltages keep the output voltage of the op-amp positive.

26.3.13   High-Precision Differential Amplifier

When $R_7 = R_6$, $R_5 = R_2$, $R_1 = R_4$, and $V_{REF1} = V_{REF2}$, Equation 26.20 reduces to Equation 26.21.

$$V_{OUT} = (V_{IN2} + V_{REF2})\left(\frac{2R_4 + R_3}{R_3}\right)\left(\frac{R_7}{R_5 + R_7}\right)\left(\frac{R_6 + R_2}{R_2}\right) -$$

$$(V_{IN1} + V_{REF1})\left(\frac{2R_1 + R_3}{R_3}\right)\frac{R_6}{R_2} + V_{REF3}\left(\frac{R_5}{R_5 + R_7}\right)\left(\frac{R_6 + R_2}{R_2}\right) \qquad (26.20)$$

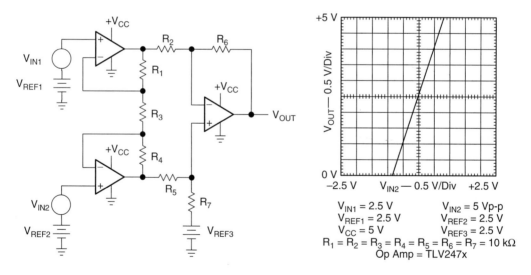

Figure 26.17: High-precision differential amplifier

$$V_{OUT} = (V_{IN2} - V_{IN1})\left(\frac{2R_1}{R_3} + 1\right)\left(\frac{R_6}{R_2}\right) + V_{REF3} \qquad (26.21)$$

In this circuit configuration, both sources work into the input impedance of a noninverting op-amp. This impedance is very high, and if the op-amps are identical, both impedances are very nearly equal. The propagation delay is still equal to two op-amp propagation delays, but the propagation delay is very nearly equal, so any distortion resulting from unequal propagation delays is minimized.

The equal resistors should be matched with more precision than is expected from the circuit. Resistor matching eliminates distortion due to unequal gains, and it reduces the common-mode voltage feed through. Resistors equal to $(R_1 \| R_3)/2$ may be placed in series with the sources to reduce errors resulting from bias currents. This differential amplifier has the unique feature that the gain can be changed with only one resistor, and if the gain setting resistor is R_3, no resistor matching is required to change gain.

26.3.14 Simplified High-Precision Differential Amplifier

When RF is set equal to R2, RG is set equal to R1, and VREF1 = VREF2, Equation 26.22 reduces to Equation 26.23.

$$V_{OUT} = (V_{IN2} + V_{REF2})\left(\frac{R_F + R_G}{R_G}\right)\left(\frac{R_2}{R_1 + R_2}\right) - (V_{IN1} + V_{REF1})\left(\frac{R_F}{R_G}\right) +$$

$$V_{REF3}\left(\frac{R_1}{R_1 + R_2}\right)\left(\frac{R_F + R_G}{R_G}\right) \qquad (26.22)$$

$$V_{OUT} = (V_{IN2} - V_{IN1})\left(\frac{R_F}{R_G}\right) + V_{REF3} \qquad (26.23)$$

Both input sources are loaded equally with very high impedances in the simplified high precision differential amplifier. This configuration eliminates three resistors, two of which are matched, but it sacrifices flexibility in gain setting capability because the gain must be set with a matched pair of resistors.

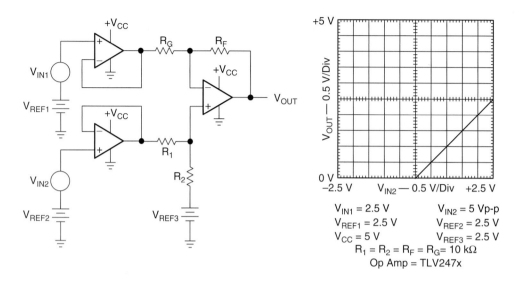

Figure 26.18: Simplified high-precision differential amplifier

26.3.15 Variable Gain Differential Amplifier

When R_1 is set equal to R_3 and R_2 is set equal to R_4, Equation 26.24 reduces to Equation 26.25.

$$V_{OUT} = V_{IN1}\left(\frac{R_2}{R_1 + R_2}\right)\left(\frac{R_3 + R_4}{R_3}\right)\left(\frac{R_G}{R_F}\right) - V_{IN2}\left(\frac{R_4}{R_3}\right)\left(\frac{R_G}{R_F}\right) \tag{26.24}$$

$$V_{OUT} = (V_{IN2} - V_{IN1})\left(\frac{R_4}{R_3}\right)\left(\frac{R_F}{R_G}\right) \tag{26.25}$$

When a function is enclosed in a feedback loop, the function acts inverted on the closed loop transfer function. Thus, the gain stage R_F/R_G ends up being an attenuator. The circuit shown in Figure 26.19 can be used with any of the differential amplifiers to change gain without affecting matched resistors. R_1, R_3 and R_2, R_4 must be matched to reduce the common-mode voltage.

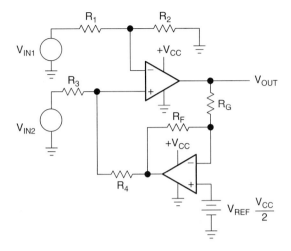

Figure 26.19: Variable gain differential amplifier

26.3.16 T *Network in the Feedback Loop*

Sometimes it is desirable to have a low-resistance path to ground in the feedback loop. Standard inverting op-amps cannot do this when the driving circuit sets the input resistor value and the gain specification sets the feedback resistor value. Inserting a *T* network in the feedback loop yields a degree of freedom that enables both specifications to be met with a low DC resistance path to ground in the feedback loop.

$$V_{OUT} = -V_{IN}\left[\frac{R_2 + R_3 + \dfrac{R_2 R_3}{R_4}}{R_1}\right] + V_{REF}\left[1 + \frac{R_2 + R_3 + \dfrac{R_2 R_3}{R_4}}{R_1}\right] \quad (26.26)$$

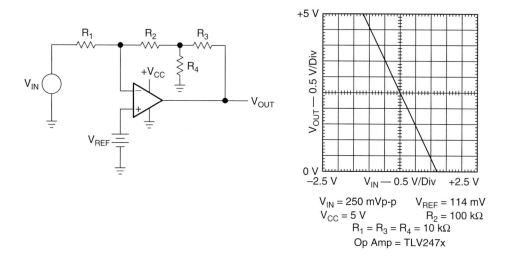

Figure 26.20: T network in the feedback loop

26.3.17 Buffer

The buffer input signal polarity must be unipolar because the output voltage swing is unipolar. When this limitation precludes the buffer, a differential amplifier with the negative input correctly biased is used, or a reference voltage is added to the buffer to offset the output voltage. R_F must be included when the op-amp inputs are not rated

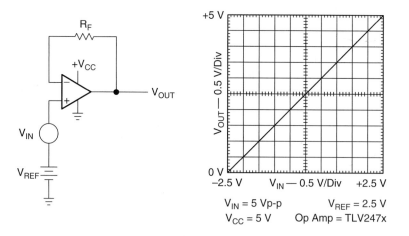

Figure 26.21: Buffer

for the full supply voltage. In that case, R_F limits the current into the op-amp inputs, thus preventing latch up. Most new op-amp inputs can withstand the full supply voltage, so they often leave R_F out as cost savings. The main attraction of the buffer is that it has very high input impedance and very low output impedance. The impedance transformation capability is why buffers are often added to the input of other circuits.

$$V_{OUT} = V_{IN} + V_{REF} \qquad (26.27)$$

26.3.18 Inverting AC Amplifier

V_{CC} and resistors R set a DC level of $V_{CC}/2$ at the inverting input. R_G is connected to ground through a capacitor, thus the circuit functions as a buffer for DC. This causes the DC output voltage to be $V_{CC}/2$, so the quiescent output voltage is the middle of the supply voltage, and it is ready to swing to either rail as the input signal commands.

The AC gain is given in Equation 26.28. R_G and C form a coupling network for the AC signal. Good coupling networks should be constant low impedance at the signal frequencies, so Equation 26.29 should be satisfied to get good low-frequency

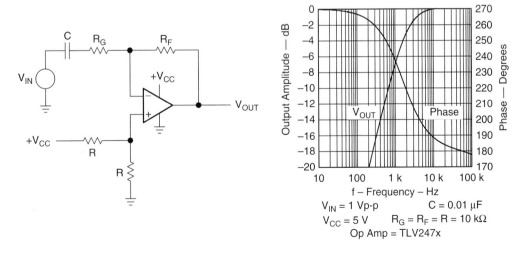

Figure 26.22: Inverting AC amplifier

performance. The lowest frequency component of the input signal, f_{MIN}, is determined by completing a Fourier series on the input signal. Then, setting $f_{MIN} = 100f$ in Equation 26.29 ensures that the 3-dB breakpoint introduced by R_G and C is two decades lower than f_{MAX}.

$$V_{OUT} = -V_{IN}\frac{R_F}{R_G} + \frac{V_{CC}}{2} \qquad (26.28)$$

$$f = \frac{1}{200\pi R_G C} \qquad (26.29)$$

26.3.19   Noninverting AC Amplifier

V_{CC} and the resistors (R) set a DC level of $V_{CC}/2$ at the inverting input. R_G is connected to ground through a capacitor, thus the circuit functions as a buffer for DC. This causes the DC output voltage to be $V_{CC}/2$, so the quiescent output voltage is the middle of the supply voltage, and it is ready to swing to either rail as the input signal commands.

The AC gain is given in Equation 26.30. R_G and C create a coupling network for the AC signal. Good coupling networks should be a constant low impedance at the signal

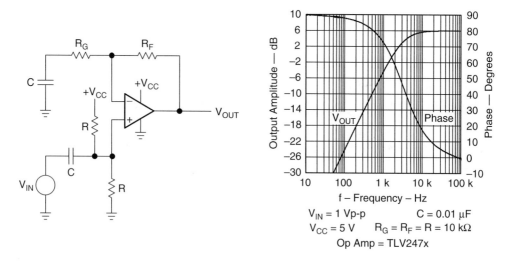

Figure 26.23: Noninverting AC amplifier

frequencies, so Equation 26.31 should be satisfied to get good low frequency performance. The lowest frequency component of the input signal, f_{MIN}, is determined by completing a Fourier series on the input signal. Then, setting $f_{MIN} = 100f$ in Equation 26.31 ensures that the 3-dB breakpoint introduced by R_G and C is two decades lower than f_{MAX}. The breakpoint for R_G and C_1 is set in a similar manner.

$$V_{OUT} = V_{IN} \frac{R_F + R_G}{R_G} + \frac{V_{CC}}{2}\left(\frac{R_F + R_G}{R_G}\right) \qquad (26.30)$$

$$f = \frac{1}{200\pi R_G C} \qquad (26.31)$$

26.4   Computing Circuits

Four versions of the inverting op-amp and four versions of the noninverting op-amp were given in the previous section. During the concept stage of the design, one of these eight op-amp circuits is selected. Specifications for the input and output voltage are the selection criteria that determines which circuit configuration is used.

There are four versions of most of the circuits given in this and following sections, but just the simplest version of any circuit is included in this chapter because of space limitations. Each circuit configuration can be modified as required to fit specific applications. Look back to the first section to determine what bias is required to fit the application, and adapt that bias to the new circuit.

26.4.1   Inverting Summer

The three input voltages are inverted and added as Equation 26.32 shows. R_B should be made equal in value to the parallel combination of R_F, R_{G1}, R_{G2}, and R_{G3} to convert the input bias current to a common-mode voltage so the op-amp can reject it. V_{REF} sets the output voltage somewhere between the supply limits, and this allows negative addition (subtraction) to take place.

$$V_{OUT} = -\left(V_{IN1}\frac{R_F}{R_{G1}} + V_{IN2}\frac{R_F}{R_{G2}} + V_{IN3}\frac{R_F}{R_{G3}} + \cdots\right) + V_{REF}\left(1 + \frac{R_F}{R_{G1}\|R_{G2}\|R_{G3}\cdots}\right)$$

$$(26.32)$$

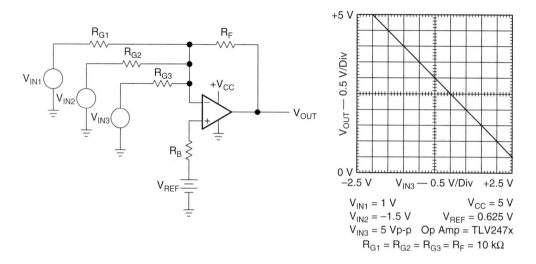

Figure 26.24: Inverting summer

26.4.2   Noninverting Summer

This circuit adds the input voltages and multiplies them by the stage gain. R_{G1}, R_{G2}, and R_{G3} in parallel should be equal to R_F in parallel with R_G to cancel the input bias current using the common-mode input voltage rejection technique. VREF is added to the circuit to enable the addition of negative values.

$$V_{OUT} = +\left(\frac{V_{IN1}(R_{G2} \| R_{G3})}{R_{G1} + R_{G2} \| R_{G3}} + \frac{V_{IN2}(R_{G1} \| R_{G3})}{R_{G2} + R_{G1} \| R_{G3}} + \frac{V_{IN3}(R_{G2} \| R_{G1})}{R_{G3} + R_{G2} \| R_{G1}}\right)$$

$$\times \left(\frac{R_F + R_G}{R_G}\right) - V_{REF}\frac{R_F}{R_G}$$

$$(26.33)$$

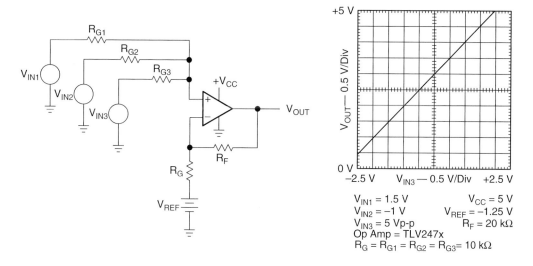

$V_{IN1} = 1.5\ V$ $V_{CC} = 5\ V$
$V_{IN2} = -1\ V$ $V_{REF} = -1.25\ V$
$V_{IN3} = 5\ Vp\text{-}p$ $R_F = 20\ k\Omega$
Op Amp = TLV247x
$R_G = R_{G1} = R_{G2} = R_{G3} = 10\ k\Omega$

Figure 26.25: Noninverting summer

26.4.3 Noninverting Summer with Buffers

V_{REF1} and V_{REF2} are added to enable the buffers to handle positive input voltages. Their output contribution to the last stage is cancelled out by V_{REF3}. This configuration uses fewer resistors at the expense of two op-amps. R_{G1}, R_{G2}, and RF in parallel should be made equal to R_B to cancel the input bias current.

$$V_{OUT} = V_{IN1} \frac{R_F}{R_{G1}} + V_{IN2} \frac{R_F}{R_{G2}} - V_{REF1} \frac{2R_F}{R_{G1}} - V_{REF2} \frac{2R_F}{R_{G2}}$$

$$+ V_{REF3}\left(1 + \frac{R_F}{R_{G1} \,\|\, R_{G2}}\right) \tag{26.34}$$

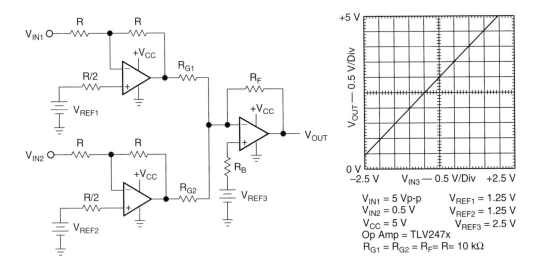

Figure 26.26: Noninverting summer with buffers

26.4.4 Inverting Integrator

The Laplace operator, $s = j\omega$, is used in Equation 26.35, and the mathematical operation $1/s$ constitutes an integration. Differentiation circuits are shown later, and the mathematical operation, s, constitutes a differentiation. The integration time constant is RC, thus the magnitude crosses 0 dB on a log plot when RC = 1. Also the phase is $-45°$ when RC = 1.

$$V_{OUT} = -V_{IN}\frac{1}{RCs} \tag{26.35}$$

This integrator is not very practical because there is no method of discharging the capacitor; hence, any leakage current will eventually charge the capacitor until the circuit becomes saturated. The positive input of the integrator is biased at $V_{CC}/2$ to center the output voltage at $V_{CC}/2$; thus allowing for positive and negative voltage swings. The bias resistors are selected as 2R so that the parallel combination equals R. This offsets the input current drawn through R.

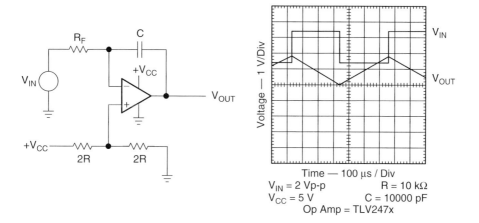

Figure 26.27: Inverting integrator

26.4.5 Inverting Integrator with Input Current Compensation

Functionally, this circuit is the same as that shown in Figure 26.27, but a current compensation network has been added to offset the input current. V_{CC}, R_1, and R_2 bias the positive input at $V_{CC}/2$ to center the output voltage at $V_{CC}/2$; thus allowing for positive and negative voltage swings.

R_1 and R_2 are selected as relatively small values because the current flowing through R_A also flows through the parallel combination of R_1 and R_2. R_A forward biases the diode with a constant current, thus the diode acts like a small voltage regulator. The diode voltage drop is temperature sensitive, and this factor works in our favor because the input transistors are temperature sensitive. The two temperature sensitivities cancel out if the diode current is selected correctly. R_B is a large-value resistor that acts like a current source, so it is selected such that it supplies the input bias current. Selecting R_B correctly ensures that no input current flows through the integration resistor, R.

This integrator is not very practical because there is no method of discharging the capacitor. Hence, any input current will eventually charge the capacitor until the circuit becomes saturated. The bias circuit drastically reduces the input current flowing through R, thus it extends the integration time. A reset circuit is needed to make the integrator more practical.

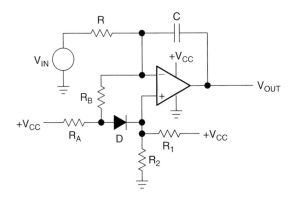

Figure 26.28: Inverting integrator with input current compensation

This bias compensation scheme is set up for an op-amp that has NPN input transistors. The diode must be reversed and connected to ground for op-amps with PNP input circuits.

$$V_{OUT} = - V_{IN1} \frac{1}{RCs} \qquad (26.36)$$

26.4.6   Inverting Integrator with Drift Compensation

Functionally, this circuit is the same as that shown in Figure 26.27, but it uses an RC circuit in the positive lead to obtain drift compensation. The voltage divider is made from a series string of resistors (R_A), and VCC biases the input in the center of the power supply.

Positive input current flows through R and C in parallel, so the positive input current drops the same voltage across the parallel RC combination as the negative input current drops across its series RC combination. The common-mode rejection capability of the op-amp rejects the voltages caused by the input currents. Much longer integration times can be achieved with this circuit, but when the input signal does not center around $V_{CC}/2$, the compensation is poor.

$$V_{OUT} = -V_{IN1} \frac{1}{RCs} \qquad (26.37)$$

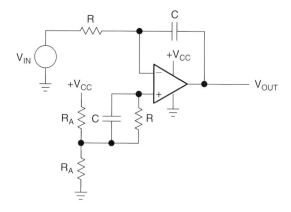

Figure 26.29: Inverting integrator with drift compensation

26.4.7 Inverting Integrator with Mechanical Reset

Functionally, this circuit is the same as that shown in Figure 26.27, but a method has been provided to discharge (reset) the capacitor. S_1 is a mechanical switch or relay and when the contacts close, they short the integrating capacitor forcing it to discharge. Some capacitors are sensitive to fast discharge cycles, so R_S is put in the discharge path to limit the initial discharge current. When R_S is absent from the circuit, the impulse of current that occurs at the first instant of discharge causes considerable noise, so the selection of R_S is also based on noise considerations. For all practical purposes, the time constant formed by R_S and C determines the discharge rate.

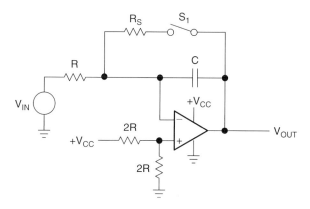

Figure 26.30: Inverting integrator with mechanical reset

One advantage of mechanical discharge methods is that they are isolated from the remainder of the circuit. Their size, weight, time delay, and uncertain actuating time offset this advantage. When the disadvantages of mechanical reset outweigh the advantages, circuit designers go to electronic reset circuits.

$$V_{OUT} = -V_{IN1} \frac{1}{RCs} \tag{26.38}$$

26.4.8 Inverting Integrator with Electronic Reset

Functionally, this circuit is the same as that shown in Figure 26.27, but an electronic method has been provided to discharge (reset) the capacitor. Q_1 is controlled by a gate drive signal that changes its state from on to off. When Q_1 is on, the gate-source resistance is low, less than 100Ω. And when Q_1 is off, the gate-source resistance is high—about several hundred $M\Omega$.

The source of the FET is at the inverting lead that is at ground, so the Q_1 gate-source bias is not affected by the input signal. Sometimes, the output signal can get large enough to cause leakage currents in Q_1, so the designer must take care to bias Q_1 correctly. Consult a transistor book for more detailed information on transistor reset circuits. A major problem with electronic reset is the charge injected through the transistor's stray capacitance. This charge can be large enough to cause integration errors.

$$V_{OUT} = -V_{IN1} \frac{1}{RCs} \tag{26.39}$$

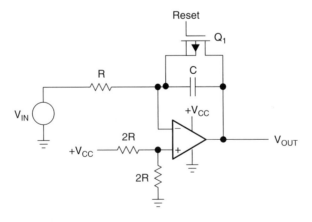

Figure 26.31: Inverting integrator with electronic reset

26.4.9   *Inverting Integrator with Resistive Reset*

This circuit differs from that shown in Figure 26.27 because it yields a breakpoint rather than a pure integration. On a log plot, the integrator slope is –6 dB per octave at the 0 frequency intercept, and the 0 dB intercept occurs when f = 1/2πRC. A breakpoint plots flat on a log plot until the breakpoint where it breaks down at –6 dB per octave. It is –3 dB when f = 1/2πRC.

R_F is in parallel with the integrating capacitor, C, so it is continually discharging C. The low frequency attenuation that is the best attribute of the pure integrator is sacrificed for the reset circuit complexity:

$$V_{OUT} = - V_{IN1}\left(\frac{R_F}{R_G}\right)\frac{1}{R_FC + 1} \tag{26.40}$$

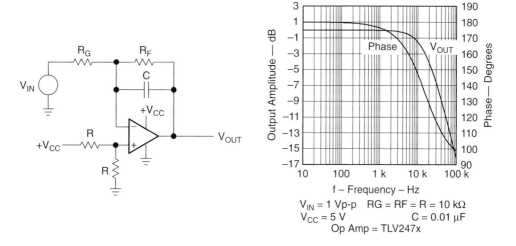

Figure 26.32: Inverting integrator with resistive reset

26.4.10   *Noninverting Integrator with Inverting Buffer*

This circuit is an inverting integrator preceded by an inverting buffer. Eliminating the signal inversion costs an op-amp and four resistors, but this is the easiest way to get true noninverting integrator performance.

$$V_{OUT} = \frac{R_A}{R_A}\left(-V_{IN}\frac{1}{RCs}\right) = V_{IN}\frac{1}{RCs} \tag{26.41}$$

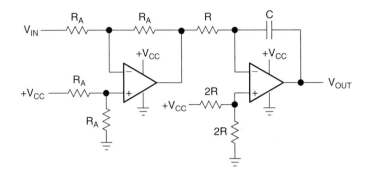

Figure 26.33: Noninverting integrator with inverting buffer

26.4.11   Noninverting Integrator Approximation

This circuit has fewer parts than the *Noninverting Integrator with Inverting Buffer* (Figure 26.33), but it is not a true integrator because there is a zero in the transfer equation. The log plot starts rolling off at a –6 dB per octave rate at low frequencies, but when f = 1/2πRC, the zero cuts in. The zero causes the log plot to flatten out because the slope decreases to 0 db per decade.

This circuit functions as an integrator at very low frequencies, but at frequencies higher than f = 1/2πRC, it functions as a buffer.

$$V_{OUT} = \frac{RCs + 1}{RCs} \qquad (26.42)$$

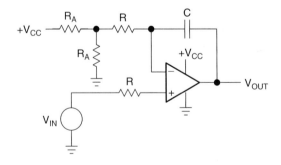

Figure 26.34: Noninverting integrator approximation

26.4.12 Inverting Differentiator

The log plot of the differentiator is a positive slope of 6-dB per octave passing through 0 dB at f = 1/2πRC. At extremely high frequencies, the capacitive reactance goes to very low values, thus the circuit gain approaches the op-amp open-loop gain. This performance emphasizes any system noise or noise generated by the op-amp. The poor noise performance of this circuit limits its application to a very few specialized situations.

This configuration has a pole in the feedback loop. If the op-amp has more than one pole, and most op-amps have several poles, this configuration can become oscillatory. The V_{CC} and R_A circuit bias the output in the center of the power supplies. $R_A/2$ should be selected equal to RG∥RF so that input currents are canceled out.

$$V_{OUT} = -V_{IN}RC_S \qquad (26.43)$$

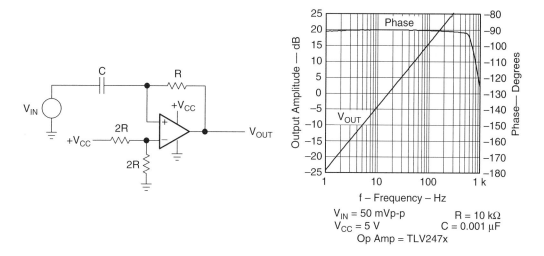

Figure 26.35: Inverting differentiator

26.4.13 Inverting Differentiator with Noise Filter

This circuit has a pure differentiator that rises at a 6-dB per octave slope from zero frequency. At f = 1/2πR$_F$C$_F$, the pole kicks in and the slope is reduced to zero. The pole has two effects. First, it stabilizes the circuit by canceling zero's phase shift. Second, it limits the circuit gain to 1 at high frequencies, so it acts like a noise filter.

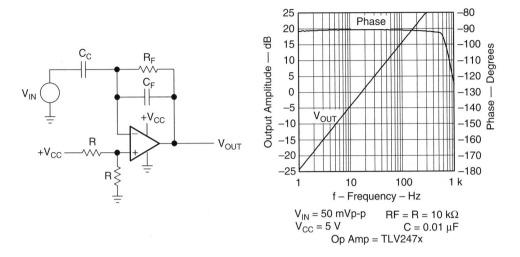

Figure 26.36: Inverting differentiator with noise filter

R/2 should equal R_F for good input current cancellation, and V_{CC} coupled with R centers the output voltage.

$$V_{OUT} = -V_{IN} \frac{R_F Cs}{R_F C_F s + 1} \tag{26.44}$$

26.5   Oscillators

This section describes some general op-amp sinewave oscillator circuits that fall under three main categories: Wien bridge, phase shift, and quadrature. A brief description and of each type is provided, along with one or two variations. Op-amp sinewave oscillators are used to create references in applications such as audio and function/waveform generators.

26.5.1   Basic Wien Bridge Oscillator

When $\omega = 2\pi f = 1/RC$, the feedback is in phase (this is positive feedback), and the gain is 1/3, so oscillation requires an amplifier with a gain of 3. When $R_F = 2R_G$ the amplifier gain is 3 and oscillation occurs at $f = 1/2\pi RC$. Normally, the gain is larger than 3 to ensure oscillation under worst case conditions.

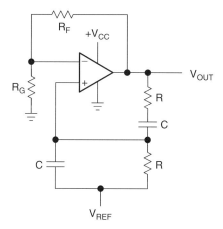

Figure 26.37: Basic Wien bridge oscillator

V_{REF} sets the output DC voltage in the center of the span.

The output sine wave is highly distorted because limiting by saturation and cut-off is controlling the output voltage excursion. The distortion decreases when the gain is decreased, but the circuit may not oscillate under worst-case low gain conditions.

$$V_{REF} = \frac{\dfrac{V_{CC}}{2}}{1 + \dfrac{R_F}{R_G}} \qquad (26.45)$$

26.5.2   Wien Bridge Oscillator with Nonlinear Feedback

When the circuit gain is 3, $R_L = R_F/2$.

Substituting a lamp (R_L) for the gain setting resistor reduces distortion because the nonlinear lamp resistance adjusts the gain to keep the output voltage smaller than the power supply voltage. The output voltage never approaches the power supply rail, so distortion doesn't occur. R_F and R_L determine the lamp current (see Equations 26.46 and 26.47).

$$I_{LAMP} = \frac{V_{OUT(RMS)}}{R_F + R_L} \qquad (26.46)$$

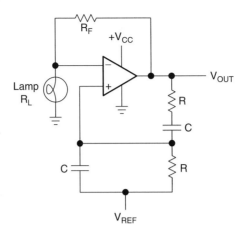

Figure 26.38: Wien bridge oscillator with nonlinear feedback

$$R_F = \frac{2(V_{OUT(RMS)})}{3(I_{OUT(RMS)})} \qquad (26.47)$$

The lamp is selected by examining lamp resistance curves until a lamp with a resistance approximately equal to $R_F/2$ at $I_{OUT(RMS)}$ is found. The output voltage swing should be less than 75% of the maximum guaranteed voltage swing, and 3 R_L must be greater than the load resistance specified for the voltage swing specification. V_{REF} should be $V_{CC}/5$.

26.5.3   Wien Bridge Oscillator with AGC

The op-amp is configured as an AC amplifier to ease biasing problems. The gain equation for the op-amp is given below. R_{G1} or R_{G2}, but not both resistors, is required depending on the selection of the Q_1.

The diode, D_1, half-wave rectifies the output voltage and applies it to the voltage divider formed by R_1 and R_2. The voltage divider biases Q_1 in its linear region, and they eventually set the output voltage. C_1 filters the rectified sine wave with a long time constant so that the output voltage stays constant. C_2 must be selected large enough to act as a short at the oscillation frequency.

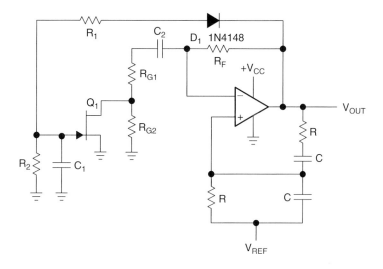

Figure 26.39: Wien bridge oscillator with AGC

As the output voltage increases, the negative voltage across the gate of Q_1 increases. The increased negative gate voltage causes Q_1 to increase its drain-to-source resistance. This results in increased op-amp gain and an output voltage decrease. When the voltage divider and FET are selected properly, the output voltage swing is less than the guaranteed maximum swing, so distortion doesn't occur.

$$ G = \frac{R_F}{R_{G1} + R_{G2} \, \| \, R_{FET}} \tag{26.48} $$

26.5.4 Quadrature Oscillator

Quadrature oscillators produce sine waves 90° out of phase, so they output sine/cosine, or quadrature waves.

When $R_1C_1 = R_2C_2 = R_3C_3$, the circuit oscillates at $\omega = 2\pi f = 1/RC$. Both op-amps act as integrators causing two poles at $1/RC$, thus the circuit oscillates when the loop gain crosses the 0-dB axis. The integrators ensure that gain is always sufficient for oscillation. There is a slight bit of distortion at the sine output, and it is very hard to eliminate this distortion.

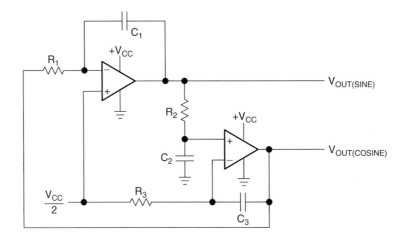

Figure 26.40: Quadrature oscillator

26.5.5 Classical Phase Shift Oscillator

Theoretically, the three RC sections do not load each other, thus the loop gain has three identical poles multiplied by the op-amp gain.

The loop phase shift is $-180°$ when the phase shift of each section is $-60°$, and this occurs when $\omega = 2\pi f = 1.732/RC$ because the tangent of $60° = 1.73$. The magnitude of β at this point is $(1/2)^3$, so the gain, $A = R_F/R_G$, must be greater or equal to 8 for the system gain to be equal to 1.

The assumption that the RC sections do not load each other is not entirely valid, thus the circuit does not oscillate at the specified frequency, and the gain required

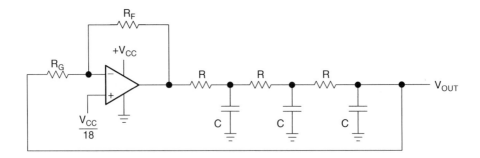

Figure 26.41: Classical phase shift oscillator

for oscillation is more than 8. This circuit configuration was very popular when an active component was large and expensive, but now that op-amps are inexpensive, small, and come quad packages, the classical phase shift oscillator is losing popularity.

The classical phase shift oscillator has an undistorted sine wave available at the output of the third RC section. This is not a low-impedance output, and the signal amplitude is smallest here, but these sacrifices have to be made to get away from distortion. An undistorted output can be obtained from the op-amp if an AGC circuit similar to the one shown in Figure 26.39 is employed. The reference voltage is set according to the equation $V_{REF} = V_{CC}/2(1 + R_F/R_G)$ to center the output voltage at $V_{CC}/2$.

$$A\beta = \left(\frac{1}{RCs + 1}\right)^3 \tag{26.49}$$

26.5.6   Buffered Phase Shift Oscillator

A noninverting op-amp buffers each RC section in this oscillator. Equation 26.49, repeated below, truly represents the transfer function of this circuit if RG >> R.

$$A\beta = \left(\frac{1}{RCs + 1}\right)^3 \tag{26.50}$$

The loop phase shift is −180° when the phase shift of each section is −60°, and this occurs when $\omega = 2\pi f = 1.732/RC$ because the tangent 60° = 1.73. The magnitude of β

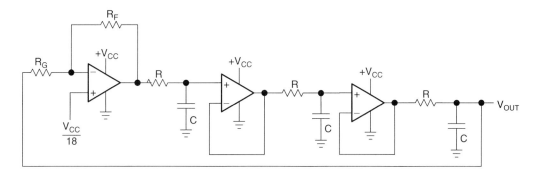

Figure 26.42: Buffered phase shift oscillator

at this point is $(1/2)^3$, so the gain, $A = R_F/R_G$, must be greater or equal to 8 for the system gain to be equal to one.

The buffered phase shift oscillator has an undistorted sine wave available at the output of the third RC section. This is not a low-impedance output, and the signal amplitude is smallest here, but these sacrifices have to be made to get away from distortion. An undistorted output can be obtained from the op-amp if an AGC circuit similar to the one shown in Figure 26.39 is employed.

There are three op-amps, so the gain can be distributed among the op-amps at the expense of a few resistors, and the distortion is reduced. Another method of reducing distortion is to limit the output voltage swing softly with external components. The limiting technique does not yield as good results as the AGC technique does, but it is less expensive. The reference voltage is set according to the equation $V_{REF} = V_{CC}/2$ $(1 + R_F/R_G)$ to center the output voltage at $V_{CC}/2$.

26.5.7 Bubba Oscillator

The Bubba oscillator is another phase shift oscillator, but it takes advantage of the quad op-amp package to yield some unique advantages. Each RC section is buffered by an op-amp to prevent loading. When $R_G \gg R$ there is no loading in the circuit, and the circuit yields theoretical performance.

Four RC sections require $-45°$ phase shift per section to accumulate $-180°$ phase shift. Each RC section contributes $-45°$ phase shift when $\omega = 1/RC$. The gain required for oscillation is $G \geq (1/0.707)^4 = 4$. Taking outputs from alternate sections yields low-impedance quadrature outputs. When an output is taken from each op-amp, the circuit delivers four $45°$ phase-shifted sine waves.

The gain, A, must equal 4 for oscillation to occur. Very low distortion sine waves can be obtained from the junction of R and R_G. When low-distortion sine waves are required at all outputs, the gain should be distributed among the op-amps. Gain distribution requires biasing of the other op-amps, but it has no effect on the oscillator frequency. This oscillator has the best $d\varphi/df$ of the phase shift oscillators, so it has minimum frequency drift. The reference voltage is set according to the equation $V_{REF} = V_{CC}/2(1 + R_F/R_G)$ to center the output voltage at $V_{CC}/2$.

$$A\beta = \left(\frac{1}{RCs + 1}\right)^4 \tag{26.51}$$

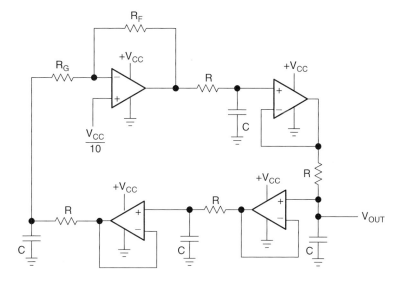

Figure 26.43: Bubba oscillator

26.5.8 Triangle Oscillator

The triangle oscillator produces triangle waves and square waves. The op-amp functions as an integrator. When the output voltage of the comparator is low, the output of the op-amp charges C until the output voltage exceeds the hysteresis voltage set by R_1 and R_F and the reference voltage ($V_{CC}/2$). At this point, the comparator output switches to a high state and the op-amp integrates the voltage in a negative direction. The triangle wave (op-amp output voltage swing) is given in Equation 26.52. The frequency of oscillation is given in Equation 26.53.

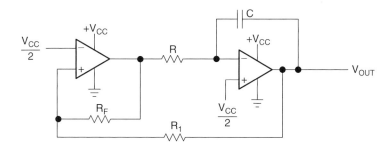

Figure 26.44: Triangle oscillator

$$V_{OUT} = \frac{V_{CC}}{2} \pm \frac{V_{CC}R_1}{2R_F} \qquad (26.52)$$

$$f = \frac{R_F}{4CRR_1} \qquad (26.53)$$

The op-amp reference voltage can be adjusted to equalize the triangle rise and fall times.

26.6　Some Favorite Circuits

Every engineer has their favorite batch of circuits and I'm no exception. There are tons of circuit cookbooks out there that show how to implement no end of cool features. There are so many that you could spend all your time searching them and never getting anything done. I suggest you develop your own favorite basic circuits that you know well and intuitively understand. This is simply an extension of the "Lego" philosophy that was discussed earlier in this book. Here are a few of my favorites. These are in addition to all the circuits used as examples up to this point; one reason they make such good examples is that they are so useful.

26.6.1　Hybrid Darlington Pair

Cool application note, using two transistors to switch a signal level Vcc PNP switched by NPN (see Figure 26.45).

This is a handy circuit that switches a higher level voltage with a lower level one. Say, for example, you have a micro with a 5V output and you need to drive a 12V load. For a reason you can't change, you have to switch the Vcc leg. In this circuit you turn on one transistor with a 5V signal, which in turn activates the other transistor switching the higher voltage to the load. This works because the transistors are current driven; when you shut off the current flow to the PNP transistor, it shuts off regardless of the voltage. Another plus is that this circuit has Darlington-like properties without one of the downsides. You won't need a lot of current to the input to switch the output and, unlike a traditional Darlington pair, the voltage drop across the output is much smaller. You don't have two series base junctions to contend with at the output.

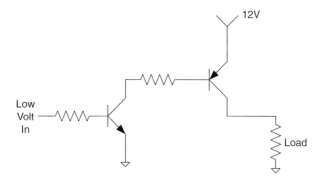

Figure 26.45: Vcc PNP switched by NPN

26.6.2   DC Level Shifter

This is really the high-pass filter that we have already studied but with a slight twist. Instead of ground, we hook the resistor to a reference voltage. Since DC has a frequency of zero, only the AC component will pass and in the process a DC bias will be applied to the signal. Make sure you don't size the cap and resistor so that the signal you want is attenuated.

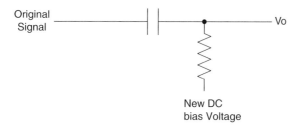

Figure 23.46: Change the DC bias on an AC signal

26.6.3   Virtual Ground

Using the voltage divider as a reference, the op-amp becomes a voltage source with the output matching the voltage at the divider. This can be very useful when you are trying to handle AC signals with only a single-ended supply circuit.

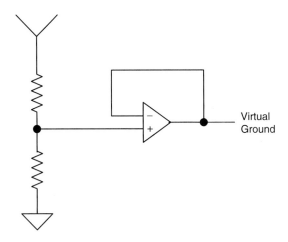

Figure 26.47: Create a "ground" at any level you want

26.6.4   Voltage Follower

This one is mighty useful when trying to measure a signal that is easily affected by load. $V_i = V_o$, but, best of all, V_i isn't loaded at all, thanks to the buffering effect of the op-amp.

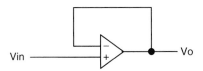

Figure 26.48: Voltage follower

This is another great circuit that works nicely amplifying AC signals with a single-ended supply. It also has the benefit of not amplifying any DC signal components, keeping things like DC offsets from making your signal rail. This happens because of the cap in the feedback circuit. Since the cap only passes AC current, DC signals see that point as disconnected. When the resistor to ground is disconnected, the op-amp acts like the voltage follower in the previous circuit.

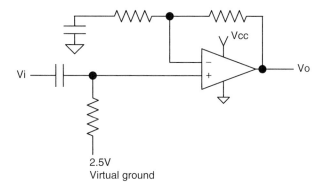

Figure 26.49: AC only amplifier

26.6.5 Inverter Oscillator

I saw this in the back of a data book years ago; I think it was a Motorola logic data book. This was way back before the internet—you used to have to turn pages to find this stuff! The way it works is based on the fact that the Schmidt trigger inverter has hysteresis built into the input. This makes the output stick at a high or low level till the cap on the input charges to the threshold voltage that trips the inverter. Output flips and everything goes the other direction, repeating indefinitely. Adding some diodes to the charge and discharge path can affect the duty cycle of the output.

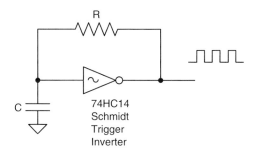

Figure 26.50: Schmidt trigger oscillator

26.6.6 Constant Current Source

Using negative feedback, the op-amp tries to maintain the voltage drop across R input. Even if the resistance of the load changes, the drop across the R input stays the same. According to Ohm's Law, keeping R and V the same will keep current the same too. Remember, though, this current control has operational limits; it can only swing the output voltage so far to compensate for load variance. Once these limits are reached, the current regulation can no longer exist.

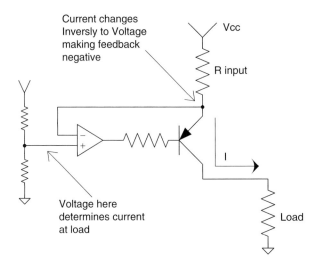

Figure 26.51: Voltage-controlled constant current source

26.6.7 Get Your Own

These are just a few of my favorites. Get your own and know them well. You will be better served knowing a few circuit concepts inside and out than knowing thousands superficially.

Thumb Rules
• Keep your own cookbook of cool circuits.   • Learn them well

References

Everyone learns from many sources. I have read many books that helped to gain that insight. Here is a list of those books, with a few comments to help you decide if you want to read them yourself.

Yellow Control Theory, Fundamentals of Automatic Control, by Robert C. Weyrick, McGraw Hill, ISBN 0-07-069493-1. Good read, helped me understand control theory.

"Pink Motor Book" – DC Motors Speed Controls Servo Systems: *The Electro-Craft Engineering Handbook*, by Reliance Motion Control, Inc. I like to call it the pink motor book due to an interesting choice of color for the cover, and I highly recommend it for anyone who is working with DC motors. It's heavy on the equations, but a good source for understanding all the complexities of motors.

Grounding and Shielding Electronic Systems, by Dr. Tom Van Doren, University Missouri Rolla, Van Doren Company, Rt. 6, Box 319, Rolla, MO 65401, Ph 314-341-4097.

Intuitive IC Op-Amps, by Tom Frederickson. This classic paperback book was originally published in 1984. The book describes how op-amps work and how they are used, from a practical, commonsense perspective. It is currently out of print. However, you may be able to find it in university libraries or by browsing the Internet. As of March 2005, the book was also available from Rector Press.

This book was written by the inventor of the most widely used op-amp in the world, the LM324. This book gave me the first hint that op-amps should be easy to use, not hard.

CHAPTER 27

Programmable Logic to ASICs

Bob Zeidman

Programmable devices have progressed through a long evolution to reach the complexity today to support an entire system on a chip (SOC). This chapter gives an approximately chronological discussion of these devices from least complex to most complex. I say "approximately" because there is definitely overlap between the various devices, which are still in use today. The chapter includes a discussion on application-specific integrated circuits (ASICs) and how CPLDs and FPGAs fit within the spectrum of programmable logic and ASICs.

The objectives of this chapter are to become aware of the different programmable devices available and how they led to the current state-of-the-art device. These objectives are summarized here:

- Learn the history of programmable devices.

- Obtain a basic knowledge of the technologies of programmable devices.

- Understand the architectures of earlier programmable devices.

- Discover the speed, power, and density limitations of earlier programmable devices.

- Appreciate the needs that arose and that were not addressed by existing devices, and that created a market for CPLDs and FPGAs.

NOTE: The ROM cell

The basic diagram for a ROM cell containing a single bit of data is shown in Figure 27.1. The word line is turned on if the address into the chip includes this particular bit cell. The metal layer is used to program the data into the ROM during fabrication. In other words, if the metal layer mask has a connection between the transistor output and the data line, the bit is programmed as a zero. When the bit is addressed, the output will be pulled to a low voltage, a logical zero. If there is no connection, the data line will be pulled up by the resistor to a high voltage, a logical one.

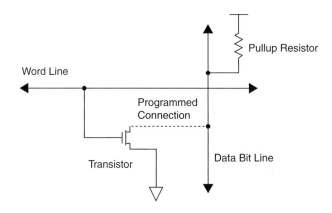

Figure 27.1: The ROM cell

27.1   Programmable Read-Only Memory (PROM)

The first field-programmable devices were created as alternatives to expensive mask-programmed ROM. Storing code in a ROM was an expensive process that required the ROM vendor to create a unique semiconductor mask set for each customer. Changes to the code were impossible without creating a new mask set and fabricating a new chip. The lead time for making changes to the code and getting back a chip to test was far too long.

PROMs solved this problem by allowing the user, rather than the chip vendor, to store code in the device using a simple and relatively inexpensive desktop programmer. This new device was called a *programmable read-only memory* (PROM). The process for storing the code in the PROM is called *programming*, or *burning* the PROM. PROMs, like ROMs, retain their contents even after power has been turned off.

NOTE: One-time programmable PROM cells
One-time programmable PROMs rely on an array of fuses and either diodes or transistors, as shown in Figure 27.2 and Figure 27.3. These fuses, like household fuses, consist of a wire that breaks connection when a large amount of current goes through it. To program a one-bit cell as a logic one or zero, the fuse for that cell is selectively burned out or left connected.

Although the PROMs were initially intended for storing code and constant data, design engineers also found them useful for implementing logic. The engineers could program state machine logic into a PROM, creating what is called *microcoded* state

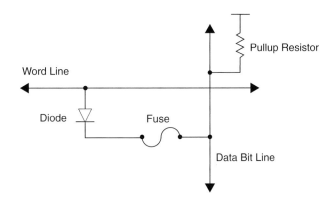

Figure 27.2: One-time programmable, diode-based PROM cell

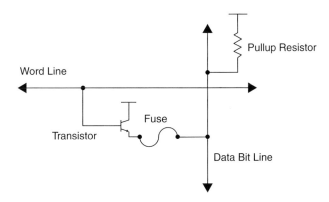

Figure 27.3: One-time programmable, transistor-based PROM cell

machines. They could easily change these state machines in order to fix bugs, test new functions, optimize existing designs, or make changes to systems that were already shipped and in the field.

Eventually, erasable PROMs were developed which allowed users to program, erase, and reprogram the devices using an inexpensive, desktop programmer. Typically, PROMs now refer to devices that cannot be erased after being programmed. Erasable PROMS include erasable programmable read only memories (EPROMs) that are programmed by applying high-voltage electrical signals and erased by flooding the devices with UV light. Electrically erasable programmable read only memories (EEPROMs) are programmed and erased by applying high voltages to the device. Flash EPROMs are programmed and erased electrically and have sections that can be erased electrically in a short time and independently of other sections within the device. For the rest of this chapter, I use the term PROM generically to refer to all of these devices unless I specifically state otherwise.

NOTE: Reprogrammable PROM cells

Reprogrammable PROMs essentially trap electric charge on the input of a transistor that is not connected to anything. The input acts like a capacitor. The transistor amplifies the charge. During programming, the charge is injected onto the transistor by one of several methods, including *tunneling* and *avalanche injection*. This charge will eventually leak off. In other words, some electrons will gradually escape, but the leakage will not be noticeable for a long time, on the order of ten years, so that they remain programmed even after power has been turned off to the device. Programming one of these devices causes wear and tear on the chip while the electrons are being injected. Most devices can be programmed about 100,000 times before they begin to lose their capability to be programmed.

PROMs are excellent for implementing any kind of combinatorial logic with a limited number of inputs and outputs. Each output can be any combinatorial function of the inputs, no matter how complex. As I said, this isn't usually how engineers use PROMs in today's designs; they're used to hold bytes of data. However, if you look at Figure 27.4, you can see how each address bit for the PROM can be considered a logic input. Then, simply program each data output bit to have the value of the combinatorial function you are creating. Some early devices used PROMs in this way to create combinatorial logic.

For sequential logic, one must add external clocked devices such as flip-flops or microprocessors. A simplified example of a state machine built using a PROM is shown

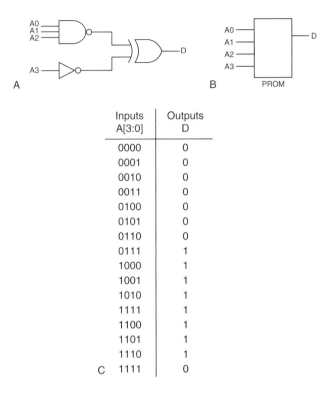

Inputs A[3:0]	Outputs D
0000	0
0001	0
0010	0
0011	0
0100	0
0101	0
0110	0
0111	1
1000	1
1001	1
1010	1
1111	1
1100	1
1101	1
1110	1
1111	0

Figure 27.4: A) combinatorial logic, B) equivalent PROM, C) logic values

in Figure 27.5. The PROM is used to combine inputs with bits representing the current state of the machine, to produce outputs and the next state of the machine. This allows the creation of very complex state machines. Microcode is often decoded within a microprocessor using this method, where the microcode for controlling the various stages of the microprocessor is stored in ROM.

The problem with PROMs is that they tend to be extremely slow—even today, access times are on the order of 40 nanoseconds or more—so they are not useful for applications where speed is an issue. These days, speed is always an issue. Also, PROMs are not easily integrated into logic circuits on a chip because they require a different technology and therefore a different set of masks and processes than for logic circuits. Integrating PROMs onto a chip with logic circuitry involves extra masks and extra processing steps, all leading to extra costs.

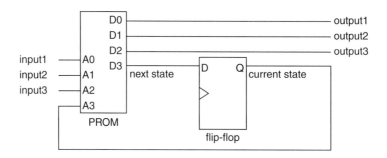

Figure 27.5: PROM-based state machine

27.2   Programmable Logic Arrays (PLAs)

Programmable logic arrays (PLAs) were a solution to the speed and input limitations of PROMs. PLAs consist of a large number of inputs connected to an AND plane, where different combinations of signals can be logically ANDed together according to how the part is programmed. The outputs of the AND plane go into an OR plane, where the terms are ORed together in different combinations and finally outputs are produced, as shown in Figure 27.6. At the inputs and outputs there are inverters (not shown in the figure) so that logical NOTs can be obtained. These devices can implement a large number of combinatorial functions, but, unlike a PROM, they can't implement every possible mapping of their input set to their output set. However, they generally have many more inputs and are much faster.

As with PROMs, PLAs can be connected externally to flip-flops to create state machines, which are the essential building blocks for all control logic.

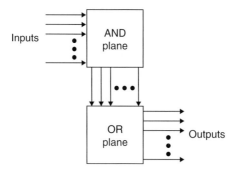

Figure 27.6: PLA architecture

Each connection in the AND and OR planes of a PLA could be programmed to connect or disconnect. In other words, terms of Boolean equations could be created by selectively connecting wires within the AND and OR planes. Simple high level languages — ABEL, PALASM, and CUPL — were developed to convert Boolean equations into files that would program these connections within the PLA. These equations looked like this:

```
a = (b & !c) | (b & !d & e)
```

to represent the logic for:

```
A = (B AND NOT C) OR (B & NOT D AND E)
```

This added a new dimension to programmable devices in that logic could now be described in readable programs at a level higher than ones and zeroes.

27.3   Programmable Array Logic (PALs)

The programmable array logic (PAL) is a variation of the PLA. Like the PLA, it has a wide, programmable AND plane for ANDing inputs together. The AND plane is shown by the crossing wires on the left in Figure 27.7. Programming elements at each intersection in the AND plane allow perpendicular traces to be connected or left open, creating "product terms," which are multiple logical signals ANDed together. The product terms are then ORed together. The Boolean equation in Figure 27.8 has four product terms.

In a PAL, unlike a PLA, the OR plane is fixed, limiting the number of terms that can be ORed together. This still allows a large number of Boolean equations to be implemented. The reason for this can be demonstrated by DeMorgan's Law, which states that a | b = !(!a & !b) or A OR B is equivalent to NOT(NOT A AND NOT B).

That means if you use inverters on the inputs and outputs, you can create all the logic you need with either a wide AND plane or a wide OR plane, but you don't need both.

Including inverters reduced the need for the large OR plane, which in turn allowed the extra silicon area on the chip to be used for other basic logic devices, such as multiplexers, exclusive ORs, and latches. Most importantly, clocked elements, typically flip-flops, could be included in PALs. These devices were now able to implement a large number of logic functions, including clocked sequential logic needed for state machines. This was an important development that allowed PALs to replace much of

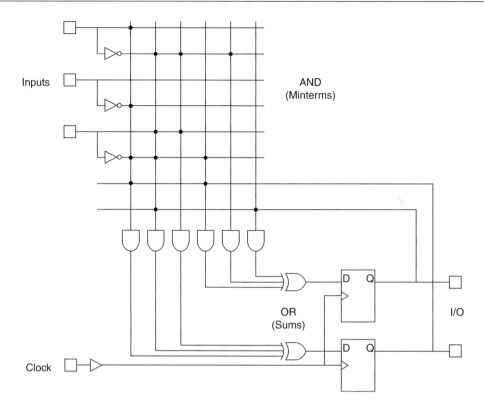

Figure 27.7: PAL architecture

```
xyz =   a1 &  b1 &  c2
    |  !a1 &  b1 & !c2
    |   a1 & !b1
    |   a1        & !c2
```

Figure 27.8: Boolean equation with four product terms

the standard logic in many designs. PALs are also extremely fast. With PALs, high-speed controllers could be designed in programmable logic.

Notice the architecture of a PAL, shown in Figure 27.7. The AND plane is shown in the upper-left corner as a switch matrix. The dots show where connections have been programmed. The fixed-size ORs are represented as OR gates. A clock input is used to clock the flip-flops. The outputs of the flip-flops can be driven off the chip, or they can be fed back to the AND plane in order to create a state machine.

The inclusion of extra logic devices, particularly flip-flops, greatly increased the complexity and potential uses of PALs, creating a need for new methods of programming that were flexible and readable. Thus the first hardware description languages (HDLs) were born. These simple HDLs included ABEL, CUPL, and PALASM, the precursors of Verilog and VHDL, much more complex languages that are in use today for CPLD, FPGA, and ASIC design.

A simple ABEL program for a PAL is shown in Listing 27.1. Don't worry about trying to understand the details—it's for illustration purposes only. Notice that the programming language allows the use of simulation test vectors in the code. The simulation vectors are at the end of the program. This simulation capability brought better reliability and verification of programmable devices, something that was critical when CPLDs and FPGAs were developed.

Listing 27.1 A simple ABEL program

```
MODULE DECODE;

FLAG '-R3' , '-T1' , '-V' , '-F0' , '-G' , '-Q2';

TITLE'
  CHIP        : Decode PAL - Version A
  DATE        : July 17, 1991
  DESIGNER    : Bob Zeidman'

" PAL to decode addresses.

decode DEVICE 'P20R6';

"CONSTANTS:
  h = 1;
  l = 0;
  c = .C.;
  x = .X.;
  z = .Z.;

"INPUTS:
  clk       PIN  1;    "System clock
  !res      PIN  2;    "System reset
  !req      PIN  3;    "Instruction/Data Request from processor
  !emacc    PIN  4;    "Emulator access
  opt0      PIN  5;    "Opt bit from processor
```

Continued onto next page

```
   opt1       PIN  6;   "Opt bit from processor
   opt2       PIN  7;   "Opt bit from processor
   a19        PIN  8;   "Address bit from processor
   a20        PIN  9;   "Address bit from processor
   a21        PIN 10;   "Address bit from processor
   a22        PIN 11;   "Address bit from processor
  !oe         PIN 13;   "Output enable
   a23        PIN 14;   "Address bit from processor
   a31        PIN 23;   "Address bit from processor

"OUTPUTS:
 !sram       PIN 15;   "SRAM select
 !dram       PIN 16;   "DRAM select
 !parallel  PIN 17;   "Parallel port select
 !leds       PIN 18;   "LEDs select
 !switch     PIN 19;   "Switches select
 !serial     PIN 20;   "Serial port select
 !config     PIN 21;   "Configuration register select
 !eprom      PIN 22;   "EPROM select

"MEMORY MAP
addr      =[ a31, a23, a22, a21, a20, a19] ;

EPROM     =[ 0, 0, 0, 0, 0, x ] ;
SRAM      =[ 0, 0, 0, 0, 1, 0 ] ;
DRAM_LO   =[ 0, 0, 0, 0, 1, 1 ] ;
DRAM_HI   =[ 0, 1, 0, 0, 1, 0 ] ;
PARALLEL  =[ 0, 1, 0, 1, 0, 0 ] ;
SERIAL    =[ 0, 1, 0, 1, 0, 1 ] ;
SWITCHES  =[ 0, 1, 0, 1, 1, 0 ] ;
LEDS      =[ 0, 1, 0, 1, 1, 1 ] ;
CONFIG    =[ 0, 1, 1, 0, 0, 0 ] ;

EQUATIONS

  eprom     = req & !emacc & !opt2 & !res & (addr == EPROM);

  sram      = req & !emacc & !opt2 & !res & (addr == SRAM);

  dram      := req & !emacc & !opt2 & !res & (addr >= DRAM_LO) & (addr <=
              DRAM_HI);

  parallel := req & !emacc & !opt2 & !res & (addr == PARALLEL);
```

```
   serial   := req & !emacc & !opt2 & !res & (addr == SERIAL);

   switch   := req & !emacc & !opt2 & !res & !switch & (addr == SWITCHES);

   leds     := req & !emacc & !opt2 & !res & !leds & (addr == LEDS);

   config   := req & !emacc & !opt2 & !res & !config & (addr == CONFIG);

TEST_VECTORS(
     [ clk, !res, !oe, !req, !emacc, opt2, opt1, opt0 ]
  ->[ eprom,!sram,!dram,!parallel,!serial,!switch,!leds,!config] );
     [ c, 0, 0, x, x, x, x, x] ->[ 1, 1, 1, 1, 1, 1, 1, 1];
     [ c, 1, 0, 1, x, x, x, x] ->[ 1, 1, 1, 1, 1, 1, 1, 1];
     [ c, 1, 0, x, 0, x, x, x] ->[ 1, 1, 1, 1, 1, 1, 1, 1];
     [ c, 1, 0, x, x, 1, x, x] ->[ 1, 1, 1, 1, 1, 1, 1, 1];

TEST_VECTORS(
     [ !res,!oe,!req,!emacc,opt2,opt1,opt0,a31,a23,a22,a21,a20,a19]
  ->[ !eprom, !sram] );
 "5"[ 0, 0, x, x, x, x, x, x, x, x, x, x, x] ->[ 1, 1];
    [ 1, 0, 1, x, x, x, x, x, x, x, x, x, x] ->[ 1, 1];
    [ 1, 0, 0, 1, 0, x, x, 0, 0, 0, 0, 0, 0] ->[ 0, 1];
    [ 1, 0, 0, 1, 0, x, x, 0, 0, 0, 0, 0, 1] ->[ 0, 1];
    [ 1, 0, 0, 1, 0, x, x, 0, 0, 0, 0, 1, 0] ->[ 1, 0];
"10"[ 1, 0, 0, 1, 0, x, x, 0, 0, 0, 0, 1, 1] ->[ 1, 1];

END;
```

Listing 27.2 shows the compiled output from this code, consisting of a map of connections within the device to program in order to obtain the correct functionality.

Listing 27.2 The compiled output

```
ABEL(tm) 3.00a FutureNet Div, Data I/O Corp. JEDEC file for: P20R6
Created on: 29-Nov-:1 06:52 PM

  CHIP       : Decode PAL - Version A
  DATE       : July 17, 1991
  DESIGNER   : Bob Zeidman*
QP24* QF2560* QV10*
L0000
```

Continued onto next page

```
111111111111111111111111111111111111111111
011010110111111111110111111011110111010
000000000000000000000000000000000000000000
000000000000000000000000000000000000000000
000000000000000000000000000000000000000000
000000000000000000000000000000000000000000
000000000000000000000000000000000000000000
000000000000000000000000000000000000000000
011010110101111111110110111011110110101
000000000000000000000000000000000000000000
000000000000000000000000000000000000000000
000000000000000000000000000000000000000000
000000000000000000000000000000000000000000
000000000000000000000000000000000000000000
000000000000000000000000000000000000000000
000000000000000000000000000000000000000000
011010110111111111110110111011101111001
000000000000000000000000000000000000000000
000000000000000000000000000000000000000000
000000000000000000000000000000000000000000
000000000000000000000000000000000000000000
000000000000000000000000000000000000000000
000000000000000000000000000000000000000000
000000000000000000000000000000000000000000
011010110111111111011011101101111011111001
000000000000000000000000000000000000000000
000000000000000000000000000000000000000000
000000000000000000000000000000000000000000
000000000000000000000000000000000000000000
000000000000000000000000000000000000000000
000000000000000000000000000000000000000000
000000000000000000000000000000000000000000
011010110111111111110010111011101111001
000000000000000000000000000000000000000000
000000000000000000000000000000000000000000
000000000000000000000000000000000000000000
000000000000000000000000000000000000000000
000000000000000000000000000000000000000000
000000000000000000000000000000000000000000
000000000000000000000000000000000000000000
011010110111111111110111011101101111001
000000000000000000000000000000000000000000
000000000000000000000000000000000000000000
000000000000000000000000000000000000000000
000000000000000000000000000000000000000000
```

```
00000000000000000000000000000000000000000
00000000000000000000000000000000000000000
00000000000000000000000000000000000000000
01101011011111111111110110111111110111001
01101011011111111111110111111101110111001
01101011011111111111110110111011111111110
01101011011111111111110111111111101111110
01101011011111111111110111111111111110110
00000000000000000000000000000000000000000
00000000000000000000000000000000000000000
00000000000000000000000000000000000000000
11111111111111111111111111111111111111111
01101011011111111111110111011011110111010
00000000000000000000000000000000000000000
00000000000000000000000000000000000000000
00000000000000000000000000000000000000000
00000000000000000000000000000000000000000
00000000000000000000000000000000000000000
000000000000000000000000000000000000000000*
V0001 C0XXXXXXXXXN0XHHHHHHHHXN*
V0002 C11XXXXXXXXN0XHHHHHHHHXN*
V0003 C1X0XXXXXXXN0XHHHHHHHHXN*
V0004 C1XXXX1XXXXN0XHHHHHHHHXN*
V0005 X0XXXXXXXXXN0XHNNNNNNHXN*
V0006 X11XXXXXXXXN0XHNNNNNNHXN*
V0007 X101XX00000N00HNNNNNNL0N*
V0008 X101XX01000N00HNNNNNNL0N*
V0009 X101XX00100N00LNNNNNNH0N*
V0010 X101XX01100N00HNNNNNNH0N*
C3B9E*
7519
```

27.4 The Masked Gate Array ASIC

An application-specific integrated circuit, or ASIC, is not a programmable device, but it is important precursor to the developments leading up to CPLDs and FPGAs. An ASIC is a chip that an engineer can design with no particular knowledge of semiconductor physics or semiconductor processes. The ASIC vendor has created a library of cells and functions that the designer can use without needing to know precisely how these functions are implemented in silicon. The ASIC vendor also typically supports software tools that automate such processes as circuit synthesis and circuit layout. The ASIC vendor may even supply application engineers to assist the

ASIC design engineer with the task. The vendor then lays out the chip, creates the masks, and manufactures the ASICs.

ASICs can be implemented using one of two internal architectures—gate array or standard cell. The differences between the two architectures are beyond the scope of this book. The standard cell architecture is not as relevant to CPLDs and FPGAs as the gate array architecture, which I describe briefly.

The gate array ASIC consists of rows and columns of regular transistor structures, as shown in Figure 27.9. Around the sides of the chip die are I/O cells containing input and output buffers along with some limited number of transistor. These I/O cells also contain the large bonding pads, shown in the figure, which are simply metal pads that are connected or "bonded" to the external pins of the chip using very small bonding wires.

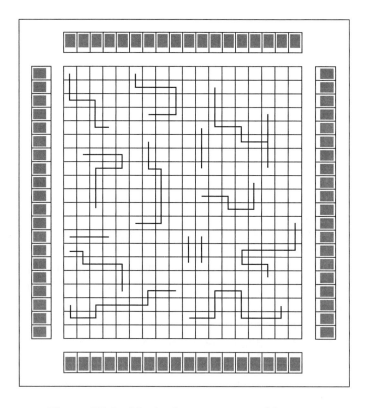

Figure 27.9: Masked gate array architecture

Within the core array are basic cells, or gates, each consisting of some small number of transistors that are not connected. In fact, none of the transistors on the gate array are initially connected at all. The reason for this is that the connection is determined completely by the design that you implement. Once given a design, the layout software figures out which transistors to connect by placing metal connections on top of the die as shown. First, the low level functions are connected together. For example, six transistors could be connected to create a D flip-flop. These six transistors would be located physically very close to each other. After the low level functions have been routed, they would in turn be connected together. The software would continue this process until the entire design is complete.

The ASIC vendor manufactures many unrouted die that contain the arrays of gates and that it can use for any gate array customer. An integrated circuit consists of many layers of materials, including semiconductor material (e.g., silicon), insulators (e.g., oxides), and conductors (e.g., metal). An unrouted die is processed with all of the layers except for the final metal layers that connect the gates together. Once the design is complete, the vendor simply needs to add the last metal layers to the die to create your chip, using photo masks for each metal layer. For this reason, it is sometimes referred to as a "masked gate array" to differentiate it from a field-programmable gate array.

The advantage of a gate array is that the internal circuitry is very fast; the circuit is dense, allowing lots of functionality on a die; and the cost is low for high volume production. Gate arrays can reach clock frequencies of hundreds of megahertz with densities of millions of gates. The disadvantage is that it takes time for the ASIC vendor to manufacture and test the parts. Also, the customer incurs a large charge up front, called a nonrecurring engineering (NRE) expense, which the ASIC vendor charges to begin the entire ASIC process. And if there's a mistake, it's a long, expensive process to fix it and manufacture new ASICs.

27.5   CPLDs and FPGAs

Ideally, hardware designers wanted something that gave them the advantages of an ASIC—circuit density and speed—but with the shorter turnaround time of a programmable device. The solution came in the form of two new devices—the complex programmable logic device (CPLD) and the field-programmable gate array (FPGA). Figure 27.10 shows how CPLDs and FPGAs bridge the gap between PALs and gate arrays. All of the inherent advantages of PALs, shown on the left of the diagram, and all

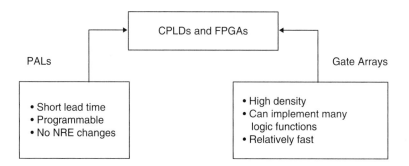

Figure 27.10: The evolution of CPLDs and FPGAs

of the inherent advantages of gate array ASICS, shown on the right of the diagram, were combined. CPLDs are as fast as PALs but more complex. FPGAs approach the complexity of gate arrays but are still programmable. CPLD architectures and technologies are the same as those for PALs. FPGA architecture is similar to those of gate array ASICs.

27.6 Summary

Several programmable and semi-custom technologies preceded the development of CPLDs and FPGAs. This chapter started by reviewing the architecture, properties, uses, and trade-offs of the various programmable devices (PROMS, PLAS, and PALs) that were in use before CPLDs and FPGAs. Later the chapter described ASICs and examined the contribution of a specific type of ASIC architecture called a *gate array*. The architecture, properties, uses, and trade-offs of the gate array were discussed. Finally, CPLDs and FPGAs were introduced, briefly, as programmable chip solutions that filled the gap between programmable devices and gate array ASICs.

References

Logic Design Manual for ASICs, Santa Clara, CA: LSI Logic Corporation, 1989.

Davenport Jr., Wilbur B, *Probability and Random Processes,* New York, NY: McGraw-Hill Book Company, 1970.

Dorf and Richard C., editor. *Electrical Engineering Handbook.* Boca Raton, FL: CRC Press, Inc., 1993.

EDA Industry Working Groups Web site, www.eda.org

Maxfield, Clive "Max." *Designus Maximus Unleashed!* Woburn, MA: Butter-worth-Heinemann, 1998.

Zeidman Bob. *Introduction to Verilog,* Piscataway, NJ: Institute of Electrical and Electronic Engineers, 2000.

Zeidman Bob. *Verilog Designer's Library,* Upper Saddle River, NJ: Prentice-Hall, Inc., 1999.

CHAPTER 28

Complex Programmable Logic Devices (CPLDs)

Bob Zeidman

Complex programmable logic devices are exactly what they claim to be: logic devices that are complex and programmable. There are two main engineering features to understand about CPLDs that separate them from their cousins, FPGAs. One feature is the internal architecture of the device and how this architecture implements various logic functions. The second feature is the semiconductor technology that allows the devices to be programmed and allows various structures in the device to be connected.

28.1 CPLD Architectures

Essentially, CPLDs are designed to appear just like a large number of PALs in a single chip, connected to each other through a crosspoint switch. This architecture made them familiar to their target market—PC board designers who were already designing PALs in their boards. Many CPLDs were used to simply combine multiple PALs in order to save real estate on a PC board. CPLDs use the same development tools and programmers as PALs, and are based on the same technologies as PALs, but they can handle much more complex logic and more of it.

The diagram in Figure 28.1 shows the internal architecture of a typical CPLD. Although each manufacturer has a different variation, in general they are all similar in that they consist of function blocks, input/output blocks, and an interconnect matrix.

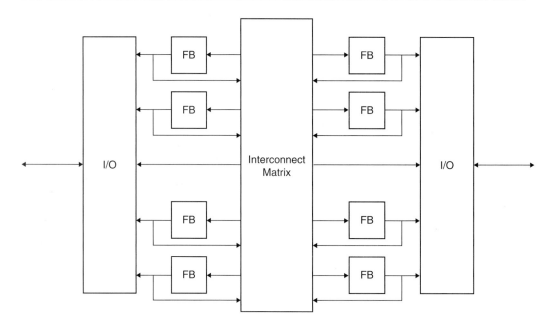

Figure 28.1: CPLD architecture (courtesy of Altera Corporation)

28.2   Function Blocks

A typical function block is shown in Figure 28.3. Notice the similarity to the PAL architecture with its wide AND plane and fixed number of OR gates. The AND plane is shown by the crossing wires on the left. The AND plane can accept inputs from the I/O blocks, other function blocks, or feedback from the same function block. Programming elements at each intersection in the AND plane allow perpendicular traces to be connected or left open, creating "product terms," which are multiple signals ANDed together, just like in a PAL. The product terms are then ORed together and sent straight out of the block, or through a clocked flip-flop. The Boolean equation in Figure 28.2 has four product terms.

```
xyz =  a1 &  b1 &  c2
    | !a1 &  b1 & !c2
    |  a1 & !b1
    |  a1       & !c2
```

Figure 28.2: Boolean equation with four product terms

There are also multiplexers in the diagram, shown as boxes labeled M1, M2, and M3. Each mux has an FET transistor beneath it, representing a programmable element attached to the select line. In other words, the mux can be programmed to output one of the inputs. M1 is the "Clear Select" because it selects the signal that is used to clear the flip-flop. The M2 mux is labeled "Clock/Enable Select" because its two outputs are programmed to control the clock and clock enable input to the flip-flop. The M3 mux is labeled "Register Bypass" because it is programmed to determine whether the output of the functional block is a registered signal (i.e., is the output of a flip-flop) or a combinatorial signal (i.e., is the output of combinatorial logic).

Many CPLDs include additional, specialized logic. This particular block includes an exclusive OR, which can be effectively bypassed by programming one input to always be a 0. An XOR can be a nice gate to have because it is otherwise difficult to implement this function in a PAL. Exclusive ORs are used to easily generate parity in a bus for simple error detection.

Though not explicitly shown in Figure 28.3, each functional block would have many OR gates, logic gates, muxes, and flip-flops. Usually, the function blocks are designed

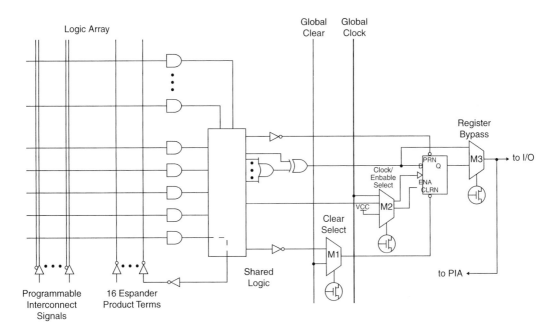

Figure 28.3: CPLD function block (courtesy of Altera Corporation)

to be similar to existing PAL architectures, such as the 22V10, so that the designer can use familiar tools to design them. They may even be able to fit older PAL designs into the CPLD without changing the design.

28.3   I/O Blocks

Figure 28.4 shows a typical I/O block of a CPLD. The I/O block is used to drive signals to the pins of the CPLD device at the appropriate voltage levels (e.g., TTL, CMOS, ECL, PECL, or LVDS). The I/O block typically allows each I/O pin to be individually configured for input, output, or bidirectional operation. The I/O pins have a tri-state output buffer that can be controlled by global output enable signals or directly connected to ground or VCC. Each output pin can also be configured to be open drain. In addition, outputs can often be programmed to drive different voltage levels, enabling the CPLD to be interfaced to many different devices.

One particularly useful feature in high speed CPLDs is the ability to control the rise and fall rates of the output drivers by using a slew rate control. Designers can configure the output buffers for fast rise and fall times or for slow transition times. An advantage

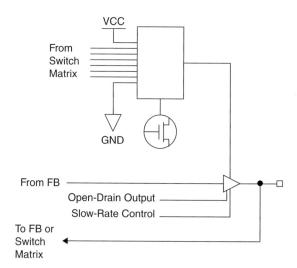

Figure 28.4: CPLD input/output block (courtesy of Altera Corporation)

of the fast speed of these devices is less delay through the logic. A disadvantage of faster transition is times that they can cause overshoot and undershoot, which can potentially damage the device that the CPLD is driving. Also, fast transitions introduce noise, which can create problems. By programming the slew rate of the output buffer to a relatively slow transition, you can preserve the small logic delays of the device while avoiding undershoot, overshoot, and noise problems.

The input signal from the I/O block goes into the switch matrix in order to be routed to the appropriate functional block. In some architectures, particular inputs have direct paths to particular functional blocks in order to lower the delay on the input, reducing the signal setup time. In most architectures, specific pins of the device connect to specific I/O blocks that can drive global signals like reset and clock. This means that only certain pins of the device can be used to drive these global signals. This is particularly important for clock signals, as described in the next section.

28.4   Clock Drivers

Synchronous design is the only accepted design methodology that will ensure that a CPLD-based design is reliable over its lifetime. In order to design synchronous CPLDs, the clock signal must arrive at each flip-flop in the design at about the same time and with very little delay from the input pin. In order to accomplish this, special I/O blocks have clock drivers that use very fast input buffers and which drive the input clock signal onto an internal clock tree. The clock tree is so named because it resembles a tree, with each branch driving the clock input of a fixed number of flip-flops. The clock driver is designed to drive the entire tree very quickly. The trees are designed to minimize the skew between clock signals arriving at different flip-flops throughout the device. Each branch of the tree is of approximately equal length, or if not, internal buffers are used to balance the skew along the different branches. It is important that clock signals are only driven through the clock input pins that connect to these special drivers.

In large devices, there may be several clock input pins connected to different clock drivers. This feature helps in designs that use multiple clocks. You need to have at least as many clock drivers in the CPLD as you need clocks in your design. Also, the different clocks must be considered to be asynchronous with respect to each other, because the CPLD vendor does not typically guarantee skew between multiple clocks. Signals clocked by one clock will need to be synchronized with the other clock before use by any logic clocked by the second clock.

28.5   Interconnect

The CPLD interconnect is a very large programmable switch matrix that allows signals from all parts of the device to go to all other parts of the device. Figure 28.5 shows the architecture of the switch matrix. The switch matrix takes the outputs of the functional blocks and is programmed to send those outputs to functional blocks. This way, the designer can route any output signal to any destination.

One advantage of the CPLD switch matrix routing scheme is that delays through the chip are deterministic. Designers can determine the delay for any signal by computing the delay through functional blocks, I/O blocks, and the switch matrix. All of these delays are fixed, and delays due to routing the signal along the metal traces are negligible. If the logic for a particular function is complex, it may require several functional blocks, and thus several passes through the switch matrix, to implement.

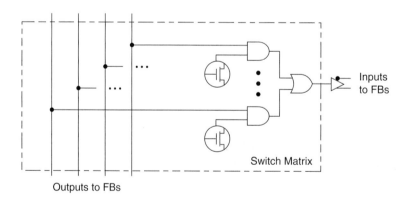

Figure 28.5: CPLD switch matrix (courtesy of Altera Corporation)

Computing Parity without Exclusive OR

The Boolean expression for generating even parity for a bus is shown in the following equation:

parity = a0 ^ a1 ^ a2 ^ a3 ^ a4 ^ a5 ^ a6 ^ a7

If we implement this equation using AND and OR logic, the result is:

```
parity = a0 & !a1 & !a2 & !a3 & !a4 & !a5 & !a6 & !a7
       | !a0 & a1 & !a2 & !a3 & !a4 & !a5 & !a6 & !a7
       | a0 & a1 & a2 & !a3 & !a4 & !a5 & !a6 & !a7
       | !a0 & !a1 & a2 & !a3 & !a4 & !a5 & !a6 & !a7
       | a0 & !a1 & a2 & a3 & !a4 & !a5 & !a6 & !a7
       | !a0 & a1 & a2 & a3 & !a4 & !a5 & !a6 & !a7
       | a0 & a1 & !a2 & a3 & !a4 & !a5 & !a6 & !a7
       | !a0 & !a1 & !a2 & a3 & !a4 & !a5 & !a6 & !a7
       | a0 & !a1 & !a2 & a3 & a4 & !a5 & !a6 & !a7
       | !a0 & a1 & !a2 & a3 & a4 & !a5 & !a6 & !a7
       | a0 & a1 & a2 & a3 & a4 & !a5 & !a6 & !a7
       | !a0 & !a1 & a2 & a3 & a4 & !a5 & !a6 & !a7
       | a0 & !a1 & a2 & !a3 & a4 & !a5 & !a6 & !a7
       | !a0 & a1 & a2 & !a3 & a4 & !a5 & !a6 & !a7
       | a0 & a1 & !a2 & !a3 & a4 & !a5 & !a6 & !a7
       | !a0 & !a1 & !a2 & !a3 & a4 & !a5 & !a6 & !a7
       | a0 & !a1 & !a2 & !a3 & a4 & a5 & !a6 & !a7
       | !a0 & a1 & !a2 & !a3 & a4 & a5 & !a6 & !a7
       | a0 & a1 & a2 & !a3 & a4 & a5 & !a6 & !a7
       | !a0 & !a1 & a2 & !a3 & a4 & a5 & !a6 & !a7
       | a0 & !a1 & a2 & a3 & a4 & a5 & !a6 & !a7
       | !a0 & a1 & a2 & a3 & a4 & a5 & !a6 & !a7
       | a0 & a1 & !a2 & a3 & a4 & a5 & !a6 & !a7
       | !a0 & !a1 & !a2 & a3 & a4 & a5 & !a6 & !a7
       | a0 & !a1 & !a2 & a3 & !a4 & a5 & !a6 & !a7
       | !a0 & a1 & !a2 & a3 & !a4 & a5 & !a6 & !a7
       | a0 & a1 & a2 & a3 & !a4 & a5 & !a6 & !a7
       | !a0 & !a1 & a2 & a3 & !a4 & a5 & !a6 & !a7
       | a0 & !a1 & a2 & !a3 & !a4 & a5 & !a6 & !a7
       | !a0 & a1 & a2 & !a3 & !a4 & a5 & !a6 & !a7
       | a0 & a1 & !a2 & !a3 & !a4 & a5 & !a6 & !a7
       | !a0 & !a1 & !a2 & !a3 & !a4 & a5 & !a6 & !a7
       | a0 & !a1 & !a2 & !a3 & !a4 & a5 & a6 & !a7
       | !a0 & a1 & !a2 & !a3 & !a4 & a5 & a6 & !a7
       | a0 & a1 & a2 & !a3 & !a4 & a5 & a6 & !a7
       | !a0 & !a1 & a2 & !a3 & !a4 & a5 & a6 & !a7
       | a0 & !a1 & a2 & a3 & !a4 & a5 & a6 & !a7
```

Designers can very easily calculate delays from input pins to output pins of a CPLD by using a few worst-case timing numbers supplied by the CPLD vendor. This contrasts greatly with FPGAs, which have very unpredictable and design-dependent timing due to their routing mechanism.

28.6   CPLD Technology and Programmable Elements

Different manufacturers use different technologies to implement the programmable elements of a CPLD. The common technologies are EPROM, EEPROM, and Flash EPROM. These technologies are versions of the technologies that were used for the simplest programmable devices, PROMs, which we discussed earlier. In functional blocks and I/O blocks, single bits are programmed to turn specific functions on and off, Figure 28.3 and Figure 28.4 show. In the switch matrix, single bits are programmed to control connections between signals using a multiplexer, as shown in Figure 28.5.

When PROM technology is used for these devices, they can be programmed only once. More commonly these days, manufacturers use EPROM, EEPROM, or Flash EPROM, allowing the devices to be erased and reprogrammed.

Erasable technology can also allow in-system programmability of the device. For CPLDs with this capability, a serial interface on the chip is used to send new programming data into the chip after it is soldered into a PC board and while the system is operating. Typically this serial interface is the industry-standard 4-pin Joint Test Action Group (JTAG) interface (IEEE Std. 1149.1-1990).

28.7   Embedded Devices

A relatively recent addition to the architecture of many CPLD devices is embedded devices, which consists of large devices integrated into the CPLD. These devices can be connected to the rest of the CPLD via the switch matrix. The availability of embedded devices brings designers closer to the concept of a system on a programmable chip (SOPC). Engineers can now move the processors, memory, and other complex standard devices that would normally be on a circuit board along with a CPLD directly into the CPLD.

The main advantages of embedded devices are cost reduction, reduced circuit board space, and often lower power consumption. A disadvantage is that it tends to

Computing Parity without Exclusive OR (Continued)

```
| !a0 & a1 & a2 & a3 & !a4 & a5 & a6 & !a7
| a0 & a1 & !a2 & a3 & !a4 & a5 & a6 & !a7
| !a0 & !a1 & !a2 & a3 & !a4 & a5 & a6 & !a7
| a0 & !a1 & !a2 & a3 & a4 & a5 & a6 & !a7
| !a0 & a1 & !a2 & a3 & a4 & a5 & a6 & !a7
| a0 & a1 & a2 & a3 & a4 & a5 & a6 & !a7
| !a0 & !a1 & a2 & a3 & a4 & a5 & a6 & !a7
| a0 & !a1 & a2 & !a3 & a4 & a5 & a6 & !a7
| !a0 & a1 & a2 & !a3 & a4 & a5 & a6 & !a7
| a0 & a1 & !a2 & !a3 & a4 & a5 & a6 & !a7
| !a0 & !a1 & !a2 & !a3 & a4 & a5 & a6 & !a7
| a0 & !a1 & !a2 & !a3 & a4 & !a5 & a6 & !a7
| !a0 & a1 & !a2 & !a3 & a4 & !a5 & a6 & !a7
| a0 & a1 & a2 & !a3 & a4 & !a5 & a6 & !a7
| !a0 & !a1 & a2 & !a3 & a4 & !a5 & a6 & !a7
| !a0 & a1 & a2 & a3 & a4 & !a5 & a6 & !a7
| a0 & a1 & !a2 & a3 & a4 & !a5 & a6 & !a7
| !a0 & !a1 & !a2 & a3 & a4 & !a5 & a6 & !a7
| !a0 & !a1 & !a2 & a3 & a4 & !a5 & a6 & !a7
| a0 & !a1 & !a2 & a3 & !a4 & !a5 & a6 & !a7
| !a0 & a1 & !a2 & a3 & !a4 & !a5 & a6 & !a7
| a0 & a1 & a2 & a3 & !a4 & !a5 & a6 & !a7
| !a0 & !a1 & a2 & a3 & !a4 & !a5 & a6 & !a7
| a0 & !a1 & a2 & !a3 & !a4 & !a5 & a6 & !a7
| !a0 & a1 & a2 & !a3 & !a4 & !a5 & a6 & !a7
| a0 & a1 & !a2 & !a3 & !a4 & !a5 & a6 & !a7
| !a0 & !a1 & !a2 & !a3 & !a4 & !a5 & a6 & !a7
| a0 & !a1 & !a2 & !a3 & !a4 & !a5 & a6 & a7
| !a0 & a1 & !a2 & !a3 & !a4 & !a5 & a6 & a7
| a0 & a1 & a2 & !a3 & !a4 & !a5 & a6 & a7
| !a0 & !a1 & a2 & !a3 & !a4 & !a5 & a6 & a7
| a0 & !a1 & a2 & a3 & !a4 & !a5 & a6 & a7
| !a0 & a1 & a2 & a3 & !a4 & !a5 & a6 & a7
| a0 & a1 & !a2 & a3 & !a4 & !a5 & a6 & a7
| !a0 & !a1 & !a2 & a3 & !a4 & !a5 & a6 & a7
| a0 & !a1 & !a2 & a3 & a4 & !a5 & a6 & a7
| !a0 & a1 & !a2 & a3 & a4 & !a5 & a6 & a7
| a0 & a1 & a2 & a3 & a4 & !a5 & a6 & a7
| !a0 & !a1 & a2 & a3 & a4 & !a5 & a6 & a7
| a0 & !a1 & a2 & !a3 & a4 & !a5 & a6 & a7
| !a0 & a1 & a2 & !a3 & a4 & !a5 & a6 & a7
| a0 & a1 & !a2 & !a3 & a4 & !a5 & a6 & a7
| !a0 & !a1 & !a2 & !a3 & a4 & !a5 & a6 & a7
| a0 & !a1 & !a2 & !a3 & a4 & a5 & a6 & a7
```

Table 28.1: JTAG signals

Signal	Description
TCK	Test Clock Input A clock signal used to shift test instructions, test data, and control inputs into the chip on the rising edge and to shift the output data from the chip on the falling edge.
TMS	Test Mode Select Serial input for controlling the internal JTAG state machine. The state of this bit on the rising edge of each clock determines which actions the chip is to take.
TDI	Test Data Input Serial input for instructions and program data. Data is captured on the rising edge of the clock.
TDO	Test Data Output Serial output for test instruction and program data from the chip. Valid data is driven out on the falling edge of the clock.
TRST	Test Reset Input (Extended JTAG only) An asynchronous active low reset that is used to initialize the JTAG controller.

tie your design into a specific CPLD offered by a single CPLD vendor because different vendors supply different embedded devices in their CPLDs, if they offer them at all.

The number and kinds of embedded devices that are being integrated into CPLDs are increasing annually. Currently, these devices include:

- SRAM memories

- Flash memories

- microcontrollers

- microprocessors

- Digital Signal Processors (DSPs)

- Phase Locked-Loops (PLLs)

- network processors

Computing Parity without Exclusive OR (Continued)

```
| !a0 & a1 & !a2 & !a3 & a4 & a5 & a6 & a7
| a0 & a1 & a2 & !a3 & a4 & a5 & a6 & a7
| !a0 & !a1 & a2 & !a3 & a4 & a5 & a6 & a7
| a0 & !a1 & a2 & a3 & a4 & a5 & a6 & a7
| !a0 & a1 & a2 & a3 & a4 & a5 & a6 & a7
| a0 & a1 & !a2 & a3 & a4 & a5 & a6 & a7
| !a0 & !a1 & !a2 & a3 & a4 & a5 & a6 & a7
| a0 & !a1 & !a2 & a3 & !a4 & a5 & a6 & a7
| !a0 & a1 & !a2 & a3 & !a4 & a5 & a6 & a7
| a0 & a1 & a2 & a3 & !a4 & a5 & a6 & a7
| !a0 & !a1 & a2 & a3 & !a4 & a5 & a6 & a7
| a0 & !a1 & a2 & !a3 & !a4 & a5 & a6 & a7
| !a0 & a1 & a2 & !a3 & !a4 & a5 & a6 & a7
| a0 & a1 & !a2 & !a3 & !a4 & a5 & a6 & a7
| !a0 & !a1 & !a2 & !a3 & !a4 & a5 & a6 & a7
| a0 & !a1 & !a2 & !a3 & !a4 & a5 & !a6 & a7
| !a0 & a1 & !a2 & !a3 & !a4 & a5 & !a6 & a7
| a0 & a1 & a2 & !a3 & !a4 & a5 & !a6 & a7
| !a0 & !a1 & a2 & !a3 & !a4 & a5 & !a6 & a7
| a0 & !a1 & a2 & a3 & !a4 & a5 & !a6 & a7
| !a0 & a1 & a2 & a3 & !a4 & a5 & !a6 & a7
| a0 & a1 & !a2 & a3 & !a4 & a5 & !a6 & a7
| !a0 & !a1 & !a2 & a3 & !a4 & a5 & !a6 & a7
| a0 & !a1 & a2 & a3 & a4 & a5 & !a6 & a7
| !a0 & a1 & !a2 & a3 & a4 & a5 & !a6 & a7
| a0 & a1 & a2 & a3 & a4 & a5 & !a6 & a7
| !a0 & !a1 & a2 & a3 & a4 & a5 & !a6 & a7
| a0 & !a1 & a2 & !a3 & a4 & a5 & !a6 & a7
| !a0 & a1 & a2 & !a3 & a4 & a5 & !a6 & a7
| a0 & a1 & !a2 & !a3 & a4 & a5 & !a6 & a7
| !a0 & !a1 & !a2 & !a3 & a4 & a5 & !a6 & a7
| a0 & !a1 & !a2 & !a3 & a4 & !a5 & !a6 & a7
| !a0 & a1 & !a2 & !a3 & a4 & !a5 & !a6 & a7
| a0 & a1 & a2 & !a3 & a4 & !a5 & !a6 & a7
| !a0 & !a1 & a2 & !a3 & a4 & !a5 & !a6 & a7
| a0 & !a1 & a2 & a3 & a4 & !a5 & !a6 & a7
| !a0 & a1 & a2 & a3 & a4 & !a5 & !a6 & a7
| a0 & a1 & !a2 & a3 & a4 & !a5 & !a6 & a7
| !a0 & !a1 & !a2 & a3 & a4 & !a5 & !a6 & a7
| a0 & !a1 & !a2 & a3 & !a4 & !a5 & !a6 & a7
| !a0 & a1 & !a2 & a3 & !a4 & !a5 & !a6 & a7
| a0 & a1 & a2 & a3 & !a4 & !a5 & !a6 & a7
| !a0 & !a1 & a2 & a3 & !a4 & !a5 & !a6 & a7
```

NOTE: JTAG interface

The JTAG interface, IEEE Standard 1149.1, is a simple serial interface specification created by the Joint Test Action Group of the Institute of Electrical and Electronic Engineers. This interface is typically used for adding boundary scan testability to a chip. Recently, though, programmers have begun using JTAG for programming CPLDs and FPGAs while the chip is in an active system. This capability is called *in-system programming*, or *ISP*.

A JTAG interface is defined as having four pins, as described in Table 28.1. Extended JTAG includes a fifth reset pin. Instructions can be serially shifted into the chip on the TDI input. The TMS input controls the stepping through internal state machines to allow the programming of the device. Internal registers and the current state of the state machine can be shifted out via the TDO pin. The TRST pin is used to asynchronously initialize the internal state machine to prepare the chip for programming.

28.8   Summary: CPLD Selection Criteria

The internal architecture and the semiconductor technology used to implement its programmable elements strongly influence how well it "fits" a particular application. When designing a CPLD you should take the following architectural and technological issues into account:

- The programming technology—PROM, EPROM, EEPROM, or Flash EPROM. This will determine the equipment you will need to program the devices and whether they can be programmed only once or many times. The ability to reprogram during development will reduce your cost for parts, though that's not usually a significant part of the entire development cost.

- In-system programmability—This feature will allow engineers to update functionality in the field. This creates many options for upgrading existing customers, including network or Internet-based upgrades and fully automatic upgrades via software. Of course, developing the software to support an in-field upgrade for a system may require a lot of effort. Sending personnel out to upgrade hardware manually may or may not be cost effective for all applications. And the CPLDs in some systems simply cannot be disabled in the field, so in-system programmability may not be an option.

Computing Parity without Exclusive OR (Continued)

      | a0 & !a1 & a2 & !a3 & !a4 & !a5 & !a6 & a7
      | !a0 & a1 & a2 & !a3 & !a4 & !a5 & !a6 & a7
      | a0 & a1 & !a2 & !a3 & !a4 & !a5 & !a6 & a7
      | !a0 & !a1 & !a2 & !a3 & !a4 & !a5 & !a6 & a7)

As you can see, this requires a large number of AND gates and OR gates. In a typical PAL or CPLD, there are many AND gates that can be used, through DeMorgan's Law, as OR gates, but we do not have the resources for a large number of both AND and OR gates. Thus, including an XOR in the functional block makes implementation of parity practical.

Note that the flip-flop in this functional block has an asynchronous preset and a synchronous clear. The preset is controlled by the logic in the functional block, whereas the reset can be controlled by the logic of the functional block or by a global clear signal used to initialize each flip-flop in the entire device. The flip-flop clock can also be generated from the functional block logic as well as from a global clock line, as is the case for the clock enable input for the flip-flop. Note that not every CPLD from every manufacturer has all of these capabilities for the flip-flops. Also note that, for reliability reasons, clocks and asynchronous inputs should only be controlled by the global signal lines and not by any internal logic, even though the CPLD may give that ability.

Consider all of these factors before deciding whether this feature is useful for the design.

- The function block capability—Although most CPLDs have similar function blocks, there are differences, for example, in the number of flip-flops and the number of inputs to each block. Try to find a function block architecture that fits your design. If the design is dominated by combinatorial logic, you will prefer function blocks with large numbers of inputs. If the design performs a lot of parity checking, you will prefer function blocks with built-in XOR gates. If the design has many pipelined stages, you will prefer function blocks with several flip-flops.

- The number of function blocks in the device—This will determine how much logic the device can hold and how easily the design will fit into it.

- The kind of flip-flop controls available (e.g., clock enable, reset, preset, polarity control) and the number of global controls—CPLDs typically have global resets that simplify the design for initializing registers and state machines. Clock enables can often be useful in state machine design if you can take advantage of them.

- Embedded devices—Does the design interface with devices like a microcontroller or a PLL? Many CPLDs now incorporate specialized functions like these, which will make your job much easier and allow you to integrate more devices into a single CPLD.

- The number and type of I/O pins—Obviously, the CPLD will need to support the number of I/O pins in your design. Also, determine how many of these are general-purpose I/O and how many are reserved for special functions like clock input, master reset, etc.

- The number of clock input pins—Clock signals can be driven only into particular pins. If the design has several clock domains (i.e., sections driven by separate clocks), you will need a CPLD that has that many clock input pins.

References

Logic Design Manual for ASICs. Santa Clara, CA: LSI Logic Corporation, 1989.

Davenport Jr., Wilbur B, *Probability and Random Processes.* New York, NY: McGraw-Hill Book Company, 1970.

Dorf, Richard C., editor. *Electrical Engineering Handbook.* Boca Raton, FL: CRC Press: Inc., 1993.

EDA Industry Working Groups Web site, www.eda.org

Maxfield, Clive "Max." *Designus Maximus Unleashed!* Woburn, MA: Butter-worth-Heinemann: 1998.

Zeidman, Bob. *Introduction to Verilog.* Piscataway, NJ: Institute of Electrical and Electronic Engineers, 2000.

Zeidman, Bob. *Verilog Designer's Library.* Upper Saddle River, NJ: Prentice-Hall, Inc., 1999.

CHAPTER 29

Field-Programmable Gate Arrays (FPGAs)

Bob Zeidman

Field-programmable gate arrays are given this name because they are structured very much like a gate array ASIC. Like an ASIC, the FPGA consists of a regular array of logic, an architecture that lends itself to very complex designs.

29.1 FPGA Architectures

Each FPGA vendor has its own FPGA architecture, but in general terms they are all a variation of that shown in Figure 29.1. The architecture consists of configurable logic blocks, configurable I/O blocks, and programmable interconnect to route signals between the logic blocks and I/O blocks. Also, there is clock circuitry for driving the clock signals to each flip-flop in each logic block. Additional logic resources such as ALUs, memory, and decoders may also be available. The two most common types of programmable elements for an FPGA are static RAM and antifuses. Antifuse technology is a cousin to the programmable fuses in EPROMs. You will learn about antifuses, along with these other aspects of FPGAs, in the following sections.

The important thing to note about the FPGA architecture is its regular, ASIC-like structure. This regular structure makes FPGAs useful for all kinds of logic designs.

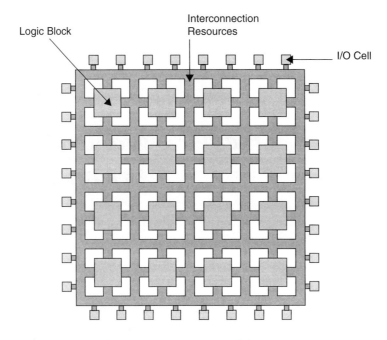

Figure 29.1: FPGA architecture

29.2   Configurable Logic Blocks

Configurable logic blocks (CLBs) contain the programmable logic for the FPGA. The diagram in Figure 29.2 shows a typical CLB, containing RAM for creating arbitrary combinatorial logic functions. It also contains flip-flops for clocked storage elements and multiplexers in order to route the logic within the block and to route the logic to and from external resources. These muxes also allow polarity selection, reset input, and clear input selection.

On the left of the CLB are two 4-input memories, also known as 4-input lookup tables or 4-LUTs. As discussed in an earlier chapter, 4-input memories can produce any possible 4-input Boolean equation. Feeding the output of the two 4-LUTs into a 3-LUT, produces a wide variety of outputs (for up to nine inputs).

Four signals labeled C1 through C4 enter at the top of the CLB. These are inputs from other CLBs or I/O blocks on the chip, allowing outputs from other CLBs to be input to this particular CLB. These interconnect inputs allow designers to partition large logic functions among several CLBs. They also are the basis for connecting CLBs in order to create a large, functioning design.

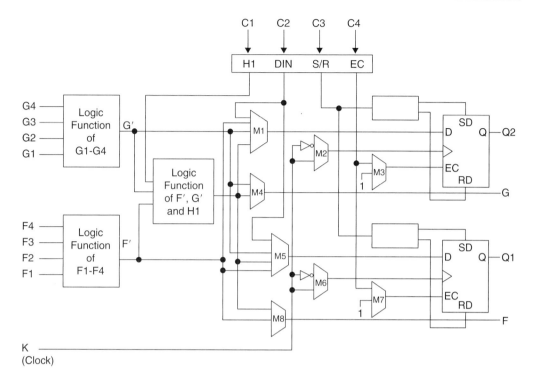

Figure 29.2: FPGA configurable logic block (CLB) (courtesy of Xilinx Inc.)

The muxes throughout the CLB are programmed statically. In other words, when the FPGA is programmed, the select lines are set high or low and remain in that state. Some muxes allow signal paths through the chip to be programmed. For example, mux M1 is programmed so that the top right flip-flop data is either input C2, or the output of one of the two 4-LUTs or the output of the 3-LUT.

Some muxes are programmed to affect the operation of the CLB flip-flops. Mux M2 is programmed to allow the top flip-flop to transition on the rising or falling edge of the clock signal. Mux M3 is programmed to always enable the top flip-flop, or to enable only when input signal C4 is asserted to enable it.

Note that the clock input to the flip-flops must come only from the global clock signal. Earlier architectures allowed flip-flops to be clocked by the outputs of the combinatorial logic. This allowed asynchronous designs that created lots of problems, and FPGA vendors eventually took that capability out of their architectures, greatly reducing their headaches and greatly increasing the reliability of their customers' designs.

Note that the logic outputs do not need to go through the flip-flops. Designers can use a CLB to create simple combinatorial logic. Because of this, multiple CLBs can, and often are, connected together to implement complex Boolean logic. This advantage of FPGAs over CPLDs means that designers can implement very complex logic by stringing together several CLBs. Unfortunately, routing delay in an FPGA is a significant amount of the overall delay. So this advantage also results in an overall decrease in the speed of the design.

Fine-Grained vs. Large-Grained CLBs

In theory, there are two types of CLBs, depending on the amount and type of logic that is contained within them. These two types are called *large grain* and *fine grain*.

In a large grain FPGA, the CLB contains larger functionality logic. For example, it can contain two or more flip-flops. A design that does not need many flip-flops will leave many of these flip-flops unused, poorly utilizing the logic resources in the CLBs and in the chip. A design that requires lots of combinatorial logic will be required to use up the LUTs in the CLBs while leaving the flip-flops untouched.

Fine grain FPGAs resemble ASIC gate arrays in that the CLBs contain only small, very basic elements such as NAND gates, NOR gates, etc. The philosophy is that small elements can be connected to make larger functions without wasting too much logic. If a flip-flop is needed, one can be constructed by connecting NAND gates. If it's not needed, then the NAND gates can be used for other features. In theory, this apparent efficiency seemed to be an advantage. Also, because they more closely resembled ASICs, it seemed that any eventual conversion of the FPGA to ASIC would be easier.

However, one key fact renders the fine grain architecture less useful and less efficient. It turns out that routing resources are the bottleneck in any FPGA design in terms of utilization and speed. In other words, it is often difficult to connect CLBs together using the limited routing resources on the chip. Also, in an FPGA, unlike an ASIC, the majority of the delay comes from routing, not logic. Fine grain architectures require many more routing resources, which take up space and insert a large amount of delay, which can more than compensate for their better utilization. This is why all FPGA vendors currently use large grain architectures for their CLBs.

In the early days of the industry several FPGA manufacturers produced fine grain architectures for their devices. Thinking like ASIC vendors, they missed the significance of the routing issues. All of these vendors have either fallen by the wayside or have abandoned their fine grain architectures for large grain ones.

29.3   Configurable I/O Blocks

A configurable I/O block, shown in Figure 29.3, is used to bring signals onto the chip and send them back off again. The output buffer, B1, has programmable controls to make the buffer three-state or open collector and to control the slew rate. These controls allow the FPGA to output to most standard TTL or CMOS devices. The slew rate control is important in controlling noise, signal reflections, and overshoot and undershoot on today's very fast parts. Slowing signal rise and fall times, reduces the noise in a system and reduces overshoot, undershoot, and reflections.

The input buffer B2 can be programmed for different threshold voltages, typically TTL or CMOS level, in order to interface with TTL or CMOS devices. The combination of input and output buffers on each pin, and their programmability, means that each I/O block can be used for an input, an output, or a bidirectional signal.

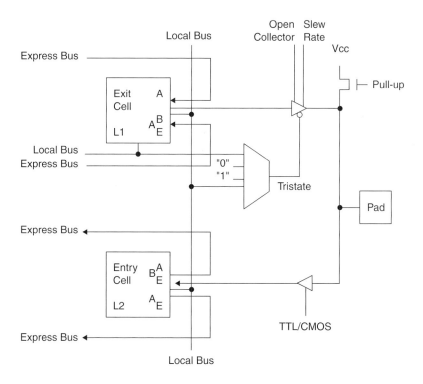

Figure 29.3: FPGA configurable I/O block (courtesy of Xilinx Inc.)

The pull-up resistors in the I/O blocks are a nice feature. They take up little space in the FPGA and can be used to pull up three-state buses on a board. Floating buses increase the noise in a system, increase the power consumption, and have the potential to create metastability problems.

There are two small logic blocks in each I/O block, labeled L1 and L2 in the diagram. These blocks are there for two reasons. First, it always makes sense to stick logic wherever there is space for it. In the I/O block, there is space for it. Second, unlike an ASIC, the routing delay in an FPGA is much more significant than the logic delay. In an ASIC, signals are routed using metal layers, leading to RC delays that are insignificant with respect to the delay through logic gates. In an FPGA, the routing is done through programmed muxes in the case of SRAM-based devices and through conducting vias in the case of antifuse devices. Both of these structures add significant delay to a signal. The muxes have a gate delay associated with them. The conducting vias have a high resistance, causing an RC delay. I discuss this in further detail in Section 29.5.

Pull-Ups, Floating Buses, and Stubbornness

I would think it's obvious to anyone who understands CMOS technology that floating buses are bad. CMOS devices dissipate power unnecessarily when the inputs are floating, and floating signals are more prone to noise. A pull-up resistor is a very simple, small, low-power, inexpensive solution to the problem of floating buses. In my career, though, I have encountered, on two occasions, a religious fervor about not putting pull-ups on potentially floating buses. I still don't completely understand the reasons.

In one case, a career marketing manager at a large semiconductor company, who still prided himself on being a top-notch engineer, did a board design that I was asked to optimize, lay out, and debug. When I saw that he had omitted any pull-up resistors, I simply put them back in the schematic. When he saw this, he became angry. He told me that in all his years, he had never seen a problem, nor had he ever encountered a metastability problem. I replied that a reliability problem like metastability might only be seen once every year on a continually running board. It's not something that can be measured. This manager went so far as to inform the layout designer to tell me that she couldn't fit the pull-up resistor pack (nine pins on a small sliver of material) on the board. I could tell she felt ridiculous about this because she was telling me that she couldn't do something that any fresh-out-of-school layout designer could accomplish with the cheapest layout software package available.

In the other case, I was brought in to do a sanity check of a board design that was nearing completion. A small startup had a contract to design some specialized network boards for Cisco Systems. A consultant had been hired to design the board, and the project manager then hired me to review the design. In my review, one of the potential problems I found

---

Pull-Ups, Floating Buses, and Stubbornness (Continued)

was, yes, no pull-up resistors on the buses. I mentioned this casually, and the board designer became irate for the same reasons as the manager I had met. There was no reason for it. They were too expensive (actually about $.01 per resistor), and they took up too much space (a pack of ten resistors takes a few square millimeters). Finally he said, "We met with those guys at Cisco and they said the same thing. They wanted those stupid, unnecessary resistors on the buses. I just won't waste my time doing it." Later, in private, I talked with the project manager. "You may not think there's a need for those resistors," I said, "and you may not trust my judgment. But if I were selling boards to Cisco and they said to spread peanut butter on the boards, I'd break out the Skippy®."

The point of these stories is that internal resistors on I/O pins of FPGAs make this problem go away. With internal resistors on the I/O pins, you can connect pull-ups to all of your buses, saving the tiny cost and area of a resistor pack, and no one will be the wiser.

Because of the large routing delays, if an input signal needed to be routed from an input buffer through internal interconnect to a flip-flop in a CLB inside the chip, as shown in Figure 29.4, there would be a large delay, labeled d, from the input pin to the data input of the flip-flop. In order to meet the hold time requirement of the internal flip-flop, labeled h, the hold time requirement for that signal with respect to the clock at the pins of the chip would be the sum of the delay d and hold time h which would be a large number and difficult to meet for devices interfacing with the FPGA. Instead, by placing flip-flops in the I/O blocks, the delay d is very small, resulting in a reasonable hold time at the pins of the chip.

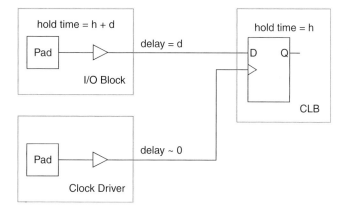

Figure 29.4: Hold time issues in FPGAs

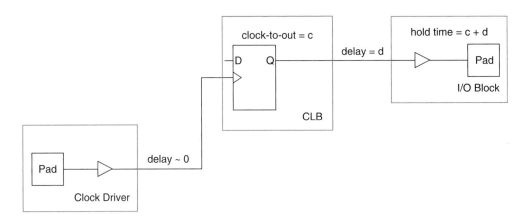

Figure 29.5: Setup time issues in FPGAs

Similarly, if an output needed to be routed from a flip-flop in a CLB inside the device directly to an output buffer, as shown in Figure 29.5, there would be a large delay from the flip-flop output to the pin of the chip. This means that the clock-to-output delay for all signals would be the delay of the flip-flop, labeled c, plus the delay of the routing, labeled d. All chips interfacing with the FPGA would need to have a very small setup time requirement, or the clock speed of the system would need to be reduced. The solution is to have the output flip-flop inside the I/O block so that it is right next to the buffers. Then the routing delay is not significant. Having input and output flip-flops in the I/O blocks allows the FPGA and the system in which it is designed to be as fast as possible.

29.4   Embedded Devices

Many newer FPGA architectures are incorporating complex devices inside their FPGAs. These devices can range from relatively simple functions, such as address decoders or multipliers, all the way through ALUs, DSPs, and microcontrollers and microprocessors. These embedded devices are optimized, and the FPGA devices that include them can offer you a very nice way of integrating an entire system onto a single chip, creating what is being called a *system on a programmable chip* or *SOPC*.

The advantage of FPGAs with embedded devices is that you can save board area and power consumption. You can usually save cost and increase system speed with these FPGAs. The embedded devices are already tested, just like a standalone chip, so that you don't need to design the circuit from scratch and verify its functionality.

The disadvantage of these devices is that you tend to tie yourself into a single FPGA from a single FPGA vendor, losing some of the portability that engineers prefer. Each vendor has specific devices embedded into their FPGAs. In the case of embedded processors, each FPGA vendor usually licenses a specific processor core from a different processor manufacturer. This is good for the FPGA vendor because once they have a design win, that design is committed to their FPGA for some time. This is also a great way for the smaller FPGA vendors to differentiate themselves from the larger ones, by providing an embedded device that is in demand or soon will be, and can also produce a nice niche market for themselves that allows them to fend off eventual annihilation by the bigger vendors.

29.5 Programmable Interconnect

The interconnect of an FPGA is very different than that of a CPLD, but is rather similar to that of a gate array ASIC. Figure 29.6 shows the CLB's hierarchy of interconnect resources. Each CLB is connected with the immediately neighboring CLBs, as shown in the top left. These connections are sometimes called *short lines*. (Note that for simplicity only the connections with CLB1 in the top left are shown. In reality, all four CLBs have connections to their nearest neighbors. These connections allow logic that is too complex to fit into a single CLB to be mapped to multiple CLBs.)

Other routing resources consist of traces that pass by a number of CLBs before reaching switch matrices. These switch matrices allow a signal to be routed from one switch matrix to another to another, eventually connecting CLBs that can be relatively far from each other. The disadvantage to this method is that each trip through a switch matrix

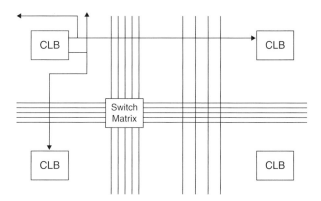

Figure 29.6: FPGA programmable interconnect (courtesy of Xilinx Inc.)

results in a significant delay. Often, in order to route signals through a chip, the routing delay becomes greater than the logic delay. This situation is unique to FPGAs and creates some design issues.

The third type of routing resource is the long line, which designers can use to connect critical CLBs that are physically far from each other on the chip without inducing much delay. These lines usually go from one end of the die to the other without connecting to a switch matrix. For critical path logic, long lines ensure that there will not be a significant delay.

The long line can also be used as buses within the chip. Three-state buffers are used to connect many CLBs to a long line, creating a bus. In an ASIC, three-state buses should be avoided because all three-state buses present a possible danger of contention or floating nodes, both of which can introduce long-term reliability problems if not designed carefully. Instead, muxes are used to combine many outputs because muxes are simple, easily expandable devices. In an ASIC, routing is not a significant problem. For an FPGA, though, muxes are not practical for connecting multiple outputs because this would require bringing outputs of different CLBs to a single CLB that contains the mux. Sometimes the CLBs producing the outputs are spread over a large section of the chip, requiring the signals to go through many switch matrices to reach the final destination. This introduces a very significant delay, slowing down the entire design.

Instead, CLBs near the long lines can directly drive the long lines with three-state drivers. The routing delay in this case is small. Of course, designers need to follow proper design techniques carefully to avoid any bus contention or floating buses.

Figure 29.7 shows a more detailed view of the routing for a single CLB. Note that the CLB has connections to local interconnect to connect it to neighboring CLBs. Also, it uses muxes to drive outputs onto the longer interconnects to connect it to devices at other parts of the chip. In this particular architecture, inputs from the interconnects go directly into the CLB, where the logic determines whether to use or ignore these signals.

Note that CLBs themselves are often used for routing. In cases of high congestion on a chip, where routing resources are used up and not all signals and CLBs are connected, CLBs can be used for routing. In this case, the logic and muxes are set up so that a signal coming in simply goes out without any logical changes. This effectively increases routing resources in a densely packed design, but of course results in additional significant delay.

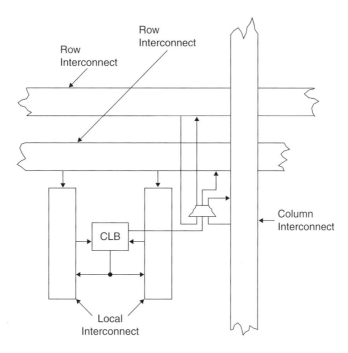

Figure 29.7: CLB programmable interconnect (courtesy of Altera Corporation)

29.6 Clock Circuitry

Special I/O blocks with special high drive clock buffers, known as clock drivers, are distributed around the chip. These buffers are connected to clock input pins and drive the clock signals onto the global clock lines distributed throughout the device in a configuration called a *clock tree*. These clock lines are designed for low clock skew times and fast clock propagation times. Synchronous design is a must with FPGAs, because absolute skew times and delay times for signals cannot be guaranteed using the routing resources of the FPGA. Only when using clock signals from clock buffers can the relative delays and skew times be small and predictable.

29.7 SRAM vs. Antifuse Programming

There are two competing methods of programming FPGAs. The first, SRAM programming, involves static RAM bits as the programming elements. These bits can be combined in a single memory and used as a LUT to implement any kind of combinatorial logic, as described earlier. Also, programmers can use individual SRAM bits to control muxes, which select or deselect particular logic within a CLB, as described in Section 29.2. For routing, these bits can turn on a transistor that connects two traces in a switch matrix, or they can select the output of a mux that drives an interconnect line. Both methods are illustrated in Figure 29.8.

The other programming method involves antifuses. A regular fuse normally makes a connection until an excessive amount of current goes through it, generating heat and breaking the connection. With an antifuse, there is a small link between two conductors that are separated by an insulator, as shown in Figure 29.9. When a large voltage is applied across the link, the link melts. As the link melts, the conductor material migrates across the link, creating a conducting path between the two conductors. This process is used to connect traces inside the FPGA.

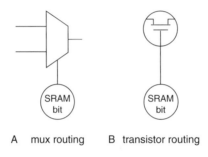

A   mux routing   B transistor routing

Figure 29.8: FPGA routing using SRAM bits

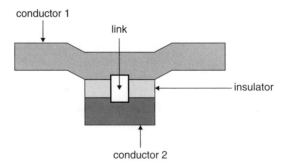

Figure 29.9: FPGA antifuses (courtesy of Quicklogic Corporation)

There are two main types of antifuses in production today. In one type, conductor 1 is polysilicon, and conductor 2 is n+ diffused silicon. The insulator is oxide-nitride-oxide (ONO) and the link is composed of silicon.

In the second type of antifuse, both conductors are metal, the insulator is amorphous silicon, and the link is composed of titanium or tungsten silicide.

SRAM-based FPGAs have the advantage of being reprogrammable. Especially as FPGAs become larger and therefore more expensive, it is a nice feature during debugging to be able to reprogram them rather than toss out a bad design. SRAM-based FPGAs can also be reprogrammed while in the system, which makes in-field upgrading very easy. Programmers can alter a communication protocol or add a feature to the FPGA by a simple software change. SRAM-based FPGAs allow you to include small memories like FIFOs in your design, though large memories inside an FPGA are not cost effective. Also, SRAM-based FPGAs can be used for reconfigurable computing, a concept whereby computers contain FPGAs and algorithms can be compiled to run in the FPGAs.

A disadvantage of SRAM-based FPGAs is that they're reprogrammable. Some applications, particularly military ones, often require that a device be nonvolatile and not susceptible to changes from radiation or power glitches. Antifuse FPGAs fit these criteria.

In theory, antifuse FPGAs are much faster than SRAM FPGAs. This is because antifuse FPGAs have a real connection between conductors for routing traces, as opposed to the logic or transistors used in SRAM-based FPGAs. Although the antifuse connections have a high resistance and therefore some RC delay associated with them, this delay should be much lower than the delay in SRAM-based FPGAs. You'll notice that I use some wishy-washy terms here. The reason is that, in practice, antifuse FPGAs are not significantly faster than SRAM-based FPGAs, despite the theory. That's because every semiconductor company in the world knows how to make SRAMs. It's a standard product using a standard technology. Even companies that do not produce SRAMs often use SRAM structures to test their new processes because the structures are so regular and their performance is predicable. And because each semiconductor company is continually trying to improve its processes, they are always making faster and smaller SRAMs. On the other hand, only a small number of semiconductor companies, those manufacturing antifuse FPGAs, know how to make antifuses. There simply aren't as many people or companies at work attempting to improve the yields, size, and speed of antifuses. For this reason, from a practical point of view, the speed difference between the two technologies is, and will probably remain, fairly small.

Antifuse FPGAs do have the advantage of lower power over SRAM-based FPGAs. Antifuse FPGAs also have an intellectual property security advantage. By this I mean that SRAM-based FPGAs must always be programmed by an external device upon power-up. It's possible, then, for some unscrupulous engineer to copy your design simply by capturing the external bit stream. This engineer can then load the bit stream into any other FPGA, making a perfect copy of your design. Because an antifuse FPGA is programmed once, the program and the design are safe from prying eyes.

A newer technology that shows promise is Flash-based FPGAs. These devices are essentially the same as SRAM-based devices, except that they use Flash EPROM bits for programming. Flash EPROM bits tend to be small and fast. They are nonvolatile like antifuse, but reprogrammable like SRAM. Flash-based FPGA routing is shown in Figure 29.10.

Table 29.1 summarizes how SRAM and antifuse programming technologies compare.

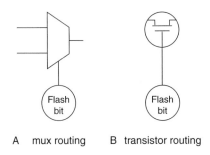

A   mux routing   B   transistor routing

Figure 29.10: FPGA routing using Flash bits

Table 29.1: Comparison of FPGA programming technolog

	SRAM	Antifuse
Volatile	Yes	No
In-system programmable	Yes	No
Speed	Fast	Somewhat faster
Power consumption	Higher	Lower
Density	High	High
IP security	No	Yes
Embedded RAM	Yes	No

29.8 Emulating and Prototyping ASICs

Designers can also use FPGAs in places where an ASIC will eventually be used. For example, designers may use an FPGA in a design that needs to get to market quickly at a low initial development cost. Later, they can replace the FPGA with an ASIC when the production volume increases, in order to reduce the part cost. Most FPGA designs, however, are not intended to be replaced by an ASIC. FPGAs tend to be used where programmability, up-front cost, and time to market are more important than part cost, speed, or circuit density.

Reconfigurable Computing

An interesting concept, reconfigurable computing, has been floating around universities for the past decade or so and has been the subject of a number of research papers. In recent years, some companies have begun offering different variations of the concept in actual products. Essentially, the concept behind configurable computing is that computations can be performed much faster using special purpose hardware than using general-purpose hardware (a processor) running software. Therefore, general computing could be sped up significantly if hardware could be dynamically reconfigured to act as specialized coprocessors. Obviously, SRAM-based FPGAs would be an ideal implementation for this hardware.

The first, and most difficult, obstacle to reconfigurable computing is that a compiler must be able to partition a general-purpose computer program into software and hardware functionality by extracting the algorithms that can be implemented in hardware. It must then compile the algorithm code into an FPGA design. Synthesis programs have a difficult enough time doing the same operation using a hardware description language, which is designed specifically for this use. And a synthesis program doesn't need to partition the code between hardware and software—all of the HDL code represents hardware. Because the process is so difficult, synthesis programs restrict the HDL code to a specific coding style to make the job easier. Also, hardware designers still need to set switches within comments in the code and change settings of the synthesis program, in order to get usable results. And after all of this, the designer often needs to tweak the HDL code to fit the design into the FPGA and to meet the required performance criteria.

The second obstacle is the relatively long time required to load a new design into an FPGA. Unless companies can speed up this process, reconfigurable computing will be limited to algorithms that are repeated in a loop in the code, so that the overhead of reprogramming the FPGA is compensated by the number of times the hardware algorithm is executed.

As companies come to market with reconfigurable computing solutions, they have been taking a more practical approach than the "Holy Grail" described above. Some of the solutions

---

Continued onto next page

Reconfigurable Computing (Continued)

include libraries of algorithms that have already been developed, tested, and synthesized and that can be called from software. Other companies have created new programming languages that combine the flexibility of C++, for example, with an HDL. Such languages make it easier for compilers to partition and synthesize the code. The disadvantage is that these new nonstandard languages represent a particularly challenging learning hurdle, because they require a knowledge of both software and hardware design techniques.

I believe that these practical, but limited solutions will eventually produce a real-world product that will have some use in specific areas. The progress of reconfigurable computing will probably parallel that of silicon compilation. In the early 1980s, some companies, such as Silicon Compilers, Inc., were touting the ability to go directly from a schematic diagram to a chip layout. The problem turned out to be bigger than these advocates thought; the algorithms needed were much more complex than they originally believed, and the computing power to execute the algorithms just wasn't available yet at a reasonable cost. So these companies all folded, but not without first producing corporate offspring and cousins, such as Synopsys Corporation, that decided they could tackle a much easier problem and still provide a solution that engineers could use. Their solution was software synthesis—software that could produce a gate level description from an RTL level description. This much less ambitious but much more achievable solution was a great success and may still eventually lead to the ultimate solution of silicon compilation. Many successful new products in the engineering design automation (EDA) industry follow this same trajectory. An ambitious first product fails, leading others to attempt smaller, less costly, more achievable products that succeed. In the same way, I believe, restricted solutions to reconfigurable computing will make their way into the marketplace and be successful, and eventually lead to more and more complex implementations.

There are two methodologies in integrated circuit chip design where FPGAs are being used to assist in the development of chips and the development of software that depends on these chips. These methodologies are known as emulation and prototyping.

29.8.1 Emulation

Several companies provide standalone hardware boxes for emulating the function of an ASIC or a custom integrated circuit. These hardware emulators can be programmed with a design for a chip. Once programmed, the emulator can be physically connected to a target circuit board where the chip would normally be connected. Then, the entire target system can be run as if the chip were actually available and functioning. You can then debug the design using real world hardware and real world software.

You can stop and start the hardware emulator in order to examine internal nodes of the design. You can capture waveform traces of internal and I/O signals for debugging. You can make changes to the design to correct mistakes or to improve performance, before the design is committed to silicon.

For example, if you are designing a new microprocessor, you could load the microprocessor design into a hardware emulator, plug the emulator into a target personal computer motherboard, and actually boot Linux or any other operating system. You could even run real applications. Of course, a hardware emulator runs at a fraction of the speed of the final chip, but it affords a form of testing that is otherwise not possible, except with prototyping.

Different hardware emulators from different manufacturers have different internal architectures. Many of them, though, use large sets of FPGAs to emulate the chip design, because FPGAs allow users to easily load and modify designs, stop the design while it is in the system, and easily examine internal nodes and external I/O.

29.8.2 Prototyping

As FPGAs become faster and denser, and ASICs become larger, prototyping has become an important alternative to emulation. Prototyping involves loading a chip design into one or more FPGAs. If the chip design fits into a single FPGA, the FPGA can be plugged into a socket or soldered into a target circuit board where the final chip will ultimately go. The board can then be powered up and tested using real data. If the design cannot fit into a single FPGA, a board can be designed that contains several FPGAs into which the chip design is partitioned. Companies now provide software that will automatically partition a chip design into multiple FPGAs.

These design-specific FPGA prototypes generally run faster than a hardware emulator because they do not have the overhead required for a general-purpose machine, and there are fewer FPGAs—only the exact number required to implement your chip design. On the other hand, they do not have the built-in diagnostic capabilities of a hardware emulator, and they do not come with application engineers to support you. FPGA prototypes are generally cheaper than hardware emulators, but you must do all of the work, including partitioning the design, designing the board to hold the FPGAs, and designing whatever debug capabilities you require.

29.9   Summary

This section summarizes the various aspects of FPGAs that we have learned in this chapter. This section also provides a list of factors to use when deciding whether to choose a CPLD or FPGA for your design.

29.9.1   FPGA Selection Criteria

Knowledge of the internal architecture of FPGAs and the semiconductor technologies used to implement the programmable elements is critical for considering which FPGA to use in your design. When making that decision, you should take into account the following architectural and technological issues:

Configurable logic blocks — Although most FPGAs have similar logic blocks, there are differences, for example, in the number of flip-flops and the width of the lookup tables. Try to find a CLB architecture that fits your design. If your design has wide combinatorial functions, choose an FPGA using CLBs with large numbers of inputs. If your design has many pipelined stages, you will prefer CLBs with several flip-flops. Newer architectures are always being developed that fit the needs of specific types of designs, such as digital signal processing.

- The number of CLBs in the device — This will determine how much logic the device can hold and how easily your design will fit into it.

- The number and type of I/O pins — Obviously, the FPGA will need to support the number of I/O pins in your design. Also, determine how many of these are general-purpose I/O and how many are reserved for special functions such as clock input, master reset, etc.

- The number of clock input pins — Clock signals can be driven only into particular pins. If your design has several clock domains (i.e., sections driven by separate clocks), you will need an FPGA that has that many clock input pins.

- Embedded devices — Does your design interface with devices such as a microcontroller or a PLL? Many FPGAs now incorporate specialized functions like these, which will make your job much easier and allow you to integrate more devices into a single FPGA.

- Antifuse vs. SRAM programming — Which technology makes sense for your design? Do you need the speed, low power, nonvolatility, and security of an antifuse device, or do you need the reprogrammability of an SRAM-based device?

Table 29.2: CPLDs vs. FPGAs

	CPLD	FPGA
Architecture	PAL-like	Gate array–like
Speed	Fast, predictable	Application dependent
Density	Low to medium	Medium to high
Interconnect	Crossbar	Routing
Power consumption	High per gate	Low per gate

- Emulating and prototyping ASICs — FPGAs can be found in off-the-shelf hardware emulators for testing the design of an ASIC in a real-world target before it goes to silicon. Or you can use FPGAs to create your own custom prototype of an ASIC for the same kind of pre-silicon real-world testing.

29.9.2 *Choosing Between CPLDs and FPGAs*

Choosing between a CPLD and an FPGA will depend on the requirements of your project. Table 29.2 shows a summary of the characteristics of each type of programmable device. You will notice that I use fuzzy terms like "low," "medium," and "high" for some of the characteristics. People often want me to give a definitive answer on, for example, the number of gates in a typical CPLD or the cost of a typical FPGA. Because these numbers change so quickly, they are wrong as soon as they leave my lips (or in this case when they reach print). For that reason, I prefer to give relative characteristics that will still be correct for a while after I give them.

References

Logic Design Manual for ASICs. Santa Clara, CA: LSI Logic Corporation, 1989.

Davenport Jr., Wilbur B. *Probability and Random Processes*. New York, NY: McGraw-Hill Book Company, 1970.

Dorf, Richard C., editor. *Electrical Engineering Handbook*. Boca Raton, FL: CRC Press, Inc., 1993.

EDA Industry Working Groups Web site, www.eda.org

Maxfield, Clive, "Max." *Designus Maximus Unleashed!* Woburn, MA: Butterworth- Heinemann, 1998.

Zeidman, Bob, *Introduction to Verilog*. Piscataway, NJ: Institute of Electrical and Electronic Engineers, 2000.

Zeidman, Bob, *Verilog Designer's Library*. Upper Saddle River, NJ: Prentice- Hall, Inc., 1999.

CHAPTER 30

Design Automation and Testing for FPGAs

Peter Wilson

30.1   Simulation

30.1.1   Test Benches

The overall goal of any hardware design is to ensure that the design meets the requirements of the design specification. In order to measure that this is indeed the case, we need not only to simulate the design representation in a hardware description language (such as VHDL), but also to ensure that whatever tests we undertake are appropriate and demonstrate that the specification has been met.

The way that designers can test their designs in a simulator is by creating a *test bench*. This is directly analogous to a real experimental test bench in the sense that stimuli are defined and the responses of the circuit measured to ensure that they meet the specification.

In practice, the test bench is simply a VHDL model that generates the required stimuli and checks the responses. This can be in such a way that the designer can view the waveforms and manually check them, or by using VHDL constructs to check the design responses automatically.

30.1.2   Test Bench Goals

The goals of any test bench are twofold. The first is primarily to ensure that correct operation is achieved. This is essentially a functional test. The second goal is to ensure that a synthesized design still meets the specification (particularly with a view to timing errors).

30.1.3 Simple Test Bench: Instantiating Components

Consider a simple combinatorial VHDL model given below:

```
library ieee;
use ieee.std_logic_1164.all;
entity cct is
      port (   in0, in1 : in std_logic;
               out1 : out std_logic
      );
end;

architecture simple of cct is
begin
      out1 <= in0 AND in1 ;
end;
```

This simple model is clearly a two input AND gate, and to test the operation of the component we need to do several things.

First, we must include the component in a new VHDL design. So we need to create a basic test bench. The listing below shows how a basic entity (with no connections) is created, and then the architecture contains both the component declaration and the signals to test the design.

```
-- library declarations
library ieee;
use ieee.std_logic_1164.all;

-- empty entity declaration
entity test is
end;

-- test bench architecture
architecture testbench of test is
      -- component declaration
      component cct
            port (   in0, in1 : in std_logic;
                     out1 : out std_logic
            );
      end component;
      -- test bench signal declarations
```

```
        signal in0, in1, out1 : std_logic;
-- architecture body
Begin
        -- declare the Circuit Under Test (CUT)
        CUT : cct port map ( in0, in1, out1 );
end;
```

This test bench will compile in a VHDL simulator, but is not particularly useful as there are no definitions of the input stimuli (signals in0 and in1) that will exercise the *circuit under test* (CUT).

If we wish to add stimuli to our test bench we have some significant advantages over our design VHDL; the most appealing is that we generally don't need to adhere to any design rules or even make the code synthesizable. Test bench code is generally designed to be "off chip" and therefore we can make the code as abstract or behavioral as we like and it will still be fit for purpose. We can use wait statements, file read and write, assertions and other nonsynthesizable code options.

30.1.4 Adding Stimuli

In order to add a basic set of stimuli to our test bench, we could simply define the values of the input signals in0 and in1 with a simple signal assignment:

```
begin
        CUT : cct port map ( in0, in1, out1 );

        in0 <= '0' ;
        in1 <= '1' ;
end;
```

Clearly this is not very complex or dynamic test bench, so to add a sequence of events we can modify the signal assignments to include numerous value, time pairs defining a sequence of values.

```
begin
        CUT : cct port map ( in0, in1, out1 );

        in0 <= '0' after 0 ns, '1' after 10 ns, '0'
           after 20 ns;
        in1 <= '0' after 0 ns, '1' after 15 ns, '0'
           after 25 ns;
end;
```

While this method is useful for small circuits, clearly for more complex realistic designs it is of limited value. Another approach is to define a constant array of values that allow a number of tests to be carried out with a relatively simple test bench and applying a different set of stimuli and responses in turn.

For example, we can exhaustively test our simple two input logic design using a set of data in a *record*. A VHDL record is simply a collection of types grouped together defined as a new type.

```
type testdata is record
      in0 : std_logic;
      in1 : std_logic;
end;
```

With a new composite type, such as a record, we can then create an array, just as in any standard VHDL type. This requires another type declaration, of the array type itself.

```
type data_array is array (natural range <>) of data_array
```

With these two new types we can simply declare a constant (of type data_array) that is an array of record values (of type testdata) that fully describe the data set to be used to test the design. Notice that the type data_array does not have a default range, but that this is defined by the declaration in this particular test bench.

```
constant test_data : data_array := ( ('0', '0'), ('0', '1'), ('1', '0'),
('1', '1') );
```

The beauty of this approach is that we can change from a system that requires every test stimulus to be defined explicitly, to one where a generic test data process will read values from predefined arrays of data. In the simple test example presented here, an example process to apply each set of test data in turn could be implemented as follows:

```
process
begin
      for i in test_data'range loop
            in0 <= test_data(i).in0;
            in1 <= test_data(i).in1;
            wait for 100 ns;
      end loop
      wait;
end process;
```

There are several interesting aspects to this piece of test bench VHDL. The first is that we can use behavioral VHDL (wait for 100 ns) as we are not constrained to synthesize this to hardware. Second, by using the range operator, the test bench becomes unconstrained by the size of the data set. Finally, the individual record elements are accessed using the hierarchical construct test_data(i).in0 or test_data(i).in1, respectively.

30.2 Libraries

30.2.1 Introduction

VHDL as a language on its own is actually very limited in the breadth of the data types and primitive models available. As a result, libraries are required to facilitate design reuse and standard data types for model exchange, reuse and synthesis. The primary library for standard VHDL design is the IEEE library. Within the IEEE Design Automation Standards Committee (DASC), various committees have developed libraries, packages and extensions to standard VHDL. Some of these are listed below:

- IEEE Std 1076 Standard VHDL Language

- IEEE Std 1076.1 Standard VHDL Analog and Mixed-Signal Extensions (VHDL-AMS)

- IEEE Std 1076.1.1 Standard VHDL Analog and Mixed-Signal Extensions – Packages for Multiple Energy Domain Support

- IEEE Std 1076.4 Standard VITAL ASIC (Application-specific Integrated Circuit) Modeling Specification (VITAL)

- IEEE Std 1076.6 Standard for VHDL Register Transfer Level (RTL) Synthesis (SIWG)

- IEEE Std 1076.2 IEEE Standard VHDL Mathematical Packages (math)

- IEEE Std 1076.3 Standard VHDL Synthesis Packages (vhdlsynth)

- IEEE Std 1164 Standard Multivalue Logic System for VHDL Model Interoperability (Std_logic_1164)

Each of these working groups are volunteers who come from a combination of academia, EDA industry and user communities, and collaborate to produce the IEEE Standards (usually revised every four years).

30.2.2 Using Libraries

In order to use a library, first the library must be declared:

```
library ieee;
```

Within each library a number of VHDL packages are defined, that allow specific data types or functions to be employed in the design. For example, in digital systems design, we require logic data types, and these are not defined in the basic VHDL standard (1076). Standard VHDL defines integer, Boolean and bit types, but not a standard logic definition. This is obviously required for digital design and an appropriate IEEE standard was developed for this purpose—IEEE 1164. It is important to note that IEEE Std 1164 is NOT a subset of VHDL (IEEE 1076), but is defined for hardware description languages in general.

30.2.3 Std_Logic Libraries

There are a number of std_logic libraries available in the IEEE library and these are:

- std_logic_1164
- std_logic_arith
- std_logic_unsigned
- std_logic_signed
- std_logic_entities
- std_logic_components
- std_logic_misc
- std_logic_textio

In order to use a particular element of a package in a design, the user is required to declare their use of a package using the USE command. For example, to use the standard IEEE logic library, the use needs to add a declaration after the library declaration as follows:

```
library ieee;
use ieee.std_logic_1164.all;
```

The std_logic_1164 package is particularly important for most digital design, especially for field-programmable gate arrays (FPGA), because it defines the standard logic types used by ALL the commercially available simulation and synthesis software tools, and is included as a standard library. It incorporates not only the definition of the standard logic types, but also conversion functions (to and from the standard logic types) and also manages the conversion between signed, unsigned and logic array variables.

30.2.4 Std_logic Type Definition

As it is such an important type, the std_logic type is described in this section. The type has the following definition:

- 'U': uninitialized; this signal hasn't been set yet

- 'X': unknown; impossible to determine this value/result

- '0': logic 0

- '1': logic 1

- 'Z': High Impedance

- 'W': Weak signal, can't tell if it should be 0 or 1

- 'L': Weak signal that should probably go to 0

- 'H': Weak signal that should probably go to 1

- '-': Don't care

These definitions allow resolution of logic signals in digital designs in a standard manner that is predictable and repeatable across software tools and platforms. The operations that can be carried out on the basic std_logic data types are the standard built in VHDL logic functions:

- and

- nand

- or

- nor

- xor

- xnor

- not

An example of the use of the std_logic library would be to define a simple logic gate, in this case a three input nand gate:

```
library ieee;
use ieee.std_logic_1164.all;

entity nand3 is
        port ( in0, in1, in2 : in std_logic;
                    out1 : out std_logic ) ;
end;

architecture simple of nand3 is
begin
        out1 <= in0 nand in1 nand in2;
end;
```

30.3   Synthesis

30.3.1   Design Flow for Synthesis

The basic HDL design flow is shown in Figure 30.1.

As can be seen from this figure, synthesis is the key stage between high-level design and the physical place and route which is the final product of the design flow. There are several different types of synthesis ranging from behavioral, to RTL and finally physical synthesis.

Behavioral synthesis is the mechanism by which high level abstract models are synthesized to an intermediate model that is physically realizable. Behavioral models can be written in VHDL that are not directly synthesizable and so care must be taken with high level models to ensure that this can take place, in fact. There are limited tools that can synthesize behavioral VHDL and these include the Behavioral Compiler from Synopsys, Inc and MOODS, a research synthesis platform from the University of Southampton.

RTL synthesis is what most designers call synthesis, and is the mechanism whereby a direct translation of structural and register level VHDL can be synthesized to individual gates targeted at a specific FPGA platform. At this stage, detailed timing analysis can

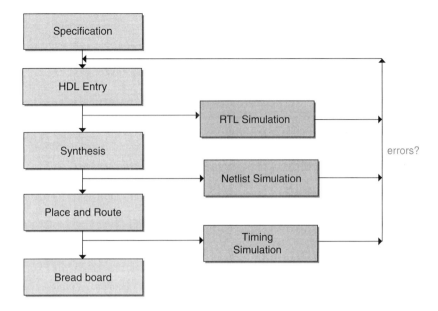

Figure 30.1: HDL design flow

be carried out and an estimate of power consumption obtained. There are numerous commercial synthesis software packages including Design Compiler (Synopsys), Leonardo Spectrum (Mentor Graphics) and Synplify (Synplicity). This is not an exhaustive list; there are numerous offerings available at a variety of prices.

Physical synthesis is the last stage in a synthesis design flow and is where the individual gates are placed (using a *floorplan*) and routed on the specific FPGA platform.

30.3.2 Synthesis Issues

Synthesis basically transforms program-like VHDL into a true hardware design (netlist). It requires a set of inputs, a VHDL description, timing constraints (when outputs need to be ready, when inputs will be ready, data to estimate wire delay), a technology to map to (list of available blocks and their size/timing information) and information about design priorities (area vs. speed)

For big designs, the VHDL will typically be broken into modules and then synthesized separately. 10K gates per module was a reasonable size in the 1990s, however tools can handle a lot more now.

30.3.3   RTL Design Flow

RTL VHDL is the input to most standard synthesis software tools. The VHDL must be written in a form that contains registers, state machines (FSM) and combinational logic functions. The synthesis software translates these blocks and functions into gates and library cells from the FPGA library. The RTL design flow is shown in Figure 30.2, in more detail than the overall HDL design flow. Using RTL VHDL restricts the scope of the designer as it precludes algorithmic design, as we shall see later. This approach forces the designer to think at quite a low level, making the resulting code sometimes verbose and cumbersome. It also forces structural decisions early in the design process, which can be restrictive and not always advisable, or helpful.

The design process starts from RTL VHDL:

- Simulation (RTL) – needed to develop a test bench (VHDL).

- Synthesis (RTL) – targeted at a standard FPGA platform.

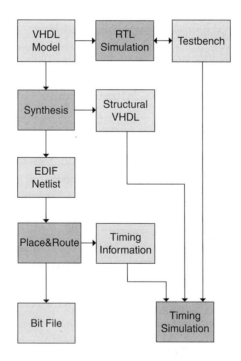

Figure 30.2: RTL synthesis and design flow

- Timing simulation (structural) – simulate to check timing.

- Place and route using standard tools (e.g., Xilinx Design Manager).

Although there are a variety of software tools available for synthesis (such as Leonardo Spectrum or Synplify), they all have generally similar approaches and design flows.

30.4   Physical Design Flow

Synthesis generates a netlist of devices plus interconnections. The 'place and route' software figures out where the devices go and how to connect them. The results are not as good as you'd like; a 40 to 60% utilization of devices and wires is typical. The designer can trade off run time against greater utilization to some degree, but there are serious limits. Typically the FPGA vendor will provide a software toolkit (such as the Xilinx Design Navigator or Altera's Quartus tools) that manages the steps involved in physical design.

Regardless of the particular physical synthesis flow chosen, the steps required to translate the VHDL or EDIF output from an RTL synthesis software program into a physically downloadable bit file are essentially the same and are listed below:

1. Translate

2. Map

3. Place

4. Route

5. Generate accurate timing models and reports

6. Create binary files for download to device

30.5   Place and Route

There are two main techniques to place and route in current commercial software, which are *recursive cut* and *simulated annealing*.

30.5.1   Recursive Cut

In a recursive cut algorithm, we divide the netlist into two halves, move devices between halves to minimize the number of wires that cross cut (while keeping the number of devices in each half the same). This is repeated to get smaller and smaller blocks.

30.6   Timing Analysis

Static timing analysis is the most commonly used approach. In static timing analysis, we calculate the delay from each input to each output of all devices. The delays are added up along each path through circuit to get the critical path through the design and hence the fastest design speed.

This works as long as there are no cycles in the circuit, however in these cases the analysis becomes less easy. Design software allows you to break cycles at registers to handle feedback if this is the case.

As in any timing analysis, the designer can trade off some accuracy for run time. Digital simulation software such as Modelism or Verilog will give fast results, but will use approximate models of timing, whereas analog simulation tools like SPICE will give more accurate numbers, but take much longer to run.

30.7   Design Pitfalls

The most common mistake that inexperienced designers make is simply making things too complex. The best approach to successful design is to keep the design elements simple, and the easiest way to manage that is efficient use of hierarchy.

The second mistake that is closely related to design complexity is not testing enough. It is vital to ensure that all aspects of the design are adequately tested. This means not only carrying out basic functional testing, but also systematic testing, and checking for redundant states and potential error states.

Another common pitfall is to use multiple clocks unnecessarily. Multiple clocks can create timing related bugs that are transient or hardware dependent. They can also occur in hardware and yet be missed by simulation.

30.8 VHDL Issues for FPGA Design

30.8.1 Initialization

Any default values of signals and variables are ignored. This means that you must ensure that synchronous (or asynchronous) sets and resets must be used on all flip-flops to ensure a stable starting condition. Remember that synthesis tools are basically stupid and follows a basic set of rules that may not always result in the hardware that you expect.

30.8.2 Floating-Point Numbers and Operations

Data types using floating-point are currently not supported by synthesis software tools. They generally require 32 bits and the requisite hardware is just too large for most FPGA and ASIC platforms.

30.9 Summary

This chapter has introduced the practical aspect of developing test benches and validating VHDL models using simulation. This is an often overlooked skill in VHDL (or any hardware description language) and is vital to ensuring correct behavior of the final implemented design. We have also introduced the concept of design synthesis and highlighted the problem of not only ensuring that a design simulates correctly, but also how we can make sure that the design will synthesize to the target technology and still operate correctly with practical delays and parasitics. Finally, we have raised some of the practical implementation issues and potential problems that can occur with real designs.

An important concept useful to define here is the difference between validation and verification. The terms are often confused, leading to problems in the final design and meeting a specification. Validation is the task of ensuring that the design is "doing the right thing." If the specification asks for a low-pass filter, then we must implement a low-pass filter to have a valid design. We can even be more specific and state that the design must perform within a constraint. Verification, on the other hand, is much more specific and can be stated as "doing the right thing *right*" In other words, verification is ensuring that not only does our design do what is required functionally, but in addition it must meet ALL the criteria defined by the specification, preferably with some head-room to ensure that the design will operate to the specification under all possible operating conditions.

References

Mano, M. Morris, *Digital Design*, Prentice Hall
This is a good background text for digital design and computer design. A particularly useful aspect for those designing embedded processors is the section of the book that discusses the difference between high level languages, assembly language and machine code and then develops that into a design methodology. For anyone starting out with processor design this is a very useful text. Mano also has a related book called *Computer System Architecture* that has more detail in this area and is equally useful.

Maxfield, Clive, Design Warriors Guide to FPGAs, Elsevier
This is an excellent introduction to the field of FPGAs. It introduces the main concepts in designing with FPGAs as the platform and does not get into low level details of VHDL or Verilog, but does have a balance between high level design issues and low level details. This is especially useful for the student who needs to know how FPGAs work and also for engineers who need a "heads up" on how FPGAs can be used in practice.

CHAPTER 31

Integrating Processors onto FPGAs

Peter Wilson

31.1 Introduction

This application example chapter concentrates on the key topic of integrating processors onto field-programmable gate array (FPGA) designs. This ranges from simple 8-bit microprocessors up to large IP processor cores that require an element of hardware-software co-design involved. This chapter will take the reader through the basics of implementing a behavioral-based microprocessor for evaluation of algorithms, through to the practicalities of structurally correct models that can be synthesized and implemented on an FPGA.

One of the major challenges facing hardware designers in the 21st century is the problem of hardware-software co-design. This has moved on from a basic partitioning mechanism based on standard hardware architectures to the current situation where the algorithm itself can be optimized at a compilation level for performance or power by implementing appropriately at different levels with hardware or software as required.

This aspect suits FPGAs perfectly, as they can handle fixed hardware architecture that runs software compiled onto memory, they can implement optimal hardware running at much faster rates than a software equivalent could, and there is now the option of configurable hardware that can adapt to the changing requirements of a modified environment.

31.2   A Simple Embedded Processor

31.2.1   Embedded Processor Architecture

A useful example of an embedded processor is to consider a generic microcontroller in the context of an FPGA platform. Take a simple example of a generic 8-bit microcontroller shown in Figure 31.1.

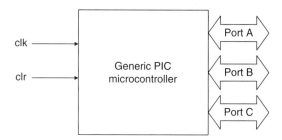

Figure 31.1: Simple microcontroller

As can be seen from Figure 31.1, the microcontroller is a "general-purpose microprocessor," with a simple clock (clk) and reset (clr), and three 8-bit ports (A, B and C). Within the microcontroller itself, there needs to be the following basic elements:

1. A control unit: This is required to manage the clock and reset of the processor, manage the data flow and instruction set flow, and control the port interfaces. There will also need to be a program counter (PC).

2. An arithmetic logic unit (ALU): a PIC will need to be able to carry out at least some rudimentary processing – carried out in the ALU.

3. An address bus.

4. A data bus.

5. Internal registers.

6. An instruction decoder.

7. A read only memory (ROM) to hold the program.

While each of these individual elements (1–6) can be implemented simply enough using a standard FPGA, the ROM presents a specific difficulty. If we implement a ROM as a set of registers, then obviously this will be hugely inefficient in an FPGA architecture. However, in most modern FPGA platforms, there are blocks of random access memory (RAM) on the FPGA that can be accessed and it makes a lot of sense to design a RAM block for use as a ROM by initializing it with the ROM values on reset and then using that to run the program.

This aspect of the embedded core raises an important issue, which is the reduction in efficiency of using embedded rather than dedicated cores. There is usually a compromise involved and in this case it is that the ROM needs to be implemented in a different manner, in this case with a hardware penalty. The second issue is what type of memory core to use. In an FPGA RAM, the memory can usually be organized in a variety of configurations to vary the depth (number of memory addresses required) and the width (width of the data bus). For example, a 512 address RAM block with an 8-bit address width would be equivalent to a 256 address RAM block with a 16-bit address width.

If the equivalent ROM is, say 12 bits wide and 256, then we can use the 256×16 RAM block and ignore the top four bits. The resulting embedded processor architecture could be of the form shown in Figure 31.2.

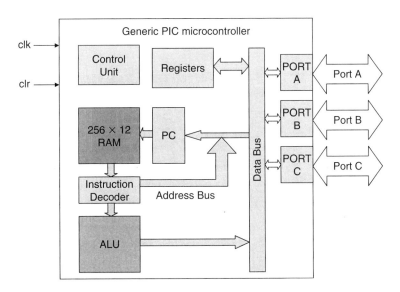

Figure 31.2: Embedded microcontroller architecture

31.2.2   Basic Instructions

When we program a microprocessor of any type, there are three different ways of representing the code that will run on the processor. These are machine code (1's and 0's), assembler (low-level instructions such as LOAD, STORE, ...) and high-level code (such as C, Fortran or Pascal). Regardless of the language used, the code will always be compiled or assembled into machine code at the lowest level for programming into memory. High-level code (e.g., C) is compiled and assembler code is assembled (as the name suggests) into machine code for the specific platform.

Clearly a detailed explanation of a compiler is beyond the scope of this book, but the same basic process can be seen in an assembler and this is useful to discuss in this context.

Every processor has a basic "Instruction Set" which is simply the list of functions that can be run in a program on the processor. Take the simple example of the following pseudocode expression:

$$b = a + 2$$

In this example, we are taking the variable a and adding the integer value 2 to it, and then storing the result in the variable b. In a processor, the use of a variable is simply a memory location that stores the value, and so to load a variable we use an assembler command as follows:

LOAD a

What is actually going on here? Whenever we retrieve a variable value from memory, the implication is that we are going to put the value of the variable in the register called the *accumulator* (ACC). The command 'LOAD a' could be expressed in natural language as 'LOAD the value of the memory location denoted by a into the accumulator register ACC'.

The next stage of the process is to add the integer value 2 to the accumulator. This is a simple matter, as instead of an address, the value is simply added to the current value stored in the accumulator. The assembly language command would be something like:

ADD #x02

Notice that we have used the x to denote a hexadecimal number. If we wished to add a variable, say called c, then the command would be the same, except that it would use the address c instead of the absolute number. The command would therefore be:

ADD c

Now we have the value of a + 2 stored in the accumulator register (ACC). This could be stored in a memory location, or put onto a port (e.g. PORT A). It is useful to notice that for a number we use the key character # to indicate that we are adding the value and not using the argument as the address.

In the pseudocode example, we are storing the result of the addition in the variable called *b*, so the command would be something like this:

STORE b

While this is superficially a complete definition of the instruction set requirements, there is one specific design detail that has to be decided on for any processor. This is the number of instructions and the data bus size. If we have a set of instructions with the number of instructions denoted by I, then the number of bits in the opcode (n) must conform to the following rule:

$$2^n \leq I \tag{31.1}$$

In other words, the number of bits provides the number of unique different codes that can be defined, and this defines the size of the instruction set possible. For example, if $n = 3$, then with three bits there are eight possible unique opcodes, and so the maximum size of the instruction set is eight.

31.2.3 Fetch Execute Cycle

The standard method of executing a program in a processor is to store the program in memory and then follow a strict sequence of events to carry out the instructions. The first stage is to use the PC to increment the program line, this then calls up the next command from memory in the correct order, and then the instruction can be loaded into the appropriate register for execution. This is called the *fetch execute cycle*.

What is happening at this point? First the contents of the PC is loaded into the memory address register (MAR). The data in the memory location are then retrieved and loaded into the memory data register (MDR). The contents of the MDR can then be transferred into the instruction register (IR). In a basic processor, the PC can then be incremented by one (or in fact this could take place immediately after the PC has been loaded into the MDR).

Once the opcode (and arguments if appropriate) are loaded, then the instruction can be executed. Essentially, each instruction has its own state machine and control path, which is linked to the IR and a sequencer that defines all the control signals required to move the data correctly around the memory and registers for that instruction. We will discuss registers in the next section, but in addition to the PC, IR and accumulator (ACC) mentioned already, we require two memory registers as a minimum, the MDR and MAR.

For example, consider the simple command LOAD a, from the previous example. What is required to actually execute this instruction? First, the opcode is decoded and this defines that the command is a 'LOAD' command. The next stage is to identify the address. As the command has not used the # symbol to denote an absolute address, this is stored in the variable a. The next stage, therefore is to load the value in location a into the MDR, by setting MAR = a and then retrieving the value of a from the RAM. This value is then transferred to the accumulator (ACC).

31.2.4 Embedded Processor Register Allocation

The design of the registers partly depends on whether we wish to "clone" a PIC device or create a modified version that has more custom behavior. In either case there are some mandatory registers that must be defined as part of the design. We can assume that we need an accumulator (ACC), a program pounter (PC), and the three input/output ports (PORTA, PORTB, PORTC). Also, we can define the IR, MAR, and MDR.

In addition to the data for the ports, we need to have a definition of the port direction and this requires three more registers for managing the tristate buffers into the data bus to and from the ports (DIRA, DIRB, DIRC). In addition to this, we can define a number (essentially arbitrary) of registers for general-purpose usage. In the general case the naming, order and numbering of registers does not matter, however, if we intend to use a specific device as a template, and perhaps use the same bit code, then it is vital that the registers are configured in exactly the same way as the original device and in the same order.

In this example, we do not have a base device to worry about, and so we can define the general-purpose registers (24 in all) with the names REG0 to REG23. In conjunction with the general-purpose registers, we need to have a small decoder to select the correct register and put the contents onto the data bus (F).

31.2.5 A Basic Instruction Set

In order for the device to operate as a processor, we must define some basic instructions in the form of an instruction set. For this simple example we can define some very basic instructions that will carry out basic program elements, ALU functions, memory functions. These are summarized in Table 31.1.

In this simple instruction set, there are 10 separate instructions. This implies that we need at least 4 bits to describe each of the instructions given in the table above. Given that we wish to have 8 bits for each data word, we need to have the ability to store the program memory in a ROM that has words of at least 12 bits wide. In order to cater for a greater number of instructions, and also to handle the situation for specification of different addressing modes (such as the difference between absolute numbers and variables), we can therefore suggest a 16-bit system for the program memory.

Table 31.1: Basic instruction set

Command	Description
LOAD arg	This command loads an argument into the accumulator. If the argument has the prefix # then it is the absolute number, otherwise it is the address and this is taken from the relevant memory address. Examples: LOAD #01 LOAD abc
STORE arg	This command stores an argument from the accumulator into memory. If the argument has the prefix # then it is the absolute address, otherwise it is the address and this is taken from the relevant memory address. Examples: STORE #01 STORE abc
ADD arg	This command adds an argument to the accumulator. If the argument has the prefix # then it is the absolute number, otherwise it is the address and this is taken from the relevant memory address. Examples: ADD #01 ADD abc

(Continued)

Table 31.1: Basic instruction set (Cont'd)

Command	Description
NOT	This command carries out the NOT function on the accumulator.

AND arg	This command ands an argument with the accumulator. If the argument has the prefix # then it is the absolute number, otherwise it is the address and this is taken from the relevant memory address.
	
	Examples:
	
	    AND #01
	
	    AND abc

OR arg	This command ors an argument with the accumulator. If the argument has the prefix # then it is the absolute number, otherwise it is the address and this is taken from the relevant memory address.
	
	Examples:
	
	    OR #01
	
	    OR abc

XOR arg	This command xors an argument with the accumulator. If the argument has the prefix # then it is the absolute number, otherwise it is the address and this is taken from the relevant memory address.
	
	Examples:
	
	    XOR #01
	
	    XOR abc

INC	This command carries out an increment by one on the accumulator.

SUB arg	This command subtracts an argument from the accumulator. If the argument has the prefix # then it is the absolute number, otherwise it is the address and this is taken from the relevant memory address.
	
	Examples:
	
	    SUB #01
	
	    SUB abc

BRANCH arg	This command allows the program to branch to a specific point in the program. This may be very useful for looping and program flow. If the argument has the prefix # then it is the absolute number, otherwise it is the address and this is taken from the relevant memory address.
	
	Examples:
	
	    BRANCH #01
	
	    BRANCH abc

Notice that at this stage there are no definitions for port interfaces or registers. We can extend the model to handle this behavior later.

31.2.6 Structural or Behavioral?

So far in the design of the simple microprocessor, we have not specified details beyond a fairly abstract structural description of the processor in terms of registers and busses. At this stage we have a decision about the implementation of the design with regard to the program and architecture.

One option is to take a program (written in assembly language) and simply convert this into a state machine that can easily be implemented in a VHDL model for testing out the algorithm. Using this approach, the program can be very simply modified and recompiled based on simple rules that restrict the code to the use of registers and techniques applicable to the processor in question. This can be useful for investigating and developing algorithms, but is more ideal than the final implementation as there will be control signals and delays due to memory access in a processor plus memory configuration, that will be better in a dedicated hardware design.

Another option is to develop a simple model of the processor that does have some of the features of the final implementation of the processor, but still uses an assembly language description of the model to test. This has advantages in that no compilation to machine code is required, but there are still not the detailed hardware characteristics of the final processor architecture that may cause practical issues on final implementation.

The third option is to develop the model of the processor structurally and then the machine code can be read in directly from the ROM. This is an excellent approach that is very useful for checking both the program and the possible quirks of the hardware/ software combination as the architecture of the model reflects directly the structure of the model to be implemented on the FPGA.

31.2.7 Machine Code Instruction Set

In order to create a suitable instruction set for decoding instructions for our processor, the assembly language instruction set needs to have an equivalent machine code

Table 31.2: Machine code instruction set

Command	Opcode (Binary)
LOAD arg	0000
STORE arg	0001
ADD arg	0010
NOT	0011
AND arg	0100
OR arg	0101
XOR arg	0110
INC	0111
SUB arg	1000
BRANCH arg	1001

instruction set that can be decoded by the sequencer in the processor. The resulting opcode/instruction table is given in Table 31.2.

31.2.8 Structural Elements of the Microprocessor

Taking the abstract design of the microprocessor given in Figure 31.2 we can redraw with the exact registers and bus configuration as shown in the structural diagram in Figure 31.3. Using this model we can create separate VHDL models for each of the blocks that are connected to the internal bus and then design the control block to handle all the relevant sequencing and control flags to each of the blocks in turn.

Before this can be started, however, it makes sense to define the basic criteria of the models and the first is to define the basic type. In any digital model (as we have seen elsewhere in this book) it is sensible to ensure that data can be passed between standard models and so in this case we shall use the std_logic_1164 library that is the standard for digital models.

In order to use this library, each signal shall be defined as of the basic type std_logic and also the library ieee.std_logic_1164.all shall be declared in the header of each of the models in the processor.

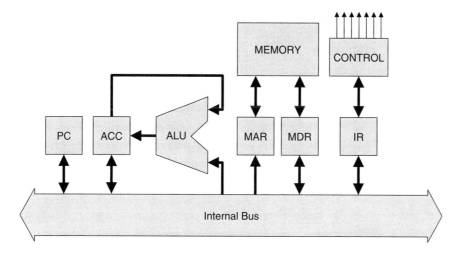

Figure 31.3: Structural model of the microprocessor

Finally, each block in the processor shall be defined as a separate block for implementation in VHDL.

31.2.9 Processor Functions Package

In order to simplify the VHDL for each of the individual blocks, a set of standard functions have been defined in a package called *processor_functions*. This is used to define useful types and functions for this set of models. The VHDL for the package is given below:

```
Library ieee;
Use ieee.std_logic_1164.all;

Package processor_functions is
    Type opcode is (load, store, add, not, and, or,
        xor, inc, sub, branch);
    Function Decode (word : std_logic_vector) return
        opcode;
    Constant n : integer := 16;
    Constant oplen : integer := 4;
    Type memory_array is array (0 to 2** (n-oplen-1)
        of Std_logic_vector(n-1 downto 0);
    Constant reg_zero : unsigned (n-1 downto 0) :=
        (others => '0' );
End package processor_functions;
```

```
Package body processor_functions is
    Function Decode (word : std_logic_vector) return
       opcode is
            Variable opcode_out : opcode;
    Begin
      Case word(n-1 downto n-oplen-1) is
            When "0000" => opcode_out : = load;
            When "0001" => opcode_out : = store;
            When "0010" => opcode_out : = add;
            When "0011" => opcode_out : = not;
            When "0100" => opcode_out : = and;
            When "0101" => opcode_out : = or;
            When "0110" => opcode_out : = xor;
            When "0111" => opcode_out : = inc;
            When "1000" => opcode_out : = sub;
            When "1001" => opcode_out : = branch;
            When others => null;
      End case;
      Return opcode_out;
  End function decode;
End package body processor_functions;
```

31.2.10   The PC

The PC needs to have the system clock and reset connections, the system bus (defined as inout so as to be readable and writable by the PC register block). In addition, there are several control signals required for correct operation. The first is the signal to increment the PC (PC_inc), the second is the control signal load the PC with a specified value (PC_load) and the final is the signal to make the register contents visible on the internal bus (PC_valid). This signal ensures that the value of the PC register will appear to be high impedance ('Z') when the register is not required on the processor bus. The system bus (PC_bus) is defined as a std_logic_vector, with direction inout to ensure the ability to read and write. The resulting VHDL entity is given below:

```
library ieee;
use ieee.std_logic_1164.all;
entity pc is
    Port (
        Clk : IN std_logic;
        Nrst : IN std_logic;
        PC_inc : IN std_logic;
```

```
            PC_load : IN std_logic;
            PC_valid : IN std_logic;
            PC_bus : INOUT std_logic_vector(n-1 downto 0)
        );
    End entity PC;
```

The architecture for the PC must handle all of the various configurations of the PC control signals and also the communication of the data into and from the internal bus correctly. The PC model has an asynchronous part and a synchronous section. If the PC_valid goes low at any time, the value of the PC_bus signal should be set to 'Z' across all of its bits. Also, if the reset signal goes low, then the PC should reset to zero.

The synchronous part of the model is the increment and load functionality. When the clk rising edge occurs, then the two signals PC_load and PC_inc are used to define the function of the counter. The precedence is that if the increment function is high, then regardless of the load function, then the counter will increment. If the increment function (PC_inc) is low, then the PC will load the current value on the bus, if and only if the PC_load signal is also high.

The resulting VHDL is given below:

```
architecture RTL of PC is
    signal counter : unsigned (n-1 downto 0);
begin
    PC_bus <= std_logic_vector(counter)
            when PC_valid = '1' else (others =>
                '2' );
    process (clk, nrst) is
    begin
        if nrst = '0' then
            count <= 0;
        elsif rising_edge(clk) then
            if PC_inc = '1' then
                count <= count + 1;
            else
                if PC_load = '1' then
                    count <= unsigned(PC_bus);
                end if;
            end if;
        end if;
    end process;
end architecture RTL;
```

31.2.11   The IR

The IR has the same clock and reset signals as the PC, and also the same interface to the bus (IR_bus) defined as a std_logic_vector of type INOUT. The IR also has two further control signals, the first being the command to load the IR (IR_load), and the second being to load the required address onto the system bus (IR_address). The final connection is the decoded opcode that is to be sent to the system controller. This is defined as a simple unsigned integer value with the same size as the basic system bus. The basic VHDL for the entity of the IR is given below:

```
library ieee;
use ieee.std_logic_1164.all;
use work.processor_functions.all;
entity ir is
    Port (
        Clk : IN std_logic;
        Nrst : IN std_logic;
        IR_load : IN std_logic;
        IR_valid : IN std_logic;
        IR_address : IN std_logic;
        IR_opcode : OUT opcode;
        IR_bus : INOUT std_logic_vector(n-1 downto 0)
    );
End entity IR;
```

The function of the IR is to decode the opcode in binary form and then pass to the control block. If the IR_valid is low, the bus value should be set to 'Z' for all bits. If the reset signal (nsrt) is low, then the register value internally should be set to all 0's.

On the rising edge of the clock, the value on the bus shall be sent to the internal register and the output opcode shall be decoded asynchronously when the value in the IR changes.

The resulting VHDL architecture is given below:

```
architecture RTL of IR is
    signal IR_internal : std_logic_vector (n-1 downto 0);
begin
    IR_bus <= IR_internal
        when IR_valid = '1' else (others => 'Z' );
    IR_opcode <= Decode(IR_internal);
    process (clk, nrst) is
    begin
```

```
            if nrst = '0' then
                    IR_internal <= (others => '0' );
            elsif rising_edge(clk) then
                    if IR_load = '1' then
                            IR_internal <= IR_bus;
                    end if;
            end if;
        end process;
    end architecture RTL;
```

In this VHDL, notice that we have used the predefined function Decode from the processor_functions package previously defined. This will look at the top four bits of the address given to the IR and decode the relevant opcode for passing to the controller.

31.2.12   The Arithmetic and Logic Unit

The *arithmetic and logic unit* (ALU) has the same clock and reset signals as the PC, and also the same interface to the bus (ALU_bus) defined as a std_logic_vector of type INOUT. The ALU also has three further control signals, which can be decoded to map to the eight individual functions required of the ALU. The ALU also contains the accumulator (ACC) which is a std_logic_vector of the size defined for the system bus width. There is also a single-bit output ALU_zero, which goes high when all the bits in the accumulator are zero.

The basic VHDL for the entity of the ALU is given below:

```
library ieee;
use ieee.std_logic_1164.all;
use work.processor_functions.all;
entity alu is
    Port (
        Clk : IN std_logic;
        Nrst : IN std_logic;
        ALU_cmd : IN std_logic_vector(2 downto 0);
        ALU_zero : OUT std_logic;
        ALU_valid : IN std_logic;
        ALU_bus : INOUT std_logic_vector(n-1 downto 0)
    );
End entity alu;
```

The function of the ALU is to decode the ALU_cmd in binary form and then carry out the relevant function on the data on the bus, and the current data in the accumulator. If the ALU_valid is low, the bus value should be set to 'Z' for all bits. If the reset signal (nsrt) is low, then the register value internally should be set to all 0's.

On the rising edge of the clock, the value on the bus shall be sent to the internal register and the command shall be decoded.

The resulting VHDL architecture is given below:

```
architecture RTL of ALU is
      signal ACC : std_logic_vector (n-1 downto 0);
begin
      ALU_bus <= ACC
            when ACC_valid '1' else (others => 'Z' );
      ALU_zero <= '1' when acc reg_zero else '0' ;
      process (clk, nrst) is
      begin
            if nrst = '0' then
                  ACC <= (others => '0' );
            elsif rising_edge(clk) then
                  case ACC_cmd is
                  -- Load the Bus value into the
                        accumulator
                  when "000" => ACC <= ALU_bus;
                  -- Add the ACC to the Bus value
                  When "001" => ACC <= add(ACC,ALU_bus);
                  -- NOT the Bus value
                  When "010" => ACC <= NOT ALU_bus;
                  -- OR the ACC to the Bus value
                  When "011" => ACC <= ACC or ALU_bus;
                  -- AND the ACC to the Bus value
                  When "100" => ACC <= ACC and ALU_bus;
                  -- XOR the ACC to the Bus value
                  When "101" => ACC <= ACC xor ALU_bus;
                  -- Increment ACC
                  When "110" => ACC <= ACC + 1;
                  -- Store the ACC value
                  When "111" => ALU_bus <= ACC;
            end if;
      end process;
end architecture RTL;
```

31.2.13 The Memory

The processor requires a RAM memory, with an address register (MAR) and a data register (MDR). There therefore needs to be a load signal for each of these registers: MDR_load and MAR_load. As it is a memory, there also needs to be an enable signal (M_en), and also a signal denote Read or Write modes (M_rw). Finally, the connection to the system bus is a standard inout vector as has been defined for the other registers in the microprocessor.

The basic VHDL for the entity of the memory block is given below:

```
library ieee;
use ieee.std_logic_1164.all;
use work.processor_functions.all;
entity memory is
        Port (
                Clk : IN std_logic;
                Nrst : IN std_logic;
                MDR_load : IN std_logic;
                MAR_load : IN std_logic;
                MAR_valid : IN std_logic;
                M_en : IN std_logic;
                M_rw : IN std_logic;
                MEM_bus : INOUT std_logic_vector(n-1
                    downto 0)
        );
End entity memory;
```

The memory block has three aspects. The first is the function that the memory address is loaded into the MAR. The second function is either reading from or writing to the memory using the MDR. The final function or aspect of the memory is to store the actual program that the processor will run. In the VHDL model, we will achieve this by using a constant array to store the program values.

The resulting basic VHDL architecture is given below:

```
architecture RTL of memory is
        signal mdr : std_logic_vector(wordlen-1 downto 0);
        signal mar : unsigned(wordlen-oplen-1 downto 0);
                    begin
        MEM_bus <= mdr
            when MEM_valid = '1' else (others => 'Z');
```

```
process (clk, nrst) is
        variable contents : memory_array;
        constant program : contents :=
        (
                0 => "0000000000000011",
                1 => "0010000000000100",
                2 => "0001000000000101",
                3 => "0000000000001100",
                4 => "0000000000000011",
                5 => "0000000000000000" ,
                Others => (others => '0' )
        );
Begin
        if nrst = '0' then
                mdr <= (others => '0' );
                mdr <= (others => '0' );
                contents := program;
        elsif rising_edge(clk) then
                if MAR_load = '1' then
                        mar <= unsigned(MEM_bus(n-oplen-
                                1 downto 0));
                elsif MDR_load = '1' then
                        mdr <= MEM_bus;
                elsif MEM_en = '1' then
                        if MEM_rw = '0' then
                                mdr <= contents(to_integer
                                        (mar));
                        else
                                mem(to_integer(mar))
                                  := mdr;
                        end if;
                end if;
        end if;
end process;
end architecture RTL;
```

We can look at some of the VHDL in a bit more detail and explain what is going on at this stage. There are two internal signals to the block, mdr and mar (the data and address, respectively). The first aspect to notice is that we have defined the MAR as an unsigned rather than as a std_logic_vector. We have done this to make indexing direct. The MDR remains as a std_logic_vector. We can use an integer directly, but an unsigned translates easily into a std_logic_vector.

```
signal mdr : std_logic_vector(wordlen-1 downto 0);
signal mar : unsigned(wordlen-oplen-1 downto 0);
```

The second aspect is to look at the actual program itself. We clearly have the possibility of a large array of addresses, but in this case we are defining a simple three line program:

$$c = a + b$$

The binary code is shown below:

```
0 => "0000000000000011",
1 => "0010000000000100",
2 => "0001000000000101",
3 => "0000000000001100",
4 => "0000000000000011",
5 => "0000000000000000",
Others => (others => '0')
```

For example, consider the line of the declared value for address 0. The 16 bits are defined as 0000000000000011. If we split this into the opcode and data parts we get the following:

```
Opcode  0000
Data    000000000011 (3)
```

In other words, this means LOAD the variable from address 3. Similarly, the second line is ADD from 4, and finally the third command is STORE in 5. In addresses 3, 4 and 5, the three data variables are stored.

31.2.14 Microcontroller: Controller

The operation of the processor is controlled in detail by the sequencer, or controller block. The function of this part of the processor is to take the current PC address, look up the relevant instruction from memory, move the data around as required, setting up all the relevant control signals at the right time, with the right values.

As a result, the controller must have the clock and reset signals (as for the other blocks in the design), a connection to the global bus and finally all the relevant control signals must be output. An example entity of a controller is given below:

```
library ieee;
use ieee.std_logic_1164.all;
use work.processor_functions.all;
entity controller is
        generic (
                n : integer := 16
        );
        Port (
                Clk : IN std_logic;
                Nrst : IN std_logic;
                IR_load : OUT std_logic;
                IR_valid : OUT std_logic;
                IR_address : OUT std_logic;
                PC_inc : OUT std_logic;
                PC_load : OUT std_logic;
                PC_valid : OUT std_logic;
                MDR_load : OUT std_logic;
                MAR_load : OUT std_logic;
                MAR_valid : OUT std_logic;
                M_en : OUT std_logic;
                M_rw : OUT std_logic;
                ALU_cmd : OUT std_logic_vector(2 downto 0);
                CONTROL_bus : INOUT std_logic_vector(n-1 downto 0)
        );
    End entity controller;
```

Using this entity, the control signals for each separate block are then defined, and these can be used to carry out the functionality requested by the program. The architecture for the controller is then defined as a basic state machine to drive the correct signals. The basic state machine for the processor is defined in Figure 31.4.

We can implement this using a basic VHDL architecture that implements each state using a new state type and a case statement to manage the flow of the state machine. The basic VHDL architecture is shown below and it includes the basic synchronous machine control section (reset and clock) the management of the next stage logic:

```
architecture RTL of controller is
        type states is
            (s0,s1,s2,s3,s4,s5,s6,s7,s8,s9,s10);
        signal current_state, next_state : states;
```

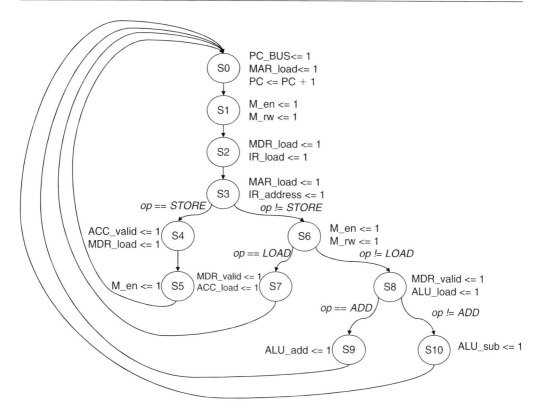

Figure 31.4: Basic processor controller state machine

```
begin
        state_sequence: process (clk, nrst) is
            if nrst = '0' then
                current_state <= s0;
            else
                if rising_edge(clk) then
                    current_state <=
                        next_state;
                end if;
            end if;
        end process state_sequence;

        state_machine : process ( present_state,
            opcode ) is
                -- state machine goes here
        End process state_machine;
end architecture;
```

You can see from this VHDL that the first process (state_sequence) manages the transition of the current_state to the next_state and also the reset condition. Notice that this is a synchronous machine and as such waits for the rising_edge of the clock, and that the reset is asynchronous. The second process (state_machine) waits for a change in the state or the opcode and this is used to manage the transition to the next state, although the actual transition itself is managed by the state_sequence process. This process is given in the VHDL below:

```
state_machine : process ( present_state,
    opcode ) is
begin
    -- Reset all the control signals
    IR_load <= '0' ;
    IR_valid <= '0' ;
    IR_address <= '0' ;
    PC_inc <= '0' ;
    PC_load <= '0' ;
    PC_valid <= '0' ;
    MDR_load <= '0' ;
    MAR_load <= '0' ;
    MAR_valid <= '0' ;
    M_en <= '0' ;
    M_rw <= '0' ;
    Case current_state is
    When s0 =>
        PC_valid <='1' ; MAR_load <='1' ;
        PC_inc <='1' ; PC_load <='1' ;
        Next_state <= s1;
    When s1 =>
        M_en <='1' ; M_rw <='1' ;
        Next_state <= s2;
    When s2 =>
        MDR_valid <='1' ; IR_load <='1' ;
        Next_state <= s3;
    When s3 =>
        MAR_load <='1' ; IR_address <='1' ;
        If opcode = STORE then
            Next_state <= s4;
        else
            Next_state <=s6;
        End if;
```

```
                    When s4 =>
                        MDR_load <='1'; ACC_valid <=' 1';
                        Next_state <= s5;
                    When s5 =>
                        M_en <= '1';
                        Next_state <= s0;
                    When s6 =>
                        M_en <='1'; M_rw <='1';
                        If opcode = LOAD then
                                Next_state <= s7;
                        else
                                Next_state <= s8;
                        End if;
                    When s7 =>
                        MDR_valid <='1'; ACC_load <='1';
                        Next_state <= s0;
                    When s8 =>
                        M_en<='1'; M_rw <='1';
                        If opcode = ADD then
                                Next_state <= s9;
                        else
                                Next_state <= s10;
                        End if;
                    When s9 =>
                        ALU_add <= '1';
                        Next_state <= s0;
                    When s10 =>
                        ALU_sub <= '1';
                        Next_state <= s0;
            End case;
    End process state_machine;
```

31.2.15 Summary of a Simple Microprocessor

Now that the important elements of the processor have been defined, it is a simple matter to instantiate them in a basic VHDL netlist and create a microprocessor using these building blocks. It is also a simple matter to modify the functionality of the processor by changing the address/data bus widths or extend the instruction set.

31.3   Soft Core Processors on an FPGA

While the previous example of a simple microprocessor is useful as a design exercise and helpful to gain understanding about how microprocessors operate, in practice most FPGA vendors provide standard processor cores as part of an embedded development kit that includes compilers and other libraries. For example, this could be the Microblaze core from Xilinx or the NIOS core supplied by Altera. In all these cases the basic idea is the same: that a standard configurable core can be instantiated in the design and code compiled using a standard compiler and downloaded to the processor core in question.

Each soft core is different and rather than describe the details of a particular case, in this section the general principles are covered and the reader is encouraged to experiment with the offerings from the FPGA vendors to see which suits their application the best.

In any soft core development system there are several key functions that are required to make the process easy to implement. The first is the system building function. This enables a core to be designed into a hardware system that includes memory modules, control functions, direct memory access (DMA) functions, data interfaces and interrupts. The second is the choice of processor types to implement. A basic NIOS II or similar embedded core will typically have a performance in the region of 100–200 MIPS, and the processor design tools will allow the size of the core to be traded off with the hardware resources available and the performance required.

31.4   Summary

The topic of embedded processors on FPGAs would be suitable for a complete book in itself. In this chapter the basic techniques have been described for implementing a simple processor directly on the FPGA and the approach for implementing soft cores on FPGAs have been introduced.

CHAPTER 32

Implementing Digital Filters in VHDL

Peter Wilson

32.1 Introduction

An important part of systems that interface to the "real world" is the ability to process sampled data in the digital domain. This is often called *sampled-data systems* (SDS) or operating in the Z-domain. Most engineers are familiar with the operation of filters in the Laplace or S-domain where a continuous function defines the characteristics of the filter and this is the digital domain equivalent to that.

For example, consider a simple RC circuit in the analog domain as shown in Figure 32.1. This is a low-pass filter function and can be represented using the Laplace notation shown in Figure 32.1.

This has the equivalent S-domain (or Laplace) function as follows:

$$L(s) = \frac{1}{1 + sCR}$$

This function is a low-pass filter because the Laplace operator s is equivalent to $j\omega$, where $w = 2\pi f$ (with f being the frequency). If f is zero (the DC condition), then the gain will be 1, but if the value of sCR is equal to 1, then the gain will be 0.5. This in dB is -3 dB and is the classical low-pass filter cut-off frequency.

In the digital domain, the s operation is replaced by Z. Z^{-1} is practically equivalent to a delay operator, and similar functions to the Laplace filter equations can be constructed for the digital, or Z-domain equivalent.

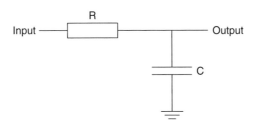

Figure 32.1: RC filter in the analog domain

There are a number of design techniques, many beyond the scope of this book (if the reader requires a more detailed introduction to the realm of digital filters, Cunningham's *Digital filtering: an introduction* is a useful starting point); however, it is useful to introduce some of the basic techniques used in practice and illustrate them with examples.

The remainder of this chapter will cover the introduction to the basic techniques and then demonstrate how these can be implemented using VHDL on field-programmable gate arrays (FPGAs).

32.2   Converting S-Domain to Z-Domain

The method of converting an S-domain equation for a filter to its equivalent Z-domain expression is done using the 'bilinear transform'. This is simply a method of expressing the equation in the S-domain in terms of Z. The basic approach is to replace each instance of s with its equivalent Z-domain notation and then rearrange into the most convenient form. The transform is called *bilinear*, as both the numerator and denominator of the expression are linear in terms of z.

$$s = \frac{z - 1}{z + 1}$$

If we take a simple example of a basic second-order filter we can show how this is translated into the equivalent Z-domain form:

$$H(s) = \frac{1}{s^2 + 2s + 1}$$

replace s with $(z - 1)/(z + 1)$:

$$H(Z) = \cfrac{1}{\left(\cfrac{(z-1)}{(z+1)}\right)^2 + 2\cfrac{(z-1)}{(z+1)} + 1}$$

$$H(Z) = \frac{(z+1)^2}{(z-1)^2 + (z-1)(z+1) + (z+1)^2}$$

$$H(Z) = \frac{z^2 + 2z + 1}{3z^2 + 1}$$

Now, the term $H(Z)$ is really the output $Y(Z)$ over the input $X(Z)$ and we can use this to express the Z-domain equation in terms of the input and output:

$$H(Z) = \frac{z^2 + 2z + 1}{3z^2 + 1}$$

$$\frac{Y(Z)}{X(Z)} = \frac{z^2 + 2z + 1}{3z^2 + 1}$$

$$3z^2Y(Z) + Y(Z) = z^2X(Z) + 2zX(Z) + X(Z)$$

This can then be turned into a sequence expression using delays (z is one delay, z^2 is two delays and so on) with the following result:

$$3z^2Y(Z) + Y(Z) = z^2X(Z) + 2zX(Z) + X(Z)$$
$$3y(n + 2) + y(n) = x(n+) + 2x(n + 1) + x(n)$$

This is useful because we are now expressing the Z-domain equation in terms of delay terms, and the final step is to express the value of $y(n)$ (the current output) in terms of past elements by reducing the delays accordingly (by 2 in this case):

$$3y(n + 2) + y(n) = x(n + 2) + 2x(n + 1) + x(n)$$
$$3y(n) + y(n - 2) = x(n) + 2x(n - 1) + x(n - 2)$$
$$y(n) + 1/3y(n - 2) = 1/3x(n) + 2/3x(n - 1) + 1/3x(n - 2)$$
$$y(n) = 1/3x(n) + 2/3x(n - 1) + 1/3x(n - 2) - 1/3y(n - 2)$$

The final design note at this point is to make sure that the design frequency is correct, for example the low-pass cut-off frequency. The frequencies are different between the

S- and Z-domain models, even after the bilinear transformation, and in fact the desired digital domain frequency must be translated into the equivalent S-domain frequency using a technique called *pre-warping*. This simple step translates the frequency from one domain to the other using the expression below:

$$\omega_c = \tan\left(\frac{\Omega_c T}{2}\right)$$

where Ω_c is the digital domain frequency, T is the sampling period of the Z-domain system and ω_c is the resulting frequency for the analog domain calculations.

Once we have obtained our Z-domain expressions, how do we turn this into practical designs? The next section will explain how this can be achieved.

32.3   Implementing Z-Domain Functions in VHDL

Z-domain functions are essentially digital in the time domain as they are discrete and sampled. The functions are also discrete in the amplitude axis, as the variables or signals are defined using a fixed number of bits in a real hardware system, whether this is integer, signed, fixed-number or floating-point, there is always a finite resolution to the signals. For the remainder of this chapter, signed arithmetic is assumed for simplicity and ease of understanding. This also essentially defines the number of bits to be used in the system. If we have 8 bits, the resolution is 1 bit and the range is −128−+127.

32.3.1   Gain Block

The first main Z-domain block is a simple gain block. This requires a single signed input, a single signed output and a parameter for the gain. This could be an integer or also a signed value. The VHDL model for a simple Z-domain gain block is given below:

```
Library ieee;
Use ieee.numeric_std.all;

    Entity zgain is
        Generic ( n : integer := 8;
                    gain : signed
        );
        Port (
                Zin : in signed (n-1 downto 0);
                Zout : out signed (n-1 downto 0)
        );
```

```
End entity zgain;
Architecture zdomain of zgain is
Begin
     p1 : process (zin)
          variable product : signed (2*n-1 downto 0);
     begin
          product := zin * gain;
          zout <= product (n-1 downto 0);
     end process p1;
End architecture zdomain;
```

We can test this with a simple testbench that ramps up the input and we can observe the output being changed in turn:

```
library ieee;
use ieee.std_logic_1164.all;
use ieee.numeric_std.all;

entity tb is
end entity tb;

architecture testbench of tb is
     signal clk : std_logic := '0';
     signal dir : std_logic := '0';
     signal zin : signed (7 downto 0) := X"00";
     signal zout : signed (7 downto 0) := X"00";

     component zgain
     generic (
          n : integer := 8;
          gain :signed := X"02"
     );
     port (
          signal zin : in signed(n-1 downto 0);
          signal zout : out signed(n-1 downto 0)
     );
     end component;
     for all : zgain use entity work.zgain;
begin
     clk <= not clk after 1 us;

     DUT : zgain generic map ( 8, X"02" ) port map (zin, zout);

     p1 : process (clk)
```

```
        begin
                zin <= zin + 1;
        end process p1;
  end architecture testbench;
```

Clearly, this model has no error checking or range checking and the obvious problem with this type of approach is that of overflow. For example, if we multiply the input (64) by a gain of 2, we will get 128, but that is the sign bit, and so the result will show −128! This is an obvious problem with this simplistic model and care must be taken to ensure that adequate checking takes place in the model.

32.3.2   Sum and Difference

Using this same basic approach, we can create sum and difference models which are also essential building blocks for a Z-domain system. The sum model VHDL is shown below:

```
        Library ieee;
        Use ieee.numeric_std.all;

        Entity zsum is
            Generic ( n : integer := 8
            );
            Port (
                    Zin1 : in signed (n-1 downto 0);
                    Zin2 : in signed (n-1 downto 0);
                    Zout : out signed (n-1 downto 0)
            );
        End entity zsum;

        Architecture zdomain of zsum is
        Begin
            p1 : process(zin)
                    variable zsum : signed (2*n-1 downto 0);
            begin
                    zsum := zin1 + zin2;
                    zout <= zsum (n-1 downto 0);
            end process p1;
  End architecture zdomain;
```

Despite the potential for problems with overflow, both of the models shown have the internal variable that is twice the number of bits required, and so this can take care of any possible overflow internal to the model, and in fact checking could take place prior to the final assignment of the output to ensure the data is correct. The difference model is almost identical to the sum model except that the difference of zin1 and zin2 is computed.

32.3.3 Division Model

A useful model for scaling numbers simply in the Z-domain is the division by 2 model. This model simply shifts the current value in the input to the right by 1 bit—hence, giving a division by 2. The model could easily be extended to shift right by any number of bits, but this simple version is very useful by itself. The VHDL for the model relies on the logical shift right operator (SRL) which not only shifts the bits right (losing the least significant bit) but adding a zero at the most significant bit. The resulting VHDL is shown below for this specific function:

```
zout <= zin srl 1;
```

The unit shift can be replaced by any integer number to give a shift of a specific number of bits. For example, to shift right by 3 bits (effectively a divide-by-8) would have the following VHDL:

```
zout <= zin srl 3;
```

The complete division by 2 model is given below:

```
Library ieee;
Use ieee.numeric_std.all;

Entity zdiv2 is
     Generic ( n : integer := 8
     );
     Port (
           Zin : in signed (n-1 downto 0);
           Zout : out signed (n-1 downto 0)
     );
End entity zdiv2;

Architecture zdomain of zdiv2 is
Begin
     zout <= zin srl 1;
End architecture zdomain;
```

In order to test the model a simple test circuit that ramps up the input is used and this is given below:

```
library ieee;
use ieee.std_logic_1164.all;
use ieee.numeric_std.all;

entity tb is
end entity tb;

architecture testbench of tb is

      signal clk : std_logic := '0';
      signal dir : std_logic := '0';
      signal zin : signed (7 downto 0) := X"00";
      signal zout : signed (7 downto 0) := X"00";

      component zdiv2
      generic (
            n : integer := 8
      );
      port (
            signal zin : in signed (n-1 downto 0);
            signal zout : out signed (n-1 downto 0)
      );
      end component;
      for all : zdiv2 use entity work.zdiv2;

begin
      clk <= not clk after 1 us;

      DUT : zdiv2 generic map (8) port map (zin, zout);

      p1 : process (clk)
      begin
            zin <= zin + 1;
      end process p1;
end architecture testbench;
```

The behavior of the model is useful to review, if the input is X"03" (Decimal 3), binary '00000011' and the number is right shifted by 1, then the resulting binary number will be '00000001' (X"01" or decimal 1), in other words this operation always rounds down. This has obvious implications for potential loss of accuracy and the operation is skewed downward, which has again, implications for how numbers will be treated using this operator in a more complex circuit.

32.3.4 Unit Delay Model

The final basic model is the unit delay model (zdelay). This has a clock input (clk) using a std_logic signal to make it simple to interface to standard digital controls. The output is simply a one clock cycle delayed version of the input.

```
Library ieee;
use ieee.std_logic_1164.all;
Use ieee.numeric_std.all;

Entity zdelay is
      Generic ( n : integer := 8 );
      Port (
            clk : in std_logic;
            Zin : in signed (n-1 downto 0);
            Zout : out signed (n-1 downto 0) := (others
               => '0' )
      );
End entity zdelay;

Architecture zdomain of zdelay is
      signal lastzin : signed (n-1 downto 0) := (others => '0' );
   Begin
      p1 : process(clk)
      begin
            if rising_edge(clk) then
                  zout <= lastzin;
                  lastzin <= zin;
            end if;
      end process p1;
End architecture zdomain;
```

Notice that the output zout is initialized to all zeros for the initial state, otherwise "don't care" conditions can result that propagate across the complete model.

32.4 Basic Low-Pass Filter Model

We can put these elements together in simple models that implement basic filter blocks in any configuration we require, as always taking care to ensure that overflow errors are checked for in practice.

To demonstrate this, we can implement a simple low-pass filter using the basic block diagram shown in Figure 32.2.

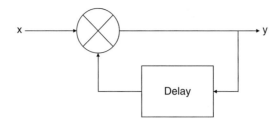

Figure 32.2: Simple Z-domain low-pass filter

We can create a simple test circuit that uses the individual models we have already shown for the sum and delay blocks and apply a step change and observe the response of the filter to this stimulus. (Clearly, in this case, with unity-gain the filter exhibits positive feedback and so to ensure the correct behavior we use the divide by 2 model zdiv2 in both the inputs to the sum block to ensure gain of 0.5 on both. These are not shown in the Figure 32.2.) The resulting VHDL model is shown below (note the use of the zdiv2 model):

```
library ieee;
use ieee.std_logic_1164.all;
use ieee.numeric_std.all;

entity tb is
end entity tb;

architecture testbench of tb is
    signal clk : std_logic := '0' ;
    signal x : signed (7 downto 0):= X"00";
    signal y : signed (7 downto 0):= X"00";
    signal y1 : signed (7 downto 0):= X"00";
    signal yd : signed (7 downto 0):= X"00";
    signal yd2 : signed (7 downto 0):= X"00";
    signal x2 : signed (7 downto 0):= X"00";
component zsum
generic (
        n : integer : = 8
);
port (
    signal zin1 : in signed(n-1 downto 0);
    signal zin2 : in signed(n-1 downto 0);
    signal zout : out signed(n-1 downto 0)
);
end component;
```

```
for all : zsum use entity work.zsum;

component zdiff
generic (
      n : integer := 8
);
port (
      signal zin1 : in signed(n-1 downto 0);
      signal zin2 : in signed(n-1 downto 0);
      signal zout : out signed(n-1 downto 0)
);
end component;
for all : zdiff use entity work.zdiff;

      component zdiv2
      generic (
            n : integer := 8
      );
      port (
            signal zin : in signed(n-1 downto 0);
            signal zout : out signed(n-1 downto 0)
      );
      end component;
      for all : zdiv2 use entity work.zdiv2;

      component zdelay
      generic (
            n : integer := 8
      );
      port (
            signal clk : in std_logic;
            signal zin : in signed(n-1 downto 0);
            signal zout : out signed(n-1 downto 0)
      );
      end component;
      for all : zdelay use entity work.zdelay;
begin
      clk <= not clk after 1 us;

      GAIN1 : zdiv2 generic map (8) port map (x, x2);
      GAIN2 : zdiv2 generic map (8) port map (yd, yd2);
      SUM1 : zsum generic map (8) port map (x2, yd2, y);
      D1 : zdelay generic map (8) port map (clk, y, yd);
      x <= X"00", X"0F" after 10 us;
end architecture testbench;
```

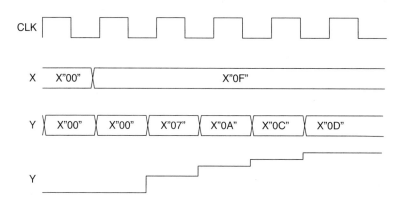

Figure 32.3: Basic low-pass filter simulation waveforms

The test circuit applies a step change of X"00" to X"0F" after 10 μs, and this results in the filter response. We can show this graphically in Figure 32.3 with the output in both hexadecimal and "analog" form for illustration.

It is interesting to note the effect of using the zdiv2 function on the results. With the input of 0F (binary 00001111) we lose the LSB when we divide by 2, giving the resulting input to the sum block of 00000111 (7), which added together with the division of the output gives a total of 14 as the maximum possible output from the filter. In fact, the filter gives an output of X"0D" or binary 00001101, which is two down from the theoretical maximum of X"0F" and this highlights the practical difficulties when using a "coarse" approximation technique for numerical work rather than a fixed- or floating-point method. On the other hand, it is clearly a simple and effective method of implementing a basic filter in VHDL.

Elsewhere in this book, the use of fixed- and floating-point numbers are discussed, as is the use of multiplication for more exact calculations and for practical filter design. Where higher accuracy is required, then it is likely that both these methods would be used. There may be situations, however, where it is simply not possible to use these advanced techniques, particularly a problem when space is at a premium on the FPGA and in these cases, the simple approach described in this chapter will be required.

There are numerous texts on more advanced topics in digital filter design, and these are beyond the scope of this book, but it is useful to introduce some key concepts at this stage of the two main types of digital filter in common usage today. These are the recursive (or Infinite Impulse Response – IIR) filters and nonrecursive (or Finite Impulse Response – FIR) filters.

32.5   FIR Filters

FIR filters are characterized by the fact that they use only delayed versions of the input signal to filter the input to the output. For example, if we take the expression for a general FIR filter below, we can see that the output is a function of a series of delayed, scaled versions of the input:

$$y = \sum_{i=0}^{n} A_i x[i]$$

where A_i is the scale factor for the ith delayed version of the input. We can represent this graphically in the diagram shown in Figure 32.4. We can implement this model using the basic building blocks described in this chapter of gain, division, sums and delays to develop block based models for such filters. As noted in the previous section, it is

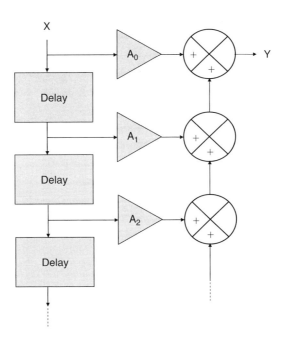

Figure 32.4: FIR Filter Schematic

important to ensure that for higher accuracy filters, fixed- or floating-point arithmetic is required and also the use of multipliers for added accuracy is preferable in most cases to that of simple gain and division blocks as described previously in this chapter.

32.6   IIR Filters

IIR filters are characterized by the fact that they use delayed versions of the input signal and fed back and delayed version of the output signal to filter the input to the output. For example, if we take the expression for a general IIR filter below, we can see that the output is a function of a series of delayed, scaled versions of the input and output:

$$y = \sum_{i=0}^{n} \frac{A_i x[i]}{B_i y[i]}$$

where Ai is the scale factor for the ith delayed version of the input and B_i is the scale factor for the ith delayed version of the output. This is obviously very similar to the FIR example previously and can be built up using the same basic elements. If we consider the simple example earlier in this chapter, it can be seen that this is in fact a simple first-order (single delay) IIR filter, with no delayed versions of the input and a single delayed version of the output.

32.7   Summary

This chapter has introduced the concepts of implementing basic digital filters in VHDL and has given examples of both the building blocks and constructed filters for implementation on an FPGA platform. The general concepts of FIR and IIR filters have been introduced so that the reader can implement the topology and type of filter appropriate for their own application.

CHAPTER 33

Microprocessor and Microcontroller Overview

Mike Tooley

Many of today's complex electronic systems are based on the use of a microprocessor or microcontroller. Such systems comprise hardware that is controlled by software. If it is necessary to change the way that the system behaves it is the software (rather than the hardware) that is changed.

In this chapter we provide an introduction to microprocessors and explain, in simple terms, both how they operate and how they are used. We shall start by explaining some of the terminology that is used to describe different types of system that involve the use of a microprocessor or a similar device.

33.1 Microprocessor Systems

Microprocessor systems are usually assembled on a single PCB comprising a microprocessor CPU together with a number of specialized support chips. These very large scale integrated (VLSI) devices provide input and output to the system, control and timing as well as storage for programs and data.

Typical applications for microprocessor systems include the control of complex industrial processes. Typical examples are based on families of chips such as the Z80CPU plus Z80PIO, Z80CTC, and Z80SIO.

33.2   Single-Chip Microcomputers

A single-chip microcomputer is a complete computer system (comprising CPU, RAM and ROM etc.) in a single VLSI package. A single-chip microcomputer requires very little external circuitry in order to provide all of the functions associated with a complete computer system (but usually with limited input and output capability).

Single-chip microcomputers may be programmed using in-built programmable memories or via external memory chips. Typical applications of single-chip microcomputers include computer printers, instrument controllers, and displays. A typical example is the Z84C.

33.3   Microcontrollers

A microcontroller is a single-chip microcomputer that is designed specifically for control rather than general-purpose applications. They are often used to satisfy a particular control requirement, such as controlling a motor drive. Single-chip microcomputers, on the other hand, usually perform a variety of different functions and may control several processes at the same time.

Typical applications include control of peripheral devices such as motors, drives, printers, and minor subsystem components. Typical examples are the Z86E, 8051, 68705 and 89C51.

33.4   Microprocessor Systems

The basic components of any microprocessor system (see Figure 33.1) are:

(a)   a central processing unit (CPU);

(b)   a memory, comprising both 'read/write' and 'read only' devices (commonly called *RAM* and *ROM*, respectively);

(c)   a means of providing input and output (I/O). For example, a keypad for input and a display for output.

In a microprocessor system the functions of the CPU are provided by a single very large scale integrated (VLSI) microprocessor chip (see Figure 33.2). This chip is equivalent to many thousands of individual transistors. Semiconductor devices are also used to provide the read/write and read-only memory. Strictly speaking, both types of memory

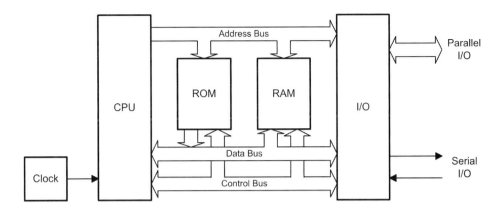

Figure 33.1: Block diagram of a microprocessor system

Figure 33.2: A Z80 microprocessor

permit "random access" since any item of data can be retrieved with equal ease regardless of its actual location within the memory. Despite this, the term *RAM* has become synonymous with semiconductor read/write memory.

The basic components of the system (CPU, RAM, ROM, and I/O) are linked together using a multiple-wire connecting system known as a *bus* (see Figure 33.1). Three different buses are present, these are:

(a) the *address bus* used to specify memory locations;

(b)   the *data bus* on which data is transferred between devices; and

(c)   the *control bus* which provides timing and control signals throughout the system.

The number of individual lines present within the address bus and data bus depends upon the particular microprocessor employed. Signals on all lines, no matter whether they are used for address, data, or control, can exist in only two basic states: logic 0 (*low*) or logic 1 (*high*). Data and addresses are represented by *binary numbers* (a sequence of 1s and 0s) that appear respectively on the data and address bus.

Many microprocessors designed for control and instrumentation applications make use of an 8-bit data bus and a 16-bit address bus. Others have data and address buses which can operate with as many as 128 bits at a time.

The largest binary number that can appear on an 8-bit data bus corresponds to the condition when all eight lines are at logic 1. Therefore the largest value of data that can be present on the bus at any instant of time is equivalent to the binary number 11111111 (or 255). Similarly, the highest address that can appear on a 16-bit address bus is 1111111111111111 (or 65,535). The full range of data values and addresses for a simple microprocessor of this type is thus:

Data	from	00000000
	to	11111111
Addresses	from	0000000000000000
	to	1111111111111111

33.4.1   Data Representation

Binary numbers—particularly large ones—are not very convenient. To make numbers easier to handle we often convert binary numbers to *hexadecimal* (base 16). This format is easier for mere humans to comprehend and offers the advantage over denary (base 10) in that it can be converted to and from binary with ease. The first sixteen numbers in binary, denary, and hexadecimal are shown in the table below. A single hexadecimal character (in the range zero to F) is used to represent a group of four binary digits (bits). This group of four bits (or single hex character) is sometimes called a *nibble*.

A *byte* of data comprises a group of eight bits. Thus a byte can be represented by just two hexadecimal (hex) characters. A group of sixteen bits (a word) can be represented by four hex characters, thirty-two bits (a double word by eight hex characters, and so on).

The value of a byte expressed in binary can be easily converted to hex by arranging the bits in groups of four and converting each nibble into hexadecimal using Table 33.1.

Note that, to avoid confusion about whether a number is hexadecimal or decimal, we often place a symbol before a hexadecimal number or add an H to the end of the number. For example, 64 means decimal "sixty-four"; whereas, $64 means hexadecimal "six-four", which is equivalent to decimal 100. Similarly, 7FH means hexadecimal "seven-F" which is equivalent to decimal 127.

Table 33.1: Binary, denary and hexadecimal

Binary (base 2)	Denary (base 10)	Hexadecimal (base 16)
0000	0	0
0001	1	1
0010	2	2
0011	3	3
0100	4	4
0101	5	5
0110	6	6
0111	7	7
1000	8	8
1001	9	9
1010	10	A
1011	11	B
1100	12	C
1101	13	D
1110	14	E
1111	15	F

Example 33.1

Convert hexadecimal A3 into binary.

Solution

From Table 33.1, A = 1010 and 3 = 0101. Thus, A3 in hexadecimal is equivalent to 10100101 in binary.

Example 33.2

Convert binary 11101000 binary to hexadecimal.

Solution

From Table 33.1, 1110 = E and 1000 = 8. Thus, 11101000 in binary is equivalent to E8 in hexadecimal.

33.5   Data Types

A byte of data can be stored at each address within the total memory space of a microprocessor system. Hence, one byte can be stored at each of the 65,536 memory locations within a microprocessor system having a 16-bit address bus.

Individual bits within a byte are numbered from 0 (least significant bit) to 7 (most significant bit). In the case of 16-bit words, the bits are numbered from 0 (least significant bit) to 15 (most significant bit).

Negative (or signed) numbers can be represented using **two's complement** notation where the leading (most significant) bit indicates the sign of the number (1 = negative, 0 = positive). For example, the signed 8-bit number 10000001 represents the denary number –1.

The range of integer data values that can be represented as bytes, words and long words are shown in Table 33.2.

33.6   Data Storage

The semiconductor ROM within a microprocessor system provides storage for the program code as well as any permanent data that requires storage. All of this data is

Table 33.2: Data types

Data type	Bits	Range of values
Unsigned byte	8	0 to 255
Signed byte	8	–128 to +127
Unsigned word	16	0 to 65,535
Signed word	16	–32,768 to +32,767

referred to as nonvolatile because it remains intact when the power supply is disconnected.

The semiconductor RAM within a microprocessor system provides storage for the transient data and variables that are used by programs. Part of the RAM is also be used by the microprocessor as a temporary store for data while carrying out its normal processing tasks.

It is important to note that any program or data stored in RAM will be lost when the power supply is switched off or disconnected. The only exception to this is CMOS RAM that is kept alive by means of a small battery. This *battery-backed memory* is used to retain important data, such as the time and date.

When expressing the amount of storage provided by a memory device we usually use kilobytes (Kbyte). It is important to note that a kilobyte of memory is actually 1,024 bytes (not 1,000 bytes). The reason for choosing the Kbyte rather than the kbyte (1,000 bytes) is that 1,024 happens to be the nearest power of 2 (note that $2^{10} = 1,024$).

The capacity of a semiconductor ROM is usually specified in terms of an address range and the number of bits stored at each address. For example, 2K × 8 bits (capacity 2 Kbytes), 4K × 8 bits (capacity 4 Kbytes), and so on. Note that it is not always necessary (or desirable) for the entire memory space of a microprocessor to be populated by memory devices.

33.7 The Microprocessor

The microprocessor *central processing unit* (*CPU*) forms the heart of any microprocessor or microcomputer system computer and, consequently, its operation is crucial to the entire system.

The primary function of the microprocessor is that of fetching, decoding, and executing instructions resident in memory. As such, it must be able to transfer data from external memory into its own internal registers and vice versa. Furthermore, it must operate predictably, distinguishing, for example, between an operation contained within an instruction and any accompanying addresses of read/write memory locations. In addition, various system housekeeping tasks need to be performed including being able to suspend normal processing in order to respond to an external device that needs attention.

The main parts of a microprocessor CPU are:

(a)   *registers* for temporary storage of addresses and data;

(b)   an *arithmetic logic unit* (ALU) that performs arithmetic and logic operations;

(c)   a unit that receives and decodes instructions; and

(d)   a means of controlling and timing operations within the system.

Figure 33.3 shows the principal internal features of a typical 8-bit microprocessor. We will briefly explain each of these features in turn.

33.7.1   Accumulator

The accumulator functions as a source and destination register for many of the basic microprocessor operations. As a *source register* it contains the data that will be used in a particular operation while as a *destination register* it will be used to hold the result of a particular operation. The accumulator (or *A-register*) features in a very large number of microprocessor operations, consequently more reference is made to this register than any others.

33.7.2   Instruction Register

The instruction register provides a temporary storage location in which the current microprocessor instruction is held while it is being decoded. Program instructions are passed into the microprocessor, one at time, through the data bus.

On the first part of each *machine cycle*, the instruction is fetched and decoded. The instruction is executed on the second (and subsequent) machine cycles. Each machine cycle takes a finite time (usually less than a microsecond) depending upon the frequency of the microprocessor's clock.

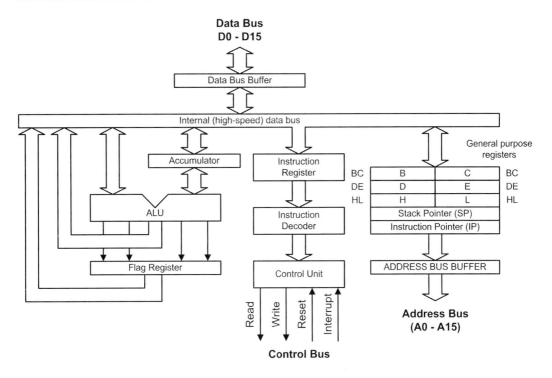

Figure 33.3: Internal architecture of a typical 8-bit microprocessor CPU

33.7.3 Data Bus (D0 to D7)

The external data bus provides a highway for data that links all of the system components (such as random access memory, read-only memory, and input/output devices) together. In an 8-bit system, the data bus has eight data lines, labeled D0 (*the least significant bit*) to D7 (*the most significant bit*) and data is moved around in groups of eight bits, or *bytes*. With a sixteen bit data bus the data lines are labeled D0 to D15, and so on.

33.7.4 Data Bus Buffer

The data bus buffer is a temporary register through which bytes of data pass on their way into, and out of, the microprocessor. The buffer is thus referred to as *bidirectional* with data passing out of the microprocessor on a *write operation* and into the processor during a *read operation*. The direction of data transfer is determined by the *control unit* as it responds to each individual program instruction.

33.7.5 Internal Data Bus

The internal data bus is a high-speed data highway that links all of the microprocessor's internal elements together. Data is constantly flowing backward and forward along the internal data bus lines.

33.7.6 General-Purpose Registers

Many microprocessor operations (for example, adding two 8-bit numbers together) require the use of more than one register. There is also a requirement for temporarily storing the partial result of an operation while other operations take place. Both of these needs can be met by providing a number of general-purpose registers. The use to which these registers are put is left mainly up to the programmer.

33.7.7 Stack Pointer

When the time comes to suspend a particular task in order to briefly attend to something else, most microprocessors make use of a region of external random access memory (RAM) known as a *stack*. When the main program is interrupted, the microprocessor temporarily places in the stack the contents of its internal registers together with the address of the next instruction in the main program. When the interrupt has been attended to, the microprocessor recovers the data that has been stored temporarily in the stack together with the address of the next instruction within the main program. It is thus able to return to the main program exactly where it left off and with all the data preserved in its registers. The stack pointer is simply a register that contains the address of the last used stack location.

33.7.8 Program Counter

Programs consist of a sequence of instructions that are executed by the microprocessor. These instructions are stored in external random access memory (RAM) or read-only memory (ROM). Instructions must be fetched and executed by the microprocessor in a strict sequence. By storing the address of the next instruction to be executed, the program counter allows the microprocessor to keep track of where it is within the program. The program counter is automatically incremented when each instruction is executed.

33.7.9   Address Bus Buffer

The address bus buffer is a temporary register through which addresses (in this case comprising 16-bits) pass on their way out of the microprocessor. In a simple microprocessor, the address buffer is unidirectional with addresses placed on the address bus during both read and write operations. The address bus lines are labeled A0 to A15, where A0 is the least-significant address bus line and A16 is the most significant address bus line. Note that a 16-bit address bus can be used to communicate with 65,536 individual memory locations. At each location a single byte of data is stored.

33.7.10   Control Bus

The control bus is a collection of signal lines that are both used to control the transfer of data around the system and also to interact with external devices. The control signals used by microprocessors tend to differ with different types; however, the following are commonly found:

READ	An output signal from the microprocessor that indicates that the current operation is a read operation.
WRITE	An output signal from the microprocessor that indicates that the current operation is a write operation.
RESET	A signal that resets the internal registers and initializes the program counter so that the program can be re-started from the beginning.
IRQ	Interrupt request from an external device attempting to gain the attention of the microprocessor (the request may be obeyed or ignored according to the state of the microprocessor at the time that the interrupt request is received).
NMI	Nonmaskable interrupt (i.e., an interrupt signal that cannot be ignored by the microprocessor).

33.7.11   Address Bus (A0 to A15)

The address bus provides a highway for addresses that links with all of the system components (such as random access memory, read-only memory, and input/output devices). In a system with a 16-bit address bus, there are sixteen address lines, labeled A0 (the least significant bit) to A15 (the most significant bit). In a system with a 32-bit address bus there are 32 address lines, labeled A0 to A31, and so on.

33.7.12   Instruction Decoder

The instruction decoder is nothing more than an arrangement of logic gates that acts on the bits stored in the instruction register and determines which instruction is currently being referenced. The instruction decoder provides output signals for the microprocessor's control unit.

33.7.13   Control Unit

The control unit is responsible for organizing the orderly flow of data within the microprocessor as well as generating, and responding to, signals on the control bus. The control unit is also responsible for the timing of all data transfers. This process is synchronized using an internal or external clock signal (not shown in Figure 33.3).

33.7.14   Arithmetic Logic Unit (ALU)

As its name suggests, the ALU performs arithmetic and logic operations. The ALU has two inputs (in this case these are both 8-bits wide). One of these inputs is derived from the Accumulator while the other is taken from the internal data bus via a temporary register (not shown in Figure 33.3). The operations provided by the ALU usually include addition, subtraction, logical AND, logical OR, logical exclusive-OR, shift left, shift right, etc. The result of most ALU operations appears in the accumulator.

33.7.15   Flag Register (or Status Register)

The result of an ALU operation is sometimes important in determining what subsequent action takes place. The flag register contains a number of individual bits that are set or reset according to the outcome of an ALU operation. These bits are referred to as flags. The following flags are available in most microprocessors:

ZERO	The zero flag is set when the result of an ALU operation is zero (i.e., a byte value of 00000000).
CARRY	The carry flag is set whenever the result of an ALU operation (such as addition) generates a carry bit (in other words, when the result cannot be contained within an 8-bit register).
INTERRUPT	The interrupt flag indicates whether external interrupts are currently enabled or disabled.

33.7.16   Clocks

The clock used in a microprocessor system is simply an accurate and stable square wave generator. In most cases the frequency of the square wave generator is determined by a quarts crystal. A simple 4 MHz square wave clock oscillator (together with the clock waveform that is produces) is shown in Figure 33.4. Note that one complete clock cycle is sometimes referred to as a T-state.

Microprocessors sometimes have an internal clock circuit in which case the quartz crystal (or other resonant device) is connected directly to pins on the microprocessor chip. In Figure 33.5(A) an external clock is shown connected to a microprocessor while in Figure 33.5(B) and internal clock oscillator is used.

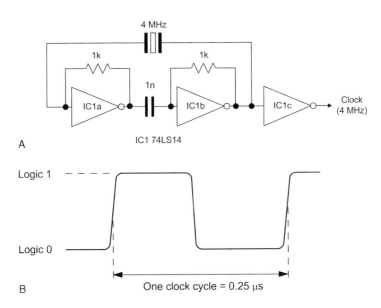

Figure 33.4: (A) A typical microprocessor clock circuit (B) waveform produced by the clock circuit

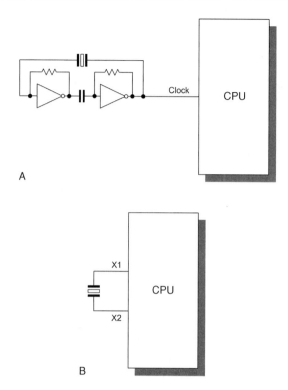

Figure 33.5: (A) An external CPU clock, and (B) an internal CPU clock

33.8   Microprocessor Operation

The majority of operations performed by a microprocessor involve the movement of data. Indeed, the program code (a set of instructions stored in ROM or RAM) must itself be fetched from memory prior to execution. The microprocessor thus performs a continuous sequence of instruction fetch and execute cycles. The act of fetching an instruction code (or operand or data value) from memory involves a read operation while the act of moving data from the microprocessor to a memory location involves a write operation—see Figure 33.6.

Each cycle of CPU operation is known as a machine cycle. Program instructions may require several machine cycles (typically between two and five). The first machine cycle in any cycle consists of an instruction fetch (the instruction code is read from the memory) and it is known as the M1 cycle. Subsequent cycles M2, M3, and so on, depend on the type of instruction that is being executed. This fetch-execute sequence is shown in Figure 33.7.

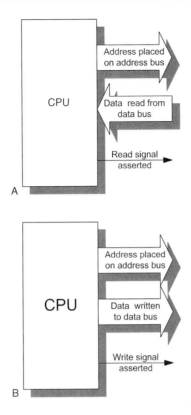

Figure 33.6: (A) Read, and (B) write operations

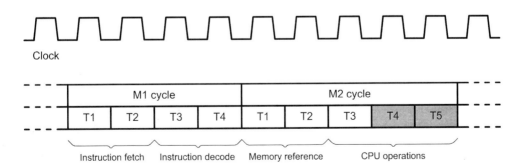

Figure 33.7: A typical timing diagram for a microprocessor's fetch-execute cycle

Microprocessors determine the source of data (when it is being read) and the destination of data (when it is being written) by placing a unique address on the address bus. The address at which the data is to be placed (during a write operation) or from which it is to be fetched (during a read operation) can either constitute part of the memory of the system (in which case it may be within ROM or RAM) or it can be considered to be associated with input/output (I/O).

Since the data bus is connected to a number of VLSI devices, an essential requirement of such chips (e.g., ROM or RAM) is that their data outputs should be capable of being isolated from the bus whenever necessary. These chips are fitted with select or enable inputs that are driven by address decoding logic (not shown in Figure 33.2). This logic ensures that ROM, RAM and I/O devices never simultaneously attempt to place data on the bus!

The inputs of the address decoding logic are derived from one, or more, of the address bus lines. The address decoder effectively divides the available memory into blocks corresponding to a particular function (ROM, RAM, I/O, etc). Hence, where the processor is reading and writing to RAM, for example, the address decoding logic will ensure that only the RAM is selected while the ROM and I/O remain isolated from the data bus.

Within the CPU, data is stored in several registers. Registers themselves can be thought of as a simple pigeon-hole arrangement that can store as many bits as there are holes available. Generally, these devices can store groups of sixteen or thirty-two bits. Additionally, some registers may be configured as either one register of sixteen bits or two registers of thirty-two bits.

Some microprocessor registers are accessible to the programmer whereas others are used by the microprocessor itself. Registers may be classified as either general-purpose or dedicated. In the latter case a particular function is associated with the register, such as holding the result of an operation or signalling the result of a comparison. A typical microprocessor and its register model is shown in Figure 33.8.

33.8.1   The Arithmetic Logic Unit

The ALU can perform arithmetic operations (addition and subtraction) and logic (complementation, logical AND, logical OR, etc). The ALU operates on two inputs (sixteen or thirty-two bits in length depending upon the CPU type) and it provides one output (again of sixteen or thirty-two bits). In addition, the ALU status is preserved in the *flag register* so that, for example, an overflow, zero or negative result can be detected.

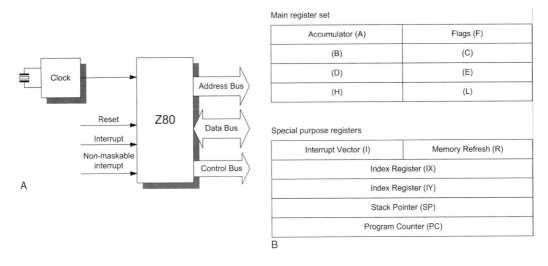

Figure 33.8: The Z80 microprocessor (showing some of its more important control signals) together with its register model

The control unit is responsible for the movement of data within the CPU and the management of control signals, both internal and external. The control unit asserts the requisite signals to read or write data as appropriate to the current instruction.

33.8.2   Input and Output

The transfer of data within a microprocessor system involves moving groups of 8, 16 or 32-bits using the bus architecture described earlier. Consequently it is a relatively simple matter to transfer data into and out of the system in parallel form. This process is further simplified by using a *Programmable Parallel I/O* device (a Z80PIO, 8255, or equivalent VLSI chip). This device provides registers for the temporary storage of data that not only buffer the data but also provide a degree of electrical isolation from the system data bus.

Parallel data transfer is primarily suited to high-speed operation over relatively short distances, a typical example being the linking of a microcomputer to an adjacent dot matrix printer. There are, however, some applications in which parallel data transfer is inappropriate, the most common example being data communication by means of telephone lines. In such cases data must be sent serially (one bit after another) rather than in parallel form.

To transmit data in serial form, the parallel data from the microprocessor must be reorganized into a stream of bits. This task is greatly simplified by using an LSI interface device that contains a shift register that is loaded with parallel data from the data bus. This data is then read out as a serial bit stream by successive shifting. The reverse process, serial-to-parallel conversion, also uses a shift register. Here data is loaded in serial form, each bit shifting further into the register until it becomes full. Data is then placed simultaneously on the parallel output lines. The basic principles of parallel-to-serial and serial-to-parallel data conversion are illustrated in Figure 33.9.

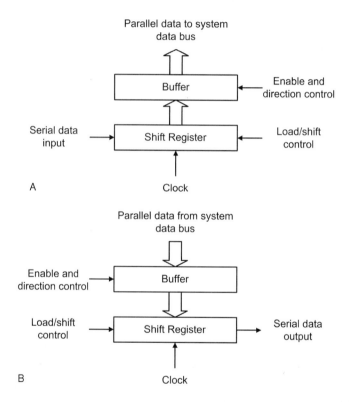

Figure 33.9: (A) Serial-to-parallel data conversion, and (B) parallel-to-serial data conversion

33.8.3   An Example Program

The following example program (see Table 33.3) is written in assembly code. The program transfers 8-bit data from an input port (Port A), complements (i.e., inverts) the data (by changing 0's to 1's and 1's to 0's in every bit position) and then outputs the result to an output port (Port B). The program repeats indefinitely.

Table 33-3: A simple example program

Address	Data	Assembly code	Comment
2002	DB FF	IN A, (FFH)	Get a byte from Port A
2002	2F	CPL	Invert the byte
2003	D3 FE	OUT (FEH), A	Output the byte to Port B
2005	C3 00 20	JP 2000	Go round again

Just three microprocessor instructions are required to carry out this task together with a fourth (jump) instruction that causes the three instructions to be repeated over and over again. A program of this sort is most easily written in assembly code which consists of a series of easy to remember mnemonics. The flowchart for the program is shown in Figure 33.10(A).

The program occupies a total of eight bytes of memory, starting at a hexadecimal address of 2000 as shown in Figure 33.10(B). You should also note that the two ports, A and B, each have unique addresses; Port A is at hexadecimal address FF while Port B is at hexadecimal address FE.

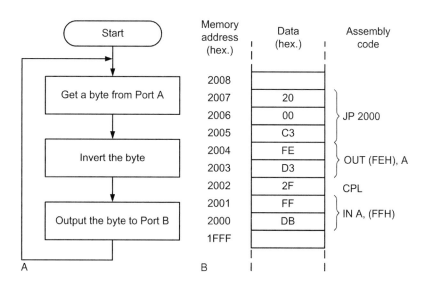

Figure 33.10: **(A) Flowchart for the example program, and (B) the eight bytes of program code stored in memory**

33.8.4 Interrupts

A program that simply executes a loop indefinitely has a rather limited practical application. In most microprocessor systems we want to be able to interrupt the normal sequence of program flow in order to alert the microprocessor to the need to do something. We can do this with a signal known as an *interrupt*. There are two types of interrupt; maskable and nonmaskable.

When a *nonmaskable interrupt* input is asserted, the processor must suspend execution of the current instruction and respond immediately to the interrupt. In the case of a *maskable interrupt*, the processor's response will depend upon whether interrupts are currently enabled or disabled (when enabled, the CPU will suspend its current task and carry out the requisite interrupt service routine).

The response to interrupts can be enabled or disabled by means of appropriate program instructions. In practice, interrupt signals may be generated from a number of sources and since each will require its own customized response a mechanism must be provided for identifying the source of the interrupt and calling the appropriate interrupt service routine. In order to assist in this task, the microprocessor may use a dedicated programmable interrupt controller chip.

33.9   A Microcontroller System

Figure 33.11 shows the arrangement of a typical microcontroller system. The sensed quantities (temperature, position, etc.) are converted to corresponding electrical signals by means of a number of sensors. The outputs from the sensors (in either digital or analog form) are passed as input signals to the microcontroller. The microcontroller also accepts inputs from the user. These user set options typically include target values for variables (such as desired room temperature), limit values (such as maximum shaft speed), or time constraints (such as "on" time and "off" time, delay time, etc).

The operation of the microcontroller is controlled by a sequence of software instructions known as a control program. The control program operates continuously, examining inputs from sensors, user settings, and time data before making changes to the output signals sent to one or more controlled devices.

The controlled quantities are produced by the controlled devices in response to output signals from the microcontroller. The controlled device generally converts energy from one form into energy in another form. For example, the controlled device might be an

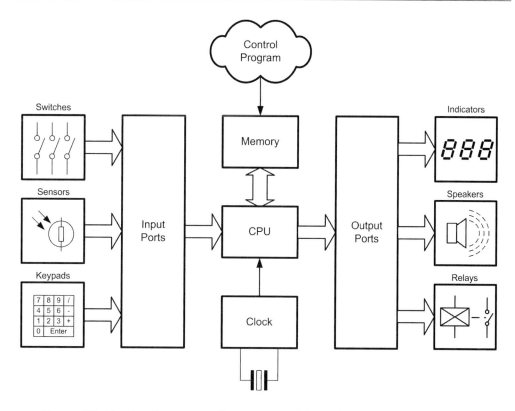

Figure 33.11: A microcontroller system with typical inputs and outputs

electrical heater that converts electrical energy from the AC mains supply into heat energy thus producing a given temperature (the controlled quantity).

In most real-world systems there is a requirement for the system to be automatic or self-regulating. Once set, such systems will continue to operate without continuous operator intervention. The output of a self-regulating system is fed back to its input in order to produce what is known as a closed loop system. A good example of a closed-loop system is a heating control system that is designed to maintain a constant room temperature and humidity within a building regardless of changes in the outside conditions.

In simple terms, a microcontroller must produce a specific state on each of the lines connected to its output ports in response to a particular combination of states present on each of the lines connected to its input ports (see Figure 33.11). Microcontrollers must also have a central processing unit (CPU) capable of performing simple arithmetic, logical and timing operations.

The input port signals can be derived from a number of sources including:

- switches (including momentary action pushbuttons);

- sensors (producing logic-level compatible outputs); and

- keypads (both encoded and unencoded types).

The output port signals can be connected to a number of devices including:

- LED indicators (both individual and multiple bar types);

- LED seven segment displays (via a suitable interface);

- motors and actuators (both linear and rotary types) via a suitable buffer/driver or a dedicated interface);

- relays (both conventional electromagnetic types and optically couple solid-state types); and

- transistor drivers and other solid-state switching devices.

33.9.1 Input Devices

Input devices supply information to the computer system from the outside world. In an ordinary personal computer, the most obvious input device is the keyboard. Other input devices available on a PC are the mouse (pointing device), scanner and modem. Microcontrollers use much simpler input devices. These need be nothing more than individual switches or contacts that make and break but many other types of device are also used including many types of sensor that provide logic level outputs (such as float switches, proximity detectors, light sensors, etc).

It is important to note that, in order to be connected directly to the input port of a microcontroller, an input device must provide a logic compatible signal. This is because microcontroller inputs can only accept digital input signals with the same voltage levels as the logic power source. The 0V ground level (often referred to as V_{SS} in the case of a CMOS microcontroller) and the positive supply (V_{DD} in the case of a CMOS microcontroller) is invariably 5V $\pm$ 5%. A level of approximately 0V indicates a logic 0 signal and a voltage approximately equal to the positive power supply indicates a logic 1 signal.

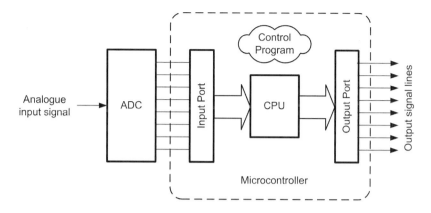

Figure 33.12: An analog input signal can be connected to a microcontroller input port via an analog-to-digital converter (ADC)

Other input devices may sense analog quantities (such as velocity) but use a digital code to represent their value as an input to the microcontroller system. Some microcontrollers provide an internal analog-to-digital converter (ADC) in order to greatly simplify the connection of analog sensors as input devices but where this facility isn't available it will be necessary to use an external ADC which usually takes the form of a single integrated circuit. The resolution of the ADC will depend upon the number of bits used and 8, 10, and 12-bit devices are common in control applications.

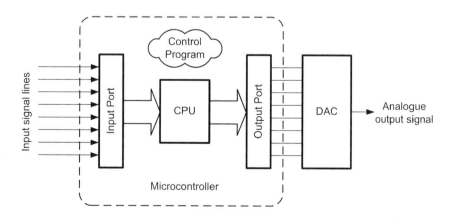

Figure 33.13: An analog output signal can be produced by connecting a digital-to-analog converter (DAC) to a microcontroller output power

33.9.2   Output Devices

Output devices are used to communicate information or actions from a computer system to the outside world. In a personal computer system, the most common output device is the CRT (cathode ray tube) display. Other output devices include printers and modems. As with input devices, microcontroller systems often use much simpler output devices. These may be nothing more than LEDs, piezoelectric sounders, relays and motors. In order to be connected directly to the output port of a microcontroller, an output device must, once again, be able to accept a logic compatible signal.

Where analog quantities (rather than simple digital on/off operation) are required at the output a digital-to-analog converter (DAC) will be needed. All of the functions associated with a DAC can be provided by a single integrated circuit. As with an ADC, the output resolution of a DAC depends on the number of bits and 8, 10, and 12-bits are common in control applications.

33.9.3   Interface Circuits

Finally, where input and output signals are not logic compatible (i.e., when they are outside the range of signals that can be connected directly to the microcontroller) some additional interface circuitry may be required in order to shift the voltage levels or to provide additional current drive. Additional circuitry may also be required when a load (such as a relay or motor) requires more current than is available from a standard logic device or output port. For example, a common range of interface circuits (solid-state relays) is available that will allow a microcontroller to be easily interfaced to an AC mains-connected load. It then becomes possible for a small microcontroller (operating from only a 5V DC supply) to control a central heating system operating from 240V AC mains.

CHAPTER 34

Microcontroller Toolbox

Stuart Ball

This chapter contains some miscellaneous topics that pull together multiple concepts from preceding chapters.

34.1 Microcontroller Supply and Reference

In many cases, you can minimize the effects that supply voltage can have by referencing your analog inputs to the supply voltage. Figure 34.1A shows a thermistor connected to an analog input and using a pull-up to a precision reference voltage. At first glance, it might appear that this is a very accurate design because the precision reference gives a repeatable voltage versus temperature at the analog input. The problem with this design is that the microcontroller is measuring the temperature using the supply voltage as a reference, so the overall accuracy is only as good as the microcontroller supply voltage.

Figure 34.1B shows the same circuit, but with the thermistor referenced to the microcontroller supply. This provides a more repeatable result. If the thermistor is 10K and R1 is 10K, for example, the analog input to the microcontroller will always sense half the supply voltage regardless of what the supply voltage actually is. This method will work only if the analog input can be made to follow the supply voltage. This essentially means that the output being measured by the microcontroller is referenced to the supply. Note that just powering the sensor or sensor circuit from the microcontroller supply may not be sufficient. If the sensor circuit has its own internal reference that controls the output value, it will produce the same output regardless of variations in the supply voltage.

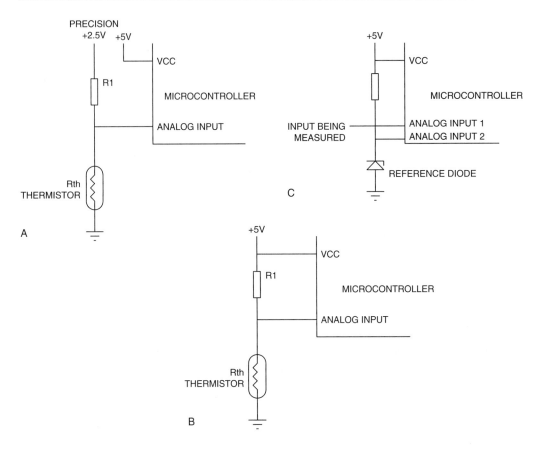

Figure 34.1: Microcontroller supply reference

An alternative compensation method, for cases in which the input is independent of the microcontroller supply voltage, is shown in Figure 34.1C. In this figure, a second analog input is used to measure the value of a precision reference diode. Of course, the microcontroller must have at least two analog inputs to take advantage of this technique. In operation, the microcontroller would use the reference diode to determine the error caused by the supply voltage. For example, if the reference diode is 2.5V and the supply voltage is 5V, the reference diode will produce a value of 80_{16} (128 decimal) when the voltage is converted (assume the internal ADCs are 8 bits). If the supply is 4.8V, the reference diode will convert to a value of 85_{16} (133 decimal). The microcontroller can use this value to correct the values from the independent input. In this case, the values read from that input will be multiplied by 128/133, or 0.96, to get the correct reading. Note, though, that the overall accuracy is only as good as the combined accuracy of the reference diode and whatever reference the external input is using. Finally, to make use of this technique, the microcontroller must have sufficient throughput to make the required calculations and memory to hold the math algorithms; this may be a problem on some small microcontrollers.

34.2   Resistor Networks

Some applications need better repeatability than you can get with standard 1% resistors, but don't need the level of precision (and cost) of going to 0.1% resistors. Sometimes you can gain an advantage by using resistor networks. Resistor networks are typically specified with the same resistance tolerances as discrete resistors: 0.1%, 1%, and 5%. However, the matching between resistors within the same network is often twice as good as the absolute resistance accuracy. If your circuit uses multiple resistors of the same value, you can often get better accuracy by using a resistor network rather than discrete parts. Note, though, that this works only for resistors in the same package; it doesn't work across packages.

Figure 34.2 shows a simple voltage divider. This circuit might be used to bring an analog input that swings between 0 and 8V down to the 0 to 5V range used by a microcontroller analog input. In the figure, both resistors are 10K. Ideally, the output voltage would be half the input voltage. However, if the resistors are 1% discrete parts, the output voltage may not be exactly half the input.

Say that R1 is high by 1% and equals 10,100 ohms, and R2 is low by 1% and equals 9900 ohms. The output is given by the following equation:

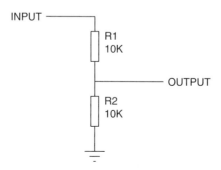

Figure 34.2: Resistor voltage divider

$$\text{INPUT} \times \frac{R2}{R1 + R2} = \text{INPUT} \times 0.495$$

which is incorrect by 1%, the tolerance of the resistors.

Now, say that R1 and R2 are resistors that are in the same resistor network package. The specified resistance is 1%, but the part-to-part variation is 0.5%. If R1 is high by 1%, it will be 10,100 ohms, as before. However, because parts within the network package can only vary by 0.5%, R2 cannot be less than 10,049.5. As a result, the output will be:

$$\text{INPUT} \times \frac{R2}{R1 + R1} = \text{INPUT} \times 0.4987$$

which is within 0.25%, of the ideal value.

34.3   Multiple Input Control

In some cases a system will have multiple inputs. For example, you might be controlling a telescope that is taking pictures of clouds. For some reason (perhaps to protect sensitive optics coatings), you don't want to aim the telescope at the sun (Figure 34.3). In this case, one of the inputs would be the position of the sun, probably determined by the date and time of day. You would also have as inputs the telescope's current position and the desired position.

This is a good example of a two-input problem, because the position of the sun is not fixed. You can't just use a table to look up the move path because the position of the sun varies. To move the telescope without crossing the path of the sun, you can take two approaches.

The first approach is to calculate the direct path, determine that it crosses the sun's path, then calculate a new path that just misses the sun. This is illustrated in Figure 34.3A and 34.3B. This calculation may be complicated and time-consuming, especially on a microcontroller or other system with limited capability.

A simpler approach is to calculate the position of the sun and then calculate a path that remains as far as possible from the sun. One way to do this is to divide the telescope range into a grid of, say, 8 or 16 regions. When calculating the move path, any region overlapped by the sun is avoided. A typical example is shown in Figure 34.3C. Figure 34.3D shows how the move area can be subdivided into 16 regions.

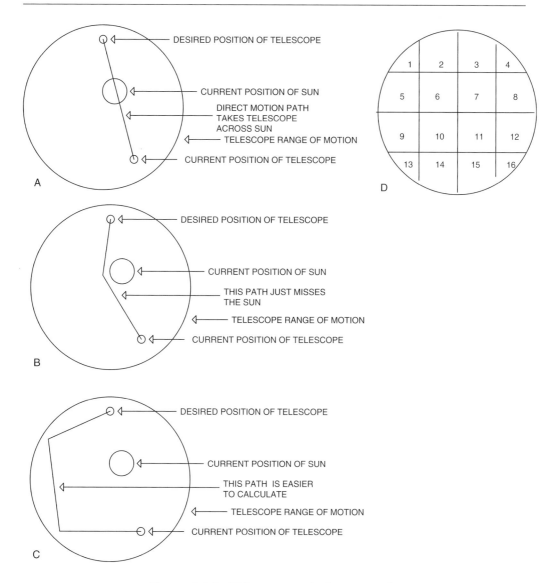

Figure 34.3: Telescope pointing example

Another alternative would be to find a sunless path to the perimeter of the motion circle, and then determine which way to move around the circle to avoid the sun. In the example shown, either direction would work, but if the sun were on the perimeter of the motion circle, the direction would have to be determined.

You could make the path decision table driven by having two move paths from any region to any other region. One path would be a straight line and the other path would avoid any region in common with the first path. In this way, you could determine a safe path by checking the straight line move for "interference" from the sun. If the straight line path goes through a region containing the sun, just pick the other path. Once in the target region, calculate a straight line move to the desired point. This method has the advantage of minimizing calculation requirements for simple processors.

The foregoing assumes that the system requirements don't include taking photos of any region containing the sun. If you did have to take photos right next to the sun, you could make the regions smaller or you could have multiple paths from any region to any other, arriving at the destination from different directions.

Although the telescope example is very specific, the general principles are applicable to many similar multiple input problems, such as:

- multihead fluid pipetting systems in which the pipettes can interfere with each other;

- a heating system in which the maximum safe heater power depends on a fluid level;

- a stepper motor with resonance points that depend on the load;

- a valve control system in which valve closing/opening speed depends on fluid viscosity and flow rate—variable closing/opening time might be required to avoid "water hammer" or similar effects; and

- a heater or cooler system in which the intent is to quickly get the target to a specific temperature and where the amount of heating or cooling applied depends on the size and initial temperature of the target.

Another example of multiple input systems is the need to adjust system parameters based on an input—for example, from a heater. In this example, the proportional (or PID) control might have an offset which had to be adjusted for varying loads and ambient conditions. A large load or very cold ambient temperature might need a larger offset to maintain the temperature. In a case like this, additional sensors may be needed to measure these parameters. The system could calculate parameters such as the offset and/or gain, or a set of tables could be used to select the values based on the values of the additional input parameters.

34.4 AC Control

Some designs require control of AC power to turn on lights, motors, heaters, or other AC devices. The simplest method of controlling such devices is with solid-state relays (SSRs), as shown in Figure 34.4. An SSR consists of an optoisolator driving an SCR or triac. The internal optoisolator is selected by the manufacturer to ensure that it will be capable of driving the SCR or triac. Some SSRs have heatsink plates on the back that need to be bolted to a metal chassis or heatsink to avoid overheating the SSR.

In many cases you need to perform *zero crossing switching*. This consists of switching the load only when the AC signal crosses zero (Figure 34.5). If the AC signal is switched when the voltage is not zero, then the load will see a sudden jump in the applied voltage instead of a smooth sine wave. This can damage some loads. In addition, the fast rising edge causes considerable EMI. Finally, in some cases, the load will draw excessive current if the AC voltage is suddenly applied and the value of the voltage isn't zero. You can get solid-state relays that have zero crossing built in. These parts include circuitry that turns on the SCR or triac only when the AC voltage is zero.

In some cases, you need to perform the zero crossing detection in software. Figure 34.5C shows a way to do this; an optoisolator is connected, with a current limiting resistor, across the AC line. Each time the AC voltage goes through zero, the optoisolator turns off and an interrupt is generated to the microprocessor. All switching of external AC loads is performed in the ISR. Typically, to get fast response, the software outside the ISR will set flags or semaphores to determine what AC outputs should be turned on. The ISR reads the flags and switches the appropriate outputs on;

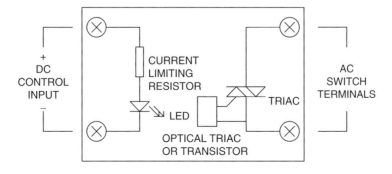

Figure 34.4: SSR

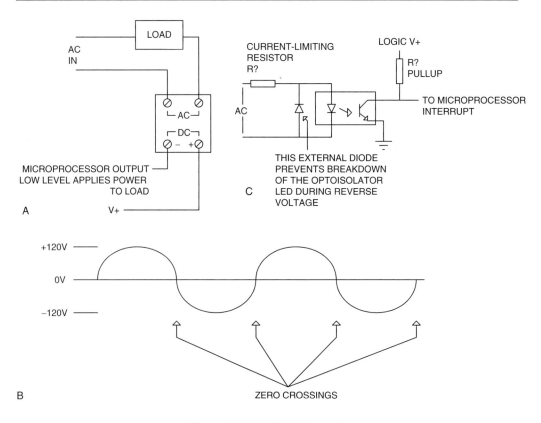

Figure 34.5: SSR Control

the ISR does not do whatever processing is required to determine what should be off or on, it just switches the outputs. This provides minimum latency between the interrupt and the output switch.

Note the external diode across the optoisolator LED. This diode conducts during the negative half of the AC cycle, preventing excessive reverse voltage across the optoisolator LED.

34.5   Voltage Monitors and Supervisory Circuits

A number of ICs are available that provide voltage monitoring functions for microprocessor circuits. An example is the Texas Instruments TL7770. The TL7770 has two voltage comparators that can monitor either of two voltage inputs. Generally, these devices work by asserting a microprocessor reset when the supply voltage reaches some predefined value (1V for the TL7770), and then remove the reset when the monitored voltage has been above a preset threshold for some predefined period of time. This ensures that the microprocessor is held in reset until the supply voltages are stable.

Although many supervisory ICs are intended for monitoring multiple microprocessor supplies, they can be used to monitor other voltages as well. Typically, one input would be used to monitor the microprocessor supply and the other would be used to monitor a higher voltage such as that used to drive a motor. In some cases, you will need to use resistive voltage dividers to bring the voltage you want to monitor within the range of the supervisory IC.

34.6   Driving Bipolar Transistors

A bipolar transistor is often used as an output driver for a microprocessor. Figure 34.6A illustrates how a transistor can be used. When the microcontroller output pin is high, it sources current into the transistor base and the transistor turns on.

When the output is low, the transistor turns off. The requirements for driving a bipolar transistor from the output of a microcontroller are:

- Voltage output from the microcontroller must be high enough to turn the transistor on, typically greater than 0.8V.

- Current into the base of the transistor must be high enough to saturate the transistor.

- Current into the base of the transistor must be limited to a value that will avoid damage to the transistor.

- Low-output voltage of the microcontroller output must be low enough to ensure that the transistor turns off. This is typically not a problem unless the output must sink significant current.

The current into the base of the transistor is calculated as:

$$\frac{\text{Logic high voltage} - \text{transistor Vbe}}{\text{Base resistor}}$$

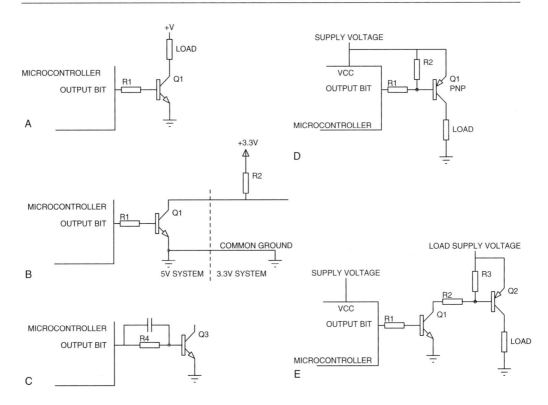

Figure 34.6: Driving bipolar transistors

(In the figure, R1 is the base resistor.) The logic high voltage is the value of the output voltage for the logic used. It may vary with the load, so a logic output that nominally swings to the supply may deliver less voltage if it cannot supply adequate current. The transistor base-emitter voltage, Vbe, is typically 0.6V to 0.8V.

Whether the transistor can pull its collector close enough to ground to function as a logic output depends on the load and the base drive. The maximum collector current is approximately equal to the base current times the current gain of the transistor, up to the point at which the transistor saturates. A signal transistor might have a gain of 100, so a few mA of base current can switch a few hundred mA of collector current. A large power transistor may only have a gain of 10 or 20, so it is difficult to drive them directly from the microcontroller outputs—there isn't enough gain to ensure that the transistor is saturated when driving a high-current load. Consequently, driving very high current loads typically requires a signal transistor driving the base of a larger power transistor.

The simplest approach to using bipolar transistors is to set the base current at half or less of the maximum rated base current. If this does not provide sufficient collector current for your application, or if you calculate the required base current and find that it exceeds what the microcontroller can produce, then you are trying to switch too much current. Use another type of driver.

Finally, remember that the gain of transistors tends to vary quite a bit from one lot to the next, so don't build a circuit that depends on very high gain transistors unless you are willing to sort them in production.

34.6.1 Logic Level Translation

Bipolar transistors provide a convenient means to pass signals between two systems at different supply voltages. Figure 34.6B shows a transistor used to connect a 5V microcontroller and a 3.3V external system. The collector of the transistor is pulled up on the 3.3V system through a resistor.

34.6.2 Switching Speed

One problem with driving bipolar transistors directly from the output of a microcontroller (or other logic) is speed. When the transistor is saturated, the base-emitter junction exhibits a characteristic known as *stored charge*. This, in effect, acts as a capacitor, making the transistor slow to turn off when the logic input goes low. In addition, the output of the transistor circuit (the collector) does not have an active pull-up to force the output high. Instead, a resistor pulls the output high when the transistor turns off. Consequently, the risetime of the output is dependent on the transistor switching speed and the capacitance in the collector circuit. If the transistor is connected to another board via a long cable, this capacitance can be significant.

The turn-on and turn-off speed of the transistor can be improved with the addition of a capacitor across the base resistor, as shown in Figure 34.6C. The capacitor is a low impedance when the logic output is changing states, rapidly charging or discharging the base circuit. A typical value for this capacitor is 220 pF, although larger values may be needed for large transistors.

The collector risetime can be reduced by reducing the value of the pull-up resistor. However, smaller resistor values increase supply current drain and transistor dissipation. Also, a smaller pull-up resistor means more base current is needed to ensure that the transistor will be saturated. These techniques will make a transistor circuit switch faster, but a discrete transistor design will never be as fast as a driver or interface IC designed for a specific application. Bipolar transistors find primary use in controlling currents or voltages beyond the capability of the microcontroller/microprocessor itself.

34.6.3   High-Side Switches

In some cases, you need to pull an output up instead of clamping it to ground. Figure 34.6D shows a PNP transistor used in this way. The PNP is wired with the emitter connected to the positive supply voltage (the NPN had the emitter grounded), so pulling the base toward ground turns the transistor on. The resistor between the base and emitter of the transistor ensures that the base goes all the way to the supply, turning the transistor completely off, in case the microcontroller output doesn't quite swing all the way.

The same considerations apply as for the NPN transistor in terms of the base current. The base current in this case is at maximum when the microcontroller output is low.

In some cases, you need to supply current from a higher supply voltage than the microcontroller is using. For instance, a 5V or 3.3V microcontroller may need to switch the 12V supply to a motor. Figure 34.6E shows how a PNP and NPN can be used together for this. The NPN transistor isolates the microcontroller from the high voltage on the base of the PNP transistor.

34.7   Driving MOSFETs

Like bipolar transistors, MOSFETs also provide a means to control voltages and currents outside the range of the microcontroller. The simplest MOSFET drive is shown in Figure 34.7A. Here, a microcontroller output directly drives the MOSFET gate. When the microcontroller output is high, the MOSFET is turned on and sinks current. When the microcontroller output is low, the MOSFET is turned off. The key points to remember in driving a MOSFET in this way are

- The output voltage of the microcontroller must be greater than the MOSFET gate-to-source threshold voltage or the MOSFET will not turn on. This is more of a problem with 3V logic than with 5V logic, but either logic voltage requires the use of a MOSFET with a logic-level threshold voltage. If necessary, a pull-up resistor can be added to the logic signal to ensure that it goes all the way to the supply voltage.

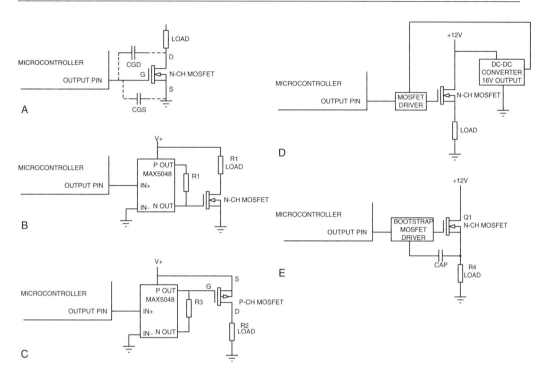

Figure 34.7: Driving MOSFET transistors

- The MOSFET has significant gate-to-source and gate-to-drain capacitance, shown as Cgs and Cgd in the figure. Generally, the larger the MOSFET, the greater this capacitance is. If the MOSFET is driving a signal that can have large voltage spikes, such as an inductive load, sufficient voltage can be coupled back into the microcontroller to damage the outputs.

- The MOSFET turn-on time is limited by the speed at which the gate-to-source voltage rises. This in turn is determined by how quickly the microcontroller can charge up the gate-source capacitance. Many microcontroller outputs have very limited output current capability. If the MOSFET turn-on time is too long and the switching frequency is high, the MOSFET will dissipate excessive power as it transitions from cut-off to saturation.

- If a pull-up resistor is needed to ensure adequate turn-on voltage, the turn-on time of the MOSFET will be limited to the risetime of the pull-up resistor in combination with the gate-source capacitance of the MOSFET. Because the current sinking capability of the microcontroller output limits the size of the pull-up resistor, the switching speed of the MOSFET is also limited by the same current sink capability.

Many of these problems can be eliminated by using a MOSFET driver IC, as shown in Figure 34.7B. In this circuit, a Maxim MAX5048 is used to drive the MOSFET. The MAX5048 provides logic level inputs and can operate on supply voltages up to 12.6V. The MAX5048 has separate sourcing (P-channel) and sinking (N-channel) outputs. In the figure, resistor R1 is not needed. If R1 were not used, the P output and N output would be tied together and to the gate of the FET. R1 in series with the P output limits the rise-time of the gate, and thereby the turn-on time of the FET. If the gate of the FET is connected to the P output instead of to the N output, then R1 will limit the fall time of the gate and thereby the turn-off time of the FET.

34.7.1   High-Side Switching

In some cases, you want to source current instead of sinking current. The simplest way to do this is with a P-channel MOSFET, as shown in Figure 34.7C. In this circuit, the MAX5048 is used to drive the P-channel output transistor. Note that the P-channel MOSFET has the source connected to the positive supply and the gate must be driven toward ground to turn the transistor on.

The problem with P-channel MOSFETs is that they tend to be more expensive than equivalent N-channel MOSFETs and they usually have a higher ON resistance, causing the transistor to dissipate more power when turned on. In some applications, the gate-to-source capacitance can couple the load voltage into the MOSFET gate, turning it on when it should be off. This typically occurs with inductive loads or when there is another transistor pulling the load to ground when the P-channel MOSFET is off. For these reasons, N-channel MOSFETs are usually preferred for high-side switching applications.

The primary difficulty in using an N-channel MOSFET for high-side switching is the gate drive voltage. To turn the N-channel MOSFET on, the gate must be driven higher than the source; because the source is connected to the load in a high side application, this means the gate must be driven higher than the positive supply voltage.

In most cases, the MOSFET is used to drive the load from the highest voltage available in the system, so there is no higher voltage available to drive the MOSFET gate. You have two choices in this case: you can use a bootstrap MOSFET driver or you can add a DC-DC converter.

The DC-DC converter is the simplest solution, as shown in Figure 34.7D. You add a DC-DC converter to the board and use a MOSFET driver IC. The output of the DC-DC converter must not exceed the maximum gate-source voltage, or the MOSFET may be damaged. In the figure, a DC-DC converter with a 16V output produces the gate drive voltage for a driver IC. The gate of the MOSFET will switch between ground and 16V, and the load will switch between ground and 12V. Note that the gate drive voltage cannot exceed the gate-to-source breakdown voltage, which is typically 18V for a MOSFET.

A bootstrap MOSFET driver IC can also drive a high-side MOSFET, as shown in Figure 34.7E. A bootstrap driver uses a capacitor (external to the IC) that is charged up to the supply voltage when the load is low. If the circuit is being used to drive a high-side switch, with no low-side driver, the capacitor charges through the load. If the circuit is being used to drive a pair of MOSFETs, one providing high-side drive and one providing low-side drive, the capacitor charges through the low-side MOSFET when it is turned on.

When the high-side driver is turned on, the bootstrap capacitor is switched so that it drives the gate of the MOSFET above the supply voltage to turn it on. Typically, the bootstrap capacitor is much larger than the gate-source capacitance of the MOSFET, so the voltage across the capacitor does not drop very far when driving the MOSFET gate. However, once the high-side MOSFET is turned on, there is no longer a charging path for the bootstrap capacitor, so it will eventually discharge. For this reason, bootstrap circuits are normally used in applications in which the MOSFET is continuously switching. If you need to turn the MOSFET on and leave it on, you will need a DC-DC converter or some similar method.

34.8 Reading Negative Voltages

Sometimes you need to read and convert a negative voltage with an ADC that operates only from ground and a positive supply. Sometimes the only way to accomplish this is to use an op-amp, powered from both positive and negative supplies, to shift the signal to a range the ADC can use.

Figure 34.8 shows a simple resistor voltage divider that will accomplish the same thing, with some limitations. In the figure, the input is a sine signal that swings

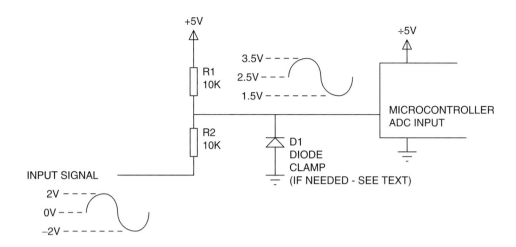

Figure 34.8: Resistive divider for reading negative input signals

between −2V and +2V, being read by a microcontroller that operates from +5V and ground. Using a voltage divider (R1 and R2) brings the signal within the 0–5V range of the microcontroller ADC input. With the values used in the figure, the signal swing is 1.5V to 3.5V. There are a few limitations on this technique:

- The voltage divider essentially acts as a resistive pull-up to the supply voltage. This may affect the input signal.

- The voltage swing is reduced; in the figure, a 4VP-P signal is reduced to 2V P-P at the microcontroller ADC input.

- Large resistors may be needed to avoid loading the input signal source. Large resistors, coupled with the input capacitance of the microcontroller, limit the speed.

- If the input can occasionally go negative enough to bring the microcontroller input below ground, the microcontroller may be destroyed. The maximum signal excursion must be known, or a diode, as shown in the figure, must be used to clamp the signal to ground.

- The actual voltage produced at the microcontroller input is dependent on both the input signal voltage and the supply voltage. Variations in the supply voltage will affect the ADC reading.

34.9 Example Control System

To illustrate some of the principles described in previous chapters, an example control system was developed. This concept system is easy to build and is useful for experimenting with control concepts. Figure 34.9 shows a block diagram of the system. The control system is simulated with an inexpensive lamp coupled to an infrared phototransistor. The lamp and phototransistor are held in place with a length of heatshrink or other opaque tubing.

A PWM circuit is used to control the current through the lamp. The prototype used for the examples here operated at about 14 kHz. An analog control could also be used, with a DAC followed by an op-amp capable of delivering sufficient current to the bulb. The ADC was an 8-bit converter, with an output value of 0 representing 0V and a value of 255 representing 5V.

The system is controlled by a PC, although the same arrangement could be controlled by a microcontroller or single-board computer. Using a PC is less precise than using a more hardware-oriented approach, because the sampling rate in a PC will vary with operating system activity. However, it is close enough to make a useful experimentation tool. For the examples used here, the code was written in Python.

This simple arrangement provides a good simulation of a control system. The lamp filament is, in effect, a heater. The lamp filament does not heat up instantly and the phototransistor is relatively slow, so the combination has many of the characteristics of a real heater or motor arrangement.

In Figure 34.9, R2 is shown with dashed connections. R2 is installed in parallel with R1 to simulate an external load, as will be described later.

Note that this is a reversed control system—a higher control value results in a lower ADC value because a hotter filament results in more phototransistor current.

Figure 34.10A shows the step response of the system. This waveform was created by starting with a PWM value of 1 (just barely turning the lamp on) and then changing to a PWM value of 250 (almost 100% on) and sampling the resulting voltage from the phototransistor once per millisecond. Note that the lamp has a short delay before it starts heating, then a rapid heating period, then a slower curve as it approaches its final temperature. This data was plotted using Microsoft Excel.

Figure 34.10B shows the reverse of the positive step. Here, the PWM value was set to 250 and the output was allowed to settle for one second. The PWM was then

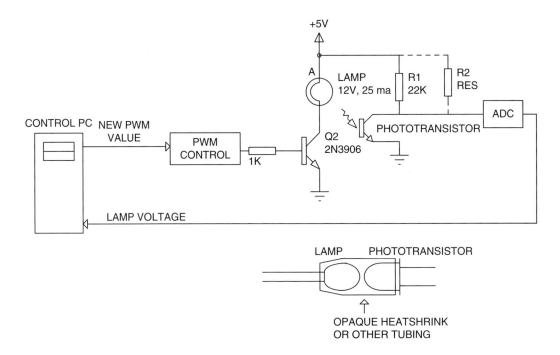

Figure 34.9: Simulation system block diagram

turned off and the output was measured once per millisecond. The result is an exponential curve as the lamp filament cools. This asymmetrical characteristic of the system is typical of many real-world environments.

Figure 34.10C shows the characterization of the system with respect to the control value. This curve was made by applying 16 equally spaced control values from 1 to 241, allowing the output to settle, and measuring the ADC result.

34.9.1 On-Off (Bang-Bang) Control

An on-off control is illustrated in Figure 34.11A. The setpoint for this example was 100, corresponding to about 1.95 V at the phototransistor collector. Note the oscillation around the setpoint—it ranges from 98 to 112, a range of 0.3 volts, or 15% of the setpoint value. The oscillation is not centered around the setpoint, but is skewed toward the high values. This occurs because the control response is not symmetrical—the filament cools down more quickly than it heats up.

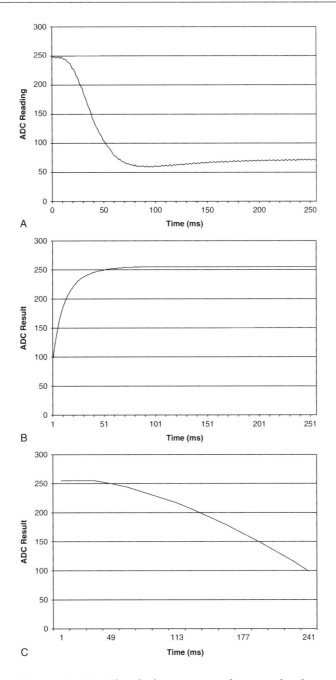

Figure 34.10: Simulation system characterization

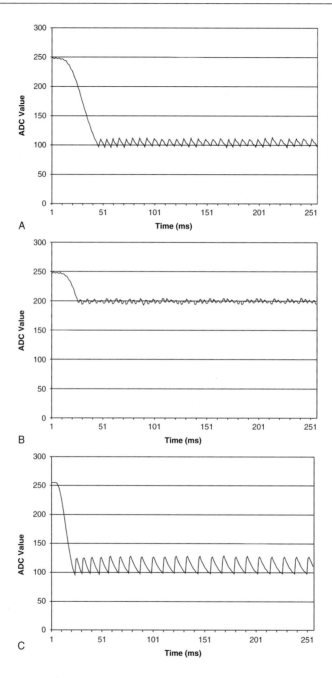

Figure 34.11: On-off control examples

Figure 34.11B shows an on-off control with a setpoint of 150. There is less oscillation at this setpoint; a control system that is not linear across its range will exhibit characteristics like this. Figure 34.11C shows an on-off control with a setpoint of 100 and a sampling interval of 4 ms. Note the size of the oscillation; the sampling interval has a significant effect on the result.

Figure 34.12 shows an on-off control starting with the PWM full on and using a setpoint of 150. Unlike the case that started with the PWM off, there is significant overshoot past the setpoint; the lamp filament cools down more easily than it heats up, so there is more momentum in that direction.

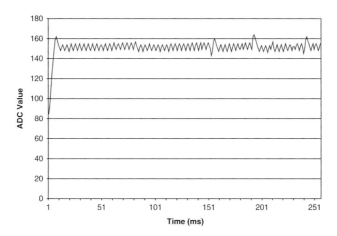

Figure 34.12: On-off control, starting with PWM 100% on

34.9.2   Proportional Control

Figure 34.13A shows a proportional control with a setpoint of 150 (about 2.9 volts) and a gain of 2. Using the input-to-output characterization curve, an offset of 200 was selected for this setpoint. The equation for the control value is

Control output = 200 + (ADC value – setpoint) × Gain
If control output > 254, control output = 254.
If control output < 1, control output = 1.

The last two statements limit the control value to the 8-bit range of the system.

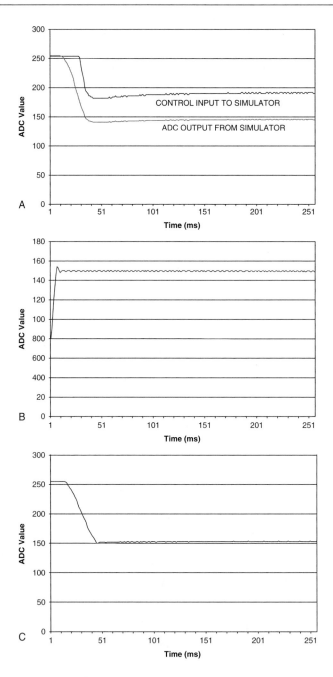

Figure 34.13: Proportional control

At a gain of 2, the system stabilizes with an output around 145. Figure 34.13B shows a proportional control, but starting at the top of the range (100% PWM) and using a gain of 20. This time the result makes it to the setpoint of 150, but with significant oscillation. Note the overshoot as the signal passes through 150; like the on-off example, this is caused by the asymmetrical nature of the heater and the additional gain. If the gain is reduced, the overshoot can be eliminated, but the result ends up below the setpoint (150). Although a graph is not shown for this condition, at a gain of 10, the waveform overshoots just slightly and then settles down to oscillate between 149 and 150.

Figure 34.13C shows a proportional system with a gain of 10, setpoint of 150, and an offset of 100. The lower offset results in a final result between 157 and 158. As you can see, the gain and offset both affect the final result in a proportional system. However, the proportional control is still better than open-loop control, because an open-loop control value of 100 results in an ADC value of 222 (see the characterization chart).

Figure 34.14 shows the proportional system with a setpoint of 150, gain of 10, and a load of 47K (R2) in parallel with the 22K collector resistor (R1). There is a small overshoot as the output passes 150, then the output settles down to oscillate between 152 and 153. Note that the addition of a load caused a permanent offset in the output—the proportional system was unable to completely compensate for the effects of the added load.

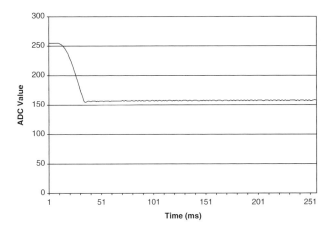

Figure 34.14: Proportional control with load

34.9.3   Pid

Figure 34.15A shows a simple PID control. The parameters are:

- Proportional gain = 2

- Derivative gain = 2

- Integral gain = 2

- Setpoint 150 = 2

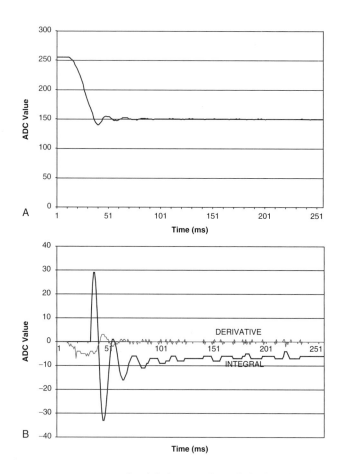

Figure 34.15: PID control with integral and derivative waveforms

To prevent integral windup, the integral is held at zero until the ADC result is within 10% of the setpoint. As you can see, there is a little overshoot and then the output settles down to values of 150 and 151. Figure 34.15B shows the integral and derivative terms. Note that changes (edges) in the integral waveform correspond to a positive and negative transition in the derivative waveform, because the derivative is measuring the amount that the error changed from the previous sample.

Figure 34.16A shows what happens if the derivative gain is set to 40; a high-frequency oscillation occurs, although it is centered on the setpoint. In Figure 34.16B, the derivative gain is set to 2 again, and the integral gain is set to 40. This condition causes an oscillation between about 135 and 172, and at a lower frequency than the

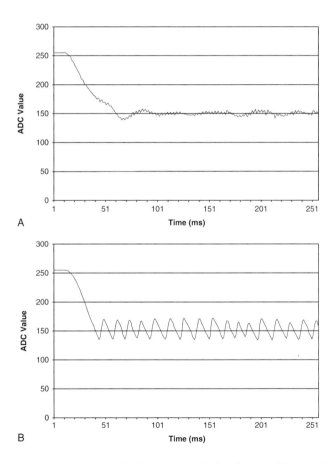

Figure 34.16: PID control with large derivative and integral values

oscillation caused by the large derivative value. This is typical of PID control systems—excessive derivative gain and excessive integral gain both cause oscillation, but the oscillation caused by the integral gain is at a lower frequency.

Figure 34.17A shows the following conditions:

- Proportional gain = 4

- Derivative gain = 2

- Integral gain = 2

- Setpoint = 150

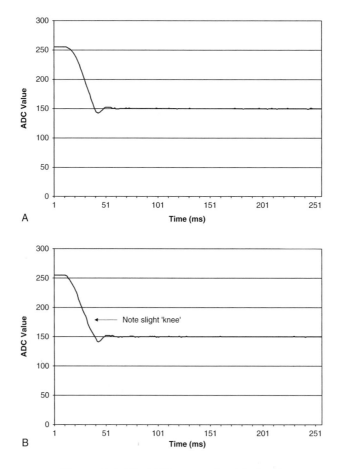

Figure 34.17: PID control with load

The result is a very smooth waveform with good control at the setpoint. The waveform in Figure 34.17B uses the same parameters, but adds a 47K resistor (R2) in parallel with the 22K resistor R1. The important thing to note here is that the final value still reaches the setpoint, although there is a "knee" in the waveform around sample 37.

34.9.4 Proportional-Integral Control

Figure 34.18 shows a proportional-integral control only, with proportional gain = 4 and integral gain = 0.1. The waveform overshoots to 140, which is past the value of 145 that was reached with the proportional-only control, but the integral eventually brings the result up to the target value of 150.

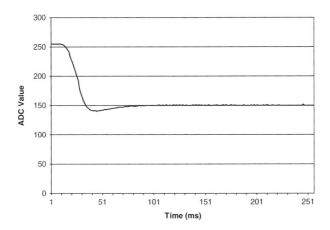

Figure 34.18: Proportional-integral control

CHAPTER 35

Power Supply Overview and Specifications

Darren Ashby
Tim Williams

35.1 Power Supplies

Whatever you do with electronics, you are going to need power to accomplish it. It will be useful to understand the basics of power supplies, as you are nearly guaranteed to deal with them at some point in your career.

35.1.1 It's All About the Voltage, Baby!

Most devices today want to keep the voltage constant. This means that current can vary as needed. In the world of power, particularly as it relates to the ubiquitous IC, it often seems that you never have the exact voltage you want.

A huge number of products run off of 120 VAC out of a wall socket. Another huge group runs off of batteries that are charged from those wall sockets, and another significant number runs off of batteries that you can buy by the caseload at any supermart. Just ask yourself, how many batteries did you buy last Christmas?

The problem is that most ICs these days want 5, 3.3, or even 1.5V DC. This is nowhere near 120V, and definitely not AC! Enter the power supply. They come in two flavors, linear and switcher.

35.1.2   Linear Power Supplies

AC rules! It is everywhere. It may seem like the world runs on batteries these days, but AC still has the majority foothold. Back when Edison and Tesla argued over what type of electrical power distribution we should have, I'll bet they had no idea of the type of integration that would occur in the world of electricity over the next 100 years.

One thing they did know about was the transformer. The basis of the transformer is AC current. Put AC into one side of the transformer and, depending on the ratio of windings, you get AC out the other side. So, put 120 VAC into a 10-to-1 ratio transformer and you will get 12 VAC out (minus heat losses due to the resistance of the windings).

The basic transformer is a very simple design. It is coils of wires on hunks of metal. That makes it robust. A transformer is a perfectly acceptable way to change the voltage of an AC signal. Transformers are used to jack the voltage way up to minimize losses over long wires, and then they are used again to bring the voltage back down to something safer to bring into your house.

They further knock the voltage down again in millions of products, but at that point they still output an AC signal. However, most of our chips want a DC signal, so what happens next?

It goes through a rectifier. There are two commonly used options, a bridge rectifier as shown in Figure 35.1, and a center tap rectifier, as shown in Figure 35.2. Note how this uses two fewer diodes and another wire to the transformer, yet the rectified output is the same.

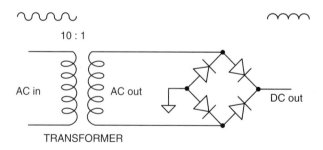

Figure 35.1: Bridge rectifier

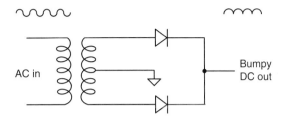

Figure 35.2: Center tap bridge rectifier

The output at this point is still too "bumpy" to be of much use to our sensitive DC circuits. The next step is to add a large filter capacitor to smooth out the bumps, as seen in Figure 35.3.

Here you need to understand the principle of output impedance. Every power supply has it. The more current you pull out of the circuit, the bigger issue the output impedance is. Remember that Ohm's Law says that, as current increases through an impedance, the voltage drop across it increases. This means that the voltage at the output will drop as load increases. To further complicate things, due to the rectifier in this circuit you will see an increased ripple voltage on the output as load increases. So there are two important things affecting the voltage on the output of this circuit: the voltage going into it (which on most AC circuits can vary 10% or more) and the amount of current being drawn, increasing the voltage drop and the voltage ripple.

This is important to know as we feed this into the next part of the circuit, called a *regulator*. The regulator is a part that adjusts its output to maintain a constant voltage in the face of a changing load and a changing input voltage.

A linear regulator typically has a voltage reference (like a zener diode) inside it running on a small current that isn't disrupted by the load. It uses this reference and a negative feedback loop to control a transistor or other part inside to maintain a constant

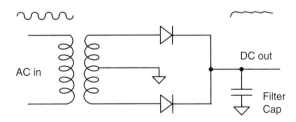

Figure 35.3: Center tap bridge rectifier with cap filter

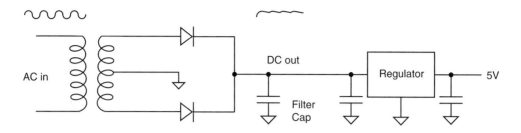

Figure 35.4: Typical linear regulated power supply

voltage at the output. This gets you to the nice DC voltage that your IC wants. The whole circuit from the wall looks something like that shown in Figure 35.4.

There are a couple of important things to know about linear regulators. They have a minimum input voltage. If the input voltage falls below this rating due to circumstances described above, the output will fall out of regulation. If this happens you can get ripple on the power supply to your chip. If it is small enough, you may not ever notice it, but if you have some high-gain circuits picking up AC noise, check out the power supply for problems first.

The other often-overlooked important spec is the power rating of the regulator. A regulator can only handle so much power, even with a heat sink. The power being dissipated by the regulator is the current times the voltage drop across the regulator, not the voltage at the output!

There are many other specs you should review in the data sheet, but these are the most important and often overlooked. Check them first. You can use linear regulators in any DC in, DC out situation. They will do very well in most cases and, to top that, hey are very simple and robust circuits. Use them whenever you can. There is nothing wrong with this technology in certain applications, but if you need more efficiency, or maybe less heat, you should consider a switcher.

35.1.3 Switcher

A type of regulator and power supply rapidly gaining footholds over the older, linear designs is called a *switcher*. As implied by the name, it regulates power to a load by switching current (or voltage) on and off. For this writing we will focus on the current method. (Don't forget, however, that current and voltage are invariably linked as

Ohm proved so well.) The secret to these supplies is the inductor and the secret to understanding an inductor for me is to think in terms of current. In the same way a capacitor wants to keep voltage across it constant, an inductor wants to keep the current flowing through it constant as well.

35.1.3.1 DC is What We Start With

Switching power supplies are DC-to-DC converters. Even those that have an AC input create a DC bus, using a rectifier circuit before implementing a switcher. You will see switchers replacing just the regulator in the circuit above working off of a DC bus voltage that has already been stepped down by a transformer. You will also see switchers that use rectified voltage right off the AC line and drop and regulate all in one step from 120V down to 5V.

The most basic current-switching supply I know of is the buck converter. A buck converter will knock a DC voltage from a higher level to a lower level. The following diagram shows the heart of a buck circuit.

First identify four main parts: the inductor, the switch, the diode, and the load.

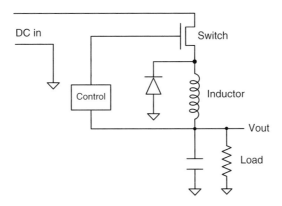

Figure 35.5: Basic switching buck converter

35.1.3.2 Flip the Switch

Let's start with the load and work our way backward. To begin with, switching supplies like to have a load. Without a load, funny things can happen, but more on that later.

What the load wants (in most cases) is a constant voltage. If I remember Ohm's Law correctly, one can control the voltage across a resistor (i.e., load) by controlling the current through it, so let's consider the flow of current in this circuit. We will begin with the switch closed. With the switch closed, current will flow through the inductor into the load. The current will rise based on the time constant of the inductor and the impedance of the load. Since the current rises, so does the voltage across the load. Assume now that we have a circuit that is monitoring the voltage across the load, and as soon as it gets too high it opens the switch.

Now what happens? First, remember this fact. Just like a capacitor resists a change in voltage, an inductor resists a change in current. When the switch opens, the inductor tries to keep the current flowing. If there is nowhere for it to go, you will see a large voltage develop across the inductor as the magnetic field collapses. In fact, at time = 0 the value of this voltage is infinite or undefined, whichever suits you. That doesn't happen in this case due to the diode and the load.

The current flows into the load, and the reason it does so is because of the diode. Consider it this way—current wants to keep flowing out of the inductor and into the other side of the inductor. Without the diode there would not be a path for this current to follow. However, with the diode, this current is pushed through the load. So now the switch is open, and current is still flowing into the load. This current starts out at the same level it was when the switch opened (an inductor wants to keep current the same, remember!) and it decays from there. As the current falls, so does the voltage. Of course we still have a circuit monitoring the voltage across the load, and as soon as it gets too low, it closes the switch again.

Voila, the process starts all over. There are two important things I noticed once the pieces fell into place in my head. The first is this control circuit I just described can be implemented with a simple comparator and a little hysteresis. Of course, that would lead to the frequency of the switcher being determined by the value of the inductor and the impedance of the load. That may or may not be a desirable trait. The other thing I realized was that when you first turned it on, the circuit would want to slam the switch shut and keep it there for a long time while current builds up in the circuit. Are you beginning to see why switchers need a load?

Luckily, others much smarter than I have dealt with these problems already. That is why you hear terms like "soft start" and "built in PWM" when you start studying switching supplies.

35.1.3.3 Some Final Thoughts on Switchers

Since designing switching supplies, getting them stable and dealing with the inductor specs can be a bit demanding, technical and tedious, all sorts of industry help has sprung up in the effort of various companies trying to get you to use their parts. You will find design guides and even web design platforms out there to help you build a switcher for your design, and I highly suggest you take advantage of them.

These days you will often find all the brains, switching components and feedback parts in one cute little package, making the design nicely compact and small. You can make switchers that boost voltage as well as the buck versions, and some that even go both ways, but ultimately they rely on the fact that the inductor wants to keep current flow the same. We will save the more in-depth review for another book on another day.

The best thing about switching supplies is the fact you can get by with relatively little copper and attain very high efficiency (meaning less heat). The reason for this is that the decay rate of the current in the inductor depends on the size of it, but if you switch it faster, the average current and thus the average voltage is still maintained. So you can get by with much less copper, especially for larger current draws at low voltages. The efficiency is good because much less power is spent heating the copper in the small inductor than in an equivalent transformer design. However, all this comes at a price. Switchers are known for their high-frequency noise that has disrupted many a sensitive analog design. But who cares about analog any more, right?

Thumb Rules

- Make sure the lowest dip on the ripple voltage doesn't go below minimum input of the regulator.
- Check your supply at $\pm15\%$ of the AC input signal.
- Linear regulators dissipate heat/power based on the current times the voltage from input to output (i.e., across it).
- Switchers exploit the fact that inductors want to keep current flowing even when the switch is open.
- Switchers are more efficient and create less heat but generally are more finicky to setup.
- Linear supplies are very quiet when it comes to EMI.
- Switchers tend to be very noisy when it comes to EMI.
- Switchers need a minimum load to work correctly.

35.2 Specifications

The technical and commercial considerations that apply to a power supply can add up to a formidable list. Such a list might run as shown in Table 35.1.

Table 35.1: Power supply specifications

• input parameters:	minimum and maximum voltage maximum allowable input current, surge and continuous frequency range, for AC supplies permissible waveform distortion and interference generation
• efficiency:	
• output parameters:	
• abnormal conditions:	output power divided by input power, over the entire range of load and line conditions
• mechanical parameters:	
	minimum and maximum voltage(s) minimum and maximum load current(s) maximum allowable ripple and noise load and line regulation transient response
• safety approval requirements	
• cost and availability requirements.	performance under output overload performance under transient input conditions such as spikes, surges, dips and interruptions performance on turn-on and turn-off: soft start, power-down interrupts
	size and weight thermal and environmental requirements input and output connectors screening

35.3 Off-the-Shelf or Roll Your Own

The first rule of power supply design is: do not design one yourself if you can buy it off the shelf. There are many specialist power supply manufacturers who will be only too pleased to sell you one of their standard units or, if this doesn't fit the bill, to offer you a custom version.

The advantages of using a standard unit are that it saves a considerable amount of design and testing time, the resources for which may not be available in a small company with short timescales. This advantage extends into production—you are buying a completed and tested unit. Also your supplier should be able to offer a unit which is already known to meet safety and EMC regulations, which can be a very substantial hidden bonus.

35.3.1 Costs

The major disadvantage will be unit cost, which will probably though not necessarily be more than the cost of an in-house designed and built power supply. The supplier must, after all, be able to make a profit. The exact economics depend very much on the eventual quantity of products that will be built; for lower volumes of a standard unit it will be cheaper to buy off the shelf, for high volumes or a custom-designed unit it may be cheaper to design your own. It may also be that a standard unit won't fulfill your requirements, though it is often worth bending the requirements by judicious circuit redesign until they match. For instance, the vast majority of standard units offer voltages of 3.3V or 5V (for logic) and ±12V or 15V (for analog and interface). Life is much easier if you can design your circuit around these voltages.

A graph of unit costs versus power rating for a selection of readily-available single output standard units is shown in Figure 35.6. Typically, you can budget for £1 per watt in the 50 to 200W range. There is little cost difference between linear and switch-mode types. On the assumption that this has convinced you to roll your own, the next chapter will examine the specification parameters from the standpoint of design.

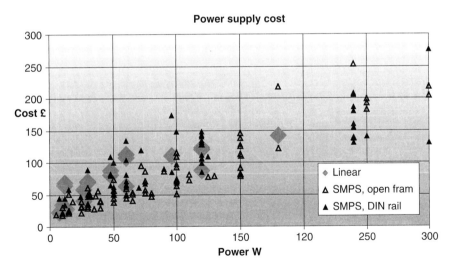

Figure 35.6: Prince vs. power rating for standard power supplies

CHAPTER 36

Input and Output Parameters

Tim Williams

36.1 Voltage

Typically you will be designing for 230V AC in the UK and continental Europe and 115V in the US. Other countries have frustratingly minor differences. The usual supply voltage variability is $\pm 10\%$, or sometimes $+10\%/-15\%$. In the UK the supply authorities are obliged to maintain their voltage at the point of connection to the customer's premises within $\pm 6\%$, to which is added an allowance for local loading effects. If the voltage tolerance is applied to the UK/Europe nominal then the input voltage range becomes 207–253V or 195–253V. This range must be handled transparently by the power supply circuitry.

To cope simultaneously with both the American supply voltage, which may drop below 100V, and the European voltages is difficult for a linear supply although it is possible to design "universal" switch-mode circuits which can accept such a wide range (see the comment at the end of section 36.5). Historically, this problem was handled by using a mains transformer with a split primary (Figure 36.1) which can be connected in series or parallel by means of a discreetly mounted voltage selector switch. This has the disadvantage that the switch may be so discreet that the user doesn't know about it, or else it may not be discreet enough and the user may be tempted to fiddle with it. This is not a real problem in the US, but applying 230V to a unit which is set for 115V will at least annoy the user by blowing a fuse, and at worst cause real damage. Universal switch-mode supplies are therefore popular.

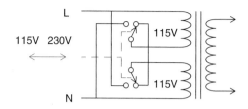

Figure 36.1: Split-primary transformer wiring

36.2   Current

The maximum continuous input current should be determined by the output load and the power conversion efficiency of the circuit. The main interest in this parameter is that it determines the rating of the input circuit components, especially the protective fuse. You have to decide whether an overload on the output will open the input circuit fuse or whether other protection measures, such as output current limiting, will operate. If the input fuse must blow, you need to characterize the input current very carefully over the entire range of input voltages. It is quite possible that the difference between maximum continuous current at full load, and minimum overload current at which the fuse should blow, is less than the fusing characteristics allow. Normally you need at least a 2:1 ratio between prospective fault current and maximum operating current. This may not be possible, in which case the input fuse protects the input circuit from faults only and some extra secondary circuit protection is necessary.

36.3   Fuses

A brief survey of fuse characteristics is useful here. The important characteristics that are specified by fuse manufacturers are the following:

- *Rated current I_N*: that value by which the fuse is characterized for its application and which is marked on the fuse. For fuses to IEC 60127 this is the maximum value which the fuse can carry continuously without opening and without reaching too high a temperature, and is typically 60% of its minimum fusing current. For fuses to the American UL-198-G standard the rated current is 85–90% of its minimum fusing current, so that it runs hotter when carrying its rated current. The minimum fusing current is that at which the fusing element just reaches its melting temperature.

- *Time-current characteristic*: the pre-arcing time is the interval between the application of a current greater than the minimum fusing current and the instant at which an arc is initiated. This depends on the over-current to which the fuse is subjected and manufacturers will normally provide curves of the time-current characteristic, in which the fuse current is normalized to its rated current as shown in Figure 36.2. Several varieties of this characteristic are available:

FF: very fast acting

F: fast acting

M: medium time lag

T: time lag (or anti-surge, slow-blow)

TT: long time lag

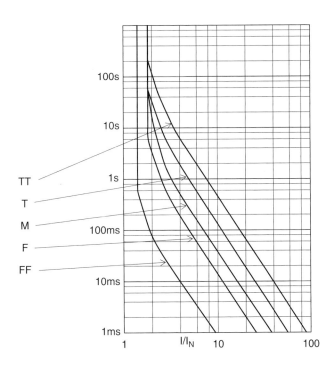

Figure 36.2: Typical fuse time-current curves

Most applications can be satisfied with either type F or type T and it is best to specify these if at all possible, since replacements are easily obtainable. Type FF is mainly used for protecting semiconductor circuits.

The total operating time of the fuse is the sum of the pre-arcing time and the time for which the arc is maintained. Normally the latter must be taken into account only when interrupting high currents, typically more than ten times the rated current.

The energy in a short-duration surge required to open the fuse depends on I^2t, and for pulse or surge applications you should consult the fuse's published I^2t rating. Current pulses that are not to open the fuse should have an I^2t value less than 50–80% of the I^2t value of the fuse.

- *Breaking capacity*: breaking capacity is the maximum current the fuse can interrupt at its rated voltage. The rated voltage of the fuse should exceed the maximum system voltage. To select the proper breaking capacity you need to know the maximum prospective fault current in the circuit to be protected – which is usually determined in mains-powered electronic products by the characteristics of the next fuse upstream in the supply. Cartridge fuses fall into one of two categories, high breaking capacity (HBC) which are sand-filled to quench the arc and have breaking capacities in the 1000s of amps, and low breaking capacity (LBC) which are unquenched and have breaking capacities of a few tens of amps or less.

36.4   Switch-on Surge, or Inrush Current

Continuous maximum input current is usually less than the input current experienced at switch-on. An unfortunate characteristic of mains power transformers is their low impedance when power is first applied. At the instant that voltage is applied to the primary, the current through it is limited only by the source resistance, primary winding resistance and the leakage inductance.

The effect is most noticeable on toroidal mains transformers when the mains voltage is applied at its peak halfway through the cycle, as in Figure 36.3. The typical mains supply has an extremely low source impedance, so that the only current limiting is provided by the transformer primary resistance and by the fuse. Toroidals are particularly efficient and can be wound with relatively few turns, so that their series

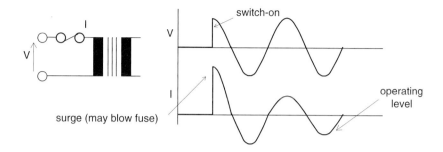

Figure 36.3: Switch-on surge

resistance and leakage inductance is low; the surge current can be more than ten times the operating current of the transformer.[†] In these circumstances, the fuse usually loses out. The actual value of surge depends on where in the cycle the switch is closed, which is random; if it is near the zero crossing the surge is small or nonexistent, so it is possible for the problem to pass unnoticed if it is not thoroughly tested.

A separate component of this current is the abnormal secondary load due to the low impedance of the uncharged power supply reservoir capacitor. For the same reason, inrush current is also a problem in direct-off-line switch-mode supplies, where the reservoir capacitor is charged directly through the mains rectifier, and comparatively complex "soft-start" circuits may be needed in order to protect the input components.

Several simpler solutions are possible. One is to specify an anti-surge or time-lag (type T or TT) fuse. This will rupture at around twice its rated current if sustained for tens or hundreds of seconds, but will carry a short overload of ten or twenty times rated current for a few milliseconds. Even so, it is not always easy to size the fuse so that it provides adequate protection without eventually failing in normal use, particularly with the high ratios of surge to operating current that can occur. A resettable thermal circuit breaker is sometimes more attractive than a fuse, especially as it is inherently insensitive to switch-on surges.

36.4.1 Current Limiting

A more elegant solution is to use a negative-temperature-coefficient (NTC) thermistor in series with the transformer primary and fuse. The device has a high initial resistance

---

[†] The effect happens with all transformers, but is more of a problem with toroidals.

which limits the inrush current but in so doing dissipates power, which heats it up. As it heats, its resistance drops to a point at which the power dissipated is just sufficient to maintain the low resistance and most of the applied voltage is developed across the transformer. The heating takes one or two seconds during which the primary current increases gradually rather than instantaneously.

NTC thermistors characterized especially for use as inrush current limiters are available, and can be used also for switch-mode power supply inputs, motor soft-start and filament lamp applications. Although the concept of an automatic current-limiter is attractive, there are three major disadvantages:

- because the devices operate on temperature rise they are difficult to apply over a wide ambient temperature range;

- they run at a high temperature during normal operation, so require ventilation and must be kept away from other heat-sensitive components; and

- they have a long cool-down period of several tens of seconds and so do not provide good protection against a short supply interruption.

36.4.2   PTC Thermistor Limiting

Another solution to the inrush current problem is to use instead a *positive-temperature-coefficient* thermistor in place of the fuse. These are characterized such that provided the current remains below a given value self-heating is negligible, and the resistance of the device is low. When the current exceeds this value under fault conditions the thermistor starts to self-heat significantly and its resistance increases until the current drops to a low value. Such a device does not protect against electric shock and so cannot replace a fuse in all applications, but because of its inherent insensitivity to surges it can be useful in local protection of a transformer winding.

A further more complex solution is to switch the AC input voltage only at the instant of zero crossing, using a triac. This results in a predictable switch-on characteristic, and may be attractive if electronic switching is required for other reasons such as standby control. Similarly, DC input supplies can use a power MOSFET to provide a controlled resistance at turn-on, as well as other circuitry such as reverse polarity protection and standby switching.

36.5   Waveform Distortion and Interference

36.5.1   Interference

Electrical interference generated within equipment and conducted out through the mains supply port was already subject to regulation for some product sectors in some countries, and with the adoption of the European EMC Directive it is mandatory for *all* electrical or electronic products to comply with interference limits. The usual method of reducing such interference is to use a radio frequency filter at the mains supply inlet, but good design practice also plays a substantial part. Switch-mode power supplies are normally the worst offenders, because they generate large interference currents at harmonics of the switching frequency well into the HF region. The size and weight advantages of switch-mode supplies are balanced by the need to fit larger filters to meet the interference limits.

36.5.2   Peak Current Summation

An increasing problem for electricity supply systems is the proportion of semiconductor-based equipment in the supply load. This is because the load current that such equipment takes is pulsed rather than sinusoidal. Current is only drawn at the peak of the input voltage, in order to charge the reservoir capacitors in the power supply. The normal RMS-to-average ratio of 1.11 for a sinusoidal current is considerably higher for this type of waveform (Figure 36.4).

The ratio of the peak load current I_{pk} to I_{rms} is called the *crest factor*, and here it depends on the input impedance of the reservoir circuit. The lower the impedance, the faster the reservoir capacitor(s) will charge, which results in lower output voltage ripple but higher peak current.

The significance of crest factor is that it affects the power handling capability of the supply network. A network of a given sinusoidal RMS current rating will show

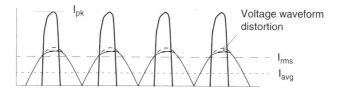

Figure 36.4: Peak input current in a rectifier/reservoir power supply

considerable extra losses when faced with loads of a high crest factor. The supply mains does not have zero impedance, and the result of the extra network voltage drop at each crest is a waveform distortion in which the sinusoidal peak is flattened. This is a form of harmonic distortion and its seriousness depends on the susceptibility of other loads and components in the network.

Large systems installations, in which there are many electronic power supplies of fairly high rating fed from the same supply, are the main threat. In domestic premises, the switch-mode supplies of TV sets are the main offenders; in commercial buildings, the problem is worst with switch-mode supplies of PCs and their monitors, and fluorescent lamps with electronic ballasts. The current peaks always occur together and so reinforce each other. A network which is dominated by resistive loads such as heating and filament-lamp lighting can tolerate a proportion of high crest factor loads more easily.

36.5.3   *Power Factor Correction*

The "peakiness" of the input current waveform is best described in terms of its harmonic content and legislation now exists in Europe, under the EMC Directive, to control this. The European standard EN 61000-3-2:2000 places limits on the amplitude of each of the harmonic components of the mains input current up to the 40th (2 kHz at 50 Hz mains frequency), and it applies to virtually all electrical and electronic apparatus up to an input current of 16A, although products other than lighting equipment with a rated power of less than 75W are exempt. The limits, although not particularly stiff, are pretty much impossible for a switched-mode power supply to meet without some treatment of the input current. This treatment is generically known as *power factor correction* (PFC).

In this context, Power Factor (PF) is the ratio between the real power, as transferred through the power supply to its load with associated losses, and the apparent power drawn from the mains: RMS line voltage times RMS line current. A purely resistive load will have a PF of unity, but since peaks increase the RMS current, one drawing a peaky waveform will have a PF of 0.5–0.75. "Correcting" the PF toward unity requires that the input current waveform is made nearly sinusoidal, so that its harmonic content is much reduced. This is done by a second switching "pre-regulator" operating directly at the mains input. The usual topology is a boost regulator, as shown in Figure 36.5.

The input rectifier supplies a full-wave-rectified half-sine voltage across C_{IN}. This capacitor is too low a value to affect the 50 Hz input current significantly, but high

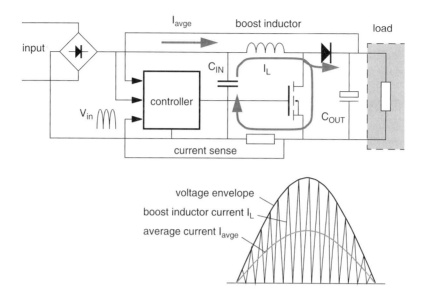

Figure 36.5: Power factor correction

enough to act as an effective reservoir at the switching frequency (typically 50–100 kHz). One sense input of the switching controller comes from this input voltage, and the controller is designed to maintain an average input current through the inductor in phase with this voltage. It does this by varying the switching pulse width or frequency as the input voltage changes. The rectified output is a DC voltage slightly higher than the highest peak supply voltage, which forms a reasonably well-regulated input to the main SMPS converter—which can of course be for any application, not just for an electronic power supply.

Naturally the addition of a second switching converter increases the cost of the total power supply, and contributes to more interference that must be filtered out at the mains input. Neither of these disadvantages are excessive, and commercial PFC power supply modules are now widely available. If you are designing your own, several IC manufacturers offer controllers specifically for the purpose, such as the L4981A/B, L6561, UC3853-5, and MC33626/33368. An extra advantage of the PFC pre-regulator is that almost by definition it will work over a wide input voltage range; so that a by-product of including it is that a single power supply will cover all worldwide markets, and will also have a uniform and predictable response to dips and interruptions.

36.6   Frequency

The UK and European mains frequency is held to 50 Hz ±1%. The American supply standard is 60 Hz. The difference in frequencies does not generally cause any problem for equipment that has to operate off either supply (provided that it's designed in Europe!), since mains transformers and reservoir circuits that perform correctly at 50 Hz will have no difficulty at 60 Hz. The sensitivity of the power supply circuits to supply voltage droops at 60 Hz should be less than at 50Hz since the ripple amplitude is only 83% of the 50 Hz figure, and the minimum voltage will thus be higher (Figure 36.6).

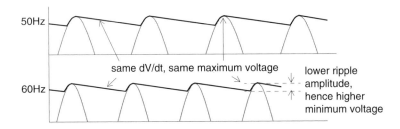

Figure 36.6: Ripple voltage vs. frequency

The ±1% tolerance on the mains frequency is slightly misleading because the supply authorities maintain a long-term tolerance very much better than this. Diurnal variations are arranged to cancel out, and this allows the mains to be used as a timing source for clocks and other purposes. If you are planning to use the mains frequency for internal timing then you will need to incorporate some kind of switching arrangement if the equipment will be used on both US and European systems.

36.7   Efficiency

The efficiency of a power supply module is its output power divided by its input power. The difference between the two quantities is accounted for by power losses in the various components in the power supply.

$$\text{Efficiency } \eta \; = \; P_{out}/P_{in} \; = \; P_{out}/(P_{out} \; + \; P_{loss})$$

The efficiency normally worsens as the load is reduced, because the various losses and quiescent operating currents assume a greater proportion of the input power. Therefore, if you are concerned about efficiency, do not use a power supply that is heavily overrated for its purpose. Linear supply efficiency also varies considerably with its input voltage, being worst at high voltages, because the excess must be lost across the regulator. Switch-mode supplies do not have this problem.

Normally efficiency is not of prime concern for mains power supplies, since it is not essential to make optimum use of the available power, although at higher powers the heat generated by an inefficient unit can be troublesome. It is far more important that a power converter for a portable instrument should be efficient because this directly affects useable battery life.

Linear power supplies are rarely more than 50% efficient unless they can be matched to a narrow input voltage range, whereas switch-mode supplies can easily exceed 70% and with careful design can reach 90%. This makes switch-mode supplies more popular, despite their greater complexity, at the higher power levels and for battery-powered units.

36.7.1   Sources of Power Loss

The components in a power supply which make the major contribution to losses are:

• the transformer:	core losses, determined by the operating level and core material, and copper losses, determined by I^2R where R is the winding resistance
• the rectifiers:	diode forward voltage drop, V_F, multiplied by operating current; more significant at low output voltages
• linear regulator:	the voltage dropped across the series pass element multiplied by the operating current; greatest at high input voltages
• switching regulator:	power dissipated in the switching element due to saturation voltage, plus switching losses in this and in snubber and suppressor components, proportional to switching frequency.

If you sum the approximate contribution of each of these factors you can generally make a reasonable forecast of the efficiency of a given power supply design. The actual figure can be confirmed by measurement and if it is wildly astray then you should be looking for the cause.

36.8   Deriving the Input Voltage from the Output

In a linear supply with a series pass regulator element, the design must proceed from the minimum acceptable output voltage at maximum load current and minimum input voltage. These are the worst-case conditions and determine the input voltage step-down required. The minimum DC input voltage is given by the minimum output voltage plus all the tolerances and voltage drops in series:

$$V_{in,dc} = V_{out(min)} + V_{tol,reg} + V_{series,reg} + V_{series,CS...}$$

where,

$V_{out(min)}$ is the minimum acceptable output voltage

$V_{tol,reg}$ is the regulator voltage tolerance, assuming it is not adjustable

$V_{series,reg}$ is the voltage drop across the regulator series pass element

$V_{series,\,CS}$ is the voltage drop across the current sense element if fitted

All the above parameters are specified at full load current. This value for $V_{in,dc}$ is then the minimum input voltage allowed for a DC input supply, or it is the voltage at the minimum of the ripple trough for a rectified and smoothed AC input supply. This is related to the transformer secondary voltage as follows:

$$V_{tx} = (V_{in,dc} + V_{ripple} + V_D)/0.92 \cdot (V_{ac(nom)}/V_{ac(min)}) \cdot 1/\sqrt{2}$$

where,

V_{ripple} is the peak ripple voltage across the reservoir capacitor

V_D is the voltage drop across the rectifier diode(s)

V_{tx} is the RMS transformer secondary voltage

$V_{ac(nom)}$ is the specified transformer input voltage

$V_{ac(min)}$ is the minimum line input voltage

All parameters at full load current

The figure of 0.92 is an approximate allowance for full-wave rectifier efficiency with a single-capacitor reservoir. It can be more accurately derived using curves published by Schade[†]. Complications set in because the current drawn through the secondary is

---

[†] O.H.Schade, Analysis of Rectifier Operation, Proc.IRE, vol 31, 1943, pp. 341–361.

not sinusoidal, but occurs at the crest of the waveform (see section 36.5). The extra peak current reduces the peak secondary voltage from its quoted value, if this value is specified for a resistive load. You can get around this either by knowing the transformer's losses in advance and allowing for the extra IR drop, or by specifying the transformer for a given circuit and letting the transformer supplier do the work for you, if you're buying a custom component. The transformer secondary RMS current rating is determined by the rectifier configuration (Figure 36.7).

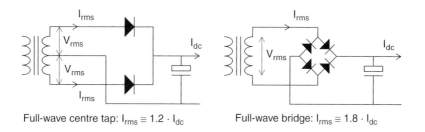

Full-wave centre tap: $I_{rms} \cong 1.2 \cdot I_{dc}$ Full-wave bridge: $I_{rms} \cong 1.8 \cdot I_{dc}$

Figure 36.7: Rectifier configuration

Take as an example a typical linear regulator circuit supplying 5V ±5% at 1A.

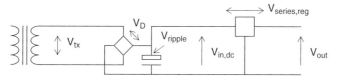

Here, $V_{out(min)}$ is allowed to be 5V − 5% = 4.75V. The regulator we shall use is a standard 7805 type with ±4% tolerance and so $V_{tol,reg}$ is 5V · 0.04 = 0.2V. Its specified minimum series voltage drop (or dropout voltage) at 1A and a junction temperature of 25°C (note the temperature restriction) is 2.5V maximum. The required minimum input voltage is:

$$V_{in,dc} = 7.45 + 0.2 + 2.5 = \underline{7.45V}$$

If the peak ripple voltage is 2V and each diode forward drop in the bridge is 1V, then the transformer voltage with a 240V nominal spec but a minimum line voltage of 195V will need to be:

$$V_{tx} = 7.45 + 2 + (2 \times 1)/0.92 \cdot 240/195 \cdot 1/\sqrt{2} = 10.83Vrms$$

From this example you can see that the secondary-side input voltage needed to assure a given output voltage is very much higher than the actual output voltage. One of the major culprits is the dropout voltage of the regulator which in this example accounts for at least 50% of the output power, although it becomes proportionally less at higher output voltages. Low-dropout voltage regulators which use a PNP transistor as the series pass element, such as National Semiconductor's LM2930 range, are popular for this reason and also where the minimum input voltage can be close to the output level, as in automotive applications.

36.8.1   Power Losses at High Input Voltage

You can also see more clearly in the above example where the power losses are which contribute to reduced efficiency. When the input voltage is increased to its maximum value the dissipation in the series-pass element is worst. In the above example with the mains input at 264V, the average value of $V_{in,dc}$ rises to 12.5V, and 7.45V of this must be lost across the regulator, which because it is passing the full load current amounts to one-and-a-half times the load power! The advantage of the switch-mode supply is that it adjusts to varying input voltages by modifying its switching duty cycle, so that an increased input voltage automatically reduces the input current and the overall power taken by the unit remains roughly constant.

36.9   Low-Load Condition

When the output load is removed or substantially reduced then the dissipation in the power supply will drop. This is good news for almost all parts of the circuit, except for the voltage rating of the components around $V_{in,dc}$. When there is a combination of low load and maximum supply input voltage, the peak value of $V_{in,dc}$ is highest. A crucial factor here is the transformer regulation. This is the ratio:

$$\text{Regulation} = (V_{sec,unloaded} - V_{sec,loaded})/V_{sec,loaded}$$

and a small or poorly-designed transformer can have a regulation exceeding 20%. If this figure is used for the transformer in the above example then the peak V_{tx} off-load at maximum input voltage will rise to 20.2V. At the same time the diode forward voltage drops at low current will be much less, say 0.6V each, so the possible maximum voltage at the reservoir capacitor could be around 19V. Thus even the common 16V rated electrolytic will not be adequate for this circuit. For higher voltage outputs, the maximum input voltage can even exceed the voltage rating of the regulator itself, and you have to invest in a pre-regulator to hold the maximum to a manageable level. Note that this condition is not the worst-case for regulator power dissipation, because the regulator is not passing significant load current.

36.9.1 Maximum Regulator Dissipation

In fact maximum series-pass dissipation does not necessarily occur at full load current, because as the current rises the voltage across the series-pass element falls. The maximum dissipation will occur at less than full output if the voltage dropped across the DC supply's equivalent series resistance is greater than half the difference between the no-load input voltage and the output voltage. Figure 36.8 shows this graphically.

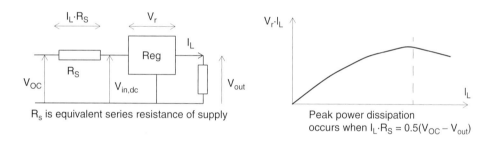

Figure 36.8: Peak power dissipation

36.9.2 Minimum Load Requirement

A further problem, particularly with switch-mode supplies, is that the stability of the regulator cannot always be assured down to zero load. For this reason some rails have to be run with a minimum load, such as a bleed resistor, to remain within specification, and this represents an unnecessary additional power drain. Many circuits, of course, always take a minimum current and so the minimum load is not then a problem.

36.10 Rectifier and Capacitor Selection

The specification of the rectifiers and capacitors is dominated by surge and ripple current concerns.

36.10.1   Reservoir capacitor

The minimum capacitor value is easily decided from the required ripple voltage:

$$C = I_L/V_{ripple} \cdot t$$

where,

I_L is the DC load current

V_{ripple} is the acceptable ripple voltage

For mains inputs, t is about 2 ms less than the AC input period, 8 ms for 50 Hz or

6 ms for 60 Hz full-wave

A more accurate value can be derived from Schade's curves (see previous footnote) which have been reprinted in numerous textbooks, but remember that the tolerance on reservoir capacitors is wide (typically ±20%) and accuracy is rarely needed.

For load currents exceeding 1A, ripple current rating tends to determine capacitor selection rather than ripple voltage. As is made clear throughout this chapter, the peak current flow through the rectifier/capacitor circuit is many times higher than the DC current, due to the short time in each cycle for which the capacitor is charging. The RMS ripple current is 2–3 times higher than the DC load. Ripple current rating is directly related to temperature and you may need to derate the component further if you need high ambient temperature and/or high reliability operation.

As an example, a load current of 2A and a permissible ripple voltage of 3V at 100 Hz suggests a 5300 μF capacitor. Typical capacitors of the next value up from this, 6800 μF, have 85°C ripple current ratings from 2 to 4A. The higher ratings are larger and more expensive. But actual ripple current requirements will be 4–6A. To meet this you will need to use either a much larger capacitor (typically 22,000 μF), or two smaller capacitors in parallel, or derate the operating temperature *and* use a slightly larger capacitor. If you don't do this, your design will become yet another statistic to prove that electrolytic capacitors are the prime cause of power supply failure.

36.10.2   Rectifiers

Although in the full-wave arrangements (Figure 36.7) the diodes only conduct on alternate half cycles, because the RMS current is 2–3 times higher than the DC load current a rating of *at least* the full load current, and preferably twice it, is necessary.

Surge current on turn-on may be much higher, especially in the higher power supplies where the ratio of reservoir capacitance to operating current is increased. This is of even greater concern in direct-off-line switch-mode supplies where there is no transformer series resistance to limit the surge, and a diode rating of up to 5 times the average DC current is needed.

The maximum instantaneous surge current is V_{max}/R_s and the capacitor charges with a time constant of $\tau = C \cdot R_s$, where R_s is the circuit series resistance. As a conservative guide, the surge won't damage the diode if τ is less than a half-cycle at mains frequency and V_{max}/R_s is less than the diode's rated I_{FSM}. All diode manufacturers publish I_{FSM} ratings for a given time constant; for example, the typical 1N5400 series with 3A average rating have an I_{FSM} of 200A. You may discover that you have to incorporate a small extra series resistance to limit the surge current, or use a larger diode, or apply the techniques discussed in section 36.4.

The rectifier's peak-inverse-voltage (PIV) rating needs to be at least equivalent to the peak AC voltage for the full-wave bridge circuit, or twice this for the full-wave centre tap. But you should increase this considerably (by 50 to 100%) to allow for line transients. This is easy for low-voltage circuits, since 200V diodes cost hardly any more than 50V ones, and does not normally make much cost difference in mains circuits. For 240V, a minimum of 600V PIV and preferably 800V PIV should be specified, even if you are using a transient suppressor at the input.

36.11   Load and Line Regulation

Load regulation refers to the permissible shift in output voltage when the load is varied, usually from none to full. Line (or input) regulation similarly refers to the permissible shift in output voltage when the input is varied, usually from maximum to minimum. Provided that the design of the input circuit has been properly considered as described above, so that the input voltage never goes outside the regulator's operational range, these parameters should be wholly a function of the regulator circuit itself. The regulator is essentially a feedback circuit which compares its output voltage against a reference voltage, so the regulation depends on two parameters: the stability of the reference, and the gain of the feedback error amplifier. If you use a monolithic regulator IC, then these factors are taken into account by the manufacturer who will specify regulation as a data sheet parameter.

36.11.1   Thermal Regulation

A monolithic regulator IC includes the voltage reference on-chip, along with other circuitry and the series pass element. This means that the reference is subject to a thermal shift when the power dissipation of the series pass element changes. This gives rise to a separate longer term component of regulation, called *thermal* regulation, defined as the change in output voltage caused by a change in dissipated power for a specified time. Provided the chip has been well-designed, thermal regulation is not a significant factor for most purposes, but it is rarely specified in data sheets and for some precision applications may render monolithic regulators unsuitable.

36.11.2   Load Sensing

No three-terminal regulator can maintain a constant voltage at anywhere other than its output terminals. It is common in larger systems for the load to be located at some distance from the power supply module, so that load-dependent voltage drops occur in the wiring connecting the load to the power supply output. This directly impacts the achievable load regulation.

The accepted way to overcome this problem is to split the regulator feedback path, and incorporate two extra "sensing" terminals which are connected so as to sense the output voltage at the load itself (Figure 36.9). The voltage drop across this extra pair

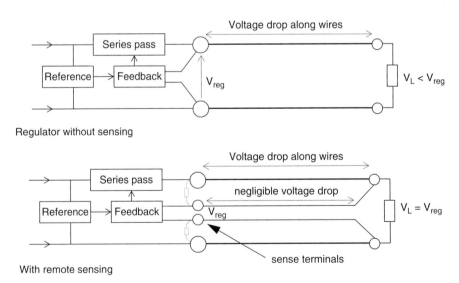

Figure 36.9: Load sensing

of wires is negligible because they only carry the signal current. The voltage at the regulator output is adjusted so as to regulate the voltage at the sensing terminals.

The minimum voltage at the regulator input must be increased to allow for the extra output voltage drop. It is wise to connect coupling resistors (shown shaded in Figure 36.9) from the output to sense terminals, so as to ensure correct operation when the sense terminals are accidentally or deliberately disconnected. Sensing can only offer remote load regulation at one point and so is not really suited when one power supply module feeds several loads at different points.

36.12   Ripple and Noise

Ripple is the component of the AC supply frequency (or more often its second harmonic) which is present on the output voltage; noise is all other AC contamination on the output. In a linear power supply, ripple is the predominant factor and is given by the AC across the reservoir capacitor reduced by the ripple rejection (typically 70–80 dB) of the regulator circuit. A figure of less than 1 mV RMS should be easy to obtain. HF noise is filtered by the reservoir and output capacitors and there are no significant internal noise sources, provided that the regulator isn't allowed to oscillate, so that apart from supply-frequency ripple linear power supplies are very "quiet" units.

36.12.1   Switching Noise

The same cannot be said for switch-mode power supplies. Here the noise is mainly due to output voltage spikes at the switching frequency, caused by fast-rise-time edges and HF ringing at these edges feeding through, or past, filtering components to the output. The ESR and ESL of typical output filter capacitors limits their ability to attenuate these spikes, while the self-inductance of ground wiring limits the high frequency effectiveness of ground decoupling anyway. Switch-mode output ripple and noise is typically 1% of the rail voltage, or 100–200 mV. In fact comparing ripple and noise specifications is the easiest way to distinguish a linear from a switch-mode unit, if there is no other obvious indication. The bandwidth over which the specification applies is important, since there is significant energy in the high-order harmonics of the switching noise, and at least 10 MHz is needed to get a true picture. Because of stray coupling over this extended bandwidth the noise frequently appears in common mode, on both supply and 0V simultaneously, and is then very difficult to control. Differential mode noise spikes can be reduced dramatically by including a ferrite bead in series, and a small ceramic capacitor in parallel with the output capacitor.

The presence of switching noise is not a problem for digital circuits, but it creates difficulties for sensitive analog circuits if their bandwidth exceeds the switching frequency. It can cause interference on video signals, mis-clocking in pulse circuits and voltage shifts in DC amplifiers. These effects have to be treated as EMC phenomena and can be cured by suitable layout, filtering and shielding, but if you have the option in the early stages to choose a linear supply instead, take it—you will save yourself a lot of trouble.

36.12.2   Layout to Avoid Ripple

Power supply output ripple is aggravated by incorrect layout of the wiring around the reservoir capacitor.

At first sight, grounds A and B in Figure 36.10 look equivalent. But there will be a potential between them of $I_R \cdot R_g$, where I_R is the capacitor ripple current and R_g is the track or wiring resistance common to the two grounds through which the ripple current flows. (The ripple current path is through the transformer, the two diodes and the capacitor.) This current is only drawn on peaks of the AC input waveform to charge the reservoir capacitor, and its magnitude is only limited by the combined series resistance of the transformer winding, the diodes, capacitor and track or wiring. If the steady-state DC current supplied is 1A then the peak ripple current may be of the order of 5A; thus, 10 mΩ of R_g will give a peak difference of 50 mV between grounds A and B. If some parts of the circuit are grounded to A and some to B, then tens of millivolts of hum injection are included in the design at no additional cost, and increasing the reservoir value to try and reduce it will actually make matters worse

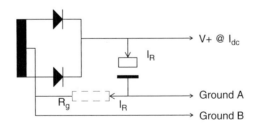

Figure 36.10: Incorrect reservoir connection

as the peak ripple current is increased. You can check the problem easily, by observing the output ripple on a 'scope; if it has a pulse shape then wiring is the problem, if it looks more like a sawtooth then you need more smoothing.

36.12.3 Correct Reservoir Connection

The solution to this problem, and the correct design approach, is to ground *all* parts of the supplied circuit on the supply side of the reservoir capacitor, so that the ripple current ground path is not common to any other part of the circuit (Figure 36.11). The same applies to the V+ supply itself. The common impedance path is now reduced to the capacitor's own ESR, which is the best you can do.

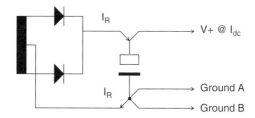

Figure 36.11: Correct reservoir connection

36.13 Transient Response

The transient response of a power supply is a measure of how fast it reacts to a sudden change in load current. This is primarily a function of the bandwidth of the regulator's feedback loop. The regulator has to maintain a constant output in the face of load changes, and the speed at which it can do this is set by its frequency response as with any conventional operational amplifier. The trade-off that the power supply designer has to worry about is against the stability of the regulator under all load conditions; a regulator with a very fast response is likely to be unstable under some conditions of load, and so its bandwidth is "slugged" by a compensation capacitor within the regulator circuit. Too much of this and the transient response suffers. The same effect can be had by siting a large capacitor at the regulator output, but this is a brute-force and inefficient approach because its effect is heavily load-dependent. Note that the 78XX series of three-terminal regulators should have a small, typically 0.1 µF capacitor at the output for good transient response and HF noise decoupling. This is separate from the required 0.33–1 µF capacitor at the input to ensure stability.

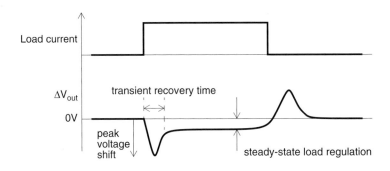

Figure 36.12: Load transient response

36.13.1   Switch-Mode vs. Linear

The transient response of a switch-mode power supply is noticeably worse than that of a linear because the bandwidth of the feedback loop has to be considerably less than the switching frequency. Typically, switch-mode transient recovery time is measured in milliseconds while linear is in the tens of microseconds.

If your circuit only presents slowly-varying loads then the power supply's transient response will not interest you. It becomes important when a large proportion of the load can be instantaneously switched—a relay coil or bank of LEDs for example—and the rest of the load is susceptible to short-duration over- or undervoltages.

Although load transient response is usually the most significant, a regulator also exhibits a delayed response to line transients, and this may become important when you are feeding it from a DC input which can change quickly. The line transient response is normally of the same order as the load response.

CHAPTER 37

Batteries

Tim Williams

Battery power is mainly used for portability or stand-by (float) purposes. All batteries operate on one or another variant of the principle of electrochemical reaction, in which anode (negative) and cathode (positive) terminals are separated by an electrolyte, which is the vehicle for the reaction. This basic arrangement forms a "cell," and a battery consists of one or more cells. The chemistry of the materials involved is such that a potential is developed between the electrodes which is capable of sustaining a discharge current. The voltage output of a particular cell type is a complex function of time, temperature, discharge history and state of charge.

The basic distinction is between primary (nonrechargeable) and secondary (rechargeable) cells. This section will survey the various types of each shortly, but first we shall make a few general observations on designing with batteries.

37.1 Initial Considerations

When you know you are going to use a battery, select the cell type as early as possible in the circuit and mechanical design. This allows you to take the battery's properties into account and increases the likelihood of a cost-effective result, as otherwise you will probably need a larger, or more expensive, battery or will suffer a reduced equipment specification. Having made the selection, you can then design the circuit so that it works over the widest possible part of the battery's available voltage range. Some of the cheaper types deliver useful power over quite a wide range, with an endpoint voltage of 60–70% of nominal, and some of this energy will be lost if the design cannot cope with it. Also, check that the battery can deliver the circuit's load current requirements over the working temperature. This capability varies considerably for different chemical systems. Rechargeable batteries can often be recharged only over a narrower temperature range than they can be discharged.

Always aim to use standard types if your specification calls for the user to be able to replace the battery. Not only are they cheaper and better documented, but they are widely distributed and are likely to remain so for many years. You should only need to use special batteries if your environmental conditions or energy density requirements are extreme, in which case you have to make special provisions for replacement or else consider the equipment as a throwaway item.

37.1.1 Voltage and Capacity Ratings

Different types of battery have different nominal open-circuit voltages, and the actual terminal voltage falls as the stored energy is used. Manufacturers provide a discharge characteristic curve for each type which indicates the behavior of voltage against time for given discharge conditions. Note that the open-circuit voltage can exceed the voltage under load by up to 15%, and the operating voltage may be significantly less than the nominal battery voltage for some of the duration.

The capacity of a battery is expressed in ampere-hours (Ah) or milliampere-hours (mAh). It may also be expressed in normalized form as the "C" figure, which is the nominal capacity at a given discharge rate. This is more frequently applied to rechargeable types. Capacity will be less than the C rating if the battery is discharged at a faster rate; for instance, a 15 Ah lead-acid type discharged at 15 amps (1C) will only last for about 20 minutes (Figure 37.2).

Three typical modes under which a battery can be discharged are constant resistance, constant current and constant power. For batteries with a sloping discharge characteristic, such as alkaline manganese, the constant power mode is the most efficient user of the battery's energy but also needs the most complex voltage regulating system to power the actual circuit.

37.1.2 Series and Parallel Connection

Cells can be connected in series to boost voltage output, but doing so decreases the reliability of the overall battery and there is a risk of the weakest cell being driven into reverse voltage at the end of its life. This increases the likelihood of leakage or rupture, and is the reason why manufacturers recommend that all cells should be replaced at the same time. Good design practice minimizes the number of series-connected cells. There are now several ICs which can be used to multiply the voltage output of even a single cell with high efficiency. It is not difficult to design a switching converter that simultaneously boosts and regulates the battery voltage.

Parallel connection can be used for some types to increase the capacity or discharge capability, or the reliability of the battery. Increased reliability requires a series diode in each parallel path to isolate failed cells. Recharging parallel cells is rarely recommended because of the uncertainty of charge distribution between the cells. It is therefore best to restrict parallel connection to specially-assembled units.

On the same subject, reverse insertion of the whole battery will threaten your circuit, and if it is possible, the user will do it. Either incorporate assured polarity into the battery compartment or provide reverse polarity protection, such as a fuse, series diode or purpose-designed circuit, at the equipment power input.

37.1.3   Mechanical Design

Choose the battery contact material with care to avoid corrosion in the presence of moisture. The recommended materials for primary cells are nickel-plated steel, austenitic stainless steel or inconel, but definitely not copper or its alloys. The contacts should be springy in order to take up the dimensional tolerances between cells. Single-point contacts are adequate for low current loads, but you should consider multiple contacts for higher current loads. The simplest solution is to use ready-made battery compartments or holders, provided that they are properly matched to the types of cell you are using. PCB-mounting batteries have to be hand soldered in place after the rest of the board has been built, and you need to liaise well with the production department if you are going to specify these types.

Rechargeable batteries when under charge, and all types when under overload, have a tendency to out-gas. Always allow for safe venting of any gas, and since some gases will be flammable, don't position a battery near to any sparking or hot components. In any case, heat and batteries are incompatible: service life and efficiency will be greatly improved if the battery is kept cool. If severe vibration or shock is part of the environment, remember that batteries are heavy and will probably need extra anchorage and shock absorption material. Organic solvents and adhesives may affect the case material and should be kept away.

Dimensions of popular sizes of battery are shown in Table 37.1.

Table 37.1: Sizes of popular primary batteries

Designation				Dimensions mm	
IEC	ANSI	Size	Voltage	Dia (or L × W)	Height
Alkaline manganese dioxid					
LR03	24A	AAA	1.5	10.5	44.5
LR6	15A	AA	1.5	14.5	50.5
LR14	14A	C	1.5	26.2	50
LR20	13A	D	1.5	34.2	61.5
6LR61	1604A	PP3	9	26.5 ×	48.5
4LR25X	908A	Lamp	6	67 × 67	115
4LR25-2	918A	Lamp	6	136.5 × 73	127
Lithium manganese dio xide - cylindrical cell					
CR17345	5018LC	2/3A	3	17	34.5
CR11108	5008LC	1/3N	3	11.6	10.8
2CR11108	1406LC	2 ×	6	25.2	13
2CR5	5032LC	2 ×	6	17 × 34	45
CR-P2	5024LC	2 ×	6	19.5 × 35	36
Lithium manganese dioxide - coin cell					
CR2016	5000LC		3	20	1.6
CR2025	5003LC		3	20	2.5
CR2032	5004LC		3	20	3.2
CR2430	5011LC		3	24.5	3
CR2450	5029LC		3	24.5	5
Silver oxide button cells mAh					
SR41	1135S0	42	1.55	7.87	3.6
SR43	1133S0	120	1.55	11.56	4.19
SR44	1131S0	165	1.55	11.56	5.58
SR48	1137S0	70	1.55	7.87	5.38

(Continued)

Table 37.1: Sizes of popular primary batteries (Cont'd)

Designation				Dimensions mm	
IEC	ANSI	Size	Voltage	Dia (or L × W)	Height
SR54	1138S0	70	1.55	11.56	3.05
SR55	1160S0	40	1.55	11.56	2.21
SR57	1165S0	55	1.55	9.5	2.69
SR59	1163S0	30	1.55	7.9	2.64
SR60	1175S0	18	1.55	6.8	2.15
SR66	1176S0	25	1.55	6.78	2.64

37.1.4 Storage, Shelf Life, and Disposal

Maximum shelf life is obtained if batteries are stored within a fairly restricted temperature and humidity range. Self-discharge rate invariably increases with temperature. Different chemical systems have varying requirements, but extreme temperature cycling should be avoided, and you should arrange for tight stock control with proper rotation of incoming and outgoing units, to ensure that an excessively aged battery is not used. Rechargeable types should be given a regular top-up charge.

In the early 90s, legislation appeared in many countries banning the use of some substances in batteries, particularly mercury, for environmental reasons. Thus mercuric oxide button cells were effectively outlawed and are not now obtainable. In Europe, this was achieved through the Batteries and Accumulators Directive (91/157/EEC).

This Directive also encourages the collection of spent NiCad batteries with a view to recovery or disposal, and their gradual reduction in household waste. In fact, what it has achieved is rather the development of alternative rechargeable technologies to NiCad, particularly NiMH and lithium. NiCads, though, are still widely used, despite the technical advantages of NiMH. The Batteries Directive is about to be updated and it is likely to propose the following changes:

- EU member states to collect and recycle all batteries, with targets of 75% consumer (disposable or rechargeable) and 95% industrial batteries

- no less than 55% of all materials recovered from the collection of spent batteries to be recycled.

In the UK in 1999, 654 million consumer batteries were sold, but the rate for recycling consumer rechargeables is a mere 5%, and less than 1% of consumer batteries are collected for recycling. On the other hand, more than 90% of automotive batteries are recycled and 24% of other industrial batteries. Clearly, for consumer batteries at least, a sea change in disposal habits is expected.

37.2   Primary Cells

The most common chemical systems employed in primary, nonrechargeable cells are alkaline manganese dioxide, silver oxide, zinc air and lithium manganese dioxide. Figure 37.1 compares the typical discharge characteristics for lithium and alkaline types of roughly the same volume on various loads.

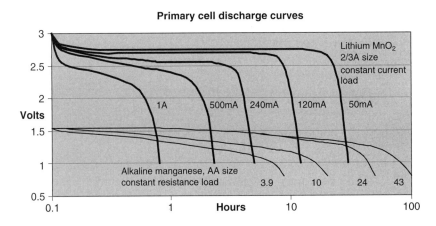

Figure 37.1: Load discharge characteristics for lithium and alkaline manganese primary cells

37.2.1   Alkaline Manganese Dioxide

The operating voltage range of this type, which uses a highly conductive aqueous solution of potassium hydroxide as its electrolyte, is 1.3 to 0.8V per cell under normal load conditions, while its nominal voltage is 1.5V. Recommended end voltage is 0.8V per cell for up to 6 series cells at room temperature, increasing to 0.9V when more cells are used. The alkaline battery is well suited to high-current discharge. It can operate between -30 and $+80°C$, but high relative humidity can cause external corrosion and should be avoided. Shelf life is good, typically 85% of stored energy being retained after 3 years at $20°C$. Standard types are now widely and cheaply available in retail outlets and it can therefore be confidently used in most general-purpose applications.

37.2.2 Silver Oxide

Zinc-silver oxide cells are used as button cells with similar dimensions and energy density to the older and now withdrawn mercury types. Their advantage is that they have a high capacity versus weight, offer a fairly high operating voltage, typically 1.5 V, which is stable for some time and then decays gradually, and can provide intermittent high pulse discharge rates and good low temperature operation. They are popular for such applications as watches and photographic equipment. Typical shelf life is two years at room temperature.

37.2.3 Zinc Air

This type has the highest volumetric energy density, but is very specialized and not widely available. It is activated by atmospheric oxygen and can be stored in the sealed state for several years, but once the seal is broken it should be used within 2 months. It has a comparatively narrow environmental temperature and relative humidity range. Consequently its applications are somewhat limited. Its open circuit voltage is typically 1.45V, with the majority of its output delivered between 1.3 and 1.1V. It cannot give sustained high output currents.

37.2.4 Lithium

Several battery systems are available based on the lithium anode with various electrolyte and cathode compounds. Lithium is the lightest known metal and the most electro-negative element. Their common features are a high terminal voltage, very high energy density, wide operating temperature range, very low self discharge and hence long shelf life, and relatively high cost. They have been used for military applications for some years. If abused, some types can be potentially very hazardous and may have restrictions on air transport. The lithium manganese dioxide ($LiMnO_2$) couple has become established for a variety of applications, because of its high voltage and "fit-and-forget" lifetime characteristics. Operating voltages range from 2.5 to 3.5V. Very high pulse discharge rates (up to 30A) are possible. Widely available types are either coin cells, for memory back-up, watches and calculators and other small, low power devices; or cylindrical cells, which offer light weight combined with capacities up to 1.5 Ah and high pulse current capability, together with long shelf life and wide operating temperature range.

Other primary lithium chemistries are lithium thionyl chloride (Li-SOCl$_2$) and lithium sulfur dioxide (Li-SO$_2$). These give higher capacities and pulse capability and wider temperature range but are really only aimed at specialized applications.

37.3 Secondary Cells

There have historically been two common rechargeable types: lead-acid and nickel-cadmium. These have quite different characteristics. Neither of them offer anywhere near the energy density of primary cells. At the same time, their heavy metal content and consequent exposure to environmental legislation have spurred development of other technologies, of which NiMH and lithium ion are the frontrunners.

37.3.1 Lead-Acid

The lead-acid battery is the type which is known and loved by millions all over the world, especially on cold mornings when it fails to start the car. As well as the conventional "wet" automotive version, it is widely available in a valve-regulated "dry" or "maintenance-free" variant in which the sulfuric acid electrolyte is retained in a glass mat and does not need topping-up. This version is of more interest to circuit designers as it is frequently used as the standby battery in mains-powered systems which must survive a mains failure.

These types have a nominal voltage of 2V, a typical open circuit voltage of 2.15V and an end-of-cycle voltage of 1.75V per cell. They are commonly available in standard case sizes of 6V or 12V nominal voltage, with capacities from 1 to 100 Ah. Typical discharge characteristics are as shown in Figure 37.2. The value "C", as noted earlier, is the ampere-hour rating, conventionally quoted at the 20-hour discharge rate (5-hour discharge rate for nickel-cadmium and nickel metal hydride). Ambient temperature range is typically from –30 to +50°C, though capacity is reduced to around 60%, and achievable discharge rate suffers, at the lower extreme.

Valve regulated lead-acid types can be stored for a matter of months at temperatures up to 40°C, but will be damaged, perhaps irreversibly, if they are allowed to spend any length of time fully discharged. This is due to build-up of the sulfur in the electrolyte on the lead plates. Self discharge is quite high—3% per month at 20°C is typical—and

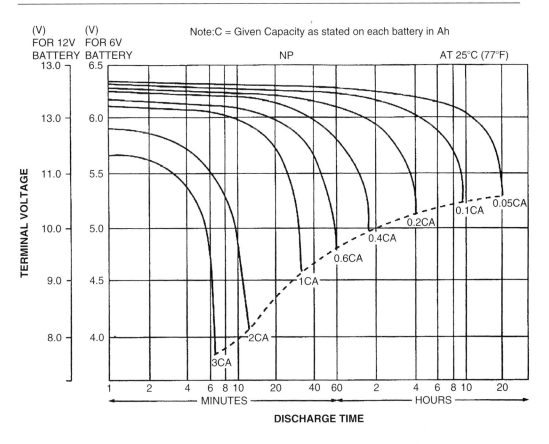

Figure 37.2: Discharge characteristics for sealed lead-acid batteries Source: Yuasa (dotted line indicates the lowest recommended voltage under load)

increases with temperature. You will therefore need to ensure that a recharging regime is followed for batteries in stock. For the same reason, equipment which uses these batteries should only have them fitted at the last moment, preferably when it is being dispatched to the customer or on installation.

Typical operational lifetime in standby float service is four to five years if proper float charging is followed, although extended lifetime types now claim up to fifteen years. When the battery is frequently discharged a number of factors affect its service life, including temperature, discharge rates and depth of discharge. A battery discharged repeatedly to 100% of its capacity will have only perhaps 15% of the cyclic service life of one that is discharged to 30% of its capacity. Overrating a battery for this type of duty has distinct advantages.

37.3.2   Nickel-Cadmium

NiCads, as they are universally known, are comparable in energy density and weight to their lead-acid competitors but address the lower end of the capacity range. Typically they are available from 0.15 to 7 Ah. Nominal cell voltage is 1.2V, with an open circuit voltage of 1.35–1.4V and an end-of-cycle voltage of 1.0V per cell. This makes them comparable to alkaline manganese types in voltage characteristics, and you can buy NiCads in the standard cell sizes from several sources, so that your equipment can work off primary or secondary battery power.

NiCads offer an ambient temperature range from –40 to +50°C. They are widely used for memory back-up purposes; batteries of two, three or four cells are available with PCB mounting terminals which can be continuously trickle charged from the logic supply, and can instantly supply a lower back-up voltage when this supply fails. Self-discharge rate is high and a cell which is not trickle charged will only retain its charge for a few months at most. Unlike lead-acid types they are not damaged by long periods of full discharge, and because of their low internal resistance they can offer high discharge rates. On the other hand they suffer from a "memory effect": a cell that is constantly being recharged before it has been completely discharged will lose voltage more quickly, and in fact it is better to recharge a NiCad from its fully discharged condition.

However, NiCads are now frowned upon because of their heavy metal content and hence the environmental consequences of their disposal to landfill. They are largely being superseded by nickel metal hydride.

37.3.3   Nickel Metal Hydride

The discharge characteristics of NiMH are very similar to those of NiCad. The charged open circuit, nominal and end-point voltages are the same. The voltage profile of both types throughout most of the discharge period is flat (Figure 37.3). NiMH cells are generally specified from –20°C to +50°C. They are around 20% heavier than their NiCad equivalents, but have about 40% more capacity. Also, they suffer less from the "memory effect" of NiCads (see above). On the other hand, they are less tolerant of trickle charging, and only very low trickle charge rates should be used if at all.

NiMH cells are available in a wide range of standard sizes, including button cells for memory back-up, and are also frequently specified in multi-cell packs for common applications such as mobile phones, camcorders and so on.

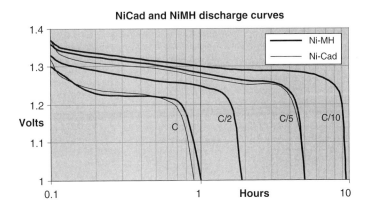

Figure 37.3: NiCad and NiMH discharge characteristics

37.3.4 Lithium-Ion

The Lithium-ion cell has considerable advantages over the types described above. Principally, it has a much higher gravimetric energy density (available energy for a given weight)—see Figure 37.5, which compares approximate figures for the three types, drawn from various manufacturers' specifications. But also, its cell voltage is about three times that of nickel batteries, 3.6–3.7V versus 1.2V. Its discharge profile with time is reasonably flat with an endpoint of 3V (Figure 37.4), and it does not suffer from the NiCad memory effect.

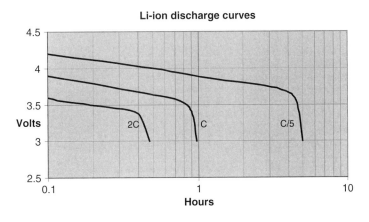

Figure 37.4: Li-ion discharge characteristics

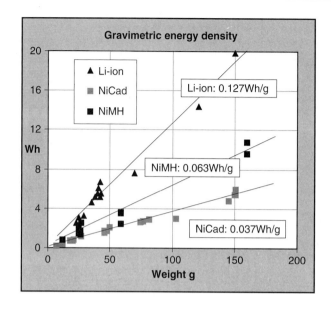

Figure 37.5: Comparison of energy density vs. weight (approximate values)

These advantages come at a price, and Li-ion batteries are more expensive than the others. Also, they are much more susceptible to abuse in charging and discharging. The battery should be protected from over-charge, over-discharge and over-current at all times and this means that the best way to use it is as a battery pack, purpose designed for a given application, with charging and protection circuits built in to the pack. This prevents the user from replacing or accidentally degrading individual cells, and gives the designer greater control over the expected performance of the battery. Since the high cost of a Li-ion battery pack makes it more suited to high value applications such as laptops and mobile phones, the extra cost of the integrated control circuitry is marginal and acceptable.

37.4   Charging

The recharging procedures are quite different for the various types of secondary cell, and you will drastically shorten their lifetime if you follow the wrong one. The greatest danger is of over-charging. Briefly, NiCad and NiMH require constant current charging, while lead acid needs constant voltage.

37.4.1 Lead-Acid

The recommended method for these types is to provide a current-limited constant voltage (Figure 37.6). The initial charging current is limited to a set fraction of the C value, generally between 0.1 and 0.25. The constant voltage is set to 2.25–2.5V per cell, depending on whether the intention is to trickle charge or recover from a cyclic discharge. The higher voltage for cyclic recharging must not be left applied continuously since it will over-cook the battery. The actual voltage is mildly temperature dependent and should be compensated by –4 mV/°C when operating with extreme variations. This charging characteristic can easily be provided by a current-limited voltage regulator IC.

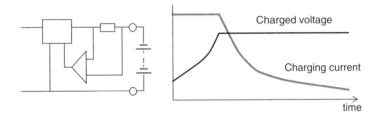

Figure 37.6: Current limited constant-voltage charging

An elegant modification to this circuit is to arrange for the output voltage to drop from the cyclic charge level to the trickle charge level when the charging current has decreased sufficiently, typically to 0.05C. This two-step charging gives the advantages of rapid recovery from deep discharge along with the benefit of trickle charging without threat to the battery's life. It does not work too well when the load circuit is connected, though.

The cheap-and-cheerful charging method is to charge the battery from a full-wave or half-wave rectified AC supply through a series resistance. This is known as "taper" charging because the applied voltage rises toward the constant voltage level as the charging current tapers away. Since it is cheap it is very popular for automotive use, but manufacturers do not recommend it because of the risk of over-charging, and the undesirable effects of the AC ripple current. Fluctuations in the mains supply can easily lead to overvoltage and the end current cannot be properly controlled. If you have to use it, include a timer to limit the overall charging time.

Lead-acid batteries *can* be charged from a constant current, typically between 0.05 and 0.2C, subject to monitoring the cell voltage to detect full charge. The technique is not often used, but can be effective for charging several batteries in series at once.

37.4.2   NiCad and NiMH

Because the voltage characteristic during charging varies substantially and actually drops when the cell is over-charged, NiCad and NiMH batteries can only be charged from a constant current. Continuous charging at up to the 0.1C rate is permissible without damage to the cell. A 0.1C charge rate will recharge the cell in 16 hours, not 10, due to the inefficiency of the charging process. An accelerated charge rate of up to 0.3C is permissible for long periods without harm, but the cell temperature will rise when charging is complete. Higher rates for a rapid charge will work, but in this case it is essential to monitor the charging progress and terminate it before the battery overheats.

A series resistor from a voltage source at a significantly higher level than the battery terminal voltage, which can rise to 1.55V per cell, is an adequate constant-current charger. For tighter control over the charging current, and especially for rapid charging, a voltage/current regulator with a battery temperature sensor is needed. Several dedicated battery charge controller ICs are available, such as the TEA1100, MC33340, and LT1510 series, and the MAX713.

Occasional overcharging, in which a partly discharged battery is put back on charge for longer than necessary, will not have much effect; but repeatedly recharging an already full battery will damage it and reduce its lifetime.

For NiMH, continuous long term "trickle" charging at 0.1C or 0.05C is not recommended. If trickle charging is necessary by design, it should be kept to C/250 or less, sufficient to replace losses due to self discharge, but not enough to severely degrade the lifetime.

37.4.3   Lithium Ion

Because of the high energy density of Li-Ion cells, charging regimes must be very carefully controlled, both to get the best out of the battery and to prevent degradation and possible serious damage. In general it is best to integrate a constant voltage/constant current controller together with over-charge, over-discharge and over-current protection, along with the battery pack.

CHAPTER 38

Layout and Grounding for Analog and Digital Circuits

Bonnie Baker

The ratio of digital designers to analog designers is increasing. This is not a news flash. Although the emphasis on digital design is providing significant advances in electronic end products, there is still and will always be a portion of circuit design that interfaces with the analog or real world. Could it be possible that analog layout differs from digital layout techniques? There is some similarity in layout strategies between these two domains. The differences can make an easy circuit layout design less than optimum if you are trying to achieve good results. In this chapter, we will discuss five topics. The first topic covers fundamental similarities and differences between analog and digital layout. Then we will talk about the hidden components (resistors, inductors, and capacitors) embedded in your PC board. The next section of this chapter will talk about how to improve your A/D converter accuracy and resolution. Getting another converter will not help here. Focusing on the interaction between the PCB and your converter will improve your results. This will be followed by the fourth section where we will discuss two-layer layout techniques. Finally, we will end with an example of how to do a poor layout and then how to fix it.

38.1   The Similarities of Analog and Digital Layout Practices

There are many similarities between analog and digital layout practices. As digital systems get faster and faster, the digital circuit looks more analog than not. When you talk about similarities between these domains, the use of bypass capacitors and power plane designs are basically the same. Differences pop up when you talk about switching noise and the location of devices on the board.

38.1.1 Bypass or Decoupling Capacitors

In terms of layout, analog devices and digital devices all require these types of capacitors. Both types of devices require that you position one capacitor as close to the power supply pin(s) as possible. A common value for this capacitor is 0.1 μF, but it is not unusual to find a 1 μF bypass capacitor (for lower frequency circuits) or a 0.01 μF capacitor in higher frequency circuits. A second class of bypass or decoupling capacitor in the system is required at the power supply source. The value of this capacitor is usually about 10 μF.

Figure 38.1 shows the position of these capacitors. The values of these capacitors can vary by being ten times higher or lower, but they are both required to have short leads. The inductance of shorter leads is smaller, reducing the chances of having a "tank" circuit. The smaller value capacitor should be as close to the device as possible and the higher value capacitor should be as close to the power supply source as possible.

The placements of the bypass or decoupling capacitors are just common sense for both types of designs, but interesting enough, for different reasons. In the analog layout

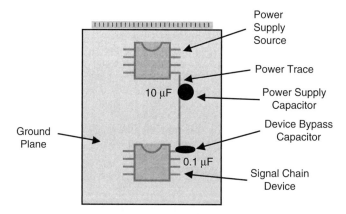

Figure 38.1: In analog and digital PCB design, you should place the bypass or decouple capacitors (0.1 μF) as close to the device as possible. You should also place the power supply decoupling-capacitor (10 μF) at the power-source or where the power-bus enters the board. In all cases, these capacitors should have short leads.

design, bypass capacitors generally serve the purpose of redirecting high frequency signals on the power supply trace. This noise would otherwise enter into the sensitive analog chip, through the power supply pin. Generally, these high frequency signals occur at frequencies beyond rejection capability of the analog device. The possible consequences of not using a bypass capacitor in your analog circuit are the addition of undue noise to the signal path or worse yet, oscillation.

For digital devices, such as controllers and processors, the decoupling capacitor on the power supply pin are required, but for a different reason. One of the functions of these capacitors is to serve as a "mini" charge reservoir. Frequently in digital circuits, a great deal of current is required to execute the transitions of the changing gate states. Because of the switching transient currents that occur on the chip and throughout the circuit board, having additional charge "on-call" is advantageous. The consequence of not having enough charge locally to execute this switching action could result in a significant dynamic and static change in the power supply voltage. When the voltage change is too large, it will cause the digital signal level to go into the indeterminate state. But more than likely, the state machines in the digital device will operate erroneously. The switching current passing through the circuit board traces causes this change in voltage. The circuit board traces have parasitic inductance. You can calculate the change in voltage results with this formula:

$$V = L\delta I/\delta t$$

where,

V = voltage change

L = board trace inductance

δI = change in current through the trace

δt = the time it takes for the current to change

So for multiple reasons, it is a good idea to bypass (or decouple) the power supply at the power supply and at the power supply pin of all of the active devices.

38.1.2   *The Power and Ground Should Be Routed Together*

When you match power and ground traces with respect to location, you lessen the opportunities for EMI. If you don't match power and ground, system loops are part of

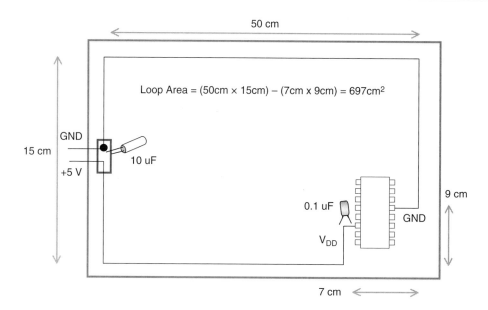

Figure 38.2: The power and ground traces are laid out using different routes to the device on this board. This mismatch opens the opportunity for EMI into the electronics of this board.

the layout. The possibility of seeing "noisy" results without explanation is real. Figure 38.2 shows an example of a PCB design with the unmatched power and ground traces.

The loop area in Figure 38.2 is 697 cm^2. This loop is a perfect antenna for noise in the area. With this board, you may be able to pick up radio signals. In the 1980s, one of the German engineers that I worked with was able to design boards of this class and "pick-up" Radio Free Europe.

Figure 38.3 shows a dramatic decrease in radiated noise off the board for induced voltages in the loop. This is because there is a decrease of radiated noise off the board and around the board.

In Figure 38.3, the signal and ground line are next to each other. This greatly reduces the loop area. A better solution would be to have a ground plane, which would be underneath the power supply trace. An even better solution would be to have a ground plane and a separate power plane.

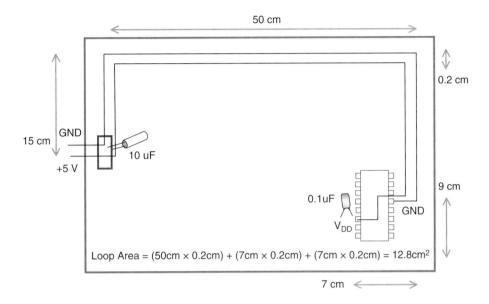

Figure 38.3: In this one-layer board, the power trace and ground trace are laid next to each other on their way to the device on this board. Figure 38.2 shows a board where the traces are better matched. The opportunity for EMI into the electronics of this board is lessened by 679/12.8 or ~54×.

38.2   Where the Domains Differ—Ground Planes Can Be a Problem

The fundamentals of circuit board layout apply to analog circuits as well as digital circuits. One fundamental rule of thumb is to use uninterrupted ground planes. This common practice reduces the effects of $\delta I/\delta t$ (change in current with time) in digital circuits. In digital circuits, the change in current with time changes the potential of ground. In analog circuits, injected noise is caused by $\delta I/\delta t$. But, when comparing digital and analog circuits, you should exercise an added precaution with analog circuits in order to keep the digital signal lines and return paths in the ground plane as far away from the analog circuitry as possible. This can be done by connecting the analog ground plane separately to the system ground connect or having the analog circuitry at the farthest side of the board—that is, at the end of the line. This is done so that signal paths have a minimal amount of interference from external sources. The opposite is not true for digital circuitry. The digital circuitry can tolerate a great deal of noise on the ground plane before problems start to appear.

38.2.1   Location of Components

In every PCB design, you should separate the noisy and quiet portions of the circuit, as mentioned above. Generally, the digital circuitry is "rich" with noise. Alternatively, digital circuitry is less sensitive to this type of noise because of the larger voltage noise margins. When you look at analog circuits, you will easily find that they are not as forgiving as the digital circuits. The voltage noise margins of the analog circuitry are much smaller. Of the two domains, the analog domain is most sensitive to switching noise. In the layout of a mixed-signal system, you should separate the two domains. Figure 38.4 shows this graphically.

The general rules of thumb are to keep the analog and digital portions of the circuit separate, with the digital circuitry closest to the connector. This is done so that the fast changing digital signals never "go past" the analog chips. A second general guideline is to place the higher frequency devices closer to the connector than the lower frequency devices. In this case, higher frequency noise will not inject into the lower frequency devices.

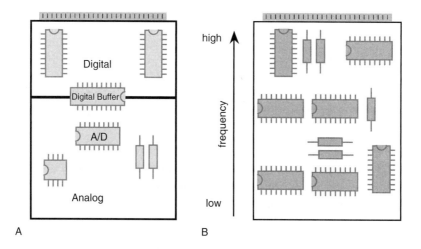

Figure 38.4: If possible, (A) the digital and analog portion of circuits should be separated in order to separate the digital switching activity from the analog circuitry. Additionally, (B) the high frequency should be separated from the low frequency where possible, keeping the higher frequency components closer to the board connector.

38.3 Where the Board and Component Parasitics Can Do the Most Damage

The major classes of parasitics generated by the PC board layout come in the form of resistors, capacitors, and inductors. For instance, you can build PCB resistors with your traces that span between components. You can build unintentional capacitors into the board with traces, soldering pads, and parallel traces. Unintentional inductors come from loop inductance, mutual inductance, and vias. All of these parasitics stand a chance of interfering with the effectiveness of your circuit as you transition from the circuit diagram to the actual PCB. You will clearly see in this section of the chapter the most troublesome class of board parasitics and see examples of where these parasitics have the most effect on circuit performance.

38.3.1 Feeling the Pain of Those Unnecessary Capacitors

You can design a capacitor into a board by simply placing two traces close to each other. This can be done by placing the two traces one on top of the other with two layers (which is harder to see) or by placing them beside each other on the same layer. Usually, you would build layout capacitors by placing two parallel traces close together. The formulas in Figure 38.5 show how you can calculate the value of this type of capacitor.

In both trace configurations, changes in voltage with time ($\delta V/\delta t$) on one trace could generate a current on a second trace. If the second trace is high impedance, the e-field creates current, which converts to voltage. Typically, you will find fast voltage transients on the digital side of the mixed signal design. If the traces that have these fast, voltage transients are in close proximity to high impedance, analog traces, this type of error will be very disruptive with analog circuitry accuracy. Analog circuitry has two strikes against it in this environment. The noise margins are much lower than digital and it is not unusual to have high impedance traces.

You can easily minimize this type of phenomena using one of two techniques. The most commonly used technique is to change the dimensions between the traces, as the capacitor equation suggests. The most effect dimension to change is the distance between the two offending traces. It should be noted that the variable, d, is in the denominator of the capacitor equation. As d is increased, the capacitance will decrease. The length of the two traces is another variable that you can change. In this case, if you reduce the length (L), this also reduces the capacitance between the two traces.

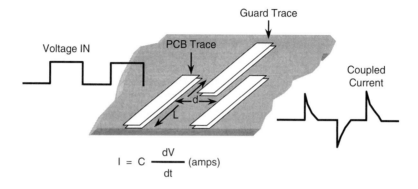

$$C = \frac{w \cdot L \cdot e_o \cdot e_r}{d} \; pF$$

w = thickness of PCB trace
L = length of PCB trace
d = distance between the two PCB traces
e_o = dielectric constant of air = 8.85×10^{-12} F/m
e_r = dielectric constant of substrate coating relative to air

Guard Trace

Voltage IN

PCB Trace

Coupled
Current

$$I = C \frac{dV}{dt} \; (amps)$$

Figure 38.5: You can easily place capacitors into a PCB by laying out two traces in close proximity. With this type of capacitor, fast voltage changes on one trace can initiate a current signal in the other trace.

Another technique used is to lay a ground trace between the two offending traces. Not only is the ground trace low impedance, but an additional trace like this will break up the E-fields that are causing the disturbance.

This type of capacitor can cause problems in mixed signal circuits where sensitive, high impedance, analog traces are in close proximity to digital traces. For example, the circuit in Figure 38.6 has the potential to have this type of problem.

To quickly explain the circuit operation in Figure 38.6, a 16-bit DAC uses three 8-bit digital potentiometers and three CMOS operational amplifiers. To the left side of this figure, two digital potentiometers (U3a and U3b) span across V_{DD} to ground with the wiper output connected to the noninverting input of two amplifiers (U4a and U4b). You program the digital potentiometers, U2 and U3 by using an SPI interface between the microcontroller, U1. In this configuration, each digital potentiometer operates as an 8-bit multiplying DAC. If V_{DD} is equal to 5V, the LSB size of these DACs is equal to 19.61 mV.

In this circuit, you connect the wipers digital potentiometers (U4a, U4b) to the noninverting inputs of two buffer amplifiers. In this configuration, the inputs to the

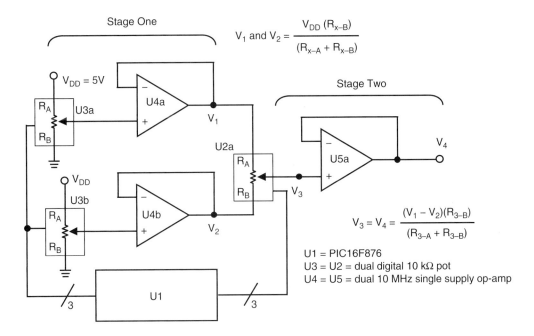

Figure 38.6: You can build a 16-bit DAC using three 8-bit digital potentiometers and three amplifiers to provide 65,536 different output voltages. If V_{DD} is 5V in this system the resolution or LSB size of this DAC is 76.3 μV.

amplifiers are high impedance, which isolates the digital potentiometers from the rest of the circuit. The output swing restrictions of the second stage of this amplifier configuration are not violated.

To have this circuit perform as a 16-bit DAC ($U2_a$), a third digital potentiometer spans across the output of these two amplifiers, U4a, and U4b. The programmed setting of U3a and U3b sets the voltage across the digital potentiometer. Again, if VDD is 5V it is possible to program the output of U3a and U3b 19.61 mV apart. With this size of voltage across the third 8-bit digital potentiometer (R_3), the LSB size of this circuit from left to right is 76.3 μV. Table 38.1 shows the critical device specifications that give optimum performance with this circuit.

You can use this circuit in two basic modes of operation. The first mode would be if you wanted a programmable, adjustable, DC reference. In this mode, you only use the digital portion of the circuit occasionally and certainly not during normal operation.

Table 38.1: From the long list of specifications that each of the devices have, there are a handful of key specifications that make this circuit more successful when it is used to provide DC reference voltages or arbitrary wave forms

Device	Specification		Purpose
Digital Potentiometers	Number of bits	8-bits	Determines the overall LSB size and resolution of the circuit.
	Nominal resistance (resistive element)	10 kΩ (typ)	The lower this resistance is the lower the noise contribution will be to the overall circuit. The trade off is that the current consumption of the circuit is high with these lower resistances.
	DNL	± 1 LSB (max)	Good Differential Non-Linearity is needed to insure no missing codes occur in this circuit which allows for a possible 16-bit operation.
	Voltage Noise Density (for half of the resistive element)	9 nV /√Hz @ 1 kHz (typ)	If the noise contribution of these devices is too high it will take away from the ability to get 16-bit noise free performance. Selecting lower resistive elements can reduce the digital potentiometer noise.
Operational Amplifiers	Input Bias Current, IB	1 pA @ 25°C (max)	Higher IB will cause a DC error across the potentiometer. CMOS amplifiers were chosen for this circuit for that reason.
	Input Offset Voltage	500 µV (max)	A difference in amplifier offset error between U4a and U4b could compromise the DNL of the overall system.
	Voltage Noise Density	8.7 nV /√Hz @ 10 kHz (typ)	If the noise contribution of these devices is too high it will take away from the ability to get 16-bit accurate performance. Selecting lower noise amplifiers can reduce amplifier noise.

The second mode would be if you used the circuit as an arbitrary wave generator. In this mode, the digital portion of the circuit is an intimate part of the circuit operation. In this mode, the risk of capacitive coupling may occur.

Figure 38.7 shows the first pass layout of the circuit. You can quickly design this circuit in your lab without attention to detail. The consequences of placing digital traces next to high impedance, analog lines were overlooked in the layout review. This speaks strongly to doing it right the first time, but to your benefit, I made this mistake and you can see how I made significant improvements.

If you take a look at this layout, it is obvious where a potential problem is. The arrow is pointing to an analog trace. This trace is from the wiper of U3a to the high impedance amplifier input of U4a. The digital trace that is pointed out carries the digital word that programs the digital potentiometer settings.

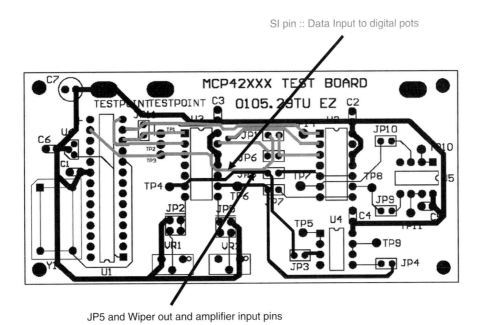

SI pin :: Data Input to digital pots

JP5 and Wiper out and amplifier input pins

Figure 38.7: This is the first attempt at the layout for the circuit in Figure 38.6. In this figure, you can see that a critical, high impedance, analog line is very close to a digital trace. This configuration produces inconsistent noise on the analog line because the data input code on that particular digital trace changes. These changes are dependent on the programming requirements for the digital potentiometer.

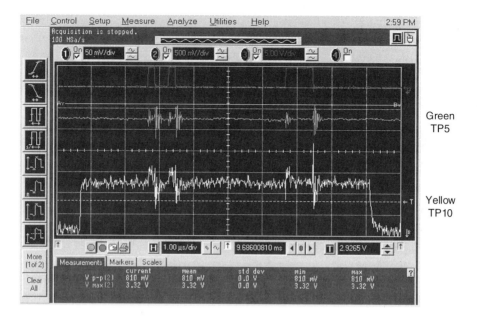

Green
TP5

Yellow
TP10

Figure 38.8: In this scope photo, the top trace was taken at JP1 (digital word to the digital potentiometers), the second trace on JP5 (noise on the adjacent analog trace), and the bottom yellow trace is taken at TP10 (noise at the output of the 16-bit DAC).

On the bench, I measured the digital signal that was coupled into the sensitive analog wiper trace. Figure 38.8 shows the scope photo.

The digital signal that is programming the digital potentiometers in the system has transmitted from trace to trace onto an analog line that is being held at a DC voltage. This noise propagates through the analog portion of the circuit all the way out to the third digital potentiometer (U5a). The third digital potentiometer is toggling between two output states.

What is the solution to this problem? Basically, you should separate the traces. Figure 38.9 shows an improved layout solution.

Figure 38.10 shows the results of the layout change. With the analog and digital traces carefully kept apart, this circuit becomes a very clean 16-bit DAC. This trace shows a single code transition of the third digital potentiometer of 76.29 μV. You may notice that the oscilloscope scale is 80 mV/div and that the amplitude of this code change is

SI pin :: Data Input to digital pots

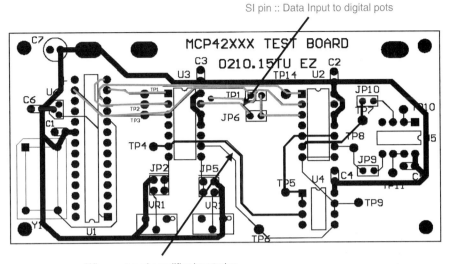

Wiper out and amplifier input pins

Figure 38.9: With a new layout, the analog lines are separate from the digital lines. This distance has eliminated the digital noise that was causing interference in the previous layout.

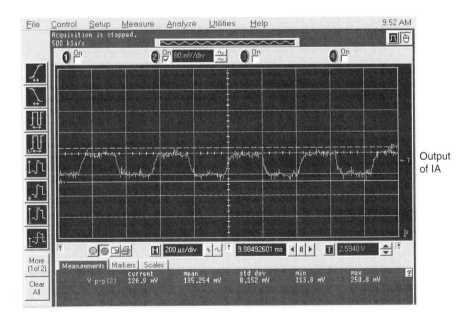

Figure 38.10: The 16-bit DAC in this new layout is showing a single code transition with no digital noise from the communication to the digital potentiometers

approximately 80 mV. In the lab, the equipment forced us to gain the output of the 16-bit DAC by 1000×.

Once again, when the digital and analog domains meet, careful layout is critical if you intend to have a successful final PCB implementation. In particular, active digital traces close to high impedance analog traces will cause serious coupling noise. You can avoid this noise coupling phenomena by putting distance between traces.

38.3.2   Inductors Designed into the PCB

The way that an inductor is designed into a board is similar to the construction of a capacitor. Again this is done by placing two traces, one on top of the other with two layers or by placing them beside each other on the same layer, as shown in Figure 38.11. In both trace configurations, changes in current with time ($\delta I/\delta t$) on one trace could generate a voltage in the same trace due to the inductance on that trace and initiate a proportional current on the second trace due to the mutual inductance. If the voltage change is high enough on the primary trace, the disturbance can reduce the voltage margin of the digital circuitry enough to cause errors. This phenomenon is

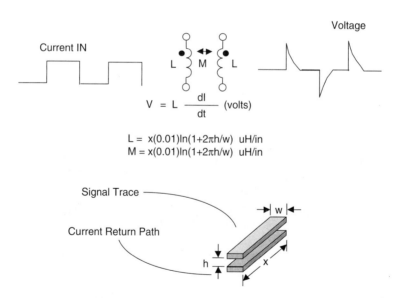

$$V = L\,\frac{dI}{dt}\ \text{(volts)}$$

$$L = x(0.01)\ln(1+2\pi h/w)\ \text{uH/in}$$
$$M = x(0.01)\ln(1+2\pi h/w)\ \text{uH/in}$$

Figure 38.11: If you pay little attention to the placement of traces, you can create line and mutual inductance with the traces in a PCB. This kind of parasitic element is most detrimental to the circuit operation where digital switching circuits reside.

not necessarily reserved for digital circuits, but is more common in that environment because of the larger, seemingly instantaneous switching currents.

To eliminate potential noise for EMI sources it is best to separate quiet analog lines versus noisy I/O ports. Try to implement low impedance power and ground networks, minimize inductance in conductors for digital circuits and minimize capacitive coupling in analog circuits.

38.4   Layout Techniques That Improve ADC Accuracy and Resolution

Initially, analog-to-digital (A/D) converters arose from an analog paradigm where a large percentage of the physical silicon was analog. As the progression of new design topologies evolves, this paradigm is shifting to where slower speed A/D converters are predominantly digital. Even with this on-chip shift from analog to digital, the PCB layout practices have not changed. Now as always, when the layout designer is working with mixed-signal circuits, you still need key layout knowledge in order to implement an effective layout. This section of the chapter will look at the PCB layout strategies required for A/D converters using successive approximation register (SAR) and sigma-delta topologies.

38.4.1   SAR Converter Layout

SAR A/D converters can be found with 8-bit, 10-bit, 12-bit, 16-bit and sometimes 18-bit resolution. Originally, the process and architecture for these converters was bipolar with R-2R ladders. But recently these devices have migrated to a CMOS process with a capacitive charge distribution topology. Needless to say, the system layout strategy for these converters has not changed with this migration. The basic approach to layout is consistent, except for higher resolution devices. These devices require more attention to the prevention of digital feedback from the serial or parallel output interface of the converter.

The SAR converter is predominantly analog in terms of circuitry and the amount of real estate dedicated to the different domains on the chip. Figure 38.12 shows a block diagram of a 12-bit, CMOS SAR converter.

These types of converters can have several pins for the ground and power connections. The pin names are often misleading in that you cannot differentiate between the analog

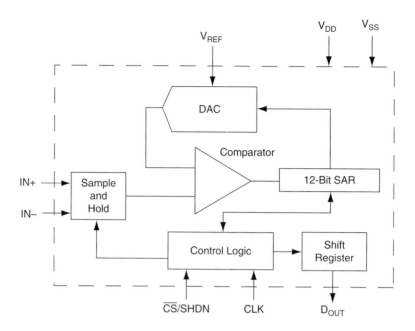

Figure 38.12: This is a block diagram of a 12-bit CMOS SAR A/D converter. This converter uses a charge distribution across a capacitive array

and digital connections with the pin label. These labels do not necessarily describe the system connections to the PCB, but rather they identify how the digital and analog currents come off the chip. Knowing this information and understanding that the primary real estate consumed on the SAR converter chip is analog, it makes sense to connect the power and ground pins on the same planes. And since the converter is primarily an analog chip, placing the pins of the device on the analog planes is very appropriate.

Figure 38.13 shows the pinout for a representative sample of 10-bit and 12-bit converters.

With these devices, the ground signal is usually directed off the chip with two pins; AGND and DGND. The power is applied to a single pin. When implementing the PCB layout, you should connect AGND and DGND to the analog ground plane. The analog and digital power pins should also be connected to the analog power plane or at least connected to the analog power train with proper bypass capacitors as close to each pin as possible. The only reason that these devices would have only one ground

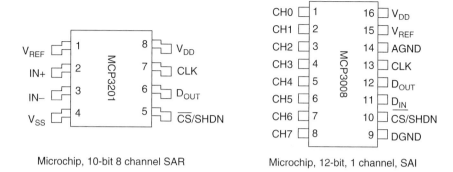

Microchip, 10-bit 8 channel SAR   Microchip, 12-bit, 1 channel, SAI

Figure 38.13: The SAR converter, regardless of resolution, usually has at least two ground connects; AGND (V_{SS}) and DGND (V_{SS}). The converters illustrated here are the MCP3008 and MCP3201 from Microchip.

pin and one positive supply pin is due to package pin limitations. However, separate grounds on the chip enhance the probability of getting good and repeatable accuracy from the converter.

With all of the converters, the power supply strategy should be to connect all grounds, positive supply and negative supply pins to the analog planes. In addition, you should connect the COM pin or IN- pin associated with the input signal as close to the signal ground as possible.

Higher resolution SAR converters (16- and 18-bit converters) require a little more consideration in terms of separating the digital noise from the quiet analog converter and power planes. When you interface these devices to a microcontroller, external digital buffers should be used in order to achieve clean operation. Although, these types of SAR converters typically have internal double buffers at the digital output, you can use external buffers to further isolate the digital bus noise from the analog circuitry in the converter. Figure 38.14 shows an appropriate power strategy for this type of system.

38.4.2 Precision Sigma-Delta Layout Strategies

The silicon area of the precision sigma-delta A/D converter is predominantly digital. In the early days, when this type of converter was first being produced, this shift in the paradigm prompted users to separate the digital noise from the analog noise by using the PCB planes. As with the SAR A/D converter, these types of A/D converters can have multiple analog-and digital-ground and power pins. Once again, the common

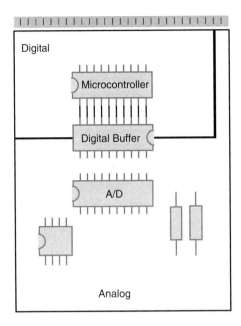

Figure 38.14: With high-resolution SAR A/D converters, you should connect the analog planes to the converter power and ground. You should then buffer the digital output of the A/D converter by using external 3-state output buffers. These buffers provide isolation between the analog and digital side, as well as high-drive capability.

tendency of a digital or analog design engineer is to try separating these pins into separate planes. Unfortunately, this is a misguided tendency, particularly if you intend to solve critical noise problems with the 16-bit to 24-bit accuracy devices.

With a high-resolution sigma-delta converter that has a 10-Hz data rate, the clock (internal or external) to the converter could be as high as 10 MHz or 20 MHz. You would use this high frequency clock for switching the modulator and running the oversampling engine. With these circuits, you should connect the AGND and DGND pins together on the same ground plane, as is the case with the SAR converter. Additionally, you should connect the analog and digital power pins together, preferably on the same plane. The requirements on the analog and digital power planes are the same as with the high-resolution SAR converters.

A ground plane is mandatory, which implies that at a minimum you need a two-layer board. On this double-sided board, the ground plane should cover at least 75% of the area if not more. You should keep interruptions in the plane to an absolute minimum. The purpose of this ground plane layer is to reduce grounding resistance and inductance as well as provide a shield against electro-magnetic interference (EMI) and radio-frequency interference (RFI). If the circuit, interconnect traces need to be on the ground-plane side of the board, they should be as short as possible and perpendicular to the ground current return paths.

You can get away without separating the analog and digital pins of low precision A/D converters, such as 6-, 8-, or maybe even 10-bit converters. But as the resolution/accuracy increases with your converter selection, the layout requirements also become more stringent. In both cases, with high-resolution SAR A/D converters and sigma-delta converters you need to connect them directly to the lower noise, analog ground and power planes.

38.5 The Art of Laying Out Two-Layer Boards

In this highly competitive marketplace, the cost objective usually dictates that a designer use two-layer boards in the design. Although the multi-layer board (4-, 6-, and 8-layers) allows the designer to build cleaner solutions in terms of size, noise, and performance, financial pressures force the engineer to rethink layout strategies with the two-layer board in mind. In this section of this chapter we will discuss the use or misuse of auto routing, the concept of current return paths with and without ground planes, and recommendations for component placement where two layer boards are concerned.

38.5.1 Pay Now or Pay Later with the Auto Router and Analog Circuits

It is tempting to use the auto router when designing printed circuit board (PCB). More often than not, a purely digital board (especially if the signals are relatively slow, and the circuit density is low) will work just fine. But as you try to lay out analog, mixed signal or high-speed circuits with the auto routing tool that is available with your layout software, there may be some issues. The probability of creating serious circuit performance problems is very high.

Figure 38.15 shows the auto routed top layer of a two-layer board. The bottom layer of this board is in Figure 38.15 and 38.16 and the circuit diagram for these layout layers is in Figure 38.17 and Figure 38.18.

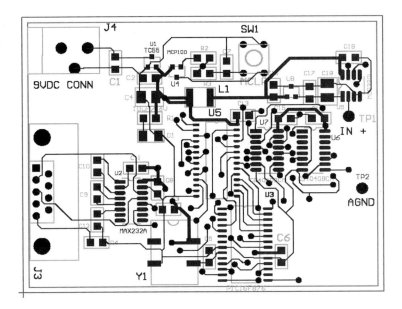

Figure 38.15: Top layer of an auto-routed layout of circuit diagram shown in Figure 38.17 and Figure 38.18

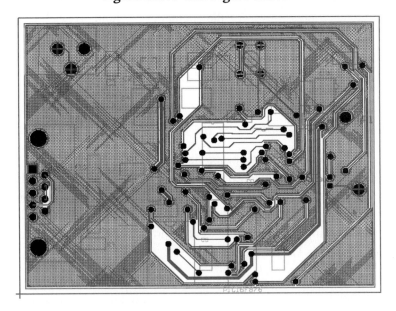

Figure 38.16: Bottom layer of an auto-routed layout of circuit diagram shown in Figure 38.17 and Figure 38.18

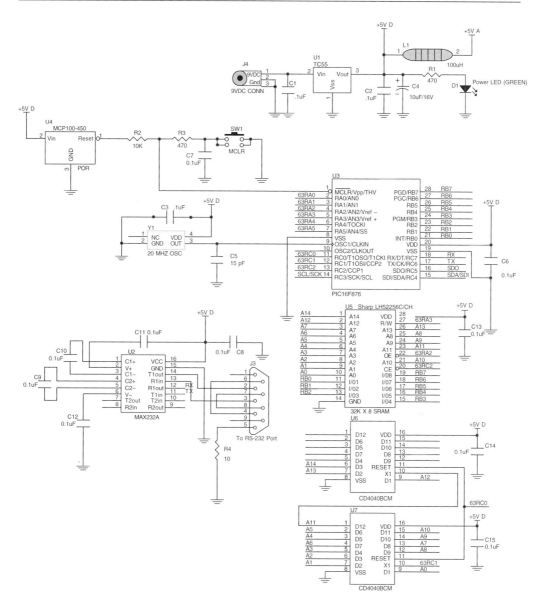

Figure 38.17: Digital section of circuit diagram for layouts in Figures 38.15, 38.16, 38.19, and 38.20. This is the circuit diagram from Microchip's MXDEV™ board, evaluation board for the 10- and 12-bit ADCs (MCP300X and MCP320x).

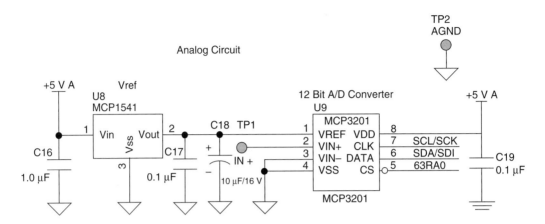

Figure 38.18: Analog section of circuit diagram for layouts in Figures 38.15, 38.16, 38.19, and 38.20. This is the circuit diagram from Microchip's MXDEV board, evaluation board for the 10- and 12-bit ADCs (MCP300X and MCP320x).

With this layout, there are several areas of concern, but the most troubling issue is the grounding strategy. If you follow the ground traces on the top layer, the traces connect every device on that layer. A second ground connection for every device uses the bottom layer with vias at the far right-hand side of the board. The immediate red flag that one should see when examining this layout strategy would be the existence of several ground loops. Additionally, horizontal signal lines interrupt the ground return paths on the bottom side. The saving grace with this grounding scheme is that the analog devices (12-bit A/D converter and 2.5V voltage reference) are at the far right-hand side of the board. This placement ensures that digital ground signals do not pass under these analog chips.

Figure 38.19 and Figure 38.20 have the manual layouts of the circuits in Figure 38.17 and Figure 38.18. For the layout of this mixed-signal circuit, the devices were manually placed on the board with careful thought to separating the digital and analog devices. With this manual layout, a few general guidelines are followed to ensure positive results. These guidelines are:

1. Use the ground plane as a current return path as much as possible.

2. Separate the analog ground plane from the digital ground plane with a break.

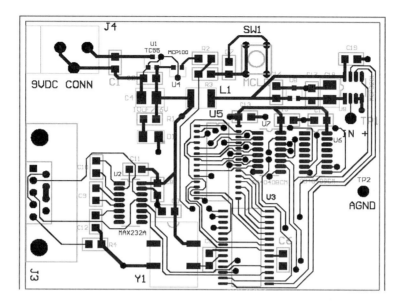

Figure 38.19: Top layer of a manual routed layout of circuit diagram shown in Figure 38.17 and Figure 38.18

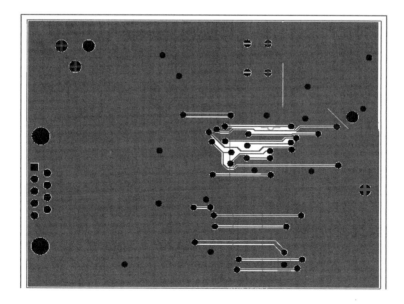

Figure 38.20: Bottom layer of a manual routed layout of the circuit diagram shown in Figure 38.17 and Figure 38.18

3. If interruptions from signal traces are required on the ground-plane side, make them vertical to reduce the interference with the ground-current, return paths.

4. Place analog circuitry at the far end of the board and digital circuitry closest to the power connects. This reduces the effects of $\delta i/\delta t$ from digital switching.

Note that with both of these two layer boards there is a ground plane on the bottom. This is only done so that an engineer working on the board can quickly see the layout when troubleshooting. You will typically find this strategy in manufacturer's demo and evaluation boards.

But more typically, the ground plane is on the top of board, thereby reducing electromagnetic interference (EMI).

At every layout-related presentation that I give in a seminar setting, the question always asked in one form or another is, "What if management tells me I can't have two layers or a ground plane, and I still need to reduce noise in the circuit? How do I design my circuit to work around the need for a ground plane?" Typically, I instruct the person asking the question to inform their management that a ground plane is simply required if they want reliable circuit performance. The primary reason for using ground planes is lower ground impedance. They also provide a degree of EMI reduction.

38.6 Current Return Paths With or Without a Ground Plane

The fundamental issues that should be considered when dealing with current return paths are:

1. If traces are used, they should be as wide as possible.

   In the event that you are considering using traces for your ground connects on your PCB, they should be as wide as possible. This is a good rule of thumb, but also understand that the thinnest width in your ground trace will be the effective width of the trace from that point to the end (where the "end" is defined as the point furthest from the power connection).

2. You should avoid ground loops.

3. If no ground plane is available, you should use a star connection strategy.

   Figure 38.21 shows a graphical example of a star connection strategy.

   With this type of approach, the ground currents return to the power connection independently. You will note that in Figure 38.21 not all of the devices have their own return path. With U1 and U2, the return path is shared. This can be done if you use guidelines #4 and #5, following.

4. Digital currents should not pass across analog devices.

   During switching, digital currents in the return path are fairly large, but only briefly. This phenomenon occurs due to the effective inductance and resistance of the ground. With the inductance portion of the ground plane or trace, the governing formula is $V = L\delta i/\delta t$, where V is the resulting voltage, L is the inductance of the ground plane or trace, δi is the change in current from the digital device and δt is the time span considered for the event. To calculate the effects of the resistance portion of the ground plane, changes in the voltage simply change because of $V = RI$. Again, V is the resulting voltage, R is the

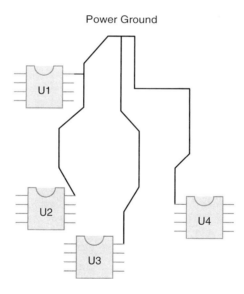

Figure 38.21: If a ground plane is not feasible, you should handle current return paths with a "star" layout strategy

ground plane or trace resistance and I is the current change caused by the digital device. These changes in the voltage of the ground plane or trace across the analog device will change the relationship between ground and the signal in the signal chain.

5. High-speed current should not pass across lower speed devices.

   Ground-return signals of high-speed circuits have a similar effect on changes to the ground plane. Again the more important formulas that determine the effects of this interference are $V = L\delta i/\delta t$ for the ground plane or trace inductance and $V = RI$ for the ground plane or trace resistance. And as with digital currents, high-speed circuits that have ground activity on the ground plane or that have a trace across the analog device change the relationship between ground and the signal in the signal chain.

6. Regardless of the technique used, you must design the ground return paths to have a minimum resistance and inductance.

7. If a ground plane is used, breaks in plane can improve or degrade circuit performance. Use with care.

But, if you are unable to win that battle with your management because of cost constraints, this book offers some suggestions. These suggestions are using star networks and current return paths, which if used properly, will give a little relief from the circuit noise.

38.7 Layout Tricks for a 12-bit Sensing System

When I started writing this chapter I thought a "cookbook" approach would be appropriate when describing the implementation of a good 12-bit layout. My assumption behind this type of approach is that I would provide a reference design, which would make the layout implementation easy. But I struggled with this topic long enough to find that this notion was fairly unrealistic.

Because of the complexity of this problem, I am going to provide basic guidelines ending with a review of issues to be aware of while implementing your layout design. Throughout this discussion I will offer examples of good and bad layout implementations. I am doing this in the spirit of discussing concepts and not with the intent of recommending one layout as the only one to use.

The application circuit that I'm going to use is a load-cell circuit that accurately measures the weight applied to the sensor, then displays the results on a LCD-display screen. Figure 38.22 shows the circuit diagram for this system. You can purchase the load cell that I used from Omega (LCL-816G). My sensor model for the LCL-816G is a four element resistive bridge that requires voltage excitation. With a 5V excitation voltage applied to the high side of the sensor, the full-scale output swing is a ±10 mV differential-signal with a 32-ounce maximum excitation. A two-op-amp instrumentation amplifier gains this small differential signal. I chose a 12-bit converter to match the required precision of this circuit. Once the converter digitizes the voltage presented at its input, the microcontroller receives the digital code by using the converter's SPI™ port. The microcontroller then uses a look-up table to convert the digital signal from the ADC into weight. Linearization and calibration activities can be implemented with controller code at this point if need be. Once this is done, the results are sent to the LCD display. As a final step, I wrote the firmware for the controller. Now the design is ready to go to board layout.

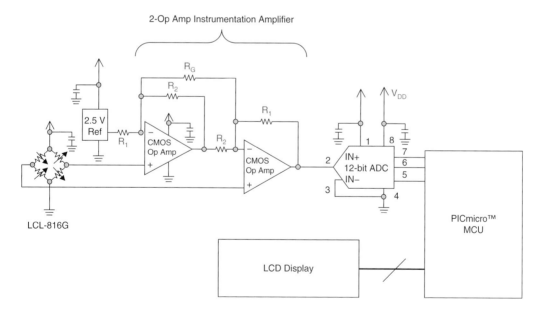

Figure 38.22: A two-op-amp instrumentation amplifier, filtered and digitized with a 12-bit A/D converter, gains the signal at the output of the load-cell sensor. The result of each conversion is sent to the LCD display.

38.8   General Layout Guidelines—Device Placement

My first step is to place the devices on the board. This critical step is done effectively because I am keeping track of my noise-sensitive devices and noise-creator devices. There are two guidelines that I use to accomplish this task:

1. Separate the circuit devices into two categories: high speed (>40 MHz) and low speed. You should place the higher speed devices closer to the board connector/power supply.

2. Separate the above categories into three subcategories: pure digital, pure analog, and mixed signal. With this delineation, you need to place the digital devices closer to the board connector/power supply.

38.9   General Layout Guidelines—Ground and Power Supply Strategy

Once I determine the general location of the devices, I was able to define my ground and power planes. My strategy of the implementation for these planes is a bit tricky.

First of all, it is dangerous for me not to use a ground plane in a PCB implementation. This is true particularly in analog and/or mixed-signal designs. One issue is that ground noise problems are more difficult to deal with than power-supply noise problems because analog signals are referenced to ground. For instance, in the circuit shown in Figure 38.22, the A/D converter's inverting input pin (MCP3201, Microchip) is connected to ground. Secondly, the ground plane also serves as a shield against emitted noise. Both of these problems are easy to resolve with a ground plane and nearly impossible to overcome if there is no ground plane.

However, with my small design, I assume that I won't need a ground plane. Figure 38.23 shows a ground plane-less, layout implementation of the circuit in Figure 38.22.

Does my "no ground plane is required" theory play out? The proof is in the pudding, or data. In Figure 38.24, 4096 samples were taken from the A/D converter and logged. There was no excitation on the sensor when this data was taken. With this circuit layout, the controller is dedicated to interfacing with the converter and sending the converter's results to the LCD display.

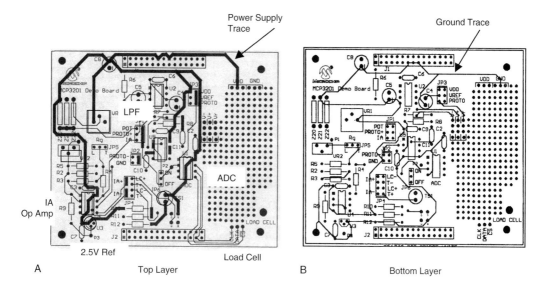

Figure 38.23: This is the layout of the top (A) and bottom (B) layers of the circuit in Figure 38.22. Note that this layout does not have a ground or power plane. Note that the power traces are considerably wider than the signal traces in order to reduce power supply trace inductance.

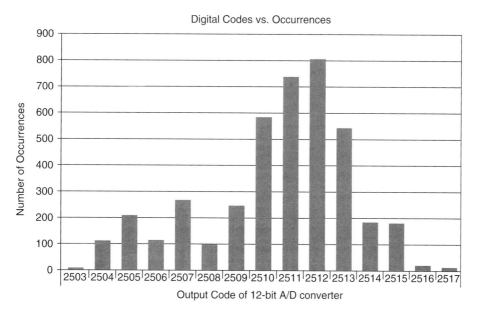

Figure 38.24: This is a histogram of 4096 samples from the output of the A/D converter. The PCB does not have a ground or power plane as shown in the PCB layout in Figure 38.25. The code of the noise from the circuit is 15 codes wide (Figure 38.23).

Figure 38.25 shows the same device layout shown in Figure 38.23, but a ground plane on the bottom layer is added. The ground plane (Figure 38.25B) has a few breaks due to signal. These breaks should be kept to a minimum. Current return paths should not be "pinched" as a consequence of these traces restricting the easy flow of current from the device to the power connector. Figure 38.26 shows the histogram for the A/D converter output. Compared to Figure 38.24, the output codes are much tighter. The same active devices were used for both tests. The passive devices were different causing a slight offset difference.

It is clear from my data that a ground plane does have an effect on the circuit noise. When my circuit did not have a ground plane, the width of the noise was ~15 codes. When I added a ground plane, I improved the performance by almost 1.5× or 15/11. You might want to know that my test set-up was in the lab, where EMI interference is relatively low.

The op-amp and absence of an anti-aliasing filter are causes of the noise shown in Figure 38.26. If my circuit has a *minimum* amount of digital circuitry on board, a single ground plane and a single power plane may be appropriate. The board designer defines

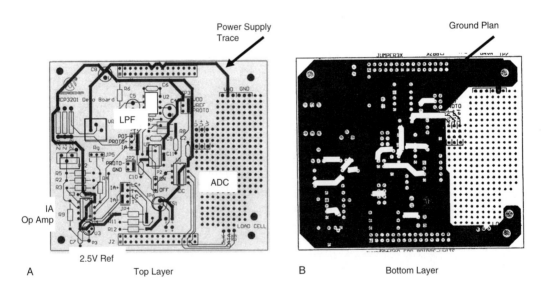

Figure 38.25: This is the layout of the top and bottom layers of the circuit in Figure 38.22. Note that this layout does have a ground plane.

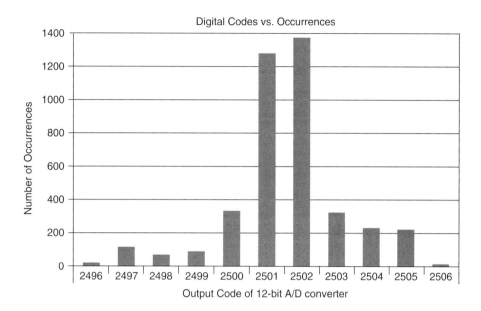

Figure 38.26: This is a histogram of 4096 samples from the output of the A/D converter on the PCB that has a ground plane as shown in the PCB layout in Figure 38.25. The code width of the noise is now 11 codes wide.

my qualifier *minimum*. The danger of connecting the digital and analog ground planes together is that my analog circuits can pick up the noise on the supply pins and couple it into the signal path. In either case, I should connect my analog and digital grounds and power supplies together at one or more points in the circuit. This ensures that my power supply, input, and output ratings of all of the devices are not violated.

The inclusion of a power plane in a 12-bit system is not as critical as the required ground plane. Although a power plane can solve many problems, making the power traces two or three times wider than other traces on the board and by using bypass capacitors effectively can reduce power noise.

38.10   Signal Traces

My signal traces on the board (both digital and analog) should be as short as possible. This basic guideline will minimize the opportunities for extraneous signals to couple into the signal path. One area to be particularly cautious of is with the input terminals of analog devices. These terminals normally have a higher impedance than the output or power supply pins. As an example, the voltage reference input pin to the A/D converter is most sensitive while a conversion is occurring. With the type of 12-bit converter I have in Figure 38.22, my input terminals (IN+ and IN−) are also sensitive to injected noise. Another potential for noise injection into my signal path is the input terminals of an operational amplifier. These terminals have typically 10^9 Ω to 10^{13} Ω input impedance.

My high impedance input terminals are sensitive to injected currents. This can occur if the trace from a high impedance input is next to a trace that has fast changing voltages, such as a digital or clock signal. When a high impedance trace is in close proximity to a trace with these types of voltage changes, charge is capacitively coupled into the high impedance trace as mentioned earlier in the chapter.

38.11   Did I Say Bypass and Use an Anti-Aliasing Filter?

Although this chapter is about layout practices, I thought it would be a good idea to cover some of the basics in circuit design. A good rule concerning bypass capacitors is to always include them in the circuit. If they are not included, the power supply noise may very well eliminate any chance for 12-bit precision.

38.12   Bypass Capacitors

Bypass capacitors belong in two locations on the board: one at the power supply (10 μF to 100 μF or both) and one for every active device (digital and analog). The value of the bypass capacitor of the device is dependent on the device in question. If the bandwidth of the device is less than or equal to ∼1 MHz, a 1 μF will reduce injected noise dramatically. If the bandwidth of the device is above ∼10 MHz, a 0.1 μF capacitor is probably appropriate. In between these two frequencies, you could use both or either one. Refer to the manufacturer's guidelines for specifics.

Every active device on the board requires a bypass capacitor. It must be placed as close as possible to the power supply pin of the device as shown in Figure 38.25.

If you use two bypass capacitors for one device, the smaller one should be closest to the device pin. Finally, the lead length of the bypass capacitor should be as short as possible.

38.13   Anti-Aliasing Filters

You will note that the circuit in Figure 38.22 does not have an anti-aliasing filter. As the data shows, this oversight has caused noise problems in the circuit. When this board has a second order, 10 Hz, anti-aliasing filter inserted between the output of the instrumentation amplifier and the input of the A/D converter, the conversion response improves dramatically. Figure 38.27 shows the resulting data.

Analog filtering can remove noise superimposed on the analog signal before it reaches the A/D converter. In particular, this includes extraneous noise peaks. Analog-to-digital converters will convert the signal that is present on its input. This signal could include the sensor voltage signal or noise. The anti-aliasing filter removes the higher frequency noise from the conversion process.

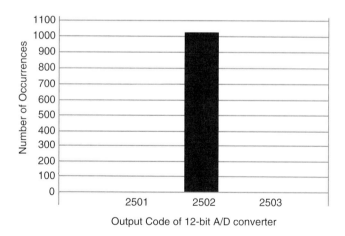

Output Code of 12-bit A/D converter

Figure 38.27: This diagram shows the conversion results of the circuit in Figure 38.22 plus a second order, anti-aliasing filter. Additionally, the board layout includes a ground plane.

38.14 PCB Design Checklist

Good layout techniques are not difficult to master as long as you follow a few guidelines:

1. Check device placement versus connectors. Make sure that high-speed devices and digital devices are closest to the connector.

2. Always have at least one ground plane in the circuit.

3. Make power traces wider than other traces on the board.

4. Review current return paths and look for possible noise sources on ground connects. Determining the current density at all points of the ground plane and the amount of possible noise present does this.

5. Bypass all devices properly. Place the capacitors as close to the power pins of the device as possible.

6. Keep all traces as short as possible.

7. Follow all high impedance traces looking for possible capacitive coupling problems from trace to trace.

8. Make sure you properly filter your signals in a mixed-signal circuit.

Analog layout and digital layout techniques differ slightly, but not completely. When it comes to the parasitic components embedded in the PCB, the analog circuits tend to show more sensitivity, but digital circuits are not completely immune. You should treat the device that straddles these two domains, such as the A/D converter, as an analog device. Two-layer boards do present some challenges, but careful manual layouts can usually work around these problems. You can fix a poor layout if you are willing to go back to the drawing board.

When the analog and digital domains meet, careful layout is critical if a designer intends to have a successful final PCB implementation. Layout strategies usually are presented as rules of thumb because it is difficult to test the success of your final product in a lab environment. So, generally speaking, although there are some similarities in layout strategies between the digital and analog domain, the differences should be recognized and worked with.

Solving signal integrity problems can take a great deal of time, particularly if you don't have the tools to tackle the tough issues. The three best analysis tools to have in your arsenal are the frequency analysis (fast Fourier transform or FFT), time analysis (scope photo), and DC analysis (histogram) tools. We used all of these tools to identify the power supply noise, external clock noise, and overdriven amplifier distortion.

References

MXDEV is a trademark of Microchip Technology Inc. in the USA and other countries.

SPI™ port is a trademark of Motorola. The Microchip name and logo, PIC, PICmicro, microID and KEE-LoQ are registered trademarks of Microchip Technology Inc. in the USA and other countries. All other trademarks are the property of their respective owners..

Noise Reduction Techniques in Electronic Systems, 2nd ed., Henry W., Ott Wiley, 1998.

Noise and Other Interfering Signals, Ralph Morrison, John Wiley & Sons, 1992.

"Circuit Layout Techniques and Tips: 6 Part," Baker, Bonnie C., First Published in *analog-Zone* (2002, 2003) and reproduced with permission.

CHAPTER 39

Safety

Tim Williams

Any electronic equipment must be designed for safe operation. Most countries have some form of product liability legislation which puts the onus on the manufacturer to ensure that his product is safe. The responsibility devolves onto the product design engineer, to take reasonable care over the safety of the design. This includes ensuring that the equipment is safe when used properly, that adequate information is provided to enable its safe use, and that adequate research has been carried out to discover, eliminate or minimize risks due to the equipment.

There are various standards relating to safety requirements for different product sectors. In some cases, compliance with these standards is mandatory. In the European Community, the Low Voltage Directive (73/23/EEC) applies to all electrical equipment with a voltage rating between 50 and 1000VAC or 75 and 1500VDC, with a few exceptions, and requires member states to take all appropriate measures:

"To ensure that electrical equipment may be placed on the market only if, having been constructed in accordance with good engineering practice in safety matters in force in the Community, it does not endanger the safety of persons, domestic animals or property when properly installed and maintained and used in applications for which it was made."

If the equipment conforms to a harmonized CENELEC or internationally agreed standard, then it is deemed to comply with the Directive. Examples of harmonized standards are EN 60065:1994, "Safety requirements for mains-operated electronic and related apparatus for household and similar general use," which is largely equivalent to IEC Publication 60065 of the same title; or EN 60950-1:2002, "Information technology equipment. Safety. General requirements," equivalent to IEC 60950-1. Proof of compliance can be by a Mark or Certificate of Compliance from a recognized laboratory, or by the manufacturer's own declaration of conformity. The Directive includes no requirement for compulsory approval for electrical safety.

39.1   The Hazards of Electricity

The chief dangers (but by no means the only ones, see Table 39.1) of electrical equipment are the risk of electric shock, and the risk of a fire hazard. The threat to life from electric shock depends on the current which can flow in the body. For AC, currents less than 0.5 mA are harmless, while those greater than 50–500 mA (depending on duration) can be fatal.[†] Protection against shock can be achieved simply by limiting

Table 39.1: Some safety hazards associated with electronic equipment

Hazard	Main risk	Source
Electric shock	Electrocution, injury due to muscular contraction, burns	Accessible live parts
Heat or flammable gases	Fire, burns	Hot components, heatsinks, damaged or overloaded components and wiring
Toxic gases or fumes	Poisoning	Damaged or overloaded components and wiring
Moving parts, mechanical instability	Physical injury	Motors, parts with inadequate mechanical strength, heavy or sharp parts
Implosion/ explosion	Physical injury due to flying glass or fragments	CRTs, vacuum tubes, overloaded capacitors and batteries
Ionizing radiation	Radiation exposure	High-voltage CRTs, radioactive sources
Nonionizing radiation	RF burns, possible chronic effects	Power RF circuits, transmitters, antennas
Laser radiation	Damage to eyesight, burns	Lasers
Acoustic radiation	Hearing damage	Loudspeakers, ultrasonic transducers

---

[†] IEC Publication 60479 gives further information.

the current to a safe level, irrespective of the voltage. There is an old saying, "It's the volts that jolts, but the mils that kills." If the current is not limited, then the voltage level in conjunction with contact and body resistance determines the hazard. A voltage of less than 50V AC rms, isolated from the supply mains or derived from an independent supply, is classified as a Safety Extra-Low Voltage (SELV) and equipment designed to operate from an SELV can have relaxed requirements against the user being able to contact live parts.

Aside from current and voltage limiting, other measures to protect against electric shock are:

- Earthing, and automatic supply disconnection in the event of a fault.

- Inaccessibility of live parts. A live part is any part contact with which may cause electric shock, that is any conductor which may be energized in normal use—not just the mains "live."

39.2 Safety Classes

IEC publication 60536 classifies electrical equipment into four classes according to the method of connection to the electrical supply and gives guidance on forms of construction to use for each class. The classes are:

Class 0: Protection relies on basic functional insulation only, without provision for an earth connection. This construction is unacceptable in the UK.

Class I: Equipment is designed to be earthed. Protection is afforded by basic insulation, but failure of this insulation is guarded against by bonding all accessible conductive parts to the protective earth conductor. It depends for its safety on a satisfactory earth conductive path being maintained for the life of the equipment.

Class II: The equipment has no provision for protective earthing and protection is instead provided by additional insulation measures, such as double or reinforced insulation. Double insulation is functional insulation, plus a supplementary layer of insulation to provide protection if the functional insulation fails. Reinforced insulation is a single layer which provides equivalent protection to double.

Class III: Protection relies on supply at safety extra low voltage and voltages higher than SELV are not generated. Second-line defenses such as earthing or double insulation are not required.

39.3   Insulation Types

As outlined above, the safety class structure places certain requirements on the insulation which protects against access to live parts. The basis of safety standards is that there should be at least two levels of protection between the casual user and the electrical hazard. The standards give details of the required strength for the different types of insulation, but the principles are straightforward:

39.3.1   Basic Insulation

Basic insulation provides one level of protection but is not considered fail-safe, and the other level is provided by safety earthing. A failure of the insulation is therefore protected against by the earthing system.

39.3.2   Double Insulation

Earthing is not required because the two levels of protection are provided by redundant insulation barriers, one layer of basic plus another supplementary; if one fails the other is still present, and so this system is regarded as fail-safe. The double-square symbol

▣   indicates the use of double insulation.

39.3.3   Reinforced Insulation

Two layers of insulation can be replaced by a single layer of greater strength to give an equivalent level of protection.

39.4   Design Considerations for Safety Protection

The requirement for inaccessibility has a number of implications. Any openings in the equipment case must be small enough that the standard test finger, whose dimensions are defined in those standards that call up its use, cannot contact a live part (Figure 39.1). Worse, small suspended bodies (such as a necklace) that can be dropped through ventilation holes must not become live. This may force the use of internal baffles behind ventilation openings.

Protective covers, if they can be removed by hand, must not expose live parts. If they do, they must only be removable by use of a tool. Or, use extra internal covers over live

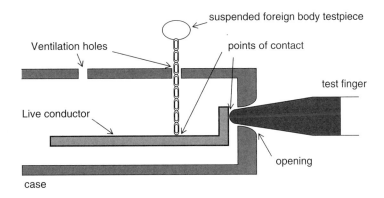

Figure 39.1: The test finger and the suspended foreign body

portions of the circuit. It is anyway good practice to segregate high-voltage and mains sections from the rest of the circuit and provide them with separate covers. Most electronic equipment runs off voltages below 50V and, provided the insulation offered by the mains isolating transformer is adequate, the signal circuitry can be regarded as being at SELV and therefore not live.

Any insulation must, in addition to providing the required insulation resistance and dielectric strength, be mechanically adequate. It will be dropped, impacted, scratched and perhaps vibrated to prove this. It must also be adequate under humid conditions: hygroscopic materials (those that absorb water readily, such as wood or paper) are out. Various standards define acceptable creepage and clearance distances versus the voltage proof required. As an example, EN 60065 allows 0.5 mm below 34V rising to 3 mm at 354V and extrapolated thereafter; distances between PCB conductors are slightly relaxed, being 0.5 mm up to 124V, increasing to 3 mm at 1240V. Creepage distance (Figure 39.2) denotes the shortest distance between two conducting parts along the surface of an insulating material, while clearance distance denotes the shortest distance through air.

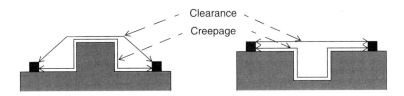

Figure 39.2: Creepage and clearance distance

Easily discernible, legible and indelible marking is required to identify the apparatus and its mains supply, and any protective earth or live terminals. Mains cables and terminations must be marked with a label to identify earth, neutral and live conductors, and class I apparatus must have a label which states "WARNING: THIS APPARATUS MUST BE EARTHED." Fuse holders should also be marked with their ratings and mains switches should have their "off" position clearly shown. If user instructions are necessary for the safe operation of the equipment, they should preferably be marked permanently on the equipment.

Any connectors which incorporate live conductors must be arranged so that exposed pins are on the dead side of the connection when the connector is separated. When a connector includes a protective earth circuit, this should mate before the live terminals and unmate after them. (The CEE-22 6A connector as an example.)

39.5   Fire Hazard

It is taken for granted that the equipment won't overheat during normal operation. But you must also take steps to ensure that it does not overheat or release flammable gases to the extent of creating a fire hazard under fault conditions. Any heat developed in the equipment must not impair its safety. Fault conditions are normally taken to mean short-circuits across any component, set of terminals or insulation that could conceivably occur in practice (creepage and clearance distances are applied to define whether a short circuit would occur across insulation), stalled motors, failure of forced cooling and so on.

The normal response of the equipment to these types of faults is a rise in operating current, leading to local heating in conductors. The normal protection method is by means of current limiting, fuses, thermal cutouts or circuit breakers in the supply or at any other point in the circuit where over-current could be hazardous. As well as this, flame-retardant materials should be used wherever a threat of overheating exists, such as for pcb base laminates.

Fuses are cheap and simple but need careful selection in cases where the prospective fault current is not that much higher than the operating current. They must be easily replaceable, but this makes them subject to abuse from unqualified users (hands up, anyone who hasn't heard of people replacing fuselinks with bent nails or pieces of cigarette-packet foil). The manufacturer must protect his liability in these cases by clear labeling of fuseholders and instructions for fuse replacement.

Thermal cutouts and circuit breakers are more expensive, but offer the advantage of easy resetting once the fault has cleared. Thermal devices must obviously be mounted in close thermal contact with the component they are protecting, such as a motor or transformer.

CHAPTER 40

Design for Production

Tim Williams

Really, every chapter in this book has been about design for production. As was implied in the introduction, the ability that marks out a professional designer is the ability to design products or systems that work under all relevant circumstances and which can be manufactured easily.

The sales and marketing engineer addresses the questions, "Can I sell this product?" and, "How much can I sell this product for?" This book hasn't touched on these issues, important though they are to designers; it has assumed that you have a good relationship with your marketing department and that your marketing colleagues are good at their job. But you as designer also have to address another set of questions, which are:

- Can the purchasing department source the components quickly and cheaply?

- Can the production department make the product quickly and cheaply?

- Can the test department test it easily?

- Can the installation engineers or the customer install it successfully?

It is as well to bear all these questions in mind when you are designing a product, or even part of one. Your company's financial health, and consequently your and others' job security, ultimately depends on it. A good way to monitor these factors is to follow a checklist.

40.1 Checklist

40.1.1 Sourcing

- Have you involved purchasing staff as the design progressed?

- Are the parts available from several vendors or manufacturers wherever possible? Have you made extensive use of industry standard devices?

- Where you have specified alternate sources, have you made sure that they are all compatible with the design?

- Have you made use of components that are already in use on other products?

- Have you specified close-tolerance components only where absolutely necessary?

- Where sole-sourced parts have to be used, do you have assurances from the vendor on price and lead time? How reliable are they? Have you checked that there is no warning, "not recommended for new designs" (implying limited availability), on each part?

- Does your company have a policy of vetting vendors for quality control? If so, have you added new vendors with this product, and will they need to be vetted?

40.1.2 Production

- Have you involved production staff as the design progressed?

- Are you sure that the mechanical and electrical design will work with all mechanical and electrical tolerances?

- Does the mechanical design allow the component parts to be fitted together easily?

- Are components, especially polarized ones, all oriented in the same direction on the PCB for ease of inspection and insertion?

- Are discrete components, notably resistors, capacitors and transistors, specified to use identical pitch spacings and footprints as far as possible?

- Have you minimized wiring looms to front or rear panels and between PCBs, and used mass-termination connections (e.g., IDC) wherever possible?

- Have you modularized the design as far as possible to make maximum use of multiple identical units?

- Is the soldering and assembly process (wave, infrared, auto-insert, pick and place, etc.) that you have specified compatible with the manufacturing capability? Will the placement machines cope with all the surface mount components you have used?

- If the production calls for any special assembly procedures (e.g., potting or conformal coating), or if any components require special handling or assembly (MOSFETs, LEDs, batteries, relays etc.) are the production and stores staff fully conversant with these procedures and able to implement them? Have you minimized the need for such special procedures?

- Do all PCBs have adequate solder mask, track and hole dimensions, clearances, and silk screen legend for the soldering and assembly process? Are you sure that the test and assembly personnel are conversant with the legend symbols?

- Are your assembly drawings clear and easy to follow?

40.1.3 Testing and Calibration

- Have you involved test staff as the design progressed?

- Are all adjustment and test points clearly marked and easily accessible?

- Have you used easily set parts such as DIL switches or linking connectors in preference to solder-in wire links?

- Does the circuit design allow for the selection of test signals, test subdivision and stimulus/response testing (including boundary scan) where necessary?

- If you are specifying automatic testing with ATE, does the pc layout allow adequate access and tooling holes for bed-of-nails probing? Have you confirmed the validity of the ATE program and the functional test fixture?

- Have you written and validated a test software suite for microprocessor-based products?

40.1.4   Installation

- Is the product safe?

- Does the design have adequate EMC?

- Are the installation instructions or user handbook clear, correct and easy to follow?

- Do the installation requirements match the conditions which will obtain on installation? For example, is the environmental range adequate, the power supply appropriate, the housing sufficient, etc.?

40.2   The Dangers of ESD

There is one particular danger to electronic components and assemblies that is present in both the design lab and the production environment. This is damage from electrostatic discharge (ESD). This can cause complete component failure, or worse, performance degradation that is difficult or impossible to detect. It can also cause transient malfunction in operating systems.

40.2.1   Generation of ESD

When two nonconductive materials are rubbed together, electrons from one material are transferred to the other. This results in the accumulation of *triboelectric* charge on the surface of the material. The amount of the charge caused by movement of the materials is a function of the separation of the materials in the triboelectric series, an example of which is shown in Table 40.1. Additional factors are the closeness of contact, rate of separation and humidity.

If this charge is built up on the human body, as a result of natural movements, it can then be discharged through a terminal of an electronic component. This will damage the component at quite low thresholds, easily less than 1 kV, depending on the device. Of the several contributory factors, low humidity is the most severe; if relative humidity is higher than 65% (which is frequent in maritime climates such as the UK's) then little damage is likely. Lower than 20%, as is common in continental climates such as the United States, is much more hazardous.

Gate-oxide breakdown of MOS or CMOS components is the most frequent, though not the only, failure mode. Static-damaged devices may show complete failure, intermittent

Table 40.1: The triboelectric series

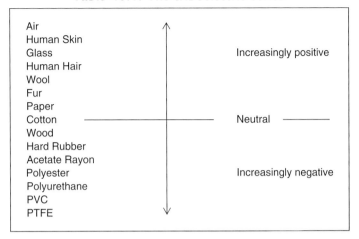

failure or degradation of performance. They may fail after one very high voltage discharge, or because of the cumulative effect of several discharges of lower potential.

40.2.2   Static Protection

To protect against ESD damage, you need to prevent static buildup and to dissipate or neutralize existing charges. At the same time, operators (including yourself and your design colleagues) need to be aware of the potential hazard. The methods used to do this are:

- package sensitive devices or assemblies in conductive containers; keep them in these until use and ensure they are clearly marked;

- use conductive mats on the floor and workbench where sensitive devices are assembled, bonded to ground via a 1 MΩ resistor;

- remove nonconductive items such as polystyrene cups, synthetic garments, wrapping film, etc. from the work area;

- ground the assembly operator through a wrist strap, in series with a 1 MΩ resistor for electric shock protection;

- ground soldering tool tips;

- use ionized air to dissipate charge from nonconductors, or maintain a high relative humidity;

- create and maintain a static-safe work area where these practices are adhered to;

- ensure that all operators are familiar with the nature of the ESD problem; and

- mark areas of the circuit where a special ESD hazard exists; design circuits to minimize exposed high-impedance or unprotected nodes.

All production assembly areas should be divided into static-safe workstation positions. Your design and development prototyping lab should also follow these precautions, since it is quite possible to waste considerable time tracking down a fault in a prototype which is due to a static-damaged device. A typical static-safe area layout is shown in Figure 40.1. BS CECC 00015: Part 1:1991 gives a code of practice for the handling of electrostatic sensitive devices.

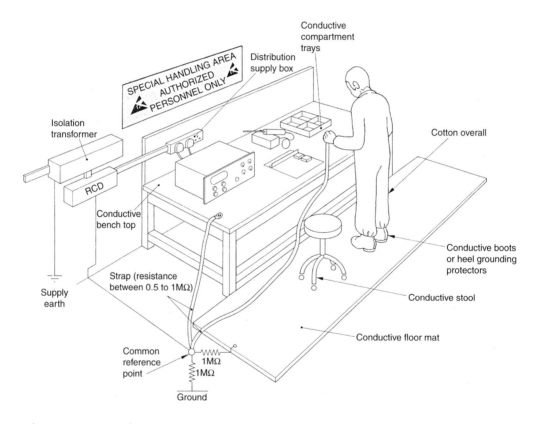

Figure 40.1: Static-safe workshop layout Extract from BS5783:1987 reproduced with the permission of BSI

CHAPTER 41

Testability

Tim Williams

The previous chapter's checklist included some items that referred to the testability of the design. It is vital that you give sufficient thought throughout the design of the product as to how the assembled unit or units will be tested to prove their correct function. In the very early stages, you should already know whether your test department will be using in-circuit testing, manual functional testing, functional testing on ATE, boundary scan, or a combination of these methods. You should then be in a position to include test access points and circuits in the design as it progresses. This is a more effective way of incorporating testability than merely bolting it on at the end.

41.1 In-Circuit Testing

The first test for an assembled PCB is to confirm that every component on it is correctly inserted, of the right type or value, and properly soldered in. It is quite possible for manual assembly personnel to insert the wrong component, or insert the right one incorrectly polarized, or even to omit a component or series of components. Automatic assembly is supposed to avoid such errors, but it is still possible to load the wrong component into the machine, or for components to be marked incorrectly. Automatic soldering has a higher success rate than hand soldering but bad joints due to lead or pad contamination can still occur.

In-circuit testing lends itself to automatic test fixture and test program generation. Each node on the PCB has to be probed, which requires a bed-of-nails test fixture (Figure 41.1) and this can be designed automatically from the PCB layout data. Similarly, the expected component characteristics between each node can be derived from the circuit schematic, using a component parameter library.

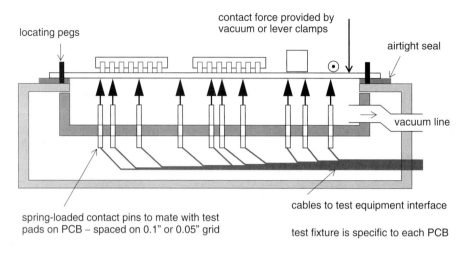

locating pegs

contact force provided by vacuum or lever clamps

airtight seal

vacuum line

cables to test equipment interface

spring-loaded contact pins to mate with test pads on PCB – spaced on 0.1" or 0.05" grid

test fixture is specific to each PCB

Figure 41.1: Principle of the bed-of-nails test fixture

An in-circuit tester carries out an electrical test on each component, to verify its behavior, value and orientation, by applying voltages to nodes that connect to each component and measuring the resulting current. Interactions with other components are prevented by guarding or back-driving. The technique is successful for discrete components but less so for integrated circuits, whose behaviour cannot be described in terms of simple electrical characteristics. Therefore, it is most widely applied on boards which contain predominantly discretes, and which are produced in high volume, as the overhead involved in programming and building the test fixture is significant. It does not of itself guarantee a working pcb. For this, you need a functional test.

41.2   Functional Testing

A functional test checks the behavior of the assembled board against its functional specification, with power applied and with simulated or special test signals connected to the input/output lines. It is often combined with calibration and set-up adjustments. For low-volume products you will normally write a test procedure around individual test instruments, such as voltmeters, oscilloscopes and signal generators. You may go so far as to build a special test jig to simulate some signals, interface others, provide monitored power and make connections to the board under test. The test procedures will consist of a sequence of instructions to the test technician—apply voltage A, observe signal at B, adjust trimmer C for a minimum at D, and so on—along with limit values where measurements are made.

The disadvantage of this approach is that it is costly in terms of test time. This puts up the overhead cost of each board and affects the final cost price of the overall unit. It is cheap as far as instrumentation goes, since you only need a simple test jig, and you will normally expect the test department to have the appropriate lab equipment to hand. Hence it is best suited to low production volumes where you cannot amortize the cost of automatic test equipment.

A further, hidden, disadvantage may be that you don't have to define the testing absolutely rigorously but can rely on the experience of the test technician to make good any deficiencies in the test procedure or measurement limits. It is common for test personnel to develop a better "feel" for the quirks of a particular design's behavior under test than its designer ever could. Procedural errors and invalid test limits may be glossed over by a human tester, and if such information is not fed back to the designer then the opportunity to optimize that or subsequent designs is lost.

41.2.1   ATE

Functional testing may more easily be carried out by automatic test equipment (ATE). In this case, the function of the human tester is reduced to that of loading and unloading the unit under test, pressing the "go" button and observing the pass/fail indicator. The testing is comprehensively de-skilled; the total unit test time is reduced to a few minutes or less. This minimises the test cost.

The costs occur instead at the beginning of the production phase, in programming the ATE and building a test fixture. The latter is similar to (in some cases may be identical to) the bed-of-nails fixture which would be used for in-circuit testing (Figure 41.1). Or, if all required nodes are brought out to test connectors, the test fixture may consist of a jig which automatically connects a suite of test instrumentation to the board under the command of a computer-based test program. The IEEE-488 standard bus allows interconnection of a desktop computer and remote-controlled meters, signal generators and other instruments, for this purpose.

The skill required of a test technician now resides in the test program, which may have been written by you as designer or by a test engineer. In any case it needs careful validation before it is let loose on the product, since it does not have the skill or expertise to determine when it is making an invalid test. The cost involved in designing and building the test fixture, programming it and validating the program, and the capital cost of the ATE itself need to be carefully judged against the savings that will be made in test time per unit. It is normally only justified if high production volumes are expected.

41.3   Boundary Scan and JTAG

Many digital circuits are too complex to test by conventional in-circuit probing. Even if bed-of-nails contact could be made to the hundreds and sometimes thousands of test pads that would be needed, the IC functions and pin states would be so involved that no absolute conclusions could be drawn from the voltage and impedance states recorded. The increased use of small sized PCBs, with surface mount, fine pitch components installed on both sides, presents the greatest problem. So a different method has been developed to address this problem, and it is known as boundary scan testing.

41.3.1   History

In 1985, an *ad hoc* group created the Joint Test Action Group (JTAG). JTAG had over 200 members around the world, including major electronics and semiconductor manufacturers. This group met to establish a solution to the problems of board test and to promote a solution as an industry standard. They subsequently developed a standard for integrating hardware into compliant devices that could be controlled by software. This was termed Boundary Scan Testing (BST). The JTAG proposal was approved in 1990 by the IEEE and defined as IEEE Standard 1149.1-1990 *Test Access Port and Boundary Scan Architecture* . Since the 1990 approval, updates have been published in 1993 as supplement IEEE Std 1149.1a-1993 and in 1995 as IEEE Std 1149.1b-1994.

41.3.2   Description of the Boundary Scan Method

Boundary scan is a special type of scan path with a register added at every I/O pin on a device. Although this requires special extra test latches on these pins, the technique offers several important benefits, the most obvious being that it allows fault isolation at the component level. A major problem driving the development of boundary scan has been the adverse effect on testability of surface-mount technology. The inclusion of a boundary-scan path in surface-mount components is sometimes the only way to perform continuity tests between devices. By placing a known value on an output buffer of one device and observing the input buffer of another interconnected device, it is easy to check the interconnection of the PCB net. Failure of this simple test indicates broken circuit traces, dry solder joints, solder bridges, or electrostatic-discharge (ESD) induced failures in an IC buffer—all common problems on PCBs.

Another advantage of the boundary-scan method is the ability to apply predeveloped functional pattern sets to the I/O pins of the IC by way of the scan path. IC manufacturers and ASIC developers create functional pattern sets for test purposes. Subsets of these patterns can be re-used for in-circuit functional IC testing, which can show significant savings on development resources.

Each device to be included within the boundary scan has the normal application logic section and related input and output, and in addition a boundary-scan path consisting of a series of boundary-scan cells (BSCs), typically one BSC per IC function pin (Figure 41.2). The BSCs are interconnected to form a shift register scan path between the host IC's test data input (TDI) pin and test data output (TDO) pin. During normal IC operation, input and output signals pass freely through each BSC. However, when the boundary-test mode is entered, the IC's internal logic may be disconnected and its boundary controlled in such a way that test stimuli can be shifted in and applied from each BSC output, and test responses can be captured at each BSC input and shifted out for inspection. External testing of traces and neighbouring ICs on a board assembly is achieved by applying test stimuli from the output BSCs and capturing responses at

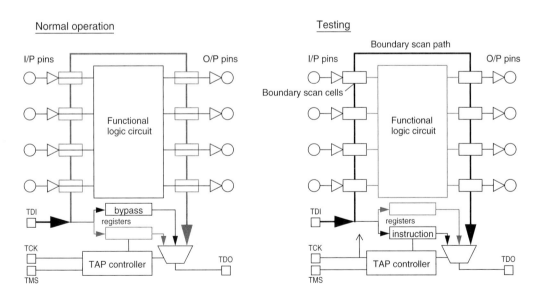

Figure 41.2: The boundary scan principle

the input BSCs. If required, internal testing of the application logic can be achieved by applying test stimuli from the input BSCs and capturing responses at the output BSCs. The implementation of a scan path at the boundary of IC designs provides an embedded testing capability that can overcome the physical access problems referred to above.

As well as performing boundary tests of each IC, ICs may also be instructed via the scan path to perform a built-in self test operation, and the results inspected via the same path. Boundary scan is not limited to individual ICs; several ICs on a board will normally be linked together to offer an extended scan path (which could be partitioned or segmented to optimize testing speed). The whole board itself could be regarded as the system to be tested, with a scan path encompassing the connections at the board's boundary, and boundary-scan cells implemented at these connections using ICs designed for the purpose.

41.3.3 Devices

Every IEEE Std 1149.1-compatible device has four additional pins—two for control and one each for input and output serial test data. These are collectively referred to as the "Test Access Port" (TAP). To be compatible, a component must have certain basic test features, but the standard allows designers to add test features to meet their own unique requirements. A JTAG-compliant device can be a microprocessor, microcontroller, PLD, CPLD, FPGA, ASIC or any other discrete device that conforms to the 1149.1 specification. The TAP pins are:

- TCK – Test Clock Input. Shift register clock separate from the system clock.

- TDI – Test Data In. Data is shifted into the JTAG-compliant device via TDI.

- TDO – Test Data Out. Data is shifted out of the device via TDO.

- TMS – Test Mode Select. TMS commands select test modes as defined in the JTAG specification.

The 1149.1 specification stipulates that at every digital pin of the IC, a single cell of a shift-register is designed into the IC logic. This single cell, known as the Boundary-Scan Cell (BSC), links the JTAG circuitry to the IC's internal core logic. All BSCs of a particular IC constitute the Boundary-Scan Register (BSR), whose length is of course determined by the number of I/O pins that IC has. BSR logic becomes active when performing JTAG testing; otherwise it remains passive under normal IC operation. A 1-bit bypass register is also included to allow testing of other devices in the scan path.

You communicate with the JTAG-compliant device using a hardware controller that either inserts into a PC add-in card slot or by using a stand-alone programmer. The controller connects to the test access port on a JTAG-compliant PCB, which may be the port on a single device, or it may be the port created by linking a number of devices. You (or your test department) then must write the software to perform boundary scan programming and testing operations.

As well as testing, the boundary scan method can be used for various other purposes that require external access to a PCB, such as Flash memory programming.

41.3.4   *Deciding Whether or Not to Use Boundary Scan*

Although the boundary scan method has enormous advantages for designers faced with the testing of complex, tightly packed circuits, it is not without cost. There is a significant logic overhead in the ICs as well as a small overhead on the board in implementing the TAP, and there is the need for your test department to invest in the resources and become familiar with the method as well as programming each product. As a rough guide, you can use the following rule[†] (relating to ASIC design) to decide on whether or not the extra effort will be cost effective:

- Designs with fewer than 10K gates: not generally complex enough to require structured test approaches. The overhead impact is usually too high to justify them. Nonstructured, good design practices are usually sufficient.

- Designs with more than 10K gates, but fewer than 20K gates: Structured techniques should be considered for designs in this density. Nonstructured, good design practices are probably sufficient for highly combinatorial circuits without memory. Structured approaches should be considered as complexity is increased by the addition of sequential circuits, feedback, and memory. Consider boundary-scan testing for reduced cycle times and high fault grades.

- Designs with more than 20K gates: The complexity of circuits this dense usually requires structured approaches to achieve high fault grades. At this density, it is often hard to control or observe deeply embedded circuits. The overhead associated with structured testability approaches is acceptable.

[†] From "IEEE Std 1149.1 (JTAG) Testability Primer," Texas Instruments, 1997.

41.4 Design Techniques

There are many ways in which you can design a PCB circuit to make it easy to test, or conversely hard to test. The first step is to decide how the board will be tested, which is determined by its complexity, expected production volume and the capabilities of the test department.

41.4.1 Bed-of-Nails Probing

If you will be using a bed-of-nails fixture, then the PCB layout should allow this. Leave a large area around the outside of the board, and make sure there are no unfilled holes, to enable a good vacuum pressure to be developed to force the board onto the probes. Or if the board will be clamped to the fixture, make sure there is space on the top of the board for the clamps. Decide where your test nodes need to be electrically, and then lay out the board to include target pads on the underside for the probes. These pads should be spaced on a 0.1" (2.5mm) grid for accurate drilling of the test jig; down to 0.05" or 1 mm is possible if the board layout is tight. It is not good practice to use component lead pads as targets, since pressure from the probe may cause a defective joint to appear good. Ensure that tooling holes are provided and are accurately aligned with the targets.

Remember that a bed-of-nails jig will connect several long, closely coupled wires to many nodes in the circuit. This will severely affect the circuit's stray reactances, and thereby modify its high frequency response. It is not really suitable for functionally testing high frequency or high speed digital circuits.

41.4.2 Test Connections

If your test department doesn't want to use bed-of-nails probing, then help them find the test points that are necessary by bringing them out to test connectors. These can be cheap-and-cheerful pin strips on the board since they will normally only be used once. The matching test jig can then take signals from these connections direct to the test instrumentation via a switch arrangement. Of course, pre-existing connectors such as multi-way or edge connectors can be used to bring out test signals on unused pins. Be careful, though, that you do not bring long test tracks from one side of the board to the other and thereby compromise the circuit's crosstalk, noise susceptibility and stability. Extra local test connectors are preferable.

41.4.3   Circuit Design

There are many design tricks to make testing easier. A simple one is to include a series resistor in circuits where you will want to back-drive against an output, or where you will want to measure a current (Figure 41.3(A)). The cost of the resistor is minimal compared to the test time it might save. Of course, you must ensure that the resistor does not affect normal circuit operation. Also, unused digital gate inputs may be taken to a pull-up resistor rather than direct to supply or ground, and this point can then be used to inhibit or enable logic signals for testing purposes only (Figure 41.3(B)).

The theme of Figure 41.3(B) can be taken further to incorporate extra logic switching to allow data or timing signals to be derived either from the normal on-board source, or from the external test equipment. This is particularly useful in situations where testing logic functions from the normal system clock would result either in too fast operation, or too slow. The clock source can be taken through a 2-input data multiplexer such as the 74HC157, one input of which is taken via a test connector to the external clock as shown in Figure 41.4. In normal operation the clock select and test clock inputs are left unconnected and the system clock is passed directly through the multiplexer.

When you are considering testing a microprocessor board, it is advantageous to have a small suite of test software resident on the main program PROM. This can be activated

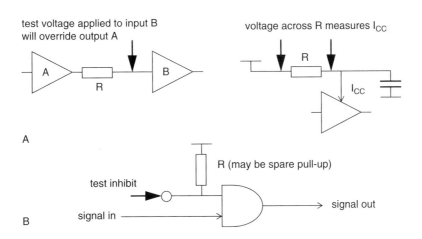

Figure 41.3: Use of an extra test resistor

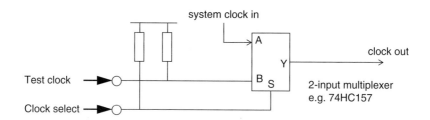

Figure 41.4: Test clock selection with data multiplexer

on start-up by reading a digital input which is connected to a test link or test probe. If the test input is set, the program jumps to the test routines rather than to the main operating routines. These are arranged to exercise all inputs, outputs and control signals continuously in a predictable manner, so that the test equipment can monitor them for the correct function. The test software operation depends of course on the core functions of the microprocessor, and its bus and control signal interconnections, being fault-free.

More complex digital systems cannot easily be tested or, more to the point, debugged, with the techniques described so far. The boundary scan methods described in section 41.3 are aimed at these applications and you need to design them in from the start, since they consume a serious amount of circuit overhead in order to function.

CHAPTER 42

Reliability

Tim Williams

The reliability of electronic equipment can to some extent be quantified, and a separate discipline of reliability engineering has grown up to address it. This chapter will serve as an introduction to the subject for those circuit designers who are not fortunate enough to have a reliability engineering department at their disposal.

42.1 Definitions

Reliability, itself, has a strictly defined meaning. This can be stated as "the probability that a system will operate without failure for a specified period, subject to specified environmental conditions." Thus, it can be quoted as a single number, such as 90%, but this is subject to three qualifications:

- Agreement as to what constitutes a "failure." Many systems may "fail" without becoming totally useless in the process.

- A specified operating lifetime. No equipment will operate forever; reliability must refer to the reasonably foreseen operating life of the equipment, or to some other agreed period. The age of the equipment, which may well affect failure rate, is not a factor in the reliability specification.

- Agreement upon environmental conditions. Temperature, moisture, corrosive atmospheres, dust, vibration, shock, supply and electromagnetic disturbances all have an effect on equipment operation and reliability is meaningless if these are not quoted.

If you offer or purchase equipment whose reliability is quoted for one set of conditions and it is used under another set, you will not be able to extrapolate the reliability figure to the new conditions unless you know the behavior of those parameters which affect it.

42.1.2   Mean Time Between Failures

For most of the life of a piece of electronic equipment, its failure rate (denoted by λ) is constant. In the early stages of operation it could be high and decrease as weak components fail quickly and are replaced; late in its life components may begin to "wear out" or corrosion may take its toll, and the failure rate may start to rise again. The reciprocal of failure rate during the constant period is known as the mean time between failures (MTBF). This is generally quoted in hours, while failure rate is quoted in faults per hour. For instance, an MTBF of 10,000 hours is equivalent to a failure rate of 0.0001 faults per hour or 100 faults per 106 hours. MTBF has the advantage that it does not depend on the operating period, and is therefore more convenient to use than reliability.

42.1.3   Mean Time to Failure

MTBF measures equipment reliability on the assumption that it is repaired on each failure and put back into service. For components which are not repairable, their reliability is quoted as mean time to failure (MTTF). This can be calculated statistically by observing a sample from a batch of components and recording each one's working life, a procedure known as life testing. The MTTF for this batch is then given by the mean of the lifetimes.

42.1.4   Availability

System users need to know for what proportion of time their system will be available to them. This figure is given by the ratio of "up-time," during which the system is switched on and working, to total operating time. The difference between the two is the "down-time" during which the system is faulty and/or under repair. Thus:

$$A = [U/(U + D)]$$

Availability can also be related to the MTBF figure and the mean time to repair (MTTR) figure by:

$$A = [MTBF/(MTBF + MTTR)]$$

The availability of a particular system can be monitored by logging its operating data, and this can be used to validate calculated MTBF and MTTR figures. It can also be interpreted as a probability that at any given instant the system will be found to be working.

42.2 The Cost of Reliability

Reliability does not come for free. Design and development costs escalate as more effort is put into assuring it, and component costs increase if high performance is required of them. For instance, it would be quite possible to improve the reliability of, say, an audio power amplifier by using massively over-rated output transistors, but these would add considerably to the selling cost of the amplifier. On the other hand, if the selling cost were reduced by specifying under-rated transistors, the users would find their total operating costs mounting since the output transistors would have to be replaced more frequently. Thus there is a general trend of decreasing operating or "lifecycle" costs and increasing unit costs, as the designed-in reliability of a given system increases. This leads to the notion of an "optimum" reliability figure in terms of cost for a system. Figure 42.1 illustrates this trend. The criterion of good design is then to approach this optimum as closely as possible.

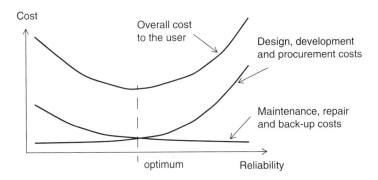

Figure 42.1: Reliability vs. cost

Of course, this argument only applies when the cost of unreliability is measured in strictly economic terms. Safety-critical systems, such as nuclear or chemical process plant controllers, railway signaling or flight-critical avionics, must instead meet a defined reliability standard and the design criterion then becomes one of assuring this level of reliability, with cost being a secondary factor.

42.3   Design for Reliability

The goal of any circuit designer is to reduce the failure rate of their design to the minimum achievable within cost constraints. The factors that help in meeting this goal are:

- use effective thermal management to minimize temperature rise;
- de-rate susceptible components as far as possible;
- specify high reliability or quality assured components;
- specify stress screening or burn-in tests;
- keep circuits simple, use the minimum number of components; and
- use redundancy techniques at the component level.

42.3.1   Temperature

High temperature is the biggest enemy of all electronic components and measures to keep it down are vital. Temperature rise accelerates component breakdown because chemical reactions occurring within the component, which govern bond fractures, growth of contamination or other processes, have an increased rate of reaction at higher temperature. The rate of reaction is determined by the Arrhenius equation,

$$\lambda = K \cdot \exp(-E/kT)$$

where,

λ gives a measure of failure rate

K is a constant depending on the component type

E is the reaction's activation energy

k is Boltzmann's constant, $1.38 \ 10^{-23}$ J K^{-1}

T is absolute temperature

Many reactions have activation energies around 0.5eV which results in an approximate doubling of λ with every 10°C rise in temperature, and this is a useful rule of thumb to apply for the decrease in reliability versus temperature of typical electronic equipment with many components. Some reactions have higher activation energy, which give a faster increase of λ with temperature.

42.3.2 De-Rating

There is a very significant improvement to be gained by operating a component well within its nominal rating. For most components this means either its voltage or power rating, or both.

Take capacitors as an example. Conventionally, you will determine the maximum DC bias voltage a capacitor will have to withstand under worst-case conditions and then select the next highest rating. Overspecifying the voltage rating may result in a larger and more costly component.

However, capacitor life tests show that as the maximum working voltage is approached, the failure rate increases as the fifth power of the voltage. Therefore, if you run the capacitor at half its rated voltage you will observe a failure rate 32 times lower than if it is run at full rated voltage. Given that a capacitor of double the required rating will not be as much as double the size, weight or cost, except at the extremes of range, the improvement in reliability is well worth having.

In many cases there is no difficulty in using a de-rated capacitor; small film capacitors, for instance, are rated at a minimum of 50 or 100V and are frequently used in 5V circuits. Electrolytics on the other hand are more likely to be run near their rating. These capacitors already have a much higher failure rate than other types because of their construction—the electrolyte has a tendency to "dry out," especially at high temperatures—and so you will achieve significant improvement, albeit at higher cost, if you heavily de-rate them.

De-rating the power dissipation of resistors reduces their internal temperature and therefore their failure rate. In low voltage circuits there is no need to check power rating for any except low value parts; if for instance you use 0.4-watt metal film resistors in a circuit with a maximum supply of 10V, you can be sure that all resistors over 500ω will be de-rated by at least a factor of 2, which is normally enough.

Semiconductor devices are normally rated for power, current and voltage, and de-rating on all of these will improve failure rate. The most important are power dissipation, which is closely linked to junction temperature rise and cooling provision, and operating voltage, especially in the presence of possible transient overvoltages.

42.3.3 High Reliability Components

Component manufacturers' reputations are seriously affected by the perceived reliability or otherwise of their product, so most will go to considerable effort not to ship defective parts. However, the cost of detecting and replacing a faulty part rises by an order of magnitude at each stage of the production process, starting at goods inward inspection, proceeding through board assembly, test and final assembly, and ending up with field repair. You may therefore decide (even in the absence of mandatory procurement requirements on the part of your customer) that it is worth spending extra to specify and purchase parts with a guaranteed reliability specification at the "front end" of production.

42.3.4 CECC

Initially it was military requirements, where reliability was more important than cost, that drove forward schemes for assessed quality components. More recently many commercial customers have also found it necessary to specify such components. The need for a common standard of assessed quality is met in Europe by the CECC[†] scheme. This has superseded the earlier national BS9000 series of standards. CECC documents refer to a "Harmonized system of quality assessment for electronic components."

Generic specifications are found in the CECC series for all types of component which are covered by the scheme. These specify physical, mechanical and electrical properties, and lay down test requirements. Individual component specifications are not found under the scheme.

[†] CENELEC Electronic Components Committee.

42.3.5 Stress Screening and Burn-in

These specifications all include some degree of stress screening. This phrase refers to testing the components under some type of stress, typically at elevated temperature, under vibration or humidity and with maximum rated voltage applied, for a given period. This practice is also called *burning in*. The principle is that weak components will fail early in their life and the failures can be accelerated by operating them under stress. These can then be weeded out before the parts are shipped from the manufacturer. A typical test might be 160 hours at 125°C. Another common test is a repeated temperature cycle between the extremes of the permitted temperature range, which exposes failures due to poor bonding or other mechanical faults.

Such stress screening can be applied to any component, not just semiconductors, and also to entire assemblies. If you are unsure of the probable quality of early production output of a new design, specifying stress screening on the first few batches is a good way to discover any recurrent production faults before they are passed out to the customer. It is expensive in time, equipment and inventory, and should not be used as a crutch to compensate for poor production practices. It should only be employed as standard if the customer is willing to pay for it.

42.3.6 Simplicity

The failure rate of an electronic assembly is roughly equal to the sum of the failure rates of all its components. This assumes that a failure in any one component causes the failure of the whole assembly. This is not necessarily a valid assumption, but to assume otherwise you would have to work out the assembly's failure modes for each component failure and for combinations of failures, which is not practical unless your customer is prepared to pay for a great deal of development work.

If the assumption holds, then reducing the number of components will reduce the overall failure rate. This illustrates a very important principle in circuit design: the highest reliability comes from the simplest circuits. Apply Occam's razor ("entities should not be multiplied beyond necessity") and cut down the number of components to a minimum.

42.3.7 Redundancy

Redundancy is employed at the system level by connecting the outputs of two or more subsystems together such that if one fails, the others will continue to keep the system working. A typical example might be several power supplies, each connected to the same power distribution rail (via isolating diodes) and each capable of supplying the full load. If the reliability of the interconnection is neglected, the probability of all supplies failing simultaneously is the product of the probabilities of failure of each supply on its own, assuming that a common mode failure (such as the mains supply to all units going off) is ruled out.

The principle can also be applied at the component level. If the probability of a single component failing is too high then redundant components can be placed in parallel or series with it, depending on the required failure mode. This technique is mandatory in certain fields, such as intrinsically safe instrumentation. Figure 42.2 illustrates redundant zener diode clamping. The zeners prevent the voltage across their terminals from rising to an unsafe value in the event of a fault voltage being applied at the input to the barrier. One zener alone would not offer the required level of reliability, so two further ones are placed in parallel, so that even with an open-circuit failure of two out of the three, the clamping action is maintained. The interconnections between the zeners must be solid enough not to materially affect the reliability of the combination.

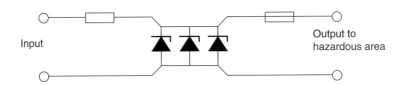

Figure 42.2: The intrinsically safe zener diode barrier

Some provision must normally be made for detecting and indicating a failed component or subsystem so that it can be repaired or replaced. Otherwise, once a redundant part has failed, the overall reliability of the system is severely reduced.

42.4   The Value of MTBF Figures

The mean time between failure figure as defined earlier can be calculated before the equipment is put into production by summing the failure rates of individual components to give an overall failure rate for the whole equipment. As discussed earlier, this assumes that a fault in any one component causes the failure of the whole assembly. This method presupposes adequate data on the expected failure rates of all components that will be used in the equipment.

Such sources of failure rate data for established component types are available. The most widely used is MIL-HDBK-217, now in its fifth revision, published by the US Department of Defense. This handbook lists failure rate models and tables for a wide variety of components, based on observed failure measurements. A failure rate for each component can be derived from its operating and environmental conditions, de-rating factor and method of construction or packaging. A further factor that is included for integrated circuits is their complexity and pinout. Another source of failure rate data, somewhat less comprehensive but widely used for telecommunications applications, is British Telecom's handbook HRD4.

The disadvantage with using such data is that it cannot be up-to-date. Proper failure rate data takes years to accumulate, and so data extracted from these tables for modern components will not be accurate. This is especially true for integrated circuits. Generally, figures based on obsolete failure rate data will tend to be pessimistic, since the trend of component reliability is to improve.

Calculations of failure rates at component level are tedious, since operating conditions for each component, notably voltage and power dissipation, must be a part of the calculation in order to arrive at an accurate value. They do lend themselves to computer derivation, and software packages for reliability prediction are readily available. Since in many cases such operating conditions are highly variable, it is arguable that you will not obtain much more than an order-of-magnitude estimate of the true figure anyway.

A published MTBF figure does not tell you how long the unit will actually last, and it does not indicate how well the unit will perform in the field under different environmental and operating conditions. Such figures are mainly used by the marketing department to make the specification more attractive. But MTBF prediction is valuable for two purposes:

- For the designer, it gives an indication of where reliability improvements can most usefully be made. For instance, if as is often the case the electrolytic capacitors turn out to make the highest contribution to overall failure rate, you can easily evaluate the options available to you in terms of de-rating or adding redundant components. You need not waste effort on optimizing those components which have little effect overall.

- For the service engineer, it gives an idea of which components are likely to have failed if a breakdown occurs. This can be valuable in reducing servicing and repair time.

42.5   Design Faults

Before leaving the subject of reliability design, we should briefly mention a very real problem, which is the fallibility of the designers themselves. There is no point in specifying highly reliable components or applying all manner of stress screening tests or redundancy techniques if the circuit is going to fail because it has been wrongly designed. Design faults can be due to inexperience, inattention or incompetence on the part of the designer, or simply because the project timescale was too short to allow the necessary cross-checking. Computer-aided design techniques and simulators can reduce the risk but they cannot eliminate the potential for human error completely.

42.5.1   The Design Review

An effective and relatively painless way of guarding against design faults is for your product development department to instigate a system of frequent design reviews. In these, a given designer's circuit is subjected to a peer critique in order to probe for flaws which might not be apparent to the circuit's originator. The critique can check that the basic circuit concept is sound and cost-effective, that all component tolerances have been accounted for, that parts will not be operated outside their ratings, and so on. The depth of the review is determined by the resources that are available within the group; the reviewers should preferably have no connection with the project being reviewed, so that they are able to question underlying and unstated assumptions. Naturally, the effectiveness of such a system depends on the resources a company is prepared to devote to it, and it also depends on the willingness of the designer to undergo a review. Personality clashes tend to surface on these occasions. Each designer develops pet techniques and idiosyncrasies during their career, and provided these are not actually wrong they should not attract criticism. Nevertheless, design reviews are valuable for testing the strength of a design before it gets to the stage where the cost of mistakes becomes significant.

CHAPTER 43

Thermal Management

Tim Williams

It is in the nature of electronic components to dissipate power while they are operating. Any flow of current through a nonideal component will develop some power within that component, which in turn causes a rise in temperature. The rise may be no more than a small fraction of a degree Celsius when less than a milliwatt is dissipated, extending to several tens or even hundreds of degrees when the dissipation is measured in watts. Since excess temperature kills components, some way must be found to maintain the component operating temperature at a reasonable level. This is known as thermal management.

43.1 Using Thermal Resistance

Heat transfer through the thermal interface is accomplished by one or more of three mechanisms: conduction, convection and radiation. The attractiveness of thermal analysis to electronics designers is that it can easily be understood by means of an electrical analog. Visualize the flow of heat as emanating from the component which is dissipating power, passing through some form of thermal interface and out to the environment, which is assumed to have a constant ambient temperature TA and infinite ability to sink heat. Then the heat source can be represented electrically as a current source; the thermal impedances as resistances; the temperature at any given point is the voltage with respect to 0V; and thermal inertia can be represented by capacitance with respect to 0V. The 0V reference itself doesn't have an exact thermal analog, but it is convenient to represent it as 0°C, so that temperature in °C is given exactly by a potential in volts. All these correspondences are summarized in Table 43.1.

Table 43.1: Thermal and electrical equivalences

Thermal parameter	Units	Electrical analog	Units
Temperature difference	°C	Potential difference	Volts
Thermal resistance	°C/W	Resistance	Ohms
Heat flow	J/s (W)	Current	Amps
Heat capacity	J/°C	Capacitance	Farads

Figure 43.1 shows the simplest general model and its electrical analog. The model can be analyzed using conventional circuit theory and yields the following equation for the temperature at the heat source:

$$T = P_D \cdot R_\theta + T_A$$

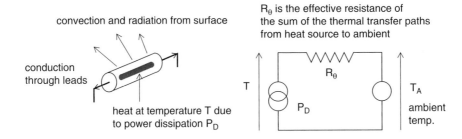

Figure 43.1: Heat transfer from hot component to ambient

This temperature is the critical factor for electronic design purposes, since it determines the reliability of the component. Reducing any of P_D, R_θ or T_A will minimize T. Ambient temperature is not normally under your control but is instead a specification parameter (but see section 43.4). The usual assumption is that the ambient air (or other cooling medium) has an infinite heat capacity and therefore its temperature stays constant no matter how much heat your product puts into it. The intended operating environment will determine the ambient temperature range, and for heat calculations only the extreme of this range is of interest; the closer this gets to the maximum allowable value of T the harder is your task. Since you are normally attempting to manage a given power dissipation, the only parameter which you are free to modify is the thermal resistance R_θ. This is achieved by heatsinking.

There are more general ways of analyzing heat flow and temperature rise, using thermal conductivity and the area involved in the heat transfer. However, component manufacturers normally offer data in terms of thermal resistance and maximum permitted temperature, so it is easiest to perform the calculations in these terms.

43.1.1 Partitioning the Heat Path

When you have data on the component's thermal resistance directly to ambient, and your mounting method is simple, then the basic model of Figure 43.1 is adequate. For components which require more sophisticated mounting and whose heat transfer paths are more complicated, you can extend the model easily. The most common application is the power semiconductor mounted via an insulating washer to a heatsink (Figure 43.2(A)).

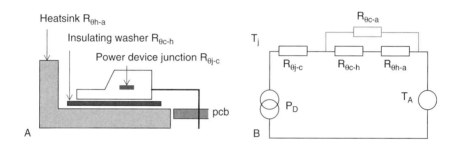

Figure 43.2: Heat transfer for a power device on a heatsink

The equivalent electrical model is shown in Figure 43.2(B). Here, T_j is the junction temperature and $R_{\theta j\text{-}c}$ represents the thermal resistance from junction to case of the device. All manufacturers of power devices will include $R_{\theta j\text{-}c}$ in their data sheets and it can often be found in low-power data as well. Sometimes it is disguised as a power derating figure, expressed in W/°C. The maximum allowable value of T_j is published in the maximum ratings section of each data sheet, and this is the parameter that your thermal calculations must ensure is not exceeded.

$R_{\theta c\text{-}h}$ and $R_{\theta h\text{-}a}$ are the thermal resistances of the interface between the case and the heatsink, and of the heatsink to ambient, respectively. $R_{\theta c\text{-}a}$ represents the thermal resistance due to convection directly from case to ambient, and can be neglected if you are using a large heatsink.

An example should help to make the calculation clear.

An IRF640 power MOSFET dissipates a maximum of 35W steady-state. It is mounted on a heatsink with a specified thermal resistance of 0.5°C per watt, via an insulating pad with a thermal resistance of 0.8°C per watt. The maximum ambient temperature is 70°C. What will be the maximum junction temperature?

From the above conditions, $R_{\theta c\text{-}h} + R_{\theta h\text{-}a} = 1.3°C/W$. The IRF640 data quotes a junction-to-case thermal resistance ($R_{\theta j\text{-}c}$) of 1.0°C/W.
So the junction temperature:

$$T_j = 35[1.3 + 1.0] + 70 = 150.5°C$$

This is just over the maximum permitted junction temperature of 150°C so reliability is marginal and you need a bigger heatsink. However, we have neglected the junction-to-ambient thermal resistance, quoted at 80°C/W. This is in parallel with the other thermal resistances. If it is included, the calculation becomes:

$$T_j = 35[2.3 \times 80]/[2.3 + 80] + 70 = 148.25°C$$

A very minor improvement, and not enough to rely on!

This example illustrates a common misconception about power ratings. The IRF640 is rated at 125W dissipation, yet even with a fairly massive heatsink (0.5°C/W will require a heatsink area of around 80 square inches) it cannot safely dissipate more than 35W at an ambient of 70°C. The fact is that the rating is specified at 25°C case temperature; higher case temperatures require de-rating because of the thermal resistance from junction to case. You will not be able to maintain 25°C at the case under any practical application conditions, except possibly outside in the Arctic. Power device manufacturers publish de-rating curves in their data sheets: rely on these rather than the absolute maximum power rating on the front of the specification.

Incidentally, if having followed these thermal design steps you find that the needed heatsink is too large or bulky, it will be far cheaper to reduce the thermal resistance of the total system. You do this by using two (or more) transistors in parallel in place of a single device. Although the thermal resistances for each of the transistors stay the same, the resultant heat flow for each is effectively halved because each transistor is only dissipating half the total power, and therefore the junction temperature rise is also half.

43.1.2 Thermal Capacity

The previous analysis assumed a steady-state heat flow, in other words constant power dissipation. If this is not a good description of your application, you may need to take

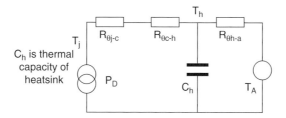

Figure 43.3: Electrical analog modified to include thermal capacity

account of the thermal capacity of the heatsink. The electrical analog circuit of Figure 43.2 can be modified according to Figure 43.3.

From this you can see that a step increase in dissipated power will actually cause a gradual rise in heatsink temperature T_h. This will be reflected at T_j, modified by $R_{\theta j-c}$ and $R_{\theta c-h}$, which will take typically several minutes, possibly hours, to reach its maximum temperature. The value of C_h depends on the mass of heatsink metal, and its heat storage capacity. Values of this parameter for common metals are given in Table 43.2. As an

Table 43.2: Thermal properties of common metals

Metal	Finish	Heat capacity J/cm3/°C	Bulk thermal conductivity W/°C/m	Surface emissivity ε (black body = 1)
Aluminium	Polished	2.47	210	0.04
	Unfinished			0.06
	Painted			0.9
	Matt anodised			0.8
Copper	Polished	3.5	380	0.03
	Machined			0.07
	Black oxidized			0.78
Steel	Plain	3.8	40–60	0.5
	Painted			0.8
Zinc	Grey oxidized	2.78	113	0.23–0.28

example, a 1°C/W aluminum heatsink might have a volume of 120 cm^3 which has a heat capacity of 296 J/°C. Multiplying the thermal resistance by the heat capacity gives an idea of the time constant, of 296 seconds.

The thermal capacity will not affect the end-point steady state temperature, only the time taken to reach it. But if the heat input is transient, with a low duty cycle to allow plenty of cooling time, then a larger thermal capacity will reduce the maximum temperatures T_h and T_j reached during a heat pulse. You can analyze this if necessary with the equivalent circuit of Figure 43.3. Strictly, the other heat transfer components also have an associated thermal capacity which could be included in the analysis if necessary.

43.1.3   Transient Thermal Characteristics of the Power Device

In applications where the power dissipated in the device consists of continuous low duty cycle periodic pulses, faster than the heatsink thermal time constant, the instantaneous or peak junction temperature may be the limiting condition rather than the average temperature. In this case you need to consult curves for transient thermal resistance. These curves are normally provided by power semiconductor manufacturers in the form of a correction factor that multiplies $R_{\theta j\text{-}c}$ to allow for the duty cycle of the power dissipation. Figure 43.4 shows a family of such curves for the IRF640. Because the period for most pulsed applications is much shorter than the heatsink's thermal time constant, the values of $R_{\theta h\text{-}a}$ and $R_{\theta c\text{-}h}$ can be multiplied directly by the duty cycle. Then the junction temperature can now be calculated from:

$$T_j \; = \; P_{Dmax} \cdot \left[K \cdot R_{\theta j-c} \; + \; \delta \; (R_{\theta c-h} \; + \; R_{\theta h-a}) \right] \; + \; T_A$$

where δ is the duty cycle and K is derived from curves as in Figure 43.4 for a particular value of $\delta \cdot P_{Dmax}$ is still the maximum power dissipated during the conduction period, not the power averaged over the whole cycle. At frequencies greater than a few kHz, and duty cycles more than 20%, cycle-by-cycle temperature fluctuations are small enough that the peak junction temperature is determined by the average power dissipation, so that K tends toward δ.

Some applications, notably RF amplifiers or switches driving highly inductive loads, may create severe current crowding conditions on the semiconductor die which invalidate methods based on thermal resistance or transient thermal impedance. Safe operating areas and di/dt limits must be observed in these cases.

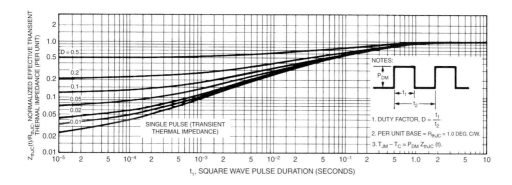

Figure 43.4: Transient thermal impedance curves for the IRF640 Source: International Rectifier

43.2   Heatsinks

As the previous section implied, the purpose of a heatsink is to provide a low thermal resistance path between the heat source and the ambient. Strictly speaking, it is the ambient environment which is the heat sink; what we conventionally refer to as a heatsink is actually only a heat exchanger. It does not itself sink the heat, except temporarily. In most cases the ambient sink will be air, though not invariably: this author recalls one somewhat tongue-in-cheek design for a 1 kW rated audio amplifier which suggested bolting the power transistors to a central heating radiator with continuous water cooling! Some designs with a very high power density need to adopt such measures to ensure adequate heat removal.

A wide range of proprietary heatsinks is available from many manufacturers. Several types are predrilled to accept common power device packages. All are characterized to give a specification figure for thermal resistance, usually quoted in free air with fins vertical. Unless your requirements are either very specialized or very high volume, it is unlikely to be worth designing your own heatsink, especially as you will have to go through the effort of testing its thermal characteristics yourself. Custom heatsink design is covered in the application notes of several power device manufacturers.

A heatsink transfers heat to ambient air primarily by convection, and to a lesser degree by radiation. Its efficiency at doing so is directly related to the surface area in contact with the convective medium. Thus, heatsink construction seeks to maximize surface area for a given volume and weight; hence the preponderance of finned designs. Orientation of the fins is important because convection requires air to move past the surface and become heated as it does so. As air is heated it rises. Therefore the best convective efficiency is obtained by orienting the fins vertically to obtain maximum air flow across them; horizontal mounting reduces the efficiency by up to 30%.

Convection cooling efficiency falls at higher altitudes. Atmospheric pressure decreases at a rate of 1 mb per 30 ft height gain, from a sea level standard pressure of 1013 mb. Since the heat transfer properties are proportional to the air density, this translates to a cooling efficiency reduction as shown in Table 43.3.

Table 43.3: Free air cooling efficiency versus altitude

Sea level	2000 ft	5000 ft	10000 ft	20000 ft
100%	97%	90%	80%	63%

The most common material for heatsinks is black anodized aluminum. Aluminum offers a good balance between cost, weight and thermal conductivity. Black anodizing provides an attractive and durable surface finish and also improves radiative efficiency by 10–15 times over polished aluminum. Copper can be used as a heatsink material when the optimum thermal conductivity is required, but it is heavier and more expensive.

The cooling efficiency does not increase linearly with size, for two principal reasons:

- longer heatsinks (in the direction of the fins) will suffer reduced efficiency at the end where the air leaves the heatsink, since the air has been heated as it flows along the surface;

- the thermal resistance through the bulk of the metal creates a falling temperature gradient away from the heat source, which also reduces the efficiency at the extremities; this resistance is not included in the simple model of Figure 43.2.

The first of the above reasons means that it is better to reduce the thermal resistance by making a shorter, wider heatsink than by a longer one. The average performance of a typical heatsink is linearly proportional to its width in the direction perpendicular to the airflow, and approximately proportional to the square root of the fin length in the direction parallel to the flow.

Also, the thermal resistance of any given heatsink is affected by the temperature differential between it and the surrounding air. This is due both to increased radiation (see below) and increased convection turbulence as the temperature difference increases. This can lead to a drop in $R_{\theta h\text{-}a}$ at 20°C difference to 80% of the value at 10°C difference. Or put the other way around, the $R_{\theta h\text{-}a}$ at 10°C difference may be 25% higher than that quoted at 20°C difference.

43.2.1 Forced Air Cooling

Convective heat loss from a heatsink can be enhanced by forcing the convective medium across its surface. Detailed design of forced air cooled heatsinks is best done empirically. Simulation software is available to map heat flow and the resulting thermal transfer in complex assemblies; most heatsink applications will be too involved for simple analytical methods to give better than ballpark results. It is not too difficult to use a thermocouple to measure the temperature rise of a prototype design with a given dissipation, most easily generated by a power resistor attached to a DC supply.

Figure 43.5 shows the improvement in thermal resistance that can be gained by passing air over a square flat plate, and of course at least a similar magnitude can be expected for any finned design. Optimizing the placement of the fins requires either experimentation or simulation, although staggering the fins will improve the heat transfer. When you use forced air cooling, radiative cooling becomes negligible and it is not necessary to treat the surface of the heatsink to improve radiation; unfinished aluminum will be as effective as black anodized.

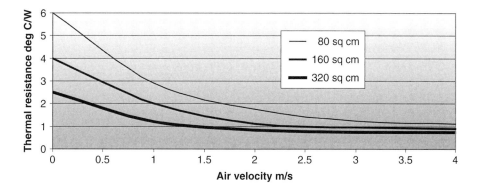

Figure 43.5: Thermal resistance vs. air velocity for various flat plate sizes

Another common use of forced air cooling is ventilation of a closed equipment cabinet by a fan. The capacity of the fan is quoted as the volumetric flow rate in cubic feet per minute (CFM) or cubic meters per hour (1 CFM = 1.7 m³/hr). The volumetric flow rate required to limit the internal temperature rise of an enclosure in which P_D watts of heat is dissipated to $\theta°C$ above ambient is:

$$\text{Flow rate} = 3600 \, P_D/(\rho \times c \times \theta) m^3/hr$$

where,

   ρ is the density of the medium

   (air at 30°C and atmospheric pressure is 1.3 kg/m3)

   c is the specific heat capacity of the medium

   (air at 30°C is around 1000J kg–1°C—1)

Fan performance is shown as volumetric flow rate versus pressure drop across the fan. The pressure differential is a function of the total resistance to airflow through the enclosure, presented by obstacles such as air filters, louvres, and PCBs. You generally need to derive pressure differential empirically for any design with a nontrivial airflow path.

43.2.2   Radiation

Radiative cooling is something of a mixed blessing. Radiant heat travels in line of sight, and is therefore as likely to raise the temperature of other components in an assembly as to be dissipated to ambient. For the same reason, radiation is rarely a significant contribution to cooling by a finned heatsink, since the finned areas which make up most of the surface merely heat each other. However, radiation can be used to good effect when a clear radiant path to ambient can be established, particularly for high-temperature components in a restricted airflow. The thermal loss through radiation is:

$$Q = 5.7 \times 10^{-12} \times \Delta T^4 \times \varepsilon$$

where,

   ε is the emissivity of the surface, compared to a black body

   ΔT is the temperature difference between the component and environment

   Q is given in watts per second per square centimeter

Emissivity depends on surface finish as well as on the type of material, as shown in Table 43.2. Glossy or shiny surfaces are substantially worse than matte surfaces, but the actual color makes little difference. What is important is that the surface treatment should be as thin as possible, to minimize its effect on convection cooling efficiency.

Poor radiators are also poor absorbers, so a shiny surface such as aluminum foil can be used to protect heat sensitive components from the radiation from nearby hot components. The reverse also holds, so for instance it is good practice to keep external heatsinks out of bright sunlight.

43.3   Power Semiconductor Mounting

The way in which a power device package is mounted to its heatsink affects both actual heat transfer efficiency and long-term reliability. Faulty mounting of metal packaged devices mainly causes unnecessarily high junction temperature, shortening device lifetime. Plastic packages (such as the common TO-220 outline) are much more susceptible to mechanical damage, which allows moisture into the case and can even crack the semiconductor die.

The factors which you should consider when deciding on a mounting method are summarized in Figure 43.6 for a typical plastic-packaged device.

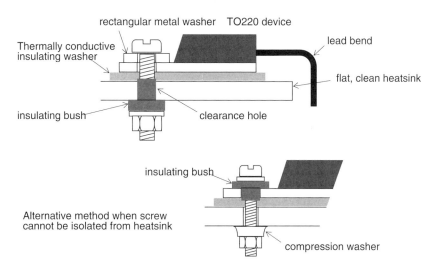

Figure 43.6: Screw mounting methods for a power device

43.3.1   Heatsink Surface Preparation

The heatsink should have a flatness and finish comparable to that of the device package. The higher the power dissipation, the more attention needs to be paid to surface finish. A finish of 50–60 microinches is adequate for most purposes. Surface flatness, which is the deviation in surface height across the device mounting area, should be less than 4 mils (0.004") per inch.

The mounting hole(s) should only be large enough to allow clearance of the fastener, plus insulating bush if one is fitted. Too large a hole, if the screw is torqued too tightly, will cause the mounting tab to deform into the hole. This runs the risk of cracking the die, as well as lifting the major part of the package which is directly under the die off the heatsink in cantilever fashion, seriously affecting thermal transfer impedance. Chamfers on the hole must be avoided for the same reason, but deburring is essential to avoid puncturing insulation material and to maintain good thermal contact. The surface should be cleaned of dust, grease and swarf immediately before assembly.

43.3.2   Lead Bend

Bending the leads of any semiconductor package stresses the lead interface and can result in cracking and consequent unreliability. If possible, mount your devices upright on the PCB so that lead bending is unnecessary. Plastic packaged devices (TO220, TO126 etc.) can have their leads bent, provided that:

- the minimum distance between the plastic body and the bend is 4 mm,

- the minimum bend radius is 2 mm,

- maximum bend angle is no greater than 90°,

- leads are not repeatedly bent at the same point, and

- no axial strain is applied to the leads, relative to each other or the package.

Use round-nosed pliers or a proper lead forming jig to ensure that these conditions are met. Metal cased devices must not have their leads bent, as this is almost certain to damage the glass seal.

When the device is inserted into the board, the leads should always be soldered after the mechanical fastening has been made and tightened. Some manufacturing departments prefer not to run cadmium plated screws through a solder bath because it contaminates the solder, and they may decide to put the screws in after the mass soldering stage. Do not allow this: insist on hand soldering or use different screws.

43.3.3   The Insulating Washer

In most devices, the heat transfer tab or case is connected directly to one of the device terminals, and this raises the problem of isolating the case. The best solution from the point of view of thermal resistance is to isolate the entire heatsink rather than use any insulating device between the package and the heatsink. This is often not possible, for EMI or safety reasons, because the chassis serves as the heatsink, or because several devices share the same heatsink. Some devices are now available in fully isolated packages, but if you aren't using one of these you will have to incorporate an insulating washer under the package.

Insulating washers for all standard packages are available in many different materials: polyimide film, mica, hard anodized aluminum and reinforced silicone rubber are the most popular. The first three of these require the use of a thermally conductive grease (heatsink compound) between the mating surfaces, to fill the minor voids which exist and which would otherwise increase the thermal resistance across the interface. This is messy and increases the variability and cost of the production stage. If excess grease is left around the device, it may accumulate dust and swarf and lead to insulation breakdown across the interface. Silicone rubber, being somewhat conformal under pressure, can be used dry and some types will outperform mica and grease.

Table 43.4 shows the approximate range of interface thermal resistances ($R_{\theta c\text{-}h}$) that may be expected. Note that the actual values will vary quite widely depending on contact pressure; a minimum force of 20N should be maintained by the mounting method, but higher values will give better results provided they don't lead to damage. When thermally conductive grease is not used, wide variations in thermal resistance will be encountered because of differences in surface finish and the micro air gaps which result. Thermal grease fills these gaps and reduces the resistance across the interface. Despite its name, it is not any more thermally conductive than the washer it is coating; it should only be applied very thinly, sufficient to fill the air gaps but no more, so that the total thickness between the case and the heatsink is hardly increased. In this context, more is definitely not better. The mounting hole(s) in the washer should be no larger than the device's holes, otherwise flashover to the exposed metal (which should be carefully de-burred) is likely.

Table 43.4: Interface thermal resistances for various mounting methods

Package type	Interface thermal resistance °C/W						
	Metal-to-metal		With insulator (1 mil = 0.001")				
Metal, flanged	Dry	Greased	2-mil mica		2-mil polyimide		6-mil silicone rubber
			Dry	Greased	Dry	Greased	
TO204AA (TO3)	0.5	0.1	1.2	0.5	1.5	0.55	0.4 - 0.6
TO213AA (TO66)	1.5	0.5	2.3	0.9			
Plastic							
TO126	2.0	1.3	4.3	3.3			4.8
TO220AB	1.2	0.6	3.4	1.6	4.5	2.2	1.8

43.3.4 *Mounting Hardware*

A combination of machine screws, compression washers, flat washers and nuts is satisfactory for any type of package that has mounting holes. Check the specified mounting hole tolerances carefully; there is a surprisingly wide variation in hole dimensions for the same nominal package type across different manufacturers. A flat, preferably rectangular (in the case of plastic packages) washer under the screw head is vital to give a properly distributed pressure, otherwise cracking of the package is likely. A conical compression washer is a very useful device for ensuring that the correct torque is applied. This applies a constant pressure over a wide range of physical deflection, and allows proper assembly by semi-skilled operators without using a torque wrench or driver. Tightening the fasteners to the correct torque is very important; too little torque results in a high thermal impedance and long term unreliability due to over-temperature, while too much can overstress the package and result in long term unreliability due to package failure.

When screw-mounting a device which has to be isolated from the heatsink, you need to use an insulating bush either in the device tab or the heatsink. The preferred method is to put the bush in the heatsink, and use large flat washers to distribute the mounting force over the package. You can also use larger screws this way. The bush material

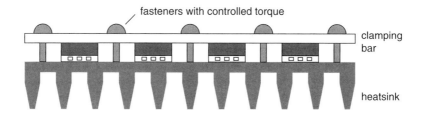

Figure 43.7: Package clamping bar

should be of a type that will not flow or creep under compression; glass-filled nylon or polycarbonate are acceptable, but unfilled nylon should be avoided. The bush should be long enough to overlap between the transistor and the heatsink, in order to prevent flashover between the two exposed metal surfaces.

A fast, economical and effective alternative is a mounting clip. When only a few watts are being dissipated, you can use board mounting or free standing dissipators with an integral clip. A separate clip can be used for larger heatsinks and higher powers. The clip must be matched to the package and heatsink thickness to obtain the proper pressure. It can actually offer a lower thermal resistance than other methods for plastic packages, because it can be designed to bear directly down on top of the plastic over the die, rather than concentrating the mounting pressure at the hole in the tab. It also removes the threat of flashover around the mounting hole, since no hole is needed in the insulating washer.

When you have to mount several identical flat (e.g., TO-220) packages to a single heatsink, a natural development of the clip is a single clamping bar that is placed across all the packages together (Figure 43.7). The bar must be rigid enough and fixed at enough places with the correct torque to provide a constant and predictable clamping pressure for each package. With suitably ingenious mechanical design, it may also contribute to the total thermal performance of the whole assembly.

43.4 Placement and Layout

If you are only concerned with designing circuits that run at slow speeds with CMOS logic and draw no more than a few milliamps, then thermal layout considerations will not interest you. As soon as dissipation raises the temperature of your components more than a few tens of degrees above ambient, it pays to look at your equipment and PCB layout in terms of heat transfer. As was shown in the last chapter, this will ultimately reflect in the reliability of the equipment.

Some practices that will improve thermal performance are:

- Mount PCBs vertically rather than horizontally. This is standard in card cages and similar equipment practice, and it allows a much freer convective airflow over the components. If you are going to do this, do not then block off the airflow by putting solid metal screens above or below the boards; use punched, louvred or mesh screens.

- Put hot components near the edge of the board, to encourage a good airflow around them and their heatsinks. If the board will be vertically mounted, put them at the top of the board.

- Keep hot components as far away as possible from sensitive devices such as precision op-amps or high failure rate parts such as electrolytic capacitors. Put them above such components if the board is vertical.

- Heatsinks perform best in low ambient temperatures. If you are using a heatsink within an enclosure without forced air cooling, remember to allow for the steady-state temperature rise inside the enclosure. However, don't position a heatsink near to the air inlet, as it will heat the air that is circulating through the rest of the enclosure; put it near the outlet. Don't obstruct the airflow over a heatsink.

- If you have a high heat density, for example a board full of high speed logic devices, consider using a thermally conductive ladder fixed on the board and in contact with the IC packages, brought out to the edge of the board and bonded to an external heatsink. PCB laminates themselves have a low thermal conductivity.

- If you have to use a case with no ventilation, for environmental or safety reasons, remember that cooling of the internal components will be by three stages of convection rather than one: from the component to the inside air, from the inside air to the case, and from the case to the outside. Each of these will be inefficient, compared to conductive heat transfer obtained by mounting hot components directly onto the case. But if you take this latter course, check that the outside case temperature will not rise to dangerously high levels.

Standards

Tim Williams

Standards are indispensable to manufacturing industry. Not only do they allow interchangeability or interoperability between different manufacturers' products, but they also represent a distillation of knowledge about practical aspects of technology—how to make measurements, what tests to use, what dimensions to specify and so on. Each standard is the result of considerable work on the part of a collection of experts in that particular field and is therefore authoritative; notwithstanding which, any standard in a fast-changing area will be subject to revision and amendment as the technology progresses.

This appendix lists a few of the more relevant British and international standards, both of those which have been referenced in the text of this book and some that the author feels to be of particular interest. The catalogs of the various standards bodies, updated yearly, give a full list of the available and current standards and are essential for the library of any development department. Although only BS and IEC publications are mentioned here, for the sake of brevity, there are of course many other standards sources which you may need to consult for a particular application.

A.1   British Standards

These are available from:

British Standards Institution
Customer Services
389 Chiswick High Road
London W4 4AL
UK
Tel +44 (0)208 996 9001 e-mail cservices@bsi-global.com
website www.bsi-global.com

Some BS standards are related to, equivalent to or identical to other European or international standards. This is indicated where appropriate. Where a British Standard's numbering starts "BS EN" this means that it is identical to the equivalent European document produced by CENELEC, which itself may well be identical to the IEC source document, although this is not assured.

BS no. **Related, equivalent or identical standards**

BS 613

Specification for components and filter units for electromagnetic interference suppression.

BS 2316 **IEC 60096**

Specification for radio-frequency cables.

BS 2488 **IEC 60063**

Schedule of preferred numbers for resistors and capacitors.

BS 2754 **IEC 60536**

Memorandum: construction of electrical equipment for protection against electric shock.

BS 4808 **IEC 60189**

Specification for LF cables and wires with PVC insulation and PVC sheath for telecommunications.

BS 5783

Code of practice for handling of electrostatic sensitive devices.

BS 6221 **IEC 60326**

Printed wiring boards.

BS 6500

Electric cables. Flexible cords rated up to 300/500V, for use with appliances and equipment intended for domestic, office and similar environments.

BS EN 13602

Copper and copper alloys. Drawn, round copper wire for the manufacture of electrical conductors.

BS EN 55014   **CISPR 14**

Electromagnetic compatibility. Requirements for household appliances, electric tools and similar apparatus.

BS EN 55022   **CISPR 22**

Information technology equipment. Radio disturbance characteristics. Limits and methods of measurement.

BS EN 60062   **IEC 60062**

Specification for marking codes for resistors and capacitors.

BS EN 60065   **IEC 60065**

Audio, video and similar electronic apparatus. Safety requirements.

BS EN 60068   **IEC 60068**

Environmental testing.

BS EN 60127   **IEC 60127**

Miniature fuses.

BS EN 60182   **IEC 60182, IEC 60851**

Basic dimensions of winding wires. Specification for maximum overall diameters of enamelled round winding wires.

BS EN 60269   **IEC 60269**

Low-voltage fuses (see also some parts of BS 88).

BS EN 60285   **IEC 60285**

Alkaline secondary cells and batteries. Sealed nickel-cadmium cylindrical rechargeable single cells.

BS EN 60431   **IEC 60431**

Specification for dimensions of square cores (RM-cores) made of magnetic oxides and associated parts.

BS EN 60529   **IEC 60529**

Specification for degrees of protection provided by enclosures (IP code).

BS EN 60617   **IEC 60617**

Graphical symbols for diagrams.

BS EN 60950   **IEC 60950**

Safety of information technology equipment.

BS EN 61951   **IEC 61951**

Secondary cells and batteries containing alkaline or other non-acid electrolytes (previously IEC 60509).

BS QC 300XXX   **IEC 60384**

Harmonized system of quality assessment for electronic components. Fixed capacitors for use in electronic equipment.

BS 9940   **IEC 60115, QC 400XXX**

Harmonized system of quality assessment for electronic components. Fixed resistors for use in electronic equipment.

A.2 IEC Standards

The IEC (International Electrotechnical Commission) is responsible for international standardization in the electrical and electronics fields. It is composed of member National Committees. IEC publications are available from the BSI, address as before, other national committees, or from the IEC webstore at www.iec.ch, postal address

IEC Central Office
3 Rue de Varembé
1211 Geneva 20
Switzerland

Most IEC publications have related British standards (see list above). A few which do not are listed below.

IEC 60479

Effects of current passing through the human body.

IEC 60647

Dimensions for magnetic oxide cores intended for use in power supplies (EC-cores).

Index